# THE ECONOMICS AND MANAGEMENT OF WATER AND DRAINAGE IN AGRICULTURE

# THE ECONOMICS AND MANAGEMENT OF WATER AND DRAINAGE IN AGRICULTURE

Edited by

**Ariel Dinar**
University of California, Davis and USDA-ERS
(Formerly University of California, Riverside and
San Joaquin Valley Drainage Program)

**David Zilberman**
University of California, Berkeley

Springer Science+Business Media, LLC

**Library of Congress Cataloging-in-Publication Data**

The Economics and management of water and drainage / edited by Ariel Dinar, David Zilberman.
    p. cm.
    Includes bibliographical references and index.
    ISBN 978-1-4613-6801-4    ISBN 978-1-4615-4028-1 (eBook)
    DOI 10.1007/978-1-4615-4028-1
    1. Irrigation—California—San Joaquin River Valley. 2. Drainage--California—San Joaquin River Valley. 3. Water-supply, Agricultural—California—San Joaquin River Valley. I. Dinar, Ariel. II. Zilberman, David, 1947-
TC824.C2E28   1991
363.73'94—dc20                                        91-16083
                                                                           CIP

**Copyright** © 1991 Springer Science+Business Media New York
Originally published by Kluwer Academic Publishers in 1991
Softcover reprint of the hardcover 1st edition 1991

All rights reserved. No part of this publication may be reproduced, stored in a retrieval system or transmitted in any form or by any means, mechanical, photo-copying, recording, or otherwise, without the prior written permission of the publisher, Springer Science+Business Media, LLC.

*Printed on acid-free paper.*

To our wives, Mati and Leorah, and children:
Roee, Shlomi, and Shira
and Shie, Eyal, and Aytan.

## Acknowledgements

*The editors would like to thank the following:*
>The San Joaquin Valley Drainage Program and the U.S. Bureau of Reclamation for funding the research, writing, editing, and design that lead to the publication of this book;
>
>The members of the book advisory committee, Margriet Caswell, Wilford Gardner, Charles DuMars, Herbert Snyder, and Henry Vaux, Jr. for providing helpful comments at the early stages of the book;
>
>The numerous reviewers who helped us with their constructive comments in the review process;
>
>Carmell Edwards of the Bureau of Reclamation's planning unit who edited and coordinated the book production;
>
>The Bureau of Reclamation's word processing unit, directed by Sammie Cervantes and Diana Savignano, for their typing efforts;
>
>Marilyn Murtos of the Bureau of Reclamation's graphics unit for the desktop publishing work and her dedication, without which this book would not have been completed;
>
>And above all, many thanks to the authors for their full cooperation. This book is theirs.

*The editors would also like to acknowledge support provided by:*
>Giannini Foundation of Agricultural Economics
>
>School of Natural Resources, University of California, Berkeley
>
>University of California Division of Agriculture and Natural Resources
>
>University of California Water Resources Center
>
>U.S. Department of Agriculture, Economic Research Service, Resources and Technology Division.

## Editorial Staff

Carmell L. Edwards, *Technical Publications Writer*
Marilyn D. Murtos, *Graphic Illustrator*
Sammie C. Cervantes, *Word Processing Supervisor*
Diana L. Savignano, *Editorial Assistant*

# Contents

DEDICATION — v

ACKNOWLEDGEMENTS — vii

PREFACE — xv
*Jan van Schilfgaarde*

## One: BACKGROUND AND SETTING — 1

1. Introduction and Overview — 3
   *Ariel Dinar and David Zilberman*

2. Irrigation, Drainage, and Agricultural Development in the San Joaquin Valley — 9
   *Lloyd J. Mercer and W. Douglas Morgan*

3. Overview of Sources, Distribution, and Mobility of Selenium in the San Joaquin Valley, California — 29
   *Robert J. Gilliom*

## Two: ENGINEERING, PHYSICAL, AND BIOLOGICAL APPROACHES TO DRAINAGE PROBLEMS — 49

4. Hydrologic Aspects of Saline Water Table Management in Regional Shallow Aquifers — 51
   *Mark E. Grismer and Timothy K. Gates*

5. Ground-water Pumping for Water Table Management and Drainage Control in the Western San Joaquin Valley — 71
   *Nigel W. T. Quinn*

6. Reuse of Agricultural Drainage Water to Maximize the Beneficial Use of Multiple Water Supplies for Irrigation — 99
   *James D. Rhoades and Ariel Dinar*

7. Land Retirement as a Strategy for Long-term Management of Agricultural Drainage and Related Problems — 117
   *Craig M. Stroh*

| 8 | San Joaquin Salt Balance: Future Prospects and Possible Solutions<br>*Gerald T. Orlob* | 143 |
|---|---|---|
| 9 | Removal of Selenium from Agricultural Drainage Water through Soil Microbial Transformations<br>*Elisabeth T. Thompson-Eagle, William T. Frankenberger, Jr., and Karl E. Longley* | 169 |
| 10 | A Conceptual Planning Process for Management of Subsurface Drainage<br>*Donald G. Swain* | 187 |

Three: **ECONOMICS OF FARM LEVEL IRRIGATION AND DRAINAGE** — 207

| 11 | Crop-Water Production Functions and the Problems of Drainage and Salinity<br>*John Letey* | 209 |
|---|---|---|
| 12 | Effects of Input Quality and Environmental Conditions on Selection of Irrigation Technologies<br>*Ariel Dinar and David Zilberman* | 229 |
| 13 | Estimation of Production Systems with Emphasis on Water Productivity<br>*Richard E. Just* | 251 |
| 14 | Increasing Block-rate Prices for Irrigation Water Motivate Drain Water Reduction<br>*Dennis Wichelns* | 275 |
| 15 | Irrigation Technology Adoption Decisions: Empirical Evidence<br>*Margriet F. Caswell* | 295 |

Four: **ENVIRONMENTAL AND PUBLIC HEALTH IMPACTS OF DRAINAGE** — 313

| 16 | Assessing Health Risks in the Presence of Variable Exposure and Uncertain Biological Effects<br>*Robert C. Spear* | 315 |
|---|---|---|
| 17 | Consideration of the Public Health Impacts of Agricultural Drainage Water Contamination<br>*Susan A. Klasing* | 327 |

CONTENTS

| | | |
|---|---|---|
| 18 | Contaminants in Drainage Water and Avian Risk Thresholds<br>*Joseph P. Skorupa and Harry M. Ohlendorf* | 345 |
| 19 | Preliminary Assessment of the Effects of Selenium in Agricultural Drainage on Fish in the San Joaquin Valley<br>*Michael K. Saiki, Mark R. Jennings, and Steven J. Hamilton* | 369 |

**Five: VALUING NON-AGRICULTURAL BENEFITS FROM WATER USE** — 387

| | | |
|---|---|---|
| 20 | Measuring the Benefits of Freshwater Quality Changes: Techniques and Empirical Findings<br>*Richard T. Carson and Kerry M. Martin* | 389 |
| 21 | Willingness to Pay to Protect Wetlands and Reduce Wildlife Contamination from Agricultural Drainage<br>*John Loomis, Michael Hanemann, Barbara Kanninen, and Thomas Wegge* | 411 |
| 22 | Valuing Environmental Goods: A Critical Appraisal of the State of the Art<br>*H. S. Burness, Ronald G. Cummings, Philip T. Ganderton, and Glenn W. Harrison* | 431 |
| 23 | Economic Value of Wildlife Resources in the San Joaquin Valley: Hunting and Viewing Values<br>*Joseph Cooper and John Loomis* | 447 |

**Six: REGIONAL ECONOMIC ANALYSIS** — 463

| | | |
|---|---|---|
| 24 | A Regional Mathematical Programming Model to Assess Drainage Control Policies<br>*Stephen A. Hatchett, Gerald L. Horner, and Richard E. Howitt* | 465 |
| 25 | The Use of Computable General Equilibrium Models to Assess Water Policies<br>*Peter Berck, Sherman Robinson, and George Goldman* | 489 |
| 26 | Analyses of Irrigation and Drainage Problems: Input-Output and Econometric Models<br>*W. Douglas Morgan and Lloyd J. Mercer* | 511 |
| 27 | Creating Economic Solutions to the Environmental Problems of Irrigation and Drainage<br>*Marca Weinberg and Zach Willey* | 531 |

| 28 | Impacts of San Joaquin Valley Drainage-Related Policies on State and National Agricultural Production<br>*Gerald L. Horner, Stephen A. Hatchett, Robert M. House, and Richard E. Howitt* | 557 |
|---|---|---|
| 29 | Cropland Allocation Decisions: The Role of Agricultural Commodity Programs and the Reclamation Program<br>*Michael R. Moore, Donald H. Negri, and John A. Miranowski* | 575 |

## Seven: DYNAMIC ASPECTS OF IRRIGATION AND DRAINAGE MANAGEMENT 597

| 30 | Optimal Intertemporal Irrigation Management Under Saline, Limited Drainage Conditions<br>*Keith C. Knapp* | 599 |
|---|---|---|
| 31 | Managing Drainage Problems in a Conjunctive Ground and Surface Water System<br>*Yacov Tsur* | 617 |
| 32 | Government Policies to Improve Intertemporal Allocation of Water Use in Regions with Drainage Problems<br>*Farhed Shah and David Zilberman* | 637 |
| 33 | Dynamic Considerations in the Design of Drainage Canals<br>*Ujjayant Chakravorty, Eithan Hochman, and David Zilberman* | 661 |
| 34 | Common Property Aspects of Ground-Water Use and Drainage Generation<br>*Lloyd S. Dixon* | 677 |

## Eight: UNCERTAINTY, ENFORCEMENT, AND POLITICAL ECONOMY OF WATER SYSTEMS 699

| 35 | Determination of Regional Environmental Policy Under Uncertainty: Theory and Case Studies<br>*Erik Lichtenberg* | 701 |
|---|---|---|
| 36 | Economic Aspects of Enforcing Agricultural Water Policy<br>*Andrew G. Keeler* | 717 |
| 37 | Organizational Failure and the Political Economy of Water Resources Management<br>*Gordon C. Rausser and Pinhas Zusman* | 735 |
| 38 | Water Market Reforms for Water Resource Problems: Invisible Hands or Domination in Disguise?<br>*Norm Coontz* | 759 |

## Nine: INSTITUTIONS, REGULATIONS, AND LEGAL ASPECTS OF WATER AND DRAINAGE PROBLEMS       779

39  Alternative Institutional Arrangements for Controlling Drainage Pollution       781
    *Alan Randall*

40  Economic Incentives and Agricultural Drainage Problems: The Role of Water Transfers       803
    *Bonnie G. Colby*

41  Water Quality and the Economic Efficiency of Appropriative Water Rights       821
    *Mark Kanazawa*

42  Institutional and Legal Dimensions of Drainage Management       841
    *Gregory A. Thomas*

43  Legal Issues Raised by Alternative Proposed Solutions to Kesterson Water Quality Problems       859
    *Charles T. DuMars*

## Ten: OUTLOOK AND DISCUSSION       875

44  Management of the San Joaquin Valley Drainage Program: The Dichotomy Between Practice and Theory       877
    *Edgar A. Imhoff*

45  Future Research on Salinity and Drainage       893
    *Henry J. Vaux, Jr. and Kenneth K. Tanji*

46  Irrigation Technology, Institutional Innovation, and Sustainable Agriculture       903
    *Vernon W. Ruttan*

47  Is the Drainage Problem in Agriculture Mainstream Resource Economics?       913
    *V. Kerry Smith*

48  Summary of Findings and Conclusions       933
    *David Zilberman and Ariel Dinar*

INDEX       941

# PREFACE

Jan van Schilfgaarde, USDA Agricultural Research Service and National Research Council Committee on Irrigation-Induced Water Quality Problems

In 1982, a startling discovery was made. Many waterbirds in Kesterson National Wildlife Refuge were dying or suffering reproductive failure. Located in the San Joaquin Valley (Valley) of California, the Kesterson Reservoir (Kesterson) was used to store agricultural drainage water and it was soon determined that the probable cause of the damage to wildlife was high concentrations of selenium, derived from the water and water organisms in the reservoir. This discovery drastically changed numerous aspects of water management in California, and especially affected irrigated agriculture. In fact, the repercussions spilled over to much of the Western United States.

For a century, water development for irrigation has been a religiously pursued means for economic development of the West. The primary objective of the Reclamation Act of 1902 was, purportedly, the development of irrigation water to support family farms which, in turn, would enhance the regional economy (Worster, 1985).

Early on, it was abundantly clear that irrigation in arid regions would bring with it salinity problems that needed special management. It was 1886 when Hilgard warned that "the evils now besetting [California's Irrigation Districts] are already becoming painfully apparent; and to expect them not to increase unless the proper remedies are applied is to hope that natural laws will be waived in favor of California" (Hilgard, 1886). Over the years, principles of soil and water management to deal with salinity were developed and applied, so that agricultural production could be maintained or enhanced. Far more recently, it became recognized institutionally that such management required drainage and that the disposal of saline drainage water had an adverse effect on the receiving water body. The passage of the Colorado River Salinity Control Act (P. L. 93-320) in 1974 illustrates the point. The discovery of selenium at Kesterson in 1982 added another dimension; to the adverse effects of salinity in drainage water was now added the potentially toxic effect of a number of minor elements--selenium, molybdenum, uranium, as well as boron. These findings led to a growing recognition of contamination problems related to irrigation drainage at other waterfowl refuges. Intensive studies were initiated by the Department of the Interior in Nevada, Utah, Wyoming, Arizona, Montana, Texas, and other areas of California in 1986-87 involving contamination of wetland habitat by irrigation return flows.

During the last two decades, the presumed and actual dominance of irrigation as the preferred (certainly, favored) user of water at the expense of other uses, was beginning to be questioned. Concern with instream uses, the public trust doctrine, competition from municipalities--these and other issues gained in importance in academic discussions as well as in the real world, in public awareness and in court decisions. The loss of birds at Kesterson in 1982 by itself might not have been of great significance had it not been related, in a broader sense, to the reduction of wetlands in California's Central Valley from over 4 million acres to less than 300,000 acres; to the concurrent degeneration of Stillwater Refuge in Nevada; to the stress on migrating birds along the Pacific Flyway; and to the increasing influence of the nonfarming population interested in fishing, hunting, and rafting.

Be this as it may, the alarm set off by Kesterson caused the State and Federal governments to react strongly and, at first, in a manner approaching panic. Ultimately, the San Joaquin Valley Drainage Program (SJVDP) was organized, a group charged with investigating all aspects of the irrigation/drainage problem in the Valley and developing a set of recommendations for resolution of the dilemma. SJVDP had participation from three Federal agencies (U.S. Bureau of Reclamation, U.S. Fish and Wildlife Service, and U.S Geological Survey) and two state groups (California Department of Water Resources and California Department of Fish and Game). SJVDP, in turn, requested advice from the National Academy of Sciences in designing its studies in ways that could be defended in a scientific forum; this led to the Committee on Irrigation-Induced Water Quality Problems of the Water Science and Technology Board, which is part of the National Research Council. This committee argued, as one of the recurring themes during its 5-year life, that SJVDP needed to look beyond the "technological fix" at the many, complex institutional constraints and possibilities for finding rational and equitable solutions, and also at economic reality.

Thus, as chair of this committee, I welcomed the initiative by SJVDP leaders to organize this volume of technical papers and essays that deal primarily with economic and social aspects of the general problem of conflict resolution in water resources, and specifically with the Valley.

As of this writing, no specific solution, or plan of action, has formally emerged as the preferred course for the Valley. This writer doesn't expect any such miracle soon, but he does anticipate the presentation of a set of options that will lead to implementation of a series of steps to help mitigate the problem, together with the realization on the part of the public and of policymakers that complex, multifaceted problems in a dynamic society, at best, lead to gradual corrective steps and incremental improvements. He hopes, and fully expects, that these steps will include adjustments in institutions, including subsidy programs, water quality regulations, water contracts, and many others.

Because of the "Kesterson scare", a great deal of activity was generated both within or through the SJVDP and within academia. New physical and biological processes were analyzed and developed and old ones reevaluated. Effects of foreign substances on wildlife were assessed, and economic models were generated. New impetus was given to evaluating nonmarket values of natural resources and questions were raised about mechanisms for water transfers.

The book that lies before you makes no claim to being the exhaustive and definitive text on institutional or economic management of water quality problems. Rather, it contains a series of chapters that illustrate the diversity and complexity of such problems, while also demonstrating the substantial body of knowledge that is available for their analysis. The book is witness to the significant amount of work done in response to the problems discovered at Kesterson in 1982, often with support of the SJVDP. It also makes clear that the current level of understanding is far from complete. It covers some old material and some new. Some of the conclusions will come as no surprise to knowledgeable readers; others may be unexpected. It certainly surprised this writer to note an assessment that enhancing clean water supplies to the Valley refuges would reap a benefit of $3 billion per year.

The title of this book stresses economics; most of the authors are economists. One might conclude that the primary issues in the Valley are economic. They are not. Yet the tools of economics, when properly blended with the knowledge of geologists, biologists, soil scientists, public health specialists, and engineers, can be used effectively to clarify the issues and to illustrate some of the consequences of choices that must be made. As mentioned, the NAS Committee expressed concern that, without in-depth economic and policy analysis, a viable plan of action was not likely to emerge. This book's emphasis on economic analysis, in conjunction with input from other disciplines, demonstrates that the leaders of the SJVDP shared the committee's concern and responded to it.

As was the purpose of the NRC Committee's report (NRC, 1989), the hope with this book is that the policymakers confronted with similar problems elsewhere in the future will be able to benefit from the experience of the 1980's in the San Joaquin Valley of California. It provided a partial, yet coherent, record of analytical techniques and methods of evaluation applied to, and often developed for, the water management problems there encountered.

## REFERENCES

Hilgard, E. W., 1886. *Irrigation and Alkali in India*. College of Agriculture, University of California, Report to the President of the University, Bulletin No. 86, California State Printing Office, Sacramento, CA, pp. 34-35.

National Research Council, 1989. *Irrigation-Induced Water Quality Problems: What Can Be Learned from the San Joaquin Valley Experience*. National Academy Press, Washington, D.C.

Worster, D., 1985. *Rivers of Empire: Water, Aridity, and the Growth of the American West*. Pantheon Books, New York.

# THE ECONOMICS AND MANAGEMENT OF WATER AND DRAINAGE IN AGRICULTURE

# Six: REGIONAL ECONOMIC ANALYSIS

# 24 A REGIONAL MATHEMATICAL PROGRAMMING MODEL TO ASSESS DRAINAGE CONTROL POLICIES

Stephen A. Hatchett, CH2M HILL, Sacramento, California;
Gerald L. Horner, Agricultural Economist, Davis, California;
Richard E. Howitt, University of California, Davis

### ABSTRACT

A regional model is described that is being used to evaluate nonstructural ways to control contaminated agricultural drainwater in California. The model links an economic decision component to ground-water hydrology and salinity in an integrated systems approach to policy evaluation. Water pricing is evaluated as a means for inducing drainage reduction in two different situations. Conclusions from this analysis plus broad conclusions from the overall modeling effort are discussed.

### BACKGROUND

Effective control of agricultural drainage requires that policymakers understand the linkages between the economic decisions of irrigators and the hydraulic response of the ground-water aquifer. Irrigation decisions are made by farmers responding to price signals, physical growing conditions, constraints on available resources, and institutional requirements. These decisions most importantly affect how much ground water is pumped and how much water and salt are added to the system via deep percolation. A systems modeling approach attempts to account for all potential interactions between the economic and physical subsystems. Described below is a major modeling effort to formulate and evaluate drainage control policies in California's San Joaquin Valley (Valley).

The Westside Agricultural Drainage Economics (WADE) model is derived from a model developed by Dudek and Horner (1980)[1] as an integrated physical-economic analytical system for studying environmental issues associated with irrigated agriculture in the Valley. Its purpose was to evaluate alternative land use and water development policies, and to develop Best Management Practices for the reduction of subsurface drainage water. The model was used to evaluate effects of mandatory tailwater recycling and irrigation scheduling on drainage quantity and quality. The model was also used to project the effect of alternative water quality policies on the distribution of income (Dudek and Horner, 1981), and to estimate the effect of water pricing on water conservation by irrigators in the Valley (Horner, Moore, and Howitt, 1983).

The model was substantially revised in 1988 and 1989 as part of a contractual agreement between the San Joaquin Valley Drainage Program (SJVDP) and CH2M Hill to analyze agricultural drainage issues on the west side of the Valley. Data was updated, the economic component was reformulated, the hydrology component was significantly enhanced, and a salinity component was added. A review of the current model and some preliminary results were presented by Hatchett et al. (1989).

The analytical system is designed to simulate the relationship of onfarm cropping and irrigation decisions to the volume and quality of subsurface drainage water. It contains three main components: The agricultural production model that simulates grower decisions, the hydrology model, and the salinity model. The system provides a basis for assessing the effectiveness of drainage control policies and for estimating benefits and costs associated with alternative planning objectives.

## USES OF THE MODEL

The economic component of the WADE model is designed to assess how agricultural production and resource use will respond to changes in hydrologic and economic conditions. The changing conditions could be simply the continuation of long-term trends or they could be abrupt shifts in Government policy. Assessing the agricultural response to abrupt changes requires a model that does more than extrapolate trends or infer future behavior from past behavior (although it could include elements of these approaches). Policymakers must try to assess the response to conditions that may be very different than the recent past. An appropriate model should describe the underlying physical and behavioral rules--rules that will still provide reasonable predictions under different conditions. A mathematical programming approach (rather than a purely empirical, econometric approach) was chosen in an attempt to satisfy these requirements.

Changes in institutional and economic parameters can be easily introduced into WADE. These can take the form of mandated land retirements, drainage water disposal limits, use changes, and changes in surface water supplies.

## USE OF POSITIVE MATHEMATICAL PROGRAMMING

Positive Mathematical Programming (PMP) is a technique that can reproduce observed activity levels in optimization models without directly constraining the levels (that is, without using devices like flexibility constraints). This allows the model to seek optimal levels based on economic criteria and not on exogenously derived right-hand side values. The PMP technique is implemented by initially constraining activity levels to observed values. This stage, called the calibration model, yields dual values that represent the marginal changes in the objective function from small changes in the activities. The dual values are positive when the constraints force a lower acreage of a particular crop than an unrestricted model would calculate (and negative when the constraints force a higher acreage). The second stage of PMP solves the first stage model again after making two important changes. First, the crop acreage constraints are removed. Then the dual values from the calibration model are used to calculate a linear marginal cost function for each crop activity. Integrating the marginal cost gives a total cost quadratic in crop acreage. The quadratic form is then appended to the stage 1 objective function. The PMP model will duplicate the crop mix from the restricted calibration model, but will also allow smooth changes in crop levels as conditions or policies change.

Previous modeling efforts have added imputed or residual costs in the objective function. Miller and Millar (1976) empirically forced a national level optimization model to reproduce observed crop acreage and production levels for a base year. Fajarado, McCarl, and Thompson (1981) used the same method to derive 1971 production levels in a national model of Nicaraguan agricultural production. In both of these papers, the process of reproducing observed activity levels without the use of constraints is not explicitly defined. They both appear to add a constant marginal cost term for each activity to assure that marginal revenue and cost are equal. Howitt and Mean (1983) developed a more flexible approach that used dual values to calculate linear marginal cost (quadratic total cost) terms. Their method, which they called PMP, allows the activity levels to adjust along a smooth, upward-sloping marginal cost curve as conditions change. Horner, Putler, and Garifo (1985) used the PMP technique described by Howitt and Mean (1983) in a modified version of the United States Mathematical Model (USMP) to analyze the role of irrigated agriculture in supplying commodities for the export market. Since then, the original USMP model has been modified to use the PMP technique (House, 1987).

The PMP procedure has also been used to analyze water transfers in the Colorado River Basin (Oamek, 1989); to assess alternative sizes and operating criteria of Bureau of Reclamation reservoirs (Oamek and Schluutz, 1989); and to analyze the effect of Common Market policies on feed ration composition (Quimby and Leuck, 1988).

## SIMPLIFICATIONS USED IN THE WADE MODEL

Because of the complexity and size of the Valley study area, a number of tradeoffs in model design had to be considered. Developers of the WADE model took great care to keep the model linear or at least convex. A convex model is usually easier to solve and has the property that any local optimum found must be the global optimum. Maintaining convexity sometimes required simplifying the model's structure. Some of the simplifying assumptions used to keep the model linear or convex include:

- Production decisions are based on single season net revenue rather than a series of net revenues. A more accurate (but much more complex) objective for farmers would be to maximize the discounted series of net revenues, accounting for the changing state of the soil over time.
- Some variables requiring nonlinear decision rules are evaluated before the optimization model is solved. This allows the model to treat them as predetermined values rather than nonlinear variables.
- Some nonlinear relationships that need to remain endogenous to the optimization model are approximated as a set of discrete points along the nonlinear curve. Although this procedure maintains linearity, it increases the dimensionality (size) of the model.
- Some nonlinear relationships are replaced with linear approximations.

## DATA AND PRELIMINARY CALCULATIONS

The WADE model currently includes five irrigated crops (plus unirrigated land), eleven irrigation systems, and three levels of water management. The crops actually represent categories of crops having similar water use patterns. They are:

- Alfalfa and irrigated pasture
- Orchards and vineyards
- Row crops (primarily cotton)
- Small grains
- Vegetables and truck crops

The number of crops can be increased or decreased as desired to strike a balance between level of detail and model size. The model requires data on land preparation and cultivation costs, harvest costs, irrigation and drainage system costs, market prices, evapotranspiration, irrigation system performance, yields, and available water supply and cost.

Data was collected from a variety of sources. Shallow ground-water and drainage information came from water districts, Bureau of Reclamation, and the California Department of Water Resources (DWR). Land use information came from water district and county reports and from the results of a survey in progress by the U.S. Geological Survey (USGS). Crop yields, prices, and production costs were taken from the most recent county and University of California extension information available. Water use and irrigation system performance was obtained from onfarm evaluations conducted in the study area (Burt and Katen, 1988 and CH2M Hill, 1988).

Soil and hydrogeologic properties were obtained from existing data sets compiled for other models. The USGS prepared a ground-water model of California's Central Valley as part of the National Regional Aquifer-System Analysis, or RASA (Williamson et al., 1985). This model contained one unconfined and three confined layers, operated on a 6-month time step, and used a 6-mile square grid. The DWR uses a ground-water and economic model for policy analysis in the Central Valley (California DWR, 1982). The ground-water section is a two-layer, finite-element model with an irregular grid. These models provided information on hydraulic conductivity, clay layer leakance, specific yield, and hydraulic head in the confined layer. Additional information was provided by the USGS from ongoing research, though these data are preliminary. Many of the hydrogeologic properties in these models were not measured, but rather were estimated using a calibration procedure. Therefore, during WADE model development these values were viewed as starting points for calibration rather than as parameters known with certainty.

The irrigation systems include surface, sprinkler, and drip technologies. Each system can be managed at three levels of water use efficiency: high, medium, and low. These levels do not imply that management quality is good or bad. Low water use efficiency may in certain cases be the most effective way to manage.

## IRRIGATION SYSTEM HYDROLOGIC CHARACTERISTICS

The irrigation systems used in the model are described in detail in a report prepared by CH2M Hill (1989). For each system, total applied water is divided into four parts, called distribution fractions: beneficial use, deep percolation, uncollected runoff, and evaporation losses. Additionally, the distribution uniformity defines the portion of deep percolation that provides effective leaching of salts. Distribution fractions were estimated for 11 common irrigation systems under low, medium, and high management levels.

The distribution fractions and evapotranspiration (ET) requirements (net of effective rainfall) are used to calculate applied water for each combination of crop and irrigation system. Applied water is divided between preseason irrigation and growing season irrigation.

## CROP YIELDS AND SOIL GROUPS

The polygon cell geometry used in the WADE model is largely based on the geometry used by the Dudek and Horner (1980) study, which attempted to define areas of uniform crop productivity based on an index of soil texture, water table elevations, and soil fertility. The basis for defining homogenous production areas for cells was soil groups which are aggregated soil associations mapped on general soil maps for each county by the U.S. Soil Conservation Service from 1966 to 1971. The original geometry has been altered by the SJVDP to account for distinctions in drainage and water supply conditions and boundaries of the major SJVDP study areas.

## CROP YIELDS, HIGH WATER TABLES, AND SALINITY

Saline high water tables have several effects on crop production. High water tables increase waterlogging of the root zone, stunting root development and plant growth. Williamson and Kriz (1970) and Arar et al. (1971) have documented the waterlogging effect. The results of Arar et al. were used by Gates (1988) to describe the relationship between depth to nonsaline water table and relative crop yield.

$$\begin{aligned} \text{RELYIELD} &= 1.0 & \text{if WTD} \geq 2.898 \text{ ft} \\ &= 0.313 + 0.237 \cdot (\text{WTD}) & \text{if WTD} < 2.898 \text{ ft} \end{aligned}$$

Below 2.898 feet, the water table had no observed waterlogging effect.

Field studies have shown that upward capillary flow from shallow ground water (sometimes referred to as upflux) can contribute a significant proportion of crop water demand. Gates and Grismer (1988) combined field and lysimeter data to estimate the percent contribution to crop ET as a linear function of depth to water table (WTD). The upper limit on upflux ranged from 35 percent in clay soils to 70 percent in sandy soils. The WADE model used this relationship, assuming a maximum contribution of 50 percent, declining linearly to 0 when water depth is greater than 8.3 feet.

$$\text{UPFLUX} = (.5 - .06 \cdot \text{WTD}) \cdot \text{ET for } 0 \leq \text{WTD} \leq 8.3$$
$$= 0 \text{ for WTD} > 8.3$$

Grimes and Henderson (1986) used field tests for several crops in California's Central Valley to estimate percent ET contribution from shallow ground water. They found both depth to water and water quality (measured as electrical conductivity, EC) significant in predicting upflux:

$$\text{UPFLUX} = 43.4 + 46.93 \cdot \text{WTD} - 18.56 \cdot \text{WTD}^2 - 7.54 \cdot \text{EC} + .13 \cdot \text{EC}^2 + 1.68 \cdot \text{WTD} \cdot \text{EC}$$

where water table depth is measured in meters and EC in $dSm^{-1}$. These two equations predict significantly different amounts of water table contribution. For example, a typical clay loam soil in the study area may have shallow ground water at an EC of 10 $dSm^{-1}$ and a depth of 5 feet. The first equation predicts 20 percent contribution to ET while the second predicts 35 percent. This kind of difference has a major effect on how much irrigation water is needed, on how much drainage to expect, and on how fast salts concentrate in the root zone. The Gates and Grismer equation generally provides more conservative estimates of upflux, and is used in the WADE model.

Salt is brought into the root zone by this upward flow and is concentrated as the water evaporates or is transpired. Additional salt is added by the irrigation water. High soil salinity increases the soil moisture tension which impairs the plant's ability to extract the moisture. Crop yield will eventually be reduced if soil salinity is not managed properly. A number of studies have been conducted in the Valley which measure the effect of root-zone salinity on crop yields (Maas and Hoffman, 1977 and Feinerman et al., 1982).

The WADE model accounts for soil and water salinity when making production decisions. An upper limit for applied water salinity can be selected for each cell (region) in the model. This limit should reflect the sensitivity of the crops grown and the quality of water sources available. The production decision model then will blend water sources to stay within this limit. The yield effect of the applied water salinity is expressed through its impact on root-zone soil salinity. The relationship described in Rhoades (1987) is used to relate soil salinity, effective leaching, and applied water salinity.

Maas and Hoffman estimated equations describing the response of various crops to soil salinity, and these equations are used in the WADE model. The equation below is the Maas-Hoffman relationship, with MHA the intercept and MHB the slope. Unique coefficients are defined for different crops. CRZ is the salt concentration (EC) of the root zone, and the .5 coefficient converts from root-zone concentration to saturated extract concentration:

$$\text{RELYIELD} = \{[100-(\text{MHB})][(0.5 \cdot \text{CRZ}) - \text{MHA}]\}/100$$

where RELYIELD can range from 0 to 100. Cotton represents about 50 percent of acreage on the Valley's west side, and is also one of the most salt tolerant crops. According to the estimates of Maas and Hoffman, the threshold at which cotton begins to show yield decline is 7.7 EC of saturation extract, or a field concentration of about 15. This implies that farmers may be able to manage for gradual salt accumulation and still grow major salt-tolerant crops for many years.[2]

## AGRICULTURAL PRODUCTION MODEL

The Agricultural Production Model uses the PMP procedure described above. The model assumes that farmers maximize their annual net return subject to constraints on technology and resources available.

### Objective Function

The model uses two versions of the objective function. The first, a linear version, is used to establish imputed crop values (the calibration portion of PMP). The second version adds additional quadratic terms to the linear function which represent the imputed costs or revenues not accounted for explicitly. As explained earlier, the quadratic terms are constructed in a way that keeps the model convex: the model maximizes a concave function over a set of linear constraints.

The objective is to select crops, irrigation technologies, water sources, and drainage disposal to maximize crop revenue minus: cultivation costs; harvest costs; annualized cost of tile drains; irrigation system cost; surface water cost; ground-water cost; fee on drain water discharged; cost of recycling drain water; and a quadratic function of imputed but unobserved costs.

### Constraints

The objective is maximized subject to the following set of linear equality and inequality constraints.

- Total crop area cannot be greater than irrigable area in each cell.
- An accounting identity is needed to assure that, for each cell and crop, total acres in each crop equals the sum over all irrigation technologies used for that crop.
- Water applied to each crop in each cell must come from either surface water, ground water, or recycled drainwater. The needed depth of applied water accounts for irrigation efficiency, effective rainfall, soil moisture storage, and contribution to ET from the shallow ground water.
- Surface water used in each cell must not exceed surface water available.
- For cells having a shallow water table and installed drains, expected drainage is estimated as expected deep percolation. Deep percolation may under- or over-predict actual drainage because of other flows in or out of the drained area. To account for this, adjustment coefficients are calculated based on the most recent prediction error. Total drainage volume is split between the amount discharged and the amount recycled.
- Applied water sources must be blended to achieve a maximum salinity of applied water. The maximum, or target, level is set by the user to prevent unreasonable water quality from being used. Relatively clean surface water is blended with each salty source to meet the target. Other sources are ground water, recycled drainwater, and water moving upward from a shallow water table. The blending weights are the deviations of the water source salinity from the target maximum allowed.
- During calibration of the model, constraints can be imposed to force results to approximate the observed mix of irrigation systems.

All acreage and volume variables are constrained to be nonnegative.

A number of policies can be explored by activating additional constraints and decision variables. These include:

- Ground water used in each cell must not exceed the ground-water limit.
- Drainwater discharged must not exceed a chosen level.
- Surface water can be traded among model cells to achieve higher net income.
- Surface water can be sold to buyers outside the study area at a user-specified price.
- Land can be removed from irrigation.
- Surface water prices can be tiered.

A much larger version of the model has been developed that allows it to choose a target applied water salinity for each crop. The larger model also allows for planned water stress of each crop.

## Transfer of Data to the Hydrology Model

The Production Model produces results on land use, crop production, irrigation systems, and water use. Some of this information is used by the Hydrology Model to estimate changes in ground-water flows and levels. Specifically, the Production Model provides estimates of ground water pumped from the three layers, total water applied to the land from all sources, the amount of ET, and the amount of subsurface drainwater recycled.

## HYDROLOGY MODEL

The hydrology model is most accurately described as a set of simultaneous difference equations calculating stocks and flows of water. It calculates ending and average stocks of water stored in the various layers of each cell, and the flows of water between cells and between layers. The model includes the following layers:

- Root zone.
- Unconfined layer extending from the root zone to a 20-foot depth.
- Semiconfined layer extending from the 20-foot depth down to confining clay.
- Confined layer below the clay.

Each of the three lower layers can potentially have both a saturated and an unsaturated zone.

All equations are specified as linear to allow the model to solve quickly and reliably. Some of the flow relationships are only linear within a feasible range - a physical boundary prevents flow from falling outside this range. In such cases a temporary, or dummy, flow is calculated and the model sets the true flow as follows:

- If the dummy flow is within the feasible range, set true equal to dummy;
- If the dummy flow is outside the feasible range, set true such that the boundary condition holds.

The objective function helps to accomplish this by minimizing the differences between dummy flows and true flows. By this technique, the model sets true value equal to dummy value when feasibility is maintained (i.e., the boundary conditions are not violated), or allows them to have different values if needed to maintain feasibility. The Hydrology Model objective function serves no other purpose than this.

The constraints of the hydrology model fall primarily into four categories:

- Feasibility constraints
- Darcy equations for calculating vertical or lateral flows
- Calculation of flow into subsurface drains
- Calculations of ending and average water table elevations

Volume of drainage is calculated as the average hydraulic head above the drains (averaged between the starting head and a driving head that adds in net deep percolation), multiplied by the effective drained area and by an effective conductivity. The value of effective conductivity is adjusted during a calibration procedure to track observed drainflows and water table elevations.

## SALINITY MODEL

The Salinity Model solves after the production and hydrology models. It is simply a solution to a set of simultaneous mass balance equations. The equations describe movements and ending levels of salt in the water. Each equation calculates the salt concentration in a volume of water by adding up the mass load of salt contributed by all flows in or out (including the starting mass load in the volume) and then dividing by the net volume of flows in or out. In most cases the water is assumed to blend homogeneously. The calculations for a given layer are summarized as follows:

- The layer starts at a homogeneous salt concentration, and the starting mass load of salt is calculated,
- The mass loads of all flows into or out of the layer (as determined in the Hydrology Model) are calculated and added to or subtracted from the starting load,
- The ending salt concentration is a homogeneous blend calculated by dividing the salt load by the starting volume of water in the layer plus the net volume flow in or out.

Drainwater concentration is calculated somewhat differently. It depends on the path of the water flowing into the drains. Three potential "sources" of water are used to estimate this concentration: Water percolating from the root zone, the existing water in the shallow layer, and the existing water in the deep, semiconfined layer. Research by Deverel (personal communication, 1989) indicates that the path of flow into subsurface drains can extend far below the 20-foot shallow layer. The proportions of drain flow drawn from these three sources are specified by the user; they can be derived from previous research or can be adjusted to duplicate the observed salinity of drainwater.

After the Salinity Model is solved, the ending values of the variables are passed on to the subsequent production model. The most important of these are ending root zone salinity, drainwater salinity, shallow layer salinity, and salinity of pumped ground water.

## MODEL CALIBRATION

The model is calibrated to 5 years of data on depth to ground water and drainage flow and salinity. Cropping patterns, rainfall, and water deliveries were held at observed values, and the values of uncertain hydrologic parameters were adjusted to track the target ground water and drainage observations.

## USING A REGIONAL MODEL FOR POLICY ANALYSIS

Two different approaches to policy analysis can be used. One approach applies policy changes to a large area comprising many cells with widely varying land use and hydrology. Results are displayed either by cell or in aggregate. This approach is appropriate for evaluating the regional impacts of prospective plans.

The second approach is more appropriate for policy formulation and sensitivity analysis. The model is applied to a fairly small number of cells chosen to represent a particular problem or situation. This approach has several advantages. The model runs much faster and more reliably (nonlinear programming always carries a penalty in solution time and the chance that the algorithm will encounter difficulty and return a suboptimal solution). The model can be restricted to an area of most complete or reliable data. This assures that the model will provide believable results even when reliable data does not exist for the entire region. The interpretation of results can focus on the areas and relationships of interest.

Both approaches can properly be used for developing regional drainage control plans, but their purposes should be kept separate. Only after individual control strategies have been analyzed so that their effects are fairly well understood should the impacts to a large region be evaluated. Model developers and analysts need to resist the pressure to produce results quickly by jumping directly to the large regional model. Rather than accelerating development, this slows it down: Solution time increases dramatically, and analysts must try to rationalize an overwhelming amount of results generated by modeling a large number of cells.

## RESULTS OF WATER PRICING POLICY ON SUBSURFACE DRAINAGE

This section presents results showing the potential for water pricing to reduce water application and subsurface drainage. The analysis compares results for two distinct situations: An area having full surface water supply at fairly low cost, and an area of restricted, and more costly, surface water supply. Results are shown for increased flat water rates, for tiered water rates, and for a plan that combines tiered rates with subsidies for improved irrigation efficiency.[3] Neither area is allowed to install new subsurface drains during these simulations.

An important consideration in designing tiered rates is whether the higher price should be triggered by the total volume of water a farmer uses or triggered separately for each crop. Model results show that both methods lead to reduced water application and drainage. The volume-based tier is somewhat more effective at reducing applied water by creating an incentive to switch to lower water use crops. The crop-specific tier, if tied to crop ET, is more effective at reducing deep percolation and drainage. It does, however, force the district to monitor cropping pattern and water diversions more closely.

Deep percolation accounts for almost all the irrigation water that is not transpired by the crop. Therefore a crop-specific water price tier triggered at or near crop ET is virtually a tax on deep percolation. This is probably fairer and more practical than a pure Pigouvian tax on drainage, given the problem of measuring upslope contribution to drain flow.

The model was used to evaluate three different crop-specific tiered rates, with the high price triggered at: ET, 10 percent below ET, and 10 percent above ET. As expected, the most restrictive was also most effective at reducing drain flow. Results are shown for the tier at 10 percent below ET.

The first set of results are for an area (denoted area A) having full water supply at a relatively low cost of $16 per acre-foot. The model was used to compare this base case to higher flat rates of $25 and $40, and to several tiered structures. Table 1 describes the pricing policies and how they are labeled in figures 1-5.

Table 1. Description of water pricing policies analyzed.

| | |
|---|---|
| ALL16: | All water is priced at $16 per acre-foot. |
| ALL25: | All water is priced at $25 per acre-foot. |
| ALL40: | All water is priced at $40 per acre-foot. |
| EM140: | A tiered water rate for Area A with the low price at $16 per acre-foot up to an amount 10 percent below the ET for the crop. Above this the price rises to $40. |
| EM14C: | Water pricing is the same as EM140. In addition, $24-40 per acre subsidies are provided for efficient irrigation systems. |
| EM160: | A tiered water rate for Area B with the low price at $16 per acre-foot up to an amount 10 percent below the ET for the crop. Above this the price rises to $60. |
| EM180: | A tiered water rate for Area B with the low price at $16 per acre-foot up to an amount 10 percent below the ET for the crop. Above this the price rises to $80. |
| EM18C: | Water pricing is the same as EM180. In addition, $24-40 per acre subsidies are provided for efficient irrigation systems. |

The reason for using tiered water pricing is to provide a pricing incentive for farmers to conserve water and reduce drainage, yet avoid the large transfers of income caused by flat rate increases. The results in figures 1-5 clearly demonstrate this. A flat rate increase to $40 (ALL40) does induce water savings and drainage reduction, but at close to a 70-percent reduction in net farm income. A simple tiered rate (EM140) achieves nearly as much drainage reduction but reduces net income only 25 percent. The total water bill paid by farmers to the water district increases under both flat and tiered rates (implying an inelastic demand for water). Because water districts are prevented from earning a profit, the extra revenue must be returned to farmers or reinvested. Rebates to farmers based on their water use clearly defeats the original purpose of inducing conservation.

Another scenario was modeled in which tiered pricing was combined with subsidies for improved irrigation efficiency. The subsidies could be wholly or partly financed by the district's extra water revenue. The scenario shown (EM14C) provided a $24 per acre subsidy for all irrigation systems greater than 70 percent efficient (measured as the ratio of beneficial use to applied water), and an additional $16 subsidy for systems more than 80 percent efficient. These amounts were estimated to exhaust the extra district revenue from the tiered rate. The subsidy achieved an additional 10-percent reduction in drainage volume yet higher farm income as compared to the simple tiered rate.[4]

# REGIONAL MATHEMATICAL PROGRAMMING MODEL

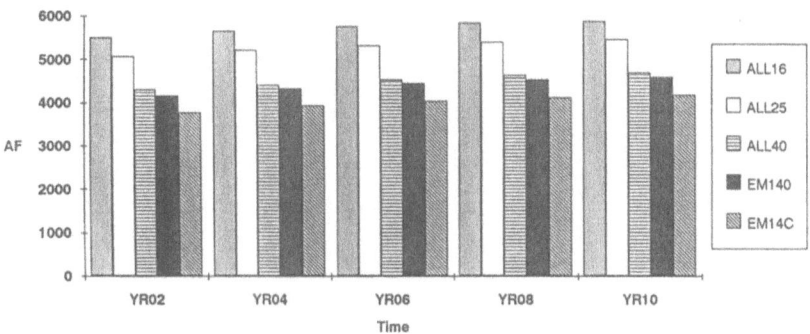

Figure 1. Drain flow in Area A.

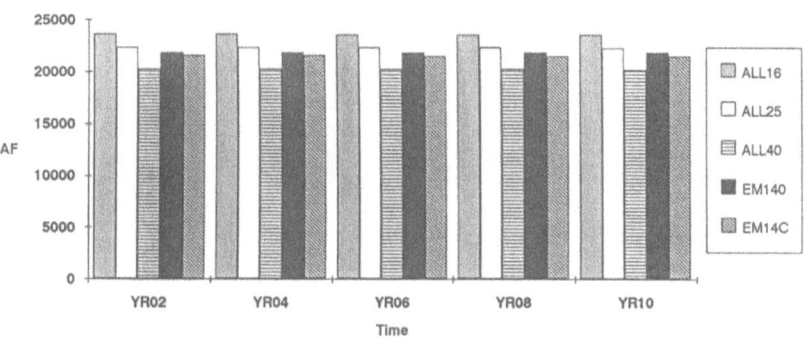

Figure 2. Surface water applied in Area A.

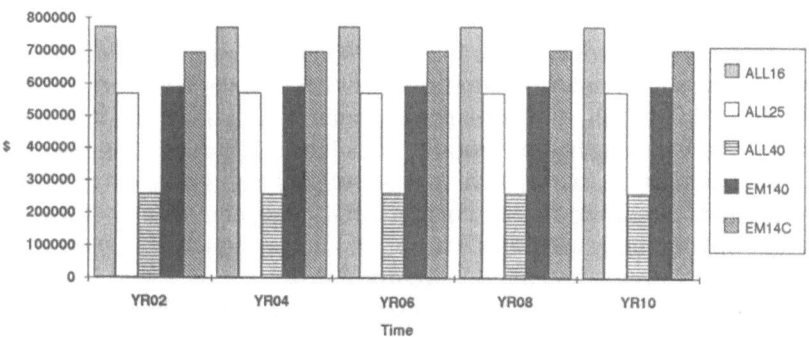

Figure 3. Net returns in Area A.

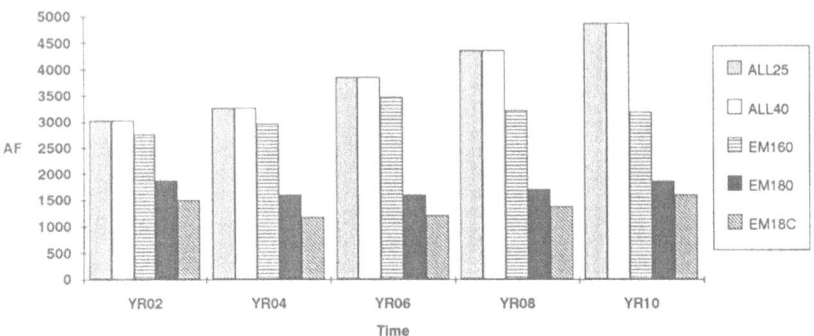

Figure 4. Drain flow in Area B.

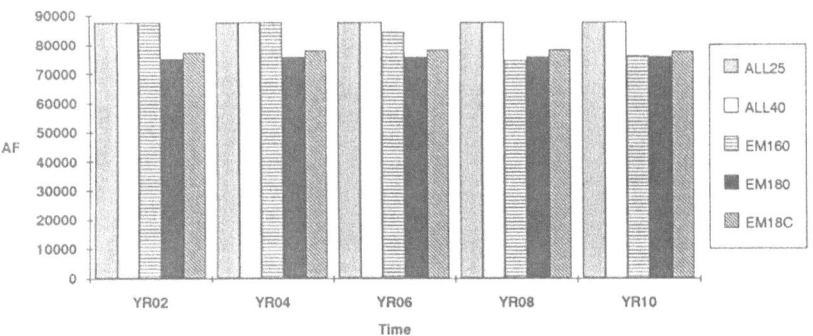

Figure 5. Surface water applied in Area B.

Subsidies for improved irrigation show some potential for reducing drain flow but have some practical limitations. Studies have shown (Burt and Katen, 1988 and CH2M Hill, 1989) that management is at least as important as hardware in achieving higher irrigation efficiency. Hardware is easy to subsidize, but management is not. Training and scheduling services can be provided at no cost, but these are only a part of improved management. A carrot and stick strategy of combining subsidies with pricing incentives may be necessary to induce water conservation and reduce drain flow.

Similar pricing policies were also tested in an area of limited surface water, where farmers supplement their supply by pumping ground water. Water is somewhat more expensive at $25 per acre-foot, but its value to the farmers (its shadow price) is substantially larger. Moderate increases in price have little or no effect on use, as seen in figure 5. Only when the price approaches that of the alternate source, ground water, is surface water use reduced. Model results show this begins to happen at $60. Significant reductions occur at $80, so the tiered price scenarios use this as the higher price. Several intriguing results should be explained.

Net returns in the base case increase over time because of increased shallow ground-water contribution to ET (upflux). Figure 8 shows the increase, which occurs as the water table rises closer to the surface. The upflux allows farmers to pump less ground water (see figure 7), hence net income rises. Upflux brings a substantial salt load into the root zone, and in time yields and crop selection will be impaired. Over the simulated 10 years, root-zone salinity increased 50 percent, although it was still acceptably low for growing most field crops. Longer simulations prepared for the SJVDP show yields and income eventually decline, and the timing and sharpness of decline depend critically on assumed starting salinity and crop salt tolerance.

As seen in figures 4 and 5, the reduction in drain flow caused by the $80 tiered price seems disproportionately large compared to the reduction in surface water use. The reason is that farmers are avoiding the high priced water by increased ground-water pumping. Total applied water changes very little but increased pumping draws down the shallow ground water thereby reducing drain flow. Because farmers substitute ground water for the high priced surface water, district revenue declines under this tiered policy. For comparison purposes, a combined tier and subsidy policy was also evaluated. Drain flow is reduced by another 12 percent compared to the simple tiered policy and income rises substantially (figure 6). This would be a very expensive policy because the subsidy could not be financed by higher revenues as it was in area A.

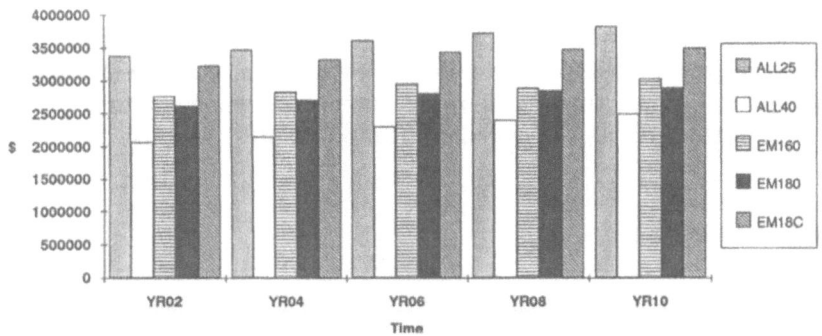

Figure 6. Net returns in Area B.

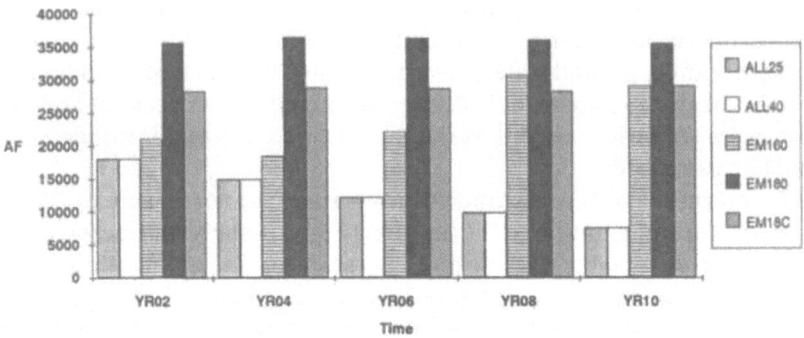

Figure 7. Groundwater Pumped in Area B

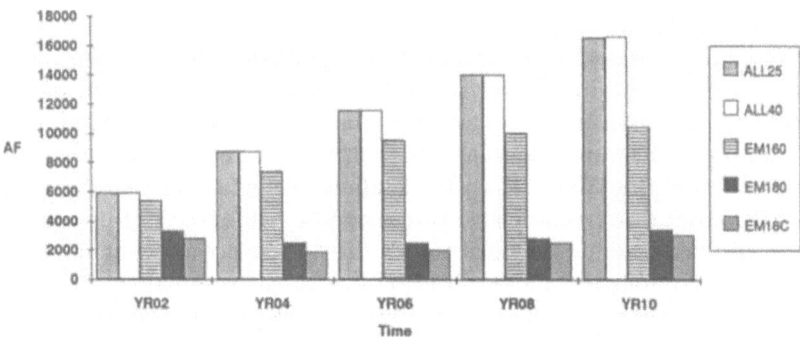

Figure 8. Volume of Upflux in Area B

These two case studies illustrate the potential for water pricing as a means of inducing water conservation and drainage reduction. They also illustrate the importance of understanding the important economic-hydrologic linkages, and the potential danger of oversimplified policy prescriptions.

## CONCLUSIONS

The following are some general observations and conclusions drawn from the WADE modeling effort. These remarks are based on both a comprehensive look at all model results and implications, and on the specific results described above.

A large number of economic and hydrologic relationships influence the WADE model policy analysis. Most of them are intuitive and are not repeated here. But a few important relationships that were not readily appreciated when the modeling effort began have had a large influence on results. Some of these are listed below:

1. Farmers have at least four potential sources of water to manage: Surface supplies, deep ground water, recycled drainwater, and upward movement from shallow ground water. The interaction between these sources and the resulting drain flow can be quite complex and difficult to predict using simplified models.
2. Ground-water pumping from above the confining clay layer can draw down shallow water levels. Therefore, any policy that results in less ground water pumped has some chance of aggravating the drainage problem. For example, planned retirement of irrigated land can increase drain flow if the surface water used on that land is not transferred out of the drainage-problem area. If that surface water is instead allocated to other lands in the drainage-problem area, it might simply replace pumped ground water. As a result, the shallow water table could rise and actually increase drainage volume. Key information needed to evaluate this potential problem includes the amount pumped, the layer from which it is pumped, the aquifer storage coefficient, and the rate of vertical ground-water flow from shallow to deep layers.
3. Farmers do not receive a full allocation of surface water in many areas, notably in the Westlands, Tulare, and Kern subareas. As a result, the shadow price of surface water is higher than the contract price. If farmers are pumping ground water to augment supply, then the cost of pumping provides a good estimate of the shadow price of surface water (ignoring the user cost of ground-water extraction, water quality differences, and other potentially limiting resources). In many areas the surface water price must rise substantially to exceed the shadow price.
4. Source control can and probably should be an important element of any plan. But the key to effective source control is reducing the deep percolation of surface water imported. Simply reducing deep percolation by adjusting irrigation or management techniques may reduce total applied water, but in many areas farmers are augmenting surface supplies with higher-cost ground water. Farmers will use less of their highest-cost source (ground water), and in some areas worsen the drainage problem. To avoid this occurrence, source control policies must either target surface water explicitly (such as tiered water pricing), or be implemented along with a plan to control shallow ground-water levels.

5. Results from analysis of tiered water pricing show that it can be a very effective way of inducing water savings and drainage reduction in areas of abundant surface water. Tiers based on crop ET seem to be more effective in reducing drain flow, while tiers based on volume of water used on all crops seem more effective in reducing applied water. As stated above, moderate water price increases (including tiered pricing) are not as effective in areas of restricted surface supply. In these areas, the price must exceed that of the supplemental source (usually ground water) before important reductions in use occur.
6. The uncertainty in available information about hydrology and economic conditions highlights the need to develop a flexible plan. Flexibility means:

    (a) The plan can be tailored to a particular area.
    (b) The plan does not rely heavily on certain economic or hydrologic-conditions being met.
    (c) The plan can be adjusted over time as more basic information is known, and as the effects of initial implementation are measured.
    (d) The plan does not lock in an expensive solution for many decades that may become obsolete in the future.

## NOTES

[1] See also Horner and Dudek (1980).

[2] Others believe, however, that under realistic growing conditions even cotton shows yield reduction at much lower concentrations. Westlands Water District (1987) presents data showing cotton yield begins to decline at an EC of 3.

[3] Many of these pricing scenarios were developed in discussions with Bruce Driver and Drainage Program staff.

[4] If the district were able to sell the water saved at, say, $50 per acre-foot, they would generate about $100,000 more that could be used for further drainage control measures.

## REFERENCES

Arar, A.; Doorenbos, J.; and Thomas, H. G., 1971. Irrigation and Drainage in Relation to Salinity and Waterlogging. Irrigation and Drainage Paper No. 7, Salinity Seminar, Baghdad. Food and Agriculture Organization of the United Nations.

Burt, C. M. and Katen, K., 1988. *Westside Resource Conservation District, 1986/87 Water Conservation and Drainage Reduction Program.* Technical Report to the California Department of Water Resources, Office of Water Conservation.

California Department of Water Resources, 1982. *Hydrologic-Economic Model of the San Joaquin Valley.* DWR Bulletin 214.

CH2M Hill, 1989. *Irrigation System Costs and Performance in the San Joaquin Valley.* Prepared for the San Joaquin Valley Drainage Program.

CH2M Hill, 1988. *On-Farm Irrigation System Hydrological Characterizations for Mathematical Modeling.* San Joaquin Valley Drainage Program, Working Paper No. 1.

Dudek, D. J. and Horner, G. L., 1980. *Integrated Physical-Economic Resource Analysis: A Case Study of the San Joaquin Valley.* Economic Research Service, USDA, Final Report, Research Agreement No. 12-17-06-8-1985-X, Robert S. Kerr Environmental Research Laboratory, U.S. Environmental Protection Agency, Ada, OK., 250 p.

Dudek, D. J. and Horner G. L., 1981. Income Distributional Impacts of Alternative Irrigation Return Flow Policies, *American Journal of Agricultural Economics*, 63(3), pp. 438-446.

Fajardo, D.; McCarl, B. A.; and Thompson, R. L., 1981. A Multicommodity Analysis of Trade Policy Effects: The Case for Nicaraguan Agriculture, *American Journal of Agricultural Economics*, 63(1), pp. 23-31.

Feinerman, E. D.; Yaron, D.; and Belorai, H., 1982. Linear Crop Response Functions to Soil Salinity with a Threshold Salinity Level, *Water Resources Research*, 18, pp. 101-6.

Gates, T. K., 1988. *Optimal Irrigation and Drainage Strategies in Regions with Saline High Water Tables.* Unpublished Ph.D. Dissertation, University of California, Davis.

Gates, T. K. and Grismer, M. E., 1988. Irrigation and Drainage Strategies in Salinity Affected Regions. Working Paper, Department of Water Science and Agricultural Engineering, University of California, Davis.

Grimes, D. W. and Henderson, D. W., 1986. Crop Water Use from a Shallow Water Table. Presented at 1986 Summer Meeting of the American Society of Agricultural Engineers, San Luis Obispo, CA.

Hatchett, S. A.; Quinn, N. W. T.; Horner, G. L.; and Howitt, R. E., 1989. Drainage Economics Model to Evaluate Policy Options for Managing Selenium Contaminated Drainage. Presented at the International Committee on Irrigation and Drainage Pan-American Conference, *Toxic Substances in Agricultural Water Supply and Drainage - An International Environmental Perspective*, Ottawa, Canada.

Horner, G. L.; Moore, C. V; and Howitt, R. E, 1983. Increasing Farm Water Supply by Conservation, *California Agriculture*, 37(11&12), pp. 6-7.

Horner, G. L.; Putler, D.; and Garifo, S., 1985. *The Role of Irrigated Agriculture in a Changing Export Market*. Natural Resource Division, Economic Research Service, U.S. Department of Agriculture, ERS Staff Report AGE 850328.

House, R. M., 1987. *USMP Regional Agricultural Model*. National Economics Division, Economic Research Service, U.S. Department of Agriculture, 46 pp.

Howitt, R. E. and Mean, P., 1983. *A Positive Approach to Microeconomic Programming Models*. Working Paper 83(6), Department of Agricultural Economics, University of California, Davis.

Maas, E. V. and Hoffman, G. J., 1977. Crop Salt Tolerance-Current Assessment, *ASCE Journal of Irrigation and Drainage Engineering*, 103, pp. 115-134.

Miller, T. and Millar, R. H., 1976. *A Prototype Quadratic Programming Model of the U.S. Food and Fiber System*. U.S. Department of Agriculture ERS, CES, and Department of Economics, Colorado State University.

Oamek, G. E., 1989. *Environmental and Economic Impacts of Interstate Water Transfers in the Colorado River Basin*. Center for Agricultural and Rural Development (CARD), Monograph No. 89-M3.

Oamek, G. E. and Schluntz, L., 1989. Sizing Multi-Purpose Reservoirs: A Methodological Approach and Applications. Presented at the WAEA Annual Meetings, Coeur d'Alene, ID.

Quimby, W. and Lueck, D. J., 1988. Analysis of Selected EC Agricultural Policies and Dutch Feed Compositions Using PMP. Presented at the 1988 American Agricultural Economics Association Summer Meeting, Knoxville, TN.

Rhoades, J. D., 1987. *Reuse of Drainage Water for Irrigation: Results of Imperial Valley Study: Hypothesis, Experimental Procedures, and Cropping Results*. U.S. Salinity Laboratory, U.S. Department of Agriculture, ARS, Riverside, CA.

San Joaquin Valley Drainage Program, 1989. *Preliminary Planning Alternatives for Solving Agricultural Drainage and Drainage-Related Problems in the San Joaquin Valley*. Final Report.

U.S. Department of Agriculture River Basin-Watershed Planning Staff, 1973. *The San Joaquin Valley Basin U.S. Department of Agriculture River Basin Study Soil Group Areas, 1972*. Berkeley, CA., 104 p.

Westlands Water District, 1985. *Water Conservation and Management Handbook*.

Westlands Water District, 1987. *Water Conservation and Management Program-Review and Evaluation 1985-1986*.

Williamson, R. E. and Kriz, G. J., 1970. Response of Agricultural Crops to Flooding, Depth of Water Table and Soil Gaseous Composition, *Transactions of the ASAE*, 13, pp. 216-220.

Williamson, A. K.; Prudic, D.; and Swain, L., 1985. *Ground Water Flow in the Central Valley, California*. U.S. Geological Survey Open-File Report 85-345.

# 25 THE USE OF COMPUTABLE GENERAL EQUILIBRIUM MODELS TO ASSESS WATER POLICIES

Peter Berck, Sherman Robinson, and George Goldman,
University of California, Berkeley

### ABSTRACT

This chapter discusses basic issues in project analysis and shows how these issues can be resolved in a computable general equilibrium (CGE) framework. The role of border prices and intersectoral linkages is explored. The CGE framework is compared to less comprehensive frameworks, including benefit-cost analysis, input-output models, multi-market models, and models based on social accounting matrices (SAM's). An illustrative CGE model of the southern portion of the San Joaquin Valley (Valley) is constructed and is used to find the effects of reducing water inputs to agriculture on aggregate Valley gross domestic product (GDP) and on sectoral output, employment, and land use. The model is also used to determine demand curves for water by the southern portion of the Valley, given alternative specifications of production technology.

## INTRODUCTION

Drainage problems and increased urban water demands have led to serious problems for agriculture in parts of the Valley. The drainage problem, in its chronic form, is that the Valley is a net importer of salt. There is extensive literature on methods of reducing this drainage problem while still engaging in agriculture. In this chapter, an alternative solution is discussed: drastic curtailment of agricultural water use. A methodology is presented to evaluate the effects of decreased water use on a region. This methodology is then applied in a preliminary and illustrative way to evaluate the "project" of withholding water from the southern portion of the Valley. The impact of this curtailment policy on regional employment, GDP, crop mix, agricultural value added, and farm income should provide an upper-bound measure of the impact of less drastic policies, such as improving residuals management. The curtailment alternative also serves as a benchmark for evaluating potential government sponsored projects, such as building a master drain.

The economic evaluation of a water-curtailment policy shares all the challenges of evaluating any type of project. The starting point is well stated by Varian (1989): "We start from a simple methodological premise: there is only one correct way to do cost-benefit analysis. First, formulate an economic model that determines the entire list of prices and incomes in an economy. Next, forecast the impact of some proposed change on this list of prices and incomes. Finally, use the utility functions of the individual agents to value the pre- and post-change equilibria. The resulting list of utility changes can then be summarized in various ways and presented to decisionmakers."

In the next section, CGE models are described. These models provide an empirical framework that incorporates "the entire list of prices and incomes in an economy." The next section discusses the relative advantages of different partial and general equilibrium approaches to project analysis. It is followed by a description of a regional CGE model developed for the southern Valley (including Fresno, Tulare, Kings, and Kern Counties). Finally, results are presented from simulation experiments performed with the model to analyze the effects of removing water from the Valley. Given the level of aggregation, difficulties in specifying alternative production technologies, and the nature of the data, the empirical results must be seen as illustrative.

## A TYPICAL CGE MODEL

A CGE model is a general equilibrium model that implements the textbook description of an economy. There are utility-maximizing consumers whose decisions determine the demand for goods and supply of labor. There are profit-maximizing producers whose decisions determine the supply of goods and the demands for primary factors (labor, capital, and land) and intermediate inputs. There is international trade. There is a government which collects taxes and tariffs; may set exchange rates; and provides transfers, subsidies, and services. Finally, there are market-clearing conditions specifying supply-demand balance, which will determine equilibrium prices. The model is a "general equilibrium" because all domestic supplies, demands, prices, and incomes are determined simultaneously within the model. It is "computable" because the model solves empirically for all endogenous variables in a highly nonlinear system of simultaneous equations.

Typically, CGE models have many sectors and factors of production. Equilibrium requires that, for each sector, supply (production) equals demand (consumption, investment, government, and exports) at market-clearing prices. The models often specify many household types, stratified by occupation or income level. Household expenditure on goods is specified as a function of household disposable income (the household's share of labor and distributed capital income less net taxes) and prices. For goods that are both traded and locally produced, the domestic market price is a function of international prices plus tariffs and producer prices. Sectoral output and demand for intermediates and labor is taken as a function of the capital stock and producer prices. All the usual neoclassical rules hold. In each sector, price equals marginal cost and wages equal the marginal value product of labor. Exchange rates and international trade flows are also usually taken to be endogenous.

The distribution of income is modeled, so profits and wages are first distributed to institutions (such as enterprises and labor) and then to households. This two-step distribution allows the inclusion of policy instruments (such as corporate and payroll taxes) and enterprise decisions about retained earnings, which affect the amount of factor income that actually ends up in households.

Changes in policy alter demand through changes in both income and prices. The wide scope of the model makes it especially useful for evaluating projects that have broad effects, changing incomes in many sectors through intersectoral linkages. When an investment project is large, generating many ripples in the economy, a general equilibrium framework is the appropriate tool of analysis (Bell and Devarajan, 1987).

Multimarket equilibrium models differ from CGE models by including fewer linkages (Braverman and Hammer, 1988). In particular, final demand for goods does not depend upon endogenous household income. Sectors that are deemed unimportant to the question at hand are also not modeled. Most econometric models fall into this category. The advantage of this approach is that the analyst can pay more attention to the remaining parts of the model, focusing on the included sectors, at the cost of introducing some bias and inaccuracy by omitting sectors and feedbacks from changes in incomes. While many projects, particularly regional projects, can be well analyzed with a multimarket model, there are also many examples of policies for which feedbacks through the omitted links are very important. For example, analyzing the impact on a developing country of pursuing an agriculture-led development strategy, requires an economywide framework. Increased agricultural incomes will result in increased demand for goods produced in the urban sector, and therefore to increased urban incomes, with further indirect effects back to the agriculture sector as well. In this case, a model in which the final demand for urban goods is independent of agricultural development will miss a crucial linkage through which the policy scenario affects economic performance (Adelman, 1984).

There is also a tradition of input-output multiplier models which focus on intermediate input flows. The "semi-input-output" model improves on the standard open Leontief model by accounting for traded goods, and has been used to evaluate large regional agricultural projects.[1] This multiplier approach has been extended to include a wider view of economic linkages through the use of a Social Accounting Matrix or SAM. The SAM extends the input-output accounts to include income flows among all agents in the economy, and also provides the data base for CGE models. A SAM is discussed below for the Valley.

In summary, the CGE framework is the most general framework in which to conduct policy analysis. Input-output analysis, SAM analysis, and multimarket analysis can all be seen as special cases of a general equilibrium model. While the CGE framework is the most general, applied CGE models tend to be more highly aggregated than models in other frameworks. A multimarket model is best seen as a subset of a CGE model, focusing on a subset of sectors and linkages. The multimarket model solves for prices in its subset of sectors and assumes all other prices are given. In particular, it assumes a fixed exchange rate. A SAM is a data framework that provides a snapshot of an economy and can easily be turned into a linear, demand-driven, multiplier model. An input-output model focuses only on intersectoral linkages and is a subset of a SAM model. Input-output and SAM-multiplier models all assume fixed prices. The price of increasing generality and economywide coverage is greatly increased demand for data and/or diminished precision.

## PROJECT EVALUATION

There is extensive literature on how to evaluate projects without using a full general equilibrium model. This section will discuss some of the general issues in project evaluation and point out how general equilibrium modeling contributes to some, but not all, of their resolution.

### Benefits

Projects are considered to be good insofar as they benefit people. The concrete expression of this principle is a "social welfare" criterion. Specifying a social welfare function, one can proceed directly to evaluating projects by maximizing this function subject to the rules of the underlying economy (for example, represented by a CGE model). The problem is that this approach requires both an explicit social welfare function and a CGE model.

The standard and familiar rule of cost-benefit analysis is to accept only those projects that have benefits in excess of their costs, to whomever those benefits may accrue. This rule reflects a particular social welfare function: the marginal social benefit of a dollar is assumed the same for all citizens. Choosing such a rule does not negate the need for a model of the economy.

Manuals of project evaluation suggest using indirect approaches for specifying the welfare criteria and the way the economy operates. Little and Mirrlees (LM, 1974), among others, suggest rules that give the same result as explicit welfare maximization in special situations and that may be more convenient to use. For example, a number of writers have argued that, when the Government has sufficient policy instruments available to channel uncommitted Government revenue to the most socially needy household or most worthy use, then uncommitted Government income is also the appropriate maximand for project analysis and the specification of a social welfare function can be avoided.

Most American water projects are undertaken because private individuals will benefit. Water projects make farmers better off and taxpayers worse off. To apply the LM methods to these projects requires evaluation of the social value of a dollar given to a farmer relative to an uncommitted dollar in Government hands. This evaluation is no easier than the direct problem of maximizing a specified social welfare function. Thus, the project manuals have no advantage over direct methods insofar as specifying benefits is concerned.

## Large and Small Projects

Small projects are those that do not change many existing prices. To evaluate a small project, one needs information on the prices (or shadow prices) of its outputs and inputs at the existing equilibrium. If the project covers costs at these prices, then it should be built. Large projects change many prices. The Aswan high dam and the California Central Valley Project (CVP) were projects big enough so that one could reasonably expect that the prices of cotton, fruit, and vegetables, as well as other prices, would change after project construction. Preproject pricing does not (generally) solve the question about building a particular large project. The technical reason why the pricing rule may not work is that the project is taken to be a discrete alternative, which cannot be built on a smaller scale. An explicit model, such as a CGE model which solves for market prices endogenously, can resolve such problems.

## Alternatives

The most difficult problem in project evaluation is the specification of alternatives. It is a problem common to all methods of project analysis. For example, the benefit-cost rule sets excess of benefits over costs as a necessary condition for the funding of a project. The rule does not guarantee that the project maximizes the difference between benefits and costs. There may well be another project (usually a smaller project when suggested by environmentalists) that has higher net benefits. In the case of a drainage project, shutting down production on some of the land would be an alternative, as would different cleanup processes. Finding these alternative projects and evaluating them is a major challenge.

In addition to physical alternatives to investment projects, there are economic alternatives. LM particularly emphasize the alternative of trade. Their border pricing rules implicitly evaluate every project against the alternative of international trade--the "make or buy" decision. When interregional or international trade is incorporated into a CGE model, the model correctly includes trade as an alternative to every project. Diamond and Mirrlees (1971) show that, in the presence of optimal commodity taxes, there are no distributional benefits of projects. Thus, taxes that separate producer prices from consumer prices are also important parts of any package of alternatives. These instruments are easy to incorporate into a CGE model--although it is hard to argue that current commodity taxes are optimal in the American economy, or in any other economy.

## Prices

Project evaluation in a developing country always runs into the problem that domestic observed prices are not reliable indicators of value. In the case of water projects, the United States is like a developing country. Observed agricultural prices cannot be trusted as indicators of social values because they are distorted by pervasive Government policies such as the loan program, deficiency payments, and export subsidies. Similarly, water prices do not reflect marginal social values because they are largely determined by Government project rules rather than the operation of free markets.

The standard solution to these problems is to choose a consistent set of prices that either equal or are based on international (border) prices. Bell and Devarajan (1987) provide the exact correspondences between the LM rules and the implementation of those rules in the CGE framework. The problem with the LM rules, in practice, is that important factors (particularly, labor) are not traded in international markets. Thus, the most difficult job for the analyst is to figure out a wage that is commensurate with the border prices that the analyst uses for traded goods. It turns out that the solution of the CGE model written in a particular form will give these hard-to-calculate prices. Thus, a CGE model, which takes border prices for tradeable goods as given, generates a set of solution prices for nontraded goods which represent their LM shadow prices.

For the Valley, border prices are crucial. Taken from the view of the United States, the border price for cotton is the world market price, which is about 72 cents per pound. A project evaluation (done from the point of view of the United States) should not, under the LM rules, include cotton deficiency payments, which are about 10.5 cents per pound. From the point of view of California, however, the appropriate border is that with the rest of the United States. If policymakers are trying to maximize California's welfare, they should certainly take the price in the United States, which is the U.S. price plus the deficiency payment (about 82 cents per pound), as the appropriate "border" price.

Another example of the border pricing rule is water. The Valley is an importer of water. The appropriate shadow price is the value of the next unit sold to the highest bidder. The East Bay Municipal Utilities District, for instance, is in the process of developing a high-cost, high-quality, water supply costing approximately $1,000 per acre-foot. Marin and Santa Barbara Counties are both giving serious consideration to building desalinization plants, yielding water at $2,000 per acre-foot. Under these circumstances, the border price is a great deal more than the $60-$70 that the water is worth if used in the Valley or the $20-$30 that is charged by Federal water projects. Both in the CGE model and following the LM project evaluation rules, water should be priced at its value in the next best alternative use.

Recent survey evidence puts the average returns (market revenues less variable costs) to growing cotton in Kern County at about $250 per acre.[2] Overhead, insurance, and such could add as much as $80 per acre for a net return per acre of $170. A border price for water of $50 per acre-foot greater than the current cost of water would make cotton farming unprofitable. Including a deficiency payment of 10 cents per pound (about $130 per acre, given average yields) would make cotton wildly profitable. In sum, evaluating a water project in the southern Valley is very dependent upon the border pricing rules.

## Secondary Benefits

No area of project evaluation is more controversial than the evaluation of secondary benefits--benefits accruing to sectors purchasing from or selling to the project sector. For example, a water project raises agricultural output, which induces increased demand for agricultural inputs, such as fertilizer. Numerous practitioners double count benefits or ignore costs. For example, by counting the additional fertilizer production as a project benefit (and not counting the natural gas used to make the fertilizer as a cost), it is possible to make almost any project seem welfare increasing. An equally egregious practice is to count the increased agricultural processing activities as "stemming" benefits and then ignore their costs.

Both the CGE methodology and the LM methods provide ways of consistently and correctly accounting for changes in the economy in sectors other than those directly affected by the project. LM concentrate on finding proper prices, taking linkage effects into account, while CGE models directly compute the effects of the project on all linked sectors. Both these methodologies correctly account for project benefits and costs in linked sectors.

## A REGIONAL CGE MODEL

The regional model used here is a special type of CGE model reflecting the smallness of the region at hand. For a "small" region, most sectoral exports face perfectly elastic demands at fixed prices, and the domestic price will be set by the export price. In the case of the Valley, it is reasonable to treat the prices of products such as grain and cotton as fixed, while a few exports (such as the fruit and nut sector) might be viewed as having an external downward-sloping demand curve. At the level of aggregation used in the model, there are exports in all sectors, so there are no pure non-traded goods.

The treatment of imports and exports also differs from most economywide models. Here, all imports are treated as "noncomparable," which means that they are not produced in the region but are consumed or used as intermediate goods. In the model, sectoral import demand, both as intermediate inputs and final demands, are given by fixed coefficients. Exports, on the other hand, are determined so as to clear the product markets in the region, given the fixed prices. For each sector, supply and demand is calculated given the fixed price. Then, net exports are determined residually to balance supply and demand. While net exports could be negative (i.e., becoming net imports), all sectors in the Valley are large net exporters in the base data.

In the Valley model, capital is assumed to be sectorally fixed and immobile. On the other hand, it is assumed that the aggregate supply of labor in the Valley is fixed and the model will solve for the market-clearing wage and the sectoral allocation of labor. Alternatively, the average wage could have been specified as fixed, allowing the model to solve for net labor migration into the Valley. In the event, the difference between these two specifications was irrelevant, given the policy changes modeled. Even the most extreme experiments yielded virtually no change in the average wage in the Valley, and hence, no change in the aggregate demand for labor.

In the regional model, the exchange rate is, by definition, fixed and set to one. This rules out problems faced in a country model of specifying how the foreign-exchange market clears. In a regional model, the capital account always adjusts to offset any balance of trade that the model yields in equilibrium. It is traditional, in the LM rules, to evaluate projects in terms of uncommitted government expenditure in local currency at international prices. Obviously, in terms of the United States, U.S. dollars will do just fine.

Hanemann et al. (1987) focuses on constraints of soil type. Such constraints have been crudely modeled here, as discussed below. Similarly, the quota constraints on dairy are modeled. The major source of data is a regional SAM for the aggregate of counties provided by the U.S. Forest Service.[3] Data on water coefficients by crop were provided by DWR. Production parameters are estimated using 1976 data on water usage, 1982 acreage data, and intermediate use from 1977 input-output data updated to 1982.

The model has 14 sectors, with 6 agricultural sectors (dairy, grazing/livestock, cotton, grains, fruits/vegetables and nuts, and other agriculture), 2 processing sectors (1 for dairy and 1 for all other agriculture), 1 other manufacturing sector, a mining sector, and 4 service sectors (trade, freight, banking, and other services).

There are five factors of production in our model: land, water, labor, capital, and intermediate inputs. Land and water are only used in the agricultural sectors. Land is taken to be in fixed aggregate supply, and there are three different types of land. Land currently growing cotton is assumed to be able to be converted to field crops or grazing, but not vice versa. Similarly, land currently growing fruits and nuts can be moved to field crops or to grazing, but not vice versa. Thus, grazing is the residual use, with land able to be converted from crops to grazing, but not vice versa. This specification captures the notion that there is a hierarchy of land qualities. Good land can be converted to "less-good" uses, but not vice versa.

Water is taken as having a fixed aggregate supply, and the experiment is simply to decrease this supply. Water coefficients are assumed fixed by sector and, hence, by crop. However, water is assumed to be freely mobile across sectors. That is, one can convert land from one crop to another (according to the hierarchy) and can also convert the water use at the same time.

Laborers are modeled as mobile between sectors within the Valley and immobile between the Valley and the rest of the world. Capital is taken as sectorally fixed. Thus, the model will solve for a single average wage which clears the labor market but will yield sectorally differentiated profit rates.

## Production Technology

Figure 1 shows the production technology for the agricultural sectors. Sectoral production is given by a nested multilevel function. Two variants of this function are specified: (1) A "high elasticity" variant and (2) a "low elasticity" variant. In both variants, domestic and imported intermediate inputs are demanded according to fixed input-output coefficients and land consists of a combination of acreage and water, with the water to land ratio given by a fixed coefficient (which differs across agricultural sectors). In the high elasticity variant, which is shown in figure 1, real value added is a Cobb-Douglas aggregation of land, labor, and capital. In the low elasticity variant, capital and land are used in fixed proportions, and labor is combined with the capital-land aggregate according to a Cobb-Douglas function.[4] In effect, capital is moved to the bottom level in figure 1.

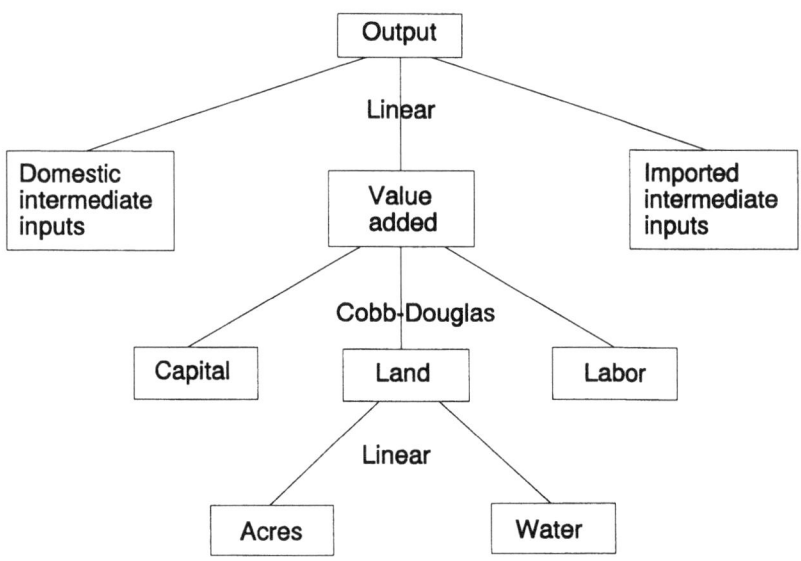

Figure 1. Production technology, high elasticity variant.

This specification of technology severely limits substitution possibilities in sectoral production. Water, land, and (in the low elasticity variant) capital are used with fixed coefficients, and there is no direct substitution between land and intermediate inputs such as fertilizer and pesticides. There are, however, substitution possibilities between land, labor, and (in the high elasticity variant) capital, with a substitution elasticity of one.[5] Thus, in response to changes in water availability and relative factor prices, yields can be changed but only by changing sectoral employment. In sum, the model probably understates substitution possibilities in sectoral production, certainly in the low-elasticity variant, although it allows adjustment by changing the cropping pattern through changes in the sectoral structure of land use and production. Given the specification of technology, one would expect the model to yield results that provide an upper bound on the impact of changes in water availability on the agricultural sectors.

The nonagricultural sectors do not use land. Thus, their technology is described by the first two-aggregation levels in figure 1, with land omitted from the value-added aggregation. The treatment of the nonagricultural sectors follows that in standard CGE models, given the assumption that all imports are noncompetitive.

## Solution Techniques

Solving CGE models numerically involves finding a general equilibrium solution with supply-demand balance in all markets.[6] In a standard model, these supply and demand equations are all written out explicitly, reflecting first-order conditions for maximization of profits by producers and utility by consumers. In the model used here, the specification of the technology for the agricultural sectors involves inequality constraints, so it is not possible to write out the factor demand equations explicitly. Instead, the explicit programming problem is written out for maximizing proprietor income (profits plus return to land) for the agricultural sectors and solved as a subproblem, thus determining product supply and factor demands for the agricultural sectors numerically. One advantage of this procedure is that the model generates the shadow price of water to the agricultural sectors, enabling the demand curve to be determined for water by running a number of experiments varying the aggregate supply of water.

## A Social Accounting Matrix

The primary data for the illustrative model comes from a SAM for Fresno, Tulare, Kings, and Kern Counties. The SAM used here was produced by the U.S. Forest Service's IMPLAN system.[7] The SAM shows the flow of income and expenditure in the four-county region.[8] Table 1 is an aggregate version of the SAM. The model distinguishes 14 sectors but, for presentation purposes, table 1 shows only 2: agriculture and nonagriculture.

The entries in the SAM are 1982 production, factor payments, transfer, trade, and final demand in dollar flows. The first sector in the aggregate SAM is agriculture. The entries down the column indicate expenditures by the

Table 1. Social Accounting Matrix for the Lower San Joaquin Valley, 1982.

| $ Millions | | | | | Expenditures | | | | | |
|---|---|---|---|---|---|---|---|---|---|---|
| | | 1 | 2 | 3 | 4 | 5 | 6 | 7 | 8 | 9 | |
| Receipts | | Agric. | Non-Ag. | Wages | Profits/ Rent/Tax | Enterprises | Households | Government | Investment | Rest of World | Total |
| 1 Agricultural | | 401 | 577 | | | | 86 | 14 | 19 | 4,009 | 5,106 |
| 2 Non-Agricultural | | 1,303 | 6,824 | | | | 6,721 | 1,637 | 1,084 | 12,117 | 29,686 |
| 3 Wages | | 645 | 8,575 | | | | | | | | 9,220 |
| 4 Profits/Rent/Tax | | 1,463 | 6,084 | | | | | | | | 7,547 |
| 5 Enterprises | | | | | 4,269 | | | | | | 4,269 |
| 6 Households | | | | 7,791 | 1,369 | 3,182 | | 2,035 | | | 13,008 |
| 7 Government | | | | 1,200 | 1,909 | 932 | 2,174 | | | | 5,675 |
| 8 Investment | | | | | | 155 | -90 | 846 | | -14 | 2,806 |
| 9 Rest of World | | 1,294 | 7,626 | 229 | | | 4,117 | 1,143 | 1,703 | | 16,112 |
| Total | | 5,106 | 29,686 | 9,220 | 7,547 | 4,269 | 13,008 | 5,675 | 2,806 | 16,112 | |

agricultural sector. The first entry is intermediate purchases by agriculture of agricultural products as intermediate inputs. The second entry is purchases of nonagricultural intermediates, followed by payments to factors of production (e.g., wages and profits). In the SAM, producing sectors make no direct payments to households. Finally, there are entries for purchase of intermediate inputs from the "rest of the world," which represents imports from the rest of the United States and other countries. The first row of the SAM is sales of the agricultural sector and describes the market for agricultural goods. Demand categories include intermediates, consumption by households, Government and investment demand, and exports to the rest of the world. The corresponding row and column sums must be equal, since we require that sales equal disbursements in every account.

The SAM treats sectors like agriculture and institutions like households symmetrically. The column for households shows their purchase of goods (domestic and imported), savings, and payments of taxes. The row for households shows that household income comes from wages, distributed profits, and transfers. The SAM captures the entire flow of funds in the Valley economy.

The SAM gives a good picture of the Valley economy. Agriculture, while important, provides only 13 percent of total value added in the Valley. Agricultural purchases of nonagricultural goods produced in the Valley are only 8 percent of Valley value added, so the "backward linkages" from agriculture through intermediate inputs produced in the Valley are not very large. Their major links are to the non-Valley economy, with intermediate imports (both agricultural and nonagricultural) equaling 53 percent of value added and exports equaling 46 percent of total sales. This underlying structure is captured in the model and largely drives the empirical results described below.

Data from the SAM provide many of the parameter estimates of the CGE model. On the production side, all the linear coefficients are taken from the SAM, while the cost shares are used as estimates of the Cobb-Douglas parameters. The demand and distributional parameters are also taken directly from the SAM. Thus the base-year solution of the CGE model exactly replicates the SAM.

## RESULTS

Starting from the base run, the simulation experiments are designed to explore the impact of removing water from agricultural use. Two sets of five experiments were run. In each set, the experiments remove water in 10 percent increments, with the last experiment forcing agriculture to use half the base-year water allocation. The two sets of experiments differ in the factor substitution elasticities assumed in the four agricultural sectors.

Figures 2 and 3 indicate the changes in water use by major crops under different assumptions about the elasticity of substitution amongst factors of production. In the high elasticity case, where factors are relatively more substitutable, water is first removed from cotton and other agriculture, and then from grains. As the cut in water use approaches 50 percent, grain acreage nearly disappears. In the low elasticity case, grains decline first, followed by other agriculture and cotton. Grain acreage nearly disappears when the cut in water reaches 30 percent. Neither dairy nor fruits and vegetables are affected at all in either case. The changes in the high elasticity case are much more gradual, with smoother changes in cropping mix than in the low elasticity case.

Agricultural sectoral production results are shown in figures 4 and 5. They closely follow the results for water. High value agriculture (such as fruits and vegetables and dairy) are not cut back as water is removed. There is increasing livestock output as irrigated crop land is diverted to dry land pasture. The decline in the output of grains actually leads to the region becoming a grain importer in the extreme experiments.

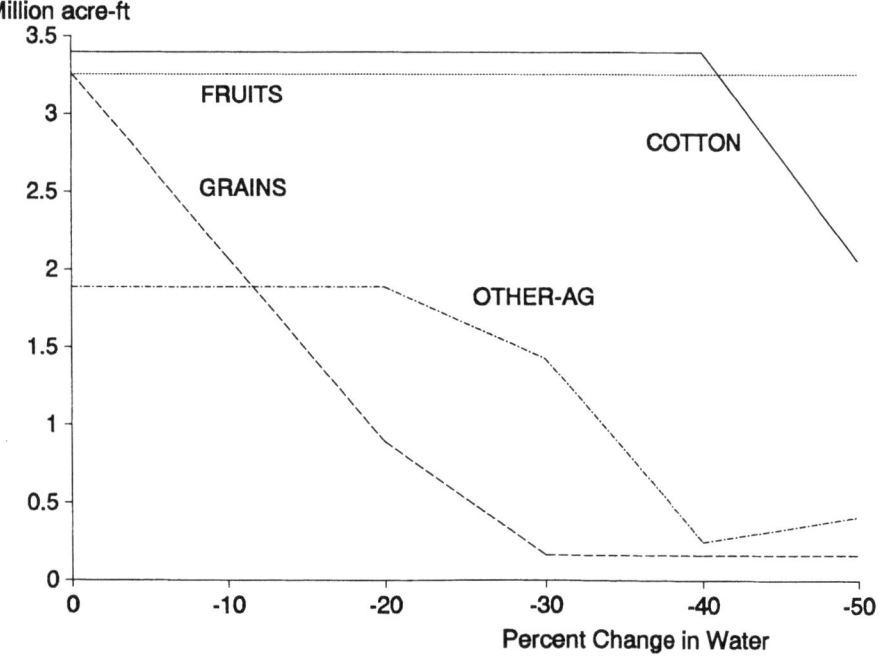

Figure 2. Water use, low substitution elasticities.

# COMPUTABLE GENERAL EQUILIBRIUM MODELS 503

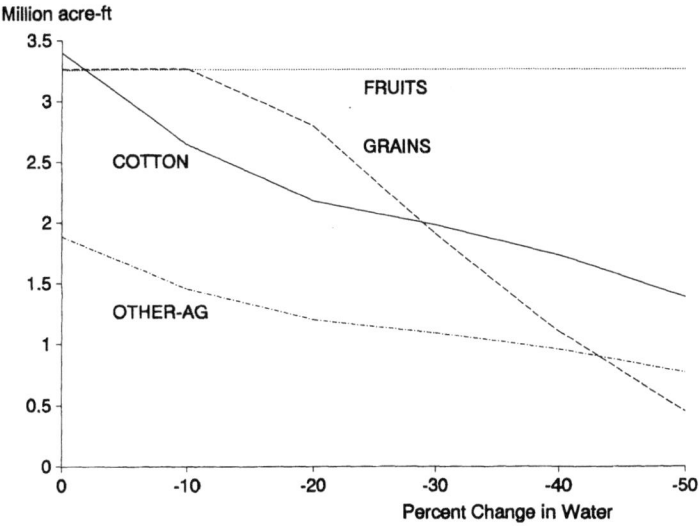

Figure 3. Water use, high substitution elasticities.

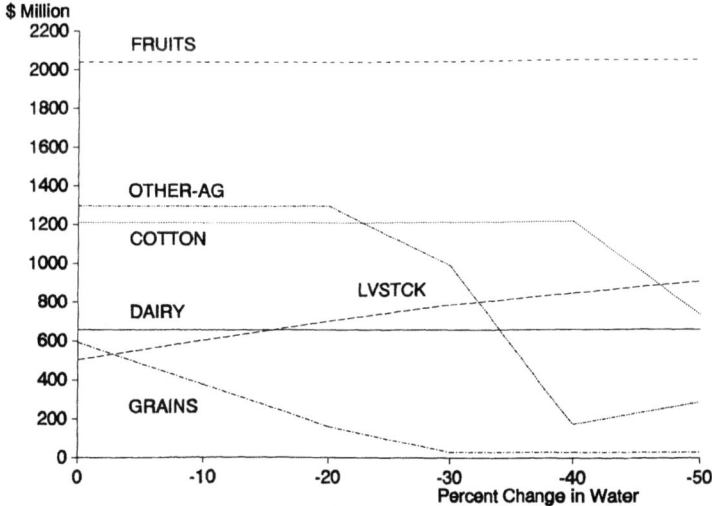

Figure 4. Output, low substitution elasticities.

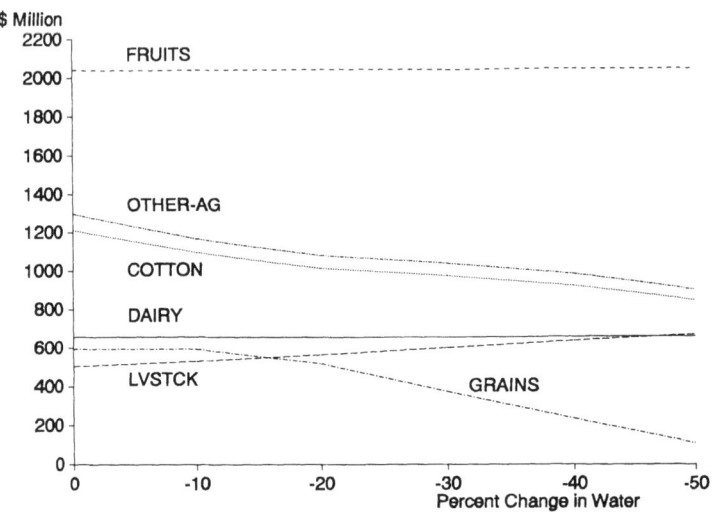

Figure 5. Output, high substitution elasticities.

The effect of water restrictions on Valley GDP, agricultural value added, and returns-to-agricultural proprietors are given in table 2. In the most extreme case, with a 50-percent cut in water and low elasticities of factor substitution, agricultural value added falls by $758 million and proprietor income falls by $401 million. Valley GDP, however, only falls by $305 million, which represents 3 percent of initial Valley GDP.

Agricultural value added includes the value of agricultural labor. When the agricultural sectors contract, labor is released to work in other sectors, thus ameliorating the impact of the water reductions on Valley GDP. The difference between the $758 million loss in agricultural value added and the $305 million loss in GDP equals $453 million, which represents the earnings of the resources shifted out of agriculture. Most of this offset is accounted for by the transfer of 22,000 laborers into other sectors (not tabulated). The losses are less extreme in the high elasticities of substitution experiments.

In the experiments, landowners are not compensated for their lost water. In the most extreme case discussed above (low elasticities, 50 percent cut in water), a payment of $67 per acre-foot of water removed per year would leave proprietor income unchanged from its base value.

The marginal value of water to proprietors (the shadow price generated by the model solution) reflects the demand for water by the agricultural sector as a whole. Plotting these marginal values against total water usage represents the demand curve for water by the agricultural sector. Figure 6 shows these demand curves for the low and high substitution elasticities cases. As expected, the demand curve is much steeper in the low substitution case. In the extreme case, the competitive price of water would rise to $88 per acre-foot, from a value of $51 in the base. In both cases, the price elasticity of demand rises above one after a 30- to 40-percent cut in water usage.

Table 2. Aggregate results.

| | | *Percent Change in Water* | | | | |
|---|---|---|---|---|---|---|
| | *Base* | *-10%* | *-20%* | *-30%* | *-40%* | *-50%* |
| *Low Elastiticities* | | | | | | |
| Valley GDP | 9,803 | 9,755 | 9,706 | 9,649 | 9,577 | 9,498 |
| Agricultural value added | 2,538 | 2,475 | 2,413 | 2,248 | 1,910 | 1,78 |
| Proprietor income | 1,515 | 1,454 | 1,394 | 1,319 | 1,218 | 1,11 |
| *High Elasticities* | | | | | | |
| Valley GDP | 9,803 | 9,770 | 9,729 | 9,685 | 9,638 | 9,586 |
| Agricultural value added | 2,538 | 2,442 | 2,353 | 2,278 | 2,192 | 2,088 |
| Proprietor income | 1,515 | 1,470 | 1,418 | 1,363 | 1,303 | 1,237 |

Note: All figures are in millions of 1982 dollars.

Table 3 presents results for multipliers with respect to changes in water usage. In the high elasticity case, the GDP multiplier for the first 10-percent cut in water usage is $28 per acre-foot. The corresponding multipliers for agricultural value added and proprietor income are $81 and $38 per acre-foot. The low elasticity multipliers are uniformly higher for proprietor income, as one would expect. The less able farmers are to adjust farming techniques, the more the withdrawal of water hurts them. There is no necessary relationship for the other multipliers between the low and high elasticity cases, and they, in fact, vary widely.

The labor multipliers show particularly wide variation, depending on the nature of the changes in cropping patterns as water is withdrawn. The largest is 13,000 workers withdrawn per million acre-feet of water withdrawn, which occurs in the low elasticity case when other agriculture is affected (see figure 6). All the labor multipliers in the high-elasticity case, and three of the five multipliers in the low-elasticity case, are under 3,000 workers per million acre-feet withdrawn.

Table 3. Water multipliers.

| | Percent Change in Water | | | | |
|---|---|---|---|---|---|
| | -10% | -20% | -30% | -40% | -50% |
| *Low elasticities* | | | | | |
| Dollars per acre-foot | | | | | |
| Valley GDP | 41 | 41 | 48 | 61 | 67 |
| Agricultural value added | 53 | 53 | 139 | 286 | 110 |
| Proprietor income | 51 | 51 | 63 | 85 | 88 |
| Workers per million acre-ft | | | | | |
| Agricultural labor | 118 | 118 | 4,684 | 12,509 | 1,387 |
| *High elasticities* | | | | | |
| Dollars per acre-foot | | | | | |
| Valley GDP | 28 | 34 | 38 | 40 | 44 |
| Agricultural value added | 81 | 75 | 64 | 72 | 88 |
| Proprietor income | 38 | 44 | 47 | 51 | 56 |
| Workers per million acre-ft | | | | | |
| Agricultural labor | 2,676 | 1,934 | 1,036 | 1,355 | 1,982 |

*Notes:* Values are dollars (or workers) lost per acre-foot (or million acre-feet) of water removed from agricultural use. Values calculated from successive 10 percent reductions in water use.

• • • • • • • • • • • • • • • • • • • • • • • • • • • • • • • • • • • • • • • • • •

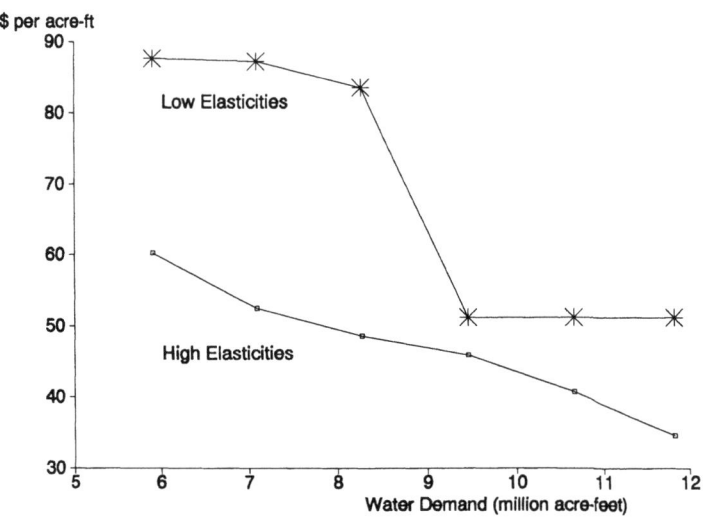

Figure 6. Demand curves for water, low and high substitution elasticities.

## CONCLUSIONS

This chapter has laid out a methodology for evaluating the economic impact of withdrawing water from agricultural use. The model presented is fairly aggregated, highly stylized, and is designed to illustrate the methodology. Two versions of the model were presented with production specifications that probably bracket the actual substitutability of labor and capital for water. Both versions are extreme in that they assume no substitution possibilities between other intermediate inputs (such as fertilizers and pesticides) and water. In the end, we believe that the aggregate results will prove to be robust and that the results from the two models will probably bracket those from a more detailed model.

These experiments indicate that removing water from the southern Valley results in a rapid decline in cotton and/or grain acreage and in an increase in acreage devoted to livestock. Coincident with this acreage shift is a decrease in Valley GDP, employment, and agricultural income. These decreases in macroeconomic indicators are much less pronounced than the acreage shift, because the released resources find alternative employment. Given that the crops withdrawn have relatively low labor intensities, the net effect of water withdrawal on agricultural employment is small. For example, the effect of withdrawing 20 percent of the water supply leads to the displacement of only 5,000 agricultural workers.

Similarly, the net effect of withdrawing 20 percent of the water supply on economic activity in the Valley as a whole is small, although it decreases proprietor incomes in agriculture by around $100-120 million. With a market in water rights, payments for water would offset these income losses. Even in the most extreme case, a payment of $67 per acre-foot of water withdrawn would compensate proprietors for their loss of income.

## ACKNOWLEDGEMENTS

The authors wish to acknowledge the input-output model and social accounting matrix provided by the U.S. Forst Service's IMPLAN system, which was used in the demonstration of the methodology.

## NOTES

[1] See, for example, Bell and Devarajan (1985), who use a semi-input-output model to analyze the impact of a large irrigation project in the Muda valley region of Malaysia.
[2] Personal communication from Richard Howitt.

[3]Alward et al. (1989) describes the method for constructing regional SAM's.

[4]In the low elasticity variant, sectoral capital in the agricultural sectors varies with land use. In the high elasticity variant, sectoral capital stocks are fixed.

[5]In the high elasticity variant, even though there are substitution possibilities between land and capital, capital is sectorally fixed. The responsiveness of output to changes in land use, however, differ between the two variants.

[6]For a survey of solution techniques used in applied models, see Ginsburgh and Waelbroeck (1981) and Dervis, de Melo, and Robinson (1982). We use a software package called General Algebraic Modelling System (GAMS), which is described in Brooke, Kendrick, and Meeraus (1988).

[7]See Alward et al. (1989). The data start from a 1977 input-output table, updated to 1982.

[8]Pyatt and Round (1985) provide a complete description of the SAM methodology.

## REFERENCES

Adelman, I., 1984. Beyond Export Led Growth, *World Development*, 12(9), pp. 937-950.

Alward, G.; Siverts, E.; Olson, D.; Wagner, J.; Senf, D.; and Lindall, S., 1989. *Micro Implan: Software Manual*. Regents of the University of Minnesota, Minneappolis/St.Paul.

Bell, C. and Devarajan, S., 1987. Intertemporally Consistent Shadow Prices in an Open Economy, Estimates for Cyprus, *Journal of Public Economics*, 32, pp. 263-285.

Bell, C. and Devarajan, S., 1985. Social Cost-Benefit Analysis in a Semi-Input-Output Framework: An Application to the Muda Irrigation Project. In: Pyatt, G. and Round, J. I. (Eds.), *Social Accounting Matrices: A Basis for Planning*, The World Bank, Washington, DC.

Braverman, A. and Hammer, J., 1988. Computer Models for Agricultural Policy Analysis, *Finance and Development*, 25(2).

Brooke A.; Kendrick, D.; and Meeraus, A., 1988. *GAMS: A User's Guide*. The Scientific Press, Redwood City, CA.

Dervis, K.; de Melo, J.; and Robinson, S., 1982. *General Equilibrium Models for Development Policy*. Cambridge University Press, New York, NY.

Diamond, P. and Mirrlees, J., 1971. Optimal Taxation and Public Production: I. Production Efficiency, *American Economic Review*, 61(1), pp. 8-27.

Goreux, L. M. and Manne, A. S. (Eds.), 1973. *Multi-Level Planning: Case Studies in Mexico*. North-Holland, Amsterdam.

Ginsburgh, V. and Waelbroeck, J., 1981. *Activity Analysis and General Equilibrium Modelling*. North-Holland, Amsterdam.

Hanemann, M.; Lichtenberg, E.; Zilberman, D.; Chapman, D.; Dixon, L.; Ellis, G.; and Hukkinen, J., 1987. *Economic Implications of Proposed Water*

*Quality Objectives for the San Joaquin River Basin*. Report to the California State Water Resources Control Board.

Little, I. M. D. and Mirrlees, J. A., 1974. *Project Appraisal and Planning for Developing Countries*. Basic Books, New York, NY.

Pyatt, G. and Round, J. I. (Eds.), 1985. *Social Accounting Matrices: A Basis for Planning*. The World Bank, Washington, DC.

Varian, H. R., 1989. Measuring the Deadweight Costs of DUP and Rent Seeking Activities, *Economics and Politics*, 1(1), pp. 81-95.

# 26 ANALYSES OF IRRIGATION AND DRAINAGE PROBLEMS: INPUT-OUTPUT AND ECONOMETRIC MODELS

W. Douglas Morgan and Lloyd J. Mercer,
University of California, Santa Barbara

## ABSTRACT

Policymakers have an interest in regional economic models which can provide answers to policy related questions. The recognition that one region has effects on adjacent regions highlights the special problems of regional analysis. This requires recognition of economic relationships between the region under study and the remainder of the economy. The two major techniques of modeling regions are input-output and econometric modeling.

Input-output models take no account of time and are static. Input-output models describe the regional flow of goods and services in a double entry accounting system. Technical coefficients by sector depict the sector's production relationships. The "ripple effects" of local spending by each sector produces the sector's multiplier. The 1977 United States National input-output table is often used as a starting point for input-output models.

A regional econometric model is a set of equations based on microeconomic and macroeconomic theory describing the economic structure of a regional economy. The parameters of the equations are estimated econometrically, usually by time series regression equations, as distincet from input-output models where parameters are based on single-point observations.

A survey of applications of regional modeling concerning water problems to the San Joaquin Valley (Valley) includes the California Department of Water Resources (DWR) Bulletin 210 which provides crop multipliers, an input-output study on the effects of salt buildup in the Sacramento-San Joaquin Delta, an input-output study on reallocating water in the Westlands Water District, and an econometric model of the impact of increased surface water prices in Fresno and Kern Counties. These studies show the kind of information obtainable from regional modeling to deal with policy issues.

## INTRODUCTION

Regional modeling of economic change is still in its infancy. Regional economic models are, in general, built for one of four main reasons: Pure economic science; economic forecasting; Government revenue forecasting; and policy analysis including impact analysis. The prime users of economic models are policymakers. Policymakers typically would like answers to the question: What are the effects of specific changes within a region on its economy and the economics of adjacent regions? Alternatively, if there is an economic change outside the region, how is the regional economy under study affected? More specific questions involve how individual sectors of the economy (e.g., finance, agriculture, Government, business, and household service providers) are affected by the change? How long do the effects last and over what geographic area will such changes be significant? In addition, what happens to relative prices, wage rates, incomes, and population growth over time are important outcomes regional modeling addresses. Current methodology can provide some answers to these questions, but no one procedure can answer all questions. Frequently the problem is simply too large to be dealt with as a whole. One can better get at an answer by dividing the problem and posing several different questions more amenable to solution. Dividing the question has the advantage that greater information is forthcoming.

There are two major techniques of modeling regions to assess the economic impact of an exogenous event on the region: input-output analysis and econometric models. This chapter examines these two alternate techniques with a discussion of their fundamental structure and operation. Empirical results of input-output and econometric regional modeling relevant to water problems in the Valley are presented. The discussion in this chapter provides a basis for conclusions regarding the applicability and usefulness of input-output and econometric modeling to dealing with the broad regional effects of water problems.

## REGIONAL ANALYSIS OVERVIEW

The schematic diagram in figure 1 illustrates the order and magnitude of regional modeling problems. The area of concern is shown as the rectangle in the upper left-hand corner and is designated region 1 for the current time period, t=0. Flows of goods, factors, and incomes between regions and over time are indicated by the connecting arrows. Assume that a specific exogenous change occurs in or is imposed on region 1 at time zero. The region is one of several similarly defined regional units indicated by additional rectangles running vertically and designated region 2, region 3, ... region N within a larger

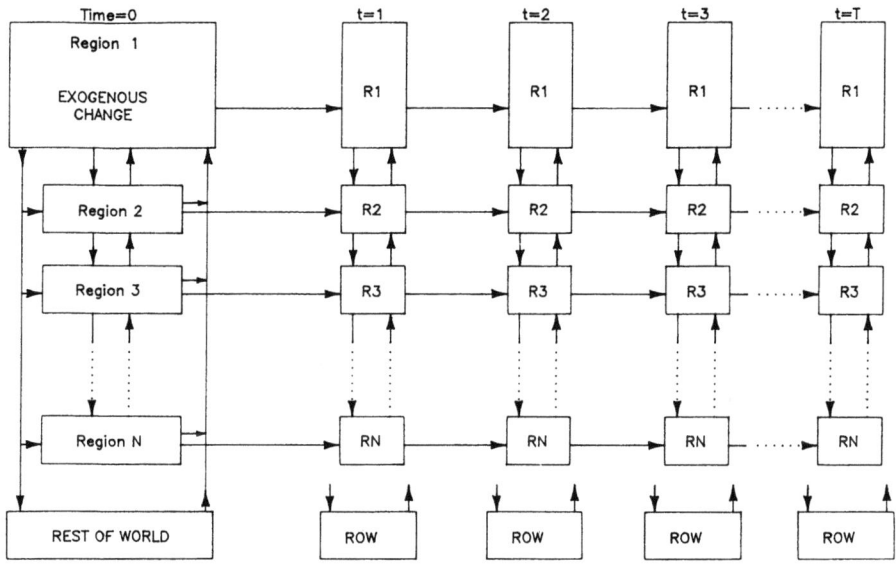

Figure 1. Schematic representation of regional effects.

economy. Some of the goods and services produced within region 1 are sold in other regions and National markets, these goods are exports of the region. Other industries exist in the region to provide local goods and services like housing and real estate and to distribute retail goods and services.

The recognition that one region has effects on adjacent regions or areas highlights the special problems of regional analysis. Most models built today truncate the analysis to a single region with all other areas absorbed into the rest-of-the-world (ROW). This dismisses the problem of regional interdependence. The problem here is illustrated by the results when governments in the coastal area of Santa Barbara County in the early 1970's adopted numerous growth control measures. These growth controls raised house prices and rents in the south coast faster than in adjacent areas. By the late 1970's the growth controls produced changes within the region which impacted employment and created some new problems. With a mobile population and good transportation routes, individuals dealt with high housing prices in the south coast by electing to work on the south coast, but live in north Santa Barbara County (Santa Ynez, Lompoc) or move south to Ventura County in order to obtain access to affordable housing. To model just the coastal area of Santa Barbara would have neglected the effects fostered on other regions by the policies adopted on the coast.

Regional firms compete for labor within the region and usually compete in the wider marketplace for financing and capital goods as inputs to the productive process. The flow of goods and factor inputs between regions is shown by the broad arrows moving both ways in the left-hand side of figure 1. In most regional analyses, individual adjacent regions are absorbed into a single mass called the rest-of-the-world (located at the bottom of figure 1). This choice reduces the information obtainable from regional analysis.

A serious problem in regional modeling is the determination of whether variables are exogenous, determined outside the model, or endogenous, determined by the model structure. The solution is generally based on the theoretical structure of the model.

Another important dimension of regional modeling, time, begins by recognizing economic relationships between the region under study and the remainder of the economy and traces the effect of a specific policy on designated economic variables over time. Schematically the development (and change) of regions over time is represented by the smaller rectangles associated with regions R1, R2, ... RN moving horizontally and designating time $t+1$, $t+2$, ... $t+T$ in the middle and right-hand side of figure 1. The imposition of an exogenous policy in region 1 can be traced out over time periods, $t+1$, to $t+T$ to measure and evaluate the dynamic or time dependent effects. The effects on a single area over time, are influenced by what is occurring in the other regions (R2, ... RN) as they also evolve and change over time. This introduces simultaneity into the relationships among regions. An exogenous decline of one type of manufacturing in region 1 may release labor, reduce the rate of growth in property values, and reduce the growth rate of retail sales and personal income in that region. But the impact on some of these variables, for example retail sales and personal income, would be much less if other regions have an increase in manufacturing output in time periods 2 to 4 to absorb some factors of production (labor) released by the decline in region 1. Of course, given enough time, most mobile factors of production (including labor) will move in response to changing economic incentives. Immobile factors, such as land, typically show the largest changes in economic value given such changes as the posited decline of manufacturing. Witness the old mining towns of the West -- when the ore ran out, mobile factors of production moved. Factors such as land and capital investments which could not move, lost all or most of their value as a result of the exogenous change, the disappearance of the ore.

This discussion illustrates two important dimensions in the development of regional models: (1) The interindustry links, both within the region under study and between the studied region and other regions and the rest-of-the-world and (2) the dynamic or time factor of how changes in the region evolve over time.

## Input-Output Modeling

Input-output (IO) models are generated without respect to time. The region is defined, the relevant economic sectors designated and the technical production processes for each sector are determined and remain the same irrespective of the production volume. The models and the economic structure of the regional economy are static. The economic structure of the regional economy is assumed to remain the same with the passage of time.

Figure 1 illustrates the limitations of the static nature of input-output analysis. An exogenous change occurs in region 1 at $t=0$ and, excluding any of the effects in intermediate time periods, the final change is reported in region 1 at $t=T$ in the upper right-hand corner of figure 1. Changes occurring outside the modeled area, in the schematic R2, R3, ... RN and rest-of-the-world, are excluded from having any effect on region 1. There is no attempt to incorporate other exogenous changes (an increase in manufacturing demand or movement of factors that may become unemployed) in other regions, although the trade interdependence (at $t=0$) with other regions is accounted for.

In this kind of analysis there are no flows of input factors (especially labor) into or out of the region in response to changing relative prices, and the technical production processes and shares of payroll remain as determined in the initial period ($t=0$). This ignores the fact that the exogenous change may produce changes in the relative prices of the factors of production (labor, capital or land) by time period $t = T$. In general, the economic changes reported from IO models tend to be <u>long-run</u> effects which do not allow for factor and output substitution. Because of the static nature of the model, IO results tend to overstate the magnitude of the effects of exogenous change.

IO models usually begin with a dollar flow table, describing the flow of sales dollars from one industry to every other and the dollar purchases of each industry from every other (see table 1). This provides a complete description of the regional flow of goods and services expressed in dollars. These relationships can be viewed as a double entry accounting system. Each industry sector is assigned a column reporting the dollar amount of purchases from all other regional industries. Purchases from other industries represent the inputs for a given industry. As calculated in the base period, this provides the fixed "mix" of goods and services (including the input labor) necessary to produce the output of each industry. This fixed technical coefficient or Leontief production function is the key to IO analysis. Each industry is assigned a row in the table reporting its sales to all other regional industries. Thus, each entry represents a purchase by the column industry from a row industry and a sale by the row industry to the column industry, producing a matrix of dollar flows.

Added to the bottom of each industry's column are one or more rows covering intermediate purchases, imports, Government and taxes and other

Table 1. Aggregated San Joaquin County Input-Output Model - 1974 Dollar-Flow Table[a].

| Industry Sector | Interindustry Sales | | | | | | | | | | Final Demand | | | Output |
|---|---|---|---|---|---|---|---|---|---|---|---|---|---|---|
| | 1-Agric. | 2-Const. | 3-Agric. Proc. | 4-Manuf. | 5-Trns. & Utilities | 6-Trade | 7-FIRE[b] | 8-Srvcs. | 9-House Expend. | 10-Int. Inputs | Regional Exports | Invest. Inventory | Govt. | |
| 1 Agriculture & Agric. Services | 68 | 0 | 155 | 0 | 0 | 0 | 3 | 0 | 15 | 226 | 273 | 0 | 0 | 520 |
| 2 Construction and Mining | 5 | 0 | 4 | 2 | 7 | 1 | 16 | 7 | 0 | 45 | 16 | 147 | | 210 |
| 3 Agricultural Processing | 10 | 0 | 81 | 0 | 0 | 0 | 0 | 1 | 34 | 95 | 807 | 1 | | 939 |
| 4 Manufacturing | 18 | 29 | 44 | 39 | 0 | 2 | 1 | 3 | 14 | 140 | 367 | 21 | | 543 |
| 5 Transport & | 11 | 7 | 49 | 21 | 23 | 12 | 5 | 16 | 65 | 149 | 15 | 19 | | 250 |
| 6 Trade | 15 | 16 | 32 | 10 | 2 | 5 | 3 | 7 | 221 | 94 | 71 | 15 | | 402 |
| 7 Finance, Insurance & Real Estate | 17 | 2 | 10 | 8 | 3 | 23 | 23 | 20 | 171 | 111 | 27 | 5 | | 316 |
| 8 Services | 7 | 6 | 8 | 5 | 13 | 10 | 10 | 10 | 138 | 72 | 87 | 20 | | 318 |
| 9 Household Income | 198 | 62 | 152 | 137 | 124 | 206 | 43 | 150 | 0 | | | | | |
| 10 Intermediate Purchases | 155 | 62 | 388 | 88 | 51 | 57 | 65 | 68 | 676 | | | | | |
| 11 Regional Imports | 106 | 68 | 338 | 246 | 35 | 52 | 42 | 60 | 343 | | | | | |
| 12 Unallocated Value Added | 60 | 17 | 60 | 71 | 40 | 86 | 165 | 40 | 0 | | | | | |
| 13 Gross Regional Outlays | 520 | 210 | 939 | 543 | 250 | 402 | 316 | 318 | 1,020 | | | | | |

[a] In millions of 19074 dollars. Row and column totals may not add exactly because of rounding.
[b] Finance, Insurance, and Real Estate

*Source:* Goldman, G., et al, 1978. *Regional Economic Impacts of Water Resource Use, an Interindustry Analysis of San Joaquin County* Berkeley, Cooperative Extension, p. 17.

components of value added. The summation of all inputs represents final gross regional outlays by the industry. Simultaneously, to the end of each row are added the sales made to regional households, total exports and other sales made to sectors not included, representing the final demand component. The final entry in each row represents the sum of all sales or the gross regional output or total sales by each industry. By accounting definitions used in the national income accounts, total purchases or outlays (the sum of each column) must equal total sales for each industry (the sum of each row).

The technical coefficients table expresses each cell as a percent of total purchases or total outlays. These coefficients, which must sum to one, depict that sector's production relationships or recipes, that is, to produce a dollar's output the construction industry buys $0.14 of its inputs from the manufacturing sector, $0.12 from local service, etc. From this table direct, indirect, and induced multipliers can be calculated. Thus, for each dollar change in a sector's sales the purchase of inputs also changes, including (for the induced effect) labor or household income. This means that other local sectors experience a change in their sales as a result of the initial change and the induced effect of the change in household incomes. It is the sum of these "rounds" or "ripple effects" of local spending of one industry that is referred to as the sector's multiplier. The greater the interaction with local industries (and households) and the greater the proportion of purchases made locally (which depends on the size of region chosen), the larger the multiplier for any industrial sector.

IO analysis at the regional level requires several decisions. First the size of the region has to be determined. The smaller the region the greater the amount of exports and imports and hence fewer local interactions and the smaller the multiplier effects. The smaller the region the more difficult and possibly the poorer the available data. Thus, usually a county unit is the smallest size estimated. Second, the kinds of problems the researcher is interested in exploring determines the degree of disaggregation of the industrial sectors.

For questions dealing with California agriculture usually the major regional crops are assigned individual sector status while manufacturing could be aggregated into a few industries. The assignment of industries in Los Angeles County would probably be just reversed, a large number of manufacturing sectors and a single or few agricultural sectors. The greater the number of "industries" the more detailed the interactions between sectors incorporated in the model. Many researchers begin with large National or state tables of technical coefficients and adjust these to the region under study, usually using supplementary information to adjust for product and input mix differences.

Nonsurvey methodology is often applied to adjust the 1977 United States National IO table with over 500 sectors. The steps involved include: (1) Determination of whether or not intermediate products needed by local sectors are produced in the region. This establishes the size of noncompetitive imports by sector; (2) the use of location quotients or other procedures to determine whether supplies of local intermediate products can satisfy full regional demand; and (3) aggregation of the model to operational size. Regional employment and output measures are frequently used for this purpose. In some recent research, more detailed surveys have been used to obtain more exact information on regional technical coefficients.

## ECONOMETRIC MODELING

Regional econometric models are a set of equations describing the economic structure of a regional economy. The theory behind regional econometric models is the standard theory of economics: microeconomic and macroeconomic theory. The set of equations of a regional model is built on this theory as a means of describing the economic structure of the region. The parameters of the equations are estimated econometrically, largely by time series regression methods using a historical period. This is distinct from an IO model in which parameters are based on single-point observations. Equations in econometric models include behavioral or stochastic, definition, and identity equations. Many regional models are "driven" by a national model or assumptions concerning economic outcomes outside the region.

Specifying regional structure is similar to specifying industrial detail. Following Klein (1969), many early models concentrated on income and its division between consumption, investment, Government and net exports. Lacking data, regional modelers concentrated on industry output, labor earnings, or employment. This permitted more detailed modeling of the labor component, either employment or earnings. Many regional models do not include capital formation (private or public) or include it only as an exogenous variable due to lack of data.

Most econometric models are short run; however, stimulation of econometric models produces dynamic (long-run) effects and is able to trace the effects of an exogenous change in the studied region over a future time period. In terms of the schematic of figure 1, economic change for a region can be evaluated for each period t=1,2,3,...,T moving horizontally across the figure, and depending on how the region is modeled, incorporate changes from other regions (R2, ... RN) or the rest-of-the-world for time periods t=1 to T. If there was an exogenous change in agricultural production in region 1, and cost conditions result in adjacent regions expanding agricultural output given total demand, then econometric models could reflect the increased production elsewhere permitting resources to move over time.

Early regional econometric models were extensions of Keynesian open macroeconomic multiplier models with full recognition of the impact of import and export sectors both on the goods and income side of the circular flow. Later model development added more complex production relationships not unlike the fixed input production functions found in IO models. Employment data became the prime mover in many models because of the availability of employment data at the subnational level. Beginning in the 1980's regional econometric models were given a new lease on life by incorporating responses to relative price and cost changes, to interaction of labor supply and demand, and to migration (see Treyz, Stevens, and Friedlaender, 1980). The result of this work was to eliminate unrealistic rigidities in the econometric framework. Other developments along these lines included changing factor proportions over time to reflect technology and consumption patterns of a changing population (see Treyz et al., 1986). The development of computable general equilibrium modeling in the 1980's has introduced more microtheory into the adjustment process. The addition of more endogenous variables and making some parameters endogenous (costs) required more equations with variables which have to be modeled. If correctly modeled, econometric models can incorporate the potential mobility of factors, especially labor to other regions. Correctly constituted, each major region (R1, R2, R3, RN) has as an input relative prices from other regions to account for interregional movements. To date work is just beginning on this aspect of regional modeling.

## APPLICATIONS TO THE SAN JOAQUIN VALLEY

The National Research Council's report titled *Irrigation-Induced Water Quality Problems* (1989) discusses four types of institutional options that might be brought to bear to reduce the Kesterson National Wildlife Refuge (NWR) contamination. These are: Price adjustments, legal changes, organizational changes, and political and social changes. Under the price option it is suggested that "adjusting the price of water so that irrigation pays the full cost of providing it" would increase efficiency.

The regulatory approach they suggest is to "restrict agricultural use that causes leaching and deep percolation." In addition, other legal or organizational changes could change crop prices, increase production costs, increase water prices, or cause acreage to be taken out of production to address the problem of water contamination. For regional analysis, one or all of these changes could be modeled as exogenous events occurring in a regional economy. Application of IO or econometric modeling could trace their impact on the industry structure or effects over time. Work has begun along these lines.

IO procedures have been applied to California and the Valley for several years. The Cooperative Extension program of the University of California developed numerous IO models in the 1970's and early 1980's which have been used for economic analysis at the county level (Goldman, 1978). In Goldman et al. (1978) a county IO model was used to assess the income effect of deteriorating water quality on crop yields in the Sacramento-San Joaquin Delta. Water quality is defined as the quantity of total dissolved salts. The outside exogenous calculation involved estimation of the decline in crop yield, and adjustment of harvest costs to yield a reduction in farm sales and income. This assumes a constant crop mix. The IO model calculates the regional multiplier effect (both direct and indirect) producing a reduction in regional sales and income "of $245,000 for a 1.0 ECw level in the south Delta region to a $31 million reduction in income and sales for a 2.0- ECw level in the north and central Delta region" (Goldman, 1978). Permitting the substitution of salt-resistant grain for higher valued corn reduces the total losses to about $19 million. These are interesting results, however, one must note the problem created by the static nature of IO analysis. We have here a slice of the economy's future without seeing the loss over time. The problem this analysis shares with other IO studies is the absence of substitution possibilities.

In 1980, the DWR published Bulletin 210, *Measuring Economic Impacts, The Application of Input-Output Analysis to California Water Resources Problems*, the results of an exhaustive study of interindustry linkages. Using the individual crop multipliers from Bulletin 210, shows that the $1,447,000 gross value earned on west-side farms in 1985 increased to about $4,544,000 as it moved through processing and trade channels. This multiplier of 3.14 includes the induced effect of additional household income generated as crop sales move through the trade paths. Given the labor requirements for processing and trade (from Bulletin 210), it is estimated that agricultural production on the west side in 1985 produced about 4,400 full-time off-farm jobs, 42 percent of these coming from the largest crop, cotton (Gaines, 1988).

An interesting application of IO analysis is the study by Wallace and Strong (April 1985) of the effects of reallocating water in the Westlands Water District (WWD). They focus on withdrawing water from 42,000 acres identified by the Bureau of Reclamation (Reclamation) as among the "key" acres contributing to the drainage problem in the Kesterson area. They ask: What are the economic effects on output and employment if these lands are restricted from receiving Reclamation water? Knowing the crops grown and average water applications, reduction in this acreage would amount to 107,000 acre-feet of water. However, WWD still has rights to this water. Application of this water to other (more marginal) lands would yield an estimated increase in agricultural value of about 15 percent of the original reduction in direct agricultural sales. The result is a net reduction of 85 percent of production using a 15-percent recovery rate, i.e., application of this water to other (marginal) lands. This reduction is estimated to produce an immediate direct loss of agricultural sales of $35.4 million (in 1984 $). The impact of this direct effect on the remainder of the economy is provided by the IO model. Farms no longer receiving water do not buy agricultural inputs, hire labor, or use transportation services. These agricultural service industries employ fewer people and buy fewer inputs themselves. Former workers do not buy retail goods either within the region and from outside.

The direct reduction in agricultural sales results in an estimated reduction of 460 jobs which (using county average household size) is equivalent to slightly over 1,000 people affected, an estimated $4 million loss in personal income and $540,000 loss in retail sales. But this is only the first stage; the reductions in people, jobs, retail sales and personal income tend to have indirect effects on the remainder of the economy. Wallace and Strong estimate an additional $28 million decline in agricultural related sales, $1.2 million in retail sales, and $5.8 million in additional personal incomes and an additional 1,000 people affected. The total effect, direct and induced, would be:

- 2,070 people affected
- 916 jobs lost
- $9.9 million reduction in personal income
- $63 million reduction in agricultural sales.

They further estimate a $7.8 million decline in residential property value, a $60 million decline in commercial property value (primarily agricultural land), and possibly 780 dwelling units "could empty" because of job losses.

These results from the IO model are interesting, but notice the information missing because of the modeling technique employed. Referring back to the schematic in figure 1, these results are equivalent to moving region 1 at time zero to region 1 at time T (the long run) without knowing anything about the time path or length of time of the transition. Also omitted are the time path and time of transition for changes taking place in regions 2,3,...,N and possibly the increase in acreage if the total demand remains constant, and how mobile factors (primarily labor) move when relative wages and other costs change.

A unique portion of this study utilizes the major changes reported from the IO models to assess their impact on the public sector, county government and the major cities of Mendota, Firebaugh, and Fresno. This involves separating Williamson Act land from the remainder of the land removed from production to determine the loss in property tax revenue and adding back in the increased values from new land placed under production as a result of water reallocation. The loss of revenues are calculated for both property and persons, using historical averages for the county, for each separate government. The people impact effects (housing, sales taxes, residential property tax) are allocated primarily to the cities while the decline in agricultural land values and its impact on property tax was allocated to the county. General fund revenue losses for the counties are estimated to be $450,000, while the losses of the westside cities of Mendota and Firebaugh are estimated at $122,000. Other cities at a distance also lose, but because of the distance, their loss is less.

The general procedures in the Wallace and Strong (1985) economic and fiscal study were modified and enlarged in the technical report by Strong, Hanemann, and Wallace (1987). Here 94,480 acres, designated the San Joaquin Drainage Study Area (DSA), located in Fresno and Merced Counties, were subjected to 10 regulatory options designed to reduce the level of selenium in the drainage water.

The options achieve different standards for discharge ranging from 2 p/b to 10 p/b. For the economic and fiscal analysis, the key variable for each option analyzed is the estimated cost per acre. These range from an increase of $21/acre for the least expensive options to a maximum of $82/acre. Direct economic impacts calculated from DWR Bulletin 210 county IO models are assumed to occur in the basin, while indirect impacts are divided two-thirds inside and one-third outside the basin.

Baseline agriculture-related jobs, population, and residential housing are affected relatively slightly, ranging from a 1 percent loss under a $21/acre option to about 10 percent loss from the baseline under the $82/acre option. This primarily represents the direct and indirect loss of employment. Agricultural property values are reduced dramatically from baseline values, as estimated rents fall by 21 percent for the least cost option, and land values decrease by 13.6 percent. The $82/acre option would force 6.5 percent of the acreage out of production but would reduce the value of agricultural acreage by an estimated 52.7 percent. The study applies these results to the revenue of the public sector similar to the original study.

Single region models fail to recognize the interconnections between regions. An exogenous change in agriculture water pricing, or institutional changes that impact more than one county could have ramifications not only within the region where the change takes place but in other counties (regions). The total economic effect could likely be larger than the sum of the regional effects. Some of these interregional linkages can be captured in an interregional input-output (IRID) model. Such models require a large amount of detailed data and require independent estimates of interregional commodity flows. Such a model is currently under development for the Valley by Cooperative Extension, University of California.

One of the early econometric models analyzing a regional water problem in the Valley is the Mercer and Morgan (1984) study of Fresno and Kern Counties. The authors were intrigued by the statement by Kahrl (1984) that with increases in the price of water: "Whole towns in the Central Valley would disappear and vast segments of our population would be thrown out into unemployment..."

The thrust of the research was to examine what would happen to the economies of Kern and Fresno Counties if water prices were adjusted upward toward cost? This was a precurser to the National Research Council's price option to improve water quality and reduce damage caused by drainage. In this study, agricultural production initially decreases due to an increase in surface water prices. Because of data availability the models focused on the county as the regional unit without any specific feedback to surrounding counties. Feedback with the remaining California and National economies was accomplished by using the relative price of housing, and unemployment rates, both California and United States. These changes in relative prices permitted factors to move into and out of the county being studied.

To determine the magnitude of the economic impact on agriculture, California Agriculture Resource Model (CARM), Howitt and Mean (1983), was run first under the scenario that average surface water prices double in both Fresno County ($7 to $14) and Kern County ($16 to $32). This pushes the price toward or slightly beyond the average unit cost of water ($23.77) from the Central Valley Project calculated on a historical accounting basis by Yvonne Levy (1982). Ground-water pumping was constrained to increase by no more than 10 percent which implies new legal and organizational changes.

The quadratic programming model of CARM was run using the new water prices to produce a set of new cropping patterns, acres, and yields. Crop acres were aggregated to show total acres removed from production and the change in the mix between field and nonfield crops to capture differing labor requirements. The reduction in acres was 122,000 acres (9.5 percent of the total) for Fresno County and 135,000 acres (15 percent) for Kern County. The change in acres was phased in evenly over a period of 4 years in the regional econometric model. A longer timeframe could be assumed which would dampen adverse effects.

The econometric portion of the model was similar in structure for both counties. All structural equations were estimated by regression over the 1967 to 1979 period and 1980 was used as the base year for 10-year simulations. The model contained a detailed demographic sector incorporating net migration as a function of relative unemployment rates (United States to California) and the level or change in total county employment to capture the encouraged/discouraged worker effect. Thus, the model explicitly permits factors of production, in this case labor, to move given changes in economic conditions within the regional economy. Most equations incorporated Koyck or distributed lag functions of independent variables to capture the lagged adjustment process and make the model dynamic in structure.

Employment for the regional economy was placed in the following classifications: Agriculture, agricultural services, State and local government, local services (trade, service, and finance), construction, and manufacturing. The model is employment based because employment was available on a time series basis and provided a better data base. Each employment sector is driven by the appropriate variables many of which are endogenous to the complete model. Employment in agriculture and agriculture service sectors are determined by the number of acres under production and the mix between field and nonfield crop acreage. State and local government employment is a function of total population levels while construction employment is determined by local housing starts which in turn is determined by population growth and the price of housing in the county relative to the cost of housing elsewhere in California. Manufacturing employment is also influenced by relative house prices between the county average and similar costs in the California coastal areas to capture the strong relative price effects first suggested by the UCLA Forecasting Project. Mining and Federal employment were assumed to be exogenous. There is no attempt to model wages, income or other financial flows for the region in this model. The entire model was simulated over the 1980 to 1990 period under a business as usual assumption. Using the same assumption but with the acreage reductions produced the alternative which was then compared to the base results to derive the regional impacts.

The results are mildly surprising...

Under the assumption of a 100-percent increase in water prices for the two counties, the hardest hit employment sector in Fresno County is agricultural services accounting for about half of the 1990 decline in county employment. In Kern County, agriculture is the hardest hit, accounting for about half of the decline in total employment. The latter is the result of the higher surface water price in Kern County. Agriculture and agricultural services together account for about 80 percent of the total decline in both counties. Local services (trade, service, and finance) is the other hardest hit sector; primarily due to the absence of population growth over the period. But even these estimates overstate the economic impact on the individual counties. Although factors (labor) are modeled as mobile through the migration function, and manufacturing employment is influenced by relative costs, the model does not capture specific economic changes in the adjacent regions, specifically increased agriculture output in adjacent counties, due to the decline in Fresno and Kern Counties.

Evaluating the overall effects on employment, the unemployment rate first rises when the land goes out of production and by the mid-80's (the middle of the simulation period) begins falling toward the base level. Table 2 reports the relevant indexes of employment, population, and unemployment rates for both counties. With this order of magnitude of acreage reductions (about 10 percent of base period totals), the economies of the counties are adversely impacted, but the impact is primarily in agriculture and related areas. To the extent that agricultural output increases elsewhere, this impact is mitigated for society as a whole. The impact beyond these sectors is fairly small. Moreover, the economy in general, but not agriculture, recovers following the shock moving back toward the 1990 base values with respect to relative aggregates like the unemployment rate. Kahrl's gloomy prognosis is untrue.

Table 2. 1990 values.

| Sector[a] | Fresno Scenario | | | Kern Scenario | | |
|---|---|---|---|---|---|---|
| | Base | 1 | 2 | Base | 1 | 2 |
| Agriculture Employment | 30,976 | 29,355 | 25,958 | 23,331 | 20,528 | 17,169 |
| Agriculture Services Employment | 14,088 | 11,488 | 6,039 | 12,896 | 10,871 | 8,444 |
| Manufacturing Employment | 25,768 | 25,906 | 26,210 | 10,560 | 10,578 | 10,594 |
| Local Services Employment | 152,362 | 151,410 | 149,454 | 95,791 | 94,914 | 94,060 |
| Construction Employment | 9,986 | 9,696 | 9,089 | 5,561 | 5,313 | 5,134 |
| Total Employment | 280,480 | 274,982 | 263,507 | 203,125 | 197,089 | 190,140 |
| Local House Price | 151.80 | 149.40 | 144.23 | 123.82 | 122.55 | 121.55 |
| Population | 549,053 | 545,623 | 538,574 | 421,059 | 417,204 | 413,452 |
| Labor Force | 304,957 | 299,923 | 288,614 | 221,657 | 215,613 | 208,364 |
| Labor Force Participation Rate (percent) | 55.5 | 54.97 | 53.59 | 52.64 | 51.68 | 50.40 |
| Unemployment Rate (percent) | 8.03 | 8.32 | 8.70 | 8.36 | 8.59 | 8.74 |

[a]Exogenous employment sectors are not shown but are included in total employment.

A second scenario was run with water prices increased 200 percent. In this scenario, the Fresno price rises from $7 to $21 and the Kern price from $16.10 to $48.30. The Fresno price is 87.5 percent of Levy's estimate of historical accounting cost while the Kern price is about double that value and approximates Levy's estimated replacement cost ($48). Fresno acres are reduced 30.5 percent (394,348) and Kern acres 35 percent (320,884) in scenario 2. The decline in acreage is in field crops with alfalfa, irrigated barley, cotton, irrigated pasture, irrigated wheat, and corn bearing the brunt. Note that these crops (or their derivatives, e.g., milk) are ones for which the Federal Government spends substantial sums to deal with "surplus production."

The employment impact in this scenario is much greater with unemployment rising to about 12 percent in both Fresno and Kern Counties. The rise in the unemployment rate is moderated by a 2-percentage point fall in labor force participation for both counties. Agriculture and agricultural services together account for almost 75 percent of the employment decline in Fresno County and over 80 percent in Kern County. As in scenario 1, agricultural services is hardest hit in Fresno and agriculture in Kern. The recovery by 1990 is to about 8 percent in Fresno County and a little under 5 percent in Kern county. The total employment ratio declines to about 94 percent in Fresno versus 98 percent for the 100-percent increase scenario and 94 percent in Kern versus 97 percent previously. The population ratio declines to about 98 percent versus 99.5 percent in Fresno County and about 98.25 versus 99.25 in Kern County. These results are produced by the fact that total employment and population rise after the water price increases, but rise more slowly than before. It is interesting that Kern County with a higher water price shows greater response than does Fresno County in terms of returning toward normal.

Tables 2 and 3 present the major empirical results from the Mercer-Morgan model. Table 2 shows total values for the base and scenarios 1 and 2 while table 3 shows the differences from the base for each scenario.

Table 3. 1990 differences from the base (scenario - base).

| Sector | Fresno Scenario 1 | Fresno Scenario 2 | Kern Scenario 1 | Kern Scenario 2 |
|---|---|---|---|---|
| Agriculture Employment | -1,620 | -5,018 | -2,803 | -6,162 |
| Agriculture Service Employment | -2,599 | -8,048 | -2,026 | -4,452 |
| Manufacturing Employment | 138 | 442 | 18 | 26 |
| Local Services Employment | -952 | -2,908 | -877 | -1,383 |
| Construction Employment | -290 | -897 | -204 | -330 |
| Total Employment | -5,497 | -16,973 | -6,036 | -12,511 |
| Local House Price | -2.44 | -7.60 | -1.27 | -1.58 |
| Population | -3,430 | -10,479 | -3,856 | -6,216 |
| Labor Force | -5,034 | -16,343 | -6,044 | -12,452 |
| Labor Force Participation Rate | -0.57 | -1.95 | -0.96 | -2.24 |
| Unemployment Rate | 0.29 | 0.67 | 0.23 | 0.55 |

The interesting result of the Mercer-Morgan study is the demonstrated resiliency of the regional economy to severe shock in the form of sharp rises in the price of surface water in an agricultural region. These results show that a move toward economic efficiency by raising water prices can be accomplished without wrecking the regional economy as suggested by Kahrl. In addition to efficiency gain, raising water prices as in this model would produce another substantial benefit. This is the reduction in drainage and drainage-associated problems. Raising water prices thus has a two-pronged benefit for society.

A review of the econometric literature suggests there is tremendous variety in model builder's objectives which is reflected in the wide variety of regional models built and estimated in the United States and elsewhere. Work is underway, usually in small experimental firms, to incorporate more completely the special effects, to incorporate the "supply side," to include more detailed description of the financial sector, and combine results of cross-section analysis in a time series format. The modeling of economic change in the Valley can employ models already constructed and modified for local conditions. A multiregional multisectoral model, based on the TFS methodology, is available from REMI (Regional Economic Models, Inc.). Both goods-market and labor-market demand and supply are explicitly modeled incorporating rapid goods market clearing and slower nonmarket clearing of the labor market based on measures of excess demand. The profit maximization hypothesis is employed to get price-cost and factor-demand equations derived from CES or Cobb-Douglas production functions. Modeling in this area is influenced by computable general equilibrium models (discussed elsewhere in this volume).

Both IO and econometric regional modeling have been applied to Valley water problems. Such modeling provides useful information for decisionmakers and is a necessary component of any regional decisionmaking process. Both general techniques (IO and econometric) are valuable and can provide useful insights and information. Because one usually needs to be concerned about the time path of change, econometric modeling is generally preferable to reliance on IO alone. Perhaps the wave of the future resides in current attempts to combine IO and econometric modeling. Decisionmakers must familiarize themselves with the basic background of modeling offered here in order to understand and profitably use the research results presented to them.

## REFERENCES

Alward, G. S.; Davis, H. C.; Despotakis, K. A.; and Lofling, E. M., 1985. Regional Non-Survey Input-Output Analysis with IMPLAN. Paper presented at the Southern Regional Science Association Conference, Washington, DC.

Batey, P. W. and Madden, M., 1986. Integrated Analysis of Regional Systems, *London Papers in Regional Science,* 15.

Bolton, R., 1985. Regional Econometric Models, *Journal of Regional Science*, 25(4).
California Department of Water Resources, 1980. *Measuring Economic Impacts, The Application of Input-Output Analysis to California Water Resources Problems.* Bulletin 210
California State Water Resources Control Board, 1987. *Regulation of Agricultural Drainage to the San Joaquin River.* Final Technical Committee Report on Order WQ 85-1.
Gaines, R. W., *1988. West San Joaquin Valley Agricultural Setting.* Prepared for San Joaquin Valley Drainage Program by Boyle Engineering.
Goldman, G.; McLeod, D.; O'Regan, M.; and McReynolds, J., 1978. *Regional Impacts of Water Resource Use, An Interindustry Analysis of San Joaquin County.* UC Cooperative Extension, Berkeley.
Howitt, R and Mean, P., 1983. *A Positive Approach to Microeconomic Programming Model.* Working Paper 83-6, Department of Agricultural Economics, University of California, Davis.
Karhl, W., 1984. Letting Price Allocate Water Would Devastate the Land, *Los Angeles Times*, Vol. 101, Part 15, p. 5.
Klein, L., 1969. The Specification of Regional Econometric Models, *Papers, Regional Science Association*, 23, pp. 105-115.
Levy, Y., 1982. Pricing Federal Irrigation Water: A California Case Study, *Economic Review*, Federal Reserve Bank of San Francisco, pp. 44-45.
Mercer, L. J. and Morgan, M. D., 1984. *A Dynamic Simulation Analysis of Regional Economic Change with Alternative Water Prices.* Final Completion Report, Water Resources Center, University of California, Davis.
Natural Research Council, 1989. *Irrigation-Induced Water Quality Problems.* National Academy Press, Washington, DC.
San Joaquin Valley Drainage Program, 1989. *Preliminary Planning Alternatives for Solving Agriculture Drainage and Drainage Related Problems in the San Joaquin Valley.* Draft Report.
Solomon, B. D., 1986. The Socioeconomic Impacts of a Regional Synthetic Fuels Industry: An Integrated Econometric Analysis. In: Batey, P.W. et al., Integrated Analysis of Regional Systems, *London Papers in Regional Science*, 15.
Strong, D; Hanemann, M.; and Wallace, L, 1987. Private and Public Sector Impacts Resulting from Regulation of Agricultural Drainage to the San Joaquin River. In: State Water Resources Control Board, Technical Committee Report, *Regulation of Agricultural Drainage to the San Joaquin River, Appendix J.*
Treyz, G. I.; Stevens, B. H.; and Friedlaender, A. F., 1980. The Employment Sector of a Regional Policy Simulation Model, *The Review of Economics and Statistics*, LXII(1).

Treyz, G. I.; Stevens, B. H.; and Friedlaender, A. F., 1985. The TFS Regional Modeling Methodology, *Regional Studies,* 19.

Wallace, L. T. and Strong, D., 1985. *Selected Economic Estimates of the Impact of Restricting Irrigation Inflows to Agriculture Lands in the Westlands Water District of California.* UC Cooperative Extension.

# 27 CREATING ECONOMIC SOLUTIONS TO THE ENVIRONMENTAL PROBLEMS OF IRRIGATION AND DRAINAGE

Marca Weinberg, University of California, Davis and
Zach Willey, Environmental Defense Fund

### ABSTRACT

The environmental problems of irrigation and drainage include surface and ground-water contamination, deterioration of fish and waterfowl habitat, public health problems, degradation of soil and land resources, and ground-water overdraft. The institutions that govern water prices and allocations are examined and found to present farmers with economic signals that do not indicate the true cost of water use. Water markets, tradable water pollution discharge permits, and liability and compensation rules for environmental damages are examined as three incentive-based policy alternatives that can be designed to correct these institutional distortions and encourage the managerial changes necessary to reduce environmental problems from irrigation and drainage. A simulation exercise is conducted to estimate the potential for water markets to address drainage reduction goals for the San Joaquin Basin. The results predict that significant reductions in total drain flows may be achieved as a by-product of a water market.

### INTRODUCTION

A number of natural resource and environmental problems are associated with irrigation and drainage. The degree to which the underlying physical processes and effects are understood varies widely. Attempts to measure resulting economic damages, particularly those that are external and non-market, are imperfect at best. All policies intended to control these damages are based on presumptions about societal values placed on the impacted environmental resources. These values are diverse, including an array of cultural, psychological, ethical, and economic interests within the society.

It is becoming apparent, as a matter of public policy, that agricultural water conservation and drainage reduction will have to be implemented as key means of controlling environmental damages. Accordingly, the institutions and incentives which govern agricultural water use are increasingly important. The institutions that currently govern agricultural water supply and quality in California are described in this chapter. Institutionally based regulations, procedures and pricing policies play an important role in the underlying framework for irrigation decisions that impact the environment and are important in the optimal design and implementation of policies to reduce that impact. Three incentive-based policies are introduced that can address drainage and other environmental problems associated with irrigated agriculture. A simulation exercise is conducted to estimate the potential for one of these policies to be effective as a solution to the agricultural drainage problem in the San Joaquin River basin. The simulation model and results of the analysis are followed by conclusions and recommendations.

## ENVIRONMENTAL PROBLEMS OF IRRIGATION AND DRAINAGE

### Surface and Ground-Water Pollution

Discharge of irrigation runoff and drainage containing salts, trace elements, or other contaminants results in surface and ground-water pollution. It is important to have a perspective on the significance of agricultural water pollution in general, and total dissolved solids (various salt compounds) and trace elements in particular, in overall surface water pollution problems. While agriculture is by far the largest nonpoint source of water pollution, salinity is a relatively minor pollutant in terms of impairing uses (Willey and Graff, 1988). In a regional context, however, total dissolved solids and trace elements are major surface water pollutants in California's Sacramento and San Joaquin River Delta (Robie, 1988) and Central Valley wetlands (Ohlendorf et al., 1986 and California SWRCB, 1987). The impact of trace elements in the Delta and San Francisco Bay is still under study -- while limited tests show no immediate toxic effects of trace elements on larval striped bass (Doroshov and Wang, 1984), such effects on the prime food source of the bass have been observed (SRI International, 1985). Selenium, which is the most studied trace element with regard to wildlife effects, may cause damage within a range of concentrations as low as under 1 p/b to about 50 p/b (Eisler, 1985).

Water pollution can also pose risks to public health. For example, significant presence in ground water of pesticide residues is an indicator of potential health risks where wells deliver drinking water supplies (Holden, 1986). Evaluation of such risks, including identification of the most hazardous substances, routes of exposure, and highly exposed population sectors, is a complicated and ongoing task (Kennedy, 1989). Further evidence of potential public health risks was presented at State hearings in 1990 on the presence of the pesticide DBCP in ground water.

## Degradation of Wildlife Habitat

Fish and waterfowl habitat have deteriorated as a result of irrigation water drainage in a number of river basins in the Western United States. The bioaccumulation of the trace element selenium and subsequent deaths and deformities among several waterfowl species at California's Kesterson Wildlife Refuge has led to a growing recognition of contamination problems at other waterfowl refuges by trace elements carried in irrigation drainage waters (U.S. Fish and Wildlife Service, 1986). Intensive studies were initiated in California, Nevada, Utah, Wyoming, Arizona, Montana, and Texas during the 1986-87 period involving waterfowl refuge contamination by irrigation return flows (U.S. Department of the Interior, 1988). In addition, 17 other possible problem refuge areas in the Western United States are under study (U.S. Fish and Wildlife Service, 1986). The significance of these studies, which will likely continue for some time, is that they resulted from a recognition during the early 1980's of the adverse environmental effects of irrigation on wetland habitats.

Surface water storage and diversion for irrigation water supplies also can degrade wildlife habitat. The adverse impacts of the Red Bluff Diversion Dam on migrating salmon populations in the Sacramento River and the loss of all salmon and steelhead fisheries in the San Joaquin River due to Friant Dam are examples.

## Degradation of Soil and Land Resources

Irrigation and drainage can cause the degradation of soil and land resources. This primarily onfarm environmental effect imposes costs on individual property owners and farmers. Buildup of soil salinity and waterlogging is of increasing concern to irrigators in many agricultural regions, including the San Joaquin Valley (Valley). Perhaps no other deleterious environmental effect of irrigated agriculture has as long a demonstrated history.

## Ground-Water Overdraft

The overdraft of ground water occurs when irrigators and other water users pump water from aquifers at a cumulative rate that exceeds replenishment from infiltration plus ground-water inflows. Overdrafts are particularly serious in parts of the Western United States (Water Efficiency Working Group, 1987) where the technical and economic resources provide the capability to pump large amounts of deep ground water. In California, ground-water overdraft provides up to 10 percent of the total average annual irrigation water supply in normal years, and much more in dry years (California Department of Water Resources, 1987).

The contribution of subsurface water sources to natural habitats and water bodies can be reduced by ground-water overdraft. In Arizona, ground-water overdraft in a number of areas has been a key factor in the decline of Sonoran desert plant communities and riparian habitats (Bryan, 1928). Instream flows--the volume and rate of stream and river flows -- have also been reduced in a number of river basins in the Southwest (Sheridan, 1981). A related effect is that the water storage capacity of some aquifer formations can be irreversibly reduced by the collapse of formerly water-filled voids in strata with a high clay content, and a concomitant subsidence in surface land elevations can occur (Willey, 1977). Overdrafts in agricultural regions can also stimulate political demands for construction of new surface water supply projects with associated environmental effects. The proposed Mid-Valley Canal in California's Tulare Basin, that would tap water supplies that maintain environmental values in other parts of the State, is one such example (U.S. Department of the Interior, 1981).

## IRRIGATION WATER SUPPLY AND POLLUTION LOAD ALLOCATION IN CALIFORNIA

Water supplies and pollution loads are distributed in the California system through a convoluted mix of laws, government agency regulations, and water district procedures and requirements. Agriculturalists' rights to use and to pollute water supplies are governed either by rigid yet unclear government regulations or effectively not at all. Agriculture, which is noted for its responsiveness to market conditions, has yet to be presented with market opportunities and constraints with regard to its water use and pollution.

### Allocating Water Supplies Without Markets

The water rights system, including the water contracts of the State and Federal water projects, is the primary mechanism governing the allocation of water supplies that are stored and/or diverted from natural waterways. Several types of usufructuary water rights have been established -- waterfront property owners can have riparian rights; landowners can have correlative rights to underlying ground water; and surface water diverters can have appropriative rights. The allocation of these rights is based on legal doctrines with little if any economic content or rationale.

California's State Constitution mandates that water rights be utilized according to the concept of "beneficial and reasonable use." The doctrine, however, has not been defined in terms of economics and is rarely applied in any meaningful way to evaluate the execution of water rights. Consequently, water rights permits are administered in an ad hoc fashion which aggravates uncertainty about the status of existing rights. In addition, although State law has been supportive of voluntary water rights transfers for nearly a decade, an active water market has not developed, due in part to the uncertain status of transferred water rights.

## Water Pricing and Distribution in California

Much of California's water is distributed by over 1,000 local water districts which have either appropriative water rights or contracts with the Federal Central Valley Project (CVP) or the State Water Project (SWP). Only about one-half of the 30 to 35 million acre-feet of irrigated water applied annually is priced on a per-water-unit usage basis, mostly on a flat-rate schedule. Approximately one-third of all irrigation water is distributed on a per-acre fee assessment basis, rather than being priced on a usage basis. These pricing procedures reflect the districts' need to generate only enough revenues to meet operating expenses and debt without making a profit. Pricing based on marginal costs is generally not applied by agricultural water districts[1].

Water pricing procedures in the Central Valley Project (CVP) and State Water Project (SWP), like those of local water districts, are not designed to encourage economic efficiency in water use decisions. While recent reforms in Federal Reclamation water contract pricing require additional operational costs in contract prices, these costs are not a substantial portion of the full marginal costs to finance and operate the CVP. This newly enhanced "cost recovery" in the CVP, which has been a more prominent feature of the SWP, does not address two key elements that distort project water prices: (1) Use of taxpayer revenues either directly to service project debt and operational expenses or indirectly through lost tax revenues from the use of tax exemption and public ratings for bonded debt and (2) application of historical average costs to the calculation of repayment in contract water prices.

The most obvious example of the use of public funds to lower water prices is the Federal practice of charging no interest to CVP contractors for the funds used in constructing project facilities. The interest payments forgone each year are probably well in excess of $50 million. Clearly, one of the original public policy goals of the Bureau of Reclamation (Reclamation) program -- to stimulate agricultural development in the West -- has come into conflict in recent decades with the goals of water conservation and pollution control.

Other cost accounting and repayment practices of the CVP and SWP result in water prices below full costs and undermine economic incentives. CVP water is delivered using Federal hydropower that is priced at a fraction of its market value. SWP water prices include interest but are below full costs due to the tax exempt status of financing bonds, the use of property tax revenues by SWP contractor districts to pay water costs and the use of SWP hydroelectric power to deliver SWP water. In addition, both the CVP and the SWP use historical average costs to determine price, wherein "old" project costs are "rolled-in" with "new" project costs.

## Pollution Control, Drainage Pricing, and Agricultural Water Districts

The link between irrigation, drainage, and water pollution has become widely acknowledged during the past decade. However, during the era of major project development in the Western States that peaked during the 1950's, drainage was viewed as an agricultural, not an environmental, problem. The threat to crop production from rising water tables, soil salinity, and waterlogging was known when California's major irrigation projects and lands were developed. In the most arid agricultural areas, where this threat is most acute, plans and some actual investments were made to remove this threat. In the Imperial Valley, subsurface drain systems convey saline water for discharge into the Salton Sea. Subsurface drain systems costing $100 to $300 per acre have been installed on over 100,000 of the Western San Joaquin Valley's approximate 1.5 million irrigated acres. The San Luis Drain was intended to discharge saline drainage water from the valley into the San Francisco Bay or Delta, but construction was halted at Kesterson Reservoir, when trace element pollution and waterfowl toxicity problems became apparent in the early 1980's. Some of the water collected in subsurface drains is conveyed to the San Joaquin River, where it mixes with seepage from irrigated lands that do not have installed drain systems. Several dozen evaporation ponds are in operation as well, mostly in the Tulare Basin. Like Kesterson, a number of these ponds have developed toxic levels of trace elements in the sediments and biota (Westcot et al., 1988).

Regulation of drainage pollution is authorized under the nonpoint source provisions of the Federal Clean Water Act. During the 1970's and 1980's, the permit system for discharge of water pollutants from point sources was not applied to irrigation. Instead, due to technical and institutional problems involving the measurement and control of nonpoint source pollution, Section 208 of the Act required extensive studies of basinwide planning and "best management practices." Some useful information, along with the encouragement of the designation of regional planning organizations to control nonpoint pollution, resulted from the 208 program before it disappeared due to funding cuts.

Nonpoint source control efforts were revitalized when the Clean Water Act was amended by the Water Quality Act of 1987. Section 319 of the amendments established the Nonpoint Source Management Program to provide states with Federal financial assistance to concentrate control efforts on nonattainment segments of rivers and streams. Regulations recently promulgated by the U.S. Environmental Protection Agency (EPA) require implementation of state plans to control nonpoint pollution in nonattainment river segments. The first step in developing a regulatory program designed to protect other water users from the damages of pollutants in agricultural drainage (surface runoff and subsurface drainwater) is to establish water quality standards for rivers, wetlands, lakes, estuaries, and bays. For river segments that have not attained instream quality standards, this first step must be followed by establishing total daily maximum loads of pollutants and allocating these loads to pollutant sources within the basin. These steps are necessary to meaningful implementation of instream standards.

In California, the Central Valley Regional Water Quality Control Board (Regional Board) is required to develop a nonpoint source control program for pollutants in the San Joaquin River. The first step was taken in 1988 when the Regional Board announced a selenium standard of 5 p/b in the river. Standards for other pollutants, including other trace elements, are also being set. Boron appears to be the next pollutant for which standards will be established. Monitoring river concentration levels for several years is required to determine its "attainment" status -- if adjustments in irrigation and drainage are sufficient to produce selenium concentration levels in the river which comply with the standard, then attainment will have been reestablished with respect to selenium. The current target year to determine attainment status is 1993.

If attainment is not achieved, then selenium load allocations may be required. The difficult problem of allocating, among discharging sources, the permissible load of selenium (and eventually other pollutants) depends upon the river's hydrology and selenium inflows. In addition to complex monitoring, measuring, and modeling tasks to document the Basin's "mass balance" of selenium, a critical problem in load allocation is heterogeneity of agricultural districts and institutions that govern the irrigated lands within the San Joaquin River basin. A patchwork of irrigation districts, water districts, canal companies, drainage districts, and individual landholdings control the distribution of the basin's irrigation water supplies and subsequent drainage and runoff. The districts vary in age and size, and thus in the type of infrastructure that has been developed to deliver water and manage drainage. Significantly, the price charged to growers for delivered water varies by district, as does the per-acre water supply allocation.

Discharge into the San Joaquin River and/or the Delta/Bay system has been the historic preference of drainers and agricultural water districts considering options for drainage management (Interagency Drainage Program, 1979). However, as the price involved in establishing rights to such discharges rises, alternatives are being considered. Drainage reduction, through water conservation and improved onfarm irrigation management, can reduce the drainage management problem (University of California Committee of Consultants, 1988). Other drainage disposal methods being considered include drainage water treatment, evaporative cogeneration, deep-well injection, vegetative uptake, ground-water management for drainage storage, and evaporation ponds.

The price of drainage will eventually depend upon some mix of drainage reduction measures and some of these management options. Questions of technical feasibility notwithstanding, institutions and policies which allow the pursuit -- within the constraints of environmental standards -- of the most economically attractive options have still not been fully implemented in California.

## POLICIES CONSISTENT WITH ECONOMIC SOLUTIONS

Conceptually simple policies exist that can allow irrigators to make the least-cost decisions consistent with environmental objectives. These policies emphasize voluntary trading in water or pollution markets with quantitative and qualitative constraints set by government regulations.

### Water Marketing and Quantitative Transfers

Water marketing and transfers can encourage shifts in scarce water supplies by providing incentives for voluntary changes in irrigation water use. Water marketing transactions can be of several types. Permanent transfers may involve outright selling of water rights or contracts in perpetuity. This can be on an intermittent basis, such as in dry years only, or on an "every-year" basis. Alternatively, leases of water rights can occur whereby, after a specified term, the rights to water use return to the lessor. A substantial variety of legal and technical arrangements characterize water marketing transactions (see Colby, this volume).

There are three ways in which water marketing transactions can benefit the environment. First, transactions can shift water from existing uses to meet new, growth-induced requirements, thereby helping to alleviate the need for new supply and diversion projects. In the Western United States, irrigation accounts for an average of approximately 80 percent of all water used and, therefore, is the prime source of supply for water marketing transactions. Where municipal and industrial growth is the primary source of new demand, water transfers from agricultural to urban uses are increasingly common.

Second, the direct acquisition of water supplies for environmental purposes, a relatively new aspect of water marketing, is being considered and implemented in several areas. Again, irrigation water supplies are the prime source for such acquisitions. As a result some of the environmental effects of irrigation can be mitigated and possibly reversed while irrigators are compensated for making adjustments in their water use.

Finally, water marketing can be an important component of an incentive-based effort to control water pollution stemming from irrigation. Reduced pollutant loading often results from decreasing irrigation water applications, which can be made economically attractive to the irrigator through water marketing options. In addition, to the degree that irrigators must pay for the right to discharge pollutants into external environments, income from water marketing can provide funding for treatment or discharge facilities, purchase of discharge permits, or compensation for liability claimants.

While development of California's water markets has been limited, substantial activity is occurring in other regions. Table 1 shows some recent water market prices in several Western States. For California, the economic potential of such trading is clear from the fact that irrigators' water prices are considerably less than urban water prices, which in turn tend to be less than the costs of new surface supply projects (see figure 1). From an economic perspective, there will likely be a substantial number of willing lessors or sellers within the irrigation community once the political, legal, and social barriers to water transfers are removed.

The costs of many irrigation water conservation practices and systems exceed water prices paid by irrigators. While there are frequently yield-related reasons for irrigators to improve their systems and practices, in many cases the value in reduced water costs does not cover the investments for water conservation. Income from water sales and leases by irrigators can underwrite improvements in irrigation systems, which would provide a real incentive for conservation investments.

Table 1. Sample water market transactions/Western States/1987.

| State | Source/Region | Amounts Per Transaction (Acre-Feet/Year) | Price ($/Acre-Foot/Year) |
|---|---|---|---|
| Arizona | Ground/Central | 500-2,000 | (50-120) |
|  | Surface/Southern | 300-400 | (200) |
| California | Ground/South Coast | 100-1,000 | (150) |
| Colorado | Surface/East | 2,000-5,000 | (190-350) |
| Nevada | Surface/West Central | 10-1,000 | (200) |
| New Mexico | Surface/North Central | 50-300 | (100) |
| Texas | Surface/Southwest | 100-2,000 | (50-65) |

*Source*: Willey and Graff, 1988

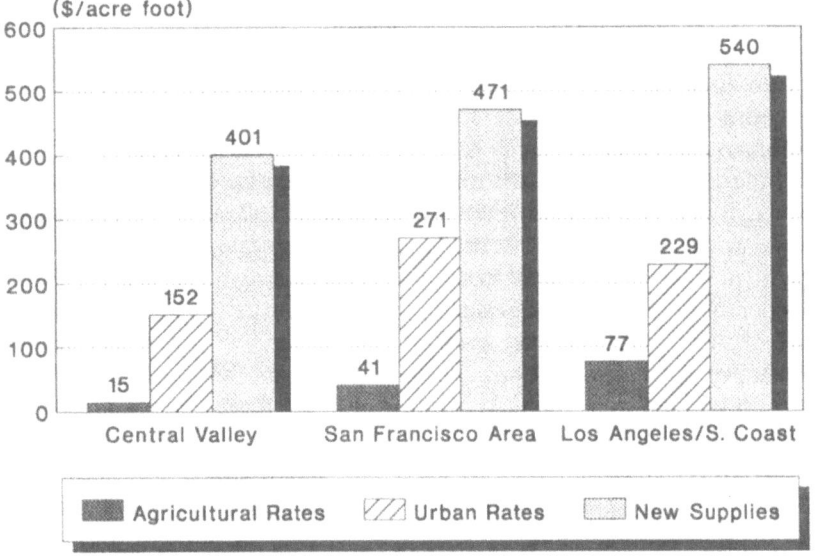

Figure 1. Retail rates and costs of new surface supplies in California's water system (1985).

*Source*: California Department of Water Resources, Bulletin 160-83; Wiley and Graff (1988).

In the midst of California's dearth of experience with actual water transfers, the 1988 agreement between the Imperial Irrigation District and Metropolitan Water District is the largest single transaction to date. The agreement transfers 100,000 acre-feet of Imperial's Colorado River water each year to Metropolitan's service area. This agreement is based on Imperial irrigation system improvements financed by Metropolitan (Smith and Vaughan, 1989). The availability of this water to Metropolitan reduces its demand for new water diversions from Northern California and consequently avoids incremental environmental damages. In addition, approximately 300,000 acre-feet of water has been estimated to be available in the Imperial system if sufficient investments are undertaken (Stavins and Willey, 1983). That is enough water to offset the water yield, for example, of the proposed Auburn Dam on California's American River or of the Animas-LaPlata Dam on Colorado's Yampa River.

Water marketing for the direct benefit of specific environmental resources is also developing. In Colorado, where rights to instream flows are allowed under State law, the Nature Conservancy is implementing an acquisition program involving irrigation water rights (Harrison and Wigington, 1987). Proposals to acquire water rights for Lahontan Valley wetlands as well as to settle the fisheries-based water rights claims of the Pyramid Lake Paiute Tribe are aimed at mitigating the environmental effects of the Newlands Project in Central Nevada (Yardas, 1989). In California, an attempt to resolve long-standing legal disputes concerning the diversion of water by the Los Angeles Department of Water and Power from streams that flow into Mono Lake includes investigation of purchase options for replacement water supplies from irrigators in several areas (Mono Lake Committee, 1987; and Conniff, 1989).

## Tradable Water Pollution Discharge Permits

Water pollution control laws in the United States have relied primarily upon a technology-based approach that sets performance standards for individual pollutant discharge points. The discharges of such "point sources" as paper processing plants, steel mills, and municipally owned wastewater treatment plants, have been regulated by the issuance of permits. Enforcement of the terms of these permits has been the responsibility of public regulatory agencies. Allowable pollutant levels in discharges are determined by various forms of "best available technology" (BAT) for control.

This approach to the control of point sources of water pollution has been criticized in several ways. The construction of many of the legally required BAT treatment facilities has been heavily subsidized by Federal taxpayers, and technological rigidities have impeded the use of the most economical control options (Tietenberg, 1985). Allowable discharge levels for individual points have often not achieved the basinwide water quality objectives of "swimmable and fishable" waters originally targeted for 1982 in the Clean Water Act of 1972. Monitoring and enforcement by EPA and state agencies is costly and, therefore, is often vulnerable to budget cuts.

Whether or not such criticisms are valid, it will be much more difficult to achieve similar water quality improvements from "nonpoint" sources. In its amendments to the Clean Water Act in 1987, Congress addressed these problems but did not formulate an effective incentive mechanism for nonpoint source controls (Ackerman and Stewart, 1988). By definition, nonpoint sources are numerous and dispersed, making it difficult to monitor them and to enforce performance under a technology-based regulatory approach. Management-intensive production technologies and practices, such as those that will be necessary to reduce drainage and other nonpoint source emissions from irrigated agriculture, are not very amenable to the BAT approach. Public subsidies, available in significant amounts during the era of point source control investments, are less available now due to chronic budget deficits.

Tradable discharge permits (TDP) offer an incentive-based system by which regulatory agencies can achieve pollutant loading targets. Under a TDP system, regulatory agencies issue exactly enough permits to attain the loading target. All sources are required to have permits equivalent to their emissions levels, but permits may be traded (sold) among sources. Because permits are valuable, an opportunity cost to hold permits exists and creates an incentive to reduce emissions. Each source is free to reduce emissions in a least cost manner, and will do so to the extent that reducing emissions costs less than buying (or not selling) discharge permits. Thus, unlike the BAT approach, TDP's allow a degree of economic flexibility and ensure a least-cost allocation of discharge reduction across sources (Baumol and Oates, 1988).

The establishment of pollutant loading targets would be within existing authorities under the Clean Water Act. Section 319(a)(4) of the Act's 1987 amendments states that nonpoint source pollutant management should "to the maximum extent practicable develop programs on a watershed-by-watershed basis." The Act does not specify what type of implementation program is necessary -- this is the states' responsibility. A "watershed basis," which involves multiple sources of pollutants, holds the possibility of pollutant trading programs. TDP's have been proposed by Congressional leaders (Heinz and Wirth, 1988). In several European countries, incentive-based water quality control programs have been utilized (Brown and Johnson, 1984 and Harrison and Sewell, 1980). Pilot projects involving potential trading between point and nonpoint sources of water pollution have been implemented in Colorado (Jaksch and Niedzialkowski, 1985).

TDP's offer great potential to actually control some of the most significant environmental effects of irrigation while minimizing the disruption of farming; however, there are significant implementation problems. While there are millions of irrigated fields, it is infeasible to issue and enforce millions of TDP's. Larger geographical units, such as irrigation or drainage districts would have to be the recipients of individual TDP's. Internal allocation of discharge rights could be determined by methods such as discharge assessments on irrigated lands, drainage fees imposed on applied water rates, and intraunit water trading. Dividing responsibility and cost for discharge loads among farms and fields would require the use of hydrogeological information and would, most certainly, require negotiated compromises among neighboring irrigators and landowners. The distribution of the initial permits, required prior to trades, is a tough political decision due to the economic value contained in such permits. Each permit would specify a given permissible load for a pollutant at a given discharge point. Equivalencies with other discharge points, and possibly with other pollutants, would have to be established by regulatory authorities. Enforcement of permit provisions would be a continuing problem. Yet, TDP's have the best potential to control irrigation pollution effects without massive and costly regulatory intrusion and constraint on irrigated agriculture.

## Liability and Compensation for Environmental Damages

Water marketing and tradable discharge permits seek to induce irrigators to make adjustments to avoid damages. Such avoidance will not always be successful, and policies to address environmental damages associated with irrigation are necessary. Policies that assign liability and require compensation for external damages can provide additional economic incentives to prevent and mitigate environmental effects. Such policies should at the same time provide a systematic means by which agriculture can protect itself from unreasonable damage claims.

The existing system of liability and compensation is administered primarily by the judiciary. During the past three decades, a complex sequence of legal opinions has created a de facto public policy in which tort litigation is increasingly used by plaintiffs to secure monetary compensation. While such litigation has been primarily focused on product liability and personal injury, various types of environmental torts appear to be on the upswing. The growth of tort litigation indicates the seriousness with which it should be viewed -- cases in which products, automobiles, and chemicals are alleged to have caused injuries increased fourfold between 1976 and 1986. Damage claims against municipalities doubled between 1982 and 1986. During a 12-month period in 1984-85, damage claims against the Federal Government increased by over 30 percent to nearly $150 billion. Plaintiffs' probability of success rose from around 25 percent in the 1960's to over 50 percent in the 1980's. During the same two decade period, the average size of award rose by a factor of five to an inflation-adjusted $250,000 (Huber, 1988).

Legal analysts have come to recognize the transformation of tort law during the 1980's. "The courts had first taken the limited legal theories of the past, meant to apply to front-yard sort of environmental mischief, and stretched them to cover the inner space of intimate contractual relations; now they extended them to the outer space of the public square, with its myriad low-level mass contact. Public risks and environmental torts, once all but excluded from the tort system, quickly became the vibrant hub of a whole new field of litigation....The environment knows no bounds. Which meant that the environmental lawsuit could not know any bounds either. The modern history of toxic-tort litigation has thus been the record of an ever-widening circle" (Huber, 1988, p. 67).

The interests of the environment and of irrigated agriculture would be served well by establishing a contractual basis for the assignment of liability risks. Negotiations leading to such contracts would require specificity of priority risk areas and issues, and opposing points of view could be expressed and evaluated outside the courtroom. A key contractual device would be direct insurance, tailored to the circumstances of individual irrigators, districts, or perhaps regional institutions.

An a priori agreement on liability for environmental risks could be helpful in implementing solutions to drainage problems. For example, a proposal for drainage discharge management was made by westside irrigators (Panoche Drainage District, 1990) to redirect drainage through Mud Slough to the San Joaquin River below the Merced River confluence (the so-called "Zahm-Samsoni" plan). But the proposal lacked clear assignment of liability for damages, thereby fostering unproductive negotiations on the need for environmental impact statements, the degree to which the draining areas are and should be engaged in water conservation, and other subjects.

Liability factors should be negotiated upfront, prior to any actual damage claims and as part of the conditions required for rights to water use and to pollution discharge. Draining entities could assume the risks of discharging into public waters or reduce drainage through improved irrigation management practices determined by their internal economics. Obviously, monitoring of water quality in receiving waters -- perhaps cost-shared by the drainers and responsible public agencies -- would be required. Implementation could involve insurance and/or the posting of bonds for agricultural water uses, and could be required to secure rights to divert water from, and to discharge pollutants to, public waters. Environmental damage assessment procedures already exist for chemical and oil spills pursuant to the Superfund legislation[2].

Reviews of methods for economic assessment of damages have been published in recent years (Yang, Dower, and Menefee, 1984).

Such a priori arrangements can be made on an individual basis and can be incorporated into routine agricultural decisionmaking. Contracts provide the bridge and would require agricultural institutions with the collateral and authority to assume liabilities. This requirement poses difficulties for the irrigation community since a determination of risk-sharing among districts, landowners, and agricultural interests is needed. On the other hand, a measure of control is moved from the courts to the agricultural community.

In addition, potential risk-sharing by public agencies might be considered, based on the public interest in protecting natural resources. Economic incentives to prevent environmental damages provided by individual assignment of risks, however, would be reduced by such public risk-sharing. For example, to the degree that public agencies such as Reclamation assume such liability, incentives for Federal water contractors will be reduced in much the same way that publicly subsidized water supplies have reduced incentives to adopt new, water-conserving irrigation technologies and practices. The dispute over payment for cleanup at Kesterson--estimated at around $50 million (U.S. Department of the Interior, 1986)-- is a significant example of this issue. This dispute stems in large part from the lack of clarity in Reclamation water supply contracts regarding liabilities. A policy of incorporation of liability provisions in future water contracts, in oversight of appropriative water rights, and in issuance of discharge permits would help avoid this ambiguity.

Where public risk-sharing may be unavoidable, explicit authorities and responsibilities of public agencies are especially important. Damage claims could apply, for example, to the U.S. Department of Agriculture for its agricultural commodity support programs; to the U.S. Army Corps of Engineers for its flood control projects; and to the U.S. Department of the Interior for its irrigation projects. Clear liability provisions applied to such public agencies would provide them with an incentive to establish the rules of risk-sharing with project beneficiaries. Compensation revenues which might be paid by these agencies could be applied to programs to avoid and mitigate environmental damages.

## A SCENARIO OF WATER MARKETING AND WATER QUALITY IMPROVEMENT IN THE SAN JOAQUIN VALLEY'S WEST SIDE

The change from current water supply institutions based on fixed allocations and low prices is for many a radical option with many unknowns. This section examines some of the impacts and responses that might be expected from implementing a water market in a drainage problem area.

A 94,000-acre drainage study area on the west side of the Valley has been targeted as the major source of the trace elements and salts in the San Joaquin River. It has been estimated that the river quality standard set by the State Water Resources Control Board could be met with approximately 30 percent reductions in drain flows from the drainage study area, and that these decreases are feasible through improved management of irrigation applications (SWRCB, 1987).

## Problem Setting and Simulation Model

Siphon tube furrow irrigation systems with half or quarter mile runs are most commonly used to irrigate cotton, tomatoes, sugarbeets, and melons in the area, while wheat and alfalfa fields are generally irrigated with border check systems. These crops represent 80-90 percent of irrigated acreage in the study area. Reclamation contracts specify water volumes to be delivered and the price per acre-foot. Prices and allocations vary by district, with prices charged to growers ranging from $0 to $36/acre-foot, and allocations from approximately 2.3 to 4+ acre-feet per acre (SWRCB, 1987).

To examine implications of introducing water markets to the drainage study area, an agricultural production model was constructed to simulate grower decisionmaking. The model reflects regional economic conditions and agronomic characteristics. For modeling purposes the region was divided into physically homogenous cells[3], which were then subdivided into areas corresponding to water district jurisdictions. Areas within a cell are thus assumed to be homogenous with respect to aggregate soil and drainage conditions but differ from each other by institutionally set parameters. Two such subareas are selected for the analysis presented here. Subarea 1 represents a district with a $16 water price and an allotment of 3 acre-feet per acre, while subarea 2 is represented by a token charge of $1 per acre-foot and a 4-foot allotment.

Observations generated with the crop water production function model presented in Letey and Dinar (1986) are used to estimate production functions for alfalfa hay, cotton, sugarbeets, tomatoes, and wheat. A production function for melons is derived from actual observations. Crop production is specified as a function of applied water, irrigation application efficiency, pan evaporation and irrigation water salinity. Output is adjusted to reflect variation in average yields achieved by the water districts represented in the model.

Changes in irrigation practices to conserve water will necessarily increase production costs. To incorporate this aspect of the problem, a frontier-estimation technique was used to estimate an irrigation technology cost-efficiency function from data on annualized capital, maintenance, and labor costs of various irrigation technologies, and the associated efficiency levels[4].

Because irrigation techniques, and thus irrigation costs, vary by crop type, crop specific cost functions were estimated. Irrigation technology costs are specified as an increasing function of irrigation efficiency. The resulting functions thus describe the lowest cost at which a given irrigation efficiency level can be expected to be attained, on each crop, by an average farming operation in the area.

The objective of the model is to maximize the sum of grower returns to land and management in each subarea. Net returns to land and management are defined as crop and water sales revenues less harvest costs, preharvest costs, water and irrigation technology costs, water sales revenues, and drain costs.

Crop budgets produced by California State Cooperative Extension are the basis for most production cost parameters. Average crop prices and yields for each district were obtained from Reclamation. Production parameters and crop prices are assumed to be constant throughout the area, while water prices, per acre water allocations, and average yields vary by district, and thus by model subarea. All prices and costs are expressed in 1988 dollars.

The optimization problem is to choose cropping patterns, irrigation efficiency, water applications and water sales to maximize net returns to land and management subject to the production and cost functions. In addition, upper bounds on crop acreage are imposed on tomatoes and sugarbeets to reflect limitations due to market conditions and processor capacity. A minimum percent of cropped acreage is constrained to be planted with wheat for rotation. Lastly, water and land constraints are specified to reflect the limited availability of these resources.

The expected volume of resulting drainwater is determined as a function of water applications and irrigation efficiency on overlying fields, soil properties, and high water table conditions. Surface runoff and evaporation losses are assumed to be 7 percent of total water applications.

## RESULTS AND ANALYSIS

The model described above is specified as a nonlinear programming problem, and solved with an appropriate optimization algorithm. For a more detailed description of the model specification and simulation results, see Weinberg et al. (1990). The results predict optimal agricultural production and water-sale decisions in response to alternative water market prices. The analysis was conducted for the two subareas in order to examine the role of heterogeneous institutional parameters, e.g., water costs and allocations, on optimal response to a water market.

## Water Market Simulation

A water "supply" curve was derived by varying the market price of water between $0 and $150 per acre-foot and is presented in figure 2. Subarea 2 enters the market when the price is $40, while subarea 1 enters at $70. As the opportunity cost of using water increases, due to increased market prices, water use is reduced on all crops. In many cases, improvements in irrigation efficiency compensate for reductions in applied water such that yields are unaffected. As water market prices rise, it becomes profitable to sacrifice some cotton and melon output, rather than incur the extra cost associated with increasing irrigation efficiency enough to compensate for further reductions in water applied on these crops.

Water may also be conserved through changes in cropping patterns. Crops with a high consumptive use will be changed to crops using less water when it becomes profitable to do so. This switch occurs between sugarbeets and cotton in subarea 2 when the water market price is $50/AF. As the price reaches $70 and $80/AF it becomes optimal to move a small percentage of the cotton acreage into melons in areas 1 and 2 respectively. Eventually, as the market price of water continues to rise, it may become profitable to take land out of production. It should be noted that a switch from crops that have high water needs to those with low consumptive requirements will free up water for sale but will not necessarily have a proportionate effect on drainage production since only that amount of water that is greater than plant needs becomes drainage.

Gradual reductions in consumptive water use result in incremental water supply response to a broad range of market prices. However, as the price approaches $140/AF it becomes suboptimal to produce cotton and a sharp jump in water sales results as the cotton acreage comes out of production.

At a market price of $80/AF, exactly 1 AF/AC is sold from subarea 2; the remaining allocation (3 AF/AC) is equal to the total allocation in subarea 1. Comparison of irrigation applications and efficiencies with the base case for subarea 1 demonstrates the equivalence of these situations. In contrast to the 1-foot sales from subarea 2, the $80 price invokes a supply response from subarea 1 of only .2 AF/AC. The difference in the initial water allocation is also reflected in the income realized in the two areas. Net revenues in subarea 2 increase by $24/acre over the base case, while the difference in subarea 1 is only $2/acre.

In subarea 1 optimal irrigation efficiencies are at least 10 percent higher when the market price for water is $120/AF than in the base case, in order to conserve .5 AF/AC for sale. The higher efficiencies might be achieved with high management levels of quarter mile furrows or border checks with tailwater return systems, or might require surge (with quarter-mile runs) or linear move sprinkler irrigation technologies. Water applications on cotton and melons are 11 percent lower than when the market price is $80, resulting in slight yield reductions. Water applications on tomatoes decreased by 8 percent in response to the $40 increase in the market price. Net returns increased by $18/acre, although returns from crop production fell by $45/acre. In subarea 2 water applications on cotton and wheat are reduced by 29 percent from base, and melon irrigations drop by 12 percent as the market price increases from $80 to $120. Water sales revenues of $163/acre easily offset $92/acre reductions in crop returns.

Figure 2 maps the responses to increasing water market prices that would be economically optimal for irrigators in subareas 1 and 2. These are the implicit water supply curves which are derived from the specific circumstances of cropping, water endowments, and other conditions within each subarea.

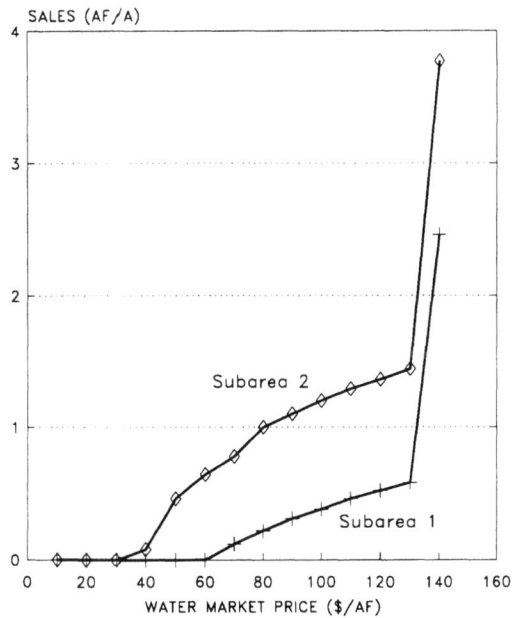

Figure 2. Predicted water sales.

## Environmental Implications

The volume of drainwater estimated to be generated under the market price scenarios is illustrated in figure 3. Water conservation in response to the water market is clearly reflected in drain flow reductions in both subareas. As expected, subarea 2 is predicted to generate more drainwater than in subarea 1 in all cases, but greater water supply responsiveness results in proportional reductions that are greater than in subarea 1.

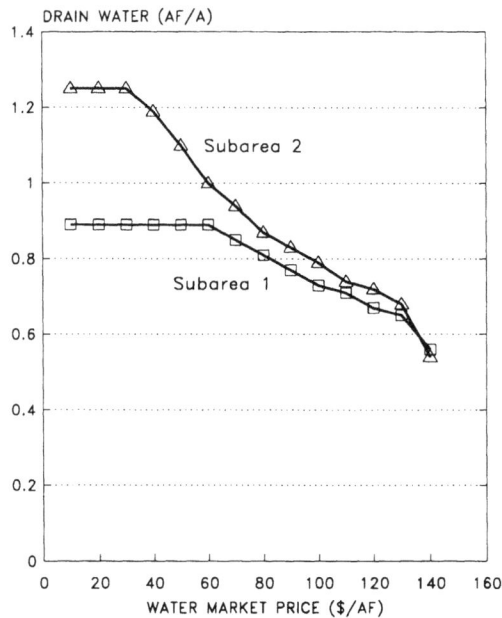

Figure 3. Predicted drainwater volumes.

In subarea 2, collected drainwater is reduced by 30 percent in response to a market price of $80/AF. Irrigation water exceeding plant needs may percolate below the root zone and enter drains; this is the portion of drain flows that a grower has direct control over. In addition, irrigation on neighboring undrained fields may contribute to a regional high water table and influence total drain flows indirectly. A market price approaching $60 in subarea 2 is sufficient to elicit a 30-percent reduction in water percolating below the root zone. In subarea 1, a 30-percent reduction in total drain flows (including grower contribution plus contribution from the high water table) is not achieved until the market price exceeds $130. A 30-percent reduction in deep percolation is realized when the water market price is $90. These results reflect the fact that the contribution of the high water table as a percent of total drain flows is higher in subarea 1. The 30-percent drainage reductions are significant in that they represent the target established by the State for San Joaquin River quality goals (California SWRCB, 1987).

## CONCLUSIONS AND RECOMMENDATIONS

Grower decisions and management practices influence surface runoff, subsurface percolation, and irrigation return flows, and are often linked to the degradation of water quality through contributions of nutrients, salts, and toxins to rivers and other aquatic systems. The environmental problems of irrigation and drainage include surface and ground-water contamination, deterioration of fish and waterfowl habitat, potential public health problems, degradation of soil and land resources, and ground-water overdraft. Yet, despite the diversity of the problems and of potential remedies, one common and central element of any mix of solutions exists -- reduced water use, and reduced drainage, in agriculture. The institutions that govern water prices and allocations have not historically presented growers with economic signals that indicate the true cost of water use.

Water markets may create an incentive to conserve water in agriculture, thus creating an important "new" water supply, and providing a flexible complement to pollution reduction policy. The simulation exercise described above examines the potential for water markets to address drainage reduction goals by modeling grower response to a range of water market prices and the drainwater changes that would result.

Water supply responsiveness is predicted for two areas with different underlying structures. Irrigation application efficiencies are increased in each area to conserve water for sale without appreciably reducing yields or cropped acreage. The results indicate that significant reductions in total drain flows may be achieved as a byproduct of a water market in which prices range from $80 to $130. These prices are in the range of those realized in Western water markets (table 1). They are also well below water prices paid by urban areas in California (figure 1), indicating a clear potential for a water transfer that is beneficial to agricultural and urban sectors, as well as to the environment.

The analysis suggests that underlying institutional factors, such as historic water allocations and delivered-water prices, can be as important as economic and hydrologic factors in conditioning water market impacts. In one of the areas considered, water allocations are relatively "tight" compared to the other, and agricultural crop decisions and irrigation practices in the base case reflect a higher shadow price of water. Participation in a water market, and expected drainage reductions, is thus more modest than in a neighboring area with a more abundant allocation.

This type of flexibility assures that the cheapest agricultural water conservation options will be developed first if water marketing opportunities are presented to irrigators. Drainage and pollution reduction will occur as benefits which are incidental to water conservation. Optimal drainage pollution reduction would, however, require that the cheapest drainage reduction, as opposed to water conservation, measures be taken to meet water quality goals. The creation of an emissions market based on tradable discharge permits is required to achieve water quality and pollution control goals efficiently. Ideally, water marketing and tradable discharge permits would operate simultaneously. These markets, strengthened by the increased certainty provided by liability contracting and insurance, would provide the foundations of a policy designed to achieve both agricultural and environmental goals.

The key obstacles to implementing these economic instruments are political and bureaucratic. The legislative authorities in State and Federal laws are probably adequate to give the relevant agencies the power to facilitate the development of these instruments. The key agencies include the State and Regional Water Quality Boards, EPA, Reclamation, and Department of Water Resources (DWR). Each has authority with regard to different and frequently overlapping functions. The waterboards have delegated water pollution control authorities. The State Board oversees the water rights system, and any water transfers must be reviewed by it. EPA provides a check on the State Board's performance with respect to water pollution control. Reclamation governs all Federal Reclamation water contracts, including any changes in place or manner of use associated with water marketing. In spite of Reclamation policies announced in 1987 to facilitate water transfers, such facilitation has not noticeably materialized in California. DWR operates the California Aqueduct to convey water to southern California. In addition, DWR controls access to substantial funds in the form of loans and subsidies.

The economic instruments discussed have not been part of the historic policy mission of any of these agencies. Their implementation would, then, constitute a substantial departure from the organizational identities extant in each. Individuals notwithstanding, each agency resists implementation. Fault is easy to find in systems which have not been "proven." Nonetheless, the potential of these systems for environmental and economic performance is so great that it is difficult to imagine that they will not be implemented in the future.

## NOTES

[1] A notable exception is the Broadview Water District, which has implemented an increasing block pricing scheme in an effort to encourage water conservation and drainage reduction. For details of the program design and implementation, see the Wichelns chapter in this volume.

[2] The Comprehensive Environmental Response, Compensation, and Liability Act of 1980, Sec. 301 (c), 42 U.S.C. Sec. 9613 (c) (1982 & Supp. IV 1986), requires that natural resource damage assessment regulations be developed. The first large financial test of these regulations involves their application to the damages from the Exxon Valdez oil spill in Alaska's Prince William Sound in March 1989.

[3] The cells indicated here correspond to those developed for the WADE model. For a more complete description see chapter by Hatchett, Horner, and Howitt this volume.

[4] Irrigation efficiency is defined as the ratio of the depth of water beneficially used (plant needs plus minimum leaching fractions) to the average depth of applied water. This information was compiled by Davids and Gohring (1989) for 11 irrigation technologies, with three management levels for each technology.

## REFERENCES

Ackerman, B. and Stewart, R., 1988. Reforming Environmental Law: The Democratic Case for Market Incentives, *Columbia Journal of Environmental Law*, 13(2), School of Law, Columbia University, New York, NY.

Baumol, W. and Oates, W., 1988. *The Theory of Environmental Policy*, Cambridge University Press, New York, NY.

Brown, G. and Johnson, B., 1984. Pollution Control by Effluent Charges: It Works in the Federal Republic of Germany, Why Not in the U.S.? *Natural Resources Journal*, 24, p. 929.

Bryan, K., 1928. Changes in Plants Associated with Changes in Groundwater Levels, *Ecology*, 9(4).

California Department of Water Resources, 1987. *California Water: Looking to the Future*. Bulletin 160-87, Chapter 4.

California State Water Resources Control Board, 1987. *Regulation of Agricultural Drainage to the San Joaquin River*. Final Technical Committee Report.

Central Valley Regional Water Quality Control Board, 1988. *Staff Report on the Program of Implementation for the Control of Agricultural Subsurface Drainage Discharges in the San Joaquin Basin*.

Conniff, R., A, 1989. Deal that Might Save a Sierra Gem, *Time Magazine*.

Davids, G. and Gohring, T., 1989. *Irrigation System Costs and Performance in the San Joaquin Valley*. Prepared for the San Joaquin Valley Drainage Program, CH2M Hill.

Doroshov, S.I. and Wang, Y. L., 1984. *The Effect of Subsurface Agricultural Drainage Water on Larval Striped Bass Morone Saxatilis (Walbaum)*. Department of Animal Science, University of California, Davis.

Eisler, R., 1987. *Selenium Hazards to Fish, Wildlife, and Invertebrates: A Synoptic Review*. Biological Report 85(1.5), U.S. Fish and Wildlife Service, Laurel, MD.

Harrison and Sewell, 1980. Water Pollution Control By Agreement: The French System of Contracts, *Natural Resources Journal*, 20, p.765.

Harrison, D. L. and Wigington, R., 1987. Water Rights: A Protection Tool for the West, Ecology Forum No. 64, *The Nature Conservancy Magazine*, Washington, DC.

Heinz, J. and T. Wirth, U.S. Senators, 1988. *Project 88: Harnessing Market Forces for the Environment*. Washington, DC.

Holden, P. W., 1986. *Pesticides and Groundwater Quality*. National Academy of Science Press, Washington, DC.

Huber, P. W., 1988. *Liability: The Legal Revolution and Its Consequences*. Basic Books, New York, NY.

Interagency Drainage Program (IDP), 1979. *Agricultural Drainage and Salt Management in the San Joaquin Valley*. Final Report.

Jaksch and Niedzialkowski, 1985. Speeding Water Cleanup While Saving Money, *EPA Journal*, Washington, DC.

Kennedy, D., 1989. Humans in the Chemical Decision Chain, *Choices*, American Agricultural Economics Association, Third Quarter.

Letey, J. and Dinar, A., 1986. Simulated Crop-Water Production Functions for Several Crops Irrigated with Saline Waters, *Hilgardia*, 54(1).

Mono Lake Committee and the Los Angeles Department of Water and Power, 1987. Press Announcement, Los Angeles, CA.

Ohlendorf, H.M.; Hothem, H.C.; Bunck, C.M.; Aldrich, T.W.; and Morre, J.F., 1986. Relationships Between Selenium Concentrations and Avian Reproduction, *Transactions of the North American Wildlife and Natural Resources Conference*, Reno, NV., pp. 330-342.

Panoche Drainage District, 1990. *Environmental Assessment and Initial Study -- Proposed Use of the San Luis Drain for Conveyance of Drainage Water through the Grasslands Water District and Adjacent Grasslands Areas*, Firebaugh, CA.

Robie, R. B., 1988. The Delta Decision, The Quiet Revolution in Water Rights, *Pacific Law Journal*, 19(4), McGeorge Law School, University of the Pacific, Sacramento, CA.

SRI International, 1985. *Chronic Toxicity of San Luis Drain Effluent to Neomysis Mercedis*. Final Report to the U.S. Bureau of Reclamation.

Sheridan, D., 1981. *Desertification in the United States*. President's Council on Environmental Quality, Washington, DC.

Smith, R. T. and Vaughan, R., (Eds.), 1989. Let's Make a Deal: The IID/MWD Water Conservation Agreement, *Water Strategist*, 2(4), Stratecon, Inc., Claremont.

Stavins, R. N. and Willey, Z., 1983. Trading Conservation Investments for Water. In: Charbeneau, R. J., (Ed.), *Regional and State Water Resources Planning and Management*, American Water Resources Association, Bethesda, MD.

Tietenberg, T., 1985. Emissions Trading: *An Exercise in Reforming Pollution Policy*. Resources for the Future, Washington, DC.

U.S. Department of the Interior, Water and Power Resources Service, 1981. *A Report on the Mid-Valley Canal Feasibility Investigation, Central Valley Project, and East Side Division*.

U.S. Department of the Interior, Bureau of Reclamation, 1986. *Final Kesterson Program Environmental Impact Statement*.

U.S. Department of the Interior, *Office of the Secretary. Interior Begins Detailed Studies of Irrigation Drainage at Four Sites*. News Release.

U.S. Fish and Wildlife Service, 1986. *Contaminant Issues of Concern-- National Wildlife Refuges*.

University of California Committee of Consultants on Drainage Water Reduction, 1988. *Opportunities for Drainage Water Reduction*. Report Number 1 in a Series on Drainage, Salinity, and Toxic Constituents, UC Salinity/Drainage Task Force, Davis.

Water Efficiency Working Group, 1987. *Water Efficiency: Opportunities for Action*. Report to the Western Governors, Western Governors' Association, Denver, CO.

Weinberg, M.; Wilen, J.E.; Kling, C.L.; and Willey, Z., 1990. Water Markets and Water Quality. Presented at the Annual Meeting of the American Agricultural Economics Association, Vancouver, British Columbia.

Westcot, D.W.; Rosenbaum, S.; Grewell, B.; and Belden, K., 1988. *Water and Sediment Quality in Evaporation Basins Used for Disposal of Agricultural Subsurface Drainage Water in the San Joaquin Valley, California*. Central Valley Regional Water Quality Control Board.

Willey, Z., 1977. *Groundwater in the California Water Quandary*. Proceedings of the Biennial Groundwater Conference, Water Resources Center, University of California, Davis.

Willey, Z. and Graff, T., 1988. Federal Water Policy in the United States An Agenda for Economic and Environmental Reform, *Columbia Journal of Environmental Law*, 13(2), pp. 325-356.

Yang, E. J.; Dower, R. C.; and Menefree, M., 1984. *The Use of Economic Analysis in Valuing Natural Resource Damages*, Environmental Law Institute and U.S. Department of Commerce, Washington, DC.

Yardas, D., 1989. Water Transfers and Paper Rights in the Truckee and Carson Basins, Indian Water Rights Section. Proceedings of the American Water Resources Association Symposium on Headwaters Hydrology, Missoula, MT.

# 28 IMPACTS OF SAN JOAQUIN VALLEY DRAINAGE-RELATED POLICIES ON STATE AND NATIONAL AGRICULTURAL PRODUCTION

Gerald L. Horner, Agricultural Economist, Davis, California;
Stephen A. Hatchett, CH2M HILL;
Robert M. House, USDA-ERS; and
Richard E. Howitt, University of California, Davis

### ABSTRACT

The chapter links National and State agricultural policy models with a regional agricultural production and ground-water hydrology model to track the effects of environmental and commodity policies. Results indicate that a policy which would eliminate subsurface drainage disposal would not significantly effect the amount of agricultural production within the 10-year time horizon assumed in the analysis. A 20-percent decrease in the demand for cotton is projected to decrease total irrigated acreage by 20 percent but subsurface drainage water is projected to double. This phenomenon is explained by decreased unconfined ground-water pumping which serves to reduce perched ground-water tables. The two policy scenarios illustrate the importance of unconfined ground-water pumping and the spatial allocation of surface water supplies in reducing the production of subsurface drainage, if disposal is restricted.

### INTRODUCTION

Integrating resource use and development policies with environmental protection goals can achieve a better allocation of natural resources. Water resource development and land-use plans can be modified if water quality problems are projected to result from development. Water quality policies and changes in commodity prices can also affect the amount and location of agricultural production.

Agricultural resource and commodity policies have never formally been integrated even though both greatly influence farm decisions. Resource policies are usually formulated and enacted at the regional level while political pressures to formulate commodity policies are Nationally orientated. Analyzing agricultural resource and commodity policies requires sufficient regional information to delineate changes in firm level resource use and resulting changes in the physical environment. National and foreign commodity markets must also be considered including evaluation of changes in the comparative advantage of competing regions in agricultural production.

## REGIONAL, STATE, AND NATIONAL MODEL DESCRIPTIONS

A traditional method of estimating the effects of National policies on resource use and environmental quality is to use National models with sufficient regional detail to identify natural resource problems. These models, however, are extremely large, very expensive to use, and contain a limited set of parameters that are of National interest (Piper et al., 1989 and Howitt, 1989). By using existing State and National models to focus on policy parameters relevant to those levels of aggregation, effort can be focused on modeling the region's physical system in sufficient detail to project the effects of policy changes on agricultural production and the environment. This approach is taken in this study. A San Joaquin Valley (Valley) agricultural production-hydrology model has been linked with a California agricultural-resources model, and a National agricultural policy model, to determine the effects of drainage policies on agricultural production and prices.

The regional effects of changes in National commodity programs and resource policies implemented at the State level can be projected on the location of regional crop production, commodity prices, and resource use. Resource and commodity policies have varying regional impacts depending on the number of alternative crops that can be grown and the supply of regional resources. Each individual policy model will project changes in commodity prices and crop production as a result of policies originating at the regional, State, or Federal level.

### Westside Agricultural Drainage Economics Model (WADE)

This model is based on Horner and Dude, (1980). It was further extended to evaluate alternative land use and water development policies, and to develop "Best Management Practices" for the reduction of subsurface drainage water from irrigated agriculture. The model was used by the San Joaquin Valley Drainage Program (SJVDP) to analyze agricultural drainage issues that existed on the west side of the Valley. The linkages between components of its analytical system were enhanced and made easier to use by rewriting all of them into the GAMS-MINOS modeling language (Brooke et al., 1988).

The analytical system is designed to simulate the relationship of onfarm cropping pattern, water use, and irrigation technology decisions to the volume and quality of subsurface drainage water. It contains three main components: the agricultural production model, the hydrology model, and the salinity model. These components are linked and represent a 6-month timeframe. The winter period extends from November-April and simulates preirrigation and subsurface drainage installation decisions. The summer period, May-October, determines optimal cropping patterns, irrigation technology use, water use, and drainage water disposal decisions. This sequence of model operations is repeated until a steady state is achieved in the results, which usually occurs after 6 to 10 years depending on the type of scenario posed.

Time is a critical parameter in this type of analysis, and a dynamic model would be desirable for analyzing policy issues. However the model is specified in a 6-month timeframe and solved sequentially rather than specified as a dynamic model. This was done for two reasons. First, a dynamic model would have to be severely limited in spatial and system detail and it was concluded that modeling the agricultural production-soil-hydrologic systems in the Valley requires substantial detail. Second, developing a dynamic model is costly and uncertain due to the limited availability of software and computer facilities required to complete such a task. However a dynamic model would definitely provide some valuable insights into deriving optimal resource and environmental policies, but models of this type are being developed within the SJVDP (Kasower, 1989). This analytical system was structured to be a positive model that simulates economic behavior and physical phenomena and provides comparisons of results based on alternative assumptions rather than deriving optimal policies. It provides a basis to assist in estimating benefits and costs associated with alternative planning objectives. A complete description of WADE is presented in a previous chapter by Hatchett et al.

Changes in institutional and economic parameters can be easily introduced into WADE. These can take the form of, but not limited to, mandated land retirements, drainage water disposal limits on quality and amounts, and changes in surface water supplies. Policy can also change how conflicts are resolved and how water is priced and allocated. Some preliminary results of agricultural drainage policy runs were presented by Hatchett et al. (1989).

## *California Agricultural Resources Model (CARM)*

This model provides a convenient method of analyzing agricultural crop production and commodity demand changes for California and it also provides a necessary aggregation to relate to the National model's production regions. CARM was developed to analyze the effects of National commodity policy, and State resource and environmental policies on shifts in the location of crop production, commodity prices, and related resource use. CARM is structured as a resource constrained quadratic programming model which is similar to the structure of WADE. Crop production activities are specified for 47 crops in 17 regions which are mostly aggregations of counties (figure 1). Some crops have multiple activities such as irrigated and dryland, and for seasonal production variations. Crops in the model account for about 95 percent of the State's total crop acreage and value. CARM is a static model representing a 1-year production period.

A demand function is specified for each commodity and the constant term is adjusted to allow for production in the rest of the United States and for demand-shift factors, such as income growth and changes in exports. The demand equations are econometrically estimated from commodity price and production data for 1969-84 (Howitt, June 1989, pp. 9-13).

Regional production cost data were derived from county level budgets available through the University of California Cooperative Extension Service (1980-85). The U.S. Department Agriculture's (USDA) commodity programs are modeled and historical data on commodity program participation comes from USDA's Agricultural Stabilization and Conservation Service (1985). This model has been used to project the effects of developing additional water supplies, declining energy supplies, and increased air pollution levels on amount and location of agricultural production (Howitt, 1989).

## United States Mathematical Programming Regional Agricultural Model (USMP)

This model is currently being used by the USDA-Economic Research Service for the analysis of alternative Federal commodity policies. USMP is a spatial model which incorporates Government agricultural commodity programs and solves for equilibrium in all major agricultural factor and product markets. USMP is formulated as a quadratic programming model and is written in the GAMS-MINOS modeling language. The nonlinear formulation and modeling system approach permit quick and elegant structuring of complex relationships, facilitate rapid-response analysis, and make spatial sector model policy analysis feasible (House, 1987).

USMP estimates equilibrium levels of factor and commodity prices and regional production as well as flows of commodity production into domestic, stock, and export demand markets. Commodity policy changes are specified in the model and results are compared to current conditions and selected baseline

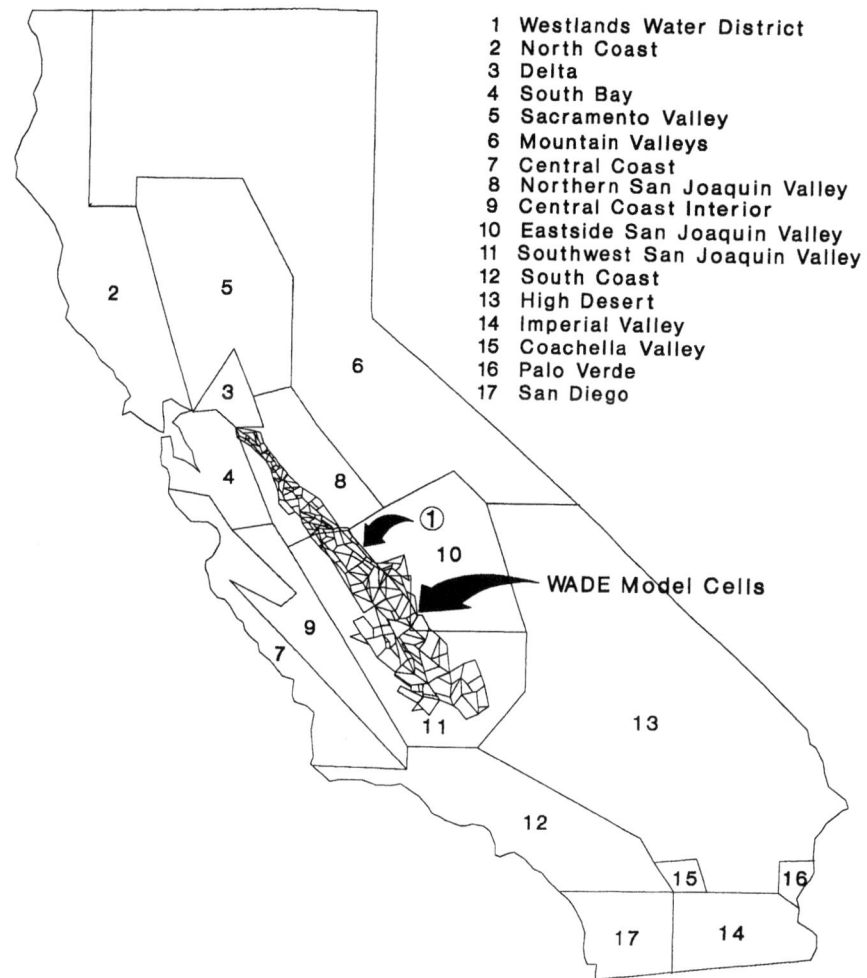

Figure 1. California agricultural resources model production regions.

performance indicators. The static, 1-year model accounts for production risk and uncertainty in response to price changes.

USMP specifies 168 crop and livestock production activities, and 5 resource supply activities for the 10 USDA farm production regions (figure 2). The model specifies activities for 36 major agricultural commodities, 55 commodity demand and supply functions, and 23 inputs on a National basis.

The U.S. Government implements a number of commodity programs to support crop prices and farm income. Programs for wheat feed grains, soybeans, cotton, rice, and dairy are modeled in USMP. Prices are supported through the Commodity Credit Corporation nonrecourse loan program. If prices fall below a specified "loan rate," producers may delay selling their crop, and take a loan from the Government, using the commodity as collateral. The producer can then default if prices remain low and the commodity is forfeited to the Government.

Figure 2. USDA Agricultural Production Regions

Incomes are supported through a target price-deficiency payment program if prices are below target levels. Payments are made to producers which are determined by the difference between the market price and the lesser of the target price or loan rate times the grower's production.

Most programs are voluntary but a variety of requirements must be met to participate. The principal requirement is that producers cannot exceed their "acreage allotment," based on historical production and, in addition, they must not plant a specified percentage of allotted acreage called "set-aside." If commodity supplies are large relative to projected demand, an acreage "diversion payment" is also made to producers for not planting additional amounts of their acreage allotments.

The dairy program supports farm milk prices through the Government purchase of processed dairy products. Farm milk prices are maintained at specified support levels.

USMP does not include other commodity programs nor cross compliance requirements, Food Security Act of 1985 (marketing loans for cotton and rice), limits on payments to a single grower, nor the Farmer Owned Reserve (FOR) program. FOR stocks are included in commercial stocks in USMP.

## LINKAGES BETWEEN WADE, CARM, AND USMP

The three models are linked by policy and institutional information flows that include crop acreage and commodity production and prices. In addition, regional definitions of disaggregated models are specified as spatial subsets of the more aggregated models.

### Policy and Institutional Informational Flows

CARM is linked to USMP by specifying regional crop production activity levels as shares of the National production and using commodity prices estimated by USMP. USMP projects the location of agricultural production based on regional comparative advantage, resource supplies, and the demand for agricultural commodities. California's share of National production and commodity prices map directly into CARM as baseline data. CARM is then solved to project the location of production within the State. The resulting Valley crop acreage are used to derive baseline production and price data for the WADE model. Solving WADE, with the revised price and production data, will project the effect of changes in commodity policy on the WADE region of the Valley.

The effect of changes in Valley commodity production on State and National production patterns can be traced from the WADE model through CARM to the USMP model by reversing the information flow. The changed regional commodity production amounts projected by WADE become fixed in CARM and solved. The CARM solution is then fixed in the USMP and solved to determine the crop shifts among the 10 USDA agricultural production regions and equilibrium commodity prices.

## Spatial Relationships

The original Valley model was spatially delineated on the basis of 66 soil groups that were specified and mapped by USDA's Soil Conservation Service (USDA River Basin-Watershed Planning Staff, 1973). A linear delineation of these areas resulted in 422 regular polygons or cells. When the WADE model was specified, just the 205 cells that were thought to affect agricultural drainage were included in the model. The smaller number of cells allowed other aspects of the drainage problem to be specified in the model. The area of interest lies to the west of the San Joaquin River in Merced and northern Fresno Counties, Westlands Water District, and most of the irrigated areas of Kings and Kern Counties. This area contains most of the Federal and State water project service areas located in the Valley. About 500,000 acres have water table depths of less than 5 feet (SJVDP, 1987).

The WADE model cells lie within four CARM regions: Westlands Water District, and the northern, eastside, and southwest areas of the Valley (figure 1). This area produces about $9 billion worth of agricultural commodities annually from about 5 million acres. About 95 percent of the Valley's agricultural land is irrigated. USDA commodity programs affect cotton and grain crop acreage in the Valley. About 1 million acres of cotton are grown in the Valley in some years. The production of feed grains and forages are also indirectly influenced by the dairy program.

## DRAINAGE POLICY ANALYSIS AND RESULTS

A WADE policy analysis would normally be conducted in three stages. First, several runs are made to calibrate the production, hydrology, and salinity models. The models are calibrated to five historical data points on crop acreage, depth of ground-water table, and salinity concentrations. A run is then made with the calibrated model assuming current economic, physical, and institutional condition for the assumed period of analysis. The period of analysis used for this set of runs was 10 years. This will be termed the "Base" run.

The second stage is to formulate a "Future Without" scenario. The "Future Without" scenario assumes the most likely set of economic, institutional, and physical conditions that are likely to exist in the absence of corrective action that could be taken by Government agencies and the private sector to solve the drainage problem. A "Future Without" alternative must be included in all Federal planning studies and serves as a baseline against which planned actions can be measured in terms of their positive or negative values (SJVDP, 1989). One possible "Future Without" scenario is to assume drainage disposal restrictions, land use changes and retirements, and restrictions on subsurface drainage installations.

The third stage is the formulation of policies to correct the drainage problem and comparing the results with the "Future Without" scenario. A program to correct drainage problems would probably include a mix of source control; ground-water management; drainage water treatment, reuse and disposal; fish and wildlife measures; and a host of institutional changes (SJVDP, 1989). The analytical procedure required to analyzed this "Future Without" scenario is very complex and too detailed to conduct and report in this chapter, and the purpose here is to demonstrate the type of analysis that is possible with the linked National-State-regional model structure. Therefore, a "Worst Case Scenario" was devised, and results from that analysis, compared with the "Base" run to demonstrate the value of the analytical system in comparing results from two very different institutional settings.

The "Worst Case Scenario" hypothesized for this analysis simply prohibits subsurface drainage water disposal except to onfarm evaporation ponds, and limits reuse as irrigation water only if, when blended with existing water supplies, it does not impair future productivity by increasing soil salinity to excessive levels. The "Worst Case Scenario" assumes that no subsurface drainage water can leave the farm, either as surface or subsurface return flows. To comply with this type of regulation, irrigators would need to construct and operate evaporation ponds that would minimize deep percolation and meet wildlife protection requirements. The preliminary cost estimate of complying with these requirements is $412 per acre-foot of drainage water (Stroh, 1990).

Results from the "Worst Case Scenario" were compared with the "Base" run, described earlier, for the assumed period of 10 years. This comparison indicates that Valley production will not be greatly affected within the 10 years. Total irrigated acreage for the WADE region does not begin to decline until the 5th year of the simulation and by the 10th year only a 3-percent decline is projected (table 1). Most of this decrease is projected to occur in grain crops and alfalfa which comprise about 25 percent of the total irrigated acreage in the WADE region of the Valley. Acreage reductions amount to 24,600 acres of grain crops, 18,300 acres of alfalfa, and 8,300 acres of tree fruits and nuts. Row crop acreage, which is mostly comprised of cotton, was projected to remain unchanged from the "Base" run. The effect of such a relatively small decrease in agricultural production in the WADE region is relatively insignificant when compared to California's production of these crops. These crops are grown in substantial amounts in the eastern Valley and the Sacramento Valley, and substantial agricultural production resources exist in those regions that would assimilate these acreage affecting commodity and resource prices only slightly. The National effects of the "Worst Case Scenario" would be imperceptible.

Table 1. Crop acreage and farm income, "Worst Case" scenario, percent change from the "Base" solution.

| | Irrigated Acreage* | | | | Farm Income |
|---|---|---|---|---|---|
| | Total | Grain | Alfalfa | Row Crop | Tree Fruits & Nuts | |
| | *(percent change from "Base" run)* | | | | | |
| Year 1 | -1 | 0 | -1 | -2 | 0 | -3 |
| Year 2 | 0 | -1 | 0 | 0 | 0 | -2 |
| Year 3 | 0 | -2 | 0 | 0 | 0 | -4 |
| Year 4 | 0 | -4 | -1 | 0 | 0 | -5 |
| Year 5 | -1 | -4 | -1 | 0 | -1 | -5 |
| Year 6 | -1 | -6 | -3 | 0 | -2 | -9 |
| Year 7 | -2 | -7 | -4 | 0 | -2 | -11 |
| Year 8 | -2 | -8 | -5 | 0 | -3 | -13 |
| Year 9 | -2 | -8 | -6 | 0 | -3 | -14 |
| Year 10 | -3 | -9 | -7 | 0 | -3 | -17 |

*Values less than 1 percent are rounded to 0.

Farm incomes decline over the 10-year simulation due to the higher costs of additional subsurface drainage installations, recycling drainage water, the small increase in evaporation ponds, and slight reductions in crop yields due to higher levels of soil salinity.

Subsurface drainage water discharge is almost eliminated under the "Worst Case Scenario" (table 2). The $412 per acre-foot cost of evaporating drainage water is not the least cost alternative and instead more drainage water is reused as irrigation water. However, recycling drainage water reduces the need for ground water, with confined pumping being reduced by 13 percent and unconfined pumping by 9 percent by year 10. Reduced unconfined ground-water pumping directly increases the levels of perched ground-water tables and the need for additional subsurface drains.

Table 2. Drainage water produced, discharged, and recycled, and irrigation water used by source, "Worst Case" scenario, percent change from the "Base" solution.

| | Subsurface Drainage Water | | | Irrigation Water | | |
|---|---|---|---|---|---|---|
| | Produced | Discharged | Recycled | Surface | Unconfined | Confined |
| | *(percent change from "Base" run)* | | | | | |
| Year 1 | -26 | -95 | 1,707 | -2 | -3 | -3 |
| Year 2 | -9 | -97 | 733 | -1 | -2 | -3 |
| Year 3 | 3 | -95 | 1,031 | -1 | -3 | -5 |
| Year 4 | 10 | -96 | 1,021 | -1 | -5 | -8 |
| Year 5 | 16 | -98 | 981 | -2 | -6 | -8 |
| Year 6 | 20 | -93 | 894 | -2 | -6 | -9 |
| Year 7 | 26 | -90 | 835 | -2 | -7 | -11 |
| Year 8 | 31 | -88 | 783 | -2 | -8 | -12 |
| Year 9 | 34 | -86 | 666 | -2 | -8 | -12 |
| Year 10 | 38 | -85 | 665 | -4 | -9 | -13 |

In summary, the effect of eliminating off-farm drainage water disposal is to hasten the rise of perched water tables. This contributes to greater drainage water disposal requirements that can only be met by recycling and the use of very expensive evaporation ponds. Although the long-term outlook is grim under this scenario, irrigators will be able to survive financially in the near term due to the use of more efficient water application and management systems and recycling drainage water. The longer-term issue is the effects of higher soil salinity levels on crop yields which will depend on crop salinity tolerance in the future and the ability of irrigators to manage water supplies with different salinity concentrations (UC Committee of Consultants on Drainage Water Reduction).

## REDUCED COTTON DEMAND AND DRAINAGE POLICY

Changes in commodity markets and Government programs can affect regional production patterns, resource use, and environmental quality. To illustrate the effect of changing commodity demands on the Valley drainage situation, a 20-percent reduction in cotton prices were simulated in the USMP. In addition to reducing domestic and international cotton demands, the cotton loan rate and target price were also reduced. USMP projected a 19.3-percent reduction in cotton acreage and 26.4 percent reduction in cotton production for the United States. However, regional crop changes were very different.

California was one of the bigger losers with a 41-percent reduction in cotton acreage. The Mountain region lost 49 percent, while Appalachia's was reduced by 38 percent, Southeast and Delta by 21 percent, and the Southern Plains by 10 percent. USMP projects changes in regional cropping patterns on the basis of profitability of alternative crops within each region, and availability and cost of water and land resources.

The assumed 20-percent reduction in cotton demand and the 41-percent reduction in California cotton production projected by USMP were specified in the CARM model to determine the regional impacts within California. Again, the regional reductions were very different with the Valley projected to lose considerable cotton acreage. Reductions projected by CARM are as follows:

| Region | % Reduction |
|---|---|
| Eastern San Joaquin Valley | 84 |
| Northern San Joaquin Valley | 9 |
| Southwestern San Joaquin Valley | 30 |
| Westlands Water District | 30 |
| Cochella Valley | 11 |
| Imperial Valley | 20 |
| Palo Verde | 14 |

The Valley cotton acreage reductions and price assumption were imposed on the WADE model and then solved to estimate changes in cropping patterns, crop yields, water use, and water table depths for each cell. It is also assumed that the cotton market will not recover within the 10 years of the analysis.

The WADE model projected substantial increases in grain crops, alfalfa, and tree fruits and nuts as a result of reduced cotton demand but total irrigated acres declined by about 20 percent. After 10 years, this resulted in a 49-percent decrease in farm income (table 3).

Projected reductions in total irrigated acreage range between 16 and 22 percent of "Base" levels over the 10-year simulation. Reduced acreage requires less irrigation water and a reduction in ground-water pumping since ground water is the most costly source of irrigation water. WADE model results indicate that about 1.6 million acre-feet of groundwater is pumped within the WADE region in an average surface water supply year. Ground-water pumping is close to being evenly split between the confined and unconfined aquifers for the WADE region as a total. However this varies considerably for individual cells. Projected reductions in unconfined and confined ground-water pumping under the "Worst Case Scenario" and reduced cotton demands range from 70 to 80 percent of "Base" levels which amounts to about 555,000 acre-feet per year (table 4).

Table 3. Crop acreage and farm income, "Worst Case" scenario with reduced cotton demand, percent change from the "Base" solution.

|  | Irrigated Acreage | | | | | Farm Income |
|---|---|---|---|---|---|---|
|  | Total | Grain | Alfalfa | Row Crop | Tree Fruits & Nuts | |
|  | *(percent change from "Base" run)* | | | | | |
| Year 1 | -22 | 17 | 5 | -49 | 4 | -22 |
| Year 2 | -19 | 26 | 14 | -48 | 11 | -19 |
| Year 3 | -17 | 33 | 21 | -47 | 6 | -19 |
| Year 4 | -16 | 36 | 25 | -47 | 5 | -21 |
| Year 5 | -16 | 35 | 28 | -46 | 5 | -27 |
| Year 6 | -17 | 32 | 28 | -47 | 4 | -34 |
| Year 7 | -18 | 26 | 31 | -47 | 5 | -38 |
| Year 8 | -19 | 20 | 29 | -47 | 5 | -43 |
| Year 9 | -19 | 19 | 29 | -46 | 5 | -45 |
| Year 10 | -20 | 15 | 29 | -47 | 6 | -49 |

Table 4. Drainage water produced, discharged, and recycled, and irrigation water used by source, "Worst Case" scenario with reduced cotton demand, percent change from the "Base" solution.

|  | Subsurface Drainage Water | | | Irrigation Water | | |
|---|---|---|---|---|---|---|
|  | Produced | Discharged | Recycled | Surface | Unconfined | Confined |
|  | *(percent change from "Base" run)* | | | | | |
| Year 1 | -28 | -98 | 1,757 | -6 | -73 | -76 |
| Year 2 | -2 | -100 | 766 | -5 | -78 | -81 |
| Year 3 | 25 | -99 | 1,208 | -5 | -77 | -82 |
| Year 4 | 48 | -98 | 1,380 | -5 | -77 | -82 |
| Year 5 | 64 | -94 | 1,381 | -6 | -75 | -81 |
| Year 6 | 80 | -86 | 1,255 | -7 | -74 | -80 |
| Year 7 | 92 | -85 | 1,224 | -9 | -73 | -78 |
| Year 8 | 87 | -85 | 1,219 | -11 | -73 | -78 |
| Year 9 | 94 | -83 | 1,002 | -11 | -71 | -76 |
| Year 10 | 99 | -81 | 951 | -12 | -70 | -75 |

The most interesting result of this scenario is the significant increase in the production of subsurface drainage water when compared to both the "Base" run (table 4) and the "Worst Case Scenario" (table 2). The doubling of subsurface drainage water production by year 10 is directly caused by the reduction of unconfined ground-water pumping by 70 percent. Under the "Worst Case Scenario," unconfined ground-water pumping declines range from only 2 to 9 percent (table 2). Decreased pumping levels cause perched water tables to rise faster which results in more irrigated acres requiring subsurface drainage and, subsequently, the production of more subsurface drainage water and the need for disposal.

These results are dependent on the sensitivity of perched water table depths to changes in hydrologic parameters. These hydrologic parameters are the amount of water being added to the aquifer from the root zone, the amount of water being removed through losses to the confined aquifer, ground-water pumping, and the specific yield of the soil profile. Of these parameters, estimates of specific yield is very important and it is not endogenous to the analytical system. Specific yield converts a pure water depth to an equivalent depth in the soil. Estimates of specific yield used in WADE were derived by Williamson et al. (1985) and Belitz (1988) for the study area and range between .14, 7.1 feet. If a soil profile has a specific yield of .07, removing 1 foot of water depth will reduce the water table depth by 14.3 feet (1/.07), and a soil with a specific yield of .14, 7.1 feet. Therefore pumping ground water from the unconfined aquifer can have a significant effect on the depths of perched water tables and the production of subsurface drainage water.

## POLICY IMPLICATIONS

One important conclusion can be drawn from these results. A policy that directly or indirectly reduces unconfined aquifer pumping will probably increase the production of subsurface drainage water. Therefore, policies that reduce irrigated acreage to reduce the need for subsurface drainage water disposal may well increase drainage water if surface water supplies are just relocated and applied to adjacent areas. Similarly, if water saving irrigation technologies and management strategies are adopted without considering the source of water savings, the desired results may not be achieved. This analysis indicates that the management of unconfined ground-water pumping and the spatial reallocation of surface water supplies may be the most important policy variables in formulating a drainage policy for the Valley.

This analysis also demonstrates the value of economic models being linked to appropriate physical models. The scale of agricultural production in the Valley is large and complex and almost everything is connected or related to some degree. By being able to estimate specific reductions in ground-water pumping with the agricultural production model and tracking those changes through the hydrology model, projections of changes in perched water table depths can be made. The linked system also serves to identify the parameters that may be important to policy formation and provide estimates of their sensitivity to ranges in input data and agricultural production practices.

These results are simple aggregations of individual cell results of the WADE model. These cells vary considerably in size, productivity, and hydrology characteristics and cell results are varied accordingly. To conclude that these results apply to all of the areas in the affected high water tables would be erroneous. Therefore results should be aggregated to regions that have the responsibility for formulating and instituting drainage policy and analyzed accordingly.

This exercise also demonstrates the value of linking several models of various spatial aggregations to estimate the importance of changes in policies formulated at the National levels. First, changes in commodity policies can change regional cropping patterns, resource use, and effect environmental quality in ways that may be counterintuitive. Second, regional environmental quality policies, even though having adverse effects on some individual producers, may not have profound effects on commodity prices if production alternatives exist outside of the region.

## ACKNOWLEDGEMENTS

The assistance of John Miranowski, Director, Resources and Technology Division, Economic Research Service, USDA, in facilitating the transfer of the USMP model is greatfully acknowledged. Views expressed are those of the authors and do not reflect the position of USDA.

## REFERENCES

Belitz, K., 1988. *Character and Evolution of the Ground-Water Flow System in the Central Part of the Western San Joaquin Valley*. U.S. Geological Survey Open-File Report 87-573.

Brooke, A.; Kendrick, D.; and Meeraus, A., 1988. *GAMS, A User's Guide*. The Scientific Press, Redwood City, CA.

Hatchett, S. A.; Quinn, N. W. T.; Horner, G. L.; and Howitt, R. E., 1989. Drainage Economics Model to Evaluate Policy Options for Managing Selenium Contaminated Drainage. Presented at the International Committee on Irrigation and Drainage Pan-American Conference, *Toxic Substances in*

*Agricultural Water Supply and Drainage-An International Environmental Perspective*, Ottawa Quebec Canada.

Horner, G. L. and Dudek, D. J., 1980. An Analytical System for the Evaluation of Land Use and Water Quality Policy Impacts Upon Irrigated Agriculture. In: Yaron, D. and Tapiero, C. (Eds.), *Proceedings of Operations Research in Agriculture and Water Resources, International Conference, Jerusalem.* North Holland, Amsterdam, pp. 537-568.

House, R. M., 1987. *USMP Regional Agricultural Model.* National Economics Division, Economic Research Service, U.S. Department of Agriculture, 46 p.

Howitt, R. E., 1989. Water Policy Effects on Crop Production and Vice Versa: An Empirical Approach. Presented at the Commercial Agriculture and Resources Policy Symposium, Baltimore, MD.

Howitt, R. E., 1989. *The Economic Assessment of California Field Crop Losses Due to Air Pollution.* Final Report prepared for California Air Resources Board, Contract No. A5-105-32, 44 p.

Kasower, S., 1989. A Dynamic Analysis of Public and Private Objectives in the Management of Saline Soils. Prospectus and a Mathematical Presentation of the Dynamic Agro-Economic Soil Salinity (DASS) Model. Preliminary Working Paper, San Joaquin Valley Drainage Program.

Piper, S.; Huang, W. Y.; and Ribaudo, M., 1989. Farm Income and Ground-Water Quality Implications From Reducing Surface Water Sediment Deliveries, *Water Resources Bulletin*, American Water Resources Association, 25(6).

San Joaquin Valley Drainage Program, 1987. *Developing Options, An Overview of Efforts to Solve Agricultural Drainage and Related Problems in the San Joaquin Valley.*

San Joaquin Valley Drainage Program, 1989. *Preliminary Planning Alternatives for Solving Agricultural Drainage and Drainage-Related Problems in the San Joaquin Valley.*

Stroh, C., 1990. Personal Communication, San Joaquin Valley Drainage Program.

University of California Committee of Consultants on Drainage Water Reduction, 1988. *Opportunities for Drainage Water Reduction.* No. 1 in a Series on Drainage, Salinity and Toxic Constituents, Sponsored by the UC Salinity/Drainage Task Force and Water Resources Center.

University of California Cooperative Extension, 1980-89. *Sample Costs of Producing Agricultural Crops in California Counties.* Department of Agricultural Economics, Davis, CA.

U.S. Department of Agriculture, Agricultural Stabilization and Conservation Service, 1985. Unpublished Commodity Program Participation Data.

U.S. Department of Agriculture, River Basin-Watershed Planning Staff, 1973. *The San Joaquin Valley Basin USDA River Basin Study Soil Group Areas, 1972,* 104 pp.

Williamson, A. K.; Prudic, D.; and Swain, L., 1985. *Ground-Water Flow in the Central Valley, California.* U.S. Geological Survey Open-File Report 85-345.

# 29 CROPLAND ALLOCATION DECISIONS: THE ROLE OF AGRICULTURAL COMMODITY PROGRAMS AND THE RECLAMATION PROGRAM

Michael R. Moore, USDA-ERS; Donald H. Negri, Willamette University, Salem, Oregon; and John A. Miranowski, USDA-ERS

### ABSTRACT

This chapter describes and analyzes the role of Federal commodity programs and the U.S. Bureau of Reclamation (Reclamation) program in California agriculture. A behavioral model is developed of Reclamation water supply and commodity program provisions as determinants of cropland allocation decisions. The model is estimated econometrically using data from irrigation districts in California. Based on the results, two simulations illustrate cropland allocation responses to generic changes in Federal program provisions. By affecting crop supply, a simulated 10-percent reduction in California's Reclamation water supply could result in 1-2 percent increases in the National market prices of rice and vegetables, while a simulated 0.05 decrease in wheat's acreage reduction program would not affect National market prices.

### INTRODUCTION

Studying Western irrigated agriculture from a public policy perspective requires more than an understanding of Western water policy. The water policy viewpoint focuses on scarcity and conservation as central themes of the new era of Western water policy. It identifies water reallocation from agriculture to alternative uses and substitution of other inputs for irrigation water as cornerstones of the new era. Two other policy arenas, however, are pertinent to the future structure and performance of Western irrigated agriculture. Environmental policy issues obviously are growing in importance. These include irrigation-related water quality problems and pesticide-related food safety issues. Federal and state governments are considering regulatory policies to address agricultural chemical use and irrigation drainage concerns. Agricultural commodity policy also constitutes an important National issue. Commodity program provisions are important elements of international trade and Federal budget deliberations, which are being conducted under the purview of the General Agreement on Tariffs and Trade and the 1990 Farm Bill process, respectively. Although commodity policy has not specified particular programs or provisions for irrigated agriculture, partial or total deregulation of commodity markets would affect irrigators directly. Further, legislative amendments linking Federal commodity and irrigation water programs were introduced in 1990. In combination, commodity policy, environmental policy, and water policy will partially determine the future characteristics of Western irrigated agriculture.

This chapter addresses policy issues associated with two Federal programs: agricultural commodity programs and the Reclamation program. Although developed independently, these programs overlap in their effect on irrigated agriculture in the West. Commodity programs restrain supply of program crops through land set-aside requirements, while (at least historically) the Reclamation program promoted expansion of irrigated agriculture in the arid and semiarid portions of the West. One theme of the chapter concerns the opportunity for policy coordination: Developing policies with compatible incentives or avoiding policies with countervailing incentives (Miranowski et al., 1989). The chapter's goals are: First, to understand the effect on individual behavior of the commodity and Reclamation programs; second, to understand the effect on aggregate behavior in California agriculture of possible changes in the programs; and, finally, to describe implications for coordinating the possible program changes.

The description and analysis focuses on the role of commodity and Reclamation programs as determinants of cropland allocation in California. The next section describes operations of the programs in California and develops a context for how program provisions affect individual behavior. In particular, it discusses prospects for renewal of water-service contracts for almost 4 million acre-feet per year of irrigation water from Reclamation's Central Valley Project (CVP). The third section of this chapter develops a behavioral model of the relationship between cropland allocation decisions and Federal program provisions. The model treats Reclamation-supplied water as a nonmarket determinant of cropland allocation decisions. Commodity program provisions also are introduced as endogenous elements of the model. The fourth section reports econometric estimates of a commodity program participation model and the cropland allocation model. The cropland allocation model is estimated using data provided by the 130 irrigation districts that receive Reclamation-supplied water in California. Using the econometric results, the fifth section simulates cropland allocation response to two generic changes in commodity and Reclamation program provisions.

## FEDERAL PROGRAMS IN CALIFORNIA

### Commodity Programs

Federal Government intervention in agricultural commodity markets, initiated in the 1930's, accomplishes three objectives. Commodity program provisions: (1) Protect producers from low market prices by setting price floors ("nonrecourse loan rates") for USDA program crops, (2) support farm income by establishing target prices and deficiency payments for program crops, and (3) provide for National food security by Government ownership of commodity reserves and provisions to guide planted acreage ("acreage reduction programs") (Langley et al., 1985). Seven major crops--barley, corn, cotton, oats, rice, sorghum, and wheat--are designated for Government support.

Because program participation is a voluntary decision, individual producers reassess the relative advantages of program participation each year.[1] The income support of deficiency payments provides one major incentive to participate. Target prices, market prices, base acreage, and base yields of program crops--each defined on a crop-specific basis--are program parameters that affect total deficiency payments received by a participant. The Government sets target prices exogenously to the producer, while a farm's historic planted acreage and yields set current base acreage and program yields for the farm. The deficiency payment rate generally is the difference between the target price and market price. Deficiency payments, then, are the product of the payment rate, crop base acreage, and program yield. They do not depend on current year production.

The minimum crop price set by nonrecourse loan rates provides a second major incentive to participate. By setting a price floor, the loan rate removes the risk of low commodity prices for participating producers. When the loan rate exceeds market price, the deficiency payment rate is defined as the difference between the target price and loan rate rather than the difference between the target price and market price. When this occurs, participating producers may repay their nonrecourse loans with actual product. This creates a de facto producers' subsidy rate equivalent to the difference between the loan rate and market price.

To qualify for income and price supports, producers must set aside a designated share of cropland. Forgone crop production from acreage reduction programs (ARP's) imposes the major cost of participation. The higher the ARP rate, the greater the participation cost.

In California, program provisions, market prices, and participation rates associated with the rice and upland cotton programs illustrate commodity program incentives. Beginning in 1983, rice program provisions were sufficiently attractive relative to the rice market price that the program appeared to dictate cropland allocated to rice from 1983-88 (table 1). In 1982, the California average producer price for rice declined to $6.44 per hundredweight[2] (cwt), a drop of roughly $4.50 per cwt from 1979 levels. Rice's target price the following year was set at $11.40 per cwt, a level 77 percent higher than the 1982 market price. Participation among California rice producers increased from 72 percent in 1982 to 98 percent in 1983--in spite of the 15-percent acreage set-aside requirement. Participation rates remained very high through 1988, fluctuating between 87 percent and 94 percent. During the period, the rice target price remained high, the rice market price remained relatively low, and the rice ARP apparently was not high enough to deter participation. Participation appeared to be a nonmarginal decision for most rice growers during the period.

Participation by California producers in the cotton program, in contrast, suggests that cotton program participation tended to be a marginal decision during 1983-88 (table 1). The 1983 average market price for cotton in California was only slightly below the target price. Participation in the cotton program decreased considerably in 1984, from 79 percent to 28 percent, likely because of this small price differential. Participation increased to 94 percent in 1986, but again decreased to the 62-68 percent range in 1987 and 1988. Based on causal empiricism, the differential between target and market price appears to explain much of the variation in participation rates. Empirical results from the commodity program participation model presented in the fourth section of this chapter demonstrate this more formally.

## The Reclamation Program

Congress originally imbued the Federal Reclamation program with a mission of establishing permanent settlement in the arid portions of the American West through expansion of irrigated agriculture. Beginning in 1902, the Reclamation Service (later the Bureau of Reclamation), an agency of the U.S. Department of the Interior, planned and constructed storage dams, diversion dams, and water conveyance systems associated with the program. Reclamation continues to operate and manage Federal water projects, administer the program's repayment contracts, and direct several other Reclamation functions, such as the Safety-of-Dams Program.

Table 1. Federal commodity programs and market prices in California - rice and upland cotton (1982-88).

| Year | Rice | | | | | Upland Cotton | | | | |
|---|---|---|---|---|---|---|---|---|---|---|
| | Target Price ($/cwt) | Loan Rate ($/cwt) | Market Price[a] ($/cwt) | ARP[b] (%) | Part. Rate[c] (%) | Target Price ($/lb) | Loan Rate ($/lb) | Market Price ($/lb) | ARP (%) | Part. Rate (%) |
| 1982 | 10.85 | 8.14 | 6.65 | 15 | 72 | 0.71 | 0.57 | 0.63 | 15 | 50 |
| 1983 | 11.40 | 8.14 | 6.96 | 15 | 98 | 0.76 | 0.55 | 0.73 | 20 | 79 |
| 1984 | 11.90 | 8.00 | 6.43 | 25 | 87 | 0.81 | 0.55 | 0.67 | 25 | 28 |
| 1985 | 11.90 | 8.00 | 5.33 | 20 | 88 | 0.81 | 0.57 | 0.61 | 20 | 50 |
| 1986 | 11.90 | 7.20 | 3.18 | 35 | 92 | 0.81 | 0.55 | 0.59 | 25 | 94 |
| 1987 | 11.66 | 6.84 | 6.72 | 35 | 93 | 0.79 | 0.52 | 0.70 | 25 | 68 |
| 1988 | 11.15 | 6.63 | 6.15 | 25 | 94 | 0.76 | 0.52 | 0.64 | 12.5 | 62 |

[a] Market price is the California average producer's market price, reported in selected years of *Agricultural Prices*, National Agricultural Statistics Service, USDA.
[b] "ARP", the acreage reduction program, is the mandatory acreage set-aside for participants in a commodity program, measured as the percentage of base acreage in the crop.
[c] "Part. Rate" measures, for the State of California, the participation rate in the respective commodity program. This is calculated as the ratio of complying base acres to total base acres in the State. The data are from unpublished Crop Compliance Reports of the Agricultural Stabilization and Conservation Service, USDA.

California receives the largest volume of Reclamation water of any state and irrigates the most acreage with Federally developed water. Reclamation provides irrigation water services to 130 irrigation districts and water districts in California (Department of the Interior, Bureau of Reclamation, 1988). These districts irrigated 3.17 million acres in 1987. This constituted 41 percent of California's 1987 harvested cropland and, in terms of the Federal program, roughly one-third of total Reclamation irrigated acreage in the 17 Western States. Reclamation water deliveries to California farms totaled 7.93 million acre-feet[3] (MAF) in 1987. Some of these farms supplement Reclamation water by purchasing water from the State Water Project or pumping ground water.

Two projects--the Boulder Canyon Project and the CVP--provide for the vast majority of Reclamation irrigated acres in California (table 2). The Boulder Canyon Project provides full water service from the Colorado River to the southern-most agricultural area of the State through the All-American Canal. In this area, over 515,000 acres were irrigated with over 2.63 MAF of Colorado River water in 1987. The CVP runs along a north-south axis through the central region of California. Composed of seven major divisions, the CVP primarily provides supplemental water service to over 2.17 million irrigated cropland acres. It delivered 3.90 MAF to agricultural producers in 1987. Major crops grown in the project service area included cotton, fruits, nuts, vegetables, alfalfa, rice and wheat.

The issue of Reclamation water price increases, a perennial topic of Reclamation policy, recently commanded public attention. Beginning in 1989, most CVP water-service contracts are eligible for renewal during the next twenty years. Contract prices for CVP water frequently do not cover operating and maintenance costs, and generally are below long-run marginal cost of water supply (table 2). The Department of the Interior (Interior), the U.S. Environmental Protection Agency (EPA), and the Council on Environmental Quality (CEQ) disagreed over the procedures governing contract renewals (Lancaster, 1989a and 1989b). Interior argued that the Federal Government was legally bound to renew the contracts at existing price and quantity terms, while the EPA and CEQ maintained that the National Environmental Policy Act required an Environmental Impact Statement (EIS) prior to contract renewal (Council on Environmental Quality). The EIS would assess the impact on irrigation water use and the natural environment of higher water prices, reduced irrigation water deliveries, and/or Reclamation water marketing. The Secretary of the Interior forged a compromise on the issue, agreeing to renew the contracts at existing terms but conditioning their final provisions on the results of an EIS. Federal policy on CVP contract renewal thus remains to be finalized.

The fourth and fifth sections of this chapter will directly address the role of Reclamation water supply and Reclamation policy in California agriculture.

Table 2. Description of Central Valley and Boulder Canyon Projects, California.

| Project | Date[a] | Irrigated Area, 1987 (acres) | Irrigation Water, 1987 (acre-ft.) | Contract Water Price[b] ($/acre-ft.) | Full-Cost Water Price[c] ($/acre-ft.) |
|---|---|---|---|---|---|
| Central Valley | 1940 | 2,166,898 | 3,901,759 | 1.50-16.50 | 7.43-68.60 |
| Boulder Canyon (All American Canal) | 1940 | 515,547 | 2,631,593 | 2.32-2.56 | 3.60-3.96 |

[a]Date of initial irrigation water service by project.
[b]Reports the range of water prices paid by irrigation districts receiving water from the project.
[c]Reports the range of full-cost water prices according to the definition of full cost contained in the Reclamation Reform Act of 1982.

*Sources:* Department of the Interior, Bureau of Reclamation (1988) and Department of the Interior, Office of the Secretary (1988).

## A BEHAVIORAL MODEL OF MULTICROP LAND ALLOCATION

This section develops a model of a multioutput firm engaged in irrigated crop production[4]. A key model component is treatment of Reclamation water supply as an exogenous determinant of cropland allocation decisions. Administrative procedures governing Reclamation water supply make this the correct approach. Long-term contracts, generally of 40 years duration, specify the volume and price of water that will be delivered to an irrigation district. Administrative procedures, rather than markets, set the water prices contained in Reclamation contracts. The prices consequently do not reflect water's marginal value in irrigation (Kanazawa, 1988 and Wahl, 1989). These nonmarket allocation mechanisms imply that water should be modeled as a fixed input rather than a variable input. Water right transfer restrictions further substantiate the argument for treating water as a fixed input. State law and Federal Reclamation law create substantial impediments to transfers (Burness and Quirk, 1980 and Wahl, 1989). Although market-oriented allocation mechanisms are now evolving, many institutional changes necessary for development of water markets remain to be completed. Thus, treating water as a fixed input in a behavioral model of Reclamation water allocation captures the institutional environment facing producers using Reclamation water.

Three assumptions guide the representation of multicrop production (Just et al., 1983): (1) Inputs are allocated to specific crop production activities, (2) production is technically nonjoint, and (3) land and surface water are fixed at the farm level but allocatable among crop production activities. In this approach, assumption (3) provides the source of jointness across crops on a farm. Although production functions can be specified for each crop, maximizing multicrop profits introduces interdependencies among crop production activities since all production activities compete for fixed land and water (Shumway et al., 1984).

A dual approach is used here to model multicrop production with fixed, allocable inputs (Chambers and Just, 1989). The approach applies the standard technique of constrained maximization. The producer chooses the allocation of land and surface water among crop production activities to maximize multicrop profit subject to the resource constraints. Formally, the problem is

$$[1] \quad \Pi(p,r,W,N) = \max_{\substack{w_1,\ldots,w_m \\ n_1,\ldots,n_m}} \left\{ \sum_{i=1}^{m} \pi_i(p_i,r,w_i,n_i) : \sum_{i=1}^{m} w_i = W \text{ and } \sum_{i=1}^{m} n_i = N \right\},$$

where p $(p_1 p_2 \ldots p_m)$ is a vector of strictly positive crop prices for the m crops; r $(r_1 r_2 \ldots r_t)$ is a vector of strictly positive variable input prices for the t inputs; W is the fixed quantity of water; w $(w_1 w_2 \ldots w_m)$ is a vector of water allocations to production of crop i; N is the fixed quantity of land; n $(n_1 n_2 \ldots n_m)$ is a vector of land allocations to production of crop i; $\pi_i(p_i,r,w_i,n_i)$ is the restricted profit function of crop i, which holds water and land allocations fixed; and $\Pi(p,r,W,N)$ is the multicrop profit function. Assuming the standard convexity, linear homogeneity, and monotonicity properties for the $\pi_i(p_i,r,w_i,n_i)$, the properties of $\Pi(p,r,W,N)$ follow as convex and linear homogeneous in p and r, nondecreasing in p, nonincreasing in r, and nondecreasing in W and N (Chambers and Just, 1989). Input and output prices are assumed to be exogenous to individual producers' decisions. A Lagrangian function, denoted L, states the constrained maximization problem as

$$[2] \quad L = \sum_{i=1}^{m} \pi_i(p_i,r,w_i,n_i) + \lambda \cdot (W - \sum_{i=1}^{m} w_i) + \mu \cdot (N - \sum_{i=1}^{m} n_i),$$

where $\lambda$ and $\mu$ are the shadow prices on the surface water and land constraints, respectively. The necessary conditions for an interior solution are

$$[3] \quad \partial L/\partial w_i = \partial \pi_i/\partial w_i - \lambda = 0 \qquad i = 1,\ldots,m,$$

[4] $\partial L/\partial n_i = \partial \pi_i/\partial n_i - \mu = 0 \qquad i = 1,...,m,$

[5] $W - \sum_{i=1}^{m} w_i = 0$ and $N - \sum_{i=1}^{m} n_i = 0.$

The necessary conditions have economic interpretation. Equation set [3] allocates surface water among crops to equate the marginal profit from each crop, while equation set [4] allocates land similarly. These establish efficient allocation of water and land within the farm production unit. The input constraints in [5] are binding since the restricted profit functions are monotonic increasing in $w_i$ and $n_i$.

Solving equations [3]-[5] yields the optimal solutions to equation [2], denoted $w_i^*(p,r,W,N)$ and $n_i^*(p,r,W,N)$. These represent the multioutput firm's production equilibrium in water and land allocations.

The equilibrium conditions provide a framework for assessing the effect of changes in water supply on crop-level land and water use. Inserting the $w_i^*$ and $n_i^*$ into the necessary conditions creates a set of identities for comparative static analysis. However, the analysis does not yield testable hypotheses. The constraints' role in land and water allocation, demonstrated by the signs of $\partial w_i^*/\partial W$, $\partial n_i^*/\partial W$, $\partial w_i^*/\partial N$, and $\partial n_i^*/\partial N$ ($i=1,...,m$), is indeterminate[5]. In words, crop-specific water allocation and land allocation may be increasing or decreasing in both the land constraint and the water constraint. Following a water-supply reduction, for instance, cropland allocated to alfalfa may decrease while cropland allocated to cotton may increase. This suggests that some positive and negative coefficients on the resource constraints in the empirical estimation of land allocation equations should be expected.

## Estimable Input Allocation Equations

To develop an estimable set of expressions for $w_i^*(p,r,W,N)$ and $n_i^*(p,r,W,N)$, specific functional forms must be adopted for the crops' restricted profit functions. Unlike conventional input demand functions, fixed-input allocation equations cannot be obtained directly using Hotelling's lemma (Shumway et al., 1984). Instead, they must be derived from the necessary conditions for multicrop profit maximization. Here it is posited that the restricted profit functions in equation [1] take a normalized quadratic form (Lau, 1976). Closed-form expressions for $w_i^*$ and $n_i^*$ are tractable using the normalized quadratic because its first derivatives are linear.

Derivation of the allocation equations begins by forming a system of (2m+2) linear equations that correspond to equations [3]-[5]. From this linear system, closed-form expressions for $w_i^*(p,r,W,N)$ and $n_i^*(p,r,W,N)$ are derived. The estimable water and land allocation equations are linear in output prices, input prices, the water constraint, and the land constraint (Moore and Negri, 1990). Output and input prices are specified relative to a numeraire price since the normalized quadratic imposes linear homogeneity on the profit function by specifying profit and prices relative to one price.

## Extending the Model to Incorporate Commodity Program Participation

The model of land allocation with fixed allocable inputs is extended to incorporate endogenous commodity program participation. For program crops, producers receive income and price supports, in the form of deficiency payments and nonrecourse loans, in exchange for acreage set asides. Producers choose to participate in a commodity program if the expected profit from participation exceeds the expected profit from nonparticipation. Thus, using a discrete choice framework (Maddala, 1983), program participation is modeled here as a function of the difference between expected profits under participation and nonparticipation. Expected profits depend on expected participation benefits (income and price supports) and expected costs (acreage reduction program rates).

Program participation directly affects land allocated to program crops through base acreage and land set-aside requirements. When a producer participates in a commodity program, the crop's base acreage is allocated to some combination of planted acreage and mandatory land set aside. For instance, 1988 planted acreage in rice could equal only 75 percent of base acreage because of the 25 percent ARP requirement. In the land allocation equations for program crops, we include the crop's acreage reduction program rate as an explanatory variable to capture acreage reduction from mandatory set asides.

Program participation also indirectly affects land allocated to nonparticipating crops and other participating crops. The indirect effect functions through the land constraint. Participation in a commodity program removes base acreage from the land available for all other program and nonprogram crops. Therefore, for nonparticipating crops, the land constraint variable equals total irrigable land minus the farm's participating base acres in all commodity programs, or

$$[6] \quad N^o = N - \sum_{j=1}^{s} B_j,$$

where $N^o$ is the new land constraint, N is the original land constraint, s is the number of participating program crops, and $B_j$ are the base acres in the jth crop. For participating crops, similarly, the land constraint variable equals the total irrigable land minus participating base acreage in other program crops. The empirical application uses the modified land constraint of equation [6] as an explanatory variable affecting cropland allocation decisions. The relationship between commodity program provisions and decisions concerning nonparticipating crops is a subtle feature of the analysis.

## EMPIRICAL APPLICATION TO RECLAMATION-SERVED CROPLAND IN CALIFORNIA

Commodity program participation rates are estimated for five program crops: Barley, wheat, grain corn, rice, and upland cotton. The remaining two program crops, sorghum and oats, are excluded because California produces minor amounts of them. Land allocation equations are estimated for 10 crops, including irrigated wheat, grain corn, barley, sugar beets, hay (primarily alfalfa), pasture, fruit and nut orchards, rice, vegetables, and upland cotton. Water allocation equations are not estimated because the data do not include crop-specific water allocations.

### Commodity Program Participation Results

California commodity program participation rates are estimated using three independent variables: Expected difference between target and market prices, ARP set-aside rates, and a dummy variable corresponding to the payment-in-kind program (PIK) in 1983. Estimates are based on annual, crop-specific data from 1979 to 1988 -- 10 observations. Annual participation rates are measured as the ratio of California base acreage on farms complying with program provisions to total California base acres. Mean participation rates range from 31 percent for corn to 88 percent for rice. Participation rates for all crops were 100 percent in 1980 and 1981 since program eligibility did not require acreage reduction in those years. The difference between target price and market price captures the benefits associated with both the deficiency payment and the price support.[6] Because land allocation and program participation decisions are made prior to the realization of market price, the income and price support variable equals the difference between current target price (established prior to the growing season) and expected market price. Here, lagged market price is used as a proxy for expected market price. This variable is truncated at zero to preclude negative benefits. The acreage reduction rate variable is the fraction of base acreage that must be set aside to qualify for program benefits. Finally, a dummy variable is included for 1983 because the 1983 PIK program differed from program provisions in other years.

Following a discrete choice modeling approach, the estimation uses a logistic transformation of the dependent variable to confine predicted participation rates to the unit interval. OLS estimates of the logit model are reported in table 3. With the exception of rice, the estimated coefficients on expected program benefits and acreage reduction requirements are statistically significant with the anticipated signs: Higher benefits in the form of either income or price supports increase participation rates while higher costs in the form of acreage reductions decrease participation rates. As reported in the second section, rice program provisions appear to be sufficiently lucrative that variation in deficiency payment rates or acreage reduction requirements did not affect participation. Although the PIK program in 1983 was a substantial departure from other years, the PIK dummy variable is not significant for any program crop.

Table 3. Commodity program participation rate estimates, California.

| Variable | Grain Corn | Barley | Crop Wheat | Rice | Upland Cotton |
|---|---|---|---|---|---|
| Intercept | 3.24 | 3.72 | 4.39 | 4.08 | 4.16 |
|  | (2.45) | (4.40) | (7.34) | (5.97) | (2.50) |
| Expected Deficiency | 5.71 | 7.03 | 2.82 | -0.08 | 0.26 |
| Payment Rate | (1.91) | (2.27) | (3.48) | (-0.15) | (2.50) |
| ARP[a] | -36.05 | -33.65 | -29.86 | -5.28 | -33.27 |
|  | (-2.81) | (-4.38) | (-6.28) | (-9.50) | (-4.32) |
| PIK[b] | 0.27 | -0.55 | -0.61 | 0.91 | 0.40 |
|  | (0.124) | (-0.39) | (-0.67) | (0.61) | (0.33) |
| R-squared | 0.57 | 0.77 | 0.88 | 0.50 | 0.82 |

[a]ARP is the land set-aside rate required by the crop's acreage reduction program.
[b]PIK is a dummy variable for 1983, the year of the payment-in-kind program.

## Cropland Allocation Results

**Data and variables.** Reclamation delivers irrigation water to 130 irrigation districts and water districts in California. Data on land allocation, total land in irrigation rotation, and water deliveries are from annual reports filed by the districts with Reclamation for 1979-1987.

In the absence of commodity programs, the land constraint measures acreage to which irrigation water could be applied with existing irrigation infrastructure. With endogenous program participation, however, participating base acreage reduces the allocatable land (as described in equation [6]). Since participation rates used to adjust the land constraint are endogenous, predicted values from the estimated participation equations are used here for participation rates to eliminate any potential simultaneous equations bias. Base acreage for the districts is constructed by inflating harvested acreage by the statewide ratio of base acreage to harvested acreage.

The water constraint measures all water delivered by an irrigation district to farms, including Reclamation project water and other district water sources. District water deliveries represent the district's nonmarket-based allocation of surface water.

District-level land allocations summarize individual production decisions of all farms served by the district. To address variation in the number of farms served by each district, we constructed observations on a per-farm basis. District values for land allocations, land constraints, and water constraints thus are divided by the number of farms in the district.

Agricultural output and input prices for California are merged with the irrigation district data. Expected output prices are used here because producers make land allocation decisions prior to the realization of output price. For program crops, the prices are the higher of current year weighted support price or lagged market price. Because output prices are highly collinear, each equation includes only the prices of two major crops, alfalfa hay and upland cotton, along with the price of the crop whose land allocation is being estimated. The current year wage rate for farm labor is the only input price.

Other independent variables include climate and soil characteristics in the counties in which the irrigation district operates. Like output prices, the climate variables reflect expected weather conditions. They include proxies for the amount of energy (average growing degree days) and rainfall (average effective rainfall) available for plant growth during the growing season. Soil variables include average soil texture, soil productivity, and soil slope for all cropland in the counties. A dummy variable measures clayey soil relative to sandy and loamy soil; a dummy variable represents a high quality land classification relative to other land classifications; and a variable measures the gradient of the soil in percentage terms. A water quality variable was available at the irrigation-district level for only 1 year. Because water quality varied between 1979 and 1987, the variable was not included in the analysis. For a detailed description of the variables used in the analysis, see Moore and Negri (1990).

Finally, program crop equations include the crop's ARP set-aside rate as an explanatory variable.

Econometric procedures and results. The cropland variables are censored dependent variables because every irrigation district does not grow every irrigated crop. Applying ordinary least squares to nonzero observations leads to biased and inconsistent estimates and eliminates information that can explain the decision to grow a crop. Tobit regression analysis, in contrast, uses all observations and produces unbiased and asymptotically efficient estimates with censored data.

While the model is derived at the farm level, estimation is based on data aggregated to the district level. Aggregation of farm-level data raises the possibility of heteroskedastic errors when the number or size of farms varies across districts. Unlike the linear model, the tobit model produces inconsistent parameter estimates in the presence of heteroskedastic errors (Maddala, 1983). To address the issue, we hypothesize that variation in the number of farms across districts produces heteroskedastic errors and estimate a heteroskedastic tobit model (Maddala, 1983).[7]

Discussion of the empirical results begins with the influence of the surface water constraint in land allocation decisions (table 4). Modeling surface water as a fixed, allocable input to reflect institutional constraints is a key dimension of the methodological approach. The water-constraint coefficients measure the marginal change in acreage allocated to the crops given a marginal change in surface water supply. For example, a 1-acre-foot increase in surface water results in an additional 0.06 acre allocated to rice production.[8]

The sign, size, and statistical significance of the water-constraint coefficients are important aspects of the results. The water constraint is a major determinant of cropland allocation. Of 11 estimated coefficients, 9 are significantly different from zero at the .01 level or better, and only hay and fallow land are not statistically different from zero. The estimated coefficients range from 0.06 for rice to -0.04 for fruit and nut crops. As discussed with the comparative static results of $\partial n_i^*/\partial w$, the mix of positive and negative coefficients was expected.

While the water-constraint coefficients are highly significant, a few signs do not conform to expectations. The expectations are based on the intuition that, as the water constraint is relaxed, farmers will reallocate land from low water-using crops to high water-using crops. Rice, vegetables, grain corn, and sugar beets, all of which have relatively large water-use intensities, have positive water-constraint coefficients. The water-constraint variable performs as ex-

## Table 4. Land allocation estimates, Reclamation-served farms in California.[a]

| Independent Variable | Alfalfa and Hay | Barley | Grain Corn | Fruit and Nuts | Irrigated Crop Pasture | Rice | Sugar Beets | Upland Cotton | Vegtbls. | Wheat | Fallow |
|---|---|---|---|---|---|---|---|---|---|---|---|
| **Prices** | | | | | | | | | | | |
| Hay ($/ton/W[b]) | 0.07 (0.43) | -1.07 (-1.30) | -0.31 (-1.05) | -0.05 (-0.24) | -0.03 (0.58) | -2.82 (-0.80) | -0.08 (-1.11) | -1.85 (-0.81) | -0.29 (-1.92) | -0.74 (-3.66) | 0.40 (2.20) |
| Cotton ($/lb/W) | -0.02 (-0.50) | 0.29 (1.15) | -0.06 (-0.82) | 0.02 (0.26) | -0.005 (-0.23) | -0.08 (-0.06) | -0.007 (-0.21) | 0.11 (0.13) | -0.06 (-1.09) | 0.05 (0.70) | -0.11 (-1.54) |
| Own Crop[c] ($/unit/W) | NA[d] | -49.32 (-1.71) | 13.89 (1.78) | 0.05 (0.60) | NA | 4.80 (0.17) | 0.27 (2.05) | NA | -0.42 (-2.09) | NA | NA |
| Wage Rate ($/hour/W) | 2.00 (1.16) | -29.57 (-2.25) | 4.22 (0.65) | -5.26 (-1.49) | -1.51 (-2.39) | -45.70 (-0.46) | -0.34 (-0.47) | -14.15 (-0.36) | 10.74 (2.60) | 1.58 (0.42) | 0.17 (0.08) |
| **Fixed Quantities** | | | | | | | | | | | |
| Surface Water (acre-feet) | 0.0008 (0.77) | 0.0099 (4.58) | 0.0022 (2.05) | -0.039 (-14.25) | -0.0017 (-3.59) | 0.059 (7.22) | 0.007 (9.61) | 0.050 (5.22) | 0.045 (18.38) | 0.016 (10.48) | -0.0007 (-0.26) |
| Irrigated Land (acres) | 0.04 (8.61) | 0.04 (5.53) | 0.004 (0.72) | 0.27 (20.43) | 0.008 (3.48) | 0.038 (1.34) | -0.0004 (-0.11) | 0.41 (16.22) | 0.16 (13.40) | 0.10 (15.68) | 0.11 (9.17) |
| **Physical Variables** | | | | | | | | | | | |
| Grow. Deg. Days[e] (days) | 0.0054 (11.54) | -0.01 (-4.54) | -0.0008 (-1.42) | 0.0026 (2.69) | -0.0005 (-2.97) | -0.08 (-5.10) | -0.0007 (-1.87) | 0.01 (1.09) | -0.0012 (-1.33) | 0.0019 (2.63) | -0.0023 (-2.36) |
| Effective Rain (inches) | -7.24 (-8.72) | -44.49 (-7.33) | -0.18 (-0.19) | 1.87 (1.03) | 4.62 (14.13) | 212.76 (8.66) | -4.87 (-4.43) | -243.86 (-6.52) | -3.38 (-2.11) | 4.97 (-3.21) | -5.48 (-2.82) |
| High Prod. Soil (dummy var.) | -5.91 (-6.39) | -1.35 (-0.39) | -1.39 (-1.30) | 20.34 (10.01) | -1.85 (-4.92) | 42.25 (2.66) | -3.76 (-5.77) | 38.70 (3.47) | -1.41 (-0.73) | -3.94 (-3.14) | -4.48 (-2.38) |
| Clayey Soil (dummy var.) | 4.33 (3.90) | 17.04 (4.03) | 4.69 (3.35) | -17.19 (-6.71) | 2.82 (6.15) | 145.66 (6.55) | 6.82 (7.56) | -18.90 (-1.21) | 2.31 (1.00) | 6.87 (4.30) | 3.95 (1.63) |
| Soil Slope 0.26 (% slope) | 0.068 (0.54) | -0.73 (-1.20) | -0.32 (-2.20) | -0.70 (-2.64) | -0.09 (-1.75) | -14.80 (-4.79) | -0.39 (-2.84) | -16.55 (-3.47) | 0.15 (0.57) | -1.11 (-4.09) | 0.35 (1.37) |
| **Other Variables** | | | | | | | | | | | |
| Own-Crop ARP (rate) | NA | 1.80 (0.053) | -11.47 (-0.64) | NA | NA | -0.92 (-0.005) | NA | -22.63 (-0.25) | NA | -23.69 (-2.99) | NA |
| GAMMA8 | 0.23 (0.54) | 28.55 (13.71) | 1.28 (1.46) | -5.08 (-11.61) | -0.46 (-3.55) | 119.65 (12.83) | -1.27 (-5.49) | 49.88 (5.05) | -5.67 (-14.44) | -1.13 (-3.57) | -5.00 (-9.49) |
| DELTA8 | 61.10 (26.61) | 29.72 (7.75) | 51.78 (14.29) | 186.98 (38.14) | 29.71 (30.06) | 43.60 (5.31) | 45.98 (25.11) | 164.41 (8.58) | 175.57 (33.97) | 87.71 (32.55) | 171.48 (35.74) |
| Intercept | -2.60 (-0.44) | 102.17 (3.14) | -6.05 (-0.82) | 9.65 (1.06) | -0.24 (-0.11) | -212.18 (-1.71) | 1.76 (0.55) | 137.19 (1.61) | 5.97 (0.78) | 5.39 (0.70) | 6.82 (0.86) |
| Likelihood Function | 4,500 | 2,916 | 2,732 | 5,421 | 3,212 | 2,564 | 2,589 | 3,211 | 4,447 | 3,977 | 5,273 |

[a]The table reports estimated regression coefficients from the tobit model and, in parenthesis, t-statistics of the estimated normalized coefficients.
[b]W is the wheat price ($/bu), which serves as numeraire to impose linear homogeneity of the profit function.
[c]Own Crop is the price of the crop whose land allocation is being estimated.
[d]NA means either not applicable (in the case of hay and cotton) or that the crop's price is not available.
[e]Growing degree days.
[f]High productivity soil, as measured by the National Resource Inventory's land classification system.
[g]GAMMA and DELTA are estimated parameters of the heteroskedastic tobit model (Maddala, p. 178-182). Number of farms in the irrigation district is the hypothesized source of heteroskedasticity. GAMMA is the coefficient on a constant and DELTA is the coefficient on the inverse of the square root of the number of farms in the irrigation district.

pected. However, the sign or relative magnitude of coefficients on fruit and nut orchards, cotton, hay crops, small grains, and fallow land differ from their established water-use intensities. Irrigation technology, a variable not in the data set, may explain the coefficients on orchards and cotton. In California, orchard crops are irrigated predominantly with sprinkler and drip systems and cotton is irrigated predominantly with gravity systems (U.S. Department of Commerce, Bureau of the Census, 1987). Although orchards require a large volume of water, their strong correlation with efficient irrigation technology may explain the large negative coefficient on the water constraint. Cotton requires relatively little water, yet its correlation with relatively inefficient irrigation technology may explain its large positive coefficient. In addition, preliminary analysis indicated that, with the remaining counterintuitive crops, the water constraint's sign and significance are sensitive to the econometric specification. For instance, correcting for heteroskedasticity or incorporating endogenous program participation in the land constraint reversed the water constraint's sign on barley, wheat, and fallow land allocations. Although a few coefficients do not conform to expectations, the overall results illustrate the direction and magnitude of changes that would occur in response to changes in water availability or commodity program parameters.

The econometric analysis yields four other general implications. First, 8 of 11 coefficients on the irrigated land constraint are significantly different from zero at the .01 level (table 4). All the significant coefficients are positive, with the largest coefficients on cotton (0.41) and orchard crops (0.27).

Second, the price variables generally perform poorly, but experience with the Reclamation data set for the 17 Western States suggested that multicollinearity among output price variables is the cause (Moore and Negri, 1990). To avoid this problem, the land allocation equations are specified here to include only the prices of two major California crops, alfalfa hay and upland cotton, in addition to the own-crop price. Nevertheless, the variables rarely are statistically significant. This may occur because the data are State level, and thus do not adequately reflect the variation in prices that exists within California.

Third, the climate and soil quality variables are generally consistent with agronomic hypotheses. They are significant at the .10 level for 38 of the 55 estimated parameters, and frequently are significant at the .01 level. To illustrate the role of these variables, consider the estimated coefficients on growing degree days (GDD). GDD serves as a proxy for the energy available for plant growth during the growing season. The results show that farms in counties with long growing seasons tend to produce those crops which require a long growing season. The positive GDD coefficients on alfalfa, fruit and nuts, and cotton, and the negative GDD coefficients on barley, rice, sugar beets, and fallow land capture the effect of season length on land allocation.[9] The consistency of the physical variables with an agronomic interpretation of the results lends credibility to the data and analysis.

Fourth, DELTA, the parameter measuring heteroskedasticity based on the number of farms in the irrigation district, is highly significant in all land allocation equations (table 4). Aggregation thus introduced heteroskedastic error terms, making a heteroskedastic tobit model necessary for consistent coefficient estimates (Maddala, 1983).

## SIMULATIONS OF CROPLAND ALLOCATION RESPONSE TO PROGRAM PROVISION CHANGES

A complete framework now exists to evaluate quantitatively the effects of policy-induced reforms in commodity and Reclamation program provisions. With Reclamation water supply, target prices of program crops, and ARP requirements for program crops specified as exogenous variables in the two econometric models, simulations are possible of cropland allocation response to changes in these variables. The simulations demonstrate the strong contribution to policy analysis of economic and econometric models that incorporate public policy variables explicitly.

Two simulations were conducted: A 10-percent reduction in Reclamation water deliveries to irrigation districts in California and a 0.05 decrease in wheat ARP set-aside rate. A 10-percent reduction in Reclamation water availability in California equals 792,000 acre-feet per year based on 1987 water deliveries. This volume of water could contribute to solving some of the major conflicts over water resources in California, such as water for ensuring wetland availability, diluting drainage water, and satisfying urban water needs. Reclamation water conservation offers one approach to solving serious water allocation issues in California.

Similarly, simulating a reduction in the wheat ARP set-aside rate addresses a realistic policy alternative. While GATT negotiations could deregulate commodity markets completely, the Federal Government appears headed toward adopting more modest changes in reauthorization of the 1990 farm program, such as small adjustments in target prices or ARP requirements.

Table 5 reports results of the two simulations for five important California crops. The simulations use 1987 harvested acres on Reclamation-served lands as base acreage.[10] The simulated 10-percent water-supply reduction increases (decreases) acreage of crops with negative (positive) estimated coefficients in table 4. Land in fruit and nut crops increases a very small amount, while the largest decrease occurs with land in cotton. Small elasticities of land allocation with respect to the water constraint generate the small absolute acreage adjustments. The elasticities are inelastic for the five crops, ranging from -0.01 for fruit and nuts to 0.44 for rice.

Table 5. Selected crops' acreage response to simulated reductions in program parameters, California.

| Crop | 1987 Reclamation Harvested Acres | Simulated Changes in Cropland Allocation | |
|---|---|---|---|
| | | 10% Reduction in Water Supply | 0.05 Decrease in Wheat ARP |
| Fruit and Nuts | 752,597 | 410 | -840 |
| Rice | 191,660 | -8,490[a] | -540 |
| Upland Cotton | 574,158 | -14,470 | -10,870 |
| Vegetables | 506,824 | -11,760[a] | -3,620 |
| Wheat | 226,216 | -3,720 | 2,340 |

[a] Acreage decreases of rice and vegetables following a 10 percent reduction in California's Reclamation water supply would create market price increases of greater than 1 percent in the National rice and vegetable markets.

The simulated 0.05 decrease in wheat ARP also generates small acreage adjustments. The simulation produces a positive change for wheat and negative changes in land allocation for other reported crops. For wheat, this occurs because reducing the set-aside requirement increases wheat harvested cropland. For other crops, decreasing the wheat ARP increases the wheat participation rate. The following sequence ensues: a higher participation rate increases acreage removed from land allocation decisions, and this decreases the operative land constraint for crops other than wheat. With fewer total acres available, land allocations decline for all crops with positive coefficients on the land constraint variable. The simulated decrease in allocations are: fruit and nuts, 840 acres; rice, 540 acres; upland cotton, 10,870 acres; and vegetables, 3,620 acres. Note that indirect effects on upland cotton and vegetables exceed in absolute value the direct effect on wheat cropland.

While the absolute changes in acreage appear small, Reclamation-served land in California represents relatively large shares of 1987 National acreage of fruits and nuts (22 percent), vegetables (21 percent), and rice (8 percent). Relatively small changes in acreage and production of these crops on California's Reclamation-served farms can affect National markets. For example, reducing Reclamation water supply can shift upward the vegetable supply curve, resulting in higher vegetable prices. The decreases in rice and vegetable acreage following a 10-percent supply reduction appear capable of increasing market prices more than one percent. These acreage decreases represent 0.4

and 0.5 percent, respectively, of National acreage in rice and vegetables. Assuming that their shares of National production are equal to these percentages, the acreage decreases translate into price increases of 1.8 percent for rice and 1.2 percent for vegetables.[11] In other words, policy-imposed changes in California's Reclamation water supply could create changes in consumer's and producer's surpluses in rice and vegetable markets. Thus, the welfare economics of Reclamation water supply policy must consider market impacts other than those limited to the allocative efficiency of water resources.

## SUMMARY AND CONCLUSIONS

Several policy proposals could directly affect irrigated agriculture, including decoupling income support from other goals of domestic commodity programs, establishing free trade in international commodity markets, reallocating a share of Western water from agriculture to other users, and internalizing agricultural chemical externalities. Each policy proposal could alter decisions concerning agricultural output and input use. In so doing, one policy reform may conflict with or complement the purpose of another policy reform. Free trade, for example, may result in some production decisions that increase environmental damage. In this setting, creating a set of policies with consistent purposes becomes a complex exercise. Policy coordination is required to integrate diverse policies and, in so doing, to achieve a variety of goals.

With this emphasis on of public policy analysis, the chapter develops a method for analyzing the relationship between cropland allocation decisions and Federal commodity and Reclamation programs. The economic and econometric models incorporate program provisions directly into the set of explanatory variables. Among other factors, cropland allocations depend on quantity of Reclamation water supply and participation decisions in commodity programs. Federal policies can determine the level of these factors. Reclamation policy on contract renewal may reduce the level of Reclamation water supplied to irrigation districts that receive water from the CVP. Commodity policy influences participation decisions by setting target prices, nonrecourse loan rates, and land set-aside requirements. Using the econometric results, a simulation demonstrates changes in cropland allocation that may occur with particular Federal policies.

Reclamation-served irrigation districts in California provide an important case study, covering water allocation, water quality, and agricultural production issues. Among the Western States, California typically experiences the greatest competition for water resources. In particular, the topic of recontracting for irrigation water from the Federal CVP has attracted National attention as a litmus test of principles governing Reclamation water allocation.

Irrigation-induced water quality problems (other than perennial salinity problems) developed recently in California's Kesterson National Wildlife Refuge. From an agricultural production perspective, California's acreage of fruits, nuts, and vegetables constitutes more than one-fifth of National acreage in the crops. National price impacts thus may occur with policy-related changes in California output. The complex array of issues in California irrigated agriculture means that, by analyzing and addressing policy issues there, methods of economic analysis of resource and agricultural policies will be available for application to other important settings.

## NOTES

[1] Many details of agricultural commodity programs are not described in this chapter. The interested reader should consult Glaser for a comprehensive account of commodity programs as defined in the Food Security Act of 1985.

[2] 1 hundredweight equals 45.36 kilogram.

[3] One acre-foot equals 1,233.5 cubic meters.

[4] This chapter uses a modeling approach originally developed in previous research (Moore and Negri, 1990). Two elements of the model and quantitative analysis are different in this chapter: discussion of how commodity programs alter the model and estimation of the model using only data on Bureau of Reclamation operations in California.

[5] See Moore and Negri for an expanded discussion of the comparative static results.

[6] The deficiency payment is the difference between target price and the higher of market price or loan rate. The price support is the difference between the loan rate and the market price when the loan rate exceeds market price. Thus, the target price minus the market price captures both benefits.

[7] In the heteroskedastic tobit model described in Maddala, the error variance assumes the form

$$\sigma^2_i = (\gamma + \delta Z_i)^2$$

In the land allocation equations, $Z_i = 1/(f_i^{1/2})$ where $f_i$ is the number of farms in the district.

Other econometric issues include problems associated with pooling of cross-sectional and time series data and the possibility of disturbance terms correlated across equations. Moore and Negri discuss these issues in detail.

[8] Coefficients obtained using a tobit model measure the effect of the explanatory variables on the model's unobserved latent variable. Multiplying by the probability of the dependent variable exceeding the threshold transforms the coefficients to reflect changes in the observed dependent variable (Maddala).

[9] In central California, alfalfa's and orchards' growing seasons are March to November, cotton's is April to October, and sugar beets' is April to September. California produces

primarily short- and medium-grain varieties of rice, which require roughly a 4-month season, and its barley primarily is planted in the fall.

[10] Other details of the simulation follow. The water-reduction simulation applies elasticities measuring the relationship between water supply and land allocation for each crop. The ARP-requirement-decrease simulation applies several elasticities because, for crops other than wheat, it operates through both the program participation and cropland allocation models. For wheat, the simulation operates directly through the ARP variable in the wheat land allocation equation. The elasticities applied in the ARP simulation measure several relationships: the effect of wheat ARP requirement on the participation rate in the wheat program; the effect of the participation rate on the land constraint facing each crop; the effect of the land constraint on land allocation to each crop; and the direct effect of the wheat ARP on wheat land allocation.

[11] The computations use price elasticities of demand for rice and vegetables of -0.2 and -0.4 respectively.

## ACKNOWLEDGEMENTS

The authors thank Daniel Hellerstein for helpful discussions on econometric issues and for use of the GRBL econometric software. The views expressed herein do not necessarily reflect those of the U.S. Department of Agriculture.

## REFERENCES

Burness, H. S. and Quirk, J. P., 1980. Water Laws, Water Transfers, and Economic Efficiency, *Journal of Law and Economics,* 23, pp. 111-134.

Chambers, R. G. and Just, R. E., 1989. Estimating Multioutput Technologies, *American Journal of Agricultural Economics,* 71, pp. 980-995.

Council on Environmental Quality, 1989. Recommendations of the Council on Environmental Quality regarding the proposal by the Department of the Interior's Bureau of Reclamation to Renew Long-Term Water Contracts for the Orange Cove and other Friant Unit Irrigation Districts of the Central Valley Project in California, *Federal Register,* 54(128), pp. 28477-28491, July 6, 1989.

Glaser, L. K., 1986. Provisions of the Food Security Act of 1985, Agricultural Information Bulletin No. 498, Economic Research Service, USDA.

Just, R. E.; Zilberman, D.; and Hochman, E., 1983. Estimation of Multicrop Production Functions, *American Journal of Agricultural Economics,* 65, pp. 770-780.

Kanazawa, M., 1988. The Economic Efficiency of the Water Supply Policies of the Bureau of Reclamation: A Reconsideration. Working Paper, Department of Economics, Carleton College, Northfield, MN.

Lancaster, J., 1989a. White House Backs Environmentalists on Water, *The Washington Post*, November 29, p. A4.

Lancaster, J., 1989b. Interior to Renew Western Water Pacts, *The Washington Post*, November 30, p. A3.

Langley, J. A.; Reinsel, R. D.; Craven, J. A.; Zellner, J. A.; and Nelson, F. J., 1985. Commodity Price and Income Support Policies in Perspective, *Agricultural Food Policy Review: Commodity Program Perspectives.* Agricultural Economics Report No. 530, Economic Research Service, USDA.

Lau, L.J., 1978. Applications of Profit Functions. In: Fuss, M. and McFadden, D. (Eds.), *Production Economics: A Dual Approach to Theory and Applications*, North-Holland, Amsterdam.

Maddala, G.S., 1983. *Limited-Dependent and Qualitative Variables in Econometrics.* Cambridge University Press, New York, NY.

Miranowski, J. A.; Hrubovcak, J.; and Sutton, J., The Effects of Commodity Programs on Resource Use. In: Just, R. and Bockstael N. (Eds.), *Commodity and Resource Policies in Agricultural Systems.* Springer-Verlag, Berlin. In Production.

Moore, M. R. and Negri, D. H., 1990. The Bureau of Reclamation, Water, and Agriculture: Irrigation Water Conservation in the American West. Working Paper, Economic Research Service, USDA.

Negri, D. H. and Brooks, D. H., 1989. Decomposing Ground Water Demand. Working Paper, *Economic Research Service*, USDA.

Shumway, C. R.; Pope, R. D.; and Nash, E. K., 1984. Allocatable Fixed Inputs and Jointness in Agricultural Production: Implications for Modeling, *American Journal of Agricultural Econ*omics, 66, pp. 72-78.

U.S. Department of Commerce, Bureau of the Census, 1986. *1984 Farm and Ranch Irrigation Survey.* U.S. Government Printing Office, Washington, DC.

U.S. Department of the Interior, Bureau of Reclamation, 1988. *1987 Summary Statistics, Vol. 1: Water, Land, and Related Data.*

U.S. Department of the Interior, Office of the Secretary, 1988. H.R. 1443, Irrigation Subsidy Legislation: Questions from the Subcommittee on Water and Power Resources of the Committee on Interior and Insular Affairs, February 24, 1988.

Wahl, R. W. 1989. *Markets for Federal Water: Subsidies, Property Rights, and the Bureau of Reclamation.* Resources for the Future, Washington, DC.

# Seven: DYNAMIC ASPECTS OF IRRIGATION AND DRAINAGE MANAGEMENT

# 30 OPTIMAL INTERTEMPORAL IRRIGATION MANAGEMENT UNDER SALINE, LIMITED DRAINAGE CONDITIONS

Keith C. Knapp, University of California, Riverside

## ABSTRACT

Dynamic optimization models are formulated to determine economically efficient irrigation practices for an individual field over a sequence of years. The models consider crop rotations, spatial variability, and investment in irrigation systems, under saline, limited drainage conditions. Crop rotations can result in significantly different decision rules for a given crop depending on its position within the rotation. A cyclical pattern of water applications and soil salinity was also found when crop rotations are considered. Nonuniform irrigation results in significantly greater water applications and significantly greater response to changing prices compared to the perfectly uniform case.

## INTRODUCTION

This chapter considers economically efficient water management strategies in irrigated agriculture where soil salinity either can or does affect crop yields, and where the resulting drainage water (deep percolation) incurs significant disposal and/or environmental costs. Management over a sequence of years is considered; however, details of within-season irrigation management are ignored in order to keep the analysis tractable. Attention is limited to management of individual fields in the philosophy that a realistic treatment of the problem at this level is a necessary prerequisite to farm and regional level models.

Previous work in this area is somewhat limited (Knapp and Wichelns, 1990, provide a review). Yaron and Olian (1973), Matanga and Marino (1979), and Dinar and Knapp (1986) formulate dynamic optimization models with soil salinity as the state variable, and irrigation volume and possibly salinity as control variables. These papers assume that soil salinity is uniform throughout the field. Bresler, Yaron, and Segev (1983) formulate a multiyear linear programming model in which heterogeneous fields result in spatially variable soil salinity. The quantity of water applied during an irrigation season is constant; however, there are multiple sources of irrigation water differing in cost and salinity and the choice of which one to use in each time period is a variable. These studies all consider constant (through time) crops and irrigation systems.

Three extensions of the existing literature are considered here:

1. Crop rotations. In most areas, the same crop is not grown continuously as assumed in the existing literature, but is instead rotated with other crops on a regular basis. As illustrated in the empirical results given later, optimal irrigation of a given crop depends, in part, on the future sequence of crops.
2. Spatial variability. The literature on soil and water shows that considerable nonuniformity in water applications and soil properties exists at the field level and that this affects the appropriate specification of cropwater production functions (see Letey, Knapp, and Solomon, 1990, for discussion and references to the literature). Several analyses in a static framework have also shown that spatial variability dramatically affects economically efficient water applications (Letey, Vaux, and Feinerman, 1984; and Dinar, Letey, and Knapp, 1985). A dynamic optimization model with spatial variability is developed here. This model takes a completely different approach from Bresler, Yaron, and Segev (1983), and allows water quantity to be a variable in the analysis.
3. Investment in irrigation systems. Irrigation systems vary in terms of water application uniformity as well as other characteristics, thus investment in new systems is a major strategy for reducing irrigation and drainage water flows. A conceptual model is developed for determining optimal irrigation investment while accounting for spatial variability in soil salinity.

Applications of the models developed here include: Efficient water use in irrigated agriculture while accounting for both scarcity of irrigation water and disposal/environmental costs, income losses from increased salinization of water sources, and analysis of grower response to various policies, including irrigation surcharges and drainage effluent fees. Also, many other resource systems are characterized by spatially varying parameters. The methods developed here for incorporating spatial variability into dynamic optimization models may be applicable to other resource systems as well.

## CROP ROTATIONS

Irrigation of an individual field over an infinite horizon is considered. The irrigation system is fixed and the crop rotation assumed to be given is n years in length. The following variables are defined:

$q_{ij}$ = depth of applied irrigation water;

$s_{ij}$ = soil salinity at the beginning of the irrigation season measured as the EC of a saturated paste extract;

$y_{ij}$ = crop yield;

$d_{ij}$ = deep percolation of water past the root zone;

where i denotes the crop rotation number and j denotes the year within the crop rotation. With this notation the calendar year may be calculated by

[1] $\quad t = (i-1)n + j$

where t denotes year.

Crop yield and deep percolation are functions of soil salinity at the beginning of the irrigation season, and the quantity of irrigation water:

[2] $\quad \begin{bmatrix} y_{ij} \\ d_{ij} \end{bmatrix} = g_j(s_{ij}, q_{ij})$

where $g_{ij}$, i=1,2, are the respective component functions for $y_{ij}$ and $d_{ij}$ and noting that the functional relationship depends on the particular crop grown in year j of a given crop rotation. The equations of motion for soil salinity are given by

[3] $\quad \begin{aligned} s_{i,j+1} &= g_{3j}(s_{ij}, q_{ij}) \quad j=1,\ldots,n-1 \\ s_{i+1,1} &= g_{3j}(s_{ij}, q_{ij}) \quad j=n \end{aligned}$

implying that soil salinity at the beginning of the irrigation season is a function of irrigation inputs and soil salinity at the beginning of the previous year. Finally, it is also assumed that soil salinity is restricted to lie within the bounds

[4] $\quad \underline{s} \leq s_{ij} \leq \bar{s}$

and applied water quantity lies within the bounds

[5] $\underline{q} \leq q_{ij} \leq \overline{q}$

where $i=1,...,\infty$ and $j=1,...,n$.

The present value of returns to land and management is given by

[6] $\sum_{i=1}^{\infty} \sum_{j=1}^{n} \alpha^t [p_j y_{ij} - h_j(y_{ij}) - w\, q_{ij} - b_j - ed_{ij}]$

where t is defined by [1], $p_j$ is the crop price in the jth year of a rotation, $h_j$ is the harvest cost function, w is the per-unit water cost, $b_j$ is all other production costs in the jth year of a rotation, and e is the per-unit disposal/environmental costs associated with drainage (deep percolation) flows. The optimization problem is to choose decision rules for the $q_{ij}$ which maximize [6] subject to [1]-[5].

The solution procedure follows the DP algorithm described in Knapp et al. (1990). Optimal value functions $V_j(s_{ij})$ are defined. These equal [6] evaluated at the optimum point for various initial conditions on soil salinity; they give the present value of future returns to land and management under optimal operation as a function of the state variable $s_{ij}$. By inspection of [6] and the constraints [1]-[5] it can be seen that the optimal value functions are the same for all rotations i, $i=1,...,\infty$; therefore, the optimal decision rules for each year j in the rotation are also the same for all rotations i, $i=1,...,\infty$. The optimal decision rules are given by the solutions to maximizing

[7]
$p_j g_{1j}(s,q) - h_j[g_{1j}(s,q)] - wq - b_j - eg_{2j}(s,q) + \alpha V_{j+1}(g_{3j}(s,q))$  $j=1,...,n-1$

$p_j g_{1j}(s,q) - h_j[g_{1j}(s,q)] - wq - b_j - eg_{2j}(s,q) + \alpha V_1(g_{3j}(s,q))$  $j=n$

with respect to q, subject to the constraints [4]-[5] and for various values of s.

The optimal value functions are calculated by a dynamic programming (DP) algorithm. Let $V_{jk}$ be the estimate of $V_j$ after the kth iteration and let $V_{10}(s) = 0$ for s satisfying [4]. The dynamic programming recursions are defined by

$V_{nk}(s) = \text{Max } pg_{1n}(s,q) - h_n[g_{1n}(s,q)] - wq - b_n - eg_{2n}(s,q) +$

$\alpha V_{1,k-1}(g_{3n}(s,q))$

[8] $V_{jk}(s) = \text{Max } pg_{1j}(s,q) - h_j[g_{1j}(s,q)] - wq - b_j - eg_{2j}(s,q) +$

$\alpha V_{j+1,k}(g_{3j}(s,q))$  $\quad\quad j=n-1,\ldots,2$

$V_{1k}(s) = \text{Max } pg_{11}(s,q) - h_1[g_{11}(s,q)] - wq - b_1 - eg_{21}(s,q) +$

$\alpha V_{2k}(g_{31}(s,q))$

where the indicated maximizations are over q subject to [4]-[5]. The DP algorithm proceeds by solving [8] in the order indicated for $k=1,\ldots,K$. Results in Bertsekas (1976) show that $V_{jk} \to V_j$ as $k \to \infty$. Error bound calculations also described in Bertsekas are used to provide a stopping criterion for the algorithm.

The empirical specification of the model is based on Knapp et al. (1990). The field is assumed to be 129.5 ha in size; however, the data and results are reported on a per-unit area basis. A 3-year crop rotation of cotton-cotton-tomatoes is assumed, and the irrigation system is furrow with 1/2-mile runs. The production functions in [2] and the soil salinity equations-of-motion [3] are specified using a modified version of the production function in Letey, Dinar, and Knapp (1985); details and parameter values are given in Appendix 1. Other parameter values are given in table 1. Harvest costs are given by 441y + .006py for cotton and 13.17 + 19.85y for tomatoes. The discount rate is assumed to be 5 percent.

Table 1. Base parameter values.

| Symbol | Description | Cotton | Tomatoes |
|---|---|---|---|
| $p_j$ | output price | $1,757.36 Mg$^{-1}$ | $56.35 Mg$^{-1}$ |
| w | water cost | $2.06 ha$^{-1}$ cm$^{-1}$ | $2.06 ha$^{-1}$ cm$^{-1}$ |
| $b_j$ | non-water production costs | $1,006.93 ha$^{-1}$ yr$^{-1}$ | $1,358.93 ha$^{-1}$ yr$^{-1}$ |
| e | drainage disposal/ environmental cost | $4 ha$^{-1}$ cm$^{-1}$ | $4 ha$^{-1}$ cm$^{-1}$ |
| c | salt concentration of the irrigation water | .67 dS m$^{-1}$ | .67 dS m$^{-1}$ |

Source: Knapp et al. (1990)

Figure 1 displays the optimal decision rules for the model, accounting only for crop rotation. As noted earlier, these give optimal water applications as a function of soil salinity at the beginning of the irrigation season. Tomatoes require less water than cotton for maximum yield; this is reflected in a generally lower optimal decision rule for tomatoes compared to cotton. More interesting is that the optimal decision rules differ for the two cotton crops. Cotton is a more salt-tolerant crop than tomatoes, thus, additional water is applied to second-year cotton to leach salts out and lower soil salinity for the following tomato crop. This illustrates the importance of accounting for both the dynamics of soil salinity in general, and crop rotations in particular. It can also be noted that the optimal decision rules first increase and then decrease as soil salinity increases. This is due to the nature of the crop water production function. As soil salinity increases, ET begins to decrease after some point. Therefore, less irrigation water is needed to maintain a given ending soil salinity.

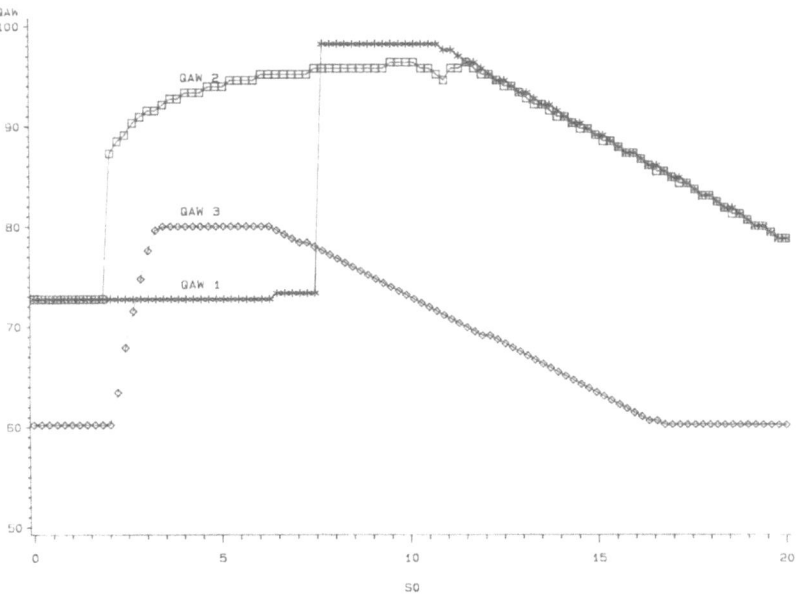

Figure 1. Optimal decision rules for the dynamic soil salinity optimization model with crop rotations. QAW1 and QAW2 are optimal water quantities for first and second year cotton, respectively; QAW3 is the optimal water quantity for tomatoes. So is soil salinity at the beginning of the growing season.

Figure 2 displays the time series plots for soil salinity and applied water when the initial soil salinity (t=0) is 2 dS/m. As can be seen, the time paths converge immediately to a limit cycle. Soil salinity is lowest for tomatoes which are a less salt-tolerant crop than cotton, and highest at the beginning of the second-year growing season for cotton. Applied water quantity is lowest for tomatoes and highest for second-year cotton. As noted earlier, applied water is greater for second-year cotton than first-year cotton in order to reduce soil salinity for the less salt-tolerant tomato crop. Additional results were generated for initial soil salinities of 8 and 14 dS/m. In both cases, additional water was applied during the initial rotation to lower soil salinity. After that, the same cyclical pattern as in figure 2 was followed for both soil salinity and applied water quantity.

## SPATIAL VARIABILITY

Irrigation water tends to be distributed nonuniformly over the field due to irrigation system characteristics and spatial variability in soil properties. As a result, soil salinity, crop yields, and deep percolation are also spatially variable. These variabilities need to be accounted for if accurate estimates of field-level yields and drainage flows are to be obtained. This section proposes an approach to incorporating spatial variability in dynamic optimization models, and generates empirical results for a specific case study. To focus on the problem at hand, it is assumed that both the crop and irrigation system are constant over time.

Let A denote the area of the field and let $\bar{q}_t$ denote the average irrigation water depth applied to the field as a whole in year t. At an individual point in the field, we assume that the irrigation water depth percolating at that point, denoted $q_t$, is given by

[9] $\quad q_t = \beta \bar{q}_t$

implying that the applied water depth at that point in the field is a constant fraction ($\beta$) of the field-average applied water depth. The crop water production functions are given by

[10]
$$y_t = g_1(s_t, q_t)$$
$$d_t = g_2(s_t, q_t)$$

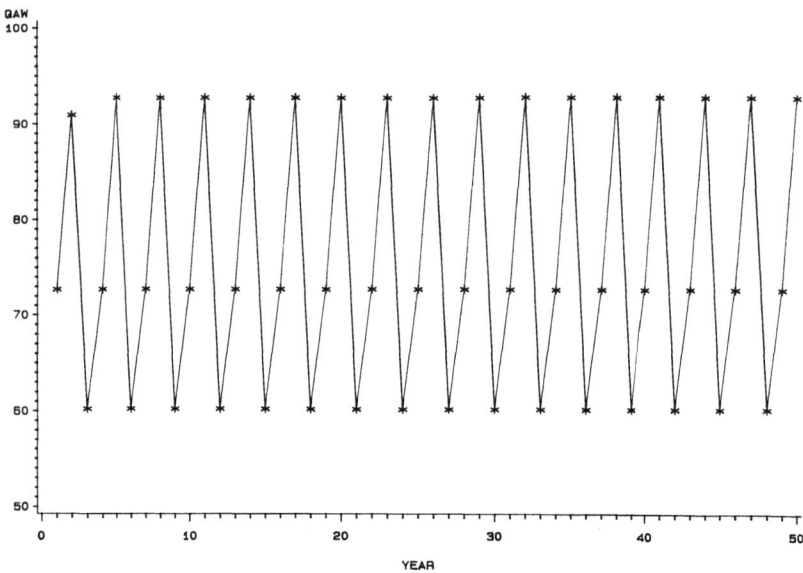

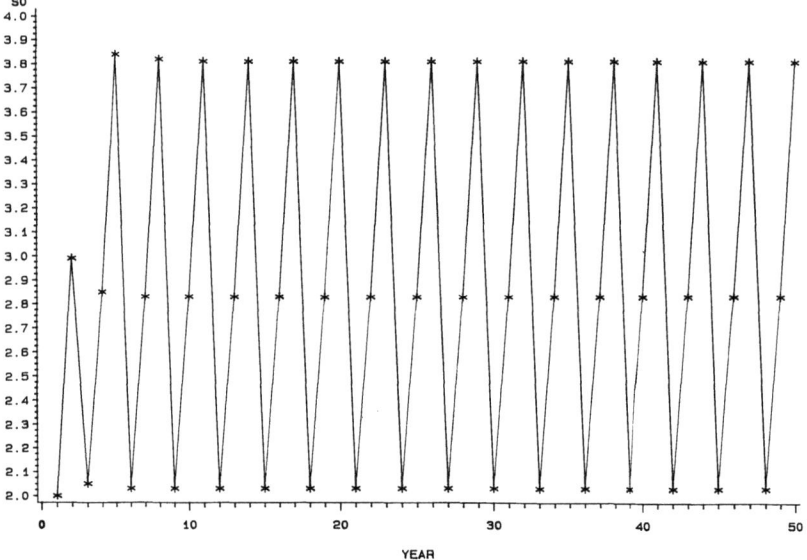

Figure 2. Optimal time paths for (a) applied water depth and (b) soil salinity for the dynamic soil salinity model with crop rotations.

where $y_t$ and $d_t$ are crop yield and deep percolation, respectively. As before, soil salinity dynamics at a given point are defined by

$$[11] \quad s_{t+1} = g_3(s_t, q_t)$$

where $s_t$ is soil salinity at the beginning of the irrigation season.

It is assumed that in any year t, $s_t$ and $\beta$ are jointly distributed over the field with density function $f_t$. Thus, the expression

$$[12] \quad \int_a^b \int_c^d f_t(s, \beta) \, ds \, d\beta$$

gives the fraction of the field in year t with a $\leq \beta \leq$ b and c $\leq s_t \leq$ d. By mass conservation

$$[13] \quad \bar{q}_t A = \int_0^\infty \int_0^\infty q_t A f_t(s, \beta) \, ds \, d\beta$$

where the left-hand side of [13] is the total volume of water applied to the field and the right-hand side of [13] is the total volume of water infiltrating into the field; obviously, these must be equal. Substituting [9] into [13] and rearranging yields

$$[14] \quad 1 = \int_0^\infty \int_0^\infty \beta \, f_t(s, \beta) \, ds \, d\beta$$

which implies $E[\beta] = 1$ by definition. Field-level yields ($\bar{y}_t$) and deep percolation depth ($\bar{d}_t$) may be calculated from

$$[15] \quad \begin{aligned} \bar{y}_t &= \int_0^\infty \int_0^\infty g_1(s, \beta \bar{q}_t) \, f_t(s, \beta) \, ds \, d\beta \\ \bar{d}_t &= \int_0^\infty \int_0^\infty g_2(s, \beta \bar{q}_t) \, f_t(s, \beta) \, ds \, d\beta \end{aligned}$$

respectively. Also, the control variable $\bar{q}_t$ is subject to the constraint

$$[16] \quad q^{min} \leq \bar{q}_t \leq q^{max}$$

where $q^{min}$ and $q^{max}$ are the minimum and maximum water quantities, respectively.

To complete the model, it remains to specify how the joint density function $f_t$ changes over time and to incorporate this into a dynamic optimization model. The approach here is to specify a particular form of the density function with a finite number of moments, and then to treat the moments as state variables in the dynamic optimization problem. The lognormal distribution is commonly used in the scientific literature to characterize spatially random variables in agricultural fields. Therefore, $s_t$ and $\beta$ are assumed to be distributed bivariate lognormal. From Mood, Graybill, and Boes (1974), the bivariate lognormal distribution is characterized by five parameters: the mean and standard deviation of both variables and the correlation coefficient. Also, the marginal distributions of both variables are lognormal. From [14], $E[\beta]=1$. Since the irrigation system is assumed to be constant over time, then the standard deviation of $\beta$ is also assumed to be constant. (Appendix 2 gives conversion formulas for the moments of a lognormal random variable and its logarithm.) This leaves the expected value and standard deviation of $\ln s_t$, and the correlation coefficient between $\ln s_t$ and $\ln \beta$ (denoted $\rho_t [\ln s_t, \ln\beta]$), as variables which can change over time depending on $\bar{q}_t$. These can be calculated in a straightforward way from:

$$E[\ln s_{t+1}] = \int_0^\infty \int_0^\infty \ln g_3(s, \beta\bar{q}_t)\, f_t(s,\beta)\, ds\, d\beta$$

$$[17]\quad E[(\ln s_{t+1})^2] = \int_0^\infty \int_0^\infty [\ln g_3(s,\beta\bar{q}_t)]^2\, f_t(s,\beta)\, ds\, d\beta$$

$$E[(\ln s_{t+1})(\ln \beta)] = \int_0^\infty \int_0^\infty [\ln g_3(s,\beta\bar{q}_t)]\ln \beta\, f_t(s,\beta)\, ds\, d\beta$$

Since the $\beta$ distribution moments are parameters in this problem, the density function $f_t$ is completely characterized once the mean and standard deviation of $\ln s_t$ and the correlation coefficient are known. Thus, equations [17] are effectively the equations of motion for the dynamic system characterizing the field.[1]

The present value of returns to land and management over an infinite horizon is given by

$$[18]\quad \sum_{t=1}^\infty \alpha^t [p\bar{y}_t - h(\bar{y}_t) - w\bar{q}_t - b - e\bar{d}_t]$$

where p is crop price, h is the harvest cost function, w is water price, b is all other production costs, and e is drainage disposal/environmental costs. The problem is to choose $\bar{q}_t$, t=1, ..., ∞ to maximize [18] subject to [15]-[17] and given the initial density function $f_1$. In this problem the control variable is $\bar{q}_t$ and the state variables are $E[\ln s_t]$, $SD[\ln s_t]$, and $\rho[\ln s_t, \ln \beta]$. This problem can be solved in a straightforward manner using dynamic programming. Also, the integrals in [15] and [17] are computed by discretizing the random variables and evaluating the functions at each of the probability mass points.

The model is applied to a continuous crop of cotton. As before, the field is assumed to be 129.5 ha. The crop-water production function described in Appendix 1 is used and the other parameter values are those in table 1. Additionally, the irrigation system is assumed to be furrow with 1/2-mile runs. Following Knapp et al. (1990), the irrigation system is assumed to have a Christiansen Uniformity Coefficient (CUC) equal to 70, implying that $SD(\beta)$ = .4.

Table 2 gives optimization results under nonuniform irrigation when the initial soil salinity is 4 dS/m throughout the field. Initially, 95 cm of water is applied. This leaches salts out so that the average salinity is 3 dS/m. After this, a constant 91 cm is applied, and all other variables reach steady-state levels in approximately 5 to 6 years.

Table 2. Optimal values with spatially variable water infiltration (CUC = 70).

| Year | $E[s_t]$ (dS/m) | $SD[s_t]$ (dS/m) | Correlation coefficient | Field-average applied water depth (cm/yr) | Field-average crop yield (Mg ha$^{-1}$ yr$^{-1}$) | Field-average deep percolation depth (cm/yr) |
|---|---|---|---|---|---|---|
| 1 | 4.00 | 0.01 | 0.00 | 95.00 | 1.51 | 22.69 |
| 2 | 3.00 | 1.90 | -0.87 | 91.00 | 1.49 | 22.62 |
| 3 | 3.07 | 2.11 | -0.90 | 91.00 | 1.49 | 22.70 |
| 4 | 3.14 | 2.22 | -0.90 | 91.00 | 1.49 | 22.74 |
| 5 | 3.18 | 2.28 | -0.90 | 91.00 | 1.49 | 22.77 |
| 6 | 3.21 | 2.32 | -0.90 | 91.00 | 1.49 | 22.79 |
| 7 | 3.23 | 2.34 | -0.90 | 91.00 | 1.49 | 22.80 |
| 8 | 3.23 | 2.36 | -0.90 | 91.00 | 1.49 | 22.81 |
| 9 | 3.24 | 2.36 | -0.90 | 91.00 | 1.49 | 22.81 |
| 10 | 3.24 | 2.37 | -0.90 | 91.00 | 1.49 | 22.81 |
| 11 | 3.24 | 2.37 | -0.90 | 91.00 | 1.49 | 22.82 |
| 12 | 3.24 | 2.37 | -0.90 | 91.00 | 1.49 | 22.82 |

Table 3 gives optimal steady-state values for the main variables under alternate water prices and drainage disposal/environmental costs. (Some of the solutions involve cyclical fluctuations in the variables; in these cases, averages over the cycle length are reported.) Imposition of a $4 ha$^{-1}$ cm$^{-1}$ drainage fee results in fairly significant reductions in applied water depth. These reductions range from 13 to 21 percent for the water prices considered. Increasing the price of irrigation water can also result in substantial water use reductions. With zero drainage disposal/environmental costs, an increase in the price of water from $1 ha$^{-1}$ cm$^{-1}$ to $2 ha$^{-1}$ cm$^{-1}$ reduces optimal applied water by 12 percent, but an increase from $2 ha$^{-1}$ cm$^{-1}$ to $3 ha$^{-1}$ cm$^{-1}$ for irrigation water only results in a 6-percent decrease in optimal water depths. The percentage reductions are smaller when drainage disposal/environmental costs are imposed. Decreasing applied water depths due to changes in water prices and drainage fees result in increased steady-state soil salinity and decreased yield and deep percolation depths as would be expected.

Table 3. Optimal steady-state values in the dynamic optimization model under nonuniform irrigation (CUC = 70).

| Water price ($ ha$^{-1}$ cm$^{-1}$) | Drainage disposal/ environ- mental cost ($ ha$^{-1}$ cm$^{-1}$) | $E[s_t]$ (dS/m) | Applied water depth (cm/yr) | Field-average yield (Mg ha$^{-1}$ yr$^{-1}$) | Field-average deep perco- lation depth (cm/yr) |
|---|---|---|---|---|---|
| 1 | 0 | 1.8 | 121 | 1.55 | 48.9 |
|   | 4 | 2.9 | 96  | 1.50 | 26.6 |
| 2 | 0 | 2.3 | 107 | 1.53 | 36.1 |
|   | 4 | 3.2 | 91  | 1.49 | 22.8 |
| 3 | 0 | 2.6 | 101 | 1.52 | 31.2 |
|   | 4 | 3.4 | 88  | 1.47 | 20.8 |

Some runs resulted in cyclical solutions, perhaps due to discretization error in the DP algorithm. In these cases, average values were computed for the variables over the cycle.

Results were also computed assuming perfectly uniform irrigation for the same water price and drainage disposal/environmental cost combinations reported in table 3. In this case there is essentially no response to changes in water costs or drainage fees. Optimal water depths ranged from 76 to 77 cm/yr for the various combinations considered, and the other variables showed even less variation. Comparing these results to the results in table 3, it can be seen that nonuniform irrigation applications result in both significantly greater applied water depths, and significantly greater response to changing economic parameters. Thus an accurate accounting of irrigation uniformity is essential if meaningful results are to be obtained.

## IRRIGATION INVESTMENT

The model in the previous section assumes a constant (over time) crop and irrigation system. The model can be extended to include exogenous crop rotations in an entirely straightforward way using the approach in the second section. This section extends the spatial variability model to include investment in new irrigation systems. The empirical results in the previous section emphasize the potential of increased water infiltration uniformity for reducing both applied water and deep percolation, and increasing yields. Different irrigation systems typically have different application/infiltration uniformities with the more expensive, capital-intensive systems typically being more uniform. A fundamental consideration in intertemporal irrigation management is, therefore, the selection of irrigation system type.

Three additional variables are defined:

$x_t$ = investment in a new irrigation system during year t,

$z_t$ = type of irrigation system existing at the beginning of year t, and

$a_t$ = age of existing irrigation system at the beginning of year t.

Let m denote the number of alternative irrigation systems available for purchase. Also, let $x_t = 0$ denote no new investment in an irrigation system in year t and $x_t = i$, $i \in \{1, ..., m\}$ denote investment in a system of type i during year t. The choice of $x_t$ is constrained by

$$[19] \quad x_t \in \begin{bmatrix} 0, ..., m & a_t < a^m(z_t) \\ 1, ..., m & a_t = a^m(z_t) \end{bmatrix}$$

where $a^m(z_t)$ is the maximum physical life of irrigation system $z_t$. This constraint guarantees that irrigation systems are replaced at or before the end of their physical life. The equations of motion for type and age of the irrigation system are:

$$[20] \quad z_{t+1} = \begin{bmatrix} z_t & x_t = 0 \\ x_t & x_t > 0 \end{bmatrix}$$

$$[21] \quad a_{t+1} = \begin{bmatrix} a_t+1 & x_t = 0 \\ 1 & x_t > 0 \end{bmatrix}$$

respectively.

As before, $E[\beta] = 1$; however, now $SD[\beta]$ is a function of the type of irrigation system used during year t. Both $E[\ln s_{t+1}]$ and $SD[\ln s_{t+1}]$ are calculated from [17] as before. If the same irrigation system is used in year t+1 as in year t, then the correlation coefficient for $\ln s_{t+1}$ and $\ln \beta$ (denoted $\rho_t[\ln s_t, \ln \beta]$) can be calculated as before from equations [17].[2] However, if the irrigation system changes, then it is likely that $\rho_t[\ln s_t, \ln \beta]$ changes as well. For example, a newly installed sprinkler system likely has a very different water distribution pattern than a previously existing furrow system; thus the correlation of $\ln s_t$ and $\ln \beta$ will change if the sprinkler system is installed. As a first approximation, it is reasonable to assume that:

$$[22] \quad \rho_t[\ln s_t, \ln \beta] = \begin{bmatrix} \rho_t'[\ln s_t, \ln \beta] & x_t = 0 \\ 0 & x_t > 0 \end{bmatrix}$$

where $\rho_t'[\ln s_t, \ln \beta]$ is calculated from equations [17]. This implies that the correlation between $\ln s_t$ and $\ln \beta$ is zero if a new system is installed.

The objective function remains as [18]. The optimization problem is then to maximize [18] subject to [15]-[17], [19]-[22] and noting that the density function $f_t$ now depends directly on $z_t$ and $x_t$ through $SD[\beta]$, as well as $E[\ln s_t]$, $SD[\ln s_t]$, and $\rho_t[\ln s_t, \ln \beta]$. The control variables in this problem are $\bar{q}_t$ and $x_t$; the state variables are $E[\ln s_t]$, $SD[\ln s_t]$, $\rho_t[\ln s_t, \ln \beta]$, $z_t$ and $a_t$. This problem can be solved in a straightforward way using dynamic programming; the model can also be extended to include crop rotations using the methods in the second section. This is likely to be a computationally demanding problem. Since a computer program has not been implemented at this point in time, empirical results are not presented here.

## CONCLUSIONS

Three extensions to the dynamic optimization models in Yaron and Olian (1973), Matanga and Marino (1979), and Dinar and Knapp (1986) have been considered. The first is inclusion of crop rotations. For the particular rotation considered (cotton-cotton-tomatoes), the empirical results illustrate that significantly greater quantities of water are applied for second-year cotton compared to first-year cotton in order to leach salts out for the more salt-sensitive tomato crop. The optimal time paths for water quantity and soil salinity thus exhibit annual fluctuations.

The second extension is inclusion of spatial variability in a dynamic optimization model. This was accomplished by defining a joint distribution function for the spatially-variable parameter and soil salinity, and then treating the moments as state variables in the dynamic optimization model. The empirical results show that spatial variability has a significant impact on both optimal water volumes and the sensitivity of optimal water use to changing water prices and drainage effluent fees. The third extension is inclusion of investment in irrigation systems. Irrigation systems vary significantly in application uniformity and are an important means of reducing water use and environmental damages from drainage effluent. The conceptual model developed here, extended to include crop rotations, can provide a realistic treatment of this problem.

## APPENDIX 1. CROPWATER PRODUCTION FUNCTION

The cropwater production function used here is based on the model in Letey, Dinar, and Knapp (1985) and Letey and Dinar (1986). That model has three equations: (1) Yield as a function of soil salinity in the root zone, (2) yield as a function of evapotranspiration (ET), and (3) soil salinity as a function of leaching fraction. These three equations are solved simultaneously to determine values for yield, ET, and soil salinity given the volume and salt concentration of applied water.

In the original model, a steady-state relation is assumed for soil salinity. To develop a dynamic model, the steady-state soil salinity relation is replaced by a dynamic equation for soil salinity which assumes piston-flow conditions. This dynamic equation is specified by

$$[23] \quad DW = \begin{cases} 0 & AW \leq ET \\ AW - ET & AW \geq ET \end{cases}$$

$$[24] \quad CDW = \begin{cases} 0 & DW = 0 \\ 2*S0 & DW \leq FC \\ \dfrac{FC(2*S0) + ECI(DW - FC)}{DW} & DW \geq FC \end{cases}$$

$$[25] \quad S1 = S0 + \frac{AW*ECI}{2*FC} - \frac{CDW*DW}{2*FC}$$

where the variables are

- AW = applied water depth
- ET = evapotranspiration
- DW = drainwater (deep percolation)
- CDW = salt concentration of the drainwater
- S0 = soil salinity at the beginning of the irrigation season
- S1 = soil salinity at the end of the irrigation season

and the parameters are

- ECI = salt concentration of the irrigation water
- FC = field capacity

Equations [23] and [24] are the piston-flow assumption; equation [25] is a mass-balance relation. Solution of these equations along with the first two equations in the original model determines values for yield, ET, DW, and S1 for given values of AW and S0. This model is then used to specify the $g_1$, $g_2$, and $g_3$ functions in the text. For this analysis, FC was assumed to be 24.72 cm for a root zone depth of 120 cm, and ECI is .67 dS/m.

## APPENDIX 2. MOMENT-CONVERSION FORMULAS FOR LOG-NORMAL RANDOM VARIABLES

Let X be distributed as a log-normal random variable and let

$$E[\ln X] = \mu$$
$$V[\ln X] = \sigma^2$$

where E and V are the expectation and variance operators, respectively. From Mood, Graybill, and Boes

$$[26] \quad E[X] = \exp[\mu + \sigma^2/2]$$

$$[27] \quad V[X] = \exp[2(\mu + \sigma^2)] - \exp[2\mu + \sigma^2]$$

and after rearranging and some algebra, it can be seen that:

[28] $\sigma^2 = \ln[(V[X] + E[X]^2) / E[X]^2]$

[29] $\mu = \ln E[X] - \sigma^2/2$

These formulas allow conversion between E[X] and V[X], and E[ln X] and V[ln X].

## NOTES

[1] Note that the assumption of a constant irrigation system enters in two ways into the model specification. First, as noted earlier in the text, this implies that the moments for $\beta$ are constant through time. Second, the assumption also implies that the $\beta$ value for each point in the field is constant through time. This is necessary for the third equation in [17] to be a valid basis for calculating the correlation between $s_{t+1}$ and $\beta$.

[2] The age of the irrigation system may also influence the distribution of $\beta$. In this case both the SD[$\beta$] and $\rho_t(s_t,\beta)$ would depend on both the age and type of the system. For simplicity, this possibility is ignored here.

## REFERENCES

Bertsekas, D. P., 1976. *Dynamic Programming and Stochastic Control*. Academic Press, New York, NY.

Bresler, E.; Yaron, D.; and Segev, A., 1983. Evaluation of Irrigation Water Quality for a Spatially Variable Field, *Water Resources Research*, 19(6), pp. 1613-1621.

Dinar, A. and Knapp, K. C., 1986. A Dynamic Analysis of Optimal Water Use Under Saline Conditions, *Western Journal of Agricultural Economics*, 11(1), pp. 58-66.

Dinar, A.; Letey, J.; and Knapp, K. C., 1985. Economic Evaluation of Salinity, Drainage and Non-Uniformity of Infiltrated Irrigation Water, *Agricultural Water Management*, 10(3), pp. 221-233.

Knapp, K. C.; Stevens, B. K.; Letey, J.; and Oster, J. D., 1990. A Dynamic Optimization Model for Irrigation Investment and Management under Limited Drainage Conditions, *Water Resources Research*, 26(7), pp. 1335-1343.

Knapp, K. C. and Wichelns, D., 1990. Dynamic Optimization Models for Salinity and Drainage Management. In: Tanji K.K. (Ed.), *Agricultural Salinity Assessment and Management*, ASCE, New York, NY.

Letey, J. and Dinar, A., 1986. Simulated Crop-Water Production Functions for Several Crops when Irrigated with Saline Waters, *Hilgardia,* 54(1), pp. 1-32.

Letey, J.; Dinar, A.; and Knapp, K. C., 1985. Crop-Water Production Function Model for Saline Irrigation Waters, *Soil Science Society of America Journal* 49(4), pp. 1005-1009.

Letey, J.; Knapp, K. C.; and Solomon, K., 1990. Crop-Water Production Functions under Saline Conditions. In: Tanji, K.K. (Ed.), *Agricultural Salinity Assessment and Management,* ASCE, New York, NY.

Letey, J.; Vaux, H. J. Jr.; and Feinerman, E., 1984. Optimum Crop Water Application as Affected by Uniformity of Water Infiltration, *Agronomy Journal,* 76, pp. 435-441.

Matanga, G. B. and Marino, M. A., 1979. Irrigation Planning: 2. Water Allocation for Leaching and Irrigation Purposes, *Water Resources Research,* 15(3), pp. 679-683.

Mood, A. M.; Graybill, F. A.; and Boes, D. C., 1974. *Introduction to the Theory of Statistics.* McGraw-Hill, New York, NY.

Yaron, D. and Olian, A, 1973. Application of Dynamic Programming in Markov Chains to the Evaluation of Water Quality in Irrigation, *American Journal of Agricultural Economics,* 55(3), pp. 467-471.

# 31 MANAGING DRAINAGE PROBLEMS IN A CONJUNCTIVE GROUND AND SURFACE WATER SYSTEM

Yacov Tsur, University of Minnesota, St. Paul

### ABSTRACT

This chapter develops a framework for the management of an irrigation and drainage system where irrigation is derived both from surface and ground-water sources. Basic principals underlying the management of a conjunctive ground and surface water system are introduced and used to derive optimal rules for managing such a system. Common property characteristics of this system and the additional drainage problem are also included in the analysis that evaluates policies aimed at enforcing irigation/drainage rules and their effects on the environment. The relevancy of the approach to the situation in the San Joaquin Valley is discussed.

### INTRODUCTION

The conjunctive use of ground water and surface water for irrigation is pervasive and has attracted much research, starting with the early work of Burt (1964a-b) followed by Brown and McGuire (1967), Cummings and Burt (1969), Burt and Cummings (1970), Cummings and Winkelman (1970), Domenico et al. (1970), Young and Bredehoeft (1972), Bredehoeft and Young (1983), Tsur (1990), and Tsur and Graham-Tomasi (1990) among others. The problem, in general terms, is that of allocating ground water over time when the demand for ground water varies according to available supply of surface water.

The term "conjunctive ground and surface water system" is applied to a number of systems; they differ according to the source. The source of surface water may consist solely of streamflows emanating from the aquifer, it may be independent of the ground-water source (rainfall) or it may be a combination of the two. The ground-water aquifer may be confined (see examples in Margat and Saad, 1985 and Issar, 1985) or replenishable, deep or shallow. The surface water source may be stable or it may stochastically fluctuate over time. Depending on the particular situation, the management problem of a conjunctive ground and surface water system can become quite involved.

In this chapter a framework is developed for the management of an irrigation and drainage system, where irrigation is derived both from surface and ground-water sources. Initially, the basic principles underlying the management of a conjunctive ground and surface water system are introduced. After deriving the optimal rules for managing such a system, a discussion of the common property nature of ground-water resources shows how market forces are unlikely to generate water use patterns which satisfy these rules. Possible policies to restore the optimal management rules are then discussed. The third section introduces the "drainage problem" (see SJVDP, 1989 for a description of drainage-related problems in the San Joaquin Valley), followed by a section on the rules governing desirable irrigation/drainage management. A section on policy distinguishes between those policies designed to enforce the optimal irrigation/drainage rules and those aimed at affecting the environment within which the management problem rests. A concluding section briefly discusses extensions relevant to the situation in the San Joaquin Valley.

## BASIC PRINCIPLES OF THE MANAGEMENT OF A CONJUNCTIVE GROUND AND SURFACE WATER SYSTEM

A conjunctive ground and surface water system consists of a surface water source (streamflows, rainfall, reservoirs), a ground-water source (aquifer) and an agriculture production process which requires water as an input. Figure 1 gives a schematic representation of such a system.

Let $F(x)$ denote the water response function, measured in dollar per hectare ($/ha), and x indicate the level of water input, measured in cubic meter per hectare $(m^3/ha)$[1]. The marginal water productivity is the change in $F(x)$ resulting from a small (marginal) change in water input x and is indicated by $F_x = \partial F/\partial x$. It plays a central role in determining the management rules. In most cases $F(x)$ increases in x at a diminishing rate, thus $F_x(x)$ is positive and decreasing in x (on different ways to estimate this function see Howitt et al., 1980 and Paris and Knapp, 1989).

The quantities of surface and ground water applied for irrigation at time t are denoted by $S_t$ and $g_t$, respectively; total water input is thus $x_t = S_t + g_t$. The amount of rainfall relevant for irrigation (during the growing season) is assumed stable at the level R and is included in $S_t$, thus $S_t \geq R$. The stock of

# CONJUNCTIVE GROUND AND SURFACE WATER SYSTEM

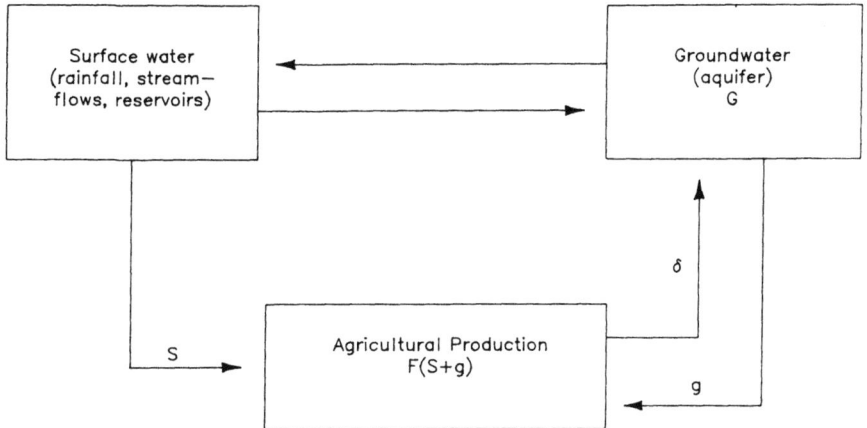

Figure 1. Schematic representation of a conjunctive ground and surface water system.

ground water at time t, denoted by $G_t$, changes over time as extraction takes place and as some of the water input (irrigation) infiltrates into the aquifer:

[1] $dG_t/dt \equiv \dot{G}_t = -(1-\delta)g_t + \delta S_t,$

where $\delta$ is a permeability parameter indicating the fraction of the water applied for irrigation that permeates into the aquifer (when the aquifer reaches its capacity level, $\dot{G}_t$ equals the minimum between the right-hand side of [1] and zero).

The cost of pumping ground water at a rate g is given by $z(G)g$, where $z(G)$ is the unit cost of ground-water extraction when the ground-water stock is at the level G. $z(G)$ is nonincreasing in G (a larger G means a higher ground-water table, a shorter distance to the surface and hence lower extraction costs). The unit cost of surface water irrigation (except for rainfall) is denoted by w. The instantaneous profit generated by $S_t$ and $g_t$ is thus given by

$F(g_t+S_t) - z(G_t)g_t - w(S_t-R).$

The amount of irrigation water may be subject to capacity constraints. Let C and B indicate these capacity limits, thus $g_t \le C$ and $S_t \le B$ for all $t \ge 0$.

A water management policy entails setting $S_t$ and $g_t$ for all time periods $t \ge 0$; it generates the benefit (the present value of the profit stream)

$$\int_0^\infty [F(g_t+S_t) - z(G_t)g_t - w(S_t-R)]e^{-rt}dt$$

where r is the time rate of discount. The policy sought is the one which maximizes this benefit.

Let V(G) be the maximum feasible benefit when the current stock of ground water is G:

[2] $$V(G) = \text{MAX} \int_0^\infty [F(g_t+S_t) - z(G_t)g_t - w(S_t-R)]e^{-rt}dt$$

subject to: Eq. [1], $0 \leq g_t \leq C$, $R \leq S_t \leq B$, $G_t \geq 0$ and $G_0 = G$.

The change in V(G) caused by a marginal (small) change in G is the unit value of the ground-water stock and is denoted by $V_G(G)$. It represents the future benefit forgone as a result of pumping a unit of ground water today and is referred to as the shadow price or the royalty value of the aquifer.

Using a dynamic programming approach, the following relation is obtained for each time period:

[3] $$rV(G_t) = \underset{g_t, S_t}{\text{MAX}} \{F(g_t+S_t) - [z(G_t)+V_G(G_t)(1-\delta)]g_t - [w-V_G(G_t)\delta]S_t + wR\}.$$

In words, the optimal conjunctive ground and surface water policy ($S^*_t$, $g^*_t$, $t \geq 0$) is the one under which the right-hand side of [3] is maximized in each time period (subject, of course, to the constraints given in [2]). The maximand on the right-hand side of [3] is the instantaneous profit corrected to account for intertemporal effects. The intertemporal effects are effects of current decisions on future profits and are represented by the shadow prices $V_G(G_t)$. Thus the cost associated with one cubic meter of ground water applied for irrigation today consists of (1) the pumping and distribution costs as given by $z(G_t)$, and (2) the effect on future profits resulting from the drop in the stock of ground water, which occurs due to higher pumping costs in the future and increased scarcity of ground water. This second cost component is represented by $V_G(G_t)[1-\delta]$ (the factor $1-\delta$ accounts for the fact that only $(1-\delta)m^3$ of each $1$ $m^3$ pumped is lost, as $\delta m^3$ leaches back into the aquifer). The economic cost of ground water is therefore given by $z(G_t)+V_G(G_t)[1-\delta]$, which is the coefficient of $g_t$ on the right-hand side of [3]. Similarly, the economic cost of surface water is $w-V_G(G_t)\delta$, which consists of the engineering cost, w, minus the contribution of surface water to future profits via its effect on the ground-water stock derived from the fraction $\delta$ of the surface water irrigation that leaches into the aquifer.

In view of [3] the characterization of the optimal policy becomes a straightforward exercise. Disregarding for the moment the capacity limits (that is assuming they are not binding) and without rainfall (i.e., $R=0$) the following management rules apply:

(i) As long as the economic cost of ground water exceeds that of surface water, i.e., $z(G_t) + V_G(G_t) > w$, only surface water is used for irrigation at a level that equates the marginal productivity of water to its cost:

$$F_x(S_t^*) = w - \delta V_G(G_t).$$

(ii) As long as the economic cost of ground water falls below that of surface water, i.e., $z(G_t) + V_G(G_t) < w$, only ground water is used for irrigation at a level that equates the marginal productivity of water to its cost:

$$F_x(g_t^*) = z(G_t) + V_G(G_t)(1-\delta).$$

(iii) When the economic costs of ground and surface water are equal, i.e., $z(G_t) + V_G(G_t) = w$, irrigation water is derived from both sources at a level that satisfies

$$F_x(g_t^* + S_t^*) = w - V_G(G_t)\delta$$

and at the mix $g_t^*/S_t^* = \delta/(1-\delta)$ such that the ground-water stock remains constant ($\dot{G}_t = 0$).

With the above interpretation of the economic costs of ground and surface water, these management rules make intuitive sense. Modifications are needed in the presence of binding capacity limits and with positive rainfall are straightforward.

The dynamic behavior of the system is depicted in figure 2. At all stock levels $G$ for which $z(G) + V_G(G)$ lies above $w$, ground water is more expensive than surface water, thus only the latter is applied for irrigation (cf. (i)). This causes the ground-water stock to increase, which in turn diminishes the pumping cost $z(G)$ and the shadow price $V_G(G)$ of ground water, as represented by the declining curve labeled $z(G) + V_G(G)$. When the ground-water stock reaches the level $\hat{G}$, the cost of ground water coincides with that of surface water and surface water is applied conjunctively with ground water so as to retain the aquifer at this stock level (cf. (iii)). For stock levels above $\hat{G}$, ground water is cheaper than surface water and irrigation water is derived solely from the aquifer (cf. (ii)). This causes the ground-water stock to decline toward $\hat{G}$. The ground-water stock level $\hat{G}$ is called the steady state; the period in which the system moves toward $\hat{G}$ is called the transition period (stage); the period in which $G = \hat{G}$ is called the steady period (stage).

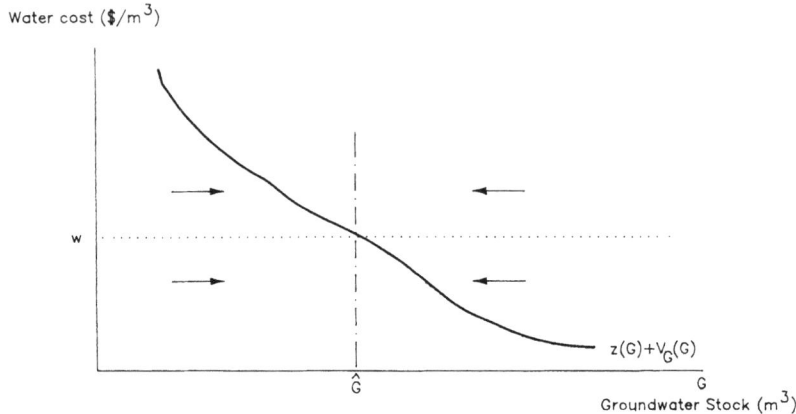

Figure 2. Dynamic behavior of the solution of Section 2.

*Policy Intervention*

The management rules (i)-(iii) differ from the myopic rules under which the instantaneous profit is maximized in each time period. The myopic rules are derived from (i)-(iii) by setting the shadow prices $V_G(G_t)$ equal to zero. A question then arises as to whether the individual growers are motivated to follow the intertemporal rules (i)-(iii) or whether they behave myopically. Unfortunately, the second possibility is more likely to prevail. The problem is similar to that of a "common property" situation (see Dasgupta, 1982 and Negri, 1989) in which the effect of each individual's extraction on the aquifer is negligible but is not at all negligible with respect to his or her own profits. Following the intertemporal rules entails giving up some present profits in return for future profits. But the future gains will materialize only if all (or most) growers follow the intertemporal rules. Now, if most growers follow the intertemporal rules, it is in the interest of the individual farmer to behave myopically because his or her effect on the aquifer is negligible and he can enjoy larger profits both in the present and in the future. On the other hand, if all other growers behave myopically then the grower should do the same, since otherwise there will be no future gains to compensate for the present losses. Realizing that this line of reasoning is not exclusive to any particular individual, the grower has good reasons to suspect that others will not follow the intertemporal rules, in which case he should not obey them either. Clearly, some regulatory policies (quota, taxes) or market mechanism (water rights) to restore intertemporal considerations are in order (on water rights see Gisser and Sanchez, 1980; Gisser, 1984; and Anderson, Burt and Fractor, 1983, among others).

Optimal tax schedule. The engineering costs of ground and surface water ($z(G)$ and w, respectively) do not reflect their economic costs ($z(G) + V_G(G)[1-\delta]$) and $w - V_G(G)\delta$, respectively). A tax schedule to correct for this discrepancy consists of taxing each cubic meter of ground water by the amount $V_G(G_t)[1-\delta]$ and subsidizing each cubic meter of surface water by the amount $V_G(G_t)\delta$. The problem with such a tax schedule is that it depends on the stock of ground water and thus must be adjusted constantly during the transition period. This might be hard to administer, since it requires constantly monitoring the aquifer level. Furthermore, it is likely to be objected to by a farmer who prefers stable water prices. An alternative scheme is therefore to impose the steady state tax schedule: a fixed tax of $V_G(\hat{G}_t)[1-\delta]$ on ground water and a fixed subsidy of $V_G(\hat{G})\delta$ on surface water. Such a tax schedule ensures a smooth transition to the steady state (though it may lengthen the transition period relative to that under the schedule described above), is easy (hence cheap) to administer, and is stable thereby facilitating compliance by growers.

Optimal water quotas. The management rules (i)-(iii) determine also the desirable quantities of ground and surface water to be applied for irrigation. During the transition period, if the aquifer stock lies below/above its steady state level $\hat{G}$, the optimal policy is to prevent the use of ground/surface water altogether; as a result only surface/ground water is applied for irrigation and the aquifer stock increases/decreases until it reaches the steady level $\hat{G}$, at which point the quota on ground and surface water is changed so as to retain the steady state, as described in (iii). The problem with this policy is that it entails a discrete jump in water policy as the system moves from the transition period to the steady stage, a jump that may require a change in the agricultural structure (crop mix) of the region. Furthermore, the option of banning the use of a particular source of water may simply be legally impossible. Such a policy, however, should be fairly simple to administer and is ensured to achieve the desirable water allocation.

A combined tax and quota schedule. A third option to be considered by water policymakers is that of a combined quota-tax schedule. Such a policy consists of setting the prices of ground and surface water at their steady levels $z(\hat{G}) + V_G(\hat{G})[1-\delta]$ and $w - V_G(\hat{G})\delta$, respectively, and at the same time regulating the quantities of the more expensive water source in order to expedite the transition to the steady stage. The tax part of such a policy is used to smooth out the transition to the steady stage whereas the quantity regulation can be used to shorten the undesirably long transition period associated with the pure tax policy.

## Policy Implementation

The minimum information required to implement a tax policy is the steady state level of the aquifer $\hat{G}$ and the shadow price $V_G(\hat{G})$ at that level. This shadow price can be derived by solving Problem [2], along the line of [3], which requires knowledge of the water response function $F(x)$ and of the permeability parameter $\delta$. A solution of Problem [2] consists of the series $S^*_t$ and $g^*_t$ and the associated stock and shadow price processes $G_t$ and $V_G(G_t)$, $t \geq 0$, and is attainable in principle (perhaps only numerically). While this is fairly easy to achieve in the simple case represented by Problem [2], it may not be so easy in more complicated and realistic cases such as those described in the next section. For such cases there exist methods that provide approximates to the optimal management rules. One such a method, which approximates the steady state solution by solving a properly defined equivalent static problem, was proposed by Burt and Cummings (1977).

# SUMMARY

Because reality is more complicated than the simple situation considered above, numerous authors have extended and applied the conjunctive use framework to particular real world situations. Young and Bredehoeft (1972), for example, considered a situation in which the only source of surface water is streamflows emanating from aquifers. Cummings and Winkelman (1970), on the other hand, analyzed a system in which surface water is independent of ground-water sources.

Tsur (1990) introduced elements of uncertainty to surface water supplies and argued that ground water, in addition to its role of increasing the supply of irrigation water, serves also as a buffer that mitigates the undesirable fluctuations in the water supply. Tsur (1990) calculated the value associated with the buffer role (the buffer value) of ground water for wheat growers in the Negev region of Isreal and found it to exceed the value associated with the increase in the water supply (the latter is the benefit to be obtained from the ground water had surface water supplies been stable at the mean). Tsur and Graham-Tomasi (1990) extended this framework to account for ground-water scarcity.

In the next section the drainage problem is incorporated within the simple case, leaving out the consideration of the above-mentioned extensions.

## THE DRAINAGE PROBLEM INCORPORATED

The drainage problem arises when two distinct processes which affect agricultural yield occur as a result of the infiltration of irrigation water into shallow aquifers. The first is the rise in the ground-water table toward the root zone as the ground-water stock G increases. The second is the deterioration in the quality of the ground water as salts and other trace elements are washed into the aquifer. Incorporating the drainage problem requires allowing the water revenue function to depend also on the ground-water stock G, which represents the ground-water table, and on a quality index Q, representing the salinity level. Avoided, for the time being, are salinity effects via the ground water applied for irrigation (for more on salinity control in ground-water management problems see Cummings, 1971 and Cummings and McFarland, 1974). Figure 3 provides a schematic presentation of such a system.

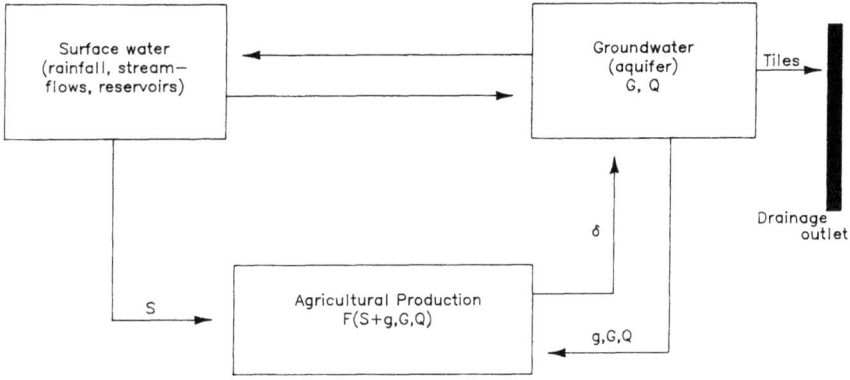

Figure 3. A conjunctive ground and surface water system with drainage.

The water response function F takes now the form

$$F(x_t, G_t, Q_t).$$

As above, F is assumed to increase at a diminishing rate with the quantity of irrigation water ($F_x > 0$ and $F_{xx} < 0$). Both G and Q, on their own, do not contribute to yield and may even cause harm ($F_G \leq 0$ and $F_Q \leq 0$). The negative effect of the one is enhanced by an increase in the quantity of the other, that is, their interaction is nonpositive ($F_{GQ} \leq 0$). Thus, as the ground-water quality deteriorates (Q increases) the negative effect of the ground water-logging is magnified ($F_G$ decreases); likewise, as the ground-water table rises (G increases) the negative effect of Q is exacerbated ($F_Q$ decreases).

Allowing for the application of drainage activities, which involves a subsurface drainage system to remove water to a drainage canal (see figure 3), the change in the aquifer stock is represented by

[4] $dG_t/dt \equiv \dot{G}_t = \delta S_t - (1-\delta)g_t - d_t$,

where $S_t$, $g_t$, and $\delta$ are as defined in the previous section and $d_t$ indicates the amount of drainage (m$^3$/ha).

The change in ground-water quality is an outcome of quite complicated hydrological processes, and may be represented implicitly as:

$$dQ_t/dt \equiv \dot{Q}_t = H(\delta x_t, G_t, Q_t).$$

The larger the amount of permeating water ($\delta x$), the greater the quantities of salts washed into the aquifer, so that H increases in $\delta x$. On the other hand, H is expected to decrease in $G_t$ (the same amount of salt changes the salinity level of a small bucket more than that of a large bucket). For the sake of concreteness, H is assumed to take the form

$$H(\delta x_t, G_t, Q_t) = q(G_t, Q_t)\delta x$$

where the nonnegative function $q(G,Q)$ translates quantities of permeating water (or of accumulated salts) into changes in the aquifer salinity level. The change in ground-water quality is thus given by

[5] $\dot{Q}_t = q(G_t, Q_t)\delta[S_t + g_t]$

A water management policy entails setting $S_t$, $g_t$, and $d_t$ for all time periods $t \geq 0$ and generates the payoff (the present value of the profit stream): ₄

$$\int_0^\infty [F(S_t + g_t, G_t, Q_t) - z(G_t)g_t - md_t - w(S_t - R)]e^{-rt}dt,$$

where $z(G_t)$, w and r are as defined in the first section and m is the unit cost of drainage activities (m is fixed and independent of the ground-water table). The policy that yields the highest payoff is sought.

## IRRIGATION AND DRAINAGE MANAGEMENT

Let V(G,Q) represent the maximum available payoff when the current stock and quality of ground water are G and Q, respectively. Formally

$$[6] \quad V(G,Q) = \text{MAX} \int_0^\infty [F(S_t+g_t,G_t,Q_t) - z(G_t)g_t - md_t - w(S_t-R)]e^{-rt}dt$$

subject to: Eqs. [4]-[5], $0 \le g_t \le C$, $R \le S_t \le B$, $0 \le d_t \le D$, $G_0 = G$ and $Q_0 = Q$,

where, as above, the parameters C and B represent respectively the capacity limits on ground and surface water supplies and D is a capacity limit on drainage activities.

The changes in V(G,Q) associated with a marginal (small) change in G or Q (the derivatives of V with respect to G or Q) are denoted by $V_G(G,Q)$ and $V_Q(G,Q)$, respectively. These quantities represent the unit value of G or Q and are thus referred to as the shadow prices of G or Q. $V_Q$ is expected to be negative (one would be willing to pay a positive amount to have Q reduced and the ground-water quality improved), while $V_G$ may be positive or negative. At low levels of G, where the ground-water table is well below the root zone, $V_G$ will be positive since the finite stock of the aquifer entails a positive royalty value (the forgone benefit of not being able to use in the future the unit of ground water pumped today). On the other hand, at high G levels where ground water has invaded the root zone, the damage to yield may outweigh the benefit of additional water, causing $V_G$ to become negative.

The Dynamic Programming equation of the present system is:

$$[7] \quad rV(G_t,Q_t) = \underset{S_t,g_t,d_t}{\text{MAX}} \{F(S_t+g_t,G_t,Q_t) - [z_t+V_{Gt} - \delta(V_{Gt}+V_{Qt}q_t)]g_t -$$

$$[w - \delta(V_{Gt}+V_{Qt}q_t)]S_t - (m+V_{Gt})d_t + wR\},$$

where $z_t = z(G_t)$, $V_{Gt} = V_G(G_t,Q_t)$, $V_{Qt} = V_Q(G_t,Q_t)$ and $q_t = q(G_t,Q_t)$. Analogous to the simpler case in the first section, the coefficients of $g_t$, $S_t$, and $d_t$ on the right-hand side of [6] represent the respective economic costs of these activities. These costs consist of the engineering costs plus terms containing the shadow prices $V_G$ and $V_Q$, which represent intertemporal effects. The economic costs of ground and surface water irrigation, compared to the simplistic ones in the first section, contain the term $-\delta V_{Qt}q_t$, which accounts for the salinity effect. Since $V_{Qt}$ is negative and $q_t$ is positive (see discussion above) this term is positive, implying that the salinization process of ground water increases the (economic) cost of irrigation.

The conjunctive ground and surface water management rules of the first section must be changed to incorporate effects of salinization of ground water and the drainage activities. In view of [7], and with no binding capacity limits on irrigation, it is straightforward to derive the following management rules:

(i') As long as the economic cost of ground-water irrigation exceeds that of surface water, $(z_t + V_{Gt} > w)$, irrigation water is derived only from surface sources at a quantity that equates the marginal productivity of water to the economic cost:

$$F_x(S^*_t, G_t, Q_t) = w - \delta(V_{Gt} + V_{Qt}q_t).$$

(ii') As long as the economic cost of surface water irrigation exceeds that of ground water, $(z_t + V_{Gt} < w)$, irrigation water is derived only from the aquifer at a quantity that equates the marginal productivity of water to its economic cost:

$$F_x(g^*_t, G_t, Q_t) = z_t + V_{Gt} - \delta(V_{Gt} + V_{Qt} q_t).$$

(iii') When the economic cost of surface water irrigation equals that of ground-water irrigation, $(z_t + V_{Gt} = w)$, irrigation water is derived from both sources at a quantity that equates the marginal water productivity to the economic cost:

$$F_x(S^*_t + g^*_t, G_t, Q_t) = z_t + V_{Gt} - \delta(V_{Gt} + V_{Qt}q_t)$$

$$= w - \delta(V_{Gt} + V_{Qt}q_t);$$

and the mix of ground and surface water is determined so as to preserve the condition $z_t + V_{Gt} = w$.[2]

(iv') Drainage activities are either applied to a full extent or not applied at all as $m + V_{Gt}$ is negative or positive, respectively:

$$d^*_t = \begin{cases} D & \text{if } V_{Gt} + m < 0 \\ 0 & \text{otherwise} \end{cases}$$

Rules (i'), (ii') and (iii') are similar in nature to their counterparts in the previous section. The main difference is in the levels of the irrigation activities, which in the present case are influenced also by the (shadow price of) salinity level of ground water. The fourth rule concerns the drainage policy. It states that drainage activities are applied only when $V_{Gt}$ falls below -m.

In view of (iii'), a steady state in this problem is characterized by the condition $z_t + V_{Gt} = w$, i.e., $z_t + V_{Gt}$ remains constant:

$$d[z(G_t) + V_G(G_t, Q_t)]/dt = z'(G_t)\dot{G}_t + V_{GG}\dot{G}_t + V_{GQ}\dot{Q}_t = 0$$

($z'(G) = dz(G)/dG$). As long as the salinity level Q affects $V_G$ (see discussion in the second section), G will not remain constant in the steady state. For suppose that the mix of ground and surface water irrigation is such that $G_t = 0$ [which can be achieved by the mix $g^*_t/S^*_t = \delta/(1-\delta)$]. Then, the irrigation water that leaches into the aquifer increases Q which, in turn, reduces $V_{Gt}$. $z(G_t)$ is unchanged (since $G_t$ is constant), thus $z_t + V_{Gt}$ falls below w. As a result, ground-water irrigation is substituted for surface water irrigation (cf. (ii')), which causes $G_t$ to fall. A similar argument can be use to rule out the possibility that $G_t$ increases. Thus, as long as $V_G(G,Q)$ decreases in Q, preserving the equality $z_t + V_{Gt} = w$ requires that the ground-water stock decreases at the appropriate rate so as to counterbalance the salinity effect on $V_{Gt}$. A constant stock level will prevail in a steady state only when the ground-water table lies well below the root zone so that changes in the salinity level cannot harm yield, that is when $V_G$ is independent of Q ($V_{GQ} = 0$).

Typically, $z(G) + V_G(G,Q)$ decreases in G. The situation $z(G) + V_G(G,Q) > w$ is therefore likely to occur at low G levels, where the ground-water table lies below the root zone. In such cases, the economic cost of ground water exceeds that of surface water and the drainage problem is not yet present; hence, it is plausible that irrigation utilizes only surface water sources (cf. (i')).

As water permeates into the aquifer, the ground-water table rises toward the root zone and its quality deteriorates. This causes both the extraction cost, $z(G)$, and the ground-water shadow price $V_G(G,Q)$ to fall. Eventually, the equality $z(G) + V_G(G,Q) = w$ holds, extraction begins and irrigation water is derived both from the aquifer and from surface sources at just the right mix so as to preserve the equality $z(G) + V_G(G,Q) = w$ (cf. (iii')).

What happens if surface water irrigation is implemented above its optimal level (say, because growers behave myopically)? Then the ground-water table and salinity continue to rise (as the stock increases and its quality deteriorates) and $V_{Gt}$ diminishes (both because ground water is less scarce and of lesser quality). As long as $z_t + V_{Gt} < w$ and $V_{Gt} > -m$, drainage activities are not required, but the situation is severe enough to warrant irrigation with ground water only and the ceasing of surface water irrigation. The situation becomes drastic when the ground-water stock achieves a level in which its shadow price, $V_{Gt}$, falls below -m; in such a case drainage activities are in order (cf. (iv)).

The dynamics of the system are characterized in figure 4. The level $\hat{G}$ is the maximum stock for which ground-water salinity does not affect the shadow price $V_G$ (at stock levels below $\hat{G}$, the ground-water table is below the root zone and its salinity cannot affect yield, i.e., $V_{GQ}(\hat{G},Q) = 0$ for all $G \leq \hat{G}$). The different curves represent the function $z(G) + V_G(G,Q)$ at different Q levels. They coincide over the interval $0 \leq G \leq \hat{G}$ (since Q is irrelevant in this interval), and for $G > \hat{G}$ they tilt clockwise as Q increases. The curves abc, abd, and abe correspond, respectively, to quality levels Q1, Q2, and Q3 with $Q1 < Q2 < Q3$. The curve ab$\hat{G}$ corresponds to the maximum possible level of ground-water salinity.

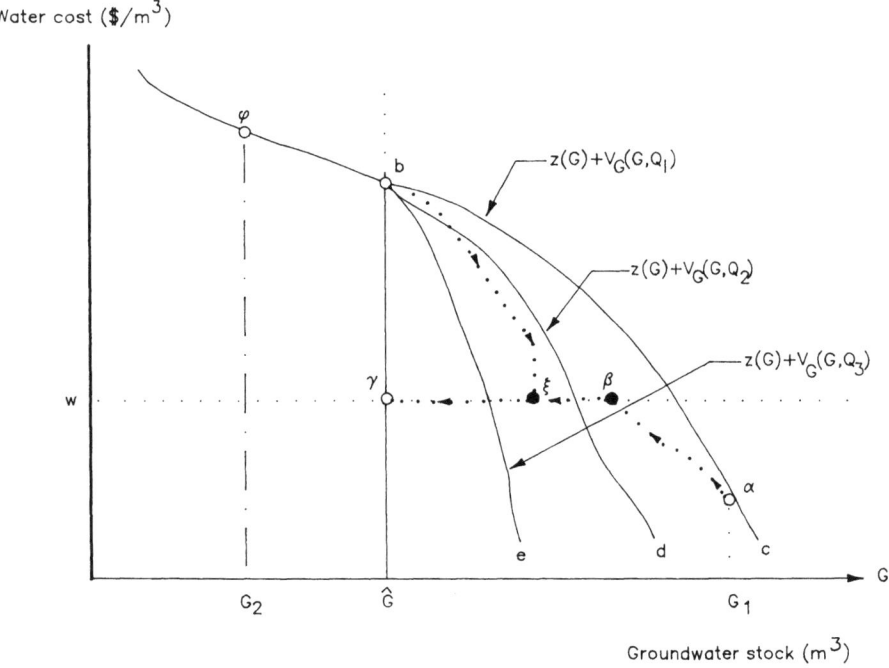

Figure 4. Dynamic behavior of the solution of section 4.

Suppose the initial stock and quality of ground water are G1 and Q1, respectively (point $\alpha$ of figure 4). Since $z(G1) + V_G(G1,Q1) < w$, irrigation water is derived solely from the aquifer. As a result G decreases, Q increases and the system moves along the line $\alpha\beta$ until it reaches the point $\beta$ where $z(G) + V_G(G,Q) = w$ holds. From there on the system progresses along the line

$\beta\gamma$ toward the point $\gamma$ (cf. (iii')) as Q increases and G diminishes at just the appropriate rate so as to preserve the equality $z(G)+V_G(G,Q) = w$. Eventually (perhaps after a very long time), the system comes to a rest at the point $\gamma$.

When the initial ground-water stock is smaller than $\hat{G}$, say at G2 (point $\varphi$ of figure 4), and $z(G2)+V_G(G2,Q) > w$, then it pays to irrigate only with surface water (cf. (i')). As a result, G increases until it reaches the level $\hat{G}$ (point b of figure 4). At this stage it is still profitable to use only surface water for irrigation, so that both G and Q increase. The system progresses along the line $b\xi$ until it reaches point $\xi$, at which stage $z(G)+V_G(G,Q) = w$ holds. From there on the system progresses along the line $\xi\gamma$ toward the point $\gamma$ as Q increases and G is reduced just at the appropriate rate to retain the condition $z(G)+V_G(G,Q) = w$.

The above management rules differ from the myopic rules under which instantaneous profit is maximized in each time period. The myopic rules are obtained by setting the shadow prices $V_{GT}$ and $V_{QT}$ equal to zero. It is clear from (iv) that, as long as drainage activities are costly (i.e., $m \geq 0$), no drainage activities are justified by the myopic rules. For reasons discussed in the first section, with no policy intervention, the individual growers are likely to behave myopically. The available policy tools include taxes and/or quotas on irrigation water as well as drainage activities. The tax and quota policies are similar in nature to those discussed in the first section; they will differ of course in the magnitudes of the taxes or quotas imposed (according to the difference between Rules (i)-(iii) and their primed counterparts). The drainage policy is unique to the present case; its implementation is characterized in (iv).

Implementing these policies requires knowledge of the shadow prices $V_G(G,Q)$ and $V_Q(G,Q)$, which can be obtained by solving problem [6], along the line of [7]. The task of solving this dynamic programming problem may turn out to be quite formidable; approximate solutions, such as the one proposed by Burt and Cummings (1977), should thus be considered.

## INVESTMENT POLICIES

It may be of interest to find out how the irrigation/drainage management rules and the associated benefit change as some of the system parameters, such as the capacity limits C, B, and D, or the water response function $F(\cdot)$ vary. A policy aimed at changing these parameters is regarded as an investment policy. A few such policies, considered to be of general interest, are discussed here.

## Extraction and Drainage Capacities

The capacity limits on ground-water extraction, C, and on drainage, D, are important components in the irrigation/drainage management rules. At the one extreme, no extraction or drainage facilities (wells, pumps, subsurface drains) are installed, (C = D = 0) so that only surface water irrigation can be applied and the region is doomed to reach a point where no agricultural production is feasible. At the other extreme, these capacities are unlimited and drainage activities can be carried out so as to instantly reduce the ground-water stock to any desirable level. Obviously, from the irrigation/drainage management point of view, unlimited capacity is preferred. However, extraction and drainage capacities entail investment costs and the benefits associated with unlimited capacities may not justify the investment.

To determine the optimal level of the extraction and drainage capacities, let V(G,Q;C,D) be the benefit of an irrigation/drainage policy when the levels of ground-water stock and salinity are G and Q, respectively, and given that extraction and drainage capacities are at the levels C and D, respectively. Let $E_c(C)$ and $E_d(D)$ be the investment costs required to achieve the capacities C and D, respectively (these technological relations depend, *inter alia*, on the hydrology, geology, and topography of the region). Then the desirable capacity levels are those that maximize $V(G,Q;C,D) - E_c(C) - E_d(D)$.

## Drainage Alternatives

It may be the case that more than one drainage alternative can be made available. Each drainage alternative entails operational costs (m in the notation of the second and third sections) and the investment cost of making it available. The latter contains direct investment costs (canals, subsurface drains, reservoirs) and possibly indirect environmental costs associated with its operation.

Suppose there are M drainage alternatives with the unit drainage cost $m_i$, i=1,2,...,M. Denote the investment and environmental costs of the i'th drainage alternative by $ID_i$, i=1,2,...,M. Let $V(G,Q;m_i)$, i=1,2,...,M, be the benefit of an irrigation/drainage policy when the unit cost of drainage is $m_i$. The desirable choice of drainage alternative is the one that generates the highest $V(G,Q;m_i)$ - $ID_i$. If a particular alternative generates prohibitive environmental effects, then the associated investment cost will be so high that it will not be selected.

## Variety or Crop Choice

Different crops, or different varieties of the same crop, respond differently to water salinity. Those which are more resistant will be affected to a lesser extent by the saline ground water. Changing the crop mix or the level of salt resistance of a particular crop entails changing the water response function $F(\cdot)$ and thereby the irrigation/drainage policy. In general, higher levels of salt resistance require less drainage activities and thus facilitate the management of the drainage problem.

## EXTENSIONS

Two extensions of the framework developed above are particularly relevant to the case under consideration. The first allows for the ground-water salinity to affect yield also through the irrigation water. The second considers the case in which a deep aquifer underlies the shallow one and water can move from the shallow to the deep aquifer. The modifications required by these two cases will be briefly outlined without going through the derivation of the management rules.

### Salinity Effects Via Irrigation

The effect of irrigation on yield depends both on the quantity and quality of the irrigation water. With quality variation, ground-water irrigation affects yield differently than surface water. The water response function takes the form

$$F(g_t, S_t, G_t, Q_t).$$

The marginal effect of irrigation depends on the source of the irrigation water and satisfies:

$$\partial F(\cdot)/\partial g \equiv F_g(\cdot) \leq F_s(\cdot) \equiv \partial F(\cdot)/\partial S.$$

Carrying out the maximization in [7] with the new $F(\cdot)$ provides a new set of management rules.

## Deep Aquifer Below the Shallow Aquifer

In some regions of the Valley, a deep aquifer underlies the shallow one. If the deep aquifer is of better quality than the shallow one, it may be worthwhile to use it for irrigation, even though pumping from the deep aquifer is more expensive. We thus have an additional activity, $n_t$--the level of irrigating with water from the deep aquifer.

The water response function takes the form

$$F(g_t, n_t, S_t, G_t, Q_t)$$

and is assumed to satisfy: $F_g(\cdot) \leq F_n(\cdot)$ and $F_g(\cdot) \leq F_s(\cdot)$. Letting $u(N_t)$ represent the unit cost of extracting from the deep aquifer at a stock level $N_t$, the benefit associated with a policy $\{S_t, g_t, n_t, d_t\}$, $t \geq 0$, is given by

$$\int_0^\infty [F(g_t, n_t, S_t, G_t, Q_t) - z(G_t)g_t - md_t - w(S_t - R) - u(N_t)n_t]e^{-rt}dt.$$

An additional equation is needed to specify the rate of change of the deep aquifer stock $N_t$ (if water can move from the shallow to the deep aquifer, equation [4] should be changed accordingly). The value function $V(G,Q,N)$ is defined as the maximum feasible benefit when the state variables are at the levels G, Q and N, allowing the Dynamic Programming equation of the new system to be derived along with the corresponding policy rules.

## NOTES

[1] $F(x)$ is derived in the following manner. Let $f(x,k)$ be an agricultural production function whose arguments are a water input, x, and a vector of other inputs, k. Given the prices of output, p, and of all inputs other than water, v, and given the level of water input, $k^*(x,p,v)$ represents the value of k that maximizes $pf(x,k) - vk$. The water response function is given by

$$F(x) = pf(x, k^*(x,p,r)) - r \cdot k^*(x,p,r).$$

where the fixed prices p and v are suppress from the notation.

[2] This mix rule is self-enforced. Suppose a nonoptimal mix is applied with too much surface water (though the quantity of irrigation water is chosen optimally). This would increase G above the level required to maintain $z_t + V_{Gt} = w$. As a result, $z_t + V_{Gt}$ falls below w so that water irrigation is derived only from the aquifer (Rule (ii')). As a result, G decreases and $z_t + V_{Gt}$ increases back toward w. Likewise, if the irrigation mix uses too much ground water, G reduces and $z_t + V_{Gt}$ rises above w, which, in turn, prompts

irrigation from surface water only (Rule (i')), causing G to increase and $z_t + V_{Gt}$ to diminish back toward w.

## ACKNOWLEDGEMENTS

The helpful comments of David Zilberman, Ted Graham-Tomasi, and William K. Easter are gratefully acknowledged.

## REFERENCES

Anderson, T. L.; Burt, O. R.; and Fractor, D. T., 1983. Privatizing Groundwater Basins: A Model and its Applications. In: Anderson, T. L. (Ed.), *Water Rights: Scarce Resource and the Environment*, Pacific Institute for Public Policy, pp. 223-248.

Bredehoeft, J. D. and Young, R. A., 1983. Conjunctive Use of Groundwater and Surface Water: Risk Aversion, *Water Resources Research*, 19, pp. 1111-1121.

Brown, G. and McGuire, C. B., 1967. A Socially Optimum Pricing for a Public Water Water Agency, *Water Resources Research*, 3, pp. 33-44.

Burt, O. R., 1964a. The Economics of Conjunctive Use of Ground and Surface Water, *Hilgardia*, 36, pp. 31-111.

Burt, O. R., 1964b. Optimal Resource use Over Time with an Application to Groundwater, *Management Science*, 11, pp. 80-93.

Burt, O. R. and Cummings, R. G., 1970. Production and Investment in Natural Resource Industries, *American Economic Review*, 60, pp. 576-590.

Burt, O. R. and Cummings, R. G., 1977. Natural Resources Management, the Steady State, and Approximate Optimal Decision Rules, *Land Economics*, 53, pp. 1-22.

Cummings, R. G., Optimum Exploitation of Groundwater Reserves with Saltwater Intrusion, *Water Resources Research*, 7, pp. 1415-1424.

Cummings, R.G. and Burt, O., 1969. The Economics of Production from Exhaustible Resources: Note, *American Economic Review*, 59, pp. 985-990.

Cummings, R. G. and Winkelman, D. L., 1970. Water Resource Management in Arid Environs, *Water Resources Research*, 6, pp. 1559-1560.

Cummings, R. G. and McFarland, J. W., 1974. Groundwater Management and Salinity Control, *Water Resources Research*, 10, pp. 909-915.

Dasgupta, P., 1982. *The Control of Resources*. Harvard University Press, Cambridge, MA.

Domenico, P. A.; Anderson, D. V.; and Case, C. M., 1970. Optimal Groundwater Mining, *Water Resources Research*, 4, pp. 247-255.

Gisser, M. and Sanchez, D. A., 1980. Competition Versus Optimal Control in Groundwater Pumping, *Water Resources Research*, 16, pp. 638-642.

Howitt, R. E.; Watson, W. D.; and Adams, R. M., 1980. A Reevaluation of Price Elasticities for Irrigation Water, *Water Resources Research*, 16, pp. 623-628.

Issar, A.; 1985. Fossil Water Under the Sinai-Negev Peninsula, *Scientific American*, pp. 104-111.

Margat, J. and Saad, K. F., 1984. Deep-Lying Aquifers: Water Mines Under the Desert? *Nature and Resources*, 20, pp. 7-13.

Negri, D. H., 1989. The Common Property Aquifer as a Differential Game, *Water Resources Research*, 25, pp. 9-15.

Paris, Q. and Knapp. K., 1989. Estimation of von Liebig Response Functions, *American Journal of Agricultural Economics*, 71, pp. 178-186.

San Joaquin Valley Drainage Program, 1989. *Preliminary Planning Alternatives for Solving Agricultural Drainage and Drainage Related Problems in the San Joaquin Valley.*

Tsur, Y., 1990. The Stabilization Role of Groundwater when Surface Water Supplies are Uncertain: The Implications for Groundwater Development, *Water Resources Research*, 26, pp. 811-818.

Tsur, Y. and Graham-Tomasi, T. *The Buffer Value of Groundwater: The Case of a Confined Aquifer.* Staff Paper, Department of Agricultural and Applied Economics, University of Minnesota, St. Paul.

Young, R.A. and Bredehoeft, J. D., 1972. Digital Computer Simulation for Solving Management Problems of Conjunctive Groundwater and Surface Water Systems, *Water Resources Research*, 8, pp. 533-556.

# 32 GOVERNMENT POLICIES TO IMPROVE INTERTEMPORAL ALLOCATION OF WATER IN REGIONS WITH DRAINAGE PROBLEMS

Farhed Shah and David Zilberman,
University of California, Berkeley

## ABSTRACT

Environmental degradation caused by current agricultural production can reduce future productivity. The generation and disposal of irrigation drainwater is an important example of this phenomenon. In this chapter, the authors model the environment's ability to absorb the polluting drainwater as exhaustible and argue that competitive farmers have every incentive to treat it like a common property resource. If such is the case, then competitive farmers use more irrigation water and generate more drainage than is socially optimal, causing production to end sooner than is optimal. A tax on drainage may be used to encourage adoption of efficient irrigation technologies, reduce drainage, and maintain a relatively high rate of production for a longer period. Whenever feasible, the development of costly drainage disposal schemes may also be used to prolong production; but this should not be expected to overcome the problem of exhaustibility or the need for policy intervention. Data from the San Joaquin Valley of California is used to show that the losses due to competitive inefficiency may be considerable, and a water-pricing strategy for correcting the problem is computed.

## INTRODUCTION

Agricultural production often generates byproducts, or has side effects, that can be a serious threat to agricultural productivity in the future. Loss of land fertility through soil erosion is one example of this phenomenon. Waterlogging of productive lands through underground accumulation of irrigation wastewater is another important example of the harm which can result over time from the cumulative effect of undesirable byproducts generated in the production process. In such cases, the basic problem is that the capacity of the productive environment to sustain damage or to absorb wastes is limited. Even when this capacity can be renewed or augmented, the associated cost may be high and substantial gains may be realized through adoption of agricultural practices which cause less damage to the environment. From a policy viewpoint, it would be highly desirable to determine the extent to which market forces may be expected to provide appropriate incentives for the adoption of conservationist agricultural practices. If the free market mechanism is likely to result in significantly inefficient outcomes, it may be worthwhile to devise interventionist measures to improve the intertemporal allocation of scarce environmental resources. This chapter presents a general framework of analysis to address the preceding issues in situations where production may cause environmental degradation. For the sake of concreteness, however, the analysis is conducted in terms of the specific problem of waterlogging, which is currently a threat to the survival of agriculture in many parts of the world.

Waterlogging occurs mainly when irrigation is practiced in regions with poorly drained soils and inadequate drainage facilities. In such circumstances, salt-laden drainwater tends to accumulate underground and has a debilitating effect on crop yields as the saline water level encroaches on the crop-root zone. In most of these situations, the development of drainage outlets is considered either too expensive or unacceptable for political or environmental reasons. This is the case, for example, in the San Joaquin Valley, where transporting drainwater to the Pacific Ocean is currently deemed infeasible and disposal in large-scale evaporation ponds has been restricted following the discovery of toxic selenium concentrations in evaporation ponds at Kesterson. Under circumstances of this type, a region's subsurface capacity for storage of drainwater may be viewed as an exhaustible resource. Some expert observers believe that if the current trends persist, more than a million acres in the Valley will become unproductive in the next century (see, for example, Frederick, 1982). Given that the storage capacity is used jointly by a large number of competitive farmers, there is every reason to suspect that the current rates of its depletion are excessive from a social point of view. In the second section of this chapter, this hypothesis is developed more formally. Regional subsurface capacity to store drainwater is modeled as a "common-property" exhaustible resource; socially optimal and competitive rates of depletion are compared; and some policy measures are suggested to correct the inefficiency due to the common pool problem.

The literature on the economics of exhaustible resources is quite extensive. The seminal paper in this area is by Hotelling (1931). The model in the next section basically extends the Hotelling model to incorporate the possibility of resource conserving technological change which may be triggered over time by increasing resource scarcity. Following earlier work by Caswell and Zilberman

(1986), technological change is modeled here as occurring in a discrete rather than continuous manner. For instance, in response to appropriate price incentives, a farmer in the model described here may decide to switch from traditional furrow irrigation to a modern drip irrigation system. In reality, such a switch may be expected to cause sizeable jumps in production as well as in resource use, and the model's results are consistent with this expectation. Aside from its obvious realism, the focus here on discrete technological choices leads to the formulation of policies for improving social welfare which are implementable. While in theory appropriate water taxes could be devised to correct for the common-property related inefficiency, in practice it is generally easier to observe and influence a farmer's choice of a discrete technology rather than the choice of a continuously variable input such as water; accordingly, the discussion focuses on the option of technology-based taxes and controls.

The third section of this chapter extends the model to include the possibility of costly drainage development and wastewater disposal. When this possibility is feasible, its consideration may be expected to prolong the productive life of an agricultural region threatened by waterlogging. However, as the Kesterson experience has shown, the basic problem of exhaustibility is difficult to overcome completely. The disposal of toxic drainwater is not only costly, but has the potential of generating new environmental problems which can also be viewed as exhaustible resource problems. Furthermore, the policymaker's role remains essentially the same as without the possibility of drainage since the common pool issue is still unresolved. The competitive rates of generation of drainage and patterns of technology use may be expected to be suboptimal from a social point of view even with the possibility of drainage. On the basis of these insights, the pure exhaustible resource model is argued to be a fair first approximation to reality. In the fourth section of this chapter, a numerical version of this model is presented using data representative of Valley cotton production. Results indicate that policy intervention could lead to substantial gains in welfare. The concluding section of the chapter presents a summary and some comments on directions for future research.

## SOCIALLY OPTIMAL AND COMPETITIVE PATTERNS OF PRODUCTION, TECHNOLOGY ADOPTION, WATER USE, AND WASTE GENERATION WITH NO DRAINAGE

### The Model

For simplification, a region is considered with a fixed amount of land of uniform quality, utilized completely in growing a single crop. Total output of the region at time t is denoted by $Y_t$ and output price at time t is denoted by $p_t$. This price is assumed to be determined exogenously. Water is assumed to be the only variable input used in production. The notation used for total water applied at time t is $A_t$ and the price of water at time t is denoted by $w_t$. This price is also assumed to be determined exogenously. Irrigation technologies (or practices) are indexed by i. It is assumed that "I" technologies are available for application of water; in other words, i takes values from 1, ..., I. Traditional furrow irrigation technology is indexed by i = 1, while more efficient technologies are indexed by higher values of i. Two coefficients are associated with each technology: $k_{it}$, irrigation efficiency coefficient (the fraction of applied water actually used by the crop at time t), and $b_{it}$, percolation coefficient (the fraction of applied water that percolates to the stock of ground water at time t). If one allows for evaporation and other losses (like surface runoff), then it is evident that $k_{it} + b_{it} < 1$. As a rule, technologies with higher coefficients of irrigation efficiency also tend to have lower percolation coefficients. For example, in certain areas of California, flood irrigation has an irrigation efficiency of 0.6 and a percolation coefficient of 0.3, while drip irrigation has an irrigation efficiency of 0.90 and a percolation coefficient of 0.05. The instantaneous fixed cost associated with the use of technology i in the hypothetical region at time t is denoted by $c_{it}$. Since the more efficient technologies are generally also more expensive, it is supposed that $c_{jt} > c_{it}$ for $j > i$.

The symbols $a_{it}$ and $y_{it}$ are employed to denote, respectively, applied water and output when irrigation technology i is used in the region at time t. Deep percolation and effective water application associated with the use of technology i in the region at time t are denoted respectively by $z_{it}$ and $e_{it}$ where $z_{it} = b_{it} \cdot a_{it}$ and $e_{it} = k_{it} \cdot a_{it}$. $S_t$ stands for the accumulated stock of saline water at time t. Salinity and effective irrigation are assumed to determine crop productivity through the production function $y = f(e, S)$, where y is output, e is effective irrigation, and S is the stock of saline ground water. It is reasonable to suppose that $f_e > 0$, $f_{ee} < 0$, and $f_s \leq 0$; in other words, marginal productivity of e is positive but diminishing, and a marginal increase in S cannot increase productivity. It is supposed, in addition, that y = 0 for S greater than $\bar{S}$, where $\bar{S}$ is the region's subsurface storage capacity for saline water.

Using the definitions given above, the production function with technology i can be written as $y_{it} = f(k_{it} \cdot a_{it}, S_t)$. Let $d_{it}$ be such that $d_{it} = 1$ if technology i is used at time t, and $d_{it} = 0$ otherwise. Then, total output in the region at time t is given by:

$$[1] \quad Y_t = \sum_{i=1}^{I} d_{it} \cdot y_{it} = \sum_{i=1}^{I} d_{it} \cdot f(k_{it} \cdot a_{it}, S_t)$$

Similarly, total applied water, total percolation, and total expenditures on technologies in the region at time t are given, respectively, by equations [2], [3], and [4]:

$$[2] \quad A_t = \sum_{i=1}^{I} d_{it} \cdot a_{it}$$

$$[3] \quad Z_t = \sum_{i=1}^{I} d_{it} \cdot z_{it}$$

$$[4] \quad C_t = \sum_{i=1}^{I} d_{it} \cdot c_{it}$$

This completes the description of the basic features of the model. Next, the discussion turns to the behavior of important variables in the model under alternative assumptions about institutional arrangements.

## The Socially Optimal Outcome

First, it is supposed that the region is managed by a social planner whose objective is to maximize the sum of discounted net benefits associated with crop production in the region over an endogenously determined and finite time horizon, T. Using the notation developed above, the planner's problem is to:

$$[5] \quad \text{Maximize} \int_0^T [p_t \cdot Y_t - w_t \cdot A_t - C_t] \cdot e^{-rt} \, dt$$

subject to [1], [2], [3], [4] and

$$[6] \quad \dot{S}_t = Z_t$$

[7] $S_t \leq \overline{S}$, $S_0$ is given

where r denotes the social rate of discount, and $\dot{S}_t \equiv dS/dt$. Equation [6] says that the increase in the saline water stock at time t equals the rate of deep percolation from irrigation at that time. The first part of [7] says that the size of water stock is bounded above by $\overline{S}$ - the region's finite underground capacity for storage of saline water; and the second part of [7] says that the initial size of the saline water stock is known to the planner.

The optimization problem in [1] - [7] is an optimal control problem involving discontinuities in the performance index and in the equation of motion (i.e., equation [6]). Attention will be restricted to an interior solution to the problem. A detailed discussion of this solution is provided in Shah, Zilberman, and Lichtenberg (1990). Only a brief description of the main results will be provided here.

The Hamiltonian function for the problem may be defined as follows:

[8] $H = [p_t \cdot Y_t - w_t \cdot A_t - C_t + q_{1t} \cdot Z_t] \cdot e^{-rt}$

where $q_{1t}$ is the shadow value of the saline water stock at time t. This shadow value measures the decline in social welfare from a marginal increase in the saline water stock. Since $q_{1t}$ takes nonpositive values, it is convenient to define $q_t = -q_{1t}$, and interpret $q_t$ as the (nonnegative) social cost of the saline water stock at time t.

Using equations [1] through [6] and the above definition of $q_t$, the Hamiltonian can be rewritten as:

[9] $H = \sum_{i=1}^{I} d_{it}[p_t \cdot f(k_{it} \cdot a_{it}, S_t) - w_t \cdot a_{it} - c_{it} - q_t \cdot b_{it} \cdot a_{it}] \cdot e^{-rt}$

$= \sum_{i=1}^{I} d_{it} \cdot B_{it} \cdot e^{-rt}$

where:

[10] $B_{it} \equiv p_t \cdot f(k_{it} \cdot a_{it}, S_t) - w_t \cdot a_{it} - c_{it} - q_t \cdot b_{it} \cdot a_{it}$ for $i = 0, ..., I$

The optimization procedure requires $B_{it}$ to be maximized by appropriate choice of $a_{it}$. The first order condition for an interior solution to this maximization problem gives:

(11) $P_t \cdot fe(k_{it} \cdot a_{it}, S_t) = (w_t + b_{it} \cdot q_t)/k_{it}$

In other words, the marginal product of effective water is equated to the social cost of effective water. Note that the social cost of applied water is the sum of its direct cost ($w_t$) and the cost it imposes on future production by increasing the saline water stock ($b_{it} \cdot q_t$). Since one unit of applied water results in $k_{it}$ units of effective water, the social cost of effective water is $(w_t + b_{it} \cdot q_t)/k_{it}$, which is greater than the social cost of applied water (as $k_{it} < 1$).

Equation [11] provides the optimal rate of applied water, $a_{it}{}^*$, as a function of the technological parameters, prices, $S_t$, and $q_t$. The maximized net benefit function for each technology is then given by:

$$[12] \quad B_{it}{}^* = pt \cdot f(k_{it} \cdot a_{it}{}^*, St) - (wt + qt \cdot b_{it})a_{it}{}^* - c_{it}$$

Equation [12] holds while technology i is being used. A necessary transversality condition requires $B_{it} = B_{jt}$ if technology switches from i to j at time t. Consequently, the maximized Hamiltonian function has to be continuous at instances when switches in technology take place, although it may well be nondifferentiable at these points in time. Since the planning horizon, T, is determined endogenously, the maximized Hamiltonian function must also go to 0 at time T.

In addition to the above conditions, the optimal control procedure requires $q_{1t}$, the shadow value of the saline water stock, to be continuous at times when technology switches and to satisfy the following differential equation at times when technology i is being used:

$$[13] \quad \dot{q}_{lt} = r \cdot q_{lt} - pt \cdot fs(e_{it}, S_t)$$

Using $q_t (\equiv -q_{lt})$ in equation [13] allows it to be rewritten as:

$$[14] \quad \dot{q}_t = r \cdot q_t + p_t \cdot f_s(e_{it}, S_t)$$

It should be noted that $r \cdot q_t \geq 0$, but $p_t \cdot fs(e_{it}, S_t) \leq 0$. Consequently, a typical social cost profile may be increasing for small values of $S_t$, but the second term is likely to dominate as time passes, causing $q_t$ to decline. A possible interpretation for this decline is that the marginal social cost of adding to the saline water stock begins to decrease after a certain amount of deterioration in soil productivity has taken place.[1] If soil productivity does not deteriorate very much until $S_t$ has reached a certain critical level and then deteriorates rapidly, one may approximate the situation by supposing that $f_s \equiv 0$ for $S < \bar{S}$ and $f(e, S) = 0$ for $S = \bar{S}$. In this important special case, $q_t$ rises at the rate r throughout the productive lifetime of the region.

Assuming that all prices and parameters remain constant (i.e., $p_t = p$, $w_t = w$, $b_{it} = b$, $k_{it} = k$), it is easy to see from equation [11] that, as long as $q_t$ is rising over time, the time paths of applied water, effective water, deep percolation, and output must decline over time while a particular technology is being used. One would certainly expect this decline in the time paths of the key variables since these variables depend critically on the social cost of using water, and this cost rises if $q_t$ rises.

Under reasonable assumptions placed on the production function, it is also possible to show that the rising social cost of applied water may trigger the adoption of successively more efficient technologies. This is another plausible result, since the more efficient technologies may be expected to reduce percolation to the saline water stock and also to increase the productivity of applied water. Indeed, the rate of deep percolation does jump down when adoption to a more efficient technology occurs. At this point in time, the rates of output and effective water show upward jumps, as may be expected when there is a discrete improvement in productive efficiency. Figure 1 illustrates a typical behavior pattern of optimal output, optimal effective water, and optimal deep percolation, under the assumption that $q_t > 0$.

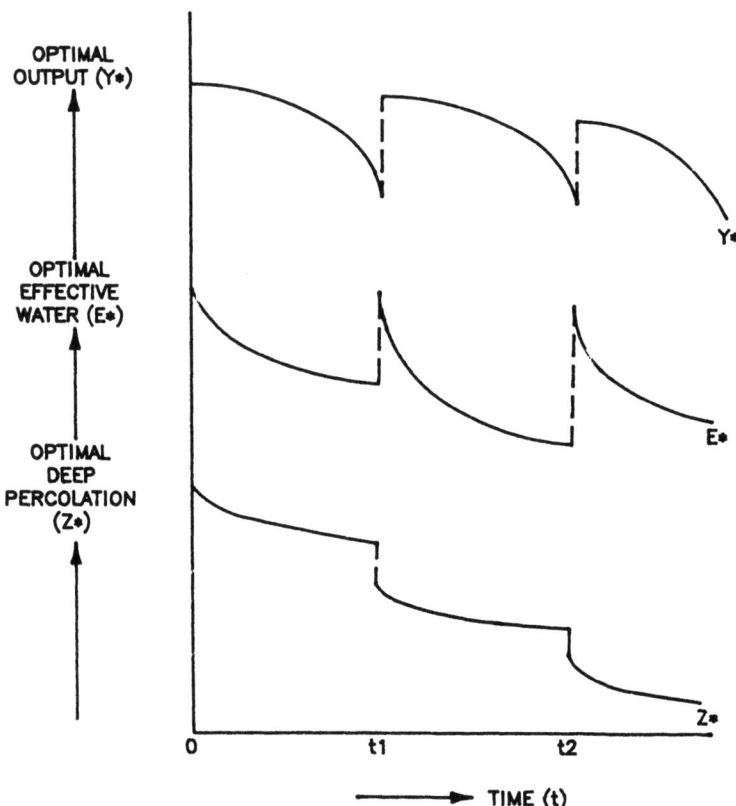

Figure 1. Time paths of optimal rates of output, effective water, and deep percolation (when q is rising).

The above discussion assumes prices and costs to be fixed. It is possible to speculate about the effect of exogenous trends in these parameters on the timing of adoption and on the manner of use of a given technology. For instance, if the price of water, $w_t$, increases over time, then the more efficient technologies would tend to get adopted earlier; however, water use with a given technology would also be lower, and this would reduce the rate of accumulation of saline water, which could have a slowing effect on adoption. Similarly, if the price of output, $p_t$, rises over time, this would encourage faster adoption of more efficient technologies since these technologies are yield increasing; but, in this case, more water would be used with a given technology, which could further accelerate adoption. Finally, if the difference between the fixed costs of a traditional and a modern technology decreases over time, then adoption of the more efficient technology would be encouraged.

## A Comparison of the Socially Optimal and the Competitive Outcomes

Instead of assuming that the region is managed by a social planner, let us now suppose that it is divided up between a large number of competitive farmers who use identical production technologies to grow the same crop. A farmer's production function is assumed to satisfy constant returns to scale with respect to land and the instantaneous fixed costs associated with the use of a particular irrigation technology are assumed to be in proportion to the area irrigated with this technology. The rest of the model is as before, and the symbols used earlier have the same meaning; however, a few additional terms need to be introduced. The proportion of the total land owned by farmer i is denoted by $h_i$; thus, if there are N farmers, then $\sum_{i=1}^{N} h_i = 1$. Farmer i's profit at time t, denoted by $\pi_{it}$, is given by

$$\pi_{it} = h_i \, (p_t \cdot Y_t - w_t \cdot A_t - C_t),$$

and his contribution to total ground-water percolation is given by $h_i \cdot Z_t$. Note that since an individual farmer cannot ensure a reduction in deep percolation through his actions alone, there is no incentive for him to adopt a technology which is more efficient than the one that maximizes his individual profits, and there is also no incentive for him to use any given technology more conservatively than is profitable from his individual point of view. In essence, therefore, this situation represents the classic common pool problem.

The total profits of the region at time t are denoted by $\pi_t$. It is obvious that:

$$\pi_t \equiv \sum_{i=1}^{N} \pi_{it} = p_t \cdot Y_t - w_t \cdot A_t - C_t$$

$\pi_t$ can be rewritten as:

$$\pi_t = \sum_{i=1}^{I} d_{it}[p_t \cdot f(k_{it} \cdot a_{it}, S_t) - w_t \cdot a_{it} - c_{it}]$$

where $d_{it} = 1$ if technology i is used, and $d_{it} = 0$ otherwise.

Since each farmer puts a shadow value of 0 on $S_t$, the farmers as a group do the same (even though this is not in their collective interest). In this situation, the farmers' profit maximization objective is achieved simply by maximizing $\pi_t$ for each t. This maximization can be done in two steps.

(a) Define the private net benefit function for technology i as:

$$G_{it} \equiv p_t \cdot f(k_{it} \cdot a_{it}, S_t) - w_t \cdot a_{it} - c_{it}$$

The first order condition for maximization of $G_i$ gives:

[15] $\quad p_t \cdot f_e(k_{it} \cdot a_{it}, S_t) = w_t/k_{it}$

Equation [15] says that the profit maximizing level of effective water is found by equating the marginal product of effective water to its private cost. Equation [15] can also be used to obtain the profit maximizing level of applied water, $\hat{a}_{it}$, which is then substituted back in $G_{it}$ to get the maximized net benefit function:

[16] $\quad \hat{G}_{it} = p_t \cdot f(k_{it} \cdot \hat{a}_{it}, S_t) - wt \cdot \hat{a}_{it} - c_{it}.$

(b) The next step is to choose $d_{it}$ to:

[17] $\quad$ maximize $\sum_{i=1}^{I} d_{it} \hat{G}_{it}$

where $d_{it} = 1$ if technology i is used, and $d_{it} = 0$ otherwise (i = 1, ..., I).

In order to make a consistent comparison with results stated in the second section, suppose that all prices and parameters are constant over time. The cases $f_s \equiv 0$ and $f_s < 0$ need to be examined separately since the time paths of the key variables in these two cases differ significantly.

### Case 1: $f_s \equiv 0$.

In this case, a farmer's production decisions are not affected by the level of underground saline water, $S_t$, until $S_t = \bar{S}$; production is no longer possible after this much saline water has accumulated. Since $f_s \equiv 0$, Equation [15] becomes:

$$[18] \quad p \cdot f_e(k_i \cdot \hat{a}_{it}) = w/k_i$$

It is apparent from equation [18] that the profit maximizing level of applied water for technology i is independent of time. Consequently, $\hat{G}_{it} = \hat{G}_i$ is constant and positive as long as $S_t < \bar{S}$ and $G_{it} = 0$ for all $t > T(i)$, where $T(i)$ is such that $S_{T(i)} = S$. The problem in [17]) is then solved simply by choosing the unique i which maximizes $G_i$ and continuing with this policy until time $T(i)$.

Now compare some aspects of the competitive (common property) and socially optimal programs. Suppose $S_0$ is the same in both cases. Let i be the technology chosen by competitive farmers and let j(t) be the socially optimal technology at time t. If ($\bar{S} - S_0$) is very large, then it is possible that the technology used in the two cases will be the same for some initial period. However, as ($\bar{S} - S_t$) declines over time, the social cost of effective water will tend to increase relative to the private cost of effective water, thereby giving greater impetus to the social planner to choose a technology which has a higher efficiency relative to the one chosen by competitive farmers. In other words, the technology choices in the two cases are likely to diverge increasingly as time passes.

Even if it happens that the socially optimal and competitive choices of technology coincide for some length of time, the socially optimal rates of output, applied water, effective water and deep percolation are likely to be lower than the corresponding competitive rates in the same period of time. This is essentially because a nonnegative wedge, $b_i \cdot q_t/k_i$, exists between the social and the private costs of effective water [compare equations [11] and [18]]. The divergence of the key variables in the two models increases as the size of this wedge increases over the time period in which the same technology is used in both models.

The above arguments imply that the competitive rate of deep percolation at time 0 cannot be less than the corresponding socially optimal rate. As stated earlier, the rate of deep percolation under the optimal program ($Z_t^*$) jumps down at every switch to a more efficient technology. In fact, as figure 1 shows, $Z_t^*$ declines monotonically over [0, T]. Consequently, the rate of deep percolation under competition is always greater than the rate of deep percolation under the socially optimal program in the time interval [0, T(i)], and the saline water stock attains its upper bound more quickly under competition than under the socially optimal program. It follows that $T(i) < T$; in other words, production under the competitive program must end earlier than under the socially optimal program.

## Case 2: $f_s < 0$.

In this case, productivity is affected adversely by every increase in the level of the underground stock of saline water. Differentiating both sides of equation [15] with respect to t and rearranging gives:

[19] $\partial \hat{a}_{it}/\partial t = -f_{es} \cdot \hat{a}_{it} \cdot b_i/(k_i \cdot f_{ee}) < 0$

If follows that, for a given i, the competitive rates of applied water, effective water, output, and deep percolation must decline continuously. Using arguments given in Shah, Zilberman, and Lichtenberg (1990), it is possible to show that, under reasonable assumptions, the competitive rates of effective water and output must jump up whenever an upward jump in technology takes place. Furthermore, the competitive rates of applied water and deep percolation must jump down at this time. It is also possible to show that, under reasonable assumptions, any technology switch under a competitive program must be from a less efficient to a more efficient technology.

This complete characterization of Case 2 can now be compared with the corresponding socially optimal version. Suppose $S_0$ is the same in both cases. The shadow cost of the saline water stock will be positive at time 0. This cost will be taken into account by the social planner, but not by the competitive producers. It follows that the social cost of effective water will exceed the private cost of effective water at time 0. Competitive behavior will, therefore, result in higher than optimal deep percolation in initial periods and use of possibly less efficient technology than is optimal in initial periods. Competitive output and profits in the initial periods may well be higher than their optimal counterparts, but soil productivity will be reduced much faster in the competitive case. Consequently, the situation with respect to output and profits is likely to be reversed in later time periods. Now, towards the end of the competitive program, there may well be a stage in which technology choice under the competitive program is more conservative than in the optimal case for the same timespan. However, this late stage adoption of efficient technologies is unlikely to prevent the "doomsday" from occurring earlier in the competitive case than in the optimal case.

A comparison of the private net benefit function, $G_{it}$, and the social net benefit function, $B_{it}$ (see second section), shows that the two functions differ only by the term $q_t \cdot b_{it} \cdot a_{it}$. This suggests that raising the price of water by $q_t \cdot b_{it}$ for users of technology i at time t would cause individual farmers to behave optimally from a social point of view. In practice, however, it may be difficult and costly to monitor water use by individual farmers, and these problems are likely to be compounded when heterogeneity among farmers is taken into consideration. Under such circumstances, a time-varying and technology based lump-sum charge could be used to promote the adoption of more efficient (but otherwise less profitable) irrigation technologies. Since irrigation technologies are discrete and easily observable, this lump-sum charge would be easier to levy than a per-unit tax on water. The magnitude of the technology charge could be based on the average amount of drainage generated by users of the particular technology. For instance, furrow irrigation generates more drainwater than drip irrigation; therefore, charging users of furrow irrigation a higher lump-sum tax than users of drip irrigation could be used to promote adoption of drip irrigation, if this were considered desirable. A social planner's problem may be set up to compute the optimal amount of the technology based lump-sum drainage charge. Shah, Zilberman, and Chakravorty (1989) have discussed in detail the solution to such a problem in a related context. It should be noted that a lump-sum drainage charge would affect only the choice of technology and would not influence the rate of water applied with a given technology. The optimal lump-sum drainage charge is, therefore, likely to be only a second-best instrument in most cases. However, empirical evidence suggests that deep percolation goes down much more drastically when technology switches than it does when a particular technology is being used and the cost of applying water is rising (see Abbot, 1984). Therefore, the optimal lump-sum charge may well achieve results which are close to the first best.

## MODELING COSTLY DRAINAGE DEVELOPMENT AND DISPOSAL

Thus far, the regional subsurface capacity to store drainwater has been considered as a nonrenewable resource. Quite often, it may be possible to augment this resource by installing tile drains at the farm level and transporting drainwater to some distant treatment or disposal facility, such as a desalination plant, an evaporation pond, wetlands, or a large body of water, such as an ocean. The last option is not always feasible, and therefore the focus here will be on the others.

Drainage has two kinds of costs associated with it: one is the cost of building the drains, and the other is the cost of transporting and disposing the drainwater. Disposal of drainwater in wetlands and evaporation ponds is often considered the least expensive alternative in the second category of costs. However, as the Kesterson experience has shown, the continued exercise of this option may cause the level of toxins in the environment to increase over the years, thereby creating a social need to resort to more expensive disposal alternatives involving treatment of drainwater and to irrigation practices which would lead to reduced generation of drainage. In such cases, one may view the basic exhaustible resource problem discussed in the second section as being "nested" within another exhaustible resource problem created by the limited capacity of the environment to safely absorb agricultural wastes. Under circumstances like this, the conservationist public policy measures discussed in the preceding section could clearly play a role in increasing social welfare and the productive lifetime of the agricultural region in question.

Suppose that the cost of increasing the region's drainage capacity by $m_t$ units of water is $u(m_t)$, where $u' > 0$ and $u'' > 0$. The cost of treating and transporting a unit of drainwater is denoted by $v_t$. It is assumed that disposal of the drainwater takes place in wetlands and that, over time, this practice causes the level of toxins in the environment to increase. The growing social awareness of this phenomenon may be expected to generate the need to carry out increasingly intensive treatments of drainage water and thereby cause $v_t$ to rise over time. Accordingly, suppose that $v_t = n(R_t)$, where $R_t$ is the stock of toxins in the environment at time t, and that $n'(R_t) > 0$.

Let $M_t$ denote the regional capacity to emit drainage at time t and let $Q_t$ denote the volume of drainage emitted at time t. Quite obviously, $Q_t$ will be less than or equal to $M_t$. $Q_t$ and $v_t$ may be expected to determine the rate at which $R_t$ increases. Since $v_t$ is itself a function of $R_t$, it is reasonable to suppose that $R_t$ increases at the rate given by the function $g(R_t, Q_t)$ where $g_Q > 0$ and $g_R < 0$. Using the other notation developed in the second section, the social optimization problem for this case may be stated as:

$$[20] \quad \text{maximize} \int_0^T [p_t \cdot Y_t - w_t \cdot A_t - C_t - n(R_t) \cdot Q_t - u(m_t)] \cdot e^{-rt} d_t$$

subject to [1], [2], [3], [4], and

$$[21] \quad \dot{S}_t = (Z_t - Q_t)$$

$$[22] \quad \dot{M}_t = m_t$$

$$[23] \quad \dot{R}_t = g(R_t, Q_t)$$

$$[24] \quad Q_t \leq M_t$$

[25] $0 \leq S_t \leq \bar{S}$, $S_0$ given

The statement of this problem differs from the social planner's problem stated in the second section in that (1) The objective functional includes two additional terms to account for the costs associated with drainage, (2) the equation of motion for saline ground-water accumulation has been modified to permit decreases due to drainage, (3) equations have been added to reflect the dynamics of drainage installation and the associated constraint of drainage, and (4) an equation of motion specifying the dynamics of toxins in the environment has been added.

Since it is relatively expensive to build drains and dispose of the drainwater, investment in these operations may be expected to be delayed until the saline water stock has risen to a certain critical level. The determination of this critical level is a matter of interest, but for the purposes of this analysis, it is supposed that the critical level has already been attained at time 0.

It should be noted that if $v_t$ is relatively low and does not rise over time, then it may become feasible at some stage to dispose of a volume of drainwater which is sufficiently large to hold $S_t$ at some desired level for an indefinite time. This "steady-state" solution would be likely, for instance, if ocean disposal was a feasible option or if very low cost desalination techniques were available. Since such options are currently not viable in most cases, it is supposed that $v_t$ has a strong positive dependence on $R_t$ and may rise high enough to cause $Q_t < M_t$ (or even $Q_t = 0$) to be optimal after some point in time. In other words, high and rising disposal costs of drainwater may force an underutilization of available drainage capacity, causing the saline water table to rise even if it is physically possible to keep it from doing so. Thus, under these assumptions, the drainage option may be used to extend the productive life of the region, but its use does not overcome the exhaustibility problem on a permanent basis.

Let $q_{2t}$, $q_{3t}$, and $q_{4t}$ be the shadow price variables associated with equations [22], [23], and [24], respectively. Then, the Hamiltonian for the problem in [20] - [25] may be written as:

[26] $H = [p_t \cdot Y_t - w_t \cdot A_t - C_t - n(R_t) \cdot Q_t - u(m_t) - q_t(Z_t - Q_t) + q_{2t} \cdot m_t + q_{3t} \cdot g(R_t, Q_t) + q_{4t} \cdot (M_t - Q_t)] \cdot e^{-rt}$

The procedures for determining optimal irrigation technology and applied water are similar to those discussed in the second section. The dynamics of $q_t$ are also given by the costate equation stated in the second section [i.e., equation [14]]. However, there are some additional necessary conditions for this problem. As long as drainage capacity is being added (i.e., $m_t > 0$), the following must hold:

[27] $u'(m_t) - q_{2t} = 0$

[28] $\dot{q}_{2t} = r \cdot q_{2t} - q_{4t}$

[29] $M_t = Q_t$

[30] $q_{4t} > 0$

[31] $q_t - n(R_t) + q_{3t} \cdot g_Q(Q, R) - q_{4t} = 0$

[32] $\dot{q}_{3t} = r \cdot q_{3t} + n'(R_t) \cdot Qt - q_{3t} \cdot g_R(Q, R)$

Equation [27] says that, at each instant, the marginal cost of adding new drainage capacity [i.e., $u'(m_t)$] must equal the marginal benefit of doing so (i.e., $q_{2t}$). Quite naturally, as long as new capacity is being added, it must be optimal to completely utilize existing capacity (i.e., $M_t$). This is the message of equation [29]. Equation [30] supports this message by saying that the existing capacity has positive social value.

Equation [31] says that the marginal social cost of adding to the saline water stock (i.e., $q_t$) exceeds the sum of the marginal current cost of cleanup [i.e., $n(R_t)$] and the marginal cost of adding to future cleanup problems [i.e., - $q_{3t} \cdot g_Q(Q, R)$] by the amount $q_{4t}$. Naturally, it pays to invest in additional drainage installation only when $q_{4t} > 0$. Now, the equations governing the dynamics of the costate variables $q_t$ and $q_{3t}$ (which is a negative number) -- that is to say, equations [14] and [32], respectively -- suggest that $q_{4t}$ must ultimately decline over time and reach zero, at which point it will no longer be desirable to utilize all of the existing drainage capacity. Obviously, additions to drainage capacity will also decline to zero no later than this point in time. In other words, $m_t$, $q_{2t}$, and $q_{4t}$, will all be equal to zero by some time $T_1$. Of course, production and drainage disposal will most probably continue beyond $T_1$, but $Q_t$ will be less than $M_t$, and its level will be determined by equation [31] with $q_{4t} = 0$. At this stage, the problem becomes very similar to a pure exhaustible resource problem: The buildup of toxins in the environment will make drainage increasingly expensive and the prospect of rising saline-water tables will make it optimal to adopt more efficient irrigation practices. The adoption of these practices may hold up productivity for a while, but ultimately, the twin menace of costly drainage disposal and rising water tables will cause production to decline and go to zero by time T.

It should be noted that, from a policy point of view, the essence of the whole problem is unchanged from the one in the second section. Given the common property nature of the subsurface aquifer, as well as of the drainage disposal facilities, it is evident that the competitive outcome will be inefficient: drainage installation under competition is likely to occur faster, and its final capacity is likely to be greater, than in the optimal situation; also, adoption of efficient technologies under competition is likely to occur in a suboptimal manner and production is likely to terminate sooner. Consequently, having the drainage option does not, by any means, eliminate the need to encourage conservationist practices through appropriate tax/subsidy and control mechanisms.

## A NUMERICAL EXAMPLE

The theoretical results of the preceding sections will now be illustrated with a specific example which is based on data from the Valley (see Hanemann et al., 1987). The analysis in Caswell and Zilberman (1986) as well as empirical work by Hexem and Heady (1978) suggests that it may be reasonable to approximate yield as a quadratic function of effective water. The data set at our disposal can be used to arrive at the following specific parameters for a quadratic form of the production function:

$$y = -1589 + 2311 \cdot e - 462 \cdot e^2$$

where y and e are yield/acre and effective water/acre, respectively.

The above relationship is representative of Valley cotton production conditions. For the sake of simplification, suppose that production is unaffected by underground saline water accumulation until the level of the stock of this water has reached a certain critical level, $\bar{S}$, at which point production must stop. Also suppose that drainage is infeasible, so that the subsurface capacity to store saline water is a nonrenewable resource. The discussion of drainage development in the preceding section makes it clear that this assumption may be considered a reasonable first approximation since the availability of the drainage option is unlikely to alter the basic nature of the problem: Sooner or later, one is bound to run into some kind of capacity constraint.

Use of up to four different irrigation technologies is considered in determining the socially optimal program. These technologies and the parameters associated with them are listed in table 1. For the purposes of this example, output price is taken as $0.75/lb, water price as $25/AF, and the social rate of discount as 5 percent per annum. The starting level of underground saline water ($S_0$) is assumed to be 1 foot. Different values of the upper limit on subsurface storage room ($\bar{S}$) are then used to compute the optimal time paths of the key variables. The values of ($\bar{S}$) are selected from the range of 6 feet to 35 feet.

The optimal initial shadow cost of the saline water stock, $q_0$, varies inversely with $\bar{S}$. Figure 2 shows this relationship for our specific example. A low value of $\bar{S}$, such as 11, results in the high value of 139.8 for $q_0$, which causes immediate adoption of technology 3 (sprinklers). The optimal $q_0$ falls exponentially as $\bar{S}$ increases. Adoption of water conserving/yield increasing technologies is delayed with higher values of S.

Table 1. Irrigation technology parameters.

| Technology | Irrigation effectiveness | Percolation coefficient | Fixed cost per acre |
|---|---|---|---|
| Furrow | 0.60 | 0.1750 | 500.00 |
| Shortened runs | 0.70 | 0.1330 | 517.00 |
| Sprinkler | 0.80 | 0.0875 | 548.00 |
| Drip | 0.95 | 0.0400 | 633.00 |

*Source:* Michael Hanemann, Erik Lichtenberg, David Zilberman, David Chapman, Lloyd Dixon, Gregg Ellis, and Janne Hukkinen. *Economic Implications of Regulating Agricultural Drainage to the San Joaquin River.* Technical Committee Report to the State Water Resources Control Board, Sacramento, California, 1987.

Figure 3 shows the optimal technology adoption pattern for $\bar{S} = 16$. In this case, $q_0 = 74.14$ and $q_t$ increases thereafter at the rate of 5 percent per year. As $q_t$ increases over time, a periodic adoption of increasingly efficient technologies takes place: Technology 2 is used up to year 7, and then technology 3 adopted; technology 3 is replaced with technology 4 in year 38, and productions ends in year 75 after the saline water stock has almost attained its maximum permissible value.

Figure 3 also shows the behavior of optimal time profiles of output ($Y_t^*$), applied water ($A_t^*$), and deep percolation ($Z_t^*$). A comparison of figures 1 and 3 indicates clearly that the empirically generated time paths of $Y_t^*$ and $Z_t^*$ correspond rather well to their theoretically predicted counterparts. For instance, note that, in both diagrams, $Y_t^*$ jumps up at the switch points, and declines smoothly between any two switch points. The behavior of $Z_t^*$ in figure 3 is also similar to its behavior in figure 1: $Z_t^*$ jumps down at the switch points and declines smoothly between any two switch points. This smooth decline is very gentle, however, and is illustrated better in table 2, which presents the optimal values of the key variables at 5-year intervals.

Figure 3 also indicates that the optimal rate of applied water ($A_t^*$) jumps down at the time of adoption of a new technology and declines continuously-while a particular technology is being used. It should be noted that the downward jump in $A_t^*$ at the two switch times could not have been predicted without a knowledge of the specific parameters of the problem.

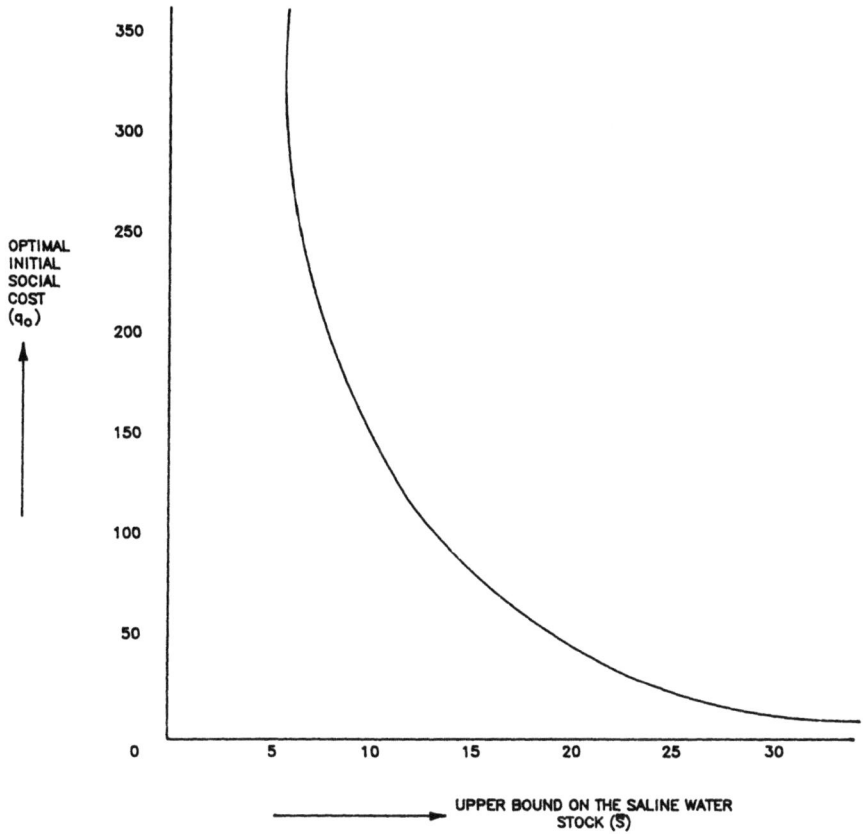

Figure 2. Behavior of the optimal initial social cost of the saline water stock as the upper bound on this stock is varied between 6 feet and 35 feet.

The preceding description of the key variables of the problem is predicated on the assumption that the subsurface drainage reservoir is managed in a socially optimal manner. If the region is owned commonly by a large number of competitive farmers, then this is unlikely to be the case. Assuming that these farmers put no value whatsoever on the depletion of the subsurface reservoir, it turns out that technology 1 is used throughout the productive life of the region. The competitive rates of per acre output, applied water, and deep percolation are all constant over time and equal to 1299.33, 4.07, and 0.71, respectively. The upper limit on drainage capacity (i.e., 16 feet) is reached in 21 years, at which point production must stop. Thus, competitive behavior with respect to the commonly owned resource reduces the productive lifetime of the region by almost 54 years. The discounted sum of profits from production in the competitive case is $5,017.41, whereas its value in the optimal case is $6,729.71. Thus, socially optimal behavior increases social welfare by approximately 34 percent. This percentage gain in social welfare increases as the upper bound on drainage capacity decreases. For instance, when $\bar{S} = 6$ feet, then social welfare goes from $3,875.07 in the competitive case to $6,242.51 in the optimal case, which is an increase of about 61 percent. It should be noted that, although the competitive rate of output in the case under consideration, as long as it is positive, is higher than the corresponding optimal rate, the difference between the two is very small (less than 0.1 percent). The optimal rates of applied water and deep percolation are, of course, always much lower than the corresponding competitive rates for as long as production is positive in the competitive situation. These observations are quite consistent with the theoretical discussion in the second section.

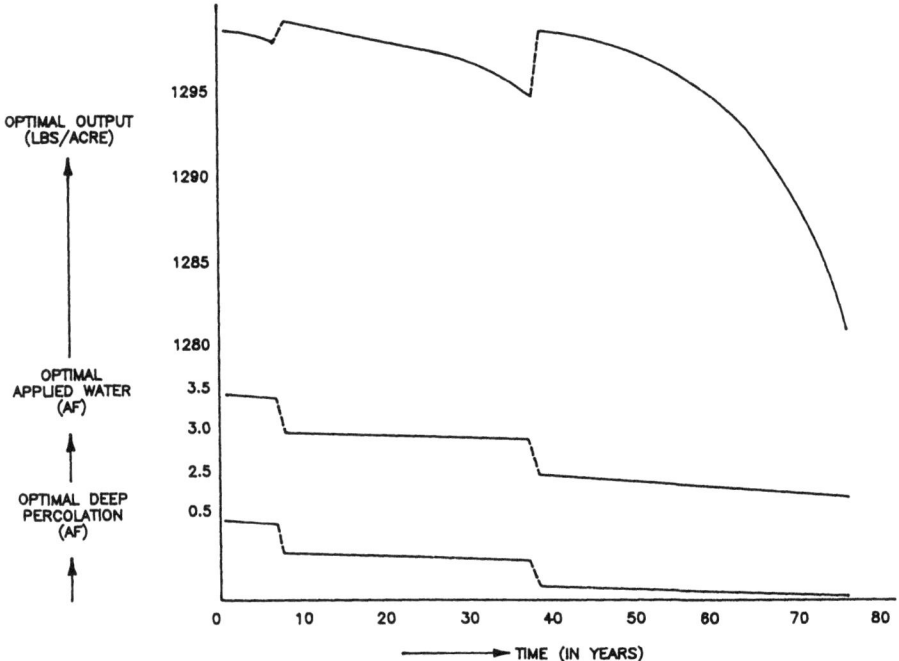

Figure 3. Time paths of optimal rates of output, applied water, and deep percolation (when $\bar{S} = 16$ feet).

Table 2. Optimal values of key variables at 5-year intervals.

| Year | Tech | Shadow cost | Output | Applied water | Percolation | Saline water level |
|---|---|---|---|---|---|---|
| 1  | 2 | 74.1432 | 1298.61 | 3.47031 | 0.461552 | 1.46155 |
| 5  | 2 | 90.1215 | 1298.31 | 3.46405 | 0.460719 | 3.30573 |
| 10 | 3 | 115.020 | 1299.15 | 3.04729 | 0.266638 | 5.02664 |
| 15 | 3 | 146.798 | 1298.85 | 3.04102 | 0.266090 | 6.35824 |
| 20 | 3 | 187.356 | 1298.43 | 3.03302 | 0.265389 | 7.68666 |
| 25 | 3 | 239.119 | 1297.83 | 3.02281 | 0.264496 | 9.01101 |
| 30 | 3 | 305.183 | 1296.98 | 3.00978 | 0.263356 | 10.3302 |
| 35 | 3 | 389.499 | 1295.75 | 2.99314 | 0.261900 | 11.6427 |
| 40 | 4 | 497.110 | 1298.85 | 2.56095 | 0.102438 | 12.4730 |
| 45 | 4 | 634.453 | 1298.30 | 2.55217 | 0.102087 | 12.9842 |
| 50 | 4 | 809.740 | 1297.49 | 2.54096 | 0.101638 | 13.4933 |
| 55 | 4 | 1033.46 | 1296.31 | 2.52665 | 0.101066 | 13.9998 |
| 60 | 4 | 1318.98 | 1294.56 | 2.50839 | 0.100336 | 14.5030 |
| 65 | 4 | 1683.39 | 1291.91 | 2.48508 | 0.099403 | 15.0020 |
| 70 | 4 | 2148.48 | 1287.88 | 2.45534 | 0.098213 | 15.4956 |
| 75 | 4 | 2742.06 | 1281.67 | 2.41738 | 0.096695 | 15.9822 |

The theoretical discussion in the second section also indicated that it should be possible to devise a water pricing strategy to induce competitive farmers to behave in a socially optimal manner. Assuming that $\bar{S} = 16$ and that the base price of water charged to farmers is \$25/AF, we can use the numbers in tables 1 and 2 to come up with the optimal pricing strategy. In year 1, for instance, charging users of technology 1 a water price of $(25 + 74.14 \times 0.175) = \$37.975$/AF, and users of technology 2 a water price of $(25 + 74.14 \times 0.133) = \$34.86$/AF, will cause users of technology 1 to abandon it in favor of technology 2, and will also cause them to use technology 2 in a socially optimal manner. Similar calculations could be performed for later years to construct a detailed water-price schedule for the full 75-year period.

## CONCLUSIONS

This chapter has presented a methodological framework for analyzing policy issues in situations where productive activity may be expected to lead to environmental degradation that may, in turn, be expected to have a negative effect on future production. Such problems may be viewed as belonging to the generic class of exhaustible resource problems. Common ownership of exhaustible resources often leads to market failures and justifies policy intervention. It was argued that the observability of discrete technological choices can play an important role in the designing of implementable corrective policies.

This analysis was conducted in terms of the important problem of waterlogging which threatens the survival of irrigated agriculture in many parts of the world. Waterlogging turns out to be a pure exhaustible resource problem if there are no outlets for drainage. However, even when drainage is possible, the safe disposal of saline drainwater may become very expensive over time, thereby reducing the situation to one of resource exhaustibility. The availability of conservationist irrigation technologies and the possibility of drainage can, of course, play an important role in prolonging a region's productive lifetime. We compared the socially optimal and the common-property (competitive) patterns of water use, production, technology adoption, drainwater generation, and drainage development. It was shown that, in the socially optimal case, the depletion of the underground capacity to store drainwater triggers the adoption of successively more efficient irrigation technologies. Common property outcomes tend to be less conservative, at least in earlier time periods, resulting in welfare losses. Efficiency can be improved in these cases through the use of appropriate public policy instruments. A "first-best" scheme of water pricing for achieving this objective was suggested, as well as a second-best scheme for situations in which water use is difficult to monitor. The observability of discrete choices of irrigation technologies plays a key role in the designing of the second-best scheme.

The "first-best" scheme relies on correct computation of the shadow cost of the underground saline water stock. Using a numerical example based on data from the Valley, it was shown that this computation can be performed and that, at least in principle, it is possible to devise a water-pricing strategy to induce individual producers to behave optimally from a social point of view.

Many simplifying assumptions were made in the theoretical analysis as well as in the construction of the numerical example. The numerical example, in particular, is purely for illustrative purposes. Several of the assumptions need to be relaxed and the model brought closer to reality before it can be put to practical use. For instance, farmers in the hypothetical region were assumed to have identical productive abilities. In reality, there is likely to be considerable variation in farmer endowments (such as land quality, general ability, etc.). Shah, Zilberman, and Chakravorty (1989) incorporate heterogeneity of this type in a model of technology adoption which treats ground water as an exhaustible resource. A similar approach could be used to model heterogeneity in the present case. The model presented here supposes that all parameters are known with certainty. In practice, there may be considerable uncertainty about prices, costs, technology coefficients, and hydrology of the saline water aquifer. The literature on the depletion of exhaustible resources under uncertainty is now fairly well developed (see Mangel, 1985, for a unified exposition of this literature), and it could be a fruitful source of ideas for appropriate extensions of this model. The assumption of a small pricetaking region could also be relaxed to incorporate a downward sloping demand curve. The number of technologies in the present model is fixed. The possible arrival of new technologies, especially back-stop, could be modeled following the approach developed by Dasgupta and Heal (1974). Undoubtedly, there are also other ways of extending the model. The authors believe that the basic problem addressed in this chapter is important and that the modelling approach presented can lead to many exciting research avenues in the future.

## NOTES

[1] In the context of mineral extraction, Solow and Wan (p. 365) interpret qt as a "degradation cost" which must start to decline at some point, and ultimately reach zero if the resource is degraded to the extent that its economic value becomes insignificant and further extraction cannot therefore cause any more degradation.

## ACKNOWLEDGEMENTS

The authors are grateful to Erik Lichtenberg for his valuable contribution to the development of the material presented in this paper.

## REFERENCES

Bryson, A. and Ho, Y., 1975. *Applied Optimal Control.* Halsted Press, New York, NY.

Carruthers, I. and Clark, C., 1981. *The Economics of Irrigation.* Liverpool University Press.

Caswell, M. and Zilberman, D., 1986. The Effects of Well Depth Land Quality on the Choice of Irrigation Technology, *American Journal of Agricultural Economics,* 68, pp. 798-811.

Dasgupta, P. and Heal, G., 1974. The Optimal Depletion of Exhaustible Resources, *Review of Economic Studies,* 41, pp. 3-28.

Frederick, K., 1982. Water Supplies. In: Portney, P. (Ed.), *Current Issues in U.S. Natural Resource Policy,* Resources for the Future, Washington, DC.

Hanemann, M.; Lichtenberg, E.; Zilberman, D.; Chapman, D.; Dixon, L.; Ellis, G.; and Hukkinen, J., 1987. *Economic Implications of Regulating Agricul-*

*tural Drainage to the San Joaquin River.* Technical Committee Report to the California State Water Resources Control Board.

Hexem, R. and Heady, E., 1978. *Water Production Functions and Irrigated Agriculture.* Iowa State University Press, Ames.

Hillel, D., 1987. *The Efficient Use of Water in Irrigation.* World Bank Technical Paper No. 64, World Bank, Washington, DC.

Hotelling, H. 1931. The Economics of Exhaustible Resources, *Journal of Political Economy*, 39, pp. 137-175.

Levhari, D. and Liviatan, N., 1977. Notes on Hotelling's Economics of Exhaustible Resources, *Canadian Journal of Economics*, 10, pp. 177-192.

Mangel, M., 1985. *Decision and Control in Uncertain Resource Systems.* Academic Press, Orlando, FL.

Shah, F.; Zilberman, D.; and Chakravorty, U., 1989. *Technological Adoption and Exhaustible Resources in Agriculture.* Unpublished Manuscript.

Shah, F.; Zilberman, D.; and Lichtenberg, E., 1990. *Choice of Techniques in the Optimal Management of Agricultural Drainage.* Unpublished Manuscript.

Seierstad, A. and Sydsaeter, K., 1987. *Optimal Control Theory with Economic Applications.* North-Holland, Amsterdam.

Solow, R. and Wan, H., 1976. Extraction Costs in the Theory of Exhaustible Resources, *Bell Journal of Economics*, 7, pp. 359-370.

Zilberman, D., 1985. Technological Change, Government Policies, and Exhaustible Resources in Agriculture, *American Journal of Agricultural Economics*, 66, pp. 634-640.

# 33 DYNAMIC CONSIDERATIONS IN THE DESIGN OF DRAINAGE CANALS

Ujjayant Chakravorty, University of Hawaii;
Eithan Hochman, Ben Gurion University, Israel; and
David Zilberman, University of California, Berkeley

### ABSTRACT

This chapter develops a dynamic framework addressing soil waterlogging and drainage problems. It suggests that optimal resource allocation may be obtained by a multifaceted policy where water conservation, pollution abatement, and disposal of drainage through a canal are induced by financial incentives and public funds are allocated to research leading to improved drainage technologies. An increase in the cost of disposal, reflecting concerns of environmental quality, may lead to a delay (or even elimination) of the use of disposal through a canal as a policy option and raise the importance of conservation and abatement prices. Higher water prices may induce conservation and reduce the severity of drainage problems.

### INTRODUCTION

Canal projects, whether for water supply or drainage, are likely to require major public expenditures and may have substantial impacts on economic development--agricultural and otherwise--as well as on environmental conditions. Yet, economic research and analyses pertaining to the design of such projects and their desirable properties have been minimal.

In most cases it has been perceived that engineers should specify design choices for the key parameters and characteristics of proposed canal projects, and economic frameworks (generally cost-benefit analysis) should be used for evaluation and comparison of a rather small number of finished proposals. Because of this "division of labor" between disciplines, design specifications have been set without recognition of their impacts on the economic environment. In particular, the impacts of canal specifications on market conditions and prices, and processes of learning and technological change that affect water use efficiencies (or drainage generation), have not been incorporated in canal design. Furthermore, segmentation of the decisionmaking process and disciplinary "division of labor" have caused the design and assessment of canal projects to be almost independent of the design and assessment of other policies within affected regions. It seems that canal projects have not been recognized as an important component of regional policy or planning but, rather, as independent entities.

This chapter presents a framework for analyzing and designing a canal project as part of a resource use plan. Canal parameters will be determined simultaneously with other resource use parameters and R&D policies to maximize the welfare of affected regions. The modeling and discussion will be accomplished within the context and conditions of the San Joaquin Valley (Valley) drainage problem, but the principles presented here apply to many other canal projects.

The analyses and discussion in most of this chapter will be conceptual and qualitative. It will identify key considerations in policy decisionmaking and establish guidelines for regional planning processes and basic relationships for project design. It will also discuss some key considerations in the quantification of the approach proposed here.

## CONCEPTUAL FRAMEWORK

To simplify the presentation, the decisionmaking system is divided into two subsystems: (1) The source region where production takes place and where drainage is generated and (2) the canal subsystem. The two subsystems are linked together interdependently, and the optimal decisionmaking model determines choices for both subsystems simultaneously.

### The Source Region Subsystem

Figure 1 depicts the key variables and relationships of the source region subsystem. Since the analysis is dynamic, this figure presents the flows within the system at any point in time.

Production is a major process in the subsystem. Output produced at time t is denoted by $q(t)$, and the inputs to the production process include applied water, $a(t)$; conservation expenditure, $c(t)$; and pollution stock, $s(t)$. In the waterlogging case considered here, the pollution stock consists of rising ground water. As this water approaches the root zone, it reduces productivity. The production function linking this variable is $q(t) = f(a(t), c(t), s(t))$.

# DESIGN OF DRAINAGE CANALS

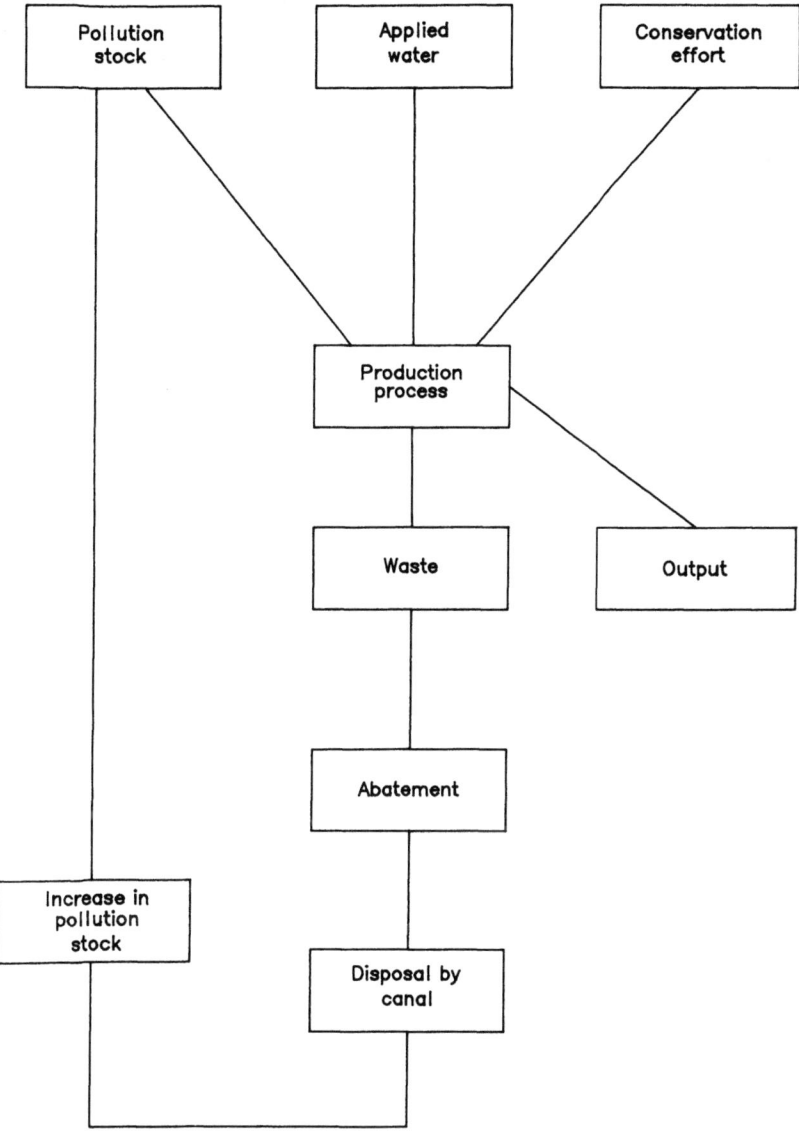

Figure 1. Relationships of the source region subsystems.

Many generalizations and simplifications are introduced in the production modeling. First, heterogeneity of inputs and environment (land quality, weather, water quality, topology) are ignored, and the production of homogeneous products is assumed. This simplification ignores many important issues such as land allocation among crops, differences in technology choices, and output among farmers and locations. The production functions presented here can also be interpreted as an aggregate production function such as the one used in economic growth modeling.

The work of Johansen (1972) presents economic foundations that give rise to aggregate relationships. The aggregate production function approach applied here is useful for the study of certain general dynamic problems. However, modeling for applied empirical analysis should be much more detailed, especially in short-run policy analysis.

Second, only two actual inputs are introduced: applied water and conservation expenditure. The analysis does not consider many other important input choice problems and concentrates only on water- and drainage-related choices. Conservation expenditures are defined as expenditures that increase irrigation efficiency. Differences in conservation expenditures may reflect differences in the cost of irrigation technologies used. Water can be applied with a wide range of technologies ranging from traditional gravitational technologies such as furrow irrigation to capital-intensive modern technologies such as drip irrigation. Modern irrigation technologies are costlier but have higher irrigation efficiency; that is, the fraction of applied water that is utilized by the crop is actually higher using modern technologies. Hence, modern technology can be viewed as input conserving (Caswell and Zilberman, 1986).

It is assumed that the marginal productivity of both applied water and conservation expenditure is positive and decreasing ($f_a > 0, f_c > 0, f_{aa} \leq 0$) and that an increase in conservation expenditure increases applied water productivity, $f_{ac} \geq 0$. It is also assumed that output is affected by contamination, $s(t)$. An increase in contamination reduces output, and the marginal effect is greater in absolute value at higher contamination levels, $f_s < 0, f_{ss} < 0$. In the case of waterlogging, contamination is measured by ground-water levels: the closer the saline ground-water level to the root zone, the higher its impact on productivity.

The production process is accompanied by a waste generation process. This waste can be interpreted as runoff water or percolating water added to the waterlogging process. The waste generated at time t is denoted by $z(t)$ and is dependent on applied water and conservation expenditure following the waste generation function, $z(t) = g(a(t), c(t))$. Conservation technologies might include land leveling, reuse of tailwater that runs off the downstream end of fields, or matching the rate of water inflow to the soil's infiltration capacity through drip or sprinkler irrigation. In the model considered here, conservation activities that increase the percentage of applied water utilized by the crops also reduce waste products. It is reasonable to assume that $g_a > 0$, $g_{aa} > 0$, $g_c < 0$, and $g_{cc} > 0$.

## The Abatement Process

Some of the waste generated by the production process can be abated through chemical treatment processes, evaporation ponds, and irrigation of salt-tolerant plants such as eucalyptus, etc. The abatement technologies are assumed to be in the early stages of development and can be improved by research and experience. Thus, let b(t) denote abatement at period t, and let the abatement cost function be denoted by h(b(t), z(t), k(t)) where k(t) is knowledge about abatement.

It is reasonable to assume that the marginal cost of abatement is positive and increasing ($h_b > 0$, $h_{bb} > 0$) and that abatement costs decline with waste levels, $h_z < 0$, and with knowledge, $h_k < 0$.

Abatement knowledge is increasing as a result of research expenditure, x(t), and the cumulative abatement level, B(t). The cumulative level of abatement is a measure of experience with abatement. It impacts on knowledge, representing the "learning by doing" effect. The importance of knowledge in explaining increases in productivity has been emphasized recently by Romer (1989). The equation of motion of knowledge is k(t) = u(B(t), x(t)), and the equation of motion of cumulative abatement is Ḃ(t) = b(t). It is reasonable to assume that marginal knowledge is increasing at a decreasing rate with both research and development (R&D) expenditures and experience, i.e., $u_B > 0$, $u_{BB} < 0$, $u_x > 0$, and $u_{xx} < 0$.

With abatement, the pollution accumulation process follows the equation of motion, ṡ(t) = z(t) - b(t) - y(t) where y(t) denotes waste disposed by the canal. Obviously, y(t) = 0 before the canal is built. The equation of motion states that waste material which is not abated or disposed through the canal is added to the pollution stock.

## The Canal Subsystem

The critical parameters in the design of a drainage canal are its construction time and structural length and diameter. Let T denote the time when the canal is built. It is assumed that it is built instantaneously, thus abstracting from choices of length of construction period, etc. The length of the canal is denoted by l. A longer canal is assumed to dispose the water to locations where it will have reduced environmental impacts. For example, for disposal of Valley water, a rather short canal may dispose drainage to the San Francisco Bay where environmental costs of the disposal are very high. A longer canal may dispose the drainage deep into the Pacific Ocean with much lower environmental costs. However, a longer canal requires higher operation and construction costs. Let r be the canal's diameter which affects the cost of water transfer. The disposal costs at time t are denoted by $n(l, y(t))$ where $y(t)$ is disposal by the canal at time t. It is assumed that the marginal cost of disposal is positive and increasing with quantity disposed, $n_y > 0$, $n_{yy} > 0$, and decreases with canal length, $n_l < 0$ and $n_{yl} < 0$. Construction costs of the canal are given by $m(r, l)$. Marginal costs of construction are assumed to be positive, $m_r > 0$, $m_l > 0$, and to increase with distance, $m_{rl} > 0$, but may decrease with diameter. The canal operation costs at time t are denoted by $v(y(t), r)l$. It is assumed that they decline with diameter and increase with water flow, $v_y > 0$, $v_r < 0$, and $v_{yr} < 0$.

## THE SOCIAL OPTIMIZATION PROBLEM

The solution to the economic welfare maximization problem determines the optimal design of canal, waste abatement, and water use policies so that the discounted net surplus from production abatement and canal activities is maximized. For simplicity, the output price-taking behavior is assumed and the discount rate is denoted by i, output price by p, and cost of applied water by w. The welfare maximization problem becomes

[1] $\max \int_0^\infty e^{-it}[pq(t)-w-c(t)-h(b(t),z(t),k(t))-x(t)]dt$

{production profit} - {abatement cost} - {R&D cost}

$- \int_T^\infty e^{-it}[v(y(t),r)l+n(l,y(t))dt-e^{-iT}m(r,l)]$

{operation and disposal cost} - {canal construction cost}

subject to

[2] $\dot{s}(t) = z(t) - b(t) - y(t)$ {waste accumulation equation}

[3] $\dot{B}(t) = b(t)$ {abatement experience}

[4] $\dot{k}(t) = u(B(t),x(t))$ {knowledge accumulation}

and given the production and waste generation functions, $q = f(a, c, s)$ and $z = g(a, c)$, and that $s(0)$ is initial pollution level, $k(0)$ is initial abatement knowledge, and $B_0$ is initial abatement experience. (Dots denote a derivative with respect to time.)

## Optimal Decision Rules

The solution for the welfare optimization problem is derived using the optimal control technique (see Appendix). The solution provides the dynamic shadow prices for the stock of ground-water pollution, $\lambda(t)$; level of abatement, $\theta(t)$; and stock of knowledge, $\phi(t)$, respectively, corresponding to conditions [2]-[4]. All these dynamic shadow prices are constructed to be nonnegative. The shadow price of pollution, $\lambda(t)$, measures the marginal welfare loss from increases in the pollution stock, $s(t)$, at time t; and $\theta(t)$ and $\phi(t)$ measure the marginal welfare benefits associated with increases in the stock of abatement experience and knowledge at time t. The solution for the optimization problem also provides values for the decision variables $a(t)$, $c(t)$, $b(t)$, $x(t)$, $l$, $r$, $y(t)$, and T. In the following discussion, the decision rules are derived from the above optimization exercise.

## Production and Conservation at Source

The choice of the optimal quantity of applied water at time t is given by the condition (time is not attached to each variable to simplify the presentation).

[5] $pf_a = w + \lambda g_a$.

The left-hand side is the marginal product value of applied water which is equated to the market price water "w" plus a term that can be interpreted as the marginal pollution cost of water. This is because the term $g_a$ is the additional water pollution caused by a unit of applied water while $\lambda$ is the shadow price of pollution. A marginal increase in the stock of pollution at time t leads to a reduction in productivity from that point on, and those productivity losses are behind $\lambda$.

Since $g_a > 0$, the term $\lambda g_a > 0$, so that, at equilibrium, the marginal product value is always greater than the marginal cost of water, and the difference reflects the social marginal cost of pollution. If the social planner were to introduce a pollution tax on each unit of water applied, it would equal $\lambda g_a$. Alternatively, a tax of $\lambda$ will be imposed on net pollution generated, which is equal to $z(t) - y(t) - b(t)$. Under a competitive market mechanism, the above pollution tax would induce optimal application of water in production.

The time path of the shadow price $\lambda$ is obtained [from condition (A10)] as

[6] $\dot{\lambda} = i\lambda + pf_s$.

Since $f_s < 0$, the sign of $\dot{\lambda}$ cannot be determined conclusively from [6]. However, if the initial stock of pollution is not large (and since $f_{ss} < 0$, the marginal effect of pollution on productivity in early periods is not very substantial), it is plausible that, at least during an initial period, the shadow price of pollution is increasing over time. Moreover, the shadow price of pollution may decline in later periods when the marginal productivity losses, because of pollution, are sufficient so that $pf_s + i\lambda < 0$.

Optimal expenditure in conservation is determined by

[7] $pf_c - \lambda g_c = 1 + h_z g_c$.

The left-hand side presents the marginal benefits from conservation. The term $pf_c$ is the marginal product value of conservation and $-\lambda g_c$ denotes the marginal benefits of conservation in reducing waste generation at the source. These marginal benefits of c are balanced against the marginal cost of conservation expenditure on the right-hand side of equation [7]. These marginal benefits consist of the unit cost of conservation (which is unity) and a term $h_z g_c$ which gives the marginal increase in abatement cost resulting from the reduction in waste generation because of conservation.

While the exact analysis of impact of changes in output and water prices is a cumbersome exercise, conditions [5] and [7] suggest that it is quite likely that increases in output price tend to increase both applied water use and conservation efforts. An increase in water price is likely to reduce water use and waste generation and lead to a reduction in optimal conservation efforts.

## Pollution Abatement and R&D

The next step is to outline rules for the choice of expenditures in pollution abatement technology and R&D and examine the behavior of their respective shadow prices over time. Expenditure in pollution abatement is governed by the equation

[8] $h_b = \lambda + \theta$

which implies that, at the optimum, the marginal cost of abatement is equal to the shadow price of pollution abated and the shadow price of abatement experience representing the benefits from learning-by-doing. Assuming increasing marginal costs of abatement, $h_{bb} > 0$, learning-by-doing increases abatement beyond what is justified by direct pollution costs.

Public expenditures in R&D are obtained from the equation

[9] $\phi u_x = 1.$

The unit cost of R&D expenditures is equated to its marginal benefit given by the shadow price of the stock of knowledge times the marginal contribution of R&D to knowledge, $u_x$. The dynamics of the shadow price of abatement knowledge is derived from condition (A12) to be

[10] $\dot{\phi} = i\phi + h_k.$

Since $h_k < 0$, the sign of $\dot{\phi}$ cannot be determined conclusively from equation [10]. However, it is quite plausible that, if the initial knowledge on abatement is very limited (small k(0)), the shadow price of abatement knowledge will decline during an initial period when the marginal effect of knowledge on abatement is of sufficient magnitude to cause $h_k + i\phi < 0$. While exact analysis of the dynamics of x is tedious, condition [9] and the assumption $u_{xx} < 0$ suggest that, during the early period where $\phi$ declines, R&D expenditure is also likely to decline. Thus, when initial knowledge on abatement is lacking, it may be desirable to establish a substantial R&D effort on abatement and reduce it over time as knowledge accumulates. The dynamics of the shadow price of the abatement stock $\theta$ is obtained from (A11) to be

[11] $\dot{\theta} = i\theta - \phi u_B.$

Again, since $u_B > 0$, the sign of $\dot{\theta}$ cannot be determined conclusively from [11]. It is plausible that, with little initial experience in abatement (small B(0)), $\theta$ will be a decreasing function of time during an early period when the learning-by-doing effect will be sufficient to cause $i\theta - \phi u_B < 0$. The decline of $\theta$ and increase of $\lambda$ during an initial period and condition [8] imply that excessive abatement (above what is required by the shadow price of pollution) for learning purposes is likely to decline over time.

## Effluent Disposal Through a Canal

The choice of canal parameters and the timing of canal construction can now be derived. From [A6],

[12] $\lambda = v_y l + n_y$.

The effluent flow in the canal at time t, y(t), is determined where the sum of its marginal disposal and operation costs is equal to the shadow price of pollution. Combining [5], [8], and [12] gives

[13] $\dfrac{pf_a - w}{g_a} = h_b \theta = \dfrac{f_a - w}{g_a} = h_b \theta = v_y l + n_y$

which implies that, at the margin, the optimal decision rule equates the benefits from pollution control from production, abatement, and disposal. The analysis can be extended to include conservation (through condition [7]).

The diameter of the canal is determined by the following equation:

[14] $e^{-iT} m_r = -\int_T^\infty e^{-it} v_r \, dt.$

Observe that $m_r > 0$ and $v_r < 0$. A larger canal diameter costs more to construct but reduces the operation costs of the canal. These elements are traded off in the determination of the canal diameter. Condition [14] suggests that the diameter is chosen at a level where marginal construction costs equal the sum of discounted marginal savings of operation costs due to larger parameters.

The length of the canal is determined according to the relation

[15] $e^{-iT} m_l = -\int_T^\infty e^{-it}[v(y,r) + n_l] \, dt.$

Rearranging [15] so that the intuition is made more clear, results in

[15'] $e^{-iT} m_l + \int_T^\infty e^{-it} v(y,r) \, dt = -\int_T^\infty e^{-it} n_l \, dt.$

Notice that $m_1 > 0$ and $\eta_1 < 0$ so that both sides of [15'] are positive. Optimal determination of the canal length involves trading off canal construction and operation costs, which increase with length, with disposal costs which decline with length. The exact optimality condition presented in [15'] states that at the optimal length is the present value of marginal construction plus the discounted marginal cost of operations which are equal to the present value of cumulative marginal disposal costs. The optimal time of construction of the canal is determined according to

[16] $\quad im(r,l) + v(y(T),r)l + n[l;y(T)] = s(T)[\lambda^-(T) - \lambda^+(T)]$.

The introduction of the canal reduces drastically the shadow price of pollution. The shadow price of the pollution prior to the construction of the canal is $\lambda^-(T)$ and, immediately after the canal introduction, it becomes $\lambda^+(T)$. Determinaton of the optimal timing for the construction of the canal involves trading off savings in pollution costs associated with the earlier construction date, $s(T) [\lambda^-(T) - \lambda^+(T)]$, with savings in the canal operation, disposal costs, and construction costs (as the construction date is delayed).

## Implementation of the Decision Rules

A range of economic policies can be adopted to correct likely market failures within a competitive setting. As suggested earlier, a Pigouvian tax equal to $\lambda$ can be levied on producers engaged in pollution-generating activity. Another component could be a subsidy scheme that induces competitive firms to employ pollution abatement technology. From condition [8], the value of this subsidy could equal the learning-by-doing effect of abatement given by $\theta(t)$. A third area for policy intervention is in the allocation of research expenditure $x(t)$, the value for which is chosen endogenously in our formulation. The construction and operation of the canal, too, will need to be undertaken by a central planning authority. Observe from condition [13] that, if the planning authority charges a fee for waste disposal, it will be equal to $\lambda(t)$, the shadow price of waste which, in turn, is equivalent to the sum of the marginal cost of operation and disposal.

## Plausible Scenarios

Several likely optimal outcomes could be achieved given specific assumptions about initial parameter values, functional relationships, and policy options.[1] From condition [16], it is easily observed that three different regimes could occur depending on whether (i) $T = 0$, (ii) $T = \infty$, or (iii) $0 < T < \infty$.

Case (iii), the most complex scenario of the three, in essence embodies the other two and may occur if the pollution level at t = 0, s(0), and the stock of knowledge at t = 0, k(0), are both small. This might lead to three distinct phases:

Phase 1 (with no abatement or disposal by canal): During this period, $\lambda(t)$ is likely to increase from an initially positive but very small $\lambda(0)$ according to condition [6]. Waste production will be taxed by $\lambda$ or water use by $\lambda g_a$. Conservation at source will be a dominant strategy for pollution control. The lack of abatement knowledge and relatively low pollution costs make abatement too costly at this stage, and it will not be performed (b = 0 as $h_b > \lambda + \theta$). No abatement will also imply that [7] is reduced to the relationship $p f_c = 1 + \lambda g_c$. However, the pollution stock will rise over time because of high levels of production triggered by a relatively low pollution shadow price in this first phase. Public expenditure in R&D will also increase in this period.

Phase 2 (with abatement and no disposal by canal): During Phase 1, $\lambda(t)$ has increased and $h_b$ has decreased sufficiently for [8] to hold at the start of Phase 2. This phase is, therefore, characterized by both conservation at source and abatement. The learning-by-doing benefits of abatement will be subsidized,[2] as discussed before. Although it is difficult to obtain precise analytical results given the large number of parameters involved, it is plausible that the subsidy ($\theta$) and research expenditures on abatement (x) will peak then decline and may even be eliminated during the second phase. The marginal benefits associated with learning-by-doing and knowledge are decreasing and, after awhile, may be exhausted. During the second phase, the stock of pollution is likely to rise continuously as well as the price of pollution $\lambda$. Conservation efforts, as well as the abatement efforts, are likely to increase during this period, but both efforts will not be sufficient to control the accumulation of pollution stocks and the increase in price of pollution until T, the time at which effluent disposal through canal becomes economically feasible.

Phase 3 (with abatement and drainage canal): This phase starts at time T. Here the dominant pollution control strategy is disposal by canal, although conservation at source and abatement might be used concurrently. The use of these policies will reduce the canal diameter. As discussed earlier, canal operation and effluent disposal are financed through a fee per unit of effluent disposed.

Case (i), where T = 0, suggests a scenario where the canal is built right at the start. This situation could arise if, for instance, s(0) is large so that $\lambda(t)$ is sufficiently high for condition [13] to hold. This case may involve simultaneous use of canal, conservation, and abatement. Furthermore, it may involve research to improve abatement. Conversely, case (ii) where the canal is never built, could occur under conditions in which the marginal cost of abatement is lower than the sum of canal operation and disposal costs, so that [13] can be written as

[13'] $\quad \lambda(t) = h_b \cdot \theta(t) < v_y l + n_y$, for $0, t \leq \infty$.

Notice that the above relationship will only hold if the system enters a steady state. This steady state occurs once the effects of learning on abatement technology improvements are exhausted and it becomes a stable technology. In this case, all the waste produced by the industry will be eliminated so that $s = 0$ and the stock of pollution is stabilized. From condition [6] for $\lambda = 0$ the steady-state pollution tax will be $-pf_s/i$.

Cases (i) and (ii) are essentially limiting cases for case (iii) and provide a better understanding of the feasible policy actions given an initial set of parameter values. For example, initial high values of $s(0)$ will cause relatively high taxes for waste generation and may increase expenditures in abatement research and accelerate the building of the canal. An increase in the price of water will reduce the amount of pollution, $s(t)$, because producers will reduce input use. Costlier water will tend to slow abatement research and delay the building of the canal.

The relative marginal costs of the alternative pollution control technologies are also important in determining their choice. If canal disposal costs were low, which could be the case, say, in sparsely inhabited areas or coastal regions (and when preservation of these environments is not valued much), the preferred policy option might be to forgo pollution abatement and build a canal at a relatively early stage. On the other hand, canal disposal and operation might be relatively costly in built-up areas, or in regions from which dumping sites are difficult to access, in which case abatement will dominate and canals will be introduced at a late stage in the program or will not be built at all. Similar tradeoffs could be discussed in terms of applying conservation technologies at source.

The above discussion essentially assumes that costs of management and enforcement are zero, so that the first-best solution can be implemented. However, in reality, this may not be the case. For instance, optimal taxation of pollution discharges may be politically infeasible or costly in terms of information requirements. An inability to tax waste will lead to higher than optimal production in earlier periods and a faster rate of waste accumulation. The canal is likely to be built more easily, and its capacity might be larger than optimal, given the increased pollution load. On the other hand, if the possibility of abatement were to be ignored, the construction time for the canal would be advanced, and its capacity would be larger. Note that a longer capacity canal implies a larger diameter and length. This, in turn, would increase the tax on waste disposal and decrease production.

## CONCLUSIONS

This chapter attempts to integrate a range of pollution control policies with the concepts of learning-by-doing and R&D, with particular reference to agricultural drainage. It proposes an analytical framework that examines policy options such as pollution control at source, abatement, and disposal in a dynamic context. It provides guidelines for technology choice and rules that determine what policies to choose in a given situation.

The model determines optimal pollution taxes on waste and subsidies to producers that incorporate the learning-by-doing effects of pollution abatement. Rules for private expenditures on pollution control at source and public investment on R&D are developed. Principles for choosing canal parameters and timing of construction are also derived. An effluent disposal tax that pays for canal operation and maintenance is obtained.

Three likely policy scenarios were identified. One involves the construction of the canal in the early stages for use in drainage disposal, the other involves not building a canal at all and relying on conservation and abatement to control pollution, and the third is a gradual three-phase scenario. In the first phase, conservation at the source, using technologies such as drip or sprinkler irrigation, may be the most feasible form of drainage control with no pollution abatement or disposal. In the second phase, abatement technologies like evaporation ponds or chemical cleanup facilities may become viable, accompanied by public expenditures in R&D, and subsidies to firms that take into account the learning-by-doing economies of pollution abatement. In the third phase, effluent disposal by drainage canal becomes the primary means of drainage control.

The likelihood of different scenarios is largely affected by societal preferences regarding the environmental quality. An increase in the social valuation of environmental amenities tends to raise disposal costs substantially, and that reduces the likelihood of using a drainage canal as the main solution to a drainage problem. With higher drainage disposal costs, it is more likely that drainage problems will be addressed by multiphase policies that provide incentives for conservation and abatement activities and which delay the construction of drainage canals (if not eliminate the need for them altogether).

The conceptual framework developed in this chapter yields insights into the dynamic character of environmental processes such as drainage. Economic prescriptions that are generated by static benefit-cost approaches or programming analyses prove to be grossly inadequate in the dynamic setting. As revealed here, the composition of economically feasible technologies and corresponding tax/subsidy schemes varies over time so that a package of technology options and economic policies that is optimal at one instant of time may be quite inappropriate at another point of time.

Although the framework is deterministic, certain relationships modeled here are prone to uncertainties that could be incorporated in future extensions of this model. The effect of R&D and abatement experience on the growth of knowledge, and the contribution of learning-by-doing to reduction in abatement costs, could be modeled as stochastic. The cost of construction and operation of the canal which is built at a future date is also a likely source of uncertainty. It is also assumed that the industry is a price taker in the output market, where a downward-sloping demand function might be more appropriate.

## APPENDIX

The problem stated in [1]-[4] is an optimal control problem with control variables $a(t)$, $c(t)$, $b(t)$, $x(t)$, $T$, $l$, $r$, and $y(t)$ and state variables $s(t)$, $k(t)$, and $B(t)$. Pontryagin's Maximum Principle can be used to construct the following Hamiltonian:

[A1] $H' = He^{it} = pq - wa - c - h(b,z,k) - x - \lambda[z - b - y] + \theta b - \phi u(B,x) - \Gamma[v(y,r)l - n(l,y)]$

where $\delta = 0$ when $t < T$ and $\delta = 1$ when $t \geq T$. Assuming the existence of an interior solution, the usual necessary conditions (see Seierstad and Sydsaeter, 1987) are given by

[A2] $H'_a = pf_a - w - \lambda g_a = 0$

[A3] $H'_c = pf_c - 1 - h_z g_c - \lambda g_c = 0$

[A4] $H'_b = -h_b + \lambda + \theta = 0$

[A5] $H'_x = -1 + \phi u_x = 0$

[A6] $H'_y = -v_y l - n_y + \lambda = 0$

The optimal determination of $l$, $r$, and $T$ are according to

[A7] $H_l = -\int_T^\infty e^{-it}[v(y),r] + n_l dt - e^{-iT} m_l = 0$

[A8] $H_r = -\int_T^\infty e^{-iT} v_r l dt - e^{-iT} m_r = 0$

$$[v(y(T),r)l + n(l,y(T))] + ie^{-iT}m(v,l)$$

[A9]  $-\lambda^-(T)s^-(T) + \lambda^+(T)s^+(T) = 0$

Note that conditions [A7]-[A9] are "salvage value" conditions. The associated co-state equations are given by

[A10]  $-\dot{\lambda} + i\lambda = -H_s = -pf_s$

[A11]  $-\dot{\theta} + i\theta = H'_B = \phi u_B$

[A12]  $-\dot{\phi} + i\phi = H'_k + h_k.$

For the purposes of this chapter it is assumed that the sufficiency conditions given, for instance in Sydsaeter and Seierstad (1987), hold and an inferior solution exists.

## NOTES

[1] A formal analysis of these situations requires extending the formal framework to allow inequalities and corner solutions. Such analysis is cumbersome and beyond the scope of this chapter. Here we present likely outcomes of an extended framework.

[2] When pollution is taxed by $\lambda$, abatement subsidy will be equal to 0, and abatement will be subsidized by $\lambda + \theta$ when water is taxed by $\lambda g_a$.

## REFERENCES

Caswell, M. F. and Zilberman, D., 1986. The Effects of Well Depth and Land Quality on the Choice of Irrigation Technology, *American Journal of Agricultural Economics*, 68(4), pp. 798-811.

Johansen, L., 1972. *Production Functions: An Integration of Micro and Macro, Short-Run and Long-Run Aspects*. North-Holland, Amsterdam.

Romer, P., 1989. *Increasing Returns and New Developments in the Theory of Growth*. National Bureau of Economic Research, Inc., Working Paper No. 3098, Cambridge, MA.

Seierstad, A. and Sydsaeter, K., 1987. *Optimal Control Theory with Economic Applications*. North-Holland, Amsterdam.

# 34 COMMON PROPERTY ASPECTS OF GROUND-WATER USE AND DRAINAGE GENERATION

Lloyd S. Dixon, The RAND Corporation

## ABSTRACT

In this chapter a game theoretic analysis of ground-water use and drainage water management is introduced. Model simulations using parameters relevant to the San Joaquin Valley (Valley) are presented to illustrate the different types of behavior that might be observed and to quantify the benefit losses of different management strategies. The implications of these models for ground-water and drainage policy are also discussed.

## INTRODUCTION

Due to incomplete property rights, many facets of agricultural water use have "common property" aspects. This is the case when many farmers pump ground water from the same aquifer to irrigate their crops. It is also the case when irrigation water percolates to a shared shallow aquifer and creates a drainage problem. In the former situation, the common property resource is the shared stock of ground water. In the latter, it is the unsaturated soil profile over which any one farmer has only incomplete control[1].

In this chapter, models of ground-water use by farmers in a common property setting are examined and the implications for farmer behavior explored. Several simulations are presented of the payoffs for different types of behavior using parameter values motivated by the ground-water extraction setting in Kern County, California. Parallels between ground-water extraction decisions and drainage water generation decisions are then discussed, and questions raised that should be addressed in formulating drainage policy.

# A FRAMEWORK FOR ANALYZING GROUND-WATER EXTRACTION

Consider two identical farmers who pump ground water from an underlying aquifer for irrigation use during periods 1, 2,.....T. Each farmer, or agent, maximizes an objective function subject to a set of constraints that characterize the aquifer, and each has no impact on agricultural prices or nonwater input costs. Pumping, however, does lower the water table and therefore affects water costs. Assume that the benefits from ground-water withdrawals are measured by the area under a linear derived demand curve for water:

[1] $MB_{it} = q - ru_{it}$   $t = 1....T$   $i = 1, 2$.

$MB_{it}$ : Marginal benefit of ground water to agent i in period t ($/acre-foot)

$u_{it}$ : ground water extracted during period t (acre-feet)

$q, r > 0$.

Each agent can pump water from the aquifer at a cost dependent on the depth-to-water at the beginning of the period:

[2] $TC_{it} = sx_{it-1}u_{it}$

$TC_{it}$ : Total cost of pumping $u_{it}$ acre-feet from depth $x_{it-1}$ ($)

$x_{it-1}$ : Depth-to-water for agent i at end of period t-1 (feet)

$s > 0$.

The objective of each agent is to maximize net discounted benefits:

[3] $\Pi_i = \max_{u_{it}} \sum_{t=1}^{T} \beta^{t-1} [qu_{it} - .5ru_{it}^2 - sx_{it-1}u_{it}]$

$\beta$: Real private discount factor $1/(1+r)$

$u_{it} \geq 0$.

The aquifer consists of two identical cells with each agent having access to one of the cells. Each agent lowers the water table in his or her cell by pumping, and the water table rises due to exogenous recharge. Water also flows between cells at a rate proportional to the difference in the ground water level in each cell (see figure 1). This leads to the following equation of motion for the depth-to-water in each cell (see Dixon, 1988, for derivation):

[4] $x_{it} = (1-\alpha)x_{it-1} + \alpha x_{jt-1} + [(1-\theta/\xi^2)]u_{it} - (1/\xi w^2)c_{it}$  $i,j=1,2; i \neq j;$
$\qquad t=1....T.$

$0 \leq x_{it} \leq z$

$x_{i0}$ given,

where

$x_{it}$: depth-to-water in cell i at the end of period t
$u_{it}$: withdrawals from cell i during period t
$c_{it}$: exogenous recharge during period t (ac-ft)
$\xi$: specific storage capability of aquifer material $(0 \leq \xi \leq 1)$
$w^2$: cross area of square aquifer cell (acres)
$z$: distance from land surface to bottom of cell (feet)
$\alpha$: leakage parameter $0 \leq \alpha \leq .5$
$\theta$: return flow coefficient $0 \leq \theta < 1$.

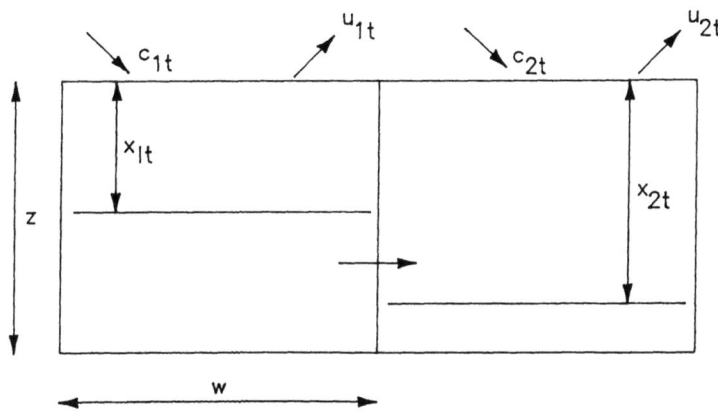

Figure 1. Water flows in and out of aquifer cells.

When $\alpha = 0$, the two cells are hydrologically separate. When $\alpha = .5$, the water levels in the cells are equalized at the beginning of each period. A two-cell aquifer model such as this is useful because it allows for the parameterization of the leakage externality. The sensitivity of the different solutions to the hydrologic connectedness of the cells can then be examined. Each farmer maximizes [3] subject to [4].

## SOLUTION CONCEPTS

Most models of farm ground-water use assume that farmers act myopically and ignore the impact of their ground-water extractions on future ground-water levels (for example, see Burt, 1964; Brown and Deacon, 1972; Bredehoeft and Young, 1970; Gisser and Sanchez, 1980; Noel, Gardner, and Moore, 1880; and Feinerman and Knapp, 1893). Farmers are assumed to pump groundwater in each period until the marginal benefit of ground-water pumping equals the marginal cost:

[5] $u_{it}^{my} = q/r - (s/r)x_{it-1}$.

This is equivalent to the solution of [3] when each agent has a discount factor equal to zero. To calculate the benefits to each agent over the time horizon $t=1.....T$, the period benefits are usually discounted and summed using the social discount factor $\beta_s$. The discounted net benefit of myopic behavior will be denoted $\Pi_i^{my}$.

The standard approach then compares the discounted net benefits of myopic behavior with the discounted net benefits of socially optimal pumping rates which fully account for the impact of pumping on future ground-water levels. In the social optimum, aquifer-wide rents are maximized using the social discount rate, and each of the two identical farmers receives one-half of the total payoff. The payoff to each farmer in the social optimum will be denoted $\Pi_i^*$. For a linear-quadratic specification, the optimal withdrawals are linear functions of the depths-to-water in the preceding period:

[6] $u_t^* = g_t^*(x_{t-1})$,

where $u_t$ and $x_{t-1}$ are vectors whose components are the individual withdrawal rates and depths-to-water. The determination of these optimal pumping rates is computationally straightforward (see Dixon, 1988).

When the private discount factor, $\beta$ equals the social discount factor, $\beta_s$, the social optimum is equivalent to the agents colluding to maximize aquiferwide net benefits and then splitting them evenly. If the private discount factor is less than the social discount factor, the withdrawal rates in the collusive solution will be different than those in the social optimum. When the period benefits of each are then discounted and summed using the social discount factor, the sum will be lower in the collusive solution. In what follows, the private discount factor will be assumed equal to the social discount factor unless otherwise noted. Thus, the collusive and social optimum solutions will be identical.

An assumption of myopic behavior by farmers may be inaccurate in many circumstances. If farmers act with foresight, the assumption of myopic behavior will lead to an overstatement of the benefit loss resulting from incomplete property rights over ground water. In recent years, some modeling studies have introduced the element of foresight into agent behavior for multiperiod resource utilization problems (for example, Negri, 1989 and Eswaran and Lewis, 1984). The following sections discuss three additional solutions to the above model in which farmers consider the impact of their pumping on future ground-water levels: Open-loop, conventional closed-loop, and trigger-strategy.

## OPEN-LOOP SOLUTION

The open-loop solution was first proposed in the ground-water literature by Wetzel (Wetzel, 1978). In the open-loop solution, each agent maximizes net present discounted value given the withdrawal path of the "opponent" (the other agent or agents). The solution is a Nash equilibrium in which each agent has no incentive to deviate from his or her withdrawal path given the path of the other agents. The solution is thus two withdrawal paths: $u_{11}....u_{1T}$ and $u_{21}....u_{2T}$. In the linear-quadratic case, the solutions are unique and computationally straightforward (see Dixon, 1988). Let the sum of the discounted payoffs to agent i of the open-loop withdrawal pattern be $\Pi_i^{ol}$.

The forward-looking behavior captured in the open-loop solution might be expected to cause $\Pi_i^{ol}$ to be greater than $\Pi_i^{my}$. However, because each agent does not take into account the effect of his or her pumping on the extraction costs of the other agent, $\Pi_i^{ol}$ is not expected to be as large as $\Pi_i^*$. This inefficiency has been labeled the "stock externality" (Negri, 1989).

While the open-loop solution may provide a somewhat more realistic description of resource extraction than the myopic solution, it has been criticized in the game theory and resource extraction literatures. In the open-loop solution, the assumption that each agent takes the opponent's extraction path as given is tantamount to assuming that the opponent will not respond to the agent's own extraction path. The impact is that "each producer effectively takes control of that part of the resource equaling the sum of its committed outputs" (Eswaran and Lewis, 1984), and has no fear of losing water to a rival

(Negri, 1989). The open-loop solution thus overlooks strategic interaction between agents to lay claim to a larger part of the resource, or put in another way, to extract the resource before his or her opponent does. The reduction in discounted payoffs caused by this strategic interaction has been called the strategic externality (Negri, 1989). In some models without extraction costs, the open-loop solution generates the counterintuitive result that noncooperative extraction of a common property resource results in no inefficiency (Kemp and Long, 1980). Nevertheless, the open-loop solution may make sense when agents learn nothing about the system as play proceeds (e.g., do not observe the state variable). In such a case, each agent has no reason to think that opponents will respond to these actions and no reason to alter his or her own actions during the course of play. The open-loop solution may also make sense where there is some institutional mechanism by which agents can commit themselves to their withdrawal paths, such as enforceable contracts between agents or the establishment of property rights to ground water.

## CONVENTIONAL CLOSED-LOOP SOLUTION

A more realistic assumption about agent behavior in most situations is that each agent will adjust behavior in response to the actions of the opponent. However, what is a reasonable conjecture of how an opponent will respond? One reasonable conjecture is that at any point in the game, each opponent will respond to an action by picking the extraction path that maximizes personal payoff for the rest of the game. In the game-theory literature, this concept is referred to as "subgame perfection." The resources literature has termed the outcome of this assumption the "closed-loop equilibrium" of the game. This conjecture will be adopted here, and for reasons that will become clear below, this outcome will be referred to as the "conventional closed-loop solution."

Closed-loop equilibrium is calculated by solving the maximization problem for each agent backward through time starting with the terminal period. A set of withdrawal rules form a closed-loop equilibrium when each is a best response to the other. The form of the solution is an extraction rule in which withdrawals during each period are a function of the depth-to-water in the proceeding period:

[7] $u_{it} = g_i(x_{t-1})$  $i = 1,2.$

Again, in the linear-quadratic case, the solutions are unique and computationally straightforward (see Dixon, 1988). Let the payoff to agent i of the conventional closed-loop withdrawal pattern be $\Pi_i^{cc1}$.

Because the closed-loop solution incorporates the strategic externality missing in the open-loop solution, one expects $\Pi_i^{cc1} < \Pi_i^{o1} < \Pi_i^*$. For many functional forms, the open-loop solution is easier to calculate than the closed-loop because it avoids the need for backward induction. Thus, it is of interest in the ensuing simulations to see how the two solutions compare.

The parameter values chosen for the above model will have an important impact on the relative payoffs of different solutions. To the author's knowledge, no simulations have been reported that might be relevant to a ground-water extraction setting. Before turning to trigger-strategy equilibria, a baseline simulation motivated by the situation in the Valley portion of Kern County will be presented and differences in the various solutions examined. The results of sensitivity analysis done for various parameters will then be reported.

## Parameterization

Baseline parameters for a two-farmer model are taken from Feinerman and Knapp's model of ground-water use in Kern County (Feinerman and Knapp, 1983); a hydrologic-economic model of the Valley developed by California's Department of Water Resources (DWR) (Department of Water Resources, 1982); and regional water budgets and cropping patterns provided by DWR. Each farmer is assumed to cultivate one acre of land and have access to one cell of the aquifer. The aquifer dimensions are scaled so that the ratio of the 2 acres to total aquifer surface area is the same as the ratio of irrigated land to aquifer surface area in Kern County in the early 1980's. The baseline set of parameter values are listed in table 1.

Table 1. Parameter baseline values and simulation ranges.

|   |   | Baseline | Range |
|---|---|---|---|
| q | Marginal Benefit Intercept | 70 | [70, 310]  $/ac-ft |
| r | Marginal Benefit Slope | 20 | [20, 170]  $/ac-ft$^2$ |
| s | Pumping Cost | .15 | $.15/ac-ft/foot lift |
| $\alpha$ | Leakage Parameter | .5 | [0, .5] |
| $\xi$ | Specific Storativity | .125 | .125 |
| $\theta$ | Return Flow Coefficient | .2 | .2 |
| w | Cell Width | 1.13 | [1, 1.414] feet |
| c | Recharge to Each Cell | .75 | [.25, 1.25] ac-ft |
| x | Initial Depth-to-Water | 225 | [175, 400] feet |
| $\beta$ | Real Discount Factor | .9524 | [.8696, .9756] ($r$ = [.15, .025]) |

There is no presumption that this two-agent, two-cell model is an accurate representation of water withdrawal in Kern County even if the results are insensitive to scaling the amount of land cultivated and aquifer size. First, in contrast to this two-farmer model, there are more like 800 farmers in Kern County (based on the number of farmers who have filed for pesticide permits with the Kern County Agricultural Commissioner in 1988). Second, there is no geophysical basis for splitting the aquifer underlying Kern County into just two interconnected cells. Rather, the Kern County estimates are used to provide plausible parameter estimates for a stylized model of a small, two-agent extraction setting. However, some inferences may be indicative of the actual Kern County situation. For example, one can see what happens to the various solutions as the number of agents increases. Since the collusive and myopic solutions are insensitive to the number of agents, as long as the results are fairly insensitive to model scaling, the myopic and collusive solutions here can be compared to those found elsewhere in the literature for Kern County.

## Simulation

To minimize the impact of the terminal period on payoffs, the length of the game, T, was selected to represent sufficient duration to assure that the payoff differences between the various solutions were fairly insensitive to any further time increase. For $\beta=.9524$ ($r=.05$), T=60 was adequate. The discounted net benefits to each farmer in the various solutions using the baseline parameter values are shown in table 2. Also shown is the approximate steady state depth-to-water in each solution, $x_{ss}$. Note that due to symmetry, the payoffs to each agent are the same as are the depths-to-water in each aquifer cell during each period. As expected, $\Pi^{my} < \Pi^{ccl} < \Pi^{ol} < \Pi^*$, and the myopic payoff is $73.84, or 14.5 percent, below the collusive payoff. At $490.47, the closed-loop payoff is 3.7 percent less than the collusive payoff, a much smaller decrease than that for the myopic solution. The 14.5 percent difference is close to the percent difference reported for the baseline simulation in Feinerman and Knapp's 1988 study of Kern County.

Table 2. Payoffs and steady-state depths-to-water for baseline parameter values in two-agent model.

|  | Collusive | Open-Loop | Closed-Loop | Myopic |
|---|---|---|---|---|
| Payoff ($\Pi$) | $509.17 | $495.89 | $490.47 | $435.33 |
| Steady-State Depth-to-Water (feet) | 249.20 | 293.93 | 302.73 | 341.67 |

The four payoffs are plotted in figure 2, and the contributions of the stock externality, strategic externality, and myopia are labeled. The difference between $\Pi^*$ and $\Pi^{ol}$ is due to the stock externality and accounts for 18 percent of the overall difference between the collusive and myopic payoffs. The strategic externality is measured by the difference between $\Pi^{ol}$ and $\Pi^{cl}$ and accounts for another 7 percent. Myopia has by far the greatest impact, accounting for 75 percent of the difference.

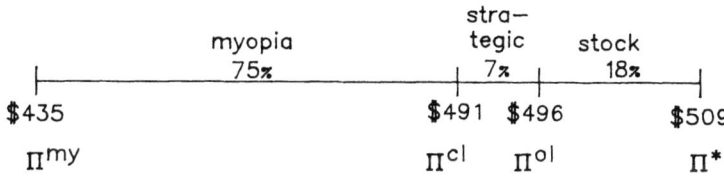

Figure 2. Payoffs of the four solutions and welfare losses due to the various factors.

What is particularly noteworthy about these results is how much smaller the benefit loss implied by the closed-loop solution is than that implied by the myopic solution. The myopic solution overstates the difference between the collusive and closed-loop solutions by 295 percent suggesting that in situations where there are a small number of agents, the myopic solution may not be a good description of agent behavior and may significantly overstate the difference between observed behavior and the social optimum. Of course, the further the private discount rate is below the social discount rate, the further both the open- and closed-loop solutions will be from the social optimum.

The open-loop solution understates the benefit loss reported by the closed-loop solution by 29 percent indicating that in situations like this, limiting attention to the open-loop solution will lead to underestimating the loss by a sizable percentage. The 29-percent difference is consistent with the results reported by Negri for a simulation of the blue whale population (Negri, 1986). Negri found that when 10 firms have access to the whole fishery the open-loop solution understates the benefit loss by 37 percent. Even though the 29 percent found here is a large percent difference, the difference in levels between the open- and closed-loop solutions is not great.

## Sensitivity Analysis[2]

Sensitivity analyses were done over reasonable ranges for various parameters. Reasonable ranges for Kern County were determined by compiling the various parameter values used in the studies previously mentioned and are listed in table 1. The percent differences between the myopic, closed-loop, open-loop, and collusive payoffs fall as the intercept of the marginal benefit function, q, rises, *ceteris paribus*, but are quite insensitive to changes in q over the range examined (q∈[70, 310]). Likewise, differences in the various steady state depths-to-water are insensitive to changes in q over this range. On the other hand, the differences between the various payoffs are quite sensitive to changes in the marginal benefit slope, r. It is only at the low end of the range of r examined (r∈[20,170]) that there are sizable differences in payoffs, and as r rises (the marginal benefit function becomes steeper), the differences fall rapidly. For example, the 14.5-percent difference between the collusive and myopic solutions in the baseline simulation (r=20) drops to 2.8 percent when r=50, *ceteris paribus*. Surprisingly, however, the differences in the steady state depths-to-water are not sensitive at all to changes in r.

The difference between the myopic and collusive solutions is unaffected by the leakage parameter $\alpha$. This is because as long as the farm operations are identical and the initial depths-to-water in the two cells are the same, there will never be any flow between the cells, and the value of $\alpha$ is immaterial. The open- and closed-loop solutions, however, approach the collusive solution as $\alpha$ falls and are identical to it when $\alpha=0$. As $\alpha$ falls, the interconnectedness of the cells decreases, and the interactions of the agents decline in importance. The sensitivity of the results to the leakage parameter emphasizes how hydrologic discontinuities in an aquifer can significantly reduce the common property problem. It also suggests that inappropriate modeling of an aquifer as single-cell can cause serious overstatement of the difference between the open- and closed-loop equilibria and the collusive solution.

When the size of each aquifer cell increases, *ceteris paribus*, both the percent and absolute differences between the various solutions drop substantially. This is consistent with the work of Gisser and Sanchez which shows that as the size of the aquifer grows the differences between the myopic and collusive solutions drops (Gisser and Sanchez, 1980). They argued that for the Ogallala Aquifer, w is large enough for there to be an insignificant difference between the two solutions. However, whether this is true in any particular setting depends on the relation between the size of the aquifer, discontinuities in the aquifer, and the marginal benefit functions of the overlying farms.

Sensitivity of the results to the number of agents will now be examined for N identical agents extracting water from the aquifer. To preserve symmetry, a single-cell model of the aquifer is used for this analysis rather than the two-cell model, but the overall size and amount of land farmed remain the same as in the previous model[3].

As would be expected, the total payoffs across all agents are independent of the number of farmers in the myopic and collusive solutions. However, total open- and conventional closed-loop payoffs rapidly approach the total myopic payoff as N grows. When there are as few as 10 agents, total payoffs in the conventional closed-loop solution is 12.3 percent below the collusive payoff, compared with 14.5 percent for the myopic. Also, the closed-loop steady-state depth-to-water quickly approaches the myopic figure--when N=2 the closed-loop steady-state depth-to-water is 35.3 feet less than the myopic, but only 5.9 feet less when N=10. These results suggest that for as few as 10 agents in this model, the myopic solution is a fairly good approximation of agent behavior, even when foresight is incorporated. Whether the number of agents is large or not in any particular setting depends on the connectedness of the aquifer. If, for example, each of ten agents had access to a separate aquifer cell that was only partially hydrologically connected to other cells, the open and conventional closed-loop solutions would be much closer to the collusive solution than if they all withdrew from a single cell.

## TRIGGER-STRATEGY EQUILIBRIA

The open- and closed-loop Nash equilibria have been proposed as two possible outcomes in ground-water extraction games with forward-looking behavior. Both these equilibria generated payoffs below those in the collusive solution with aquifer cells leakage. But if farmers all do better in the collusive solution, why do they not implement it? The problem is that in the ground-water extraction game there is a multiperiod version of the "prisoner's dilemma" at work. Even though each agent would be better off if they all play the collusive solution, any individual agent would be even better off if everyone else played the collusive solution while he or she extracted water at a higher rate. Thus, there is an incentive for each agent to defect from the collusive solution which means that the collusive solution is not a Nash equilibrium. One possible remedy is for agents to enter into binding contracts and agree to limit ground-water pumping to collusive rates. However, in most ground-water settings, property rights to ground water are ill-defined and an agreement such as this may well prove unenforceable.

Noncooperative game theory has explored situations in which collusive outcomes are expected when there is a prisoner's dilemma but no possibility of binding contracts. The results are usually based on the assumption that agents use trigger strategies. The basic idea of a trigger strategy is that all farmers agree to pump ground water at the collusive rate unless one farmer deviates by pumping more. This deviation will then trigger a punishment phase. Trigger strategies are thus closed-loop since an agent's actions depend on what has happened previously in the game, but are more complicated than the conventional closed-loop strategy considered thus far. Trigger strategies can be designed so that, when they form a Nash equilibrium, there is a self-enforcing agreement among the agents to play the collusive solution.

The possibility of trigger-strategy equilibria in the two-agent ground-water extraction model will now be examined. The existence of trigger-strategy equilibria will raise the possibility of collusive behavior even when binding agreements are not possible. If such were the case, there would be no inefficiency losses due to aquifer leakages, and outside intervention to correct the externality would be unnecessary. Until recently, there has been little work on trigger strategies in resource extraction settings. The only papers this author has come across in the resources literature are those by Hämäläinen, Haurie, and Kaitala on the exploitation of fisheries (Hämäläinen, Haurie, and Kaitala, 1984 and 1985). They show that collusive behavior is plausible in the exploitation of fisheries; however, the equilibria they consider are not subgame-perfect. Thus, it may not be in the interest of the agents to carry out the promised punishments if someone defects from the collusive solution. If this were the case, the agreement to play the collusive solution would not be self-enforcing.

## Design of a Trigger-Strategy

If agents are to use trigger-strategies, they must first agree on what to play in the periods prior to any deviation--the so-called cooperative phase. To make trigger strategies worthwhile, the payoffs in the cooperative phase must be higher than agents would otherwise receive. In this ground-water extraction game, a reasonable assumption of what they would otherwise receive is the payoff in the conventional closed-loop equilibrium. It will be assumed that agents decide on collusive behavior for the cooperative period; however, this is only one of many possibilities. The agents must also decide on responses during the punishment phase as well as the duration of the punishment phase. It will be assumed that they will choose the conventional closed-loop strategies as threats. These threats are credible in the sense that they will actually be carried out if someone deviates. To see this, remember that in each period of the game, the conventional closed-loop equilibrium strategies form a Nash equilibrium for the rest of the game, and thus punishment entails reversion to a Nash equilibrium of the game. Since in a Nash equilibrium each player is doing the best possible given the actions of the other players, if everyone else threatens punishment with the closed-loop equilibrium strategies, individual agents will also want to use this strategy. For the purposes of this chapter, the punishment phase will last for the remainder of the game. Punishments that last for only a finite number of periods are possible but less effective in deterring deviation. Moreover, the assumption of either finite or infinite punishment periods does not change the lessons to be drawn from the analysis. Finally, the players must be able to observe deviations to impose punishment. In this model, agents are assumed to observe both the actions of their opponents in each period as well as the depth-to-water (state variable). In a game of certainty, retaliation can be triggered by either a wayward action or the departure of the state variable from its agreed upon path. Either option is acceptable here.

Consider the trigger strategy, $u^{TR}$, where each agent plays the collusive solution, $u^*_{it}$, as long as all involved have done likewise in the proceeding period. If any agent does not play the collusive solution in a given period, all other players will play the conventional closed-loop equilibria, $u^{ccl}_{it}$, beginning with the next period for the remainder of the game. This can be represented as

$$[8] \quad u^{TR}_{it} = \begin{cases} u^*_{it} & \text{if } t=1 \\ u^*_{it} & \text{if } u_{t-1} = u^*_{t-1},\ t=2\ .....T \\ u^{ccl}_{it} & \text{otherwise.} \end{cases}$$

Remember that both the collusive solution and closed-loop solution are functions of the depth-to-water in the preceding period with the function dependent on time.

The set of trigger strategies $\{u^{TR}_1, u^{TR}_2\}$ forms a subgame-perfect Nash equilibrium if for each player j the best response is to play $u^{TR}_{jt}$ given that all the other players play $u^{TR}_{it}$ $i \neq j$ in each period t. Note that this response must take into account not only the actions of other players during period t, but also predicted future play. If the set of trigger-strategies forms a Nash equilibrium, then collusive play throughout the game will be an equilibrium outcome.

## *Failure of Subgame-Perfect Trigger-Strategies in Finite-Horizon Games*

In a finite-horizon game, backward induction can be used to show that an equilibrium in trigger strategies is impossible if there is a unique equilibrium for the game played over periods t....T for t=1...T. The reasoning is as follows: Consider play in the final period T. In a game such as [3], the payoff in T depends on $x_{T-1}$, but not directly on previous actions or states. Thus given $x_{T-1}$, the players can choose their strategy ignoring the particular pattern of past moves. Suppose that there is a unique Nash equilibrium to the period T game. Now consider the two-period game that consists of play in periods T-1 and T.

The players know what the play and payoffs in period T will be for any $x_{T-1}$, and again their payoffs for the two-period game depend only on past play as summarized in $x_{T-2}$. Each player must take into account the effect of period T-1 play on period T play through $x_{T-1}$, but each knows what will happen in period T for any $x_{T-1}$. If the Nash equilibrium in period T-1 is unique, the argument can be repeated for T-2. As before, this uniqueness will rule out trigger-strategy equilibria.

As discussed above, the open- and conventional closed-loop equilibria of the linear-quadratic finite-horizon game are both Nash equilibria, so it appears that there is not a unique solution and trigger-strategy equilibria are possible. However, the conventional closed-loop equilibria is the unique subgame-perfect equilibrium to this game. If attention is restricted to subgame-perfect equilibria, then subgame-perfect trigger will not exist in this game.

## Infinite-Horizon Trigger-Strategies

If there is no terminal time period, there is no point at which backward induction can start. Thus, it may be fruitful to consider games which continue forever: $T \to \infty$. In such infinite-horizon games, it has been shown that the collusive solution converges to some limiting decision rule: $u^T_{it} = g^T_{it}(x_{t-1})$ converges to $u_i = g_i(x_{t-1})$ as $T \to \infty$ where $u^T_{it}$ is the period t action in a T-period game (Kumar and Varaiya, 1986). Proofs of convergence do not exist for the finite-horizon conventional closed-loop solution as $T \to \infty$. Nevertheless, simulations suggest that the conventional closed-loop solution does converge, and it will be assumed to do so here. Derivations of the infinite horizon withdrawal rules in an N-agent, single-cell aquifer setting for the collusive and closed-loop solutions are provided in Dixon, 1988. A set of infinite-horizon trigger strategies forms a Nash equilibrium if in each period t, each player's best response is to play the trigger strategy, given that all other players play the trigger strategy.

For given values of q, r, c, and $\eta$, a set of infinite-horizon trigger strategies are expected to form an equilibrium if the discount factor $\beta$ is high enough. If the discount factor is sufficiently high, each player will not deviate from the collusive solution because the one-period gain from deviating is outweighed by the discounted value of lower payoffs associated with the closed-loop solution from the next period on. Examples where $\beta$ is high enough to support a trigger-strategy equilibrium will be examined.

## Simulation of Trigger-Strategies

Figure 3 shows whether it is in agent 2's interest to deviate in period t of an infinite horizon game when $q=200$, $r=50$, $s=.15$, $c=1$, $\eta=4$, and $x_0=0$ for several different values of $\beta$. The total discounted payoffs to each player when these strategies are played are labeled $\Pi^*$ and $\Pi^{\infty 1}$. Let $\Pi^{TR}_2(t)$ be the payoff for the entire game to agent 2 if he or she deviates in period t while agent 1 is playing $u^{TR}_1$. If $\Pi^{TR}_2(t)$ stays below $\Pi^*$, agent 2 has no incentive to deviate, while if it is above $\Pi^*$ for some t, agent 2 would rather deviate than continue playing the collusive strategy. $\Pi^{TR}_2(t)$ is plotted in figure 3 for $=.99, .8, .55, .45$, and $.10$.

First note in figure 3 that for $\beta = .55, .8$, and $.99$ that $\Pi^{TR}_2(t)$ stays below $\Pi^*$ over the 40 deviation periods checked (t=1...40). One expects $\Pi^{TR}_2(t)$ to approach $\Pi^*$ asymptotically. The longer agent 2 waits to deviate, the more the withdrawal paths for the two players look like the withdrawal paths in the collusive solution. With a discount factor less than one, deviations far in the future will have a negligible affect on the discounted payoffs and the resulting payoffs will be arbitrarily close to the collusive payoff. Given that the discount factor falls in a smooth exponential pattern from period to period, one also expects $\Pi^{TR}_2(t)$ to be monotonic. It is thus reasonable to conclude for $\beta=.99$, .8, and .55, that $\Pi^{TR}_2(t)$ is monotonically increasing and asymptotically approaches $\Pi^*$ from below and that agent 2 will not want to deviate for any t. By symmetry, neither will agent 1, so for this set of parameter values and discount rates there are trigger-strategy equilibria. This demonstrates that it is possible for agents to set up a self-enforcing agreement to play the collusive solution. The agents no longer have an incentive to deviate from the collusive solution, and the "tragedy of the common" is not necessarily the outcome in a noncooperative ground-water extraction setting.

In contrast, when $\beta=.45$ or $.1$, $\Pi^{TR}_2(t)$ lies above $\Pi^*$ for all t tested, and P2 will want to deviate. In these cases, the proposed trigger strategies do not form equilibria, and it is not in each player's individual interest to keep playing the collusive solution when his or her opponent is using a trigger strategy. Note, however, that at these extremely low discount factors (corresponding to discount rates of 122 percent and 900 percent, respectively) there is very little difference between the conditional closed-loop and collusive payoffs anyway. It appears that there is a single critical value of $\beta$ (call it $\beta_c$) below which the sets of trigger strategies do not form equilibria. A grid search is done to approximate this value, and for these particular parameter values, the trigger strategies do not form equilibria when $\beta \leq .53$ ($r \geq .89$).

A discount factor of .53 is very low--few would argue that the social discount factor is this low. Remember that the private discount factor is assumed equal to the social discount factor. If the private discount factor is significantly below the social discount factor, the collusive solution will differ from the social optimum.

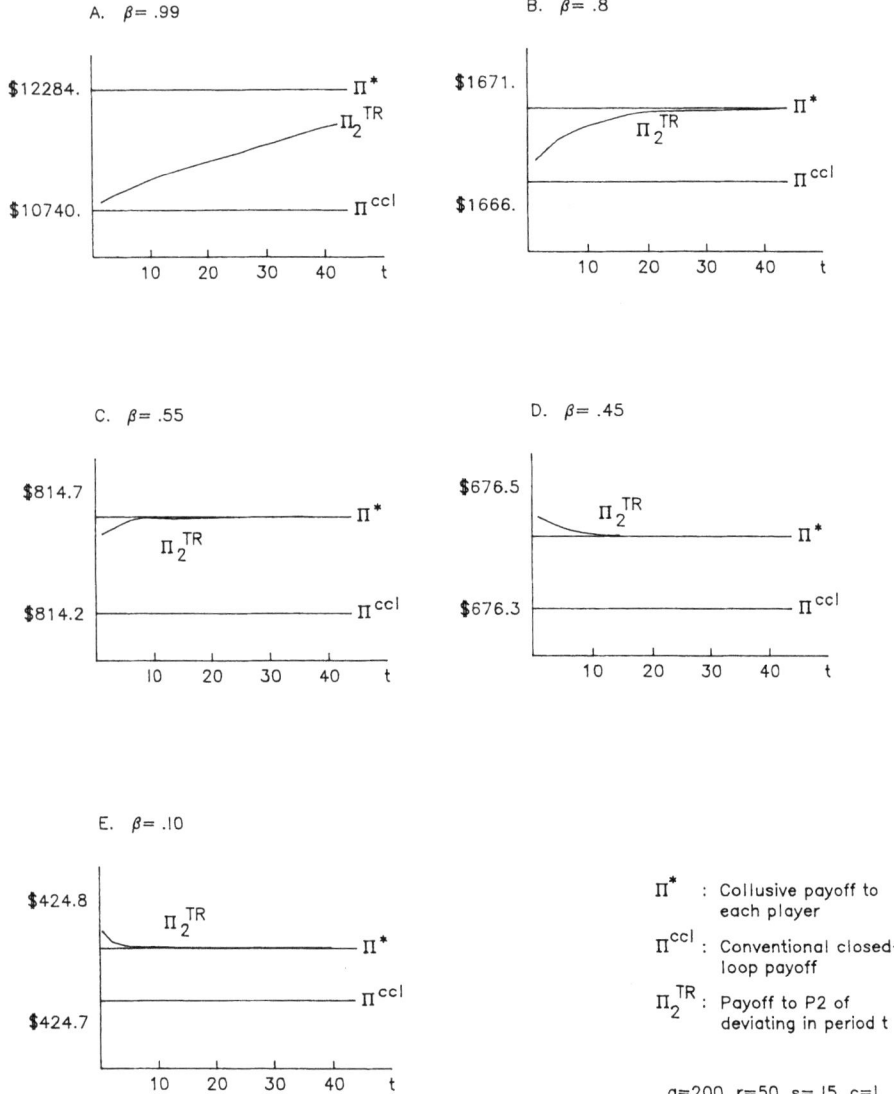

Figure 3. Benefits from deviating in period t of an Infinite-Horizon Game.

## Sensitivity Analysis

The critical values of $\beta$ were approximated over a wide range of parameter values. $\beta_c$ is determined by the difference between the collusive and conventional closed-loop strategies during the contemplated punishment phase and the one-period gain from deviating. The larger the difference between the collusive and closed-loop solutions, the smaller $\beta$ would need to be to discourage deviation, given the one-period gain from deviating. But as $\beta$ and the parameter values change, so do both the difference between the two solutions and the one-period gain from deviating, making predictions of how $\beta_c$ will change difficult.

$\beta_c$ rises as q rises and falls as r rises, but over quite a large range of values (q $\in$ [70,310] and r $\in$ [20,170]), it remains quite small and stable. For example, when q=200, r=50, s=.15, c=1, $\eta$=4, and $x_0$=0, $\beta_c$=.53, and when r=100, *ceteris paribus*, $\beta_c$=.51. For very small values of r, $\beta_c$ rises substantially. For example, when r=1, *ceteris paribus*, $\beta_c$=.93 (r=.075). Thus in this case, the trigger strategy combination will work only if the discount factor is greater than .93.

As aquifer size (w) rises, $\beta_c$ falls, but again the values do not change much and remain low over a wide range of parameter values. An interesting result from the simulations is that as the initial depth-to-water, $x_0$, rises (water level falls), $\beta_c$ drops slightly. The fact that the value of $x_0$ influences $\beta_c$ supports statements made by other authors that given a discount factor, trigger strategies might only work once the state variable passes some threshold (Benhabib and Ferri, 1987). Here, as the water level falls, the requirement on the discount rate is less stringent.

While simulations so far have only been done for two players, one can speculate on how the critical discount factor would change as the number of players increases. In the collusive solution, aggregate ground-water withdrawals and payoffs remain the same as the number of agents increases. However, as reported above, the conventional closed-loop equilibrium diverges further and further from the collusive solution. This means that the punishment phase of the trigger-strategies becomes increasingly harsh as the number of agents grow. This might be expected to lower the discount factor needed to support an equilibrium in trigger strategies. However, this may not necessarily be the case since it is not clear what happens to the single-period gain from deviating as the number of players grows.

## Evaluation of Trigger-Strategy Equilibria

Collusive behavior has been shown to be an equilibrium outcome in a noncooperative setting in infinite-horizon games when the agents use trigger-strategies. It should be noted that collusive behavior is not the only possible outcome of a trigger-strategy. Other candidates are withdrawal rates that produce payoffs between the conventional closed-loop and collusive payoffs.

While trigger-strategy equilibria are an appealing concept from the perspective of efficiency in ground-water extraction settings, there are some fairly stringent requirements to implement this type of equilibrium. First, the agents must be able to agree on what the withdrawal rates should be in the cooperative period. This may not be easy in heterogeneous settings. Second, agents must be able to observe enough about the system to detect deviation. This may be accomplished by observing opponents' behavior or the depth-to-water. In many ground-water withdrawal settings, however, agents may have little information on what others are doing and hydrologic uncertainties may make it difficult to distinguish between deviations and random fluctuations in depth-to-water. Finally, the agents must also have sufficiently high discount factors. Even if the private discount factor is high enough to support the collusive solution, if the private discount factor is less than the social discount factor, the collusive solution will not be the social optimum.

Trigger-strategy equilibria may be difficult to implement in many situations. However, there are some ground-water use settings where they are conceivable--for example, when there are a small number of forward-looking agents facing parameter values that imply significant gains to coordination. When making statements about the extent of benefit loss in a ground-water extraction setting, therefore, the possibility of trigger-strategy equilibria should be considered.

## PARALLELS WITH DRAINWATER GENERATION

As in an interconnected aquifer, water that percolates downward from one farmer's fields may flow beneath the land of another farmer. If this is the case, the first farmer does not bear all the costs of generating percolation in the same way that he or she does not bear all the costs of pumping ground water. The common property resource in such situations is the unsaturated soil profile over which any one farmer may have only incomplete control.

A model of farmer drainage water generation would look similar to one for ground water. The private benefit of applying irrigation water could be specified similarly. However, the cost of applying this water would have to be modeled differently. In the ground-water context, the cost is due to pumping water to apply for irrigation and is directly proportional to the depth-to-water. In the drainage problem, costs are due to a rising shallow aquifer and are inversely proportional to the depth-to-water. Smaller depths-to-water might negatively affect crop productivity (for example when the water table encroaches into the root zone) or may mean that more subsurface water must be collected, treated, and disposed. The model of the shallow aquifer in the drainage setting would look similar to one in the ground-water setting. Such a model would have to specify the impact of percolation on water levels and the degree to which water levels in different areas are related.

The above exploration of ground-water extraction decisions can be used to motivate the important questions that should be asked in analyzing drainage issues. First, one needs to see if the parameters of the particular setting imply a significant difference between the social optimum and the myopic behavior. It may well be the case that the common property aspect of the problem is not important. Remember that only costs that are internalized by the farmers as a whole have been examined here. Costs that are externalized to other segments of society, such as the free disposal of toxic drainwater into rivers, must be brought into farmer objective functions for farmer behavior to produce a social optima. Two particularly important parameters in common property situations are the social and private discount factors. One must ask what is the private discount rate of the farmers in the region of concern. Is it near the social discount rate? If it is low, all the various solutions examined above will approach myopic behavior, and some type of policy intervention may be warranted. If the private discount rates are high, the open-loop, conventional closed-loop, and trigger strategy equilibria may generate outcomes close to the social optimum. Second, the interconnectedness of the shallow aquifer must be considered in the drainage context. How easily does water flow from one part of the shallow aquifer to another? If there are major barriers to lateral flow in the shallow aquifer, the common property problem may not be major. A third relevant question is the number of farmers who utilize the same group of interconnected aquifer cells. Results from the ground-water model presented here show that the open and conventional closed-loop solutions approach the myopic solution as the number of agents increases. Also, as the number of agents increases, the coordination needed for trigger-strategy equilibria would be expected to be more cumbersome. Finally, degree of cooperation and communication between farmers in the affected area should be assessed. Their relationship may suggest the possibility of trigger-strategy equilibria and the potential for ameliorating the common-property problem.

## NOTES

[1] In the drainage situation there are also incomplete property rights associated with the treatment of drainage. Farmers frequently can dispose of toxic drainage water in rivers or other places for free and ignore the social costs of this disposal. In this chapter, these aspects of incomplete property rights will not be addressed.

[2] Extensive analyses of the sensitivity of the solutions to changes in parameter values are reported in Dixon, 1988. Here, the discussion is limited to some of the more interesting results.

[3] Even if agents were added in equal numbers to the land above each cell in the two-cell model, symmetry would be lost because interactions between two agents withdrawing from the same cell would not be the same as interactions between two agents withdrawing from different cells. The N-agent, single-cell model is presented in Dixon, 1988.

## ACKNOWLEDGEMENTS

This chapter is based on Chapters 1 and 2 of my Ph.D. dissertation (Dixon, 1988). I am indebted to Michael Hanemann, Tom Rothenberg, Larry Karp, and Eddie Dekel for their valuable input. All errors and omissions, of course, remain the responsibility of the author.

## REFERENCES

Bredehoeft, J. D. and Young, R. A., 1970. The Temporal Allocation of Ground Water--A Simulation Approach, *Water Resources Research*, 6(1), pp. 3-21.

Brown, G. and Deacon, R., 1972. Economic Optimization of a Single-Cell Aquifer, *Water Resources Research*, 8(3), pp. 557-564.

Burt, O., 1964. Optimal Resource Use Over Time with an Application to Ground Water, *Management Science*, 11(1), pp. 80-93.

California Department of Water Resources, 1982. *The Hydrologic-Economic Model of the San Joaquin Valley*. Bulletin 214.

California Department of Water Resources, 1982. *The Hydrologic-Economic Model of the San Joaquin Valley, Appendix C.* Bulletin 214.

California Department of Water Resources, 1982. *Estimated Components of the San Joaquin Hydrologic-Economic Modeling Study*.

Dixon, L. S., 1988. *Models of Ground-Water Extraction with an Examination of Agricultural Water Use in Kern County, California*. Ph.D. Dissertation, Department of Economics, University of California, Berkeley.

Eswaran, M. and Lewis, T., 1984. Appropriability and Extraction of a Common Property Resource. *Econometrica*, 51, pp. 393-400.

Feinerman, E. and Knapp, K., 1983. Benefits from Ground-Water Management: Magnitude, Sensitivity, and Distribution, *American Journal of Agricultural Economics*, pp. 703-710.

Gisser, M. and Sanchez, D., 1980. Competition Versus Optimal Control in Ground-Water Pumping, *Water Resources Research*, 16(4), pp. 638-642.

Hamalainen, R. P.; Kaitala, V.; and Haurie, A., 1984. Bargaining on Whales: A Differential Game Model with Pareto Optimal Equilibria, *Operations Research Letters*, 3(1), pp. 5-11.

Hamalainen, R. P.; Kaitala, V.; and Haurie, A., 1985. Equilibria and Treats in a Fishery Management Game. *Optimal Control Application and Methods*, 6, pp. 315-333.

Kumar, P. R. and Varaiya, P., 1986. *Stochastic Systems: Estimation, Identification, and Adaptive Control*. Prentice Hall, Englewood Cliffs, NJ.

Negri, D., 1986. *The Common Property Resource as a Dynamic Game*. Ph.D. Dissertation, Department of Economics, University of Michigan, Ann Arbor.

Negri, D., 1989. Stragedy of the Commons, *Water Resources Research*, 25(1), pp. 9-16.

Noel, J.; Gardner, D.; and Moore, C., 1980. Optimal Regional Conjunctive Water Management, *American Journal of Agricultural Economics*, pp. 489-498.

Wetzel, Bruce, 1978. *Efficient Water Use in California: Economic Modeling of Ground-Water Development with Applications to Ground-Water Management*. RAND Corporation, R-2388-CSA/RF.

# Eight: UNCERTAINTY, ENFORCEMENT, AND POLITICAL ECONOMY OF WATER SYSTEMS

# 35 DETERMINATION OF REGIONAL ENVIRONMENTAL POLICY UNDER UNCERTAINTY: THEORY AND CASE STUDIES

Erik Lichtenberg, University of Maryland, College Park

## ABSTRACT

Uncertainty about environmental effects is a key factor shaping environmental policy decisions. This chapter reviews alternative approaches to incorporating uncertainty into formal decision methodologies. Such alternative approaches include "conservative" environmental damage estimates; expected utility based on multiattribute decision analysis or revealed preference estimation; and cost-benefit, risk-benefit, or cost-efficiency analysis using safety rules. Safety rules are the most appealing in an empirical context and correspond to the legal framework guiding environmental regulation. Three case studies involving agricultural drainage and runoff are presented that use the safety rule approach, focusing on the impact of incorporating uncertainty, modeling behavioral responses to policy, the role of heterogeneity in production, and the relative importance of long-run versus short-run distributional effects.

## INTRODUCTION

Uncertainty is ubiquitous in environmental policy problems. One reason is simply the complexity of environmental problems which typically have multiple causes and are mediated by a multitude of factors. Some of these factors are observable, others are not; thus, science can account for part of observed variations in environmental outcomes. In addition, scientific knowledge is usually limited: There are many things about adverse environmental effects that are not fully understood in a theoretical or empirical sense. For example, little is known about the long-term effects of synthetic organic chemicals on human beings and other animal species. The aim of policy is to prevent avoidable damage. At the same time, many adverse environmental effects are quite subtle and are, therefore, reliably detectable only in cases of extreme damage. Policymakers must thus rely on estimates of adverse effects that depend heavily on assumptions made in simulation modeling, adding an extra layer of uncertainty. This preventive posture also constrains policymakers to issue decisions in a timely manner, so that data collection is often not as thorough as might be desired.

The evidence suggests that the public is quite sensitive to these uncertainties. The work of psychologists indicates that the public perceives as more hazardous effects that have greater uncertainty associated with them (for a summary see Slovic, Fischoff, and Lichtenstein, 1980). The recent furor over pesticide residues on foods (e.g., Alar on apples) bears this notion out. The best data available suggest that roughly 85 percent of fresh produce in the marketplace have no detectable residues and that almost all of the remaining cases involve residue levels that are extremely small and well below what the U.S. Environmental Protection Agency (EPA) considers to be the maximum safe levels. Yet much of the U.S. public believes that pesticide residues on foods pose a serious threat to public health.

Policymakers are also quite sensitive to these uncertainties in part because of public demands for taking uncertainty into account in making regulatory decisions; in part (perhaps) because mistakes are the most visible indicator of poor performance. Moreover, much of the legislation governing policy formulation directs decisionmakers to take uncertainty into account, by requiring policies to safeguard environmental quality with an adequate margin of safety. To be truly useful in aiding policy determination, then, quantitative decision methodologies should take uncertainty into account explicitly. Cost-benefit or risk-benefit analyses based on expected values are inadequate in this regard, since they make no adjustment for uncertainty. This chapter discusses the applicability of several approaches to uncertainty adjustments in quantitative decision methodologies, notably: (1) Cost-benefit analysis using "conservative" environmental damage estimates, as practiced by EPA and other State and Federal regulatory agencies, (2) expected utility analyses, specifically multiattribute decision analysis, and (3) cost-benefit, risk-benefit, or cost-efficiency analysis using safety rules. The safety rule approach is illustrated in the final section using problems of agricultural drainage and runoff management, specifically, river discharge of potentially toxic drainage water, groundwater contamination by agricultural pesticide use, and shellfish contamination by livestock waste runoff.

## ALTERNATIVE APPROACHES TO UNCERTAINTY ADJUSTMENT

Consider a region in which a productive activity creates a byproduct that is believed to have detrimental side effects. Irrigated agriculture in areas with perched water table problems, for example, generates drainage flows that are highly saline and may contain naturally occurring toxic elements such as selenium, arsenic, and boron as well as residues of applied chemicals such as pesticides. Surface runoff typically contains fertilizer and pesticide residues as well. Disposal of surface and subsurface runoff into rivers, lakes, or artificially created receiving waters may have adverse effects on vegetation, wildlife, and human health. The degree to which these adverse effects occur will depend on random factors such as weather that govern the amount of receiving water, breakdown or immobilization of toxic chemicals, uptake of toxic chemicals by vegetation, wildlife population sizes, chemical uptake, and so on. Estimates of the causal linkages between disposal of runoff and these adverse effects will be influenced by errors in model specification and estimation due to incomplete knowledge about the causal processes and incomplete data on causal factors, and will thus exhibit greater randomness than the effects themselves.

A decision methodology that takes uncertainty into account must thus begin with an environmental impact assessment that incorporates randomness explicitly. Two different approaches have been taken: (1) Adjusting the estimates used to ensure that they contain a suitable margin for error, and (2) building an explicitly probabilistic model of environmental impacts. The former has been standard operating procedure for the EPA, the Food and Drug Administration (FDA), the Fish and Wildlife Service (FWS), and other Federal and State agencies. Its attraction is practicality: Margins for error can be taken from existing engineering rules of thumb. This advantage is its main weakness: Margins for error are derived in an arbitrary way, with no reference to the randomness appearing in the case at hand. These margins for error have no statistical basis and therefore no real meaning. Standardization of protocols for making estimates in this way does not ensure that the resulting estimates provide the same margin for error because level of error (and inherent randomness) varies from case to case. In addition, margins for error derived in this way cannot be compared in different cases--or even for different policies designed to address a single environmental impact--in a rigorous way, making it difficult to evaluate policy alternatives. The latter approach (the explicitly probabilistic model) is more difficult to implement. It is more subject to specification error, in the sense that omission of relevant factors can bias the estimates obtained. It does make it possible to ascribe statistical meaning to any adjustments for error, however, and thus makes alternative policy options comparable. It has been growing in popularity--at least for estimating human health risks--for precisely this reason.

One implication of using probabilistic environmental impact assessments in decisionmaking is that the separation of economic from environmental impact analysis cannot be maintained. It becomes important to incorporate the effects of alternative policy options into ecological models in a complex manner, since effects on estimated outcomes and on the randomness of these estimates are both important. Thus, an interdisciplinary modeling process is a necessity to implement a more sophisticated approach to policy analysis.

## Cost-Benefit Analysis with "Conservative" Damage Estimates

The approach taken by the EPA and other regulatory agencies to adjust for uncertainty in environmental impacts is to make "conservative" estimates of potential damage under alternative policy scenarios that incorporate margins for error using engineering rules of thumb. These estimates are then provided as a rough form of "certainty-equivalent" data for cost-benefit or risk-benefit assessments. As argued elsewhere (Lichtenberg, forthcoming), this procedure does more than bias policy toward more stringent standards, as is intended. It may also bias the type of policy chosen in favor of setting stricter standards and against increased monitoring and enforcement, as indicated in the following example.

Consider the case of a rice-growing region located upstream of an urban area that uses river water for drinking. Suppose that rice growers use an herbicide believed to pose a human health risk. To control temperature, rice growers find it necessary to lower water levels in their fields. On infrequent occasions (say, a small fraction of the time $\alpha$) this occurs shortly after applying the herbicide. For convenience, assume that all rice growers discharge simultaneously, so that exposure to the herbicide in drinking water, when it does occur, is always the same. Let the risk from exposure to the herbicide be R, expressed as the number of cases occurring in the population, so that urban residents face an expected health risk $\alpha$R from exposure to the herbicide in their drinking water. One possible policy is to ban use of the herbicide. Let the social cost of banning this pesticide be $C_B$. An alternative policy is an enhanced monitoring program that detects the herbicide in time for the city water department to shut off intake until the contaminated water has passed downstream. Suppose that the monitoring program has a cost $C_M$. If only expected values matter, the pesticide should be banned as long as $C_B < C_M$. A "conservative" risk estimate of the type used by EPA treats the exceptionally high residue levels as normal occurrences and inflates the estimated risk to R. The cost per case avoided under a ban will be $C_B/R$, while the cost per case avoided under the monitoring program will remain $C_M/\alpha R$, so that the ban will be preferred as long as $C_B < C_M/\alpha$. Thus, whenever $C_M < C_B < C_M/\alpha$, the use of a "conservative" risk estimate will erroneously indicate the superiority of the ban.

## Expected Utility and Multiattribute Decision Analysis

Expected utility has long been the preferred paradigm in economics for treating issues of choice under uncertainty. Recent criticisms of this approach have focused on its inability to capture some common aspects of individuals' actual choice behavior, that is, its performance as a descriptive model (Machina, 1987). It remains attractive as a normative model, although some argue that it sets too strict a standard for rationality.

Empirical applications of expected utility depend on estimation of multiattribute utility functions describing preferences over relevant outcomes, to be combined with estimated outcome probabilities. Multiattribute utility functions can be estimated in two ways. The first involves eliciting utility function parameters by questioning a crucial decisionmaker, the second uses the revealed-preference approach to estimate parameters from past decisions.

Eliciting the preferences of a key decisionmaker has been used successfully in a number of business applications (see for example Keeney and Raiffa, 1976). Such an approach is problematic in a public policy context because it is not at all clear that any single decisionmaker can or should speak for the body politic.

An alternative is to derive information on public preferences by analyzing past decisions. Several studies have employed such a revealed preference approach to estimate the relative social welfare weights on producer welfare, consumer welfare, and similar outcomes in cases involving agricultural policies, trade policies, and highway construction (for a survey see Rausser, Lichtenberg, and Lattimore, 1983).

A number of difficulties arise in using this approach, especially for environmental policies. First, information on key variables involved in environmental policy decisions may not be available. Second, public preferences regarding policy outcomes such as environmental quality and agricultural income may change over time, so that past decisions are poor indicators of current welfare weights. For example, policy decisions in California have historically favored agriculture over urban income in cases such as water subsidies. Recent decisions appear to have reversed the trend, as evidenced by the defeat of the Peripheral Canal, the imposition of stringent standards for water quality from agricultural drainage, and the imposition of strict pesticide-use reporting requirements. Third, public preferences regarding policy outcomes may also vary from case to case, so that decisions from one situation will give erroneous information about preferences in another. Decisions about development of a

Yosemite Valley or a Glen Canyon may have little bearing on situations involving agricultural drainage. Finally, theory and empirical evidence suggest that past decisions are in large measure determined by the relative political clout exercised by different sets of agents active in political markets. Revealed preference approaches thus tend to conflate public preferences and past relative political power. It is by no means clear that the parameters estimated in this way can or should be interpreted as expressions of true social preferences.

In sum, it appears that practical difficulties in deriving estimates of parameters expressing social preferences make application of the expected utility framework to public policy issues quite questionable.

## Safety Rules

A third alternative is to assess tradeoffs between productivity losses and environmental quality using safety rules to adjust for uncertainty, as proposed by Lichtenberg and Zilberman (1988a). Such an approach has several advantages. First, it is essentially a way of deriving a "conservative" estimate of risk that has formal statistical meaning. As a result, it is likely to be appealing to regulators and scientists accustomed to dealing with "conservative" estimates while bringing some rigor to the definition of "conservative," so that the criticisms previously noted do not apply. Second, the safety rule approach conforms closely to the stricture contained in much environmental legislation that posits a goal of providing adequate protection of public health and/or the environment with a sufficient margin of safety, as well as corresponding to a "disaster avoidance" approach that is often felt to characterize bureaucratic decisionmaking. In other words, it corresponds to public preference structures codified in law and in regulatory practice. Third, it can be thought of as an extension of the Baumol and Oates (1971) standards-and-charges approach to cases involving uncertainty. Finally, safety rules have been used in a variety of economic applications; they are well understood and have been shown to give good approximations of expected utility decisions in several empirical contexts (Thomson and Hazell, 1972).

This approach views the government as having two objectives: Maximizing net market benefits and minimizing environmental damage. Net market benefits consist of the real income of producers and consumers derived from production and consumption of items affected by regulation, less government expenditures. To account for uncertainty about environmental damage estimates, the environmental quality objective is defined as an upper limit that is not exceeded with a certain degree of confidence, for example, the level below which environmental damage is estimated to occur, say, 95 percent of the time.

This corresponds to the use of confidence intervals from classical statistics to adjust for uncertainty and addresses the need to allow a margin for error.

The tradeoffs between these two objectives can be estimated by solving a constrained optimization problem of maximizing net market benefits subject to the constraint of the environmental quality objective. Solving the problem while varying the constraint repeatedly yields a set of tradeoffs between market welfare and environmental quality and an associated set of policies.

Formally, let X be a vector indicating the extent of use for the policies to be considered. For example, $X_1$ may be the level of a tax on discharge of toxic elements into a body of water, $X_2$ may indicate the severity of restrictions on pesticide use, etc. Net market benefits are a function of these policies B(X). Environmental quality is similarly a function of these policies R(X) and is a random variable. Let $R_0$ be the desired environmental quality level and P be the desired margin for error. The optimization problem is

$$\max_X B(X)$$
$$\text{s.t. } \Pr\{R(X) < R_0\} > P.$$

The solution is an optimal policy vector $X^*(R_0, P)$ that is a function of the environmental quality target and the desired confidence level, which measures the margin for error. Substituting into the net market benefits function gives the maximum net market welfare attainable given the environmental quality objective and confidence level $B(X^*) = B^*(R_0, P)$. By varying $R_0$, one obtains the set of tradeoffs with a given confidence level P. Varying the confidence level as well gives a complete set of tradeoffs between market welfare, environmental quality, and the reliability of attaining the acceptable risk level.

A key measure derived from this equation is the uncertainty premium, the absolute value of $dB^*/dP$, the reduction in net market benefits associated with a small increase in the confidence level. It indicates the additional cost required to increase reliability in meeting the environmental quality standard, and can be considered as similar to the risk premium derived from expected utility theory.

The information generated by this methodology can be used to determine policy using a variety of decision criteria, including cost-benefit and risk-benefit criteria. In cost-benefit analysis, the optimal policy equates the marginal cost of risk reduction $|dB^*/dR_0|$ with the monetary value of improved environmental quality.

## APPLICATIONS OF THE SAFETY RULE APPROACH

The preceding discussion stressed that decision methodologies to address agricultural drainage and runoff problems should: (1) Incorporate uncertainty and (2) correspond to the legal and regulatory framework that governs policy. It was argued that the safety rule approach fits these needs better than other available alternatives. This section reviews some recent applications of the safety rule approach to problems of agricultural drainage and runoff to illustrate its use and insights that can be gained from explicit consideration of uncertainty.

The discussion of these empirical applications also highlights the importance of three additional features: (1) Modeling behavioral responses of economic agents, (2) providing distributional information, and (3) modeling heterogeneity. First, as many economists have noted, economic agents seldom remain passive in the face of an altered regulatory landscape. In fact, changes in regulation typically bring forth changes in producer and consumer behavior that, if not taken into account in formulating a policy, may in large measure negate its intended effects. Thus, decision methodologies should incorporate behavioral models of producer and consumer responses. Second, the existence of political activity which surrounds proposed regulation, along with general notions of justice or fairness, indicate the distribution of costs and benefits is a key consideration in policy formulation. Thus, decision methodologies should provide this kind of information. Third, heterogeneity among agents is often critical in determining the actual effects of policies as well as shaping the distribution of gains and losses. This suggests the importance of modeling quantitatively key dimensions of heterogeneity.

### River Discharge of Agricultural Drainage

The first case study involves river discharge of agricultural drainage water (Hanemann et al., 1987). In 1983, it was established that selenium in agricultural drainage water was responsible for a variety of reproductive problems in waterfowl and other aquatic fauna in the Kesterson Reservoir (Kesterson), a repository for agricultural drainage flows emanating from the Westlands Water District on the west side of the San Joaquin Valley, California. In 1985, the California State Water Resources Control Board initiated a process to set standards for selenium and other heavy metals (boron, molybdenum) in the San Joaquin River, affecting growers cultivating 94,000 acres in four water districts to the north of Westlands that had been discharging drainage water into the San Joaquin River.

Farms in the affected area differed in terms of land quality (and therefore cropping patterns and deep percolation coefficients) and water charges (which varied according to irrigation district). Estimates of acreages and yields of crops on different soil types were obtained from soil surveys and combined with estimates of production costs and variable and fixed (per acre) water charges for each irrigation district to form distributions of quasi-rents for all possible production patterns.

The standard irrigation technique presently used with all crops in the area is furrow irrigation with half-miles runs. Subsurface drainage per acre per month was determined for this technology by combining estimates of annual drainage per acre for all districts with estimated monthly drainage distribution patterns for the one district with available monthly data. Surface runoff was estimated by subtracting estimated subsurface drainage discharges from total flows recorded in the drains of each irrigation district. Water application rates for furrow irrigation were set as equaling the average values reported in the literature.

Crop rotations, rather than individual crops, were the units of analysis. Rotation frequencies were determined by combining expert opinion on standard operating practices in the area with data on crop acreages in each district. Because the area provides a small fraction of output of all crops considered, price effects were assumed to be negligible and prices were assumed to remain constant at the most recent available average prices. The profitability of each rotation under furrow irrigation was then calculated for each quality of land in each district as the weighted average of crop profitabilities, with weights derived from the rotation frequencies. For each district, the distribution of current per-acre quasi-rents under furrow irrigation was then estimated via linear programming. Land allocations were selected to maximize quasi-rents in each district subject to the constraints that: (1) Total land allocated to each crop equaled the average level in the most recent year and (2) total land of each quality allocated to all crops equaled the estimated amount. Differences in rotational profitabilities were sufficiently large and differences in crop water requirements were sufficiently small to rule out shifts in cropping patterns in response to technology changes or cost increases.

Two possible approaches were considered for meeting selenium standards: source reduction via installation of water conserving irrigation technologies and selenium removal via water treatment. Four alternative irrigation technologies were selected for analysis: Furrow irrigation with runs shortened to one quarter-mile, installation of tailwater recovery systems, sprinkler irrigation, and drip irrigation. The parameters describing irrigation efficiency, deep percolation, and surface runoff were chosen to be broadly representative of the estimates in the literature. Reductions in water application, deep percolation, and surface runoff and increases in per-acre production costs were estimated relative to the baseline (parameter) estimates. The cost function for selenium removal consisted of three components: A cost of selenium removal, a cost of removing suspended solids (applicable when combined surface and subsurface drainage flows were treated), and a cost of storing drainwater to balance monthly treatment requirements. In addition, the minimum cost strategy always involved delivering drainage water into the San Joaquin River at a point upstream of the Merced River, to take advantage of the additional dilution capacity of the Merced. This approach required construction of a canal.

Four year types were used to characterize precipitation, and thus riverflow, patterns. The years 1978/79 and 1983/84 were chosen as representative of normal years, 1984/85 was selected as representative of a dry year, and 1980/81 was selected as representative of a critically dry year. The confidence level associated with setting standards designed to be achieved in each year was estimated using the historical distribution of riverflows reported by the California Department of Water Resources. By this criterion, 1978/79 corresponds to a 43.9-percent confidence level, 1983/84 to a 53.7-percent confidence level, 1984/85 to a 76.8-percent confidence level, and 1980/81 to an 81.7-percent confidence level. For each year, the optimal treatment capacity for each technological alternative was chosen by minimizing total treatment cost subject to the constraint of meeting selenium concentration standards of 2, 5, and 10 parts per billion (p/b) in the San Joaquin River during every month. The total treatment cost plus the investment in irrigation technology was then calculated for each technological alternative.

The analysis indicated that the choice of a control strategy depended critically on the confidence level selected; that is, the choice of policy instrument depended on the adjustment made for uncertainty. Source reduction via water conservation appeared increasingly important for more stringent selenium standards and for greater margins for error. In normal years, a standard of 10 p/b could be met entirely through dilution under the existing irrigation technology. A 76.8-percent confidence level made it optimal to construct a small treatment plant for the combined surface and subsurface flows, but implied no change in irrigation technologies. Shortened runs combined with small storage and treatment facilities became the optimal way to meet a 10-p/b standard with an 81.7-percent confidence level or to meet a 5-p/b standard under any of the safety margins considered here. Drip irrigation was optimal for meeting a standard of 2 p/b under any of these safety margins. In each of these cases the adoption of the water conserving irrigation technology reduces drainage flows sufficiently to afford substantial savings in storage and treatment costs.

The adjustment made for uncertainty had a substantial effect on the total cost of achieving most of these selenium standards. The average uncertainty premium per 1 percent increase in the confidence level ranged from zero to 1.13

percent for a standard of 10 p/b, from 0.74 to 5.83 percent for a standard of 5 p/b and from 1.32 to 3.45 percent for a standard of 2 p/b. In almost every case it increased as the selenium standard became more stringent and as the confidence level increased.

Growers were assumed to have two sorts of behavioral responses to the imposition of selenium standards: Long-run land retirement and short-run financial distress. It was assumed that land would be retired permanently whenever the cost of meeting selenium standards, spread equally among all acreage remaining in production, exceeded current quasi-rents. Short-run financial distress was assumed to occur when the per-acre costs of meeting selenium standards exceeded the debt carrying capacity of the land, which was estimated by combining estimates of the distribution of debt/asset ratios of California farmers with the estimated distribution of land values derived from the estimates of quasi-rents.

The long-run effects of any of these standards were quite small. Meeting a standard of 2 p/b under any confidence level would force retirement of only about 3.5 percent of the crop land in the area, all of which was of low productivity. With any other standard, production would remain profitable on all land currently cropped. The short-run financial effects of imposing selenium standards are quite substantial. Meeting a standard of 10 p/b would induce financial distress on 1 to 2 percent of the crop land in the area, meeting a standard of 5 p/b would induce financial distress on 5 to 7 percent, and meeting a standard of 2 p/b would cause financial distress on 17 to 28 percent. In other words, short-run financial difficulties outweighed long-run productivity effects and were likely to constitute the main incentive for political opposition to the proposed standards. This suggests that directed credit programs may often be of critical importance in making environmental quality enhancement programs both equitable and politically feasible.

## Ground-water Contamination by a Pesticide

The second case study involved residues of the nematicide 1,2-dibromo-3-chloropropane (DBCP) found in drinking water wells in Fresno County, California (Lichtenberg, Zilberman, and Bogen, 1989). DBCP had been used as a soil fumigant for orchard crops, but was banned for all agricultural uses by EPA in 1979 after having been implicated in adverse reproductive effects in chemical plant operators and oncogenesis in mice and rats. Because DBCP was no longer in use, the study focused on tradeoffs between excess gastric cancer risk and the cost of developing clean drinking water supplies.

Monte Carlo simulation was used to construct probabilistic quantitative risk assessment of the excess cancer risk faced by an individual drawn at random from the population of the county. The simulation included a multiplicative combination of the concentration of DBCP in drinking water, error in measuring that concentration, lifetime consumption of water, an interspecies dose equivalence factor, and a carcinogenic potency parameter. The distribution of DBCP concentrations in well-based water systems and the error in measuring DBCP concentrations were constructed from California State Department of Health Services data. The data presented by the International Commission of Radiological Protection were used to estimate a distribution of lifetime water consumption. The distribution of the dose-equivalence factor was estimated under the assumption that the two main hypotheses (calibrating dose on the basis of surface area versus body weight) were equally likely to be correct. The distribution of the carcinogenic potency parameter was estimated using maximum likelihood estimation of a multistage dose-response model using data from a feeding study of mice.

An element of heterogeneity was introduced by the fact that costs of developing new water supplies differed between rural and urban areas. Drilling new wells was less costly for large systems, while installing filtration devices was cheaper for individual wells. Residential areas within the county thus differed in two ways: Average DBCP concentrations in drinking water and cost of remediation. Least-cost strategies for meeting a risk standard for an individual drawn at random from the county population were derived for the entire feasible range of standards using an algorithm derived from the methodology described above. For ease of analysis, the relationship between risk standards and remediation costs were smoothed using a second-order polynomial regression of cost on the natural logarithms of the risk standard and confidence level.

Increasing the confidence level entailed substantial increases in cost. A 1-percentage point increase in the confidence level raised the total cost of meeting any given risk standard by $3-4 million, or 2 to 10 percent. Making allowance for uncertainty in this way thus had notable effects on risk-benefit tradeoffs.

Urban and rural areas differed significantly in terms of the costs of remediation. The cost of assuring clean water from individual wells in rural areas was about 2.5 times as great as the cost for community water systems in urban areas. Because of these differences, the cost-efficient strategy involved more stringent standards in urban areas and more lax ones in rural areas. In other words, heterogeneity in the population at risk implied the desirability of heterogeneity in regulation.

The marginal cost of reducing risk on average was 21 to 26 percent higher than the marginal cost with a 95-percent confidence level, and 23 to 29 percent higher than the marginal cost with a 99-percent confidence level. Making allowance for uncertainty thus reduced the marginal cost, or slope of the tradeoff curve, substantially. Economists evaluating existing health and safety regulations using cost-benefit analysis applied to estimates of average risk have typically found that marginal costs exceed marginal benefits by significant amounts. This suggests that these policies are excessively stringent. When allowance is made for uncertainty, however, marginal costs and benefits will be closer. The results obtained here indicate that the adjustment can be significant, suggesting that allowances for uncertainty account for a significant share of the observed discrepancies.

## Shellfish Contamination by Livestock Wastes

The third case study involved a shellfishery located in an estuary affected by dairy runoff (Lichtenberg and Zilberman, 1988b). During rainstorms, wastes from dairies were washed into the estuary, resulting in microbial contamination of the oysters growing there and a concomitant risk of severe gastroenteritis for anyone consuming them. The analysis centered on source reduction because open access to the fishery ruled out fishery closure as an effective means of risk reduction.

Rainfall was assumed to be the only random element affecting the risk of acute gastroenteritis, which was modeled as a multiplicative combination of parameters describing microbial contamination in runoff per cow, microbial uptake in oyster population, the probability of contracting acute gastroenteritis upon consumption of contaminated oysters and the number of cows contributing to runoff. Microbial contamination in runoff per cow was estimated from maximum fecal coliform counts observed around oyster beds in the estuary. The fraction of oysters contaminated was estimated by applying regression analysis to data in a study examining the usefulness of fecal coliform counts as an indicator of bacterial contamination of oysters. The probability of contracting acute gastroenteritis after consuming contaminated oysters was derived from epidemiological studies. The number of cows contributing to runoff in any size rainfall event equaled the number of cows at dairies with runoff control facilities of insufficient capacity for the amount of rainfall. The probability distribution of rainfall events was derived from data on local rainfall.

The dairies in the watershed differed in terms of topography and therefore in terms of the cost of constructing runoff control facilities adequate for any given size rainfall event. Data on these costs for each dairy in the region were obtained from a detailed engineering study. Least-cost patterns of runoff control facility construction and tradeoffs between gastroenteritis risk and source reduction expenditures were estimated using an algorithm derived from the methodology described above.

The optimal policy involved building holding ponds only at dairies with the lowest marginal costs. The optimal capacity at each dairy was determined by the confidence level required, and the total number of dairies subject to undertaking source reduction measures was determined by the risk standard. Because topography, and therefore cost, differed markedly at different sites, different dairies received markedly different treatment under this policy. Runoff control facilities were required at only a few sites to meet lax risk standards. As the risk standard became more stringent, the number of sites investing in source reduction grew. The optimal set of standards thus implied marked inequities among dairies, with some dairies required to undertake substantial investments in source reduction while others continued with unregulated emissions.

Economists have long argued that pollution control can be achieved by levying taxes rather than of imposing standards. In the case at hand, the per-cow tax required to meet any desired risk standard with a given confidence level equaled the marginal cost of installing runoff control facilities of the requisite capacity at the most expensive site needed. Holding pond construction patterns remained the same, but dairies not required to invest in source reduction had to pay taxes on runoff generated. The result was a much more equitable set of losses. When the risk target was lax, very few dairies found it less costly to build runoff control facilities than pay the tax, so tax payments accounted for almost all runoff control expenditures. As the risk target became more stringent and the optimal tax increased, more and more dairies found it less costly to build.

## CONCLUSIONS

Decision methodologies for addressing regional environmental policy issues should incorporate several key features characterizing these issues. The first is uncertainty, which is prevalent in ecological problems because of their complexity, because of limits on fundamental scientific knowledge and because data collection is often necessarily limited by time constraints. Existing legislation and regulatory practices have mandated that uncertainty be addressed in formulating policy; specifically, they typically require that decisionmakers provide an adequate margin for error. Second, political sense as well as most notions of fairness dictate that policymakers care about the distribution of the costs and benefits of alternative policies as much as efficiency effects, i.e., net benefits. Heterogeneity is often important in determining both the actual effects of proposed policies and the distribution of these effects across groups of economic agents and should thus also be taken into account. Finally, decision models must recognize that economic agents typically react to new policy environments, so that producer and consumer behavioral responses must be incorporated into policy models.

This chapter has argued that the safety rule approach proposed by Lichtenberg and Zilberman (1988a) allows policy analysts to make adjustments for uncertainty in a way that corresponds to existing legislative and regulatory frameworks. Three recent case studies employing this approach were discussed to examine the effects of adjusting for uncertainty and to demonstrate how heterogeneity, behavioral responses, and distributional concerns can be addressed at the same time. The case studies show that adjustment for uncertainty is feasible and that it can have significant effects on several aspects of policy, including: (1) The total cost of meeting a given environmental quality goal, (2) optimal environmental quality goals implied by any given level of marginal benefits, (3) the cost efficient choice of policy instruments, and (4) the distribution of costs among producers in the short and long run. Short-run distributional effects were shown to be substantially greater than long-run efficiency effects in some cases; in others, the distributional effects of different policy approaches differed markedly. In both sorts of cases, the analysis was able to pinpoint factors likely to determine political responses among groups of growers to proposed environmental quality goals. The results underscore the notion that failure to address these key features will result in policy analyses that fail to meet the real needs of policymakers.

## REFERENCES

Baumol, W. J. and Oates, W. E., 1971. The Use of Standards and Prices for the Protection of the Environment, *Swedish Journal of Economics*, 73, pp. 42-54.

Hanemann, M.; Lichtenberg, E.; Zilberman, D.; Chapman, D.; Dixon, L.; Ellis, G.; and Hukkinen, J., 1987. *Economic Implications of Regulating Agricultural Drainage to the San Joaquin River*. Report to the California State Water Resources Control Board, Western Consortium for Public Health, Berkeley, CA.

Keeney, R. L. and Raiffa, H., 1976. *Decisions with Multiple Objectives: Preferences in Value Trade-offs*. John Wiley and Sons, New York, NY.

Lichtenberg, E. Conservation in Risk Assessment for Food and Safety Policy. In: *Caswel, S.A. (Ed.), The Economics of Food Safety*. Iowa State University Press, Ames.

Lichtenberg, E. and Zilberman, D., 1988a. Efficient Regulation of Environmental Health Risks, *Quarterly Journal of Economics*, 49, pp. 167-178.

Lichtenberg, E. and Zilberman, D., 1988b. Regulation of Marine Contamination Under Environmental Uncertainty: Shellfish Contamination in California, *Marine Resource Economics*, 4, pp. 211-225.

Lichtenberg, E.; Zilberman, D.; and Bogen, K. T., 1989. Regulating Environmental Health Risks Under Uncertainty: Ground-water Contamination in California, *Journal of Environmental Economics and Management*, 17, pp. 22-34.

Machina, M., 1987. Choice Under Uncertainty: Problems Solved and Unsolved, *Journal of Economic Perspectives*, 1, pp. 121-154.

Rausser, G. C.; Lichtenberg, E.; and Lattimore, R., 1983. New Developments in Theory and Empirical Applications of Endogenous Governmental Behavior. In: Rausser, G. C. (Ed.), *New Directions in Econometric Modeling and Forecasting in U.S. Agriculture*. Elsevier North-Holland, New York, NY.

Slovic, P.; Fischoff, B.; and Lichtenstein, S., 1980. Facts and Fears: Understanding Perceived Risk. In: Schwing, R. C. and Albers, W. A., Jr. (Eds.), *Societal Risk Assessment: How Safe is Safe Enough?* Plenum Press, New York, NY.

Thomson, K. J. and Hazell, P. B. R., 1972. Reliability Using the Mean Absolute Deviation to Derive Efficient E, V Farm Plans, *American Journal of Agricultural Economics*, 54, pp. 503-506.

# 36 ECONOMIC ASPECTS OF ENFORCING AGRICULTURAL WATER POLICY

Andrew G. Keeler, University of California, Berkeley

## ABSTRACT

This chapter presents an overview of the important issues of enforcing public policy toward water allocation and the pollution levels in drainage water. This overview is based on the emerging literature on the enforcement of environmental policies. The focus is on the decisions made by firms whether or not to comply with the law and how regulatory actions affect these decisions. The chapter discusses the importance of defining and perceiving penalties very broadly in order to take account of the importance of cooperative values among water users and identification with the goals of water quality regulation.

## INTRODUCTION

This chapter deals with a complication commonly ignored in most of the literature on water use--the imperfect or incomplete enforcement of the regulations and mechanisms called for in law. It is clear from experience that all actors will not automatically abide by a water allocation rule, pay the prescribed cost on the entire quantity they use, or limit the residuals content of their drainage water to the maximum concentration allowed by law simply because that is official policy. Regulatory agencies currently spend significant resources monitoring and enforcing the array of policies which cover the use of water in California agriculture, and there remains a wide discrepancy between policy and reality.

These problems are likely to become even more severe as both the quantity and quality of water in California become more of a constraint to agricultural production. When water moves in an open canal, monitoring is extremely difficult, and it is hard to enforce the rights of downstream farmers. The incentives to ignore regulation and to actively subvert water metering schemes, and the necessity to monitor and to pursue effective enforcement actions, will grow as the costs of complying with regulation become greater. This will be a particularly important factor for regulatory strategies which include the ability of agricultural users to buy and sell the rights to water use.

There are three good reasons to consider incomplete enforcement of water allocation and drainage policies. One is to improve policy formation: The difficulty of enforcement can affect the optimal stringency of regulation and the choice of policy instruments to best meet regulatory goals. Understanding enforcement problems will also be necessary to design and implement fair and efficient monitoring and penalty mechanisms. A second reason is that it is necessary to understand the actual incentive structure facing individual farmers to accurately describe or predict their behavior in the face of changing market conditions or regulatory policy. A third reason is that incomplete enforcement affects farmers' views about the fairness and reasonableness of regulatory policy. Making difficult adjustments to increasingly stringent water quality standards or smaller allocations will be a far less popular and successful policy if those who evade its consequences are seen to benefit greatly. On the other hand, failure on the part of regulators to compromise in the face of exceptional difficulty in meeting regulatory standards will also be perceived as unfair.

This chapter presents an overview of the important issues of enforcing public policy toward water allocation and the pollution levels in drainage water. This overview is based on the emerging literature on the enforcement of environmental policies. Until recently, the economic literature on the identification of optimal solutions to externality problems or the selection of efficient instruments did not include problems of implementation. It was implicitly assumed that whatever policy was selected would be assiduously followed by regulated firms. An economic literature based on the analytical framework of the economics of crime (Becker, 1968) has begun to remedy this oversight. The focus is on the decisions made by firms whether or not to comply with the law and how regulatory actions affect these decisions. This chapter discusses the importance of defining and perceiving penalties very broadly in order to take account of the importance of cooperative values among water users and identification with the goals of water quality regulation. It is concluded that monitoring technologies have an important part in the design of water allocation systems. In general, enforcement parameters should be included in the design of drainage water quality regulations.

## REGULATORY OBJECTIVES AND ACTIONS

Water allocation policy in California is the result of a complex process involving Federal, State, and local government entities and a variety of interest groups. There are many institutions which set and further define policy before it is shaped into a form which directly affects the choices of farmers making drainage and water use decisions. The very fact that enforcement is incomplete is evidence that there are some financial or authority constraints on regulators; these are at least implicit choices made through the political process about how to organize and carry out regulation.

The enforcement priorities of the State Water Project toward individual farmers will not be the same as those of the local water districts. The regulator whose objectives are discussed in this chapter is the one who is responsible for monitoring, correcting, and deterring the behavior of individual water users. It should be kept in mind, however, that other parts of the government play an important role not only in setting the policy which must be enforced, but in determining key parameters of the enforcement process. These include the enforcer's budget (the legislature), the ability to obtain sanctions and withstand appeals (the courts), and the degree of independence from political interference (the executive and legislative branches).

## Regulatory Goals

The most obvious and in many ways the primary goal of the regulator concerns the behavior which the regulation is designed to govern. In the case of water quality, the goal is to minimize the damage done by polluted water, or at least to minimize some index of the pollution level. The regulator may formulate goals in terms of a complex function which gives pollution damage as a function of the distribution of pesticide and fertilizer contamination over time and space, or may be a very simple one where the regulator simply tries to minimize the deviation between the amount of water a farmer uses, and the amount paid for. The choice of a simple measure like the operation of a piece of control equipment may obscure the regulator's true goal (improved water quality) and discourage innovative actions on the part of farmers which might improve this quality. The regulator might also insist on the use of very expensive means of control even when circumstances make the marginal contribution to the regulator's goal very small (Bardach and Kagan, 1984).

The enforcement authority must also concern itself to some extent with the costs imposed on agriculture by compliance with its regulations. Meeting a more stringent standard of water quality or paying a higher price for water will reduce the profits of farmers. Although the costs imposed by regulation are a basic concern of public policy, it can be argued that the relevant institutions have already made a determination of how to balance these costs against regulatory benefits in the process of setting standards and procedures. However, regulatory policy is inexactly worded and subject to varying interpretations, and is often not really defined until enforcement takes place. Also, the ability to show flexibility in the face of special circumstances is generally regarded as fair and good policy; these special circumstances will most frequently be the very high cost of complying with regulations in the short term. As expense of meeting regulations rises, the severity of the penalties necessary to ensure compliance will also rise. Enforcement strategy will depend on how onerous compliance is to the farmer.

The regulator's choice of how to balance these two objectives--minimizing the damage from the regulated behavior and minimizing the costs imposed on farmers--is a key determinant of enforcement policy. At the most general, regulators can be thought of as minimizing the sum of the social damage of pollution and the abatement costs of firms. This corresponds to the use of economic methodology to determine an "optimal" level and allocation of externality generation by quantifying and balancing the costs and benefits of policy alternative (Veljanovski, 1984; Jones and Scotchmer, 1988; and Lee, 1984). Regulators consider imperfect enforcement for two reasons: it is costly in and of itself and it affects the balance between damages and benefits by restricting the regulator's ability to mandate any particular outcome. The polar extreme is a regulator who wishes to minimize pollution, or minimize it relative to some externally determined target, and whose inability to perfectly enforce regulations becomes part of the problem of determining how best to achieve this goal. This regulator's objective concerns the effect of its actions on pollution levels and ignores its effects on the costs faced by economic agents in meeting its regulations.

Either of these assumptions is justified by the existing institutional structure of environmental regulation. It is certainly true that tradeoffs between abatement costs and social damage take place not just in the legislature, but also in the executive branches of the Federal and State governments responsible for environmental regulation. These agencies have some responsibility for setting standards, but only within the guidelines of legislative and higher executive authority, and only within the bounds of oversight by the courts and political pressure and litigation from environmental groups and from industry. It is also true that by the time an actual code of behavior has been determined for individual farmers or water districts, the body in charge of enforcing that code is not necessarily concerned with this tradeoff nearly as much as achieving compliance or an acceptable compromise.

It is useful to think of the regulator responsible for enforcement as putting some weight on the firms compliance costs as a fraction of the weight it puts on the social damage of pollution. In models that are concerned with positive explanations of the regulator's actions, the weight is equal to that given to the externality damage caused by the pollution (Lee, 1984; Veljanovski, 1984; and Jones, 1989). Research which focuses on the mechanisms of monitoring and enforcement and the relationship between stringency and compliance has taken the regulator as being concerned, either explicitly or implicitly, only with the social costs of pollution; the compliance costs are implicitly given zero weight. The abatement costs firms incur in order to achieve compliance have mattered only insofar as they contribute to the regulator's understanding of firm behavior in response to penalty structures (Harford, 1978 and 1987; Viscusi and Zeckhauser, 1979; Jones, 1989; Russell et al., 1986; Harrington, 1989; and Scholz, 1984). Realistically, it can be expected that compliance costs, and therefore a broad definition of social efficiency, will be given some weight between zero and one relative to that given social damages; the higher the weight the more the regulator will deliberately refrain from enforcing the law. The weight which applies in any particular situation will reflect a variety of political and economic considerations.

## The Importance of Fairness as a Regulatory Goal

It is also necessary to stress the fairness with which policy is regarded by the regulated population. If the regulatory focus is on a systemwide view of efficiency, measured strictly in aggregate costs and benefits of water use and drainage water quality, then enforcement policy will center on the sources where compliance can be achieved most cost effectively. This may leave advantageously placed farmers in positions to profit greatly by continuing to ignore the law. For example, if water use is carefully metered in one area where infrastructure exists but remains basically unregulated in adjoining districts where metering is unfeasible, the result will tend to be that farmers in the second district use more water and pay less for it. This will make the farmers of the first district much more likely to bypass metering devices or exert as much political pressure as possible to change the system; if the farmers thought that everyone was in the same boat then this activity would be much less likely. Fairness, or at least the widespread perception of fairness, is in and of itself an important goal of regulatory policy.

A related reason why enforcement is important is its role in successful cooperative management of water allocation and drainage resources. While it may seem paradoxical that the authority to punish is helpful in fostering cooperation, this is undoubtedly the case for at least two reasons. The first is that some fear of retribution is necessary to keep individuals from gaining advantage by breaking cooperative agreements (Wade, 1987). Otherwise, a few deviant individuals can destroy a functioning cooperative system by making the other users feel like "suckers." The other reason is that an external authority which can effectively hold a water district responsible for the quality of its drainage water can create a tremendous impetus to cooperation and mutually agreed upon limits among a group of farmers who previously faced no responsibility for their own or their neighbors' actions.

## Monitoring Expenditures

Gaining information about the activities of the regulated population is difficult and costly. The choice of monitoring expenditure is both a part of the regulator's objectives (since minimizing enforcement costs is desirable) and more importantly one of its choice variables. The more the agency spends on monitoring, the greater the likelihood of any firm being monitored in any given interval. These costs are considerable, and are the primary reason why monitoring probabilities are relatively low. Even for large industrial sources of water pollution, the probability of being monitored once in a given year seems to be considerably less than one (Russell et al.,1990). As a first approximation, these probabilities may be equal for everyone in the regulated population (Harford, 1978 and 1987; Viscusi and Zeckhauser, 1979; and Jones, 1989). In reality, this is unlikely to be the case; the regulator will at least make use of the results of previous monitoring efforts to allocate its scarce monitoring resources (Russell et al., 1990 and Scholz, 1984). The use of such information has been formalized in a multiperiod model to show how limited fines and monitoring resources can be leveraged into more significant penalties perceived by firms (Harrington, 1988). These results emphasize that compliance is a process and enforcement is a continuing relationship between the water or environmental authority and the farmer.

Both parts of the enforcement relationship have some element of randomness. Farmers have imperfect control over both their water intake and the quality of their discharges; pipes leak, soil moisture retention depends on meteorological conditions, and so on. Monitoring is also imperfect; gauges malfunction and chemical testing equipment is only so precise. In addition, the regulator will make errors in extrapolation from a discrete-time sample to a level of continuous compliance. These factors can lead to the existence of both false negative and false positive results as well as true readings (Russell et al., 1990). The regulator must take into account the possibility of these mistakes in setting policy over fines and monitoring probabilities. This is particularly important when fairness is a criterion by which regulatory policy is judged.

## Enforcement Expenditures

Even when the regulator knows that a farmer is exceeding his legal water allocation or returning overly polluted water through drainage, it is both expensive and problematic to change his behavior by leveling penalties or taking legal action. Enforcement activities, as distinct from monitoring, are the actions undertaken by the regulator to levy penalties and to otherwise change the farmers' behavior. This class of expenditure has generally been ignored or lumped together with monitoring in the economic literature. It includes resources spent on writing enforcement orders; negotiating with firms; preparing and carrying out administrative hearings, civil litigation, and appeals; and preparing and disseminating information about the polluting firms' activities. Like monitoring, it enters into the regulators objectives because it is costly, but its primary importance is as a key choice variable which influences firm behavior by helping determine the firm's expected penalty for noncompliance. In the economic literature, penalties per unit of violation are modeled as a given: If they are not large enough to ensure compliance (given the probability of monitoring), then the problem for the regulator is how to influence the probability of monitoring. More sophisticated work views the overall penalty faced by the firm as an increasing function of the regulator's expenditure (Scholz, 1984). Even more than with monitoring, the actions a regulator takes to levy penalties on farmers will be judged not only in terms of effectiveness, but also of fairness. Penalties which are too severe will be regarded as arbitrary and capricious and will be much less likely to stand up if litigation ensues. Penalties which are so minor as to be perceived as a "slap on the wrist" will undermine the confidence of all farmers in the regulator's fairness and resolve, and will have impacts on compliance throughout the system.

There is a wide range of behaviors and compliance-producing options available to regulators (Scholz, 1984; DiMento, 1986; and Hawkins, 1984). Pursuing any of these options is costly in terms of personnel time. If the regulator follows a procedure of fining the firm or requiring costly changes in the firm's operations, then there is a good possibility that litigation will ensue. This is costly for both sides, and may require the regulator to convince other agencies to support its position. Given a monitoring reading, the penalties that the regulator can impose on the firm are an increasing function of resources devoted to enforcement over a significant range. In characterizations of optimal enforcement, the regulator pursues monitoring and enforcement activities to the point where their marginal expense is equal to the marginal gain they bring about in regulatory objectives. In reality, however, enforcement authorities have fairly rigid upper ceilings on the resources they can use for monitoring and enforcement; their task in this situation is to allocate their limited resources as effectively as possible. This allocation takes place both among the farmers in the regulated population and between monitoring and enforcement.

Because of the ability of farmers to appeal most sanctions, it is not enough for regulators to know that a violation has taken place; they must be able to prove it to the court's satisfaction. Hawkins' study of water pollution enforcement corroborates that when the regulator wants to "make a case," or pursue severe sanctions against a violator, then the evidence requirements of the monitoring process are much greater. Thus there is an element of quality to the regulator's monitoring which is important; higher quality monitoring data and other proof of infractions impose additional expense. The quality of the regulator's case could also depend on the firm's history; sanctions would be more severe if the firm was a habitual violator than for a first-time offense, holding other variables constant. It could reflect factors like recent legal precedents or the past opinions of a particular judge or circuit court.

## THE FARMER'S OBJECTIVES

Most economic research views the decision to comply with a law or regulation as the solution to the problem of minimizing the sum of compliance costs and expected penalties. In addition, the farmer can expend resources on concealing his violations from the regulator's monitoring effort and on reducing his realized penalty through litigation, negotiation, or publicity. The farmer is implicitly seen as an amoral individual with no concern about the law or the consequences of his actions except those which affect his financial status. His legal allocation of water and the pollution control regulations he faces affect him only insofar as their violation will be detected and penalized.

This extreme behavioral assumption of opportunism is an oversimplification of the motives of California water users. Research on reaction to environmental regulation by sociologists and political scientists has found this pure focus on profits to be too narrow a view of reality; the majority of individuals do not act this way in all circumstances. Chester Bowles is frequently quoted from his experience in administering regulatory price controls during World War II as saying that 20 percent of the population would automatically comply with any regulation, 5 percent would attempt to evade it, and the other 75 percent would go along with it as long as the 5 percent were caught and punished (Kagan and Scholz, 1984). We could think, following Kagan and Scholz, of a threefold categorization of opportunistic behavior. The first is the amoral calculator--this is the pure profit-maximizing firm modeled in the economic literature. Noncompliance here results from pure economic calculation. The second is the firm as corporate citizen. The regulated population complies as a matter of course with regulations it perceives as fair and reasonable, regardless of the cost. If it feels the rules are arbitrary or unequally applied, then it may not observe them out of principled disagreement or some feeling that the regulator has not lived up to its part of the implicit contract between business and industry. The third view is of the firm as organizationally incompetent--environmental violations are a result of organizational failure. Regulations, especially new ones, are difficult to interpret, and changes in processes or technologies tend to cause mistakes. In addition, farmers may simply be careless. Farmers' behavior is motivated by one of these three characterizations, although the same entity may act in different ways in different circumstances.

There are three approaches to incorporate these views of business attitudes toward environmental compliance in a framework where the regulated population optimizes some objective function. One continues to view the farmer as a pure profitmaking entity, and expands the notion of penalties to cover a much wider range of things than fines. For example, firms may avoid violations not just because of fines but because adverse publicity hurts sales. The maintained assumption is that farmers maximize profits; if they choose to comply when the penalties they face seem less costly than abating their polluting activities, then it must be assumed that there are effects on expected firm profits that have not been included. This view incorporates a kind of revealed profit function argument--if farmers did not violate legal entitlements then it must be because the perceived penalties are greater than the perceived rewards. If farmers obey pollution control laws, then it must be because they feel that it will maximize their long-run profits and for no other reason.

Another alternative is to give farmers a more complex objective function which is defined over profits, conformity to institutional and employee values and prestige, and conformity to existing laws and regulations. The firm balances the effect of compliance or violation on these objectives according to its own preferences and decides how best to meet them. This is analogous to giving firms a utility function, only one of the arguments of which is profits. The obvious disadvantage of this approach is that specifying such a function is next to empirically impossible and not very useful for prediction or even explanation.

A third approach is to allow nonmonetary goals to be incorporated into a profit-maximizing objective in terms of their monetary equivalent. A farmer may be willing to make less profits if he avoids releasing toxic runoff into nursery school drinking supplies, even if he has no fear that it will ever adversely affect profits. He reveals through his decisions how much these nonmonetary factors mean to him in monetary terms through the loss in profits they cause. This approach is logically the equivalent of the corporate-utility function approach, but incorporating it into a profit-maximization framework allows observation (at least theoretically) of penalty functions for individual farmers.

This discussion indicates that firms are influenced by more than the amount of fines collected by the regulator, and possibly by more diverse considerations than simple profit-maximization calculations. In individual water districts the farmer's attitude toward the justice of the allocation rule or drainage quality standard and his perception of how well his neighbors and his competitors are complying will be an extremely important determinant of his own compliance. It should be expected that the decision to accept the decisions of the regulator, and to what extent to follow its rules, will be made not individually but among the whole population of users. This group decision or consensus will be a key determinant of how well the regulator can achieve its objectives. For example, if all the farmers in a district join forces in a suit to protest a rule governing the selenium content of their drainage water, it will be extremely difficult for the regulator to achieve short-term compliance. If only a few farmers elect a legal challenge, their costs will be much higher and their chances of success much smaller.

These considerations also imply that the penalties for violating regulations will vary among farmers for the same level of violation because of the institutional character of agricultural enterprises. In understanding how penalties differ among firms, a useful distinction can be made among three types of penalties. One is the elements which are common to all firms in a regulated population--the fines and regulatory hassles which any farmer will face for a given type and level of violation. Another is the financial penalty that differ across farmers in the regulated populations for the same type of violation. For example, if ability to pay affects the fine, smaller and less profitable firms will face less of a violation for the same behavior, holding everything else constant. Similarly, a farm that sells a recognizable consumer product for which ready substitutes exist has much more to fear from publicity about its violation than one which supplies a homogenous commodity. A third type of penalty is that which is regarded subjectively--the extent to which acting in an environmentally damaging manner inherently reduces the farmer's (or his employees') welfare in monetary terms, or how he values the social stigma resulting from such violations.

This distinction is useful because the balance between expenditure on enforcement and monitoring may depend on which category of penalty is more significant for different firms. It may also affect how regulatory expenditure affects the composition of the penalty in terms of financial transfers to the regulator and nontransfer penalties like sales losses or production down-time. It is plausible to think that regulators have more influence over penalties which have direct consequences on profitability than on those which adversely affect firm prestige and employee satisfaction.

## The Penalty Function in Economic Models

In examining incomplete enforcement from an economic point of view, the key determinant of farmers' behavior over which the regulator has any control is ordinarily modeled through a penalty function. The expected penalty is the product of the probability of detection (the result of monitoring) and the penalty experienced by the firm if found to be in violation (which depends on enforcement actions). The fact that enforcement is imperfect is manifested in economic models as a combination of low monitoring probabilities and insufficiently high penalties. This limitation on penalties is a necessary assumption because of well-established results (Becker, 1968 and Stigler, 1970) that arbitrarily large fines can compensate for arbitrarily low detection probabilities in deterrence. The regulator has been modeled as fining the firm as much as possible in all circumstances up to the externally determined limit of its authority. The fact that penalties are not severe enough to bring about compliance is in large part due to the lack of political and social consensus about the desirability of rigid enforcement. Even when the regulator's goal is full compliance, its ability to achieve this is limited by the attitudes expressed through legal and political institutions. Hawkins found that a widespread feeling that environmental enforcement of water quality regulations should not cause job losses limited the ability of enforcement personnel to impose sanctions. Melnick (1983) found the same consideration strongly influenced court decisions on industry appeals of EPA injunctions. Silbey (1984) stresses that regulatory agencies react to the perceived attitude of their public constituencies. In situations where behavior is clearly harmful, deliberate, and outside the support of political and social forces favoring economic production over environmental quality, the penalties are quite severe and the probability of detection is of paramount importance. This is true, for example, for non-manifest disposal ("moonlight dumping") of hazardous wastes (Russell, 1990). In regulating point sources of water pollution, however, the environmental goals embodied in regulations do not command the unequivocal support of the broad range of institutions which, in addition to the regulator modeled here, have some say over the penalties faced for noncompliance. These institutions include the courts, executive agencies like the Justice Department and State Attorney General, the Office of Management and Budget, and legislators. Because both agricultural and environmental advocacy groups influence penalties through their own litigation and lobbying activities with parts of the executive and the legislature, they too have a say in determining how the actions of the regulator are translated into the individual violator's perceived penalties. If there is a genuine shift in the underlying attitudes of the public on the relative merits of environmental protection and the economic costs of achieving this protection, and if this is not reflected in changes in the law, then it will very likely be reflected through changes in the penalties for violations experienced by farmers. This could happen through changes in the legal environment, increases in the cost of bad publicity, or the internal pressures to comply with environmental regulations.

## The Consequences of Violating Environmental Regulations

Before discussing the specification of penalty functions in more detail, it is useful to elaborate what penalties are actually used in Federal environmental regulation. The most easily understandable penalty levied is a fine, or a financial transfer, from the firm to the Government. This fine can occur as part of a criminal, civil, or administrative sanction. Under a criminal sanction, the fine is likely to be only a part of the penalty. Under civil or administrative sanction, it may be the entire penalty. In any of these cases, it will take considerable resources for the regulator to levy and collect the fine.

The most frequent first step taken by environmental regulators is to issue notices of violation. This notice is frequently enough to induce a return to compliance. Whether or not this is the case, negotiation over the actions the firm will take and the penalty that the regulator will levy are quite often negotiated between the firm and the agency. Notice and negotiation do not ordinarily carry significant penalties in and of themselves, but they are a step in determining the actual penalty experienced by the firm and they also use a part of the resources available to the regulator for enforcement.

The penalty which most closely fits the "pay a fine as a result of violation" kind of penalty modeled in most of the literature is the administrative sanction. This is a penalty that the regulator has statutory authority to impose without going through the courts. EPA policies leave scope for including penalty components based on the "potential for harm," which must bear a close relationship to social damage, this indicates that, at least as a matter of policy, some balancing of social damages goes into the penalty calculation and overall enforcement strategy. Under Federal Water Pollution Control Act (FWPCA) violations, the assessed penalty is designed to cover the cost of cleaning up or otherwise remedying the damage done by the firm's activities; this penalty can also be expected to bear some strong correspondence to the social damage. Under the FWPCA, a part of the penalty is also based on the firm's financial benefit from the polluting activity. In theory, at least, this financial benefit should be equal to the abatement cost of the violation. In addition to direct financial transfers to the Government, administrative sanctions may include the withholding of Federal funds for (relatively unrelated) environmental projects to which the firm would otherwise be entitled. Under some circumstances the regulator may also withhold operating licenses the firm needs or order some specific cleanup activity which may not be the least-cost way for the firm to reach an acceptable level of compliance. Less formally, these and other sanctions can be accompanied by hassling and nitpicking which imposes high costs on the firm. Administrative sanctions can still be challenged in court by firms, leading to the same weaknesses as civil penalties. Since administrative procedures must be levied in accordance with the procedures and safeguards in the Administrative Procedures Act, they can be time consuming and costly to levy.

A related class of penalties is the civil sanction. These can be statutory or based in the common law doctrines of liability. In this case the regulator brings a civil action against the violator, requesting some specific set of penalties and the court sets the final penalty as the result of civil procedure. In addition to fines, civil procedures can result in a judicial order of injunctive relief, by which the court instructs the violator to cease and desist any processes at once which violate environmental standards. Following these injunctions can be extremely costly for firms, and may be more expensive than having complied fully in the first place because of the costs imposed by extremely rapid change; this could even require a shutdown of some productive activities. The results are subject to appeal. On the one hand, as Melnick has documented, civil sanctions which threaten severe economic hardship (particularly involving lost jobs) tend to rarely be upheld both in the civil procedure and on appeal. On the other hand, citizen groups can bring civil procedures without the concurrence of the regulator and this can interfere with the outcome of negotiations and long-term planning between the regulator and the polluting firm (DiMento, 1986). Like administrative sanctions, the process of pursuing civil sanctions is costly and time consuming for the regulator.

The regulator has the option in many cases of pursuing criminal sanctions against the firm or its executives. The deterrent effect of criminal prosecution is particularly effective for those who have not had exposure to the criminal justice system (Fisse and Braithwaite, 1983). Although these sanctions can be effective, they are thought by regulators to be most appropriate for deliberate, willful violation of codes covering actions with severe consequences. If used routinely, they lose their moral sanction; further, the probability of conviction for anything except serious, willful offenses is not great.

## INSIGHTS FROM ECONOMIC MODELS OF ENFORCEMENT

One important insight of the economic literature is the relationship between the severity of the pollution standard and the actual amount of pollution. This is important in situations where the regulator has the power to set standards and wishes to take account of the reality of incomplete enforcement in order to reach a normative objective. Viscusi and Zeckhauser (1979) show that when the fine for noncompliance was independent of the size of the violation, tightening the standard will decrease pollution (or more generally, increase the quality objective sought by the regulator) up to the point where the expected loss of compliance surpassed the fixed fine; at this point that firm would switch discontinuously to the unregulated quality level and pay the fine.

Overall pollution and the choice of an optimal standard depended on the distribution of compliance costs in the regulated population. The interesting policy implication of this research was that at some point more stringent standards would actually produce more pollution. This same consideration applies to enforcing a water allocation decision; the more stringent the allocation or higher the sale price, the more incentives the farmer faces to cheat. Harford showed that when penalties were not independent of the size of the violation, the shape of the marginal penalty curve became critical in the relationship between stringency and effluent levels. Jones linked these results to Viscusi and Zeckhauser's findings and showed that when marginal penalties were increasing, stricter standards decreased pollution but increased violation size for each firm and for the population of firms. Under declining marginal penalties stricter standards increased overall pollution and the size of firms' violations; loosening standards accomplished the opposite. Jones found that Viscusi and Zeckhauser's recommendation for restraint in setting standards with incomplete enforcement only holds when marginal penalties are declining or the penalty is independent of violation size; she argues convincingly that independence is relatively unlikely to be the case.

Another important insight is the value of efficient instruments or regulatory schemes in inducing compliance. Efficient regulation places lower costs on complying firms than inefficient regulation; in addition to being good in and of itself, these makes farmers more likely to comply at any given level of monitoring and enforcement. In general, the more fair and rational a regulatory scheme is, the more closely it will be adhered to and the more noncompliance will be regarded as deviant and unacceptable behavior.

## ISSUES IN WATER ALLOCATION

The economic logic which calls for charging farmers for the actual amount of water they use is well understood. In many parts of the world, including California, farmers are frequently assessed charges based instead on how many acres they plant or for some prorated share of the water district's total usage. In other locations charges may be based on some historical pattern or on self-report data. The most important reason that this occurs is that monitoring the amount of water used by farmers requires the installation of expensive metering equipment when pipes are used, and may not be feasible at all when irrigation water comes in ditches. It is similarly difficult to monitor groundwater pumping. The difficulties are not only technical; it will frequently be easy for the farmer to bypass or disconnect a metering device when his financial liability for water use is a direct result of its reading. Under these circumstances the cost of monitoring and enforcing unit pricing may outweigh its efficiency benefits as an allocation mechanism.

Under the above circumstances it will frequently be desirable to reduce monitoring requirements by adopting a pricing mechanism which uses an observable signal as a proxy for actual water use. While this will have costs to the system's efficiency in terms of a departure from exact pricing for actual use, these costs must be balanced against the savings from the reduced monitoring requirements. The most frequently used signals have been acreage and crop choice. Acreage is an easily observable but relatively poor signal; it fails to give farmers the incentive for conservation that is the fundamental concern for worrying about water pricing in the first place. Since water use is correlated with crop choice, this is a somewhat better signal and will influence farmers to grow less water-intensive crops when the marginal cost of water is high. It still does nothing to encourage farmers to use water only to the point where marginal benefits equal the marginal cost and not until the marginal benefit equals zero. A better signal, suggested by Shah and Zilberman (1990) is the use of irrigation technology. Once a farmer installs more efficient irrigation technologies, the optimal amount of water to use even in the absence of unit pricing is reduced; in this sense the signal is not subject to manipulation as false information to the regulator. The lower cost of water with more efficient technologies gives farmers incentives to adopt these technologies and reduces monitoring requirements at the same time. It must be noted, however, that using this kind of combination of crop and technology choice as the basis for water use and drainage regulation creates rules which are much more complex that those now is existence. This complexity carries not only expense but the possibility of mistakes and much greater demands on regulatory personnel.

As water becomes more scarce relative to demand and the marginal cost of use rises correspondingly, the pressures to charge farmers according to their actual water use will increase. At some point the efficiency gains will outweigh monitoring costs and it will be worthwhile for a regulator to install the pipes, gauges, and meters necessary to institute such a pricing scheme. As the price of water rises, enforcement agencies will gain in both monitoring budgets and the authority and legal support to levy significant penalties for cheating. Noncompliance will come to be seen less as a farmer getting around "know-nothing bureaucrats" and more as a criminal and antisocial act. This will be even more true when water markets become a reality. If farmers own property rights and elect to sell them, it will be essential that a system exists to verify the sale and ensure that the sale of water rights corresponds to the actual sale of water. It will also be even more important to guarantee the security and efficiency of conveyance systems.

## DIRECTIONS FOR FUTURE RESEARCH

More information is needed about the goals and the structure of the enforcement institutions in California water allocation. How much freedom do they have to make de facto policy, and how much do they simply administer regulations as written? What is the relationship between the enforcement authority and the executive and judicial institutions on which it depends for credibility in levying sanctions? More also needs to be known about how the stringency of regulation or the effective price of water faced by farmers affects their choice to ignore or actively evade the laws and rules governing water use.

A more immediate and practical need is to develop more information about the cost, accuracy, and reliability of the kinds of monitoring systems for water deliveries that are now available, and to anticipate technological developments which might improve monitoring performance in the near future. Higher effective water prices and institutional arrangements depending in part on water markets will create a demand for better monitoring. The need is to determine what kind of improvement this demand can induce and to include these considerations as part of the design and implementation of new institutional structures for water allocation and drainage management in California.

## ACKNOWLEDGEMENTS

The author would like to thank Robert Wade and David Zilberman for helpful comments on earlier drafts of this chapter.

## REFERENCES

Bardach, E. and Kagan, R. A., 1984. *Going by the Book: The Problem of Regulatory Unreasonableness.* Temple University Publishers, Philadelphia, PA.

Baumol, W. and Oates, W., 1988. *The Theory of Environmental Policy.* Cambridge University Press, New York, NY.

Becker, G., 1968. Crime and Punishment: An Economic Approach, *Journal of Political Economy,* 76, pp. 1669-217.

DiMento, J. F., 1986. *Environmental Law and American Business.* Plenum Press, New York, NY.

Fisse, B. and Braithwaite, J., 1983. *The Impact of Publicity on Corporate Offenders.* SUNY Press, Albany, NY.

Harford, J. D., 1987. Firm Behavior Under Imperfectly Enforceable Pollution-Standards and Taxes, *Journal of Environmental Economics and Management*, 5, pp. 26-43

Harford, J. D., 1987. Self-Reporting of Pollution and the Firm's Behavior Under Imperfectly Enforceable Regulations, *Journal of Environmental Economics and Management*, 14, pp. 293-303.

Harrington, W., 1988. Enforcement Leverage When Penalties Are Restricted, *Journal of Public Economics*, 37(1), pp. 29-53

Hawkins, K., 1984. *Environment and Enforcement: Regulation and the Social Definition of Pollution*. Claredon Press, Oxford.

Jones, C. A. and Scotchmer, S., 1988. *The Social Cost of Uniform Regulatory Standards in a Hierarchical Government*. Unpublished Manuscript, University of California, Berkeley.

Jones, C. A., 1989. Standard Setting with Incomplete Enforcement Revisited, *Journal of Policy Analysis and Management*, 8(1), 72-87.

Kagan, R. A. and Scholz, J. T., 1984. The Criminology of the Corporation and Regulatory Enforcement Strategies. In: Hawkins, K. and Thomas, J. M. (Eds.), *Enforcing Regulation*. Kluwer-Nijhoff Publishing, Boston, MA.

Lee, D. R., 1984. The Economics of Enforcing Pollution Taxation, *Journal of Environmental Economics and Management*, 11 pp. 147-160.

Melnick, R. S., 1983. *Regulation and the Courts: The Case of the Clean Air Act*. Brookings, Washington, DC.

Russell, C. S.; Harrington, W.; and Desirable, W. J., 1986. *Enforcing Pollution Control Laws*. John Hopkins University Press, Baltimore, MD.

Russell, C. S., 1990. Monitoring and Enforcement. In: Portney, P. R. (Ed.), *Public Policies for Environmental Protection*. Resources for the Future, Washington, DC.

Scholz, J. T., 1984. Cooperation, Deterrence, and the Ecology of Regulatory Enforcement, *Law and Society Review*, 18(2).

Shah, F. and Zilberman D., 1990. *Water Rights Doctrines and Technology Adoption*, University of California, Berkeley.

Silbey, S. S., 1984. The Consequences of Responsive Regulation. In: Hawkins K. and Thomas, J. M. (Eds.), *Enforcing Regulation*. Kluwer-Nijhoff Publishing, Boston, MA.

Stigler, G., The Optimal Enforcement of Laws, *Journal of Political Economy*, May 1970, pp. 526-536.

Veljanovski, C. G., 1984. The Economics of Regulatory Enforcement. In: Hawkins, K. and Thomas, J. M. (Ed.), *Enforcing Regulation*. Kluwer-Nijhoff Publishing, Boston, MA.

Viscusi, W. K. and Zeckhauser, R. J., 1979. *Optimal Standards with Incomplete Enforcement*. Public Policy 27, No. 4.

Wade, R., 1987. The Management of Common Property Resources: Collective Action as an Alternative to Privatisation or State Regulation, *Cambridge Journal of Economics*, 11, pp. 95-106.

# 37 ORGANIZATIONAL FAILURE AND THE POLITICAL ECONOMY OF WATER RESOURCES MANAGEMENT

Gordon C. Rausser and Pinhas Zusman,
University of California, Berkeley

### ABSTRACT

This chapter presents a political-economic interpretation of the legislative history of governmental intervention in Western water resource systems. In the context of regulatory behavior, a theory is presented explaining certain behavioral regularities exhibited by the political-economic forces underlying water resource management. The basis for organizational failures that often has led to serious waterlogging problems or severe damage to aquifers is developed. It is argued that these organizational failures may also explain deterioration of soil and water quality and other harmful environmental impacts.

## INTRODUCTION

Collective action, whether voluntary or through governmental intervention, is pervasive in water resource systems. In the case of the United States, collective action has occurred in water resources at both the legislative and administrative levels. A number of reasons account for this observed regularity including, *inter alia*: (1) The existence of politically powerful groups that benefit from state intervention; (2) the existence of strong nonconvexities in water resource utilization, mostly in the form of indivisibilities and sizable economies of scale; (3) the existence of strong externalities, most of which result from drawing water from a common aquifer; and (4) the desire of many governments to pursue noneconomic goals (e.g., increase settlement in particular arid regions).

In the design of Federal legislation governing Western water resource development, the above reasons have played a crucial role. They have also played an important role in the public administration or regulation of California water resource programs. In this chapter a political economic interpretation is presented of the legislative history for governmental intervention in Western water resource systems. In the context of regulatory behavior, a theory will be presented explaining certain behavioral regularities exhibited by political economies of water resource management. In particular, the basis for organizational failures that often has led to serious waterlogging problems or severe damage to aquifers is structured. The theory of organizational failure presented here may also be used to explain deterioration of soil and water quality and other harmful environmental impacts.

The history of U.S. water resource development reveals the importance of political power and influence in the legislative process. Even though the original intent of any legislation might be to serve the broader public interest, the existence of concentrated benefits invites various agents or groups to pursue their own self-interest. Moreover, although the initial public investments in water resource developments may have promoted economic growth (i.e., increased the size of the pie), vested interests emerged to redirect subsequent legislation in their favor. This general phenomenon has also occurred in the implementation or administration of the enabling water resource and land reclamation legislation. These administrative features will be examined over the balance of this chapter in terms of the regulatory structure for water resource management. A general conceptual framework will be presented which attempts to explain the behavioral regularities arising in all collective action political economies. Such political economies, whether in a legislative or administrative context, reflect both the role and importance of the public interest as well as vested interest of particular agents and groups.

## THE REGULATORY STRUCTURE OF WATER RESOURCE MANAGEMENT

The water-resource regulatory structure is a system where the principal economic and engineering decisions concerning resource management are collective choices. Group choices usually apply to such items as resource development programs (design, scheduling, investment levels, funding sources, etc.); the allocation of water among users; water quality; water pricing; operation regimes; environmental protection measures; etc. Obviously, such decisions have far-reaching allocation and distribution implications, and are likely to create considerable conflict among participants.

The water-resource political economy operates within a given physical, legal, economic, and political environment which imposes constraints and affects choices within the political economy. Thus, water allocation, in great measure, is circumscribed by existing water rights laws and water availability. The political power structure is strongly influenced by the prevailing social value system and the external, legal, political, and economic relations. The external environment thus profoundly affects the policy choices and the organizational structure of the political economy. Nevertheless, any attempt at developing a theory involving endogenous policy must combine the political power structure, the economic structure, and the hydrological system within an integrated model, thus endogenizing many political, economic, and engineering variables.

The water resource system to be considered is a simplified version of a conjunctive water-use system; it is highly schematic and ignores water quality and environmental problems. Water is the system's only output, and it is used solely for irrigation purposes. The analyzed system is thus unrealistically simple, but it does elucidate several theoretical principles at work in water resource political economies in general.

## THE PHYSICAL WATER-RESOURCE SUBSYSTEM

The physical water-resource subsystem considered is comprised of the following components (figures 1 and 2):

A. A central water supply project (CWP) which collects available water from a source located at the northern part of the country and delivers it to n districts located throughout the country: The total amount available annually at the northern source is $Z_o$, of which $x_o$ ($x_o \leq Z_o$) is collected by the CWP. No water distribution losses are incurred. The amount of water delivered by the CWP to the ith district is denoted by $x_i$. Hence, the CWP water balance relationship is

[1] $\sum_{i=1}^{n} x_i = x_o \leq Z_o$

B. n irrigation districts indexed by i (i = 1, 2, ..., n): The amount $Z_i$ of surface water is locally available at no cost at the ith district. The locally available surface water is combined with the water delivered by the CWP and the amount of locally pumped ground water to be used in irrigation, $F_i$. Hence, the amount of irrigation water used in the ith region, $I_i$, is

[2] $I_i = Z_i + x_i + F_i$      i = 1, 2, ..., n

The share of irrigation water percolating below the crop's root zone and into the underground aquifer is k (0 < k < 1), and 1 - k is the share of irrigation water

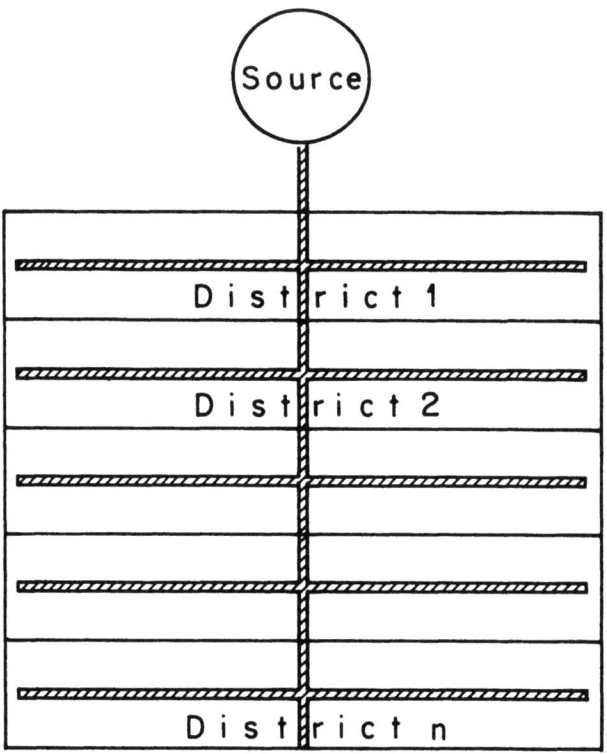

Figure 1. The physical water resource system: geographic set-up.

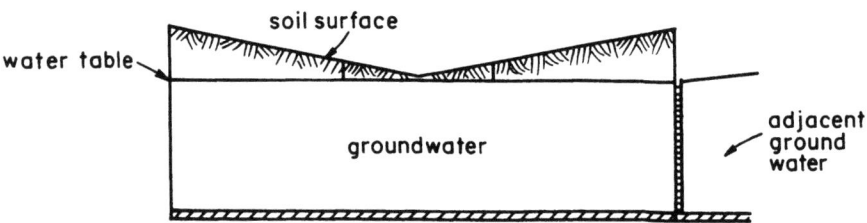

Figure 2. The physical water resource system: irrigated land and water aquifer cross section.

lost in evapotranspiration; k is assumed to be constant over all districts. Consequently, the annual addition to the amount of underground water due to the pumping and irrigation activity in district i is

[3] $G_i = k I_i - F_i = k(Z_i + x_i) - (1 - k)F_i$

$G_i$ may be negative, implying net water subtraction from the aquifer.

C. An underground aquifer spanning the entire country with perfect water conductivity within the aquifer is assumed, so that the groundwater level is equal in all districts. The elevation of the ground-water table is proportional to the total amount of water in the aquifer, Q, and may be measured by it. The evolution of ground-water level over time is given by

[4] $Q_{t+1} = Q_t + \sum_{i=1}^{n} G_i - \alpha(Q_t - H)$

where the term, $-\alpha(Q_t - H)$, refers to water inflow/outflow to areas outside the country; $\alpha$ is a positive parameter proportional to water conductivity between the aquifer and the adjacent areas; and H is a parameter such that $Q_t - H$ is proportional to the hydrostatic head determining the hydraulic flow gradient between the aquifer and adjacent ground water. When the depth of the water table relative to the soil surface is less than a specified value, waterlogging occurs and the affected land has to be withdrawn from cultivation. Hence, the amount of cultivatable land, $A_i$, in each district is a monotone decreasing function of the underground water denoted by $A_i(Q)$ level up to a certain depth. Below the critical level, $Q_c$, all land is cultivatable. That is, $A_{iQ} < 0$ for $Q \geq Q_c$ and $A_{iQ} = 0$ otherwise. Note that $A_{iQ}$ is the derivative of $A_i(Q)$ with respect to Q.

## THE ECONOMIC STRUCTURE

It is assumed that the CWP incurs two types of cost: a fixed cost denoted by $C_o$ and variable cost. The variable cost of delivering $x_i$ units of water to district i, located at a distance of $d_i$ miles from the source, is $\xi_0 d_i x_i$. Water is sold by the CWP to district i at a price $p_i$. The CWP is a nonprofit, closed-accounting unit. That is, its total cost must be exactly equal to water sales plus government net subsidy, S. Hence,

[5] $C_0 + \xi_0 \Sigma d_i x_i = \Sigma p_i x_i + S.$

$S < 0$ implies net tax.

The cost of pumping ground water denoted by $C_i(F_i, Q)$ increases at an increasing rate with the amount pumped and decreases with ground-water level, $Q$.

The marginal costs of pumping also decrease with the ground-water level. Only users located inside the district are served by the district. In the subsequent analysis each district is treated as a single, fully integrated, decision unit.[1]

The production technology in district i is described by the production function $f^i = f^i(A_i, I_i, y_i)$ where $y_i$ denotes the level of other inputs. By appropriate choice of units of output and of $y_i$, the given constant prices of output and $y_i$ are normalized to $\rho_f = \rho_y = 1$. $f^i(\cdot)$ is assumed to have the ordinary nice properties: It is monotone increasing, twice differentiable, and concave in all inputs, etc.

It is first assumed that $x_i$ is not rationed and not constrained by existing water rights, so that the amount of water imported to district i is entirely at the district's discretion. Each district selects values of $I_i$, $F_i$, $x_i$, and $y_i$ to maximize its net income $\pi_i$. Let $\pi_i(\rho_i; Q)$ be the indirect district's net income function. The optimization problem of the district is:

$$\pi_i(\rho_i; Q) = \max_{I_i, x_i, F_i, y_i} [f^i(A_i, I_i, y_i) - C_i(F_i, Q) - \rho_i x_i - y_i]$$

$$I_i = Z_i + x_i + F_i.$$

For simplicity, hereafter the index i which would otherwise appear as a subindex will be eliminated from $\rho_i$, $F_i$, $F_i$, $\hat{C}_i$, $\tilde{C}_i$, and $x_i^r$.

As is well known, $\pi_i(\rho_i; Q)$ is nonincreasing and convex in $\rho_i$. By Hotelling's lemma (Varian, 1984, Chapter 1),

$$\pi_{i\rho} = -x_i(\rho_i, Q) < 0$$

It is assumed that there are many districts (i.e., n is large), each of which is sufficiently small to ignore the effects of its own decisions on the ground-water level. That is, each district regards $Q$ as given. Recall, however, that the district's choice of its control variables and the resulting net income depend on the level of $Q$. Figure 3 provides a geometric description of the district water economy for two levels of ground water--$Q_0$ and $Q_1$, where $Q_1 > Q_0 \geq Q_c$. The price of water supplied by the CWP is maintained constant at $\bar{\rho}$. Note that, as the ground-water level rises from $Q_0$ to $Q_1$, the loss of cultivatable land shifts the demand for irrigation water from $I(\rho, Q_0)$ down to $I(\rho, Q_1)$. (In figure 3 the index, i, is suppressed in all terms.) The marginal cost of pumping is also shifted down from $C_F(F, Q_0)$ to $C_F(F, Q_1)$.

Consequently, the amount of water demanded for irrigation declines from $I(\bar{p}, Q_0)$ to $I(\bar{p}, Q_1)$ while ground-water pumping increases from $F(\bar{p}, Q_0)$ to $F(\bar{p}, Q_1)$. The demand for imported water accordingly declines from $x(\bar{p}, Q_0)$ to $x(\bar{p}, Q_1)$.

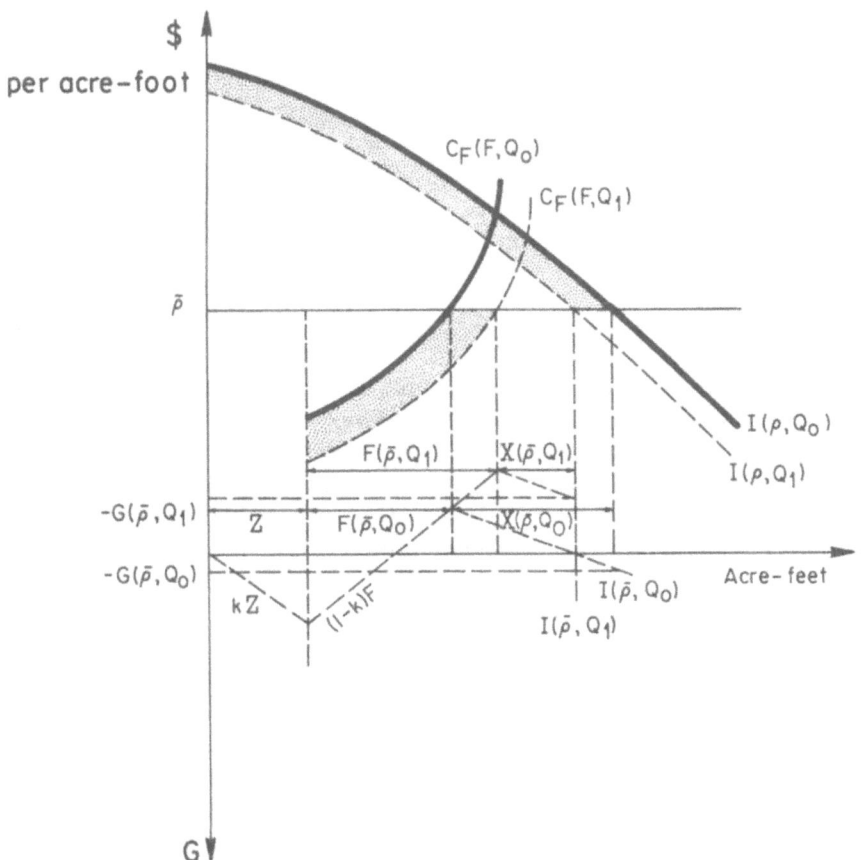

Figure 3. The district's economy ($Q_c < Q_0 < Q_1$).

The gross loss in net district income due to loss of cultivatable land is represented by the dotted area while the gross addition to net income due to savings in pumping cost is represented by the shaded area. The net change in income depends on the relative sizes of the two gross changes in the district's income.

It is also worth noting that the increase in ground-water level from $Q_0$ to $Q_1$ reduces the net addition to ground water from $G(\bar{\rho}, Q_0) > 0$ to $G(\bar{\rho}, Q_1) < 0$. Intuitively, this is due to the decrease in the amount of irrigation water and the attendant increase in pumping. Hence, $G_{iQ} < 0$.

How does a change in $\rho$, holding Q constant, affect the optimal values of the district's decision variables? Zusman and Rausser (1990b) have shown that

[6]  $F_i \rho_i = 1/C_{iFF} > 0$ when $\rho_i > C_{iF}(0, Q)$

and is zero otherwise, and that

[7]  $I_{i\rho} = x_{i\rho} + 1/C_{iFF} < 0$

and that

[8]  $G_{i\rho} = kx_{i\rho} - (1-k) F_{i\rho} < 0$

Hence, an increase in the price of imported water unambiguously reduces district i's net addition to ground water and vice versa.

Consider next a system controlled by prices, i.e., the CWP sets water prices $\rho = (\rho_1, ..., \rho_n)$ which all districts treat as parametrically given when making their economic decisions. Obviously, to be feasible, $\rho$ must be chosen to satisfy $\Sigma x_i(\rho_i, Q) \leq Z_0$. A stationary level of ground water, $Q^s$, is defined by

[9]  $\Delta Q = \Sigma G_i(\rho_i, Q^s) - \alpha(Q^s - H) = 0$.

Hence, $Q^s = Q^s(\rho)$. Is $Q^s$ stable?

It shall be assumed that $Q^s$ is globally stable.[2] Moreover, given any set of feasible constant water prices, $\rho$, the ground-water level will converge monotonically to the stationary level $Q^s = Q^s(\rho)$. As the state of the system depends on $\rho$ and Q alone, the entire water-resource system converges to a universal stationary state which is fully determined by the CWP-set water prices, $\rho$. Zusman and Rausser (1990b) have shown that the relationship between water prices and the stationary ground-water level is given by

[10]  $Q^s_{\rho i} = -G_{i\rho}/\Sigma G_{iQ} - \alpha < 0$

Thus, higher water prices imply lower stationary ground-water levels.

## THE POLITICAL POWER STRUCTURE

There are n + 2 players in the political economy: n districts, the CWP, and the government. It has been assumed, hitherto, that water prices are set by the CWP. Presumably, that is the CWP's legal authority. However, since water prices have profound effects on the well being of all other parties, price setting is essentially a political issue to be decided in the political arena in accordance with the participants' political power. In particular, water prices determine S, which is legally a governmentally controlled instrument. Hence, water prices cannot be decided without the full consent of the government. To understand the power relationships, one must correctly identify the interests of the participating parties and examine their power bases. In the following it is assumed that each group's objective function is comprised of two components: a component, $u_i$, defined exclusively over the policy space, and a component representing the value and cost of the means of power to the group. The sum of the two components, $U_i$, is designated "the extended objective function." Each group shall now be considered in turn.

### The Central Water Project (CWP)

Organizations such as the CWP are usually established as nonprofit, closed-accounting legal entities. Hence, unlike the classical capitalistic firm, the CWP does not pursue profits; and its performance is ordinarily judged by the cost efficiency of its operations since no other party is in a position to control these costs. One also expects that concern for cost efficiency may develop into an interest in economic efficiency in general.

It shall therefore be assumed that the CWP objective function is

$$[11] \quad u_0 = V(\rho, Q) = \sum_{i=1}^{n}[f^{\,i}(A_i(Q), I_i, y_i) - C_i(F_i, Q) - y_i] - \xi_0 \sum_{i=1}^{n} d_i x_i - C_0$$

However, decision agents in the CWP have other, more personal, interests as well. They usually seek recognition and sympathy from the other parties and abhor public expressions of dissatisfaction with the CWP or their personal performances. These individuals may develop political aspirations, desires to win interagency rivalries, interest in personal promotions, and material well being. To advance their interests, they must gain the other parties' support and avoid being censured. However, this is not a one-sided relationship as CWP decision agents are able to reward and penalize the other parties, primarily through their legal control over water pricing and also through loyalty and support of politicians in government. Following Zusman (1976), these relationships are introduced into the model by the device of strength functions which describe the penalties and rewards extended to decisionmakers in the CWP and the government by each of the other participants as function of the cost of power to the latter. The extended objective function of the CWP is then

[12]  $U_0 = u_0 + \sum_{i=1}^{n+1} s_i(c_i^0, \delta_i) - c_0^{n+1}$

where $s_i$ is the strength of the ith interest group power over the CWP, $c_i^0$ is the cost of power to the ith group in influencing the CWP choices, and $\delta_i$ is an indicator variable such that

$\delta_i = \begin{cases} \alpha_i^0 \text{ when i adopts a reward policy towards the CWP} \\ \beta_i^0 \text{ when i adopts a penalizing policy towards the CWP} \end{cases}$

$C_0^{n+1}$ is the cost of power to the CWP in influencing the government choices when the CWP employs means of power other than water prices. Note that the n districts (indexed by i = 1, 2, 3, ..., n) and the government (indexed by i = n + 1) all exert their influence over the CWP choices.

## The District

The objective function of the ith district is identified with its net income, i.e.,

[13]  $u_i = \pi_i(\rho_i; Q)$.

Recall that, while the ground-water level, Q, affects the district's net income, Q is viewed by each district as an exogenously given collective good/bad. The district ignores the effects of its own decisions on $G_i$, and thereby on Q. In this respect the district is narrowly rational. As asserted earlier, the individual district can contribute or detract from the welfare of decision agents in the CWP and the government. Districts may provide political rewards by contributing to election funds, by public pronouncements of support, and by denouncing opposing individuals and groups. They may mobilize goodwill toward decision agents in the CWP or the government, support their causes, and assist them in bureaucratic and political infights. They may also impose political penalties by supporting the opposition and criticizing the performance of incumbent decision agents.

But whatever the district does, whether rewarding or penalizing another party, it incurs cost--the cost of power. Hence, the extended objective function of the ith district is

[14]  $U_i = u_i - c_i^0 - c_i^{n+1}$

where $c^0_i$ is the cost to district i of influencing the CWP and $c_i^{n+1}$ is the cost of influencing the government.

## The Government

The objective function of the government is more difficult to specify. Different elements in government ordinarily pursue different and often conflicting goals; the sweeping view of government as a single entity with a well-defined goal is clearly a myth. In extant economic literature dealing with political economies, policymakers' interests are often portrayed as exclusively personal; politicians pursue purely selfish goals while political parties support particular policies not because of the policy's perceived intrinsic value but in order to maximize the likelihood of being elected (Magee et al., 1989).

We do not subscribe to this cynical view of politics and presume instead that politicians pursue both selfish and unselfish "public interest" goals. In the present context, we adopt a rather narrow interpretation of the unselfish goal of government and identify the government's objective function with the government's net revenue from the CWP (the negative value of the net subsidy to water users). That is,

[15] $u_{n+1} = -S$

where S is the water subsidy cost defined by the CWP zero profit constraint. The government thus represents taxpayers, or other claimants on the states' financial resources. This interpretation of the government interest by definition identifies the government with those responsible for the state's fiscal policy.[3]

Government decision agents also have personal, political, and economic interests which render them amenable to the influences of interest groups. The government-extended objective function may thus be formulated as follows:

[16] $U_{n+1} = u_{n+1} + \sum_{i=0}^{n} S_i(c_i^{n+1}, \eta_i) - c^0_{n+1}$

where $S_i$ is the strength of the ith group's power over the government; $c_i^{n+1}$ is the cost of power to the ith group over the government; $\eta_i$ is an indicator variable analogous to $\delta_i$ above with $\eta_i = \alpha_i^{n+1}$ if i adopts a reward policy and $\eta_i = \beta_i^{n+1}$ if i adopts a penalizing policy; and $c^0_{n+1}$ is the cost of power to the government in influencing the CWP. Note that the strength of the CWP power over the government, $S_o(\ )$, is now included in [16].

The political solution is invariant under positive linear transformation of the groups' objective functions. Accordingly, the units of the $U_i$ may be arbitrarily chosen. In the present analysis, a money metric is adopted so that each objective function is expressed in terms of an equivalent money income without loss in generality.[4]

The strength and cost of power functions are assumed to reflect the cost minimizing choice of the means of power needed to achieve a given strength. Also, the strength functions--$s_i(c_i^0, \alpha_i^0)$; $-s_i(c_i^0, \beta_i^0)$; $S_i(c_i^{n+1}, \alpha_i^{n+1})$; and $-S_i(c_i^{n+1}, \beta_i^{n+1})$--are concave in the cost of power. In other words, we assume diminishing marginal productivity in the exertion of power.

The economic and political resources at the control of each group constitute the group's base of power. There exists a variety of power bases, and we shall mention just a few. First and foremost is legitimate power, i.e., the set of actions which each party is authorized to do by general social consent. Thus, under a general acceptance of the legal system and given that the CWP is authorized by law to set prices and the government is legally authorized to determine S (within certain constitutional limits), these legal prerogatives serve as legitimate power bases of the CWP and the government, respectively. Within certain constitutional limits, the government enjoys coercive power to enforce certain laws and regulations (e.g., tax payments). Yet, legitimate power should not be interpreted as exclusively legally based. All potential group actions consistent with the beliefs, norms, and moral values prevailing in the particular society constitute that group's legitimate base of power. Control over economic resources transferable to other parties in the political process constitute the economic base of the group's power. A group's political power base consists of such political resources as the group's ability to affect appointments to various public positions, a strong representation in political party caucuses, an appeal to various constituencies and electorates, etc.

## THE HYDROLOGICAL-POLITICAL-ECONOMIC EQUILIBRIUM

Following Harsanyi (1962a and 1962b) and Zusman (1976), the political conflict among the various interest groups is viewed as a bargaining game. The resolution of the political conflict is identified with the solution to a bargaining game.

It has been shown by Zusman (1976) that the values of the policy instruments constituting a solution to the bargaining game may be obtained by solving the maximization problem

$$[17] \quad \max_{x_0 \in X_0} W = u_0(x_0) + \sum_{i=1}^{n} b_i u_i(x_0)$$

where $x_0$ denotes the vector of policy instruments, $X_0$ is the set of feasible instrumental variables, and $b_i$ is the power coefficient of the ith interest group. More specifically,

$$[18] \quad b_i = S_{ic}(\hat{c}_i, \alpha_i) = S_{ic}(\bar{c}_i, \beta_i) = (U_0^* - \bar{U}_0)/(U_i^* - \bar{U}_i) \geq 0$$

where $\hat{c}_i$ and $\bar{c}_i$ are the equilibrium costs of power under a cooperative (agreed upon) solution and under disagreement, respectively; $U_i^*$ and $\bar{U}_i$ are the values of the extended objective function of the ith group under a cooperative (agreed upon) solution and under disagreement, respectively. Notice that while the $b_i$'s are variables in the bargaining problem, they are treated as given constants in the corresponding maximization problem. It is shown in Zusman and Rausser (1990a) that the maximization problem [17] and [18] also corresponds to the bargaining problem arising in the bicentric political economic system described in the preceding section.

In the present model of a water resource system controlled by water prices, the relevant policy instruments are identified with the water prices, $\rho$, and the net subsidy, S. Note that water prices must be nonnegative and must satisfy the CWP water availability constraint, $\Sigma x_i(\rho_i, Q) \leq Z_0$, while $\rho$ and S are interdependent through equation [5]. Adopting the long-term view, we focus on the stationary states of the system ignoring the transients.[5]

The hydrological-political-economic equilibrium water prices are those maximizing:

$$[19] \quad W = u_0 + \sum_{i=1}^{n} b_i u_i + b_{n+1} u_{n+1}$$

$$= V(\rho, Q^s) + \sum_{i=1}^{n} b_i \pi_i(\rho_i; Q^s) - b_{n+1} S(\rho, Q^s)$$

where $Q^s = Q^s(\rho)$, while the narrowly rational individual districts regard $Q^s$ as exogenously given collective goods/bads. Assuming interior solution, the first order conditions (FOC) for maximum W with respect to $\rho$ are:

$$[20] \quad W_\rho = V_\rho + V_Q^s Q_\rho^s + \Sigma b_i \pi_{i\rho} - b_{n+1}[S_\rho + S_Q^s Q_\rho^s] = 0$$

Are equilibrium water prices economically efficient? To answer this question it is first noted that the following two conditions together assure efficiency.

(A) Power is uniformly distributed (i.e., $b_i = 1$ for all i).

(B) All individual districts take into account the full effects of their own decisions on the ground-water level (districts' full rationality).

Given the definition of $\pi_i(\rho;Q^s)$ and substituting equations [5], [11], [13], and [15] into [19] we get, under (A) and (B),

$$[21] \quad W = 2V(\rho, Q^s(\rho)) = 2 \left[ \sum_{i=1}^{n} [f^i - C_i - y_i] - \xi_0 \sum_{i=1}^{n} d_i x_i - C_0 \right]$$

so that maximizing W also maximizes V, the net social surplus from the water resource system. When condition (B) does not hold, the effects of each districts' choices on the ground-water level is fully externalized and the district is narrowly rational. Note also that even if (B) holds so that no externalities exist, economic efficiency still requires a uniform distribution of power.

We now explore the effects of the districts narrow rationality on the hydrological-political-economic equilibrium. To this end, let $\rho^*$ be the equilibrium vector of water prices; i.e., $\rho^*$ is the equilibrium solution obtained from [20] when condition (B) is violated. Let $\rho_B$ be the vector of equilibrium water prices when (B) holds (districts' full rationality). That is,

$$[22] \quad W_\rho(\rho_B) + \Sigma b_i \pi_{iQ^s} Q^s_\rho(\rho_B) = 0$$

In Zusman and Rausser (1990a), it is demonstrated that for an appropriate choice of weights, $a_1, ..., a_n$ $(0 \leq a_i \leq 1, \Sigma a_i = 1)$, the equilibrium water price indices $\bar{\rho}_B = \Sigma a_i \rho_{Bi}$ and $\bar{\rho}^* = \Sigma a_i \rho_i^*$ satisfy the inequality

$$[23] \quad \bar{\rho}_B > \bar{\rho}^*.$$

That is, water prices under (B) are higher than under districts' narrow rationality.

Assuming that all $\rho_i$ are positively correlated with $\bar{\rho}$, one achieves the inequality

$$[24] \quad Q^s(\rho^*) > Q^s(\rho_B).$$

The political pressure of narrowly rational districts is thus conducive to lower water prices and higher stationary ground-water levels. Losses due to waterlogging are thereby exacerbated. Essentially, this system's failure is due to the Olsonian "free riding" phenomenon (Olson, 1965) operating through political influence rather than defection.[6]

Note also, that if convergence to the stationary level, $Q^s$, is monotone, as is highly likely, the transient path leading to $Q^s(\rho^*)$ is above the transient path leading to $Q^s(\rho_B)$. Namely, commencing from any given initial level, the time path of the ground-water level under the prevailing district's narrow rationality is higher, and presumably less efficient than the time path under the hypothetical full rationality [i.e., when condition (B) is met].

In conclusion, water prices associated with the political-economic-equilibrium are inefficient. This is because setting water prices has both distribution and allocation implications, and participants in the political process exert political influence to achieve higher income at the expense of an inefficient allocation.[7] The distortion is due to the unequal distribution of power in the political-economy.

No less important is the efficiency loss due to narrowly rational districts who treat ground-water levels as collective good/bad, ignoring the effects of each district's own choices on ground-water level. The resulting failure may be manifested only after some time has elapsed. In the present case, waterlogging problems may occur only after ground-water levels have risen above the damage threshold.

## CONJUNCTIVE WATER USE WITH SHORT WATER SUPPLY

The water resource system explored next is similar to the one analyzed in the preceding sections, with one important exception: the availability of water at the northern source, $Z_0$, is highly restricted. In particular, it is assumed that $Z_0$ is so small that the CWP water balanced constrained $\Sigma x_i(\rho_i, Q^s) \leq Z_0$ entails politically prohibitively high water prices so that all parties concerned agree to substitute quantitative control for price control. Under this system control regime, the amount of water delivered to each district is rationed by the CWP. Water prices, $\rho$, now constitute pure distribution instruments devoid of an allocation effect.

Let $x^r = (x^r_1, x^r_2, ..., x^r_n)$ be the vector of delivered water quantities.
The indirect net district income function is now

[25] $\Pi_i(x^r_i; Q) = \max_{I_i, F_i, y_i} [f^i(A_i, I_i, y_i) - C_i(F_i; Q) - \rho_i x^r_i - y_i]$

It is assumed that the associated ground-water level is so low that no waterlogging occurs, i.e., $Q < Q_c$ and $A_{iQ} = 0$.

The districts' water economy is depicted in figure 4, where the index $i$ is again suppressed. The demand function for irrigation water is represented by the curve, $I(\rho; Q)$, and a rise in ground-water level produces no change in demand. The marginal cost of pumping ground water, described by $C_F$, falls as the ground-water level rises from $Q_0$ to $Q_1$; the amount of pumped water is consequently increased from $F(Q_0)$ to $F(Q_1)$. The shaded area enclosed between the two marginal cost of pumping curves represents the accompanying increase in the net district income. Note also, that while the set price of imported water, $\bar{p}$, affects the net district's income, it has no allocation effects whatsoever. Water will be imported to the district if its marginal product value exceeds its price, i.e., $f_{li}^i - \rho_i > 0$.

Zusman and Rausser (1990b) have shown that an increase in volume of imported water will reduce the amount of ground water pumped $F_{ix}r < 0$. Higher ground-water level (Q) will tend to increase net income, $\pi_{iQ} > 0$, and increase pumping, $F_{iQ} > 0$, as can be verified from figure 4.

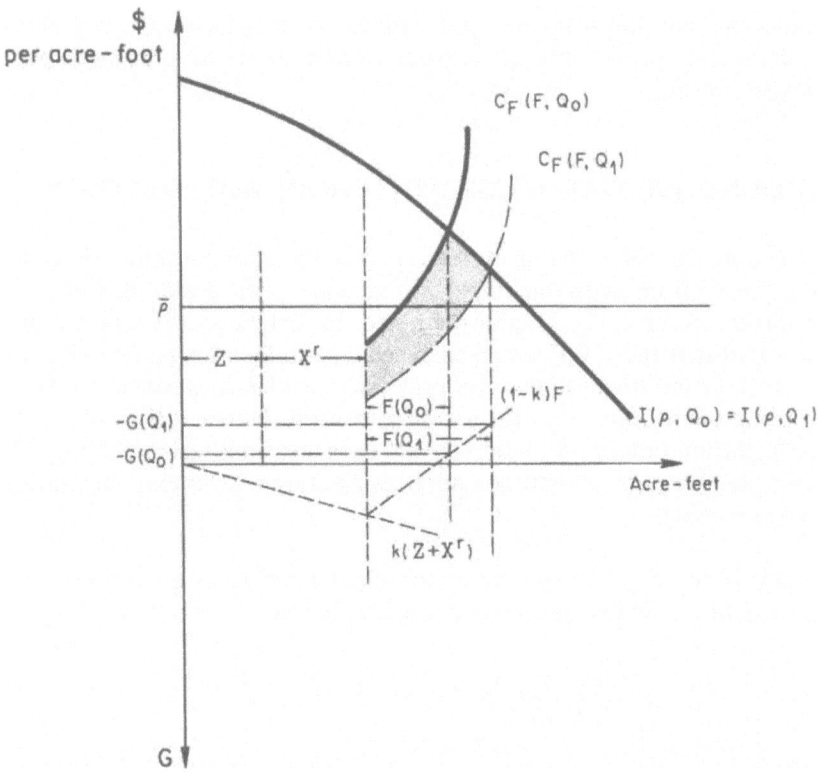

Figure 4. The district water economy under rationed imported water $(Q_o < Q_1 < Q_c)$.

The above figure and [3] suggest that increase in the amount of water imported to the ith district and/or a decline in ground-water level raises the district's net contribution to ground-water level, i.e., $G_{ix^r} > 0$ and $G_{iQ} < 0$.

The stationary ground-water level is now characterized by

[26] $\Sigma G_i(x_i^r, Q^s) - \alpha(Q^s - H) = 0$.

Total differentiation of [26] suggests that $Q^s_{x^r} > 0$ and thus by choosing a water allocation scheme, $x^r$, it is possible, to some extent, to affect the stationary ground-water level $Q^s(x^r)$. However, inasmuch as $Z_0$ is given, the water balance constraint restricts the range of possible choices of $Q^s$. The nature of this restriction can be elucidated by inserting [3] into [26] which yields the following condition for $Q^s$

[27] $(1-k) \sum_{i=1}^{n} F_i + k \sum_{i=0}^{n} Z_i - \alpha(Q^s - H) = 0$.

Suppose the stationary level, $Q^s$, is so low that the system suffers severe losses due to high pumping costs and/or progressive damages due to ground-water mineralization brought about by the penetration of salty water from adjacent areas to the aquifer. Two remedies may be considered under the circumstances:

(A) Invest heavily in developing the northern source in order to expand $Z_0$ and thereby $x^r$ and $Q^s(x^r)$.

(B) Limit the amounts of pumped ground water, which will also curtail irrigation.

If water prices could somehow be maintained at their current levels, it is in the districts interest that course of action (A) be selected, with the government shouldering the additional fixed and variable costs entailed by larger water deliveries. The districts would still prefer alternative (A) even if some increase in water prices is required. Which course of action is ultimately selected depends on the economic structure and the power structure. However, the crux of our argument is that no matter which remedy is selected, the resulting stationary ground-water level will be suboptimal due to district's narrow rationality.

Suppose it is agreed that ground-water pumping is to be rationed [course of action (B)]. Two problems then arise: How much pumping should be allowed, and how should total permissible pumping be allocated among districts?

Suppose that an equiproportional cut in current pumping levels is considered fair, then

$$F_i^r = \lambda F_i^0, \quad i = 1, 2, ..., n$$

where $F_i^r$ and $F_i^0$ denote the rationed and current unrationed amounts of pumping in district i, respectively, and $\lambda$ is a constant factor of proportionality. Since $(F_1^0, ..., F_n^0)$ are historically given parameters, and as rationing ($\lambda < 1$) imply that every district fully utilizes its pumping quota, $F_i^r$, from equation [27] we have

[28] $Q^s(\lambda) = H - \lambda(1-k) \sum_{i=1}^{n} F_i^0/\alpha + k \sum_{i=0}^{n} Z_i/\alpha.$

Hence, $Q^s_\lambda x^r = 0$ and $Q^s_\lambda < 0$

Again, the hydrological-political-economic equilibrium, $(\bar{x}^r, \bar{\lambda}, \bar{\rho})$, is that maximizing

[29] $W(x^r, \lambda, \rho) = V(x^r, \lambda, Q^s) + \sum_{i=1}^{n} b_i \Pi_i(x_i^r, \lambda, \rho_i; Q^s) - b_{n+1} S(x^r, \rho)$

subject to the constraint, $\Sigma x_i^r \leq Z_0$, and taking into account narrow rationality of districts.

As the amount of water available to the CWP for delivery to the districts is short; it is reasonable to view the water availability constraint as binding and, thus, a strict equality. We then form the Lagrangian function,

$$L(x^r, \lambda, \rho, \mu) = W(x^r, \lambda, \rho) + \mu(Z_0 - E'x^r)$$

where $E'$ is an n-vector of ones, $[E' = (1, 1, ..., 1)]$, and $\mu$ is a Lagrangian multiplier.

The vector $(\bar{x}^r, \bar{\lambda}, \bar{\rho}, \bar{\mu})$ is the equilibrium solution under districts' narrow rationality. Let $(x_B^r, \lambda_B, \rho_B)$ denote the equilibrium value of the policy instruments under district's full rationality, (i.e., condition (B) holds). Comparing the equilibrium values of the policy instruments and the stationary states of the system under districts' narrow rationality and under districts' full rationality. Zusman and Rausser (1990b) have shown that $Q^s(\lambda_B) > Q^s(\bar{\lambda})$ and that $\lambda^B < \bar{\lambda}$.

Hence, compared to the unrealistic districts' full rationality condition, the more likely districts' narrow rationality is conducive to more liberal ground-water pumping quotas and to lower stationary ground-water levels. As convergence to stationary levels is likely monotone, the time path of the ground-water level associated with districts' narrow rationality is also lower, and social losses due to higher pumping costs and damages to the aquifer are higher.

## THE DYNAMICS OF THE WATER RESOURCE POLITICAL ECONOMY

In the foregoing analysis it has been demonstrated that the political-economic equilibrium is likely to entail severe system inefficiency. In part, the inefficiency is due to nonuniform distribution of power. But more importantly, poor performance should be attributed to interest groups' narrow rationality which imposes free riding considerations on policy choices. The more powerful the narrowly rational interest groups relative to formal policymaking centers, the greater the efficiency losses. Understanding the process of power formation and distribution is, therefore, crucial; and our analytical focus is now shifted to the forces shaping the political power structure which, hitherto, was treated as given. In particular, we seek to develop a descriptive theory of structural power changes in response to internal and external developments. To this end the following heuristic propositions are posited:

1. The political power of participants in the political-economy is an increasing function of each group's bases of power.
2. A group's bases of power critically depend on the ability of a particular subset of individuals (the so-called "political entrepreneurs," the group's leadership) to mobilize the group. Effective group mobilization requires an appropriate organization capable of providing individual "selective incentives" or applying other measures designed to overcome group members' proclivity to "free ride" (Olson, 1965). Hence, to be effective in forming and maintaining the group power bases, group leaders should have their own second order power bases (Elster, 1989).
3. There are several types of power bases of which legitimate authority, economic resources, and organizational-political resources feature highest.
4. The dynamics of change in power bases involves strong, positive feedback loops. When a group is sufficiently powerful to induce favorable policy choices, it also reinforces its own power bases. Some of the political gains may be directed to augment the group's economic and political power bases. Manifest political success along with more abundant economic resources may be used in campaigns designed to foster the group's legitimate power base by promoting beliefs, values, and behavioral norms consistent with the group interests.
5. Widely perceived successful performance of policies advocated by the group contributes to the group's legitimate power. Conversely, perceived failures of these policies detract from the group's legitimate power.
6. Beyond a certain point, diminishing returns in power formation set in; that is, marginal additions to the group's power, induced by equal incre-

ments to the group's power bases, decline progressively as power bases expand.

7. Finally, not unlike capital stocks in general, power bases deteriorate in the absence of maintenance efforts. Economic and political resources must be continuously expended just to keep the power bases at their current level. Efforts must continuously be made to keep the group mobilized, and promotion campaigns must be permanently sustained to just offset public forgetfulness and negative promotion efforts by opposing groups.

Based on these propositions, we offer a particular relationship between the current group's power, $b_i(t)$, and the change in the group's power over time, $db_i(t)/dt$.

In figure 5, this relationship is presented graphically using a state-space description. Consider first the solid curve in figure 5. There are two stable equilibrium power levels, $A_1$ and $A_2$, and one unstable power level, R. The stable equilibria, $b_i = A_1$ and $b_i = A_2$, are point attractors; the unstable equilibrium, R, is a repeller. Commencing from any initial power, $b_i(0)$, to the right of R; $b_i(t)$ converges to $A_2$; while from any initial power to the left of R, $b_i(t)$ converges to $A_1$. Hence, if feasible levels of power are restricted, i.e., $B_1 \leq b_i \leq B_2$ with $B_1 < A_1$ and $B_2 > A_2$, then the interval $[B_1, R)$ constitutes the low power basin (i.e., $A_1$'s basin), while the interval $(R, B_2]$ constitutes the high power basin (i.e., $A_2$'s basin). This structure follows from the propositions stated above. Thus, the positive feedback loops operate in the interval $[A_1, A_2]$. If the group's power happens to equal R [i.e., $b_i(0) = R$], then a small increase in power will induce successive augmentations of group i's power until $b_i(t) = A_2$. However, if $b_i(0) = R$ and a slight decline in the group's power occurred, $b_i(t)$ will continue to decline until $b_i(t) = A_1$ is reached. Note that when group i power is in the neighborhood of $A_2$, $db_i(t)/dt$ is a monotonic decreasing function of $b_i(t)$ because the diminishing return effect in the formation of power becomes progressively more pronounced, and the cost of gross investment in power bases is no more justified by anticipated net future returns to the group from its incremental political power. To the right of $A_1$, even maintaining the power base is no more cost-effective. The behavior of $db_i(t)/dt$. in the neighborhood of $A_1$ reflects a similar relationship for groups opposing the ith group.

If, for some reason, the power bases of group i diminish, the dynamic relationship is shifted downward from the solid curve to the broken curve. The envisaged decline in group i power bases yields two important outcomes: The stable power equilibria move to the left from $(A_1, A_2)$ to $(A_1', A_2')$, and the power of group i diminishes; and the low power basin expands while the high power basin contracts and the likelihood of the fast weakening or the persisting weakness of group i consequently increases. It is worth noting that should $b_i(t)$ happen to be in the interval (R, R') when the decline in group i power bases occurs, $b_i(t)$ will swiftly move to $A_1'$ thus implying the steep fall from power of group i.

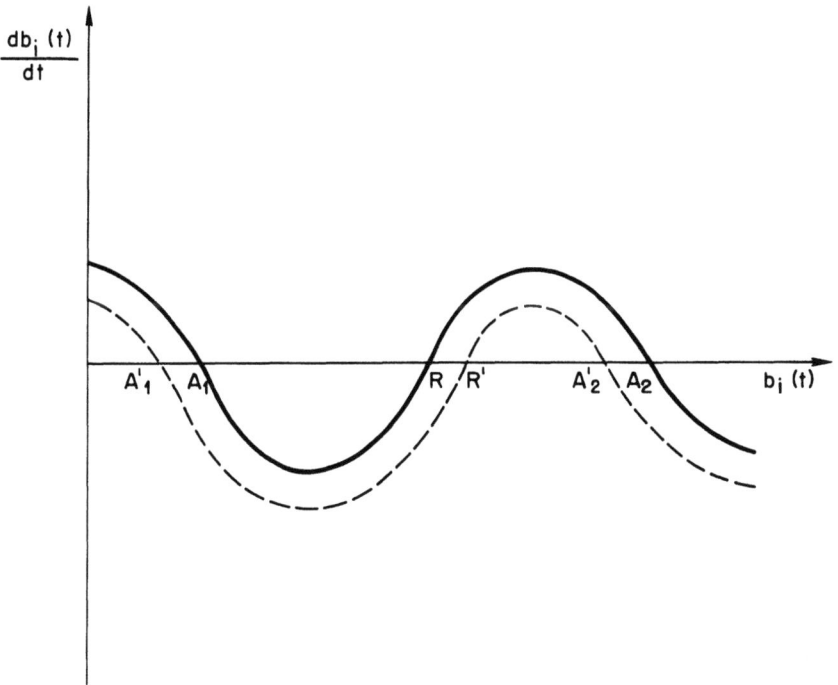

Figure 5. The dynamics of power (a state space description).

A change in a group's power bases may result from two principal causes: external environmental changes and endogenous effects. As indicated earlier, the political economy is strongly influenced by the external physical, social, political, and economic environment. Environmental changes are, therefore, likely to significantly impact the political economy in general and the political power structure in particular.

From the dynamic analytic viewpoint, endogenously induced changes in the power structure are especially interesting. Thus, if policies advocated by a group are widely perceived as failures on efficiency and/or distribution grounds, the legitimate power of the group is eroded. As was demonstrated in the case of water resource systems, districts' narrow rationality is often conducive to severe economic failures. While this process may be protracted, it eventually leads to the weakening of the districts' power and to major policy changes. The corresponding changes in the political economy can be quite dramatic.

## CONCLUDING REMARKS

This chapter has sought to apply a comprehensive theory of a political economy to a water resource management system. The theory endogenized and combined three subsystems: the physical water resource system, the economic structure, and the political power structure. The integrated model yielded several important testable predictions concerning equilibrium policies, their genesis, nature, and performance. It also predicted certain dynamic behavioral patterns.

The model expounded is rather simple, although it seems to highlight the principal forces at work in such a system. Indeed, one suspects that some of the operating principles are generic to a broader class of political economies. There can be no doubt that many of the issues encountered in the present analysis are universal.

It is also evident that concrete systems are much more complex than the hypothetical one explored here. Thus, the assumed hydrological system is unrealistically simple. In particular, water quality aspects, unavoidable in realistic empirical analyses, were completely ignored. Yet, the simplifying assumptions did not detract from the empirical usefulness of the present approach. Any analysis of real world political economies will require fuller modeling of the underlying physical system as well as a more detailed analysis of the economic and political structure. In all likelihood, recourse to simulation and numerical techniques will be unavoidable.

## NOTES

[1] In reality, every district consists of many water users, each of which constitutes an autonomous decision unit. Users in the district are usually served by a local water supply organization which is often incorporated as a nonprofit legal entity. A more realistic analysis should take into account the district's actual organizational structure. The present model ignores these complications in the interest of simplicity and brevity.

[2] Zusman and Rausser (1990b) developed conditions for a locally stable system.

[3] A broader interpretation of the government's goal may identify it with both V and S. Such a formulation, while not unreasonable, would assign the CWP a purely passive political role. Alternatively, we could interpret the "CWP" and the "government,"

respectively, as "the group interested in overall economic efficiency" and "the group interested in lower net government expenditures."

[4]Also see Zusman (1976).

[5]Analyses which take into account the entire system's path may be performed using optimal control techniques. We opted for a simpler presentation focusing on stationary states alone.

[6]This form of organizational failure is explored in Zusman and Rausser (1990a).

[7]It is worth noting that misallocation due to nonuniform social power could be removed by allowing lump-sum transfers (side payments). This device decouples distributional from allocational considerations. In terms of the present formulation, side payments are introduced by setting $s_i(c^0_i, \alpha^0_i) = c^0_i$. For further details, see Zusman and Rausser (1990a). Unfortunately, side payments are ordinarily unacceptable for good constitutional reasons (Zusman, 1990).

## ACKNOWLEDGEMENTS

While writing this chapter, the authors benefited from useful discussions with various persons. In particular, the views expressed by Mr. Habib Fetini convinced us as to the importance of dynamic processes in political economies, especially those related to changes in the power structure.

## REFERENCES

Elster, J., 1989. *The Cement of Society: A Study of Social Order*. Cambridge University Press, New York, NY.

Harsanyi, J. C., 1962a. Measurement of Social Power, Opportunity Cost, and the Theory of Two-Person Bargaining Game, *Behavioral Science*, II, pp. 67-80.

Harsanyi, J. C., 1962b. Measurement of Social Power in n-Person Reciprocal Power Situations, *Behavioral Science*, VII, pp. 81-91.

Magee, S. P.; Brock, W. A.; and Young, L., 1989. *Black Hole Tariffs and Endogenous Policy Theory*. Cambridge University Press, New York, NY.

Olson, M., 1965. *The Logic of Collective Action*. Harvard University Press, Cambridge, MA.

Varian, H. R., 1984. *Microeconomic Analysis*. Second Edition. Norton and Co., New York, NY.

Zusman, P., 1976. The Incorporation and Measurement of Social Power in Economic Models, *International Economic Review*, 17, pp. 447-462.

Zusman, P., 1990. Constitutional Selection of Collective Choice Rules in a Cooperative Enterprise, *Journal of Economic Behavior and Organization*. Forthcoming.

Zusman, P. and Rausser, G. C., 1990a. *Organizational Equilibrium and the Optimality of Collective Action*. Working Paper No. 528, Department of Agricultural and Resource Economics, University of California, Berkeley.

Zusman, P. and Rausser, G. C., 1990b. *Equilibrium States and Dynamic Behavior of a Political Economy: Applications to Water Resource Management*. Working Paper No. 531, Department of Agricultural and Resource Economics, University of California, Berkeley.

# 38 WATER MARKET REFORMS FOR WATER RESOURCE PROBLEMS: INVISIBLE HANDS OR DOMINATION IN DISGUISE?

Norm Coontz, U.S. Bureau of Reclamation and San Joaquin Valley Drainage Program

### ABSTRACT

Many neoclassical economists account for water resource scarcity and related problems as a distortion of economic processes by political institutions, which, they claim, allow and encourage individuals to exercise arbitrary and capricious power. Arguing that competitive water markets can replace and thereby neutralize existing structures and distributions of power, these economists promote politically oriented policy prescriptions to reform water rights. It follows that neoclassical theory, which claims to explain the origins and development of political and economic interests, must account for the political support and opposition found for its own reform proposals. Neoclassical theory does not, however, account for the current experience in California with water right reform proposals. A case study of the development of private property rights to Kings River water demonstrates that, contrary to neoclassical theory, water rights institutionalize relations of power; preserving proprietary relations of power may supersede other economic interests in development when power is challenged.

### INTRODUCTION

In 1776, Adam Smith's *Wealth of Nations* firmly established the idea that power over economic processes generally obstructs collective interests in the efficient production and distribution of wealth. According to Smith, collective interests are most effectively met when individuals are free to pursue self-interests in competitive markets with a minimum of political interference. Although an individual always "intends only his own gain," in free markets power is neutralized by competition, and individuals are "led by an invisible hand to promote an end which was no part of (their) intention." Building on

Smith's work, neoclassical economics makes the neutralization of power in economic development one of its central problems.

Public water policies are currently being reevaluated in response to claims that many Western water problems are caused by the inefficient development and utilization of water resources. For example, water quality problems associated with subsurface agricultural drainage are often attributed to the excessive application of low-cost subsidized water provided to farmers by public water projects. One principal reason for inefficient development and allocation of water resources given by neoclassical analyses of Western water problems is the distorting influence of power (see, for example, Anderson, 1983)[1].

Neoclassical economists generally agree that a fundamental source of this power (the ability of one actor to realize his/her will in the face of opposition from others) is the structure of water rights; Western water rights lend politicians, bureaucrats, and special interest groups power to exploit control over water resources in their own self-interests, distorting costs and benefits faced by others, and obstructing an efficient allocation of resources in society. While recognizing that both economic and political arrangements are elements of all allocation processes, neoclassical theorists argue that competitive markets neutralize power and enable the "invisible hand" to direct individual self-interested behavior into achieving an efficient, equitable allocation of water resources. Consequently, neoclassical economists have urged that water rights be reformed to promote a greater reliance on the market for allocating water resources.

Neoclassical theory holds that individuals will support or oppose reform proposals according to whether their economic interests are promoted or challenged. Therefore, a crucial test of the theory is its ability to account for the support and opposition found for its own recommendations. This theory leads to the expectation that politicians, special interest groups, and bureaucrats would oppose power-neutralizing water market reforms; yet the most visible support for market oriented reforms appears from government water bureaucracies (DWR and USBR, 1986), politicians and their staffs (Assembly Office of Research, 1982 and 1985), and special interest groups (Graff, 1986)[2]. More puzzling is the passive silence of individuals who, according to the theory, would be expected to promote water market reforms: the owners of strong, well-defined, but difficult to exchange water rights. If water resources are inefficiently allocated, then opportunities should exist for water right holders to sell their property to higher value uses and they should expect opportunities to reap larger profits from freely marketable water rights. Why aren't they actively supporting neoclassical water right reform proposals?

This chapter argues that private property institutions are social structures of power which enable owners (such as water right holders) to control and dispose

WATER MARKET REFORMS 761

of a valued resource in the face of opposition. Other interests, such as profit maximization, may easily become subordinated to interests in preserving power when property rights are threatened. Consequently, support for or opposition to reform proposals depend on who stands to win or lose by altering relations of power. Furthermore, the capacity to oppose or impose reforms is largely determined by the relative distribution of power among interested parties. Rather than neutralizing power, property right reforms express developing relations of power. Theory that claims property right reforms neutralize power can only disguise the extent and nature of power and provide ideological defense for the dominant interests in reform movements. The inability of neoclassical theory to account for the support or opposition to its reform proposals derives from the failure to recognize that social relations of power are the essence of private property rights.

This chapter first presents a case study of the development of property rights to water resources on California's Kings River[3]. Power over scarce water resources was first institutionalized by the creation of strong property rights to river water; other interests in the development of river water resources then became subordinated to the preservation and extension of the social relations of power institutionalized in those property rights. The chapter concludes that neoclassical theory must recognize that property rights always structure relations of power.

## CASE STUDY

The waters of the Kings River irrigate an area of over 1 million acres in Fresno, Kings, and Tulare Counties in California's San Joaquin Valley (Valley), counties that consistently lead the Nation in agricultural productivity (figure 1). River runoff has ranged from 391,700 acre-feet to 4,476,400 acre-feet between 1895 and 1986, averaging 1,713,600 acre-feet (KRWA, 1987). Rights to Kings River water are claimed by the 28 organizations that constitute the membership of the Kings River Water Association (Association), a nonprofit unincorporated organization that holds water right claims in trust for its members and administers the waters of the river according to private contracts: a water right indenture and an administrative agreement (KRWA, 1972).

Pine Flat Dam, located near where the river discharges from the Sierra Nevada Mountains onto the Valley floor, is capable of storing a million acre-feet of water. Built in the early 1950's by the U.S. Army Corps of Engineers (Corps), the dam provides conservation storage, flood control, and hydroelectric power generation. Kings River Conservation District, with virtually the same boundaries as the Association, is a special district which administers

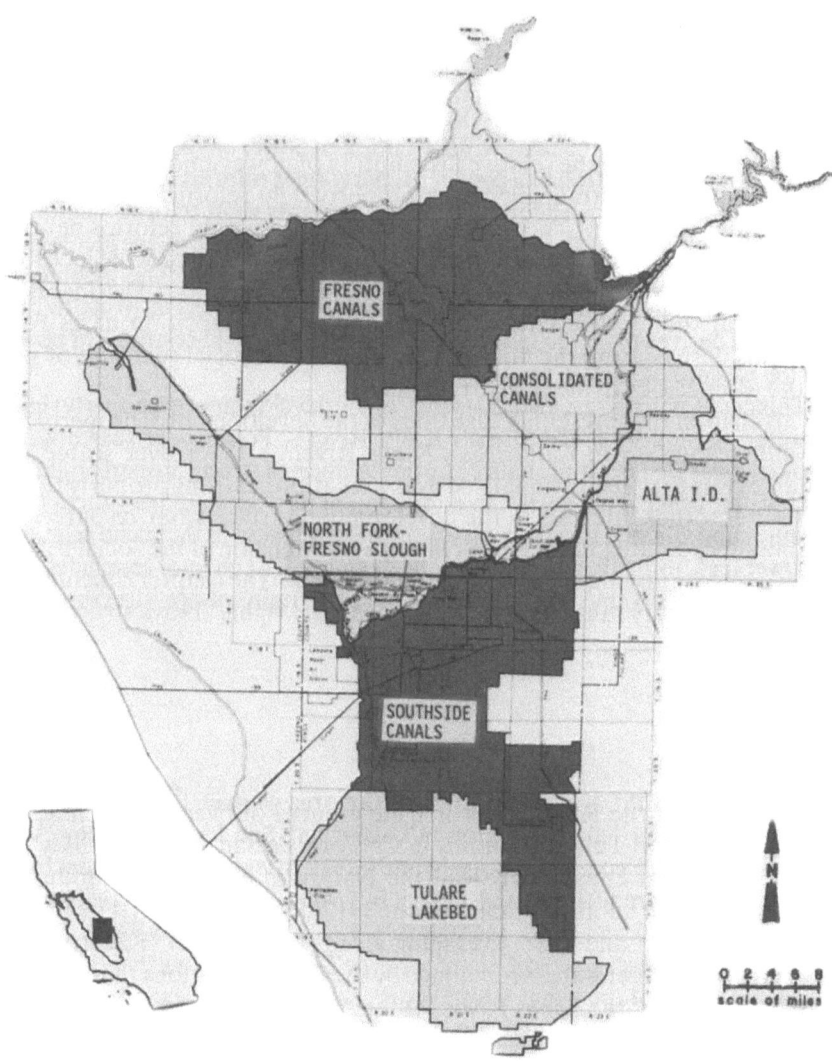

Figure 1. Kings River Service Area.

power generation and flood control at Pine Flat Dam in cooperation with the Corps.

Two phases of development characterize the institutionalization of political and economic interests in Kings River water. First, over the period 1870-1927, unstable and limited relations of power were transformed into strong private property rights to river water resources. During this period contractual property rights displaced water right claims as power was institutionalized. In the second period, 1927-63, preservation of relations of power institutionalized in contractual property rights became the fundamental interest of water right claimants in the development of the hydroelectric power generation, conservation storage, and flood control facilities on the river. During this period the ability of contractual rights to define and enforce property rights in the face of opposition by powerful government agencies was decisive in determining the course of water resource development.

## 1870-1927

Natural overflows irrigated pasture land adjacent to Kings River before minor diversions began in the late 1850's. Besides grazing, agricultural development was largely confined to dryland wheat farming on huge tracts of land until about the 1870's. Changing markets, increasing land values and taxes, and depletion of the land by continual wheat production contributed to a relative decline in grazing and grain farming. Large-scale irrigation diversions on Kings River began about 1870, contributing to the rapid diversification of agricultural production in the region. The coming of the railroad and post-Civil War land speculation encouraged agricultural ventures dependent upon irrigation (Preston, 1981).

After the brief period of small-scale irrigation, irrigation projects were developed on Kings River by canal companies to support agricultural land settlement. Companies were organized by land speculators as commercial utilities to sell water to settlers in conjunction with colony land development schemes. Other companies originated as nonprofit cooperative enterprises to convey water from a natural waterway or a commercial utility's facilities among farmers developing their own land.

Under the colony system, land development proved to be extremely profitable for developers, who concentrated their activities in upstream areas in what are now the Fresno, Consolidated, and Alta Irrigation Districts. Large tracts of unimproved land were acquired for development from Mexican land grants, public grants to railroads, purchase of land in the public domain from the Federal Government with college script, and purchase of swamp and other land from the public domain. Colonies were developed and sold as farms by subdividing large tracts of land into smaller parcels (usually 20 acres), constructing roads (and sometimes community and agricultural facilities), claiming water rights and building water conveyance facilities, advertising widely, and providing purchasers with liberal credit terms. The commercial canal operations were organized to support profitmaking from land development rather than to generate profits by themselves. Water contracts between companies and developers or colonizers tied water to specific parcels of land. Costs for water service were fixed over the term of a contract, typically established for a 50-year period. Developers such as Moses Church, William Chapman, Issac Friedlander, James Haggin, and Lloyd Tevis gained much wealth and notoriety through land development schemes (Maass and Anderson, 1978).

While large blocks of land were initially purchased at very low cost, successful development of colonies was expensive and quickly came to depend on outside financing. Absentee investors, banks, and foreign insurance companies were attracted by the spectacular returns some early developers realized. Often involved in a number of risky undertakings, land developers frequently overextended themselves. Foreclosures usually resulted in creditors taking control of undeveloped tracts of land, canal companies, and water right claims. Banks and other creditors then found themselves holding assets they had little interest in administering and which promised uncertain returns on investment without further development. Liquidating these questionable assets required patience and perserverance; frequently, a creditor's alternative was to enter into new speculative deals, possibly with the same developer, with eventual reimbursement tied to the success of yet another development scheme (Maass and Anderson, 1978).

Nevertheless, initial development proceeded quickly, and claims to water in the late 1870's and early 1880's soon exceeded the normal flow of the river. Disputes among water right claimants increased dramatically, especially during lower stages of riverflow and during dryer years. Creditors, developers, and colonizers found their prospects for profits increasingly tied to the strength of their property rights in water. Hundreds of lawsuits were filed as claimants turned to the courts to define and enforce those property rights.

Early litigation contested rights to the most valuable river water: the lowest and most reliable flows, about the first 1,900 cubic feet per second (ft$^3$/s). Flows in excess of 1,900 ft$^3$/s were either floodflows, which allowed all riverwater claimants to divert as much water as possible without depleting the river, or rose or fell too rapidly to make much difference.

Attempts to resolve disputes over low flow claims through litigation proved costly and inconclusive. Every company justified its claims to property rights in riverwater upon pre-1914 water right doctrines related to riparian ownership of land, prior appropriation, and/or prescription through adverse possession (Barnes, 1918). However, little reliable information on Kings River flows and water use, crucial to judicial decisions, was available to resolve these disputes. Riverflow measurements were practically unavailable before 1895, and flows were subject to extreme annual, seasonal, and even diurnal variations (Conkling and Kaupke, 1923). Furthermore, early speculators, intent upon land development before riverwater was scarce, had neglected to measure diversion volumes or to keep adequate records of beneficial use of diverted water. Canal managers fabricated and exaggerated data for the courts' consideration; fear that accurate measurements could be used to support the claims of an adversary helped to prevent the compilation of accurate information.

Superior courts in three counties made numerous and oftentimes contradictory judgments and decrees, sometimes contradicting even their own, earlier decisions. According to Barnes (1918), water right litigation on Kings River resulted in an increasingly incomprehensible body of conflicting and contradictory judgments and decrees:

"... (L)itigation does not seem to leave anything permanently settled. Cases are still pending in the superior courts of Kings, Tulare, and Fresno Counties, before the State Supreme Court and the Appellate Court, and no sooner is a point apparently settled than it is attacked from some different angle... In Fresno County we see Emigrant Ditch Company given a right as against Laguna De Tache Ranch and the whole world, to divert 190 second feet from the river. Some years later, in the Kings County court, Peoples, Last Chance, and Lower Kings River ditch companies are all given rights superior to Emigrant Ditch, and both Kings County and Tulare County courts give Laguna de Tache Ranch rights to water ahead of any other diversion on the river. In the Tulare County court, Fresno Canal and Irrigation Company was enjoined from diverting any water, but some years later was decreed a right to 1,000 second-feet by the Fresno courts. At one time we see Centerville and Kingsburg Irrigation Company enjoined from diverting any water at all, and another time being given a right to 600 second-feet."

Because one judicial decision could hardly be enforced without violating others, officials exercised a great deal of discretion in interpreting and enforcing complex and contradictory court decisions. The vast majority of the population of the region, residing in the upstream service areas being settled through colonization, exercised a disproportionate influence over elected officials with responsibilty for enforcing water rights. Judges and county sheriffs were understandably reluctant to enforce water rights that threatened the interests of their constituencies. The net result was that problematic decisions over the earliest claims, covering the most valuable water, were virtually unenforceable.

Unenforceability worked to the advantage of the upstream interests who, by virtue of their physical location, held a clear advantage over downstream diverters; they simply took as much of the low flows from the river as they wanted. Diversion works were constructed at several places along the river to make surveillance more difficult, and threats of physical violence were occasionally employed (Morison, 1988). Unenforceability and upstream advantages provided a foundation of power over downstream claimants in early distribution of the riverwater under low flow conditions.

Early power, while sufficient to take water, was unstable and insecure because it did not diminish opposition by competitors. Furthermore, it was limited to the low flow conditions. As developers continued to subdivide land, they expanded diversion capacity and diverted larger volumes of water to colony developments. Disputes arose between colony regions, politicizing nonenforcement. Expanded diversions were opposed with increasingly more precise and complete records. Litigation limited diversions based purely upon upstream advantage, especially after 1886 when the State Supreme Court unexpectedly determined that riparian claims (which on Kings River were strongest midstream) to be superior to other claims. As a result, litigation, which was still unenforceable, became increasingly effective in opposing the expansion of established power to control riverwater. Developers', colonizers', and investors' interests in land development tied to property interests in water became insecure. By 1887 colony development had come to a halt.

An interim strategy to overcome these constraints and obtain more defensible legal claims to riverwater for expanding upstream colony development was to purchase land with strong riparian claims and use them for upstream development. The 64,000-acre Laguna de Tache Ranch, widely thought to have the strongest riparian claim on the river, was purchased by developers in the Fresno area through the sale of $1 million in canal company bonds to an English insurance company. Acquiring this powerful claim eliminated obstructing litigation; the loan also supplied substantial capital to expand river diversion capabilities. This touched off a spectacular land boom in the developing Fresno area colonies that didn't bust until the depression of 1892-94. In 1894, when developers began going broke during the depression, the English creditors took control over the Fresno area canal companies, land, and the Laguna Ranch through foreclosure.

The Peoples Ditch Company, Last Chance Water Ditch Company, and the Lower Kings River Canal Company (now the Lemoore Canal and Irrigation Company), held some of the earliest and strongest appropriative water right claims on the river. These companies were mutual-benefit, nonprofit corporations. In contrast to the commercial canals, the mutual canals were not constructed with bank loans or bond offerings; farmer/owners, usually short on cash, obtained equity in a company in exchange for their labor in construction of company facilities. These companies, nowhere near as powerful as upstream developers, obstructed development through litigation by uniting as the Southside Group to pool their resources and their claims in opposition to the upstream diversions. Unable to prevent the colony developers from diverting lower riverflows, the Southside Group exercised powerful opposition in its ability to constrain expanded upstream development through litigation. Additionally, in 1893 Kings County split off from Tulare County, providing a direct political base sensitive to the interests of midstream Kings County diverters.

In the face of litigious challenges undermining its control of riverwater, the Fresno Group used its power to protect its economic interests by initiating and guiding the transformation of unenforceable water right claims into strong private property contract rights to the lower flows of riverwater. Power to sustain this process stemmed from several sources. First was its upstream location; only the Alta Irrigation District diverted water at a higher diversion point. Second, the Fresno Group had acquired several opposing claims based on a variety of pre-1914 water right doctrines. Ownership of riparian, appropriative, and prescriptive claims up and down the river enabled it to launch an endless barrage of legal assaults against its opponents. Third, the Fresno Group was by far the most affluent claimant, able to outspend competitors in increasing conveyance capacity as well as for litigation. Finally, the Fresno Group conveyed water to the populous Fresno area, which helped to influence the definition and enforcement of water right claims by Fresno County officials, who held jurisdiction over the upper reaches of the river.

Contract property rights to riverwater were first established between the Southside and Fresno Groups in 1897, creating an institutional basis to define and enforce property rights in Kings River water. The agreement suspended the rights of the signatories to litigate water right claims between themselves and defined property rights to riverwater according to a formal schedule, based upon riverflows up to 1,900 ft$^3$/s. The Fresno group initiated the transformation of water right claims to strenghten and stabilize its control over river diversions, enabling it to define and enforce its insecure property rights as it developed and disposed of its assets. By entering into the agreement, the southside ditches obtained contract property rights to low flows which could be enforced against the Fresno diverters. Shortly after signing the agreement, each member of the Southside Group signed a similar agreement to suspend litigation and distribute their contract entitlement among themselves.

Questions of fact regarding the extent and nature of a water right became irrelevant as claimants were bound together by contract relations that institutionalized existing distributions of power. Suspending litigation and transforming unenforceable water right claims into contract property rights strengthened the preexisting distribution of power by altering the structure of power among claimants. Instead of relying on unenforceable claims to river water, low flow claimants now held contract property rights formally sanctioned by competitors, whose interests were also tied to power structured by those rights.

As the turn of the century neared, the construction of a storage reservoir was becoming critically important to the continuing development of the region. Variable and unpredictable flows in the river ensured periodic shortages and droughts on the one hand, and damaging floods on the other. Without reservoir storage to conserve higher flows for later use, claims to riverwater exceeded normal flows under uncontrolled conditions. Ground-water pumping was becoming common and helped to supplement unpredictable riverflows, but seepage from canals was extremely important for recharging a declining water table that was increasingly expensive to pump. Even before the U.S. Geological Survey (USGS) conducted a hydrographic survey of Kings River water resources in 1900, the benefits of storage conservation were obvious to developers and irrigators whose interests were dependent upon controlling flows from a variable and unreliable river.

The USGS report and subsequent investigations led to recommendations for a 600,000 acre-foot storage project at Pine Flat. Control and receipt of stored water would depend on the power to define and enforce rights to the higher riverflows to be stored behind the dam. Investigations into a storage project aggravated conflicting interests in rights to those flows. Extending power over higher flows was opposed by water right litigation; but litigation alone was still unenforceable and too weak to overcome opposition by other claimants. Between 1900 and 1918, various water right claimants filed over 100 lawsuits. Nothing was settled.

The Fresno Group increased its holdings and power by acquiring the upstream Consolidated Canal Company in 1901. It now owned nearly all of the river claims and related facilities in Fresno County, and served an area of approximately 500,000 acres (Teilman and Shafer, 1943). In 1909 the Fresno Group filed the first claim with the U.S. Land Office for use of the Pine Flat reservoir site. Four competing interest groups now significantly influenced the development of rights to Kings River water: the Fresno Group, the Southside Group, the Alta Irrigation District (Alta), and the Tulare Lakebed interests.

Alta was incorporated in 1888 in order to purchase the facilities and water right claims of the 76 Land and Water Company. Having disposed of nearly all of its land, the company sold its claims and conveyance facilities to Alta in 1890. Increased riverwater supplies to sustain and expand colony development was very much on the minds of the members of Alta's Board of Directors, some of whom were developers themselves. Although its water right claims were based on appropriative doctrine, filed later than most claims and therefore relatively weak, Alta enjoyed the highest diversion point on the river and the strongest upstream advantage. After incorporating as an irrigation district, Alta was able to sell bonds to raise substantial sums of money to rapidly expand its diversion capacity and litigate in the courts. The population within the district, significant and growing rapidly, also lent support to the district's influence.

The lakebed interests diverted water from the river at the lowest point and held claims to only the higher riverflows. As the development of the region continued, upstream diversions made use of higher and higher flows. Like the other three interest groups, the lakebed interests wanted a storage project to schedule release of riverflows and to increase the available water supply. Because lands in the lakebed were much more vulnerable to periodic flooding, the lakebed diverters were determined to expand the size of the storage project from 600,000 acre-feet to 1,000,000 acre-feet to provide flood control protection. This interest group was situated to litigate against all other claimants, effectively challenging the power of all other water right claimants.

The years 1912 and 1913 were very dry. Insufficient riverflows and declining water tables made pumping by farmers both more necessary and expensive, generating unrest in the densely populated colony settlements. Well attended public meetings in 1914, 1915, and 1916 signaled a popular movement for progress on construction of the storage project (Maass and Anderson, 1978). Although water right claimants neither initiated nor actively participated in the agitation by colonist irrigators (Kaupke, 1957), the popular movement to develop the storage project began to pose a threat to the power and, thus, economic interests of water right claimants.

Meanwhile the State Railroad Commission had acquired greatly expanded regulatory power over public utilities through progressivist reforms of 1911 to 1913. Public utilities included commercial canal companies, but not mutual canal companies or irrigation districts. The water right contracts between the colonists and the companies held by the Fresno Group were to expire in 1921. Requests for large increases in water charges by canal companies, and a number of questions regarding the rights and obligations of both companies and colonists were unsettled. Colonists were preparing to use the Railroad Commission to oppose rate increases when contracts expired, they also appealed for Railroad Commission intervention to order changes in commercial canal maintenance and administrative practices (Maass and Anderson, 1978). However, while Railroad Commission actions had not yet seriously affected water management and development, extensive litigation, not only in the courts but also with the Railroad Commission's administrative rules and procedures, threatened to weaken power in property rights to riverwater.

Popular agitation and State regulation were not the only challenges to power structures encompassing the economic interests of water right claimants. In 1902 the newly organized U.S. Bureau of Reclamation (Reclamation) began to consider a Kings River storage project for hydropower and conservation storage. Reclamation's chief engineer spent several months trying to obtain support for a Reclamation storage project, but was unsuccessful and abandoned the project, at least temporarily (Conkling and Kaupke, 1920).

The San Joaquin Light and Power Company (SJLPC) began exploring the potential for storage projects in the Kings River watershed in 1912. In 1914, the State Water Commission was formed by ballot initiative to administer any subsequent water rights claims, including storage rights on the Kings River. In 1917, SJLPC applied to the Commission for a permit water right to store high flows. However, because permit rights were inferior to pre-1914 rights (still subject to unenforceable litigation), permit rights for storage could not be defined and enforced. The utility company tried to sidestep the permit process by negotiating an agreement with various riverwater claimants to provide conservation storage space behind its proposed dam in exchange for rights to generate hydroelectric power from the stored water. Meanwhile, Reclamation continued to compete for the project as well, submitting a report to riverwater claimants on specifications and costs of a storage project, with hydroelectric facilities to be operated by the utility. The proposals were stymied by the usual water right claim disputes (Kaupke, 1957).

In 1920-21, the city of Los Angeles' Department of Water and Power, infamous for its diversions of water from the Owens Valley through the Los Angeles Aqueduct, was conducting surveys of the Kings River watershed to assess the hydropower potential of a storage project. By 1929 Reclamation, working with Los Angeles' hydropower interests, was investigating a Kings River storage project that might increase electricity generation Statewide by as much as one-third (Reclamation, 1930).

Competition for river water resources from powerful hydroelectric interest groups, from State regulatory agencies, and from a popular movement to develop the storage project threatened to displace existing power structures. Transformation of water right claims into strong property rights to higher flows and storage took on an added sense of urgency. In 1916 representatives of pre-1914 water right claimants formed a committee to explore strategies to develop the storage project.

The committee initially settled on a strategy of forming a regional special district under local control to administer construction and operation of the Pine Flat project. After an attempt to incorporate a regional water storage district failed in 1917 the river interests drafted, sought, and succeeded in passing the California Water Conservation District Act in 1923 (Adams, 1929). Under this law, a regional district would represent a consortium of local special districts with water right claims. The distribution of storage costs and storage rights was to be determined before a vote on incorporation could be taken; each local district could independently decide whether to become part of the regional district and participate in the storage project (Maass and Anderson, 1978).

In 1917 the committee of water right claimants, dominated by the big four interest groups and led by Alta requested the State Water Commission to assign an engineer to take the first accurate measurements of canal diversions in preparation for developing strong property rights to the river. An arrangement gave the engineer authority to measure all canal diversions for 1918, temporarily suspended water right litigation among claimants, and prohibited use of the measurement data in water right litigation. In 1919 the engineer began to administer the 1897 contract rights to low flows, but continuing differences between the main four interest groups prevented settlement of disputes over higher flows.

In the meantime the Fresno Group began liquidating assets threatened by public agitation and Railroad Commission intervention. The Fresno, Consolidated, Laguna, and Riverdale irrigation districts were formed between February 1920 and August 1921 to purchase water right claims and conveyance facilities covered by colony contracts (Tielman and Shafer, 1943). The districts were formed to participate in a regional Pine Flat storage project and to avoid Railroad Commission intervention in settling pricing and administration disputes (Maass and Anderson, 1978). The value of these properties was directly dependent on the power to enforce and transfer water right claims reinforced by the 1897 contracts, diversion measurements, and engineer administered diversions. The Fresno Group's economic interests, as well as those of the purchasing districts, would be substantially enhanced by the extension of the contracts to the higher riverflows.

To "encourage" agreement on a schedule extension, the Fresno Group threatened the fundamental economic interests of the other three groups and their constituent members, initiating a joint lawsuit on behalf of the downstream Stinson Canal Company and the neighboring Cresent Canal Company against a midstream diverter. Litigation would undermine the structure of power over riverwater, obstruct progress toward a regional storage project, and render water right claimants more vulnerable to the general public, State regulators, and outside interests.

As the trial date approached, resistance to a temporary extension of the schedule, administered by the State's engineer, dissolved. By September 1921, 35 claimants, covering 95 percent of the river's water, agreed to allow the engineer to develop, interpret, and administer a temporary schedule covering the first 10,000 ft$^3$/s of flow (Kaupke, 1957). Beginning in 1922, the agreement was automatically renewed each year unless explicitly rescinded the preceding year. Litigation, the only apparent alternative, was suspended so long as the parties remained bound by this arrangement. The Fresno Group's sale of claims and facilities was completed between May 1921 and July 1922 for an approximate total of $2,863,000 (Adams, 1929).

Work toward a permanent settlement of rights to the river's water resources progressed, and a proposed schedule defining the first 10,000 ft$^3$/s for each month of the year was submitted to the Pine Flat Project Committee in May 1926. The proposal was unacceptable to the lakebed interests, and for an entire year settlement was delayed. In May 1927, 19 claimants from the other 3 interest groups signed permanent contractual agreements defining relative rights and establishing the Association to administer the contracts. Although the lakebed interests rejected the allocation schedules and refused to join the

Association, they did continue to pay for the administrative services of the Association's watermaster, who replaced the State-appointed engineer.

## 1927-1963

The contractual agreements signed in 1927 represented the culmination of the development of strong, stable relations of power necessary to create well-defined, enforceable property rights. The 1927 contracts preserved and strengthened power in four ways. First, strong contract rights displaced weak water right claims as the basis of power to control the first 10,000 ft$^3$/s of riverflow; all subsequent claims to higher flows would be inferior to these rights. Second, contract rights significantly expanded property rights to flows by defining rights for each month of the year, critical for storage operations. Third, the creation of the Association structured a permanent organization to administer the river and protect the structure of power supporting interests depending on riverwater. Finally, virtually every economic interest in riverwater resources became tied to these contracts which displaced water right litigation and upstream advantages. The preservation of contractual relations of power now became the paramount interest in Kings River water resource development for water right claimants.

Three factors intervened at this point to impede local development of the Pine Flat project: Disputes over reservoir size, competition for rights to stored waters and storage space, and the Great Depression. First, the lakebed interests refused to be bound by the permanent agreements and join the Association until differences over rights to stored waters and storage space were resolved. Second, the lakebed interests also objected to construction of a smaller reservoir without flood control capabilities and continued to threaten water rights litigation--a threat they occasionally exercised. The third problem was that the Great Depression was making itself felt in the agricultural economy; local funding of a project dependent on the prosperity of agriculture was increasingly problematic.

The storage project was stalled until the Flood Control Act (Act) of 1936 made construction of flood control projects on navigable streams the obligation of the Federal Government. The Corps was directed by the Act to investigate a flood control project in the Kings River watershed. In the interim, Reclamation had assumed responsibility for construction and operation of the Central Valley Project (CVP) from the State in 1934. In connection with the then proposed Kings Canyon National Park, Reclamation launched its own investigation of the water and power resources of the Kings River watershed. In 1940 both Federal agencies released reports on the project. Chances for a regionally administered project dwindled rapidly as the probability of a Federally constructed project grew.

The Corps' report found flood control to be the principal benefit of a storage project at Pine Flat, with incidental conservation and hydroelectric generation benefits. Under a Corps' construction project, riverwater users would be obligated to repay the United States for that portion of the project attributed to incidental benefits, and their control over the river would be subordinated only to flood control operations (Maass and Anderson, 1957).

Reclamation's report, on the other hand, found that flood control and hydroelectric generation were secondary benefits, and that conservation storage was the primary benefit of a Pine Flat storage project. On this basis, Reclamation pursued Congressional authorization of the project to be constructed and operated by Reclamation, financially and operationally integrated into the CVP, and subject to the provisions of Reclamation law. A Reclamation project would be developed and administered in support of Reclamation's State and National interests and priorities, regardless of how this corresponded with the interests of local developers and irrigators. Riverwater and power resources could be transferred to other units of the CVP instead of supporting local and regional development interests; the Association's schedule and administrative structure to control riverwater resources would be displaced by contracts negotiated and administered by Reclamation; and acreage limitation would force the breakup of larger farms, mainly in the lakebed, and empower Reclamation to oversee its redistribution (Maass and Anderson, 1979).

After intense political maneuvering by local interests, the Corps, and Reclamation, Congress authorized the Pine Flat project as a Corps project in the Flood Control Act of 1944. However, in May 1946 President Truman directed Reclamation to represent the United States in negotiating repayment contracts. Reclamation at first attempted to treat riverwater users as it would any contractor for Federal project water, ignoring their claims to the river supported by strong private property rights. Reclamation did not submit a draft contract until January 1950, after construction had already begun (Maass and Anderson, 1978 and Kaupke, 1957).

Riverwater claimants closed ranks in opposition to Reclamation's assault on the property structures of power which supported their economic interests. In 1949 the lakebed interests joined the Association and ratified a new allocation schedule. The Association had also fostered alliances with a powerful outside interest that supported its struggle against Reclamation. After adoption of the 1927 contracts, the San Joaquin Light and Power Company (SJLPC) concluded an agreement with the Association for rights to a streamflow plant, with waterflows administered by the Association's watermaster. The Pacific Gas and Electric Company (PG&E) subsequently acquired SJLPC and the streamflow plant, and continued to develop its hydropower interests in conjunction with the Association. Eventually, PG&E obtained additional hydroelectric generating capacity in the watershed, as well as the right to market the electricity generated at Pine Flat Dam.

After 1950 Reclamation repayment negotiations dragged on for several years. Construction progressed to the point that by 1953 storage operations could begin. The superior power of property rights wielded by the Association now became apparent. Reclamation was too weak to overcome Kings River water right claimants. Water right litigation which had undermined power among water right claimants under conditions of unenforceability was eliminated by mutually agreed upon contract rights. State administered permit storage rights were subordinated to pre-1914 water right claims, and local and regional interests were first to apply for permits to storage and power flows not covered by pre-1914 claims. The Corps' flood control interests did not compete with the Association-controlled contract rights.

Reclamation had little choice but to sign a temporary contract with claimants in 1953 to allow storage behind the dam. The contracts allowed the Association's watermaster to administer storage and release water according to the Association's contracts, for $1.50 per acre-foot. Most of this fee was applied toward the irrigator's repayment obligations once a permanent agreement was reached. The temporary contracts were renewed until 1963, when a permanent contract was signed. Between 1936 and 1963 the Association successfully resisted Reclamation's attempt to take control of Kings River water resources (Maass and Anderson, 1978).

## CONCLUSIONS

Neoclassical theory claims that power causes individuals to obstruct efficient economic development by promoting self-interests in ways that inevitably distort efficient economic decisionmaking. Many neoclassical economists therefore recommend institutional reforms to neutralize power and allow the "invisible hand of the market" to direct economic development. But reforming existing institutions to define and enforce marketable water rights obviously requires power to alter existing structures and distributions of power over water resources. This creates a paradox for neoclassical theory; how can power neutralizing reforms be instituted if efficient, equitable reform depends upon the acquisition and application of distorting power? Clearly, the inevitable effects of power cannot be eliminated through the application of power. Failure to identify, much less resolve, this paradox lies at the heart of the inability to explain the behavior of powerful actors, such as Kings River water rightholders, with direct economic interests in water right reform proposals.

Power is a pervasive dimension in all human relationships, property and markets included (Weber, 1968). Property reform proposals, irrespective of their structure or intent, will affect relations of power among the interested parties. Even a cursory survey of the interminable struggles over California's water resources suggests that reforming water rights to allocate water resources in competitive markets will be highly politicized, requiring considerable power to carry through.[4] Intensifying competition for developed water supplies will mobilize special interest partisans to launch reform campaigns in the State's perennial water wars.

Neoclassical analyses that obscure the reconstruction of power may provide appealing ideological weapons for urban, agricultural, environmental, or other special interest groups competing for control of the State's scarce water resources. Is it any wonder that Kings River water rightholders have not pursued potential opportunities to maximize profits by altering Kings River water rights? To rephrase Proudon, property is power--the power to overcome the opposition of competitors, power to defend control over a valued resource, power to shape competition for scarce resources, power to define and enforce economic and other interests in competitive processes.

## NOTES

[1]Externalities and bounded rationality are important concepts used in neoclassical property reform strategies, but should not be seen as offering alternative theoretical accounts of water resource problems. The concept of externalities is used to justify power in economic processes in which some costs or benefits arising from a decision do not accrue to the decisionmaker, leading to market failure. Bounded rationality describes situations where decisionmakers, do not have full knowledge of their alternatives, the consequences of their actions, or the validity of claims made by others in the market, and where obtaining such knowledge either is not cost effective or is impossible to obtain, again causing the market to fail to generate economic efficiency. These concepts are applied by neoclassical economists to determine the limits of market institutions in achieving economic efficiency. Disputes among neoclassical economists arise over whether and to what degree government intervention into economic processes is justifiable; the competitive, power neutralizing market is the normative ideal that guides these analyses.

[2]Apart from the support of professional economists, who would seem to have few direct economic interests in water markets.

[3]Data for this case study were derived from Federal Government and State of California records, Federal, State and local government reports and other documents, and the existing literature on the region and related issues. Field notes compiled from first-hand observations and interviews of more than 30 persons involved in agriculture and water management in the Kings River service area also provided information. Interviewees included public officials, area farmers and their technical advisors, managers

of special districts and mutual water companies, regional watermasters, and officials of local, regional, State, and Federal government agencies.

[4]Competition for control over California's thinly stretched water resources will intensify as water becomes increasingly scarce. Contamination and overdrafting of groundwater resources threatens to limit the production of usable water from wells. Yields from existing and potential surface water projects are being compromised by increasingly restrictive environmental constraints. The most economically feasible projects in the State are already developed and appropriated; new water projects are difficult for private interests and local governments to finance; prospects for State and Federal government financial support for new water projects have dwindled. Meanwhile, demand from a growing population and related development is rising.

## REFERENCES

Adams, F., 1929. *Irrigation Districts of California*. Bulletin No. 21, California Department of Public Works, Division of Engineering and Irrigation, Sacramento, CA.

Anderson, T. L. (Ed.), 1983. *Water Rights: Scarce Resource Allocation, the Bureaucracy, and the Environment*. Ballinger Publishing Company, Cambridge, MA.

Assembly Office of Research, 1982. *A Marketing Approach to Water Allocation*. California State Assembly, Sacramento, CA.

Assembly Office of Research, 1985. *Water Trading: Free Market Benefits for Exporters and Importers*. California State Assembly, Sacramento, CA.

Barnes, H., 1918. *Use of Water from Kings River*. Bulletin No. 7, California State Department of Engineering.

California Department of Water Resources and U.S. Bureau of Reclamation, 1986. *Buying and Selling Water in California*. Santa Monica, CA.

Conkling, H. and Kaupke, C. L., 1923. *Kings River Investigation*. Bulletin No. 2, California Department of Public Works, Division of Water Rights.

Coontz, N. D., 1990. *Organizations and Institutions: Agricultural Drainage-Related Water Management in the Kings River Region*. San Joaquin Valley Drainage Program.

Graff, T., 1986. Reflections on Water, *Aqueduct*, 4, pp. 2-3.

Kaupke, C. L., 1957. *Forty Years on Kings River*. Hume Publishing and Lithography Company, Fresno, CA., pp. 1917-1957.

Kings River Water Association, 1972. *Kings River Agreements and Pine Flat Contracts*.

Kings River Water Association, 1987. *Watermaster Report for the Year 1985-86*.

Maass, A. and Anderson, R. L., 1978. *And the Desert Shall Rejoice: Conflict, Growth, and Justice in Arid Environments*. MIT Press, Cambridge, MA.

Morrison, W., 1988. *The Alta Empire*. Alta Irrigation District, Dinuba, CA.
Preston, W. L. 1981. *Vanishing Landscapes: Land and Life in the Tulare Basin*. University of California Press, Berkeley and Los Angeles.
Smith, A., 1937. *The Wealth of Nations*. Random House, Inc., New York, NY.
Smith, Z. A., 1984. Centralized Decisionmaking in the Administration of Groundwater Rights: The Experience of Arizona, California, and New Mexico and Suggestions for the Future, *Natural Resources Journal*, 24(3) pp. 641-688.
Tielman, I. and Shafer, W. H., 1943. *The Historical Story of Irrigation in Fresno and Kings Counties in Central California*. Williams and Son, Fresno, CA.
U.S. Bureau of Reclamation, 1930. *Report to the Federal Power Commission on the Storage Resources of the South and Middle Forks of Kings River, California*. U.S. Government Printing Office, Washington, DC.
Weber, M., 1968. *Economy and Society*. Bedminster Press, New York, NY.

# Nine: INSTITUTIONS, REGULATIONS, AND LEGAL ASPECTS OF WATER AND DRAINAGE PROBLEMS

# 39 ALTERNATIVE INSTITUTIONAL ARRANGEMENTS FOR CONTROLLING DRAINAGE POLLUTION

Alan Randall, Ohio State University, Columbus

## ABSTRACT

Inefficiency is endemic to the arrangements that govern most of irrigated agriculture. Particular attention is paid in this chapter to two problems: (1) Public demands for instream flows may be subordinated to farmers' demands for water withdrawals, and (2) drainage water from irrigated land may be excessive in quantity and pollutants.

A total value framework is presented for benefit cost analysis and methods of estimating nonmarketed use and existence value are discussed. Assuming that the benefits of water quantity and quality in instream and wetlands uses can be estimated, the chapter considers what kinds of institutional arrangements can be developed to ensure that these benefits are fully addressed in water management and policy. Commonly suggested solutions include regulation, administrated efficient prices, and property rights to facilitate water markets; more recently, the possibility of stable cooperative solutions to common property problems has been suggested. These alternative approaches are examined at the conceptual level. Many of the inefficiencies of the water economy could be eliminated via water markets based on transferable property entitlements. However, this solution alone is unlikely to provide adequately for instream and wetlands uses. Literature on principal-agent problems is discussed which suggests some promising approaches.

## INTRODUCTION

The quantity of drainage water from irrigated agriculture is likely to be excessive and its quality suboptimal for further use, so long as governing institutions fail to provide adequate incentives for efficient irrigation and drainage. While inefficiency is endemic throughout irrigated agriculture, this chapter will emphasize nonfarm demands for instream flows and downstream water of acceptable quantity and quality. Particular attention will be paid to estimating the value of water in recreational and ecosystem support uses, and designing institutions that bring these values to bear on irrigators' water application and drainage decisions.

Consider an irrigation district serving a few hundred independent farmers, somewhere in the Western United States. The district is a nonprofit entity that buys water from a Federal agency, say the U.S. Bureau of Reclamation (Reclamation), at a price well below its real cost and sells it to member farms. The district operates and maintains its internal delivery system, and establishes the institutional arrangements among its members with respect to water rights, allocation of water excess to satisfying its obligations under water rights, delivery schedules, and the conditions for water transfer (or lease of water rights) among members.

It is likely that the district pays anywhere from one-quarter to one-twelfth of the real cost of providing water to it, the remainder being borne by citizens all across the Nation (U.S. Department of the Interior, 1980). In addition, much of the irrigated land is devoted to crops that are subsidized via the U.S. Department of Agriculture's (USDA) commodity programs, resulting in a "double subsidy" (Moore and McGuckin, 1988). Many of the farms in the district would not be viable economic entities in the absence of these subsidies, a fact that has obvious implications for the pattern of land settlement in the West. On the other hand, subsidization of water in service of farm viability objectives likely leads also to excessive application.

The district is in a relatively low rainfall zone and, in common with many arid area irrigation projects throughout the world, occupies land with relatively high concentrations of various salts in the soil. These salts are dissolved by irrigation water, which would benefit crop yields if drainage was adequate. Eventually, however, water tables rise, bringing excessive concentrations of salts back to the root zone. Like many Western irrigation projects, the farms in the district are experiencing salinity problems. In addition, the quality of ground water -- which is used conjunctively with surface water in the district -- is reduced, and some of the salt-polluted drainage waters return to the surface streams, reducing their quality. The salinity problem is partially internal to the district as farmers bear costs of drainage problems to agricultural productivity, and external as downstream users of surface and ground water suffer from reduced water quality. Downstream users include other irrigators, urban users, and fish and wildlife in the wetland that is the ultimate downstream sink. Many plants and aquatic creatures are sensitive to ordinary salts and thus damaged when salt concentrations rise. In addition, elements such as selenium, that are toxic in trace quantities, occur in drainage water from the district.

This stylized irrigation district is involved in a number of conflicts:

- Nationwide taxpayer interests are opposed to subsidizing the economic viability of the irrigation district.
- Western urban interests are concerned about water supplies for continued urban growth, and observe that irrigation farmers typically enjoy not only heavily subsidized water but senior rights in a prior appropriation system of water rights.
- Outdoor recreationists and environmentalists are concerned about maintaining instream flows for recreation and ecosystem support services and sufficient tailwaters to maintain wetland sinks, whereas such uses have traditionally held low priority or none at all in the hierarchy of water rights.
- Urban, recreational, and environmental uses have specific quality requirements that may conflict with the farmers' preference for regular water applications to flush salts from the root zone.
- Farmers in the district may suffer from surface waters and aquifers polluted by drainage and return flows from up stream districts, and/or may cause such problems for districts downstream.
- Farmers within the district share surface and ground-water sources, so that excessive use or polluted drainage due to the actions of one will adversely affect others. Without constructive arrangements to counteract it, nonexclusiveness is likely to be the norm with respect to drainage and ground-water pumping within the district.

Briefly, then, subsidized water, high priority for agricultural uses, and nonexclusive drainage waters lead to problems ranging from salinization of farmland within the district to selenium contamination in the wildlife reserve at the downstream end of the hydrological system. These problems are interrelated and the effectiveness of solutions to, say, the drainage problem will depend on what, if anything, is done about institutions that assign large quantities of water to agriculture at heavily subsidized prices. Some aspects of the overall problem may resist solution for technical, economic, or political reasons, introducing a classical second-best situation (Lipsey and Lancaster, 1957) in which the best attainable arrangements throughout the system may depart from what would seem optimal in principle.

While attempting to avoid the dangers inherent in piecemeal solutions to holistic problems, this chapter will focus mainly on the conflict between irrigators and the general public who demand instream flows and downstream waters of adequate quantity and quality. Two problems will be discussed in some depth: (1) Estimating the value of water in recreational and ecosystem support uses and (2) developing institutional arrangements that induce irrigators to consider these values when making decisions that affect on farm water use and the quality of drainage water.

## ESTIMATING THE VALUE OF WATER IN RECREATIONAL AND ECOSYSTEM SUPPORT USES

Valuation takes place within a benefit cost framework. First, a general framework will be introduced for benefit cost analysis, followed by the development of an approach to value estimation for recreation and ecosystem support services.

### The Benefit Cost Framework

Consider a complex environment, E, producing a vector of services, $x(t)$, over time. These services (or goods or amenities) are likely to be diverse -- for example, support services for human, animal, and plant life; esthetic services, landscape amenities, and diversity of flora and fauna; recreation opportunities; and waste disposal services -- and many of them are likely to be nonmarketed. The availability of each of these services at any time is a function -- uniquely determined by geological, hydrological, atmospheric, and ecological relationships -- of the attributes, $a(t)$, of the environment and the human-controlled inputs, $w(t)$, that are combined with them:

[1] $x(t) = f[a(t), w(t)]$.

People enter the system not only as producers of services but also as modifiers of resource attributes. They may do this directly, for example, by reassigning land to other uses, diverting water, removing vegetation, or disturbing soil for mining. They may also modify the resource as a side effect (expected or unexpected) of some other activity, for example, diverting and storing water for irrigation, and allowing drainage flows to enter aquifers and surface waters. Defining $n(t)$ as a vector of "natural" systems input, for example, geological, hydrological, atmospheric, and ecological,

[2] $a(t) = g[n(t), w(t)]$.

Interactions between n and w are likely. For example, the attempt to enjoy high productivity from irrigated crops involves large applications of water, which may affect the quantity and quality of water in aquifers and downstream wetlands.

People have preferences related to ordinary goods and services such as might be bought in the shopping center, $z(t)$, and environmental services, $x(t)$. Some environmental services are raw materials for producing goods in the z vector; and so they are valued and demanded indirectly, through direct demands for z. Preferences are specified in the form of an individual's utility function. For an individual, j,

[3] $u_j(t) = u_j[z(t), x(t)]$.

Each individual chooses a combination of goods and services so as to minimize the expenditures needed to attain a given level of utility, i.e., satisfaction. This process identifies a series of compensated demands, each specific to a given level of utility. For benefit cost analysis, a particular compensated demand schedule must be chosen which represents a specific reference level of utility. It is conventional to choose the individual's initial level of utility. Demands referenced at the initial utility level are characterized as those that would (1) Motivate voluntary exchange, buying, and selling, if markets were in operation and (2) satisfy the Kaldor-Hicks compensation test, also called the potential-Pareto-improvement (PPI) test, for public policy proposals.

The expenditure minimization process yields individual valuations of environmental services, $V_j[x_j(t)]$, for each individual in $j = 1,...,J$.

The present value of the stream of current and future environmental services is found by summing, across individuals, the discounted present value of that stream:

[4] $PV(E) = \sum_{j=1}^{J} \int_{t_o}^{\infty} V_j[x_j(t)] e^{-rt} dt,$

where r is the rate of interest.

As the notation, PV(E), suggests, the value of the environment E as an asset is identically equal to the net present value of its anticipated stream of services. Thus the environment, E, is seen as a capital good acquiring value to the extent that the services it provides are valued by people. Those services are determined by the environment's attributes, which are themselves determined by the characteristics of the natural system and by the activities of people. If that environment were to be disturbed -- that is, if the $x(t)$ vector of human-controlled inputs were to be modified -- its attributes could change, changing the $x(t)$ vector of the services it provides and its capital value.

Consider a project or policy $\Delta$, which would change $w(t)$ to $w^\Delta(t)$, thus converting the environment, E, to some "with project" state, $E^\Delta$, at some conversion cost, $C^\Delta(t)$. The proposed project would replace the "without project" stream of services, x(t), with some "with project" stream $x^\Delta(t)$, . The net present value of such a project is

[5] $PV(\Delta) = PV [ E^\Delta - C^\Delta(t) - E ]$,

where present values are discounted as in equation (4). If $PV(\Delta)$ is greater than zero, the project will be a PPI and will therefore pass the benefit cost test.

## A Total Value Framework for Environmental Services

Now, we consider step (3), the valuation process, in more detail. First, assume that the system is deterministic; assume especially, that production and demands are certain.

<u>Total Value in a Deterministic Framework.</u> Consider an individual with the utility function,

[6] $u = u(z, x_e, x_s, x_1, x_2, Q)$

where z: a vector of ordinary goods and services; x: environmental services; Q: the state or condition of the environment; and the subscripts denote, respectively, e: existence, s: ecosystem esthetics, 1: activity 1, and 2: activity 2. This formulation of use values includes the customary explicit onsite use activities such as fishing and birding, and also $x_s$, ecosystem esthetics. An example of $x_s$ would be the enjoyment elsewhere (and perhaps incidental to some other activity) of more diverse birdlife attributable to a wetland within the environment under study. At this stage, $x_e$, $x_s$, $x_1$, and $x_2$ are each single elements; however, the analysis could easily be extended to consider a vector of existence services, a vector of site experience services, and n activities rather than just two.

The solution to the problem,

min pz s.t. $u(\bullet) \geq u^\circ$,

where p is a vector of prices, is the expenditure function

[7] $e = e(p, x_e, x_s, x_1, x_2, Q, u^\circ)$.

Consider a proposed policy that would modify money income, m, and p, $x_e$, $x_s$, $x_1$, $x_2$, and Q from baseline levels (denoted by a superscript $^\circ$) to alternative levels (denoted by a superscript $^1$). The economic benefit of such a policy (if it is preferred to the baseline situation) or the economic damage from the policy (if preferred less than the baseline) is

[8] $TV(p^1,x^1_e,x^1_s,x^1_1,x^1_2,Q^1,m^1 | p^o,x^o_e,x^o_s,x^o_1,x^o_2,Q^o,m^o)$

$= e(p^o,x^o_e,x^o_s,x^o_1,x^o_2,Q^o,u^o) - e(p^1,x^1_e,x^1_s,x^1_1,x^1_2,Q^1,u^1) + m^1 - m^o.$

This holistic total value of the alternative policy accounts for the welfare impacts of changes in ordinary prices, money income, and existence and use services of the environment conditioned on an altered quality of the environment. It may be broken down into component values provided that the components are evaluated in a valid sequence. First, we simplify the notation in the following ways: assume m does not change; let p be exogenous and implicit in $e(\cdot)$ and interpret each x variable as $x|_Q$, i.e., the level of x conditioned on the level of environmental quality, so that, for example, $x^o_e \equiv x_e|_{Q^o}$ and $x^1_e \equiv x_e|_{Q^1}$. In this example, existence services become a continuous variable conditioned on the existing level of environmental quality. Now, [8] reduces to

[9] $TV(x^1_e,x^1_s,x^1_1,x^1_2 | x^o_e,x^o_s,x^o_1,x^o_2) = e(x^o_e,x^o_s,x^o_1,x^o_2,u^o) - e(x^1_e,x^1_s,x^1_1,x^1_2,u^o).$

Implementing a valid valuation sequence, [9] can be disaggregated into

[10.1] $TV(x^1_e,x^1_s,x^1_1,x^1_2 | x^o_e,x^o_s,x^o_1,x^o_2) = e(x^o_e,x^o_s,x^o_1,x^o_2,u^o) - e(x^1_e,x^o_s,x^o_1,x^o_2,u^o)$

change in existence value

[10.2] $+ e(x^1_e,x^o_s,x^o_1,x^o_2,u^o) - e(x^1_e,x^1_s,x^o_1,x^o_2,u^o)$

change in value of ecosystem esthetics

[10.3] $+ e(x^1_e,x^1_s,x^o_1,x^o_2,u^o) - e(x^1_e,x^1_s,x^1_1,x^o_2,u^o)$

change in activity-1 value

[10.4] $+ e(x^1_e,x^1_s,x^1_1,x^o_2,u^o) - e(x^1_e,x^1_s,x^1_1,x^1_2,u^o)$

change in activity-2 value,

i.e., into existence value and (in this case) three kinds of use values.

Note that TV is unique whether measured holistically [9] or in a valid sequenced piecewise structure [10.1-10.4], and is path-independent whereas the component values are sequence-dependent. Note also that there is a frequently used practice of estimating the various component values independently and then aggregating them to calculate total value. This practice, called independent piecewise aggregation to distinguish it from sequenced piecewise aggregation, generates a conceptually invalid total value, called IPV (i.e., independent piecewise total value), defined as follows:

[11.1] $\quad \text{IPV}(x^1_e, x^1_s, x^1_1, x^1_2 | x^o_o, x^o_s, x^o_1, x^o_2) = e(x^o_e, x^o_s, x^o_1, x^o_2, u^o) - e(x^1_e, x^o_s, x^o_1, x^o_2, u^o)$

[11.2] $\quad\quad\quad\quad\quad\quad\quad\quad\quad + e(x^o_e, x^o_s, x^o_1, x^o_2, u^o) - e(x^o_e, x^1_s, x^o_1, x^o_2, u^o)$

[11.3] $\quad\quad\quad\quad\quad\quad\quad\quad\quad + e(x^o_e, x^o_s, x^o_1, x^o_2, u^o) - e(x^o_e, x^o_s, x^1_1, x^o_2, u^o)$

[11.4] $\quad\quad\quad\quad\quad\quad\quad\quad\quad + e(x^o_e, x^o_s, x^o_1, x^o_2, u^o) - e(x^o_e, x^o_s, x^o_1, x^1_2, u^o)$

In general, the relationship of IPV to TV (>, =, or <), depends on the number of value components, the budget share devoted to the total policy, and the presence of complementary and substitute relationships among the components. The error becomes systematic (e.g., benefits of a preferred policy are systematically overstated by IPV), as the number of components becomes very large (Hoehn and Randall, 1989).

Total Value Under Uncertainty. It is likely that baseline conditions, the impacts of policy, and demands for environmental services may be uncertain. This observation led Weisbrod (1964) to propose an additional category of value, option value, to account for the willingness of uncertain future users to pay for assurance that the resource would still be available if and when they eventually demand use.

Smith (1987) developed a conceptual framework for use value in an ex ante context, i.e., before the uncertainty about demand and/or supply has been resolved. Randall (in press) extended these concepts to develop an ex ante total value model. Analytically, the procedure modifies the deterministic total value model (equations [6]-[10]) by introducing a vector of probabilities for various states of the world and vectors of state-conditional market-good prices and nonmarket-good quantities. State-conditional expenditures are minimized subject to a constraint that expected utility be maintained at the baseline level, to generate the planned expenditure function (Simmons, 1984 and Helms, 1985).

Using this framework, ex ante total value is defined holistically in a manner analogous to equation [9] and can be separated into its component values analogous to equation [10]. Ex ante existence value and various kinds of ex ante use values emerge. There is no separate category for option value; rather, the uncertainty is included in the ex ante existence and use values.

## Estimation Methods

The standard approaches to valuing nonmarket services and amenities include contingent valuation, weak complementarity methods such as the travel cost method of estimating recreation values, and hedonic price analysis.

Contingent valuation (Randall et al., 1974 and Mitchell and Carson, 1989) analyzes the self-reported willingness to pay or behavioral responses (buy/not buy; vote yes/no) of survey respondents or experimental subjects confronted with baseline and alternative policy scenarios constructed by the researcher. It has the advantage that the researcher can control the scenarios and thus evaluate a broad domain of alternative policies. Many researchers and commentators worry that self-reported valuations or contingent behavioral responses might be tainted in some way -- perhaps because respondents invest too little effort in formulating their responses, or because they make false reports for strategic reasons -- but the literature provides little evidence that these problems are severe (Hoehn and Randall, 1987 and Mitchell and Carson, 1989).

Weak complementarity methods and hedonic price analysis use data from actual transactions, which is clearly an advantage over contingent valuation methods. Where travel goods are a weak complement for recreation site quality (Bradford and Hildrebrandt, 1977), expenditures on travel goods contain information on the value of site quality. Where the market price of a non-homogeneous good, such as land, depends on its characteristics including local environmental quality, econometric analysis of land sales or rental data may reveal willingness to pay for environmental quality (Rosen, 1974). While the basic data for these methods are generated by actual transactions, the values calculated for nonmarketed environmental goods may be influenced by arbitrary analytical decisions (e.g., Smith et al., 1986). Further, the domain of these methods is limited to situations where the analytical assumptions hold and transactions data can be obtained.

The total value framework developed above has some clear implications for value estimation strategy and the choice of methods. Any and all of these value concepts -- total value, existence value, and use values, in a deterministic framework; and ex ante total value, ex ante existence value, and ex ante use values, when uncertainty is a concern -- can be estimated de novo (i.e., in an exercise that starts at the very beginning, or "from scratch") via the contingent valuation method (CVM). Scenarios can be constructed to elicit total value holistically, or total value and component values in a valid piecewise sequence.

Where uncertainty is involved, scenarios can be constructed to communicate that uncertainty and obtain ex ante value data.

However, there might be other questions and concerns that militate against relying on contingent valuation studies designed and executed from scratch. Is it important to include in the research design some opportunities for using travel cost or hedonic methods to estimate use values? Since new studies designed from scratch tend to be expensive, do we wish to use for at least some components the "typical" unit values that studies such as that of Sorg and Loomis (1984) have compiled? If the answer to either of the last two questions is positive, two additional problems must be resolved.

First, component values estimated by travel cost or hedonic methods, or taken from compilations of "typical" unit values, are usually in IPV form, i.e., estimated independently rather than in a total value context. To use them within a valid total value framework, it would be necessary to find a method of approximating a valid piecewise valuation using independent piecewise value estimates.

Second, when uncertainty is a concern, the total value framework calls for ex ante values, but travel cost, hedonic, and "typical" unit values are usually considered ex post, i.e., values estimated from decisions made after the uncertainty has been resolved. An eclectic strategy for valuation under uncertainty seems to require procedures for translating between ex ante and ex post values. However, Smith (1987) argues that ex ante and ex post values are fundamentally noncomparable, like apples and oranges. If ex ante benefit cost analysis requires ex ante values, actual transactions reveal ex post values, and the two kinds of values are basically noncomparable, the prospects seem bleak for using weak complementarity and hedonic price methods when uncertainty is a concern.

Things may not be quite so problemmatic. First, not all actual transactions reveal ex post values. For example, advance purchase of airplane tickets and vacation packages would seem to reveal ex ante rather than ex post values. Surely, uncertainties about one's health and the weather on site at vacation time remain unresolved at the time of advance purchase.

Second, there appear to be a variety of opportunities for the providers of projects or policies -- entrepreneurs or agencies -- to relieve demanders of any demand uncertainty and to hedge the acquired risk via purchased insurance or self-insurance. Meier and Randall (in press) show that the expected value of ex post willingness to pay will be a lower bound for ex ante benefits under such circumstances.

The situation with respect to valuation of nonmarketed environmental services may be summarized:

- Holistic total value and existence value, whether ex ante or ex post, may be measured with contingent valuation. Other methods are unavailable.
- Ex ante use value may be measured with CVM. It cannot be measured with revealed value methods and ex post transactions data. However, some actual transactions are themselves ex ante, and revealed value methods may be used to measure ex ante use value in these cases.
- Provider assumption of demand uncertainty, with efficient market insurance or self-insurance, is a sufficient condition for the expected value of ex post consumers surplus to serve as a lower-bound estimate of ex ante use value when demand is uncertain.

## INSTITUTIONAL POSSIBILITIES

Suppose that the total value of environmental services has been measured for the baseline situation in the irrigation district and for a variety of alternative policies. The analyses show that a PPI would be generated by a reallocation of water from agriculture to instream flow and wetlands uses, and a reduction in the load of ordinary salts and trace toxics due to drainage from irrigated land. What kinds of institutional arrangements might be proposed to achieve this result?

Economists would agree that the present situation is inefficient and is due, at least in part, to some kind of failure of market and/or government institutions. While this much is agreed, there is quite vigorous debate among economists as to the source of the failure and what should be done about it. Perhaps a majority of economists still adhere to what shall be labeled the "market failure, government fix" school of thought, rooted in the economics of Pigou (1932) and Bator (1958) and the political theories of the progressive era (Nelson, 1987). A vocal minority, who might be called "property rights libertarians," argue that government is more nearly the problem than the solution and that the inefficiencies of the irrigation economy should be eliminated by extending the scope of private property rights. An emerging group of economists takes yet a third view, based on recent work in game theory and related disciplines, that private property and central government hardly exhaust the institutional possibilities.

### "Market Failure, Government Fix" Approaches

In the conventional wisdom of the "market failure, government fix" paradigm, there are four kinds of circumstances in which even a fundamentally competitive economy would experience market failure. These phenomena are externality, public goods, common property resources, and natural monopoly. For three of these phenomena, the conventional solutions call unambiguously for government action: to tax or regulate externalities, to raise revenue for public provision of public goods, and to regulate the pricing policies of natural monopolies. For common property resources, the range of endorsed solutions is broader. Regulation and taxation may be suggested, but it is also frequently suggested that the government specify private property rights and then stand aside as emerging markets restore efficiency.

These concepts and solutions would find direct application to the problems of instream flows and drainage pollution. Ecosystem support and existence values would be considered public goods; on site recreation uses are nonmarketed and the demands for such uses are not translated into incentives facing irrigators; and drainage pollution is an externality to the irrigators. Given the difficulty of attributing polluted ground water and return flows to any individual farmer, a common property problem persists. Benefit cost analysis could be used to determine the efficient levels of instream flow and drainage water quality. Government could harness the farmers to internalize these externalities, solve the common property problem, and provide the desired public goods, by regulating their irrigation and drainage practices or, perhaps less likely, by establishing administered prices that would achieve efficiency.

Since Coase's classic paper (1960), the "market failure, government fix" paradigm has been in retreat. Coase, Cheung (1970), and Dahlman (1979) established that externality had little analytical content, in that inefficient externality cannot persist unless there are some additional impediments to trade among the parties involved. Ciriacy-Wantrup and Bishop (1975) pointed out that the "common property resources" analysis is really applicable only to pure nonexclusiveness. The "tragedy of the commons" analysis is misleading, if applied to the myriad common property institutions that have been developed to handle resource management problems in various traditional and modern societies. Various authors have established that "the public goods problem" of conventional analyses is really two distinct problems: nonexclusiveness and nonrivalry, which may occur separately or together. Randall (1983) argued that whatever valid content exists in the market failure concepts of externality, common property resources, and public goods can be captured, without all the confusion, by the concepts of nonexclusiveness and nonrivalry.

The implicit government activism of the "market failure, government fix" approach has also come under attack. The Coasian analysis of externality focused on nonattenuated property rights as a sufficient condition for efficiency. It drew attention to the possibility of market-like behaviors in many domains of human interaction beyond conventional markets. What at first-glance might appear to be market failure may in fact be an efficient market solution. Thus, the burden of proof was switched to those who would claim market failure in any particular case. As the Coasian tradition developed, it was argued with increasing generality that attenuation of rights was endemic in the public sector itself. Government failure may be an even more pervasive problem than market failure. That, of course, took the argument one rather large step further. A sustained posture of government activism in control of market failure was not merely unnecessary, it was undesirable.

## The Property Rights Libertarian Approach

This approach took its cue from the mid-1950's analyses of Samuelson (1954) and Gordon (1954) and the voluminous literature that followed Coase's seminal paper (1960).

Samuelson and Gordon did not merely show that rivalry and nonexclusiveness, respectively, were substantial impediments to Pareto-efficiency in a decentralized economy. Their analyses predicted the total collapse of the nonrival and nonexclusive economic sectors unless government stepped in, coercively, to save the day. The Coasian tradition, as we have seen, ridiculed the claim that fundamentally flawed government institutions could be expected to rectify market failures. Rather, the Coasian analysis identified nonattenuated property rights as the first best hope.

From these premises developed the conventional wisdom of the "property rights libertarian" approach. So-called market failures were mainly caused by attenuated property rights, and nonexclusiveness was far and away the greatest part of that problem. Privatization was the appropriate policy response to diagnosed inefficiencies. Thus, Anderson and Hill (1976) argued, essentially, that the economic history of the United States could be characterized as a triumphal march of private property institutions from east to west with the predictable result of prosperity unparalleled in other times and places. Schmid (1977) raised the argument (originated by Ciriacy-Wantrup and Bishop, 1975) that Anderson and Hill had ignored a whole universe of institutional possibilities, some of them quite serviceable, between the extremes of exclusive private property and open access. The Anderson and Hill (1977) response was scathing: the possibilities to which Schmid referred were essentially uninteresting, since any efficiency properties these institutions possessed must surely be attributable to some degree of exclusiveness inherent in them. Further, incomplete exclusiveness implied incomplete efficiency; why not go all the way? In this, Anderson and Hill were faithfully reflecting the libertarian mindset: most of the issues raised by the old-fashioned notion of market failure can be addressed with a simple dichotomy between exclusive private property, which promotes efficiency, and nonexclusiveness, which leads to the collapse of the economic sectors it afflicts.

For this simple analysis, nonrivalry poses a difficulty, since ordinary exclusion is not sufficient to restore a nonrival goods sector to efficiency. Some proponents of the property rights approach (Anderson and Hill, for example) tend to play down the issue of nonrivalry. Others (Buchanan, 1977, for example) confront nonrivalry directly, favoring voluntary taxation schemes in the tradition of Lindahl (1958) and Wicksell (1958) and endorsing the modern search for incentive-compatible collective decision mechanisms.

## Game-Theoretic and Related Approaches

Perhaps no long-established prediction of economics has been so thoroughly refuted as Samuelson's and Gordon's prediction of total collapse in the nonrival of nonexclusive sectors. There is evidence all around us that these sectors are seldom efficient, which supports Samuelson's and Gordon's predictions with respect to efficiency. But there is also ample evidence that, despite their predictions, these sectors have not totally collapsed.

The Samuelson-Gordon tradition left an escape route: Government could coercively regulate or tax and thereby provide what citizens will not provide through markets or other endogenous institutions. But this escape route is unsatisfactory. More contemporary analyses (reflecting the Coasian tradition and a variety of other influences) treat government itself as endogenous. From this perspective, government is not a wise external force capable of disciplining an unruly society. Rather, government emerges, warts and all, from society. How, then, can government (which is endogenous to society) impose upon society that which society cannot agree to impose on itself? Once the endogeneity of government is conceded, it is impossible to reconcile Samuelson's and Gordon's prediction of collapse with the observation that the nonrival and nonexclusive sectors seem to do no worse than limp along and often perform passably well. Clearly, Samuelson's and Gordon's theory of market failure is inadequate and misleading.

In the past two decades several novel and related approaches have emerged to shed new light on the possibilities for collective action. These approaches include game theory formulations of the nonrivalry and nonexclusiveness problems (Sen, 1967 and Runge, 1981), resource allocation mechanisms (Hurwicz, 1973), the theory of teams (Marshak and Radner, 1971), incentive-compatible mechanisms (Groves and Ledyard, 1980), and principal-agent models (Arrow, 1986).

An early and influential game-theoretic formulation was the prisoners' dilemma, a game in which individuals unable to communicate with each other must each choose either a cooperative or noncooperative strategy. In a two-person game, the pay off matrix for individual A is structured, from the highest down: A defects while B cooperates, both cooperate, both defect, B defects while A cooperates. In a one-shot prisoners' dilemma, the "both defect" solution always emerges, despite the fact that it is Pareto-inferior to "both cooperate."

By the 1960's it was widely held that the Samuelson-Gordon analyses of market failure could be reformulated as single-period $n$-person prisoner's dilemmas. Such reformulation would, of course, reconfirm Samuelson's and Gordon's prediction of total collapse in the nonrival nonexclusive sectors.

The single-period prisoner's dilemma was only the beginning, however. It was soon realized that the prisoner's dilemma is not necessarily the proper specification for nonrivalry and nonexclusiveness problems (Sen, 1967 and Dasgupta and Heal, 1977). As Shubik (1981) observed, games of pure opposition have many uses in, for example, military tactics but relatively few applications in economics. In many economic contexts, cooperative behavior is the individually preferred alternative and all that is required for stable cooperative solutions is credible assurance that other players will not defect.

When these various games are repeated, there is opportunity for individuals to observe some things about the other players in the game. Aggregate performance of the group is often observable, and players can discern if a substantial number of defections have occurred on previous rounds. Sometimes, individual performances on previous rounds can be observed, leading to the possibility of retaliation against defectors. Players may be encouraged to cooperate, not merely to avoid retaliation but because having a reputation for being cooperative may bring rewards.

Stochastically repeated prisoners' dilemmas (the players do not know at which round the game will end) may produce stable cooperative equilibria. Axelrod (1982) has conducted computer simulations of repeated prisoners' dilemmas and found that a simple tit-for-tat strategy -- I will cooperate on this round if the group aggregate result for the previous round suggests that everyone cooperated, but I will defect if I have reason to expect there was significant defection -- was preferred to many other plausible strategies and led frequently to stable cooperative solutions. When any of the following -- the game is not one of pure opposition, individual contributions are to some degree observable, and reputations may be developed -- holds, the results become more favorable for stable cooperative solutions.

This kind of thinking is useful in amending both the "market failure, government fix" and the "property rights libertarian" approaches. Game theory no longer confirms the Samuelson-Gordon collapse thesis for the nonrival and nonexclusive economies. Stable cooperative solutions are at least a possibility in a variety of circumstances, and some insights have been developed concerning the factors that work in favor of stable cooperation. As well as providing insights for particular cases, the work on stable cooperative solutions has more general implications for the possibility of endogenous government. These results take us some distance beyond the idea that individual actions lead to market failures that only exogenous government fixes can cure. Similarly, they tend to deny the "property rights libertarian" dichotomy that damns all institutional arrangements except private property rights.

The demonstration that, for several relevant classes of games, coordinated strategies permit stable, Pareto-efficient cooperative solutions is not entirely comforting. Coordination is likely to be a costly activity, and complete coordination, if it requires consultation among all participants, may be prohibitively costly. Private (that is, rival and exclusive) goods markets work well because prices convey, in simple signals, sufficient information and incentives to accomplish coordination and neither centralized management nor direct consultation among all market participants is necessary. Perhaps signaling devices can be developed for adequate and cost-effective coordination so that cooperative arrangements in large organizations dealing with nonrival and nonexclusive goods are reasonably stable and efficient. This is the working hypothesis that motivates research on principal-agent models.

For principal-agent models, the following situations are typical. Total costs of loss and damage may be reduced if insured parties have some incentives for loss-avoiding behavior; can insurance policies with appropriate incentives be designed? If the work effort of individual agents cannot be monitored directly, what incentives can the manager devise to encourage agent efficiency without incurring excessive turnover of agents? If the effluents from individual polluters cannot be monitored fully, can the pollution control authority devise incentives for reasonably efficient pollution control?

Each of these problems is characterized by hidden action (the agent can take some actions unobserved by the principal) or hidden information (the agent has some information the principal does not have). An interesting variant is the problem of a single principal and many agents, where the principal can observe the combined output of all agents but not the individual output of any one of them. The relevance of this kind of thinking to nonexclusiveness and nonrivalry problems is obvious.

The literature on principal-agent problems is substantial and often highly mathematical. No attempt at careful review and evaluation is offered here, but some impressions can be conveyed. Considerable progress has been made in modeling information requirements and group performance, given various combinations of problems and incentives. Results about information requirements provide indirect evidence about the transactions costs associated with various arrangements. While principal-agent models reconfirm the efficiency of price signals in a neoclassical competitive economy, they offer no support for the "private property or total collapse" thesis of the libertarians. A wide variety of workable arrangements, with outcomes falling between Pareto-efficiency and collapse, can be identified for diverse problems exhibiting aspects of nonexclusiveness or nonrivalry.

## Application to the Irrigation District

At the outset, a stylized irrigation district was depicted with endemic inefficiency due to excessive application of underpriced water. Drainage waters bearing ordinary salts and trace toxics pollute the surface waters and aquifers. Instream and wetlands uses of water are increasing in value but poorly served -- with respect to both water quality and quantity -- by an incentive structure that fails to signal these demands to the water allocation process.

The "market failure, government fix" paradigm would suggest that some reasonably efficient allocation of resources be identified by microeconomic analysis and (perhaps) operations research. Incentives to implement this efficient allocation would be designed, and it is likely that Pigovian taxes and efficient administered prices would be preferred in principle; private property rights might be extended, to solve simple nonexclusiveness problems; and command-and-control regulations would provide a last resort in the event that the first two approaches encounter technical and/or political difficulties. In the natural course of events it seems typical that much of the burden falls to this "last resort" strategy. A small number of candidate policy proposals would be selected for careful evaluation, and benefit cost analysis would play a prominent role in the ultimate policy choice.

The "property rights libertarian" paradigm would counsel extending the domain of nonattenuated property rights as far as possible, and would be suspicious of government fixes for any remaining inefficiencies.

Both of these approaches have some merit. There may in fact be roles for administered prices and command-and-control regulations. But there are obvious problems with hidden information, incomplete signaling, etc., that make it difficult for central government to effectively implement these instruments. The institutional menu is not restricted to these options and there may be large potential gains from looking beyond the government fix.

There is a large and important potential role for private property rights to mitigate the inefficiency endemic to irrigated agriculture. In 1981, I suggested a major role for transferable water entitlements (TWE's) in the Australian irrigation complex. TWE's would discourage excessive water applications and encourage transfers of water to more efficient uses. Further, if TWE's were initially given to irrigation farmers on the basis of historical water use, they would directly attack the major impediment to irrigation policy reform: many irrigators could not pay an efficient price for water and remain in business. Under TWE's, water would tend to be reallocated away from such farms, but the exiting irrigators would have a saleable asset in the water entitlements themselves. Thus, TWE's would avoid major diminutions in irrigators' wealth as a result of policy reform. My TWE proposal addressed the drainage problem to some degree: transferable entitlements should extend to drainage waters so that a farmer who could collect drainage from his fields and ensure a prospective purchaser of its quality would be rewarded. For the record, I have not changed my mind since 1981 about the advantages of TWEs, and look with favor on the development of water markets in the United States (Saliba and Bush, 1987).

Nevertheless, the foci of this chapter -- instream flows and drainage waters -- are the very problems for which nonattenuated property rights seem least promising. Drainage waters cannot all be captured on farm. Problems of hidden information remain. It is perhaps inequitable to assign TWE's at the outset to the irrigation sector on the basis of historical use when instream and wetlands uses traditionally have been slighted. Finally, many of the technologies suggested to treat polluted drainage water -- e.g., ocean disposal, deep-well injection, desalinization (National Research Council, 1989) -- are beyond the reach of individual farmers; some degree of collective action is needed.

These considerations lead me to look toward the game-theoretic approaches for institutional possibilities that plug the rather obvious holes left by the government fix and property rights approaches.

Suppose the general public is considered as the principal and each farmer in the irrigation district an agent. The principal can establish goals with respect to the quantity and quality of water delivered to the various uses: farming, urban, instream, and wetlands. Benefit cost analysis should play a role in establishing these goals, and state-of-the-art methods should be used to ensure that nonmarketed environmental uses be fully represented in the benefit cost accounts. The principal cannot only set goals, but also monitor the aggregate performance of the district with respect to water quantity and quality.

The problem is then to develop institutions to achieve these goals given that the agents have some information that the principal cannot discover cheaply and can take some actions hidden from the principal. In other words, effective monitoring of farming practices, water application procedures, and drainage quantity and quality at the individual farm level is beyond the power of central government acting on behalf of the general public.

This is exactly the kind of environment to which the literature on repeated principal-agent games is addressed. The principal can monitor aggregate performance of the group. The agents (farmers) are neighbors, usually of long standing, a situation in which reputations develop and count for something. The irrigation district exists as an entity. It purchases water from agencies of the central government, which suggests that -- subject to existing water law and the feasibility of amending it -- the principal has a rather direct avenue of applying sanctions at the group level. The district can facilitate communication among its members. Further, it has some powers to allocate water among its members, permit and facilitate internal water markets, and construct and operate capital works. It seems clear that conditions are favorable for stable cooperation among the agents, given clear collective goals set by the principal and backed up by sanctions that could be imposed on the district as a whole.

Given water quality and quantity goals set by the principal, the district could choose among, or combine, a variety of strategies, including reallocation of water deliveries, encouragement of improved on farm technologies and management practices, retiring irrigated land with very high concentrations of salts and toxic trace elements (with TWE's to provide a saleable asset for farmers on the way out), and building capital works for treatment or disposal of polluted drainage.

Of course, the basic motivation for these approaches -- the principal sets water quality and quantity goals for instream and wetlands uses, monitors the aggregate performance of the district, and can impose sanctions at the district level -- requires a series of political decisions in an environment where irrigator, urban, and environmental constituencies are represented. At the outset, the status quo -- in terms of both law and implicit contractual obligations (Paarlberg, 1989) -- largely favors the irrigators. Yet, improvements require fundamental changes in the relationships between irrigators and the rest of society. Institutional innovators should not overlook the strategic importance of including mechanisms to compensate, at least in part, the existing irrigation community for changes that will be costly to them.

## REFERENCES

Anderson, T. L. and Hill, P. J., 1976. The Role of Private Property in the History of American Agriculture, 1776-1976, *American Journal of Agricultural Economics*, 58, pp. 937-945.

Anderson, T. L. and Hill, P. J., 1977. The Role of Private Property in the History of American Agriculture, 1776-1976: Reply, *American Journal of Agricultural Economics*, 59, pp. 590-591.

Arrow, K. J., 1986. Agency and the Market. In: Arrow, K.J. and Intrilligator, M.D. (Eds.), *Handbook of Mathematical Economics*. North-Holland, Amsterdam.

Axelrod, R., 1982. The Emergence of Cooperation Among Egoists, *American Political Science Review*, 75, pp. 306-318.

Bator, F. M., 1958. The Anatomy of Market Failure, *Quarterly Journal of Economics*, 72, pp. 351-379.

Bradford, D. F. and Hildebrandt, G. G., 1977. Observable Preferences for Public Goods, *Journal of Public Economics*, 8, pp. 111-131.

Buchanan, J. M., 1977. *Freedom in Constitutional Contract*. Texas A&M University Press, College Station.

Cheung, S. N. S., 1970. The Structure of a Contract and the Theory of a Non-Exclusive Resource, *Journal of Law and Economics*, 13, pp. 49-70.

Ciriacy-Wantrup, S. von. and Bishop, R. C., 1975. Common Property, as a Concept in Natural Resources Policy, *Natural Resources Journal*, 15, pp. 713-727.

Coase, R. H., 1960. The Problem of Social Cost, *Journal of Law and Economics*, 3, pp. 1-44.

Dahlman, C., 1979. The Problem of Externality, *Journal of Law and Economics*, 22, pp. 141-162.

Dasgupta, P. S. and Heal, G. M., 1977. *Economic Theory and Exhaustible Resources*. Cambridge University Press, New York, NY.

Gordon, H. S., 1954. The Economic Theory of a Common Property Resource: The Fishery, *Journal of Political Economy*, 62, pp. 124-142.

Groves, T. and Ledyard, J., 1980. The Existence of Efficient and Incentive Compatible Equilibria with Public Goods, *Econometrica*, 48, pp. 1487-1506.

Helms, L. J., 1985. Expected Consumer's Surplus and the Welfare Effects of Price Stabilization, *International Economic Review*, 26, pp. 603-617.

Hoehn, J. P. and Randall, A., 1987. Satisfactory Benefit Cost Indicator from Contingent Valuation, *Journal of Environmental Economics and Management*, 14, pp. 226-247.

Hoehn, J. P. and Randall, A., 1989. Too Many Proposals Pass the Benefit Cost Test, *American Economic Review*, 79, pp. 544-551.

Hurwicz, L., 1973. The Design of Mechanisms for Resource Allocation, *American Economic Review*, 63, pp. 1-30.

Lindahl, E., 1958. Just Taxation: A Positive Solution. In: Musgrave, R. A. and Peacock, A. T. (Eds.), *Classics in the Theory of Public Finance*. St. Martin's Press, New York, NY.

Lipsey, R. G. and Lancaster, K., 1957. The General Theory of Second Best, *Review of Economic Studies*, 24(1), pp. 11-33.

Marshak, J. and Radner, R., 1971. *The Economic Theory of Teams*. Yale University Press, New Haven, CT.

Meier, C. E. and Randall, A., (in press). Use Values Under Uncertainty: Is There a "Correct" Measure? *Land Economics*.

Mitchell, R. C. and Carson, R. T., 1989. *Using Surveys To Value Public Goods: An Assessment of the Contingent Valuation Method*. Resources for the Future, Washington, DC.

Moore, M. R. and McGuckin, C. A., 1988. Program Crop Production and Federal Irrigation Water, *Agricultural Resources: Cropland, Water and Conservation Situation of Outlook Report*, USDA-ERS.

National Research Council, 1989. *Irrigation-Induced Water Quality Problems*. National Academy Press, Washington, DC.

Nelson, R. H., 1987. The Economics Profession and the Making of Public Policy, *Journal of Economic Literature*, 25, pp. 49-91.

Paarlberg, R., 1989. The Political Economy of American Agriculture: Three Approaches, *American Journal of Agricultural Economics*, 71, pp. 1157-1164.

Pigou, A. C., 1932. *The Economics of Welfare*. Macmillan, New York, NY.

Randall, A., 1981. Property Entitlements and Pricing Policies for a Maturing Water Economy, *Australian Journal of Agricultural Economics*, 25, pp. 195-220.

Randall, A., 1988. Market Failure and the Efficiency of Irrigated Agriculture. In: O'Mara, G. (Ed.), *Efficiency of Irrigation*. The World Bank, Washington, DC., pp.21-30.

Randall, A. Valuing Nonuse Benefits. In: Braden, J. B. and Kolstad, C. D. (Eds.), *Measuring Environmental Benefits*. North-Holland, Amsterdam. In Press.

Randall, A.; Ives, B.; and Eastman, C., 1974. Bidding Games for Valuation of Aesthetic Environmental Improvements, *Journal of Environmental Economics and Management*, 1, pp. 132-149.

Rosen, S., 1974. Hedonic Prices and Implicit Markets: Produce Differentiation in Pure Competition, *Journal of Political Economy*, 82, pp. 34-55.

Runge, C. F., 1981. Common Property Externalities in Traditional Grazing, *American Journal of Agricultural Economics*, 63, pp. 595-606.

Saliba, B. C. and Bush, D. B., 1987. *Water Markets in Theory and Practice*. Westview, Boulder, CO.

Samuelson, P. A., 1954. The Pure Theory of Public Expenditure, *Review of Economics and Statistics*, 36, pp. 387-89.

Schmid, A. A., 1977. The Role of Private Property in the History of American Agriculture, 1776-1976: Comment, *American Journal of Agricultural Economics*, 59, pp. 587-89.

Sen, A. K., 1967. Isolation, Assurance, and the Social Rate of Discount, *Quarterly Journal of Economics*, 81, pp. 112-24.

Shubik, M., 1981. Game Theory Models and Methods in Political Economy. In: Arrow, K. S. and Intrilligator, M. D., (Eds.), *Handbook of Mathematical Economics*. North-Holland, Amsterdam.

Simmons, P. J., 1984. Multivariate Risk Premia with a Stochastic Objective, *The Economic Journal*, 94, pp. 124-133.

Smith, V. K., 1987. Nonuse Values in Benefit Cost Analysis, *Southern Economic Journal*, 54, pp. 19-26.

Smith, V. K.; Desvouges, W.; and Fisher, A., 1986. A Comparison of Direct and Indirect Methods of Estimating Environmental Benefits, *American Journal of Agricultural Economics*, 68, pp. 280-290.

Sorg, C. and Loomis, J. B., 1984. *Empirical Estimates of Amenity Forest Values: A Comparative Review*. General Technical Report RM-107, Rocky Mountain Forest and Range Experiment Station, U.S. Forest Service, Fort Collins, CO.

U.S. Department of the Interior, 1980. *Draft Environmental Impact Statement, Acreage Limitation, Westside Report Appendix G.*

Weisbrod, B. A., 1964. Collective-Consumption Services of Individual-Consumption Goods, *Quarterly Journal of Economics*, 78, pp. 471-477.

Wicksell, K., 1958. New Principle of Just Taxation. In: Musgrave, R. and Peacock, A. T. (Eds.), *Classics in the Theory of Public Finance*. St. Martin's Press, New York, NY.

# 40 ECONOMIC INCENTIVES AND AGRICULTURAL DRAINAGE PROBLEMS: THE ROLE OF WATER TRANSFERS

Bonnie G. Colby, University of Arizona, Tucson

## ABSTRACT

The opportunity to transfer water creates incentives for farmers to consider the value of water in off-farm uses when making farm management decisions. Water transfers can complement other policy approaches to agricultural drainage problems by prompting farmers to use less water for irrigation, recognizing its opportunity cost, and by providing water to mitigate drainage-related contamination and to replace and restore damaged wetlands.

## INTRODUCTION

Economic incentives have played a central role in contributing to drainage-related water quality problems, and incentives generated by the potential for water transfers can be an important contributor to solutions. Historically, the low subsidized cost of water in Federal irrigation projects has had several relevant effects. First, it has encouraged farmers to use more water than would be economical at unsubsidized water prices, creating increased agricultural runoff and decreased reserves of water for other current uses and for future uses. Second, subsidized water has made it profitable to farm lands that are marginal due to slope, soil quality and other characteristics--lands which would be unprofitable to farm if farmers paid the full cost of water provision (National Research Council, 1989, p. 5). Finally, subsidized water has created a political and institutional context in which the costs of agricultural input decisions are not fully borne by the agricultural sector--sending a message that it is acceptable for taxpayers, other water users, and the environment to bear costs resulting from farm management decisions. This makes it politically difficult to require farmers to account for these costs, including the costs of cleaning up drainage water.

The possibility for transferring water creates alternative incentive structures which can cause irrigators to consider the value of water in nonirrigation uses. Policies that send appropriate signals to farmers are a microlevel tool which can reinforce regulations and other policies that encourage farmers to incorporate drainage costs in their water use and farm management decisions. New pricing policies for Federal project water are one way of creating new incentives for farmers to adjust their water use, and these are discussed elsewhere in this volume (Willey and Weinberg, and Wichelns). This chapter evaluates the role of water transfers, including market transactions and other voluntary, negotiated water use arrangements.

Possible transfer arrangements include payments to retire particular fields from irrigation, payments to adopt different farm management practices or cropping patterns, conservation easements, and leases or purchases of land and water. Developed by the U.S. Bureau of Reclamation (Reclamation), U.S. Fish and Wildlife Service (FWS), and California Department of Fish and Game (DFG), and approved by the California Regional Water Quality Control Board, the recent mitigation plan for Kesterson Reservoir (Kesterson) includes purchase and transfer of land and water. The mitigation plan calls for creation of additional wetlands to replace losses due to selenium contamination. It would use water and land acquired from private owners and from existing Central Valley Project water supplies (Water Intelligence Monthly, 1990). Thus, voluntary transfers are already being incorporated into drainage management policies.

The institutional setting is crucial in facilitating voluntary water transfer arrangements. First, criteria and procedures for formal approval, implementation, and enforcement of agreements must be provided. Second, laws and policies can provide farmers and other parties with the incentive to negotiate, sending clear signals about the consequences of failure to adopt voluntary arrangements to mitigate drainage problems.

## MARKET TRANSACTIONS—AN OVERVIEW

"... (E)ngineer the forces of the market place into our environmental programs, using economic incentives (and disincentives) to make the everyday economic decisions of individuals, businesses, and the government work effectively for the environment ... Market forces can supplement the regulatory power of the government and create a setting for private sector innovation and initiative in the pursuit of environmental quality."

Project 88 Harnessing Market Forces to Protect Our Environment: Initiatives for the New President, pp. 1-2, 1988.

The virtue of the competitive market system as a mechanism for coordinating economic activities and allocating society's resources is a time-honored theme among economists. Adam Smith described the market process as an "invisible hand" which uses price signals to guide self-interested individuals and profit-maximizing firms to buy, sell, and pursue those activities in which they have a comparative advantage. Thus, Smith argued, the value of output is maximized, all participants are better off, and resources are allocated and used efficiently[1].

A market consists of the interactions of buyers and sellers of rights to use resources (land and/or water) either for a limited period of time or into perpetuity. Conservation easements and payments to alter specific land and water management practices are also potential market transactions. Markets allocate economically scarce resources by compelling buyers to evaluate the benefits of acquiring additional quantities of water or land at the expense of forgoing something else of value. The chance to sell forces rightholders to consider the opportunity cost of their land and water uses. Markets can provide flexibility and security in land and water rights since rightholders are permitted to participate in transactions but are not required to do so. The voluntary nature of market transfers and their reinforcement of existing property rights enhances their political acceptability.

Reliance on market processes is consistent with the belief that individuals are the best judge of their own well being and have the right to make economic decisions in pursuit of their own self-interest. Transactions are fair in the sense that buyers and sellers will only participate if they believe they have something to gain. Markets disperse the capacity to make resource allocation decisions among individuals who control resources. Markets also cause individuals to reveal their values for goods, services, and amenities and to reveal information on changes in values over time.

The motivating force behind markets is a perception that economic gains may be captured by transferring water to a place or purpose of use in which it generates higher net returns than under the existing use patterns[2]. A perfectly functioning market would ensure that transfers occur whenever the net benefits from a transfer are positive[3].

There is evidence that farmers are responsive to economic incentives in making water use and farm management decisions. Caswell and Zilberman (1985) found that adoption rates for irrigation technologies among California Central Valley growers reflect the economic advantages associated with adoption. Soil characteristics, crop mix, access to marketing networks for new technology and water costs all affect adoption of water conserving technology. Caswell and Zilberman also predicted that substantial increases in use of drip and sprinkler irrigation would occur in response to increased water costs in the southern counties of the Central Valley.

The hypothesis that farmers are responsive to economic incentives is also supported by previous water transfer activity. In areas where the legal and political framework facilitates sale and lease of water rights used in irrigation, farmers have been regular participants in the water market. In Arizona, during the 1980's, tens of thousands of irrigated acres were purchased by urban interests in order to acquire water for transfer to growing cities. These water transfers, motivated by urban growth, require land acquisition under Arizona case law (Salt River Users Association v. Kavocovich). In other states, farmers sell water rights and shares in water districts while retaining their land. In Colorado, farmers routinely sell and lease their rights to receive water from the Colorado Big Thompson project to cities and other farmers within the project's boundaries. However, considerable controversy has arisen over sales to entities located outside of the project area (Howe et al., 1986). New Mexico farmers sell surface water rights for use by growing cities in the Rio Grande basin, with the senior rights being highly valued by urban areas seeking reliable supplies. Similar transactions, involving appropriative water rights held under state law and shares in irrigation districts, are occurring in Utah, Nevada, Wyoming, Texas, Montana, and other western states. Farmers have been responsive to market incentives, though there are numerous examples of market negotiations failing for political, economic, and other reasons.

Farmers' responsiveness to changing costs of irrigation water and to market signals indicates that water transfers could create incentives useful in addressing agricultural drainage problems. However, a market-oriented approach, with its many strengths, also involves complex policy questions.

## Institutional Considerations

While prospects for increasing economic benefits are the driving force behind markets, laws and policies affect the cost of market transactions and the attractiveness of market transfers relative to other means of accomplishing objectives. The legal and political setting determines the transactions costs associated with market transfers. Transactions costs are incurred in identifying legal and physical characteristics of land and water rights (priority date, return flow obligations, deed restrictions, etc.); in negotiating price, financing, and other terms of transfer; and in satisfying state laws and transfer approval procedures. State and local laws impose transactions costs on market participants in the form of approval requirements for changing the purpose and place of use of a water right, and in zoning changes related to land use (Colby et al., 1989b).

Ambiguous institutional arrangements can impose costs by creating uncertainty regarding who owns the water right, how much water can be transferred and for what purposes. Costs stemming from legal ambiguity are particularly prevalent when environmental considerations are involved, since many states have not yet developed criteria and procedures for evaluating wetland and instream water use applications. Transactions costs influence the economic viability of proposed transfers and can, therefore, affect the level of market activity.

Voluntary transfers require a well-defined set of property rights and "rules of the game" to provide a starting point for negotiations. Transferable property rights in water are often absent or ambiguous, as are criteria for implementing a proposed transfer. Most western states have developed some procedures to govern transfers of appropriative water rights (that is, rights held under the doctrine of prior appropriation) among off-stream uses. However, water uses not conducted under an appropriative right (which may include water-based recreation, wetland protection, Federal projects, tribal water uses, and groundwater pumping in some states) are typically not subject to the same transfer procedures. In some states, no clear transfer policies have been developed for these types of water use. While transfers do occur, even where ambiguities remain, the transactions costs are high and markets are likely to be less active and provide less flexibility and responsiveness than otherwise (Tregarthen).

Assignment of clear property rights and development of transfer procedures and criteria is likely to be politically charged. For instance, one of the arguments against making shares in Federal project water transferable is that irrigators should not be able to reap a profit through selling water developed with Federal subsidies. Property rights are controversial because they determine who must initiate negotiations and make payments and who receives bids for their land and water, accepting or rejecting offers. In each western state that has attempted to facilitate water transfers, considerable controversies have arisen over assigning transferable property rights, the distribution of profits from market transfers, and the third-party costs that may accompany transfers.

Liability rules are a form of property rights and send important signals to water users. If water users are liable for damages created by their management practices, there are clear incentives to alter those practices to avoid impacts on water quality and wildlife. On the other hand, if damages are typically cleaned up or contained using public funds there is little incentive to alter management practices in consideration of off-farm impacts.

One of the crucial policy decisions related to water transfers involves determining who has a voice in the review process. Obviously the buyer and seller have important decisionmaking powers since either can veto further progress in voluntary negotiations. However, what third parties have a role and how significant should their power be? In some western states only other water rightholders may formally object to a proposed water transfer. In other states, public interest case law and statutes leave room for environmental groups, recreationists, fish and wildlife interests and local communities to object to a transfer and to influence the transfer approval process to varying degrees. In many areas, contractual rights to receive water supplied by a state or Federal water project may not be transferred without the consent of the project or irrigation district governing board.

Local governments in the area of origin and residents who do not hold water rights typically cannot obtain standing to oppose water transfers under state law; thus, their interests frequently are not taken into account. Negative effects tend to be most serious when transfers involve retiring irrigated land and moving water from one region to another. Fiscal impacts include loss of property tax base and local government bonding capacity, tighter spending limitations, and reduced revenue sharing. Environmental effects associated with the retirement of irrigated land include soil erosion, blowing dust, and tumbleweeds that arise after crop production ceases.

When farmland is retired from agriculture, loss of farm sector jobs and income often follows. Businesses that provide goods and services to farmers are affected and future economic growth in the area of origin can be inhibited. As the tax base shrinks and local services decline, the area of origin becomes less attractive to new businesses. Also, water and land resources needed by new local development may become unavailable as a result of water exports. Economic losses suffered by areas of origin may be insignificant in the context of a statewide economy and may be inconsequential relative to the benefits which accrue to the new users of the water. Area of origin losses, however, can seriously impair the viability to small, rural communities which may lack the economic strength and diversity to recover.

The breadth of third-party participation in the transfer approval process determines who can play in the market "game." To participate in the market process, one must either: (1) Own something of value that others want access to, such as water rights or strategically located land, (2) have money to acquire land or water, or (3) have the power to impose transactions costs, time delays, vetoes on progress and other inconveniences on those seeking to transfer water. These powers may arise from statutes, case law, political importance, media attention, and public support. The California Environment Quality Act (CEQA), for instance, changed the "rules of the game" in California. The CEQA is viewed by some development interests as giving rural areas and environmentalists "veto power" over water transfers. The CEQA does provide a different basis for participation in the transfer approval process and a different "starting point" for negotiations than existed prior to the CEQA. Similarly, public interest clauses in western state water laws stimulate a different type of negotiation over water transfer impacts than would occur without those clauses (Colby et al., 1989a).

## Transfers of Federal Project Water

Reclamation supplies water for about 20 percent of the irrigated acreage in the 17 Western States. In addition to controlling a significant percentage of agricultural supplies, Reclamation is central to water transfer possibilities because it operates key water storage and conveyance facilities in many areas of the West. Transfers routinely occur involving water developed by Reclamation projects within the service area of the project and for the project purposes originally authorized. Transfers of Federal project water outside of the service area, or to uses not included in the project authorization, are not yet common, and policies to govern transfers involving Reclamation water and facilities are still evolving. A thorough discussion on such transfers is provided in Wahl (1989).

As a general rule, transfer to Reclamation project water is subject to state transfer procedures and also must be approved by a Reclamation contracting officer. Voluntary transfers of water between districts in a Reclamation project would normally not be actual sales of water rights but rather leases or sales of contractual deliveries, without no water rights changing hands. Such assignments of contractual deliveries can be either short-term leases, annual rentals, long-term leases, dry-year option agreements, or permanent (Wahl, 1989).

Voluntary transfers of water from Reclamation facilities are not new. Water rentals on Idaho's Upper Snake River stretch back to the 1930's and were explicitly recognized in Reclamation contracts with the water users. In 1972, the Utah Power and Light Company obtained 6,000 acre-feet of water from two irrigation companies in the Federal Emery County project for powerplant cooling. The city of Casper, Wyoming, is paying the nearby Casper-Alcova Irrigation District (a Federal project) for canal lining to reduce seepage and to provide the city with 7,000 acre-feet of additional water. In California, during the 1976-77 drought, Reclamation operated a waterbank in which over 45,000 acre-feet of water changed hands. In southern California, the Metropolitan Water District (MWD) has reached an agreement with the Imperial Irrigation District (IID) to fund conservation measures that would salvage 100,000 acre-feet of water annually for municipal and industrial uses in the MWD service area. For additional discussion concerning these and other examples (see Wahl, 1989; Wahl and Osterhoudt, 1986; Engels, 1986; and Wahl and Davis, 1986).

The Department of the Interior has issued a set of principles to govern transfer approvals (U.S. Department of Interior, 1988) and Reclamation is developing more detailed guidance for interested water users. While Reclamation will not burden a water transfer by imposing unnecessary costs on those seeking to transfer water, it must comply with existing Reclamation law and must be in the same or better position financially as a result of the transfer. In addition to protecting other authorized project water uses, instream rights, and other established water rights, water transfers involving Reclamation facilities would have to comply with the National Environmental Policy Act.

## Examples of Market Acquisitions for Environmental Objectives

Although purchase or lease of water appears to be a logical approach for assuring adequate quantity and quality of water for wildlife, there are relatively few examples of market acquisition to maintain streamflows or lake levels for environmental purposes. Lander County, Nevada, purchased senior irrigation rights in order to maintain a stable shoreline for fishing and boating on a new county reservoir (Water Market Update, Vol. 1, No. 5, 1987). The Montana Fish, Wildlife, and Parks Department has purchased water to be released from reservoirs to enhance the survival of trout fisheries during unusually dry summers (Water Market Update, Vol. 1, No. 8, August 1987). The Nature Conservancy has purchased water to support flows on a stretch of Colorado's North Poudre River in cooperation with an irrigation district that stores and releases water from a reservoir above the Conservancy's Phantom Canyon Preserve (Water Market Update, Vol. 3, No. 8, September 1989). Nevada waterfowl interests have purchased water rights to enhance the Stillwater wetlands in western Nevada (Water Market Update, Vol. 3, No. 9, October, 1989). In addition, the Nature Conservancy has purchased a ranch with water rights that it hopes to apply to restoring the Stillwater wetlands (Water Market Update, December 1989).

Diverse interests cooperated in the purchase of water to support riparian habitat near Sacramento, California. Several years of drought threatened a rare bird species, other wildlife, and fisheries along Putah Creek. County and city governments, along with several water districts, contributed money and water to assure adequate releases from an upstream reservoir and to cover the costs of substituting ground water for diversions from the creek (Water Market Update, Vol. 3, No. 8, September 1989).

California's DFG and Grasslands Water District purchased 30,000 acre-feet to maintain wildlife and fish in the drought-stricken San Joaquin River Basin (Water Market Update, Vol. 3, No.9, October 1989). The Upper Snake Water Bank in Idaho, in cooperation with the Nature Conservancy, leased water from irrigators to provide adequate water levels for trumpeter swans (Water Market Update, Vol. 3, No. 9, October 1989).

Other market acquisitions for environmental objectives are being contemplated around the West. Congress approved, as a part of the 1988 Federal budget, $1 million for acquiring water rights to protect endangered fish species in the upper Colorado River basin of Wyoming, Colorado, and Utah (Water

Market Update, Vol. 2, No. 2, February 1988). In 1989, the California legislature established a special fund for water management to provide environmental benefits, including purchases of water to preserve riparian areas and to improve water quality (Water Market Update, Vol. 3, No. 9, October 1989).

While wildlife habitat and water quality can be enhanced through market transactions, water rights for instream flows and other environmental objectives are typically recent appropriations and have low priority relative to other water rights. Those lakes, wetlands, and rivers not protected by a water right or represented only by junior rights are vulnerable to dry-year shortages and to impairment by more senior consumptive water uses.

## *Why So Few Environmentally Oriented Transactions?*

Environmental interests have yet to become well represented in Western water markets for several reasons. First, those wishing to protect wetlands, lakes, and rivers do not have legal access to water rights on the same terms as farmers, cities, and industry. Some Western States do not recognize instream flow and water quality maintenance as a beneficial use and so water rights may not be held for these purposes. Of the western states, only Alaska and Arizona allow a private party to hold a water right for the purpose of maintaining instream flows (MacDonnell, 1989). Markets could better incorporate instream flow values if State laws permitted appropriation, purchase, and seasonal leasing of water rights for wetland and streamflow maintenance by both public and private organizations.

A second reason why there have not been more market transactions is that the transactions costs for environmentally oriented acquisitions are likely to be higher than for water rights purchased for offstream uses. Organizations wishing to use water rights to maintain wetlands and streams often face opposition by neighboring water users who fear the flexibility of their own rights will be constrained. High costs can be incurred in overcoming objections to new environmental uses of water rights. Further, many state agencies have little experience in handling applications for change in purpose of use of a water right from irrigation, for instance, to wetland, stream, or water quality maintenance. New procedures and criteria often have to be developed, resulting in delays, uncertainty, and additional costs for the applicant.

Even if obstacles to acquiring water rights for environmental objectives were abolished, wildlife, water quality, and recreation have public-good characteristics which make it difficult to translate collective values into dollars to bid for water rights in the market place. Those who benefit from environmental enhancement are a large, but generally unorganized, constituency. The term "public good" refers to resources characterized by nonexcludability, meaning it is difficult or impossible to exclude those who do not pay from enjoying the benefits of the resource. Many individuals who do place a positive value on wildlife, wetlands, and rivers may be "free riders," enjoying these amenities but making no payments--since payments are not required. Funds raised to purchase water for stream, lake, and wetland maintenance, will not generally represent total willingness to pay by all potential beneficiaries due to the free-ridership phenomenon, and the difficulty of collecting contributions from all who will benefit. Additionally, there is little incentive to voluntarily contribute since those who do not contribute cannot easily be prevented from enjoying improved water quality, recreation, and wildlife habitat.

In spite of these obstacles, environmental groups have successfully organized fund raising and donations to acquire water rights. For instance, the Nature Conservancy has received donations of water rights which they intend to use for fish and wildlife enhancement on the Gunnison River in Colorado and on Aravaipa Creek in southeastern Arizona (Water Market Update, Vol. 1, No.3, March 1988). Donations of water rights and money were also crucial in implementing several of the market acquisitions described earlier.

The public sector is becoming more active in acquiring water for environmental objectives, as illustrated by the Montana Fish, Wildlife, and Parks Department acquisition, the acquisition by a county Government in Nevada and the 1988 Congressional appropriation previously described. Reclamation announced it was altering Shasta Reservoir releases into the Sacramento River of Northern California in order to enhance the Chinook salmon fishery, at the expense of $1 million in forgone hydropower revenues (Water Market Update, Vol. 1, No. 9, September 1987). The Colorado Water Conservation Board (CWCB) has appropriated water for junior instream flow rights on over 7,000 miles of streams and numerous lakes. CWCB has also participated in the acquisition of some senior rights which would protect streamflow during drought years.

## INNOVATIVE TRANSFER ARRANGEMENTS

The most common types of water market transactions thus far involve sale or lease of water rights held under State law and transfers of Federal project water within project boundaries. In order for water markets to become a key strategy for mitigating the effects of agricultural drainage problems, transfers other than the outright purchase and lease of water rights need to be considered. Innovative transfer arrangements promote efficient and flexible water use, as do traditional purchases and leases, but often have a less severe impact on areas from which water is taken and thus may be more easily negotiated, providing quicker response to environmental needs. Several different types of transfer arrangements could be relevant to agricultural drainage problems.

Water use options and conditional leasebacks, negotiated before an environmental crisis becomes acute, are two ways to ensure that water could be made available quickly. The difference between these two approaches is the degree of security and long-term control over water provided by each. Under a water use option, ownership of the water right remains with the original water user. The new user, who might be an environmental organization or a State or Federal agency, enters into an agreement to allow them use of the water under specific conditions which could be tied to flows for fish migration, waterfowl conditions, or other environmental needs. For wildlife or recreation areas that need reliable supplies, this type of arrangement provides a backup source of water for dry years. Examples to date involve urban areas desiring drought protection rather than environmentally oriented arrangements. In one instance, a central Utah city paid a nearby farmer $25,000 upfront for a 25-year dry-year option and agreed to provide, in any year the option was exercised, $1,000 and 300 tons of hay to maintain the farmer's livestock. The option was exercised 3 out of the first 25 years the option was in place (Clyde, 1986).

Although promising, water use options can be unattractive to farmers who desire more certainty when planning their farming operations[4]. A number of issues need to be addressed when water use options are considered. One of these involves defining the conditions under which the option will be exercised. Reservoir and streamflow levels or water quality and wildlife conditions can be specified as a basis for activating the environmental water use option. Additionally, it is necessary to ensure that farmers be compensated for the losses they incur--not simply for the lost revenues of forgone crops, but also for the disruption of farm planning and land use patterns and for any production and marketing expenses they may incur prior to being notified that land would not be irrigated that season. The terms and timing for notification are important issues to irrigators.

Under conditional leasebacks, water rights are purchased by the entity desiring long-term control of the water, and are leased back to the farmer so that farming can continue except when the water is needed for specific environmental needs. The new water rightholder could be a state or Federal wildlife agency or an environmental group, and the leaseback conditioned on the need for water to support wetlands or streamflows for recreation, fish, and wildlife during dry seasons and years. Conditional leasebacks are attractive because they set aside water that can be called on either during droughts or for newly recognized environmental needs.

There have been several leaseback arrangements implemented in Arizona and Colorado, all initiated by growing cities. The city of Mesa purchased 11,606 acres of farmland in an adjacent county, planning eventually to use the ground-water rights associated with those lands to supply water to the city's expanding service area. Meanwhile, the city is leasing the land back to farmers and the land continues to be irrigated (Kolhoff, 1988). The city of Phoenix purchased 14,000 acres of farmland in Western Arizona in 1986. The city plans to retire the land and transfer the associated ground water to urban uses. Phoenix agreed to keep the farmland in production for the short term, employing at least 25 local farmers and postponing some of the local economic impacts of farmland retirement. Growing cities along Colorado's Front Range have purchased irrigated lands in the Arkansas and South Platte River Valleys and, in many instances, are leasing those lands back to the sellers so that crop production can continue until the water is actually needed to support urban growth.

Exchanging water right priority can help wildlife and recreation areas to secure reliable supplies for drought years. Under such arrangements, the senior water rightholder agrees to share shortages with the junior appropriator (a wildlife refuge or recreation area). Exchanges of priority have substantial potential with Indian reserved rights, since the priority date of tribal rights goes back to when the reservation was established. There have been some agreements to defer tribal seniority so that junior right holders have more reliable water supplies. One arrangement involves the Navajo Nation, which has a senior claim on the San Juan River. In exchange for Congressional approval of the Navajo Indian Irrigation Project, the Nation agreed to defer its seniority during dry years and to share water shortages proportionately with non-Indians. This gives downstream users in the Rio Grande Basin more reliable supplies (Back and Taylor, 1980).

Possibilities also exist for exchanging priorities in the Colorado River Basin where several Indian tribes have high priority rights to the Colorado River. A number of Arizona and California cities have considered negotiating with tribes located along the Colorado River to obtain more reliable water supplies (Water Market Update, 1987-88). Exchanges of priority could be a useful strategy in protecting wetlands and wildlife habitat from drought and poor water quality in this region.

Waterbanking is another strategy for enhancing water use flexibility. Water supplies used in water banking have included treated effluent, surplus streamflows, and Federal and state project water. Waterbanking involves storing excess water in reservoirs or in underground aquifers and maintaining "savings accounts" of stored water. When water is needed for environmental purposes, withdrawals are made from stored supplies and the accounts debited accordingly.

The 1977 Federal Emergency Drought Act established a waterbank to assist California water users receiving Central Valley Project water. By the end of the drought, the waterbank had facilitated the transfer of more than 45,000 acrefeet of water (State of California, 1989). Idaho waterbanks operate to give irrigators the opportunity to rent annual excesses of contracted water from Federal projects in the Snake and Boise River Basins (Water Market Update, 1988). Waterbanking has provided much needed flexibility during recent dry years in Idaho.

California's Kern County Water Agency is utilizing a waterbanking approach, as is MWD and the California Department of Water Resources (DWR). DWR purchased 19,000 acres of land for a recharge and water banking project. Plans include conveying 1 million acre-feet of water to the site (which has a total storage capacity of 5 million acre-feet) through the State Water Project (SWP). In dry years, the SWP will pump out 140,000 acre-feet annually to offset low flows (Water Market Update, 1987-88).

Other transfer approaches that could improve water availability for environmental needs involve incentives for water conservation, salvage, and reduced consumptive use. Most western states historically have taken a hard line against transfers of conserved water, arguing that the portions of a water right "salvaged" through conservation measures become available to new or junior appropriators rather than to those taking the conserving action. California and Oregon are exceptions, having passed statutes encouraging transfer of conserved water. There are a number of policy approaches a state can take to facilitate the transfer of conserved water. A first step is to provide the statuary incentive and authority by explicitly allowing transfer of conserved water and by protecting water rights not being exercised due to conservation from loss through forfeiture and abandonment proceedings.

Even after enabling statutes are in place, a number of difficult technical and hydrologic issues remain in determining the quantity of salvaged water that actually can be transferred. Oregon legislation (1987) states that the only salvaged water that may be transferred is that which in the absence of the conservation measure would otherwise have been irretrievably lost to the system and thus unavailable to other water users (Oregon Senate Bill 24, 1987). Capture of substantial irretrievable losses probably will not come from improvements in irrigation efficiency, however, since most salvaged water previously reentered the system as return flows. Transferable water could potentially come from switching from a higher to a lower consumptive use crop. Other measures which decrease the amount of water irretrievably lost through evaporation and deep percolation include lining earthen canals, better field drainage, and improved onfield water management. Allowing farmers who reduce consumptive use, perhaps through new crop rotations, to use the additional water on other land, or to sell or lease the water can provide strong conservation incentives. Laws in the western states on use and transfer of salvaged or conserved water vary considerably, with protection of other rightholders being the primary constraint on new uses and transfers (Colby, 1989a).

There are a few examples to date of successful transfers of conserved irrigation water. The city of Casper, Wyoming, paying for canal lining on the nearby Casper-Alcova Irrigation District's conveyance system, and the MWD and Imperial Irrigation District preliminary agreement were described earlier (Wahl, 1989 and Water Market Update, 1989). Interestingly, both examples involve water provided by Reclamation projects rather than water rights held under state law by individual appropriators.

## SUMMARY AND POLICY RECOMMENDATIONS

A number of innovative transfer arrangements have been implemented or proposed to address environmental needs. While little information is available on the quantities of water that could be made available, transfers from low-value annual crop irrigation to high-value environmental uses could reduce agricultural drainage problems and could yield reliable water supplies for wetlands and streams during dry years. The key is successful negotiation between current water users and environmental interests, and clarification of state and Federal administrative processes that allow for timely evaluation and implementation of transfer proposals.

Policymakers face a number of interesting questions when transfers are being considered. Should the parties to the agreement be able to make whatever pricing arrangements they can negotiate? Is it "fair" for those who sell or lease their water to make a profit on environmental demands for water? If public project storage and conveyance facilities are involved in conveying water to new uses, who has the highest claim on these and how can these claims be altered by voluntary arrangements?

Market transfers occur more readily where property rights in water are well defined and the procedures for transferring those rights are clear and unambiguous. State law generally provides well-defined rights and procedures for water used under appropriative rights. Rights to use water for streams, wetlands, and water quality enhancement are still ill defined in many states. Rights to transfer water provided under contract from a water supplier, such as an irrigation district, also tend to be ambiguous.

Markets function best when information about water supplies for sale or lease is readily available to potential buyers and lessees. In areas with well-developed markets, brokers provide these services just as they do for other property. In the early stages of market formation, state agencies, public water purveyors, and water user organizations can serve as a clearinghouse where potential market participants can get information.

Finally, markets operate best where low-cost and well-defined processes are available to resolve conflicts regarding proposed transfers and obtain official approval for negotiated transfer arrangements. As Colby et al. (1989b) document, in some areas of the West, applicants for a transfer, and often objectors as well, can count on spending tens of thousands of dollars to resolve conflicts and obtain formal approval. These high transaction costs can discourage transfers, except in cases where environmental benefits are expected to be large enough to justify the high costs.

Conducive institutional arrangements are particularly important in encouraging transfers responsive to agricultural drainage problems. Such transfers will be for environmental rather than commercial purposes and may sometimes be temporary or intermittent. There are not likely to be large revenues generated by the new environmental use. Low cost and timely review of transfer proposals are needed if the market mechanism is to be useful in addressing environmental needs. While transfers for environmental restoration can sometimes be accommodated within the framework of existing state and Federal law, institutional innovations outlined in this paper could facilitate more active and responsive markets.

## NOTES

[1] The concept of efficiency was refined by the Italian economist Velfredo Pareto around the turn of the century. A change in resource allocation is said to be "Pareto efficient" if the reallocation can improve at least one individual's well being without decreasing the well being of anyone else. Many resource transfers involve tradeoffs; they make some individuals better off and leave others worse off, and so cannot be evaluated using the Pareto efficiency concept. Pareto efficiency has, therefore, been modified to extend its relevance. The Kaldor-Hicks compensation criterion is a widely used modification which states that a reallocation is efficient if it represents a potential Pareto improvement--that is, if the gainers from the reallocation would be able to compensate fully the losers for their sacrifice in well being and still be better off themselves. This definition of efficiency, the conceptual foundation of benefit cost analysis, requires that benefits from any resource transfer must exceed all costs. This is how the term "efficiency" will be used in this discussion.

[2] Three conditions must be satisfied for a buyer and seller to consummate a market transfer:

    a. The buyer must expect the benefits of the acquisition (which may be contributions to some production process, investment returns or environmental improvements) to exceed all costs associated with the acquisition including the price paid to the seller, and legal or engineering costs incurred to implement the transfer.

b. The seller must receive a price offer that equals or exceeds the return forgone and that covers any costs incurred due to the transfer. A farmer selling water, for instance, must consider future net returns to water in irrigation, any decreases in the value of property due to reduced water available for irrigation and expected appreciation in the value of the water right over time.

c. The buyer must view a market purchase as an economically attractive method of accomplishing objectives relative to other possibilities--such as contracting for public project water or hooking up to a water service organization.

[3] For a market to provide efficient allocation, use and supply of land, and water and environmental resources, all parties must behave as price takers. No single individual or organization can strategically affect negotiated market prices. Instead, simultaneous exchanges among many buyers and sellers jointly determine market price.

Second, all parties have access to complete information on legal and physical characteristics of land and water resources, risks associated with environmental impacts and the costs of alternative means of obtaining environmental objectives.

Finally, property rights must be completely specified and enforced, exclusive so that the benefits and costs associated with transfer decisions accrue to the decisionmakers (buyers, sellers, and rightholders), not to third parties; and transferable so that land and water resources can gravitate to their highest value uses.

[4] The following example illustrates this point. During the 1980's the Metropolitan Water District (MWD) of southern California has been attempting to negotiate a dry-year option with the Palo Verde Irrigation District (PVID). MWD has offered Palo Verde farmers varying amounts per acre up front at the time they register acreage in the dry-year option from during the seasons land is retired from irrigation (Water Market Update, 1987-89). Given water demand projections and hydrological conditions in southern California, MWD expected to call that acreage into retirement once about every 7 years in order to firm up municipal supplies. Under such arrangements, farmers face substantial uncertainty in planning their crop rotations, their marketing strategies, equipment leases, and purchases of inputs.

## REFERENCES

Back, W. D. and Taylor, J. S., 1980. Navaho Water Rights: Pulling the Plug on the Colorado River, *Natural Resources Journal* pp. 70-90.

California Department of Water Resources, 1989a. Monthly Reports to California Water Commission by David N. Kennedy.

California Department of Water Resources, 1989b. *Drought Contingency Planning Guidelines for 1989.*

Caswell, M. and Zilberman, D., 1985. The Choices of Irrigation Technologies in California, *American Journal of Agricultural Economics*, pp. 224-234.

Clyde, E., 1986. Legal and Institutional Aspects of Drought Management. *Drought Management and Its Impact on Public Water Systems.* National Academy Press, Washington, DC.

Colby, B. G.; McGinnis, M.; Rait, K.; and Wahl, R., 1989a. *Transferring Water Rights in the Western States: A Comparison of Policies and Procedures.* National Resources Law Center, University of Colorado, Boulder.

Colby, B. G.; Rait, K.; Sargent, T.; and McGinnis, M., 1989b. *Water Transfers and Transactions Costs: Case Studies in Colorado, New Mexico, Utah, and Nevada.* Research Report, Department of Agricultural Economics, University of Arizona, Tuscon.

Davis, G., 1988. Personal Interviews with Bonnie Colby, Office of Planning, City of Albuquerque, NM.

Engles, D., 1986. Augmenting Municipal Water Supplies through Agricultural Water Conservation. *Western Water: Expanding Uses/Finite Supplies.* National Resources Law Center, University of Colorado, School of Law, Boulder.

Folk-Williams, J., 1984. *What Indian Water Means to the West.* Western Network, Santa Fe, NM.

Howe, C. W.; Schurmeier, D. R.; and Snow, W. D., 1986. Innovative Approaches to Water Allocation: The Potential for Water Markets, *Water Resources Research*, 22, pp. 431-445.

Kolhoff, K., 1988. Personal communication with Water Resource Management Coordinator, City of Mesa, Arizona, Personal Communication.

Lee, C. T., 1977. *The Transfer of Water Rights in California.* Staff Paper No. 5, Governor's Commission to Review California Water Rights, Sacramento, CA.

MacDonnell, L. J. (Ed.), 1989. *Instream Flow Protection: Law and Policy in the Western States.* University of Colorado Press, Boulder.

Metropolitan Water District of Southern California, 1987. *Summary Report to Board of Directors - Water Problems Committee: An Update on Increasing Water Supplies.*

Moss, R., 1989. Personal Communication, Friant Water Users Authority.

National Commission on Air Quality, 1981. *To Breath Clear Air.* U.S. Government Printing Office, Washington, DC.

National Research Council, 1989. *Irrigation Induced Water Quality Problems.* Washington, DC.

Oregon Senate Bill 24, 1987. Conserved Water--Transfer and Instream Flow Provisions.

Price, M. and Weatherford, G., 1976. Indian Water Rights in Theory and Practice, *Law and Contemporary Problems*, p. 118.

Project 88, 1988. *Harnessing Market Forces to Protect Our Environment: Initiatives for the New President.* Sponsored by Senators Wirth and Heinz, Washington, DC.

Quinn, T., 1989. Senior Economist, Metropolitan Water District, Quoted in *Los Angeles Times*, p. A3.

Saliba, B. C. and Bush, D. B., 1987. Water Markets in Theory and Practice, *Market Transfers, Water Values and Public Policy.* Westview Press.

Salt River Users' Association v. Kavocovich, 1966. 3 Arizona App., 28, 411, p. 2d 201.

Shupe, S. J. (Ed.) *Water Market Updated,* Vol. 1, 1987; Vol. 2, 1988; and Vol. 3, 1989. Shupe and Associates, Santa Fe, NM.

Smith, A., 1904. *Wealth of Nations.* 2 Vols., Methuen and Company, Ltd., London.

Smith, R. (Ed.) *Water Intelligence Monthly,* Vol. 1, 1987; Vol. 2, 1988; and Vol. 3, 1989. Stratecon, Claremont, CA.

Thomas, A., 1988. *Personal Communication,* Attorney Retained by/for Anheuser-Bush Brewing Company, Van Neys, CA.

Tregarthen, T. D., 1983. Water in Colorado: Fear and Loathing of the Market Place. In: Anderson T. (Ed.), *Water Rights,* Pacific Institute of Public Policy Research.

U.S. Department of the Interior, 1988. *Principals Governing Voluntary Water Transactions that Involve or Affect Facilities Owned or Operated by the Department of the Interior.*

Vaux, H. and Howitt, R., 1984. Managing Water Scarcity: An Evaluation of Interregional Water Transfers, *Water Resources Research,* 20, pp. 785-792.

Wahl, R. W., 1987. Promoting Increased Efficiency of Federal Water Use. Discussion Paper, National Center for Food and Agricultural Policy, Resources for the Future, Washington, DC.

Wahl, R. W., 1989. *Markets for Federal Water: Subsidies, Property Rights, and the Bureau of Reclamation.* Resources for the Future, Washington, DC.

Wahl, R. W. and Simon B. A., 1988. Acquiring Title to Bureau of Reclamation Water Facilities. In: Schupe, S. J. (Ed.), *Water Marketing 1988: The Move to Innovation.* College of Law, University of Denver, CO.

Wahl, R. W. and Davis, R. K., 1986. Satisfying Southern California's Thirst for Water: Efficient Alternatives. In: Frederick, K. D. (Ed.), *Scarce Water and Institutional Change,* Resources for the Future, Washington, DC.

Weatherford, G. D. and Shupe, S. D., 1986. Reallocating Water in the West, *American Water Works Association Journal,* 78, pp. 63-71.

# 41 WATER QUALITY AND THE ECONOMIC EFFICIENCY OF APPROPRIATIVE WATER RIGHTS

Mark Kanazawa, Carleton College and
University of California, Berkeley

## ABSTRACT

Many economists have studied the economic efficiency of appropriative water rights under conditions of common access to surface water. The studies in this literature often focus on the issue of how to define appropriative rights in order to promote economic efficiency. An important lack in existing studies is that they typically assume constant water quality, despite the fact that declining water quality is associated with the consumptive use of water from many western rivers. This chapter extends existing analyses by introducing variable water quality into an economic model of appropriative rights, and derives some results involving the efficient definition of appropriative rights under conditions of varying water quality. In addition, the feasibility of integrating water quality considerations into the administration of appropriative rights in California are discussed through a detailed examination of that State's rights-granting institutions and procedures.

## INTRODUCTION

In the Western United States, water rights are governed primarily by the so-called appropriative doctrine. A sizable economic literature analyzes the economic efficiency of appropriative rights to surface water[1]. Most economists agree that appropriative rights have some attractive features which go some way towards encouraging efficient surface water use (Milliman, 1959; Burness and Quirk, 1979, p. 25). Equally agreed, however, is that inefficiencies can arise under appropriative law for a variety of reasons; for example, because appropriative law allocates risk unequally among appropriators, or because appropriators lack proper incentives to invest in storage capacity (Burness and Quirk, 1979 and 1980).

An additional potential source of inefficiency, and one which will be the focus of this chapter, involves the common-pool aspects of surface water use.

Many studies have noted that water diversions from a river can generate potentially important third-party effects on downstream users (Hartman and Seastone, 1970; Meyers and Posner, 1971; Burness and Quirk, 1980; and Johnson and others, 1981). A common focus of these studies is how to define appropriative rights in order to internalize these third-party effects and thus promote efficient use of the waterway. In the absence of transactions costs, the solution is relatively straightforward: simply define rights both to the quantity diverted and to any return flows to the river, and allow market transactions to achieve an efficient allocation (Meyers and Posner, 1971, pp. 27-9; Burness and Quirk, 1980, p. 130; and Johnson, et al., 1981, p. 279). However, even if significant transactions costs exist for return flows, efficiency can be approached if rights are more narrowly defined to encompass water diversions net of return flows -- what is commonly termed consumptive use -- and transfers are unrestricted. In this case, downstream appropriators would be unaffected by upstream transfers, thus eliminating third-party effects (Johnson et al., 1981, pp. 279-80).

Interestingly, the studies in this literature typically overlook issues related to the water quality of the surface waterway. They therefore ignore the fact that appropriations of surface water can adversely affect downstream water quality in at least two ways: by discharging significant amounts of dissolved solids back into the river in return flows, and/or simply by reducing the volume of water in the river. As a result of these factors, upstream appropriations can impose significant costs on downstream users. The best known example of this is the Colorado River, which for decades has been experiencing salinity buildup, particularly in its lower basin, resulting in a variety of economic costs (see, for example, Miller et al., 1986). However, declining water quality associated with consumptive use of surface waters is a more general phenomenon which characterizes many western rivers[2].

This chapter extends the analyses in the appropriative rights literature by explicitly incorporating certain types of water quality impacts into existing economic models of appropriative water use. The next section provides a general discussion of water quality issues involving consumptive water use of a surface waterway. The third section constructs a simple model of surface water use which allows for declining water quality and derives the important theoretical results of the chapter. The main conclusion is that some of the standard results regarding the efficiency of appropriative water rights need to be modified to account for water quality impacts. The fourth section examines how water quality impacts are treated under appropriative law in the state of California, and discusses the California experience in dealing with water quality through granting and modifying legal entitlement to appropriative water. The final section summarizes the conclusions and discusses potentially valuable areas for future research.

## WATER QUALITY IN WESTERN RIVERS

Declining water quality in Western rivers has been of major concern in recent years, as continuing economic growth has increased demands on virtually every river system. It is widely recognized that a leading cause of declining Water quality is the use of rivers as repositories for the waste products of many Western industries: e.g., industry, mining, agriculture, and municipal use. Less widely recognized, but no less important, is the fact that declining water quality can be the direct result of the physical use of the waters themselves. Many uses of Western rivers involve physical withdrawals of water which is diminished in quantity and quality by consumptive use before being returned to the river. Ambient water quality within the river can be considerably diminished as a result, with adverse economic consequences for other users[3]. This chapter will focus explicitly on surface waters used for agricultural purposes, though much of the discussion may be usefully applied to other water-consuming uses as well. Water quality is a vector of many variables. However, for the purposes of the analysis in this chapter, only the salinity component of water quality will be considered.

To illustrate declining surface water quality due to consumptive use by farmers, figure 1 provides a simple model of a river. Irrigating farmers are situated at points A, B, and C. For simplicity, we assume that all natural recharge occurs upstream from A. Each farmer diverts water from the river into local water conveyance facilities, transports it to his fields, and applies it to his lands. Some fraction of the water is consumed in crop production or evaporates, with the remainder eventually finding its way back to the river. In the process, the water mobilizes dissolved salts in the soil and returns to the river more saline than when first diverted. As a result, ambient river salinity steadily increases moving downstream, so that C experiences greater salinity than B, who experiences greater salinity than A.

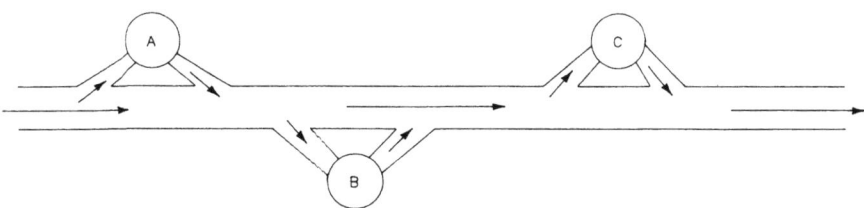

Figure 1. A stylized river.

The process of increasing river salinization may be usefully divided into two component subprocesses (Miller et al., 1986, p.4). Dissolved salts in recharge to the river directly adds to ambient salinity, or what has been termed salt loading. In addition, ambient salinity may result from reductions in river volume relative to given amounts of dissolved salts; for example, due to evaporation or withdrawals of relatively pure water (say, by farmers near the river source). This latter process has been termed salt concentration. It should be mentioned that increases in ambient salinity along the river can occur even in the absence of human intervention, for both of these reasons. Both salt loading and salt concentration will be important in the theoretical model presented in the next section.

Salinity buildup in a river can have important consequences for irrigation-based agricultural production. Many commercial crops are salt-sensitive, with production yields significantly reduced by the presence of dissolved salts in irrigation water (Miller et al., 1986, p. 9; U.S. Department of the Interior, 1980, p. 3). To counter this difficulty farmers may need to construct extra drainage facilities, take measures to apply water more evenly (e.g., switching to sprinkler irrigation or leveling the land surface), apply greater amounts of water to leach salts from the soil, and/or engage in output substitution towards salt-tolerant crops. In any case, the net result is likely to be increased costs and reduced income (U.S. Department of the Interior, 1980, pp. 6-7).

The Colorado River provides a real-world indication of the potential magnitude of the salinity buildup phenomenon. Figure 2 provides a simple schematic of the Colorado River system, with data on ambient salinity levels at various points along the river[4]. Notice that salinity levels in the upper basin states (Utah, Colorado, and Wyoming) are in general considerably lower than in the lower basin states (Arizona, New Mexico, Nevada, and California). Furthermore, although a considerable amount of the salinity buildup results from natural processes, economic activities also play an important role. For example, the U.S. Bureau of Reclamation has estimated that, while natural sources are responsible for nearly one-half of the ambient salinity in the Colorado River at Hoover Dam, irrigation practices in the Colorado River basin also account for a considerable amount--about 37 percent (Miller et al., 1986, p. 5).

The economic costs associated with increased salinity in the Colorado River are also thought to be considerable. For example, the Bureau of Reclamation has estimated the total 1-year economic costs of Colorado River salinity to be nearly $100 million in 1983, of which about 30 percent was incurred by agriculture. Furthermore, these figures were expected to nearly triple by the year 2010 (Miller et al., 1986, p. 10). Consequently, Congress has directed the Department of the Interior, in conjunction with various other Federal and state agencies, to institute a costly salinity control program. A recent plan proposed

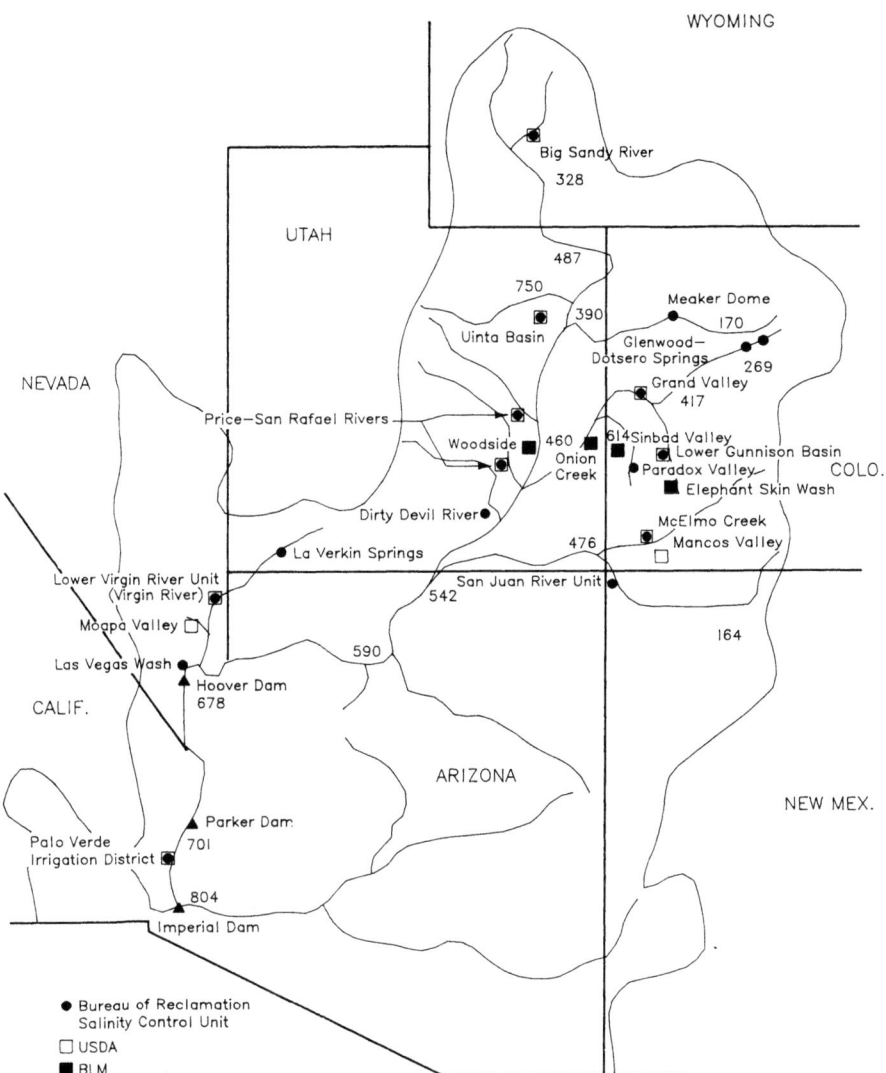

**Figure 2. Salinity along the Colorado River.**

*Source:* Miller et al., map insert.

jointly by the U.S. Departments of the Interior and Agriculture includes constructing desalting works, implementing various drainage control measures, continued monitoring of river salinity conditions, and overseeing research and development and educational activities. The plan is estimated to decrease ambient salinity in the river at Imperial Dam by less than 10 percent and would cost nearly half a billion dollars (U.S. Department of the Interior, 1986, pp. 1-3).

## A SIMPLE MODEL OF WATER QUALITY AND APPROPRIATIVE RIGHTS

Existing studies of appropriative rights efficiency focus on the effects of consumptive use of surface waters on water quantities available to other users. Declining water quality adds an extra dimension to the problem, and modifies the nature of the third-party effect. In this section, previous analyses are extended by constructing a model of a surface waterway which allows for reductions in water quality resulting from consumptive use. In the spirit of Johnson, Gisser, and Werner, it is implicitly assumed that markets in return flows are characterized by large transactions costs, so that private contracting cannot be relied upon to eliminate resulting inefficiencies. Under these conditions, specifying consumptive use as the basis for the appropriative right will not entirely eliminate third-party effects if consumptive use degrades water quality.

For the sake of clarity, much of the notation used in previous studies for modeling appropriative rights along a surface waterway is retained. Assume that there are N users (farmers) located along a river. These farmers divert water from the river for irrigation, part of which returns to the river in the form of recharge. Let $S_i$ be the water quantity diverted by the ith farmer, in acre-feet per unit of time; and let $R_i$ represent the fraction of that water which returns to the river as recharge. Therefore, $S_i(1-R_i)$ represents this farmer's consumptive use of water. Without loss of generality, we shall order the farmers from 1 to N such that $j > i$ implies that j is downstream from i.

Let $S_o$ denote the quantity of water available at the source of the river, in acre-feet per unit of time. For simplicity, assume that no other natural sources of replenishment exist (as from tributaries along the river). Following Johnson, Gisser, and Werner, assume also that there is a compact which requires some minimum amount of water to be left in the river downstream from the last user (this assumption can be very easily relaxed). Denote this quantity $\underline{S}$, also defined in acre-feet per unit of time. Then, in order to fulfill the requirements of the compact, the following inequality constraint must be satisfied:

$$[1] \quad S_o - \sum_i S_i(1 - R_i) \geq \underline{S}$$

where the summation is taken over all N farmers. That is, the total quantity of water consumed along the river must not exceed the total quantity available minus the amount which must be left in the river to satisfy the compact.

The value of water derives from its value as an input into productive use. Assume that there is some well-behaved crop production function, which may vary across farmers. Denote the value of marginal product of water for farmer i as:

$$[2] \quad VMP_i = f_i(S_i)$$

where $f < 0$. For the sake of simplicity, it shall be assumed that diversion costs are zero in the following discussion. Under reasonable assumptions regarding positive diversion costs, the results are not qualitatively affected (Johnson et al., 1981, p. 277). Economic efficiency then requires that we perform the following constrained optimization:

$$[3] \quad L = \text{Max} \sum_i \left\{ \int_0^{S_i} f_i(\theta_i) \, d\theta_i \right\} + \tau[(S_o - \underline{S}) - \sum_i S_i(1 - R_i)]$$

where maximization is with respect to the diversions $S_i$. To incorporate the river flow constraint [1] in strict equality form, it is presumed that surplus maximization from water consumption will require that the minimum possible flow which satisfies the compact be left in the river. From the first-order conditions, the following condition for water use efficiency may be derived:

$$[4] \quad \tau = \frac{f_1}{(1 - R_1)} = \frac{f_2}{(1 - R_2)} = \ldots = \frac{f_N}{(1 - R_N)}$$

where $\tau$ is the Lagrange multiplier. Deriving this result ignores the possibility of binding streamflow constraints examined by Johnson, Gisser, and Werner. This is the standard condition that efficiency requires equating the value of marginal product of consumptive water use for all farmers. As has been noted elsewhere, this outcome may be approached by specifying consumptive use as the legal basis for the individual appropriative right, and then allowing market transactions to achieve an efficient allocation (Johnson et al., 1981, p. 279).

## CONSUMPTIVE USE AND DIMINISHING WATER QUALITY

The preceding analysis has ignored the water quality dimension. The next step is to allow varying water quality to enter the model by defining a water quality indicator $\alpha_i$, which may vary across farmers. Consistent with the earlier discussion of salinity buildup along a river, $\alpha_i$ will vary with the amount of salt loading which occurs upstream, and the amount of salt concentration, which is determined by river volume, denoted $F_i$. It will be assumed that salt loading is directly associated with the volume of upstream diversions $X_i$. These effects may be represented with the following general functional form:

[5] $\alpha_i = \alpha(X_i, F_i)$

where $X_i = \sum_j S_j$, and $F_i = S_o - \sum_j S_j(1 - R_j)$, and where j is summed from 1 to (i-1) in both cases. That is, $X_i$ is defined simply as the sum of all upstream diversions, and $F_i$, the total river volume enjoyed by farmer i, equals the source quantity minus the sum of upstream consumptive use. The following will be assumed about $\alpha$: $\alpha(0, \infty) = 0$; $\alpha(X,F) < 0$, for $X,F > 0$; $\alpha_X < 0$; $\alpha_{XX} > 0$; $\alpha_F > 0$; $\alpha_{FF} < 0$; and $\alpha_{XF} < 0$.

In words, water quality diminishes at a decreasing rate with increases in upstream diversions, and improves at a decreasing rate with increasing river flow. It merits mentioning that the volume of river flow is implicitly assumed to be sufficient to ignore river flow constraints of the kind examined by Johnson, Gisser, and Werner. In other words, F is large compared to S. Notice also that the $\alpha$ function is unsubscripted, which means that it is implicitly assumed that the effect of any given amount of upstream diversions on productive value (either through salt loading or salt concentration) does not vary by location along the river. Relaxing this assumption unnecessarily complicates the analysis.

To model the water quality impact on crop production, it will be assumed for tractability that $\alpha$ adversely affects the productive value of water in the following manner:

[6] $VMP_i = f_i(S_i) + \alpha(X_i, F_i)$

In this case, the constrained optimization problem becomes the following, more complicated expression:

$$[7] \quad L = \text{Max} \sum_i \left\{ \int_0^{S_i} [f_i(\theta_i) + \alpha(\sum_j S_j, (S_o - \sum_j S_j(1-R_j)))] \, d\theta_i \right\}$$

$$+ \tau[(S_o - \underline{S}) - \sum_i S_i(1-R_i)]$$

where the equivalent expressions for $X_i$ and $F_i$ have been substituted. The condition for water use efficiency then becomes:

$$[8] \quad \tau = \frac{f_1 + \sum_j S_j[\alpha_X - \alpha_F(1-R_1)]}{(1-R_1)} = \ldots = \frac{f_N}{(1-R_N)}$$

where j is summed from 2 to N or more generally, from i+1 to N. This result is derived in Appendix 1 for the simple case of three users.

This result states that with declining water quality, efficiency no longer dictates equal VMP of consumptive use for all farmers. Indeed, such a policy will result in water use inefficiency in the presence of water quality impacts of the type modeled here. Since the entire summed expression in the numerator is negative, as are all of its individual arguments, equation [8] says that the VMP of consumptive use should be greater for upstream users. Given declining marginal productivity, this directly implies that upstream appropriators should receive less water than in the absence of water quality effects. Furthermore, this effect becomes larger further upstream. These results make intuitive sense, since upstream appropriators are now imposing externalities on downstream users, and appropriators further upstream are affecting larger numbers of downstream users.

Further interpreting equation [8], notice that this result is accentuated when diversions have a strong adverse effect on water quality through salt loading (large $\alpha_x$). That is, when recharge from diversions contains large amounts of dissolved salts, upstream users should receive less water. On the other hand, it is interesting that this result holds even when salt loading is absent ($\alpha_x = 0$). In this case, equation [8] becomes:

$$[9] \quad \tau = \frac{f_1 - \sum_j S_j \alpha_F (1-R_1)}{(1-R_1)} = \ldots = \frac{f_N}{(1-R_N)}$$

where, again, j is summed from 2 to N. As before, the VMP of consumptive use increases going upstream, so that again, upstream appropriators should receive less water than in the absence of water quality effects. The reason for this perhaps counterintuitive result is that, even when subsequent recharge contains no salts, water diversions may nevertheless diminish downstream water quality simply by reducing river volume. That is, salt concentration may be operative even in the absence of salt loading. Alternatively stated, the standard result that the VMP of consumptive use should be equated for all users holds only in the absence of both salt loading and salt concentration.

It may be directly inferred from this discussion that defining the appropriative right on the basis of consumptive use will not necessarily eliminate third-party effects. For example, suppose that an upstream appropriator were permitted to export from the river only the amount which he consumptively uses. Although downstream river volume would remain unchanged, downstream water quality may well be improved by this exportation (if $a_x$ is not zero) because the amount left in the river and not applied to crop production would contain fewer dissolved solids. From an efficiency viewpoint, upstream appropriators would tend to export too little from the river, unless downstream users get together to subsidize the exportation. However, it is easy to envision such an action failing due to free rider problems, particularly if many downstream users would benefit. More generally, defining appropriative rights purely on the basis of consumptive use will maintain a system of third-party impacts from upstream to downstream users. This conclusion is supported by the fact that transactions costs in return flows are likely to be even higher than in the absence of water quality impacts, because of increased measurement and enforcement difficulties.

This analysis suggests that defining appropriative rights to surface water without considering the potential water quality impacts of consumptive use may lead to inefficiencies. Moreover, efficiency dictates a reallocation of water away from upstream appropriators, who are more likely to impose adverse third-party effects. Unfortunately, upon entering a world in which water quality degradation is directly tied to consumptive use, there is no simple rule for efficiently allocating water rights among users. In such a world, administrative flexibility in issuing appropriative rights is likely to be of added importance. Obvious questions then arise. How does appropriative law deal with water quality impacts in practice? What administrative procedures are used in issuing appropriative rights, and how do water quality concerns enter into the process? Since standard expositions of appropriative law often do not discuss water quality issues in any comprehensive fashion, these questions merit some discussion. The next section examines these issues as they are treated under California law.

## APPROPRIATIVE WATER RIGHTS AND WATER QUALITY IN CALIFORNIA

The following discussion is not meant to be an exhaustive overview of either water quality or water rights administration in California. Such undertakings are beyond the scope of this chapter and indeed, have been admirably accomplished elsewhere (Attwater and Markle, 1988). Instead, this section will focus on the relation between water quality and appropriative rights, examining specifically how water quality concerns enter into the process of granting and modifying appropriative rights. A central issue involves administrative flexibility; or more specifically, the ability and willingness of the state to consider water quality in issuing appropriative rights.

Since 1913 the state of California has administered the distribution of appropriative rights to surface water through various state agencies. In that year the State legislature passed the Water Commission Act, which established the basic procedures for issuing water rights which survive to this day (Robie, 1972, p. 697). Currently, individuals wishing to obtain an appropriative right in California must submit an application to the State Water Resources Control Board (Board), which possesses broad statutory authority to permit or deny such applications. Importantly, the Board also enjoys ultimate jurisdiction over state water quality planning, overseeing the activities of nine regional water quality control boards whose primary mandate is to develop and implement plans to protect water quality within their respective regions. The fact that these functions have been merged into one administrative body is no mere happenstance: the merger was undertaken during the late 1960's as part of a conscious effort to improve water resource allocation within the state (Robie, 1972, pp. 698-99; California Legislature, 1966). There is little doubt that it has facilitated consideration of water quality impacts when the State issues appropriative rights.

A key issue which the Board must consider is whether granting an appropriation will serve a loosely defined and elastic welfare ideal commonly known as the public interest (Robie, 1972, p. 699). California law specifically requires the Board to reject applications for appropriations which it deems to be counter to the public interest (California Water Code, S1255, Deering, 1977). Furthermore, the Board is empowered to impose various conditions on the exercise of an appropriative right as necessary to protect the public interest (California Water Code, S1257, Deering, 1977; Brandt, 1987, p. 720). If these conditions are violated, the right may be subsequently revoked. Alternatively, the Board may issue cease-and-desist orders or seek injunctive relief to enforce conditions (Brandt, 1987, p. 720; Dunning, 1982, p. 30).

An important component of whether an appropriation serves the public interest involves its effect on third parties. Before granting an appropriative right, the Board is required to consider the effect of the proposed diversion on a wide variety of other beneficial uses of the water, including domestic, agricultural, municipal, and industrial users, as well as instream uses such as fish and wildlife, and to weigh the relative benefits to the applicant and other users (California Water Code, S1257, Deering, 1977). The Board must also consider water quality control plans established by the regional boards, and may impose conditions on use of the water "as it finds are necessary to carry out such plans" (California Water Code, S1258, Deering, 1977; Schneider, 1978, pp. 85-6).

The California Environmental Quality Act (CEQA) of 1970 confers additional statutory authority upon the Board to consider water quality impacts in granting appropriative rights (California Public Res. Code, S21000, West, 1977). CEQA broadly mandates state agencies to regulate economic activities in order to protect and promote environmental quality. One important feature of CEQA is that it requires the Board to provide an Environmental Impact Report (EIR) for all projects with a potentially "significant" impact on the environment (Robie, 1972, p. 702). Although the Board did not initially interpret CEQA as requiring an EIR whenever it granted a water right, subsequent judicial decisions more broadly construed the CEQA provisions to apply in principle whenever a permit is required for a private action (Robie, 1972, pp. 708-09). Since state administrative procedures for granting appropriative rights fall into this category, it appears that EIR's are in fact required under CEQA for proposed appropriations with significant environmental impacts, although they appear not to be a factor in most cases (Robie, 1972, pp. 705-06). Overall, the likely effect of CEQA has been to broaden the scope of the public interest as applied by the Board in acting on applications to appropriate water (Robie, 1972, pp. 703-04; Schneider, 1978, p. 107).

A significant development in a closely related area involves the recent application of the public trust doctrine to environmental protection in California[5]. In 1983, the California Supreme Court handed down an important ruling in the struggle over the waters to Mono Lake between environmental groups and the city of Los Angeles (National Audubon Society v. Superior Court, 33 Cal 3rd. 419). At issue was whether Los Angeles would be allowed to retain rights to Mono Lake water when such diversions were causing severe environmental damage to the lake (Sax and Abrams, 1986, pp. 467-68). The court invoked the public trust doctrine in calling for a reconsideration of the rights to the water flowing into Mono Lake. In particular, it held that the Board must consider public trust values in acting upon appropriations and "must attempt to prevent or minimize any harm to public interests," even if this meant reallocating existing rights (Spangler, 1988, p. 1579). This decision further expanded the powers of the Board in considering water quality impacts when acting upon applications for appropriative water rights.

The preceding discussion indicates that the Board possesses broad-based legal authority to consider water quality impacts in deciding whether to issue an appropriative right. In practice, however, the Board has been rather selective in exercising this authority. In the aforementioned case involving Mono Lake, the suit to enjoin diversions was brought by an environmental group when the Board refused to act (Sax and Abrams, 1986, p. 470). One observer has noted that in the vast majority of cases, the Board considers only whether the proposed appropriation is beneficial and whether unappropriated water is available (Robie, 1972, pp. 705-06). On the other hand, in certain instances the Board has handed down water rights decisions in which water quality has played a key role.

The most famous of these decisions has involved waters in the Sacramento-San Joaquin Delta. Since the 1960's there has been concern about the effect of water diversions from rivers flowing into the Delta on the water quality within the Delta. The controversy has centered primarily on the appropriative rights of two large users: the Central Valley Project (CVP) and the State Water Project (SWP), which divert massive amounts of water into upstream reservoirs. In a variation on the earlier discussion, the resulting reduced flows into the Delta were held to diminish Delta water quality by increasing sea water intrusion. In two decisions handed down in 1971 and 1978, the Board declared that all beneficial uses of the Delta must be protected regardless of prior vested rights, and reduced the appropriative entitlements of the CVP and SWP in order to meet water quality standards set by the regional board (Robie, 1988, pp. 1125-31; Brandt, 1987, pp. 722-24). Although an appellate court subsequently ruled that the Board's decision unfairly discriminated against the CVP and SWP by ignoring the water quality impact of all other upstream diversions, it nevertheless upheld the Board's power to modify water rights in order to achieve water quality standards (United States v. State Water Resources Control Board, 227 Cal Rptr. 161; Robie, 1988, pp. 1132-40; and Brandt, 1987, p. 713).

Another very different decision, handed down by the Board in 1984, involved the return flows from the Imperial Irrigation District (IID) which recharged the Salton Sea. The Salton Sea is a large inland sea in southern California which is considerably saltier than the ocean. The IID was allowing massive amounts of return flows to enter the Salton Sea, which were actually reducing ambient salinity in the Sea, thus maintaining the viability of a local fishery. On the other hand, these return flows were causing flooding to neighboring farmlands. The Board ruled that allowing such massive return flows constituted an unreasonable use of water and imposed the condition that the IID undertake water conservation measures (Attwater and Markle, 1988, pp. 1026-27; Spangler, 1988, p. 1584). In this case, the Board decided that maintaining water quality in the Salton Sea was a less beneficial use that flood prevention. This decision was fully upheld by an appellate court, which ruled that the Board enjoys broad adjudicatory powers in administering California water law (Imperial Irrigation District v. State Water Resources Control Board, 231 Cal Rptr. 283; Spangler, 1988, p. 1585).

It is significant that each of these cases involves an obvious water quality impact imposed by large, easily identifiable appropriators. The ability of the Board to control water quality degradation through administration of water rights may depend greatly on how costly it is to monitor return flows and their water quality impacts. Several authors have noted inherent difficulties in using water rights administration to control nonpoint source pollution (such as agricultural runoff and drainage), where it is difficult to physically monitor discharges (Robie, 1970, p. 18; Johnson, 1989, pp. 490-91). Recognizing these difficulties, regional boards have historically not required many agricultural operations even to report waste discharges (Report of the Assembly Committee on Water on the Porter-Cologne Act, cited in Robie 1970, p. 18).

A related difficulty which may hinder the Board in addressing water quality impacts is a procedural one. One author has pointed out, for example, that administrative agencies are constrained to consider only factors in the record currently being considered (Robie, 1972, p. 705). As a result, water quality considerations were historically often omitted from the rights-granting process simply because they were never brought up, either by the applicant or by protestors. This has probably been particularly true of return flows from irrigation. Recent developments which have broadened the scope of the public interest, however, require the Board to consider a much broader range of factors, particularly third-party effects, in administering appropriative rights. This suggests that active information-gathering by the Board will be increasingly important in the future.

A third set of difficulties involve possible political pressures on the Board because of the redistributional impact of its policies. Reallocating appropriative rights in order to curb third-party impacts would likely involve a transfer of in-basin quasi-rents from upstream to downstream users. One would thus expect upstream users to be politically opposed to such reallocations, and willing to pay (such as for lobbying state legislators, court costs, and/or bribes) up to the amount of their expected losses under this policy to have it defeated. This implies of course that the Board will experience the most political pressure precisely when the third-party impact is most severe (and therefore, when the efficient policy calls for the largest adjustment in the upstream appropriation). It is perhaps not coincidental that the Board decisions handed down in both of the above-cited Delta and IID cases were subsequently challenged in court. Future attempts to apply this policy more broadly may encounter more broad-based opposition.

## CONCLUSIONS

This chapter has attempted to introduce water quality considerations systematically into the economic discussion of the efficiency of appropriative water rights. The main conclusion is that existing economic models of appropriative water use which omit the possibility of water quality degradation may ignore an important source of inefficiency. The analysis indicates that redefining water rights along surface waterways to account for water quality effects may improve water use efficiency, but derives no simple definitional rule which eliminates third-party effects. Consequently, an administrative agency with broad discretionary powers to issue and adjust appropriative rights to account for water quality impacts would be required.

It appears that California has developed the necessary administrative machinery and procedures to effectively address water quality issues through the appropriative rights-granting process. Most importantly, the state has combined the functions of water rights administration and water quality planning into one agency which enjoys very broad statutory authority to consider water quality impacts in granting and modifying appropriations. Furthermore, this statutory authority has been consistently affirmed by the courts in cases where the extent of the Board's authority was in question. This may have contributed to an increasing inclination by the Board to actively intervene to promote improved water allocation policies (Spangler, 1988, p. 1567; Robie, 1972, p. 723).

Questions nevertheless remain concerning the Board's practical ability to effectively integrate water quality and water rights concerns in all cases where it is necessary. Monitoring return flows from nonpoint sources such as agricultural runoff will probably require significant amounts of additional resources. Furthermore, better understanding of the link between diminishing surface water quality and the costs imposed on downstream appropriators is probably required as well. Future research should attempt to quantify this link, and derive estimates of the potential gains from redefining water rights along a particular surface waterway.

An additional set of questions, not directly addressed in this chapter, involve the effect which a powerful state agency may have on private incentives to develop and exchange water. In particular, imbuing the Board with broad discretion to adjust appropriations once granted may tend to promote a general perception that appropriative rights are not secure. This could discourage local investment in water development facilities, particularly under rapidly changing conditions related to the nature and magnitude of costs associated with declining water quality. To some extent, it may also discourage movement towards greater reliance on water transfers if potential buyers are uncertain how much water they will ultimately be legally entitled to. Finally, it raises a serious legal question as to what, if anything, constitutes a vested right to water. These issues should also be examined in future research.

## APPENDIX 1: DERIVATION OF WATER USE EFFICIENCY CONDITIONS FOR THE CASE OF THREE USERS

Call the three users Upstream, Middle, and Downstream, and denote them as U, M, and D, respectively. Then rewrite the Lagrangean equation [7] as follows:

$$[7] \quad L = \text{Max} \left\{ \int_0^{S_U} f_U(\theta_U) \, d\theta_U + \int_0^{S_M} [f_M(\theta_M) + \alpha[S_U, S_O - SU(1 - RU)]] \, d\theta_M \right.$$
$$\left. + \int_0^{S_D} [f_D(\theta_D) \, d\theta_D + \alpha[(S_U + S_M), S_O - S_U(1 - R_U) - S_M(1 - R_M)]] \, d\theta_D \right\}$$
$$+ \tau \left[ (S_O - \underline{S}) - S_U(1 - R_U) - S_M(1 - R_M) - S_D(1 - R_D) \right]$$

Derive the first-order conditions by partially differentiating the Lagrangean with respect to the diversion amounts SU, SM, and SD, and setting these expressions equal to zero.

$$\frac{dL}{dS_D} = f_D(S_D) - \tau(1 - R_D) = 0$$

$$\frac{dL}{dS_M} = f_M(S_M) + S_D[\alpha_X - \alpha_F(1 - R_M)] - \tau(1 - R_M) = 0$$

$$\frac{dL}{dS_u} = f_U(S_U) + S_M[\alpha_X - \alpha_F(1 - R_U)] + S_D[\alpha_X - \alpha_F(1 - R_U)] - \tau(1 - R_U) = 0$$

Rearranging these expressions yields:

$$\tau = \frac{f_D}{(1-R_D)} = \frac{f_M + S_D[\alpha_X - \alpha_F(1-R_M)]}{(1-R_M)}$$

$$= \frac{f_U + (S_M + S_D)[\alpha_x - \alpha_F(1-R_U)]}{(1-R_U)}$$

## NOTES

[1] See, for example, J. Milliman, "Water Law and Private Decision-Making: A Critique," 2 J. Law & Econ. 41 (1959); L. M. Hartman & D. Seastone, Water Transfers: Economic Efficiency and Alternative Institutions (1970); C. J. Meyers & R. A. Posner, "Market Transfers of Water Rights: Towards an Improved Market in Water Resources," National Water Commission Legal Study No. 4 (1971); H. S. Burness & J. P. Quirk, "Appropriative Water Rights and the Efficient Allocation of Resources," 69 Amer. Econ. Rev. 25(1979); H. S. Burness & J. P. Quirk, Water Law, Water Transfers, and Economic Efficiency: The Colorado River," 23 J. Law & Econ. 111(1980); R. N. Johnson, M. Gisser, and M. Werner, "The Definition of a Surface Water Right and Transferability," 24 J. Law & Econ. 273(1981).

[2] Much of the discussion centers around the effect of withdrawals on instream uses such as fish and wildlife. See, for example, A. J. Schneider, "Legal Aspects of Instream Water Uses in California: Background and Issues, "Governor's Commission to Review Water Rights Law Staff Paper No. 6 (1978)." The potential importance of this phenomenon has been recognized for some time by water experts, as seen in the following statement by the Deputy Attorney General of California over 30 years ago:

"Most water uses need water of a certain minimum quality. On the other hand, most water uses tend to degrade quality of the water and thus affect its suitability for subsequent use. Hence, if multiple and successive uses of the same water are to be protected, the maintenance of water quality and prevention of water pollution become as vital as the development of sufficient water quantities."

See A. Moskovitz, "Quality Control and Reuse of Water in California: 45 Cal. Law Rev. 586 (1957).

[3] See, for example, Miller et al. for a discussion involving the Colorado River.

[4] Data on salinity levels were taken from map insert contained in Miller et al.

[5] The public trust doctrine has been the subject of considerable interest recently. For discussions of the public trust doctrine as it applies to water allocation policies in California, see J. L. Sax & R. H. Abrams, Legal Control of Water Resources (1986), pp. 466-75; R. W. Johnson, "Water Pollution and the Public Trust Doctrine," 19 Environmental Law 485(1989); S. S. Spangler, "Imperial Irrigation District v. State Water

Resources Control Board: Board as Arbiter of Reasonable and Beneficial Use of California Water," 19 Pac. Law J. 1565 (1988).

## ACKNOWLEDGEMENTS

This chapter was written while the author was on leave from Carleton College as Ciriacy-Wantrup postdoctoral fellow in natural resource economics at University of California. The author acknowledges financial support from the Graduate Division at University of California, Berkeley and partial support from Carleton College.

## REFERENCES

Attwater, W. R. and Markle, J., 1988. Overview of California Water Rights and Water Quality Law, *Pacific Law Journal*, 19, pp. 957-1030.

Brandt, A. W., 1987. United States v. State Water Resources Control Board: A Comprehensive Approach to Water Policy in California, *Ecology Law Quarterly*, 14, pp. 713-40.

Burness, H. S. and Quirk, J. P., 1979. Appropriative Water Rights and the Efficient Allocation of Resources, *American Economic Review*, 69, pp. 25-37.

Burness, H. S. and Quirk, 1980. Water Law, Water Transfers, and Economic Efficiency: The Colorado River, *Journal of Law and Economics*, 23, pp. 111-34.

California Legislature, 1966. *A Proposed Water Resources Control Board for California.* Assembly Interim Committee on Water.

Dunning, H. C., 1982. *Water Allocation in California: Legal Rights and Reform Needs.* Institute of Governmental Studies Research Paper, University of California.

Hartman, L. M., and Seastone, D., 1970. *Water Transfers: Economic Efficiency and Alternative Institutions.* Johns Hopkins University Press, Baltimore, MD.

Johnson, R. W., 1989. Water Pollution and the Public Trust Doctrine, *Environmental Law*, 19, pp. 485-513.

Johnson, R. N.; Gisser, M.; and Werner, M., 1981. The Definition of a Surface Water Right and Transferability, *Journal of Law and Economics*, 24, pp. 273-88.

Meyers, C. J. and Posner, R. A., 1971. *Market Transfers of Water Rights: Towards an Improved Market in Water Resources.* National Water Commission Legal Study No. 4, Arlington, VA.

Miller, Taylor O. and others, 1986. *The Salty Colorado.* The Conservation Foundation, Washington, DC.

Milliman, J., 1959. Water Law and Private Decision-Making: A Critique, *Journal of Law and Economics*, 2, pp. 41-63.

Moskovitz, A., 1957. Quality Control and Reuse of Water in California, *California Law Review*, 45, pp. 586-603.

Robie, R. B., 1970. Water Pollution: An Affirmative Response by the California Legislature, *Pacific Law Journal*, 1, pp. 2-35.

Robie, R. B., 1972. Some Reflections on Environmental Considerations in Water Rights Administration, *Ecology Law Quarterly*, 2, pp. 695-731.

Robie, R. B., 1988. The Delta Decisions: The Quiet Revolution in California Water Rights, *Pacific Law Journal*, 19, pp. 1111-42.

Sax, J. L. and Abrams, R. H., 1986. *Legal Control of Water Resources. Cases and Materials*. West Publishing Co., St. Paul, MN.

Schneider, A. J., 1978. *Legal Aspects of Instream Water Uses in California: Background and Issues*. Governor's Commission to Review California Water Rights Law Staff Paper No. 6, Sacramento, CA.

Spangler, S. S. 1988. Imperial Irrigation District v. State Water Resources Control Board: Board as Arbiter of Reasonable and Beneficial Use of California Water, *Pacific Law Journal*, 19, pp. 1565-96.

U.S. Department of the Interior, Bureau of Reclamation, 1986. *Joint Evaluation of Salinity Control Programs in the Colorado River Basin*.

U.S. Department of the Interior, Water and Power Resources Service, 1980. *Colorado River Salinity: Economic Impacts on Agricultural, Municipal, and Industrial Users*.

West Publishing Co., 1977. *California Public Resources Code*.

# 42 INSTITUTIONAL AND LEGAL DIMENSIONS OF DRAINAGE MANAGEMENT

Gregory A. Thomas, Natural Heritage Institute

## ABSTRACT

This chapter discusses the drainage problem in California's western San Joaquin Valley and the difficulties created by current institutional arrangements (within water and drainage districts, between districts and State and Federal agencies, etc.). A proposed institutional solution package is suggested based on similar problems elsewhere. The anatomy of the solution is presented and discussed. The proposed package is applied to a small region. A case study analysis is used to demonstrate the advantages of the proposed institutional solution. The conclusion section addresses obstacles to be removed if such solution is to be implemented.

## INTRODUCTION

On the west side of California's San Joaquin Valley (Valley), the soils are naturally permeated with salts and trace elements such as selenium and rich in clays that slow the percolation of water. The salts and trace elements leach into the shallow ground-water table, where they concentrate, eventually to toxic levels. If left in place, saline drainage invades the root zone, imperiling crops. If collected and discharged, drainage water can threaten the receiving streams, ponds, or wetlands and the fish and wildlife that inhabit them. At the present rate of increase, by the year 2000, high ground-water levels may adversely affect about 1 million acres of irrigated land, or about 40 percent of irrigable land in the Valley study area (SJVDP, 1990). Much of this land would be forced out of production.

The drainage problem in the Valley--and its solutions--are a function of how irrigation water is applied and managed and how subsurface drainage is collected and disposed in the environment. These water management decisions are ultimately made by growers at the farm level, but are influenced by a variety of governmental institutions and laws at every level.

Analysts and planners have suggested technically and economically feasible measures to ameliorate the problem and extend the life of the farmland, if not to solve the problem over the longer term. However, the institutional and legal aspects, which in some cases are predominant, have often been ignored. This chapter will discuss institutional responses that entail minimal changes in the existing legal structure.

## BACKGROUND AND STATUS

The Federal Central Valley Project (CVP), managed by the U.S. Bureau of Reclamation (Reclamation), and the State Water Project (SWP), managed by the Department of Water Resources (DWR), are the major importers of water to the Valley (collectively "the Projects"). To remove drainage, the Federal project called for a master drain to the Sacramento-San Joaquin Delta. This drain was never completed beyond the Kesterson Reservoir (Kesterson), which imported drainage between 1975 to 1985 (SJVDP, June 1990). In 1985, the Secretary of the Interior stopped the discharge of drainage to Kesterson in the wake of migratory bird deaths and deformities induced by high levels of selenium in the water (USDI-OIG, 1985). In 1988 the Central Valley Regional Water Quality Control Board (Regional Board) adopted a moratorium on expansion or construction of new subsurface drainage water collection facilities in high selenium areas until water quality standards are met (CVRWQCB, 1988).

The volume of drainage water that must be disposed is estimated to reach 314,000 acre-feet per year by the year 2000 (SJVDP, June 1990). It currently threatens four important resources:

(1) Cropland productivity. Agricultural lands are threatened by a buildup of saline water in the root zone of crops. If the current trend continues, an estimated 529,000 acres of irrigated land will have to be taken out of production within 50 years (SJVDP, June 1990).

As a private resource, farmland productivity is arguably not a legitimate governmental concern. From another view, the public's investment in water projects made these desert lands into a remarkably productive source of food and fibre; thus the public has a stake in the return on its investment.

(2) Surface waters. Agricultural drainage discharged to the San Joaquin River, and ultimately the San Francisco Bay, threatens surface water quality, a resource protected by State and Federal law.

(3) Ground water. One physical solution to drainage accumulation in the shallow water table is a planned depletion of the deeper ground water, within the semiconfined aquifer, to accelerate the downward percolation of drainage. This strategy could cause contamination of usable aquifers, in violation of the nondegradation policy of the State Water Resources Control Board (State Board) (see State Board Resolution 68-16).

(4) Fish and wildlife. Evaporation ponds in the southern portion of the Valley, like Kesterson, have been found by the U. S. Fish and Wildlife Service (FWS) to threaten migratory birds, a potential violation of the Migratory Bird Treaty Act (MBTA).[1] Moreover, agricultural return flows that formerly provided a source of water for waterfowl refuges have been rendered unsuitable by commingling with drainage. Substitute water supplies for refuges are important to mitigate drainage.

## INSTITUTIONAL RESPONSES

A satisfactory institutional response to the drainage problem would be designed to protect each of the four natural resources of public value and would be efficacious to reduce the generation of drainage, assure collection and disposal of the residual drainage, and protect and restore degraded habitat.

### Reducing the Volume of Drainage - Source Control

Approximately 90 percent of the 2.4 million irrigable acres on the Valley's western side are producing irrigated crops at any time. Currently, 6.5 million acre-feet of water is used annually for this irrigation. Improved irrigation efficiency could cut the volume of problem drainage by up to 60 percent in some areas (SJVDP, June 1990).

By reducing the generation of drainage, water conservation would also reduce the cost of managing and disposing of drainage. Conservation would also reduce the transboundary impacts of drainage between farms and between districts,[2] and would make water available for transfer to other beneficial uses, including environmental restoration.

Two economic incentives (in preference to regulatory controls) are proposed to induce conservation: inverted block (tiered) rates for water deliveries from districts to growers, and institutional reforms to facilitate transfer of conserved water.[3]

Conservation-Inducing Rates. A recent National Research Council Report considers the water rate to be "the most pervasive economic issue contributing to irrigation-related water quality problems and affecting the choice and success of solutions . . .[T]he subsidized cost of water results in more water being used, encourages farmers to cultivate less desirable lands, and leads to increased agricultural runoff" (NRC, 1989).

Water districts already possess the authority to design conservation-inducing water rates,[4] and several have done so.[5] Inverted or tiered block rates that incorporate the cost of drainage management in the tail block(s), would apportion these costs more accurately than the existing rate structure.[6] Such rates would reward efficiency and penalize inefficiency without changing the aggregate cost of irrigation water plus drainage management in the area. Those growers who apply irrigation water greatly in excess of evapotranspiration and minimal leaching requirements would pay the greatest proportion of drainage costs.[7]

To finance drainage management and internalize the costs, volume sensitive drainage fees would be desirable. However, variations in evapotranspiration, gradients, soil permeability, and other geohydrologic factors render accurate monitoring of contribution to drainage extremely difficult. By contrast, the volume of applied irrigation water represents an easily monitored surrogate that is reasonably, though not precisely, related to contribution to drainage. It is therefore proposed that drainage management costs be incorporated in tail blocks of tiered water rates.

The correlation between volume of water applied and the contribution to drainage problems is imprecise as a result because of variability in the constituents of the soils and, hence, of the drainage water, which may directly affect the degree of hazard and the costs of effecting solutions. Uncertainties regarding the movement of drainage along the gradient also complicate the relative contribution to the drainage problem water. These variables are problematic only to the extent that they exist <u>within a district</u>. Variations in soils and gradients between districts do not matter to the extent that each district is a closed system for purposes of allocating the costs and benefits of eliminating drainage water.

Water Transfers. Sale of conserved water[8] could induce and finance conservation and move excess water out of the study area. Despite the declaration of the California Water Code that it is the established policy of this State to facilitate the voluntary transfer of water and water rights where consistent with the public interest in the place of export and the place of import (California Water § 109(a)), there have been few attempts to transfer water in California. The State Board, which must approve any transfer that alters the point of diversion, place or purpose of use (California Water § 386), received only 23 applications to transfer water between 1981 and 1989 (Gray, 1989).

To be sure, much of the transfer within the Valley takes place without the approval of the State Board. Given the broad scope of permits held by Reclamation, transfers between Federal contractors in the Valley do not change point of diversion, or place of use (the factors which trigger oversight by the State Board). In this way over 3 million acre-feet of water was transferred within the CVP system during the 1980's (Gray, 1989). However, unless the water is transferred out of the drainage problem area, it may not contribute to drainage reduction.

The main barrier to water transfers is uncertainty in Federal Reclamation law and State water law. Upon close inspection, the potential legal impediments may preclude transfers in limited instances only. Among the legal complications are: Questions regarding the transferor's entitlement to profits; uncertainty as to the nature of the legal interests that can be conveyed;[9] area of use restrictions on project-supplied water; nature of use restrictions; third-party approval requirements;[10] and in general, high transaction costs that may negate potential profits.[11] In addition, assurances beyond economic benefits may be necessary to overcome the psychological reticence of farmers to transfer water out of agricultural usage, sometimes referred to in California as the "Owen's Valley Syndrome."

<u>Land Retirement.</u> Retiring the problem lands that contribute most to the salt and selenium burden is another source control option. Water transfers could stimulate some voluntary retirement of marginal lands from irrigated agricultural use. Where voluntary retirement is not a sufficient response, water districts could withhold deliveries to the problem lands and provide compensation to the landowner. The costs of land retirement to the district should then be distributed to the other growers as a part of the cost of drainage management.

## Drainage Management

Source control can only reduce drainage, not eliminate it.[12] Substantial residual drainage water will have to be managed, in part because the soils within the study area generally require periodic leaching of the salts in order to remain productive.

<u>Water Districts to Manage the Drainage Problem.</u> The California Constitution (Article X, Section 2) and statutory law (California Water § 275) condition water rights upon reasonable and beneficial use of the water, and the State

Board has held that failure to implement conservation measures may constitute unreasonable use (SWRCB, D-1600, 1984). Furthermore, a California Court of Appeals held that the State Board may use its water rights authority to enforce a water quality requirement.[13]

Water districts already possess the authority to manage drainage,[14] but their charters impose no clear duty to do so.[15] Some districts, most notably the Westlands Water District, have developed aggressive programs for assisting growers in drainage reduction and in experimenting with improved disposal options.

Clearly there is much that the districts can do to promote drainage solutions. Moreover it is the local water agencies that possess the knowledge and expertise to tailor solutions to diverse soil and geohydrologic conditions and to the preferences of individual farmers. Where drainage districts exist, the water or irrigation districts may choose to contract with them for drainage services, as a factory would contract with a private waste management firm to handle its hazardous wastes. Where no such entity exists, the water or irrigation districts may form local or regional drainage management districts, acting either alone, or in concert with other districts.

Regional drainage management entities may be formed by the districts through Joint Powers Agreements (JPA's) (Natural Heritage Institute, 1990). A JPA may exercise the powers common to the districts involved (California Government § 6502), and has the advantage of providing a regional solution and economies of scale.

External Standards - Existing and Proposed Legal Regimes. Although the districts are best suited to manage drainage, they are unlikely to act in the absence of external stimuli. To protect the four public resources at risk from drainage, enforceable performance standards with penalties externally imposed by the regulatory agencies (the State and Regional Water Boards) are necessary. Standards for surface and ground-water quality and for waterfowl protection are already in place. Standards do not exist, however, to protect farmland productivity. Uniform standards of this nature may be desirable eventually to ensure that one district's or grower's efforts to reduce drainage will be reciprocated by neighboring districts and growers, since drainage respects neither district nor farm boundaries. Private actions for nuisance or trespass and their attendant problems of proof are too expensive and uncertain to force action.

To protect all four natural resources at risk, performance standards (as contrasted with prescriptive requirements) are preferable. Performance standards specify an acceptable level of protection or control without specifying the particular method to be used.[16] This leaves the districts the maximum discretion to choose the most cost-efficient measures, best suited to local conditions.

Existing legal regimes to protect surface and ground water and fish and wildlife are exercised by the State Board,[17] which oversees the allocation of all water rights and the protection of water resources in the State, governed by the reasonable and beneficial use requirements.

Instead of regulating agricultural drainage via Waste Discharge Requirements (WDR's), the Regional Board recently adopted amendments to the San Joaquin Basin Plan[18] that set water quality objectives for agricultural constituents and required districts to develop Drainage Operation Plans (DOP's) which specify the voluntary measures they will take to meet these objectives.[19]

The Regional Board's basin plan recommends that drainers adopt "best management practices" (BMP's), however, they are not mandatory. Although the State Board has authority to modify water rights for failure to meet water quality standards,[20] the recent basin plan amendments indicate that the Regional Board would only request the State Board to exercise this authority as a last resort.

The DOP requirements are aimed at managing residual drainage, the "back-end" of the problem. Both DWR and Reclamation already possess authority to address the "front-end" of the problem, drainage generation, by requiring districts to prepare water conservation plans. This planning process has so far yielded little improvement in water management. If desired, the process could be upgraded to resemble that applied to coal mining or oil and gas production under Federal leases.[21] In the drainage context, DWR and Reclamation could condition the right to receive publicly developed water upon the contractor submitting a water conservation plan meeting certain levels of sufficiency prescribed by the agency, and conditioning eligibility for renewal of current water contracts upon compliance. If reasonably applied, such a sanction is consistent with the authority that the Federal and State agencies possess to police the beneficial use of publically developed water.

Options for Disposal of the Residual. In choosing among methods to dispose of residual drainage, the districts will encounter certain legal complications with some options, including use of a master drain to the Delta, various methods of concentrating the volume of residual drainage, and ground-water storage schemes.

- Master Drain. Conducting drainage to a discharge point below the confluence with the Merced River and below the runoff from the Grasslands refuges would facilitate disposal by making use of the increased assimilative capacity of the San Joaquin River. Dilution of discharge in this manner to meet Federal and State water quality standards would be problematic if the Regional Board should decide to regulate the mass emissions of persistent toxic contaminants that bioaccumulate, such as selenium,[22] rather than regulating only the concentration of target pollutants in the receiving water as it does now. At present, the Regional Board supports the concept of a Valleywide drain to the Delta to carry salts generated by agricultural irrigation out of the Valley as the best technical solution to the water quality problems of the San Joaquin River and Tulare Lake Basin (CVRWQCB, 1988).

- Evaporation Ponds. There are roughly 22 ponds covering 6,800 acres currently in operation in the Valley. The Regional Board expects that an additional 3,000 acres of ponds will be constructed within 15-20 years.[23] Some of the evaporation ponds pose a threat to migratory birds and may violate the MBTA. Although the MBTA provides for both civil and criminal penalties for violations,[24] the FWS has not yet imposed penalties or requested the prosecution of violators (CVRWQCB, 1989). The Regional Board has prohibited expansion and new construction of ponds until a Valleywide waterfowl impact assessment is completed and mitigation measures are developed. If the Board adopts mitigation measures currently being considered, such as development of alternative wildlife habitat, total evaporation pond acreage could be reduced to as little as 5,000 total acres.[25]

The threat to birds aside, there are numerous regulatory barriers to disposal in evaporation ponds. Under the State Hazardous Waste Management Act (Health and Safety § 25179 et. seq.), disposal of hazardous waste in an evaporation pond is permitted only when that is the best demonstrated and available technology for disposal, and when that will not present an immediate or significant long-term risk to public health or the environment.

The State Toxic Pits Control Act (Health and Safety § 250208 et. seq.), prevents discharge of waste exceeding specified concentrations of constituents having persistent and bioaccumulative properties, such as selenium and arsenic, to ponds near drinking water sources. The legislature granted the Regional Board authority to exempt evaporation ponds containing agricultural drainage from certain Toxic Pits Control Act siting, design, and operation requirements until January 1, 1993, if the ponds are designed and operated to prevent the migration of contaminants and the endangerment of wildlife, as well as other standards. Renewal of this exception (due in January 1992) is uncertain. The Regional Board will not allow the operation of new ponds under this exception until an assessment of wildlife is complete.

The Federal Resource Conservation and Recovery Act (RCRA), in contrast to these State regulations, exempts "irrigation return flows" from its hazardous waste disposal requirements (42 U.S.C. § 6903(27)).

Ground-Water Basin Storage. If not drained, the saline drainage water builds up in the root zone, preventing salt leaching and endangering crop productivity. A coordinated program of pumping the deeper ground water of usable quality within the semiconfined aquifer, could provide storage capacity for--and accelerate the deep percolation of--the shallow ground water. The correlative rights of overlying landowners to extract ground water do not readily accommodate a ground-water depletion strategy orchestrated by water districts or a regional management entity. Moreover, the option entails a planned degradation of usable ground water.

Currently the right of an overlying landowner to pump ground water is limited only by the requirement of reasonable and beneficial use.[26] Surplus ground water, and imported water which is abandoned may be appropriated.[27] Unlike surface water, appropriated ground water is not subject to the California permit system. No State agency has authority to regulate the ground-water resource as the State Board regulates surface water usage. The State Board is only authorized to regulate to prevent ground-water overdraft,[28] and to protect ground-water quality.[29] There is no authority to require an overlying landowner to pump.[30]

Contractual arrangements with overlying landowners may be the best tool for districts or regional authorities to implement a comprehensive groundwater program for drainage management. Such contracts could give these authorities the prerogative to dictate the location and scheduling of ground-water extraction, and to coordinate the conjunctive use of ground and surface water. Experience under the Arizona Groundwater Management Act[31] teaches the virtue of local implementation to allow adaptation to variations in soils, hydrology, and farm practices.

The primary obstacle to ground-water storage of drainage may be acceleration of the inevitable degradation of the ground-water quality in the deeper aquifer. Deep percolation of saline water as the result of pumping is not covered by the enforcement provisions of the Porter-Cologne Act.[32] However, the State Board may commence an adjudication "to restrict pumping, or to impose physical solutions, or both, to the extent necessary to prevent destruction of or irreparable injury to the quality of [ground] water" (California Water § 2100). Reconciling ground-water storage of drainage with State groundwater quality objectives would require a policy determination by the State legislature accompanied by delegation of authority to the State Board to consider the overall benefit to both surface and ground water and cropland productivity in deciding to allow drainage storage in a given ground-water basin.

## Environmental Protection

Fish and wildlife are protected by laws implemented by FWS and Department of Fish and Game (DFG). These laws alone may not be sufficient to protect wildlife from drainage impacts. Additional measures are necessary to mitigate these impacts and restore degraded wildlife habitat. Therefore, the problem of drainage management in the Valley cannot be regarded as solved without addressing the need for freshwater to replace or dilute toxic drainage in wildlife refuges, wetlands and the San Joaquin River. Ideally, this component of the solution would also mitigate some of the historic adverse effects that the CVP and SWP have had on environmental resources. This may entail a new framework for transferring some water from irrigation to habitat maintenance and restoration.

Water transfers provide an opportunity for reallocating water to environmental enhancement. Transfers to new uses or locations require approval by the Reclamation for the CVP and DWR for the SWP and by the State Board. Approval of such transfers could be made contingent on the transferor contributing a percentage of the net proceeds to a trust fund, administered by the State or Federal fish and wildlife agencies, and dedicated to the purchase of water specifically for environmental enhancement. This approach has the advantage of allowing water to be purchased wherever in the State environmental enhancement is most needed. It returns to the public a share of the profits from marketing publicly developed water supplies, and routes these monies to the agricultural sector, through the purchase of surplus water, where they may be used to defray drainage management expenses.

## CONCLUSIONS

The justification for governmental intervention in solving the Valley drainage problem is the fact that each district's and each grower's application and drainage management decisions (or nondecisions) affect neighboring farmers and districts, and the environment. It must be appreciated that each grower and district is both a potential perpetrator and a potential victim of drainage. It is a problem that is unlikely to be solved through individualized and voluntary action. Instead, uniform and mandatory requirements are in the best interests of all stakeholders.

The institutional solutions suggested here do not depend upon significant infusions of new subsidies or other public resources. The suggested approach would distribute the costs of drainage management and compensatory fish and wildlife measures on the drainage generators--the growers themselves.[33] In economic terms, this scenario represents a cost internalization approach.

The key political calculation is that the cost of not solving the drainage problem in all its dimensions is simply too high for all stakeholders to make the status quo acceptable. The common ground is the imperative that the problem be solved, equitably and expeditiously.

## ACKNOWLEDGEMENTS

Assistance provided by Christine Leas (J. D. 1991) and Barbara Cosens (J. D. 1990), Hastings College of Law.

## NOTES

[1] 16 U.S.C. §§ 703 et. seq. See U.S. Fish and Wildlife Service, Information Bulletin No. 88-49, July 1988, "Deformed Waterbird Embryos Found Near Agricultural Drainage Ponds in the Tulare Basin:" 50 CFR 10.13; U.S. Fish and Wildlife Service, Memorandum from Drainwater Studies Coordinator, Sacramento, California, to Deputy Regional Director and Regional Director - Region 1, Portland, Oregon, "Applicability of MBTA to Operation of Evaporation Ponds in the San Joaquin Valley," July 31, 1988; Testimony of James J. McKevitt, U.S. Fish and Wildlife Service, Field Supervisor, Sacramento Office, Presented to the Central Valley Regional Water Quality Control Board Workshop, "Waterfowl Impacts Occurring from Agricultural Evaporation Basins," June 2, 1989; Harry M. Ohlendorf and Joseph P. Skorupa, "Selenium in Relation to Wildlife and Agricultural Drainage Water," paper presented at Fourth International Symposium on Uses of Selenium and Tellurium; Banff Alberta, May 7-10, 1989.

[2] Although the magnitude of the lateral flow in most regions of the study area is not well documented (due in part to the prohibitive cost of monitoring), in various areas it is a significant component of the drainage problem. One area that has been extensively studied is the Broadview Water District. It is estimated that, on an average, some 27 percent of the volume of drainage does not correlate with applied water and may be attributable to lateral flows. Figures in the rest of the study area would probably vary substantially, depending on soil and subsoil conditions, gradients, irrigation application methods and quantities, and the efficiency of collector drains.

[3] Water subject to transfer would also include surplus water that becomes available from retirement of marginally productive lands with high selenium and salt and additional ground water pumped to accelerate deep percolation of drainage.

[4] 43 U.S.C. § 485h(d)(2).

⁵Broadview Water District, Central California Irrigation District and Pacheco Water District.

⁶Transferring drainage costs to growers in the form of volume sensitive charges would allocate to those who contribute more to the problem a proportionally larger share of the costs of solutions. No grower would subsidize the drainage contribution of another. Nor would the public be subsidizing improvident choices by growers.

⁷One study estimates the social cost of water to be roughly $200 per acre-foot for Westlands Water District (cost of delivery = $75, lost opportunity cost = $100, environmental cost of return flow = $33). At present, agricultural water rates in the study area range from $16 to $50 per acre-foot. (See Kenneth D. Frederick, December 12, 1989, "Qualitative Assessment of Policy Tools to Alter Water Use in the San Joaquin Valley," prepared for the Natural Heritage Institute, at 4.)

⁸The excess or salvaged water that contributes to subsurface drainage is the target for salvage and transfer. Water that currently leaves the field as tailwater or operational spill comprises an important source of supply for other growers (particularly in the unincorporated areas of the study area), for refuges (when it is not commingled with drainage water), and for dilution water in the San Joaquin River.

⁹The U.S. Supreme Court has indicated that landowners hold equitable title to the project water which can be transferred subject only to the Bureau's reasonable regulation. Fox v. Ickes, (1939) 300 U.S. 82, 95.

¹⁰Transfer of water under contract requires approval by the Bureau for contract modification of the original contract and formation of the new contract. 43 U.S.C. § 423e. See NHI document by Richard Roos-Collins, May 29, 1990, "Impact of Reclamation Law on Water Management and Environmental Mitigation in Federal Projects," p. 23.

¹¹The State legislature has been at pains to remove legal barriers to transfers. The Water Code has been amended to provide that conserved water may be sold without jeopardizing the water right, and that "[t]he sale, lease, exchange or transfer of water or water rights, in itself, shall not constitute evidence of waste or unreasonable use." DWR must allow bonafide transferors access to unused aqueduct capacity at a fair rate to transport water.

¹²Improved irrigation management through scheduling and technologies such as shorter furrows and tailwater return systems are predicted to reduce drainage production by only 40 percent in most zones. Even with additional techniques such as drip systems, lining of canals, and better tillage practices, substantial quantities of drainage water will still have to be managed. See SJVDP, June 1990 at 5-35.

¹³United States v. SWRCB (1986) 182 Cal. App. 3d 82, 125. In theory, this authority permits the State Board to require Reclamation and DWR, or their

contractors, or the growers directly, to take steps to reduce the generation of subsurface drainage caused by excessive water use. In practice, however, the Board has never used this power to address the drainage problem, and its exercise is sufficiently discretionary and judgmental that it is unlikely to provide a reliable solution to the problem.

[14]In general, water districts possess the power to manage drainage, levy fees, construct, operate and maintain drainage works, exercise eminent domain, and acquire, hold and dispose of real property.

[15]The consolidated settlement of <u>Westlands Water District v. U.S.</u>, (E.D. Cal. 1986) No. CV-F-81-245-EDP, and <u>Barcellos & Wolfsen, Inc. v. Westlands Water District</u>, (E.D. Cal.) No. CV-79-106-EDP (settlement order filed Dec. 30, 1986) places responsibility for the construction of drainage facilities and 65 percent of their cost on the Federal Government. A recent U.S. Court of Claims case states that:

> [i]f the evidence is correct that every irrigation canal will eventually cause high ground water and drainage problems, then the problem becomes a foreseeable one, and more importantly, the duty of the Federal Government to correct. (<u>Baker v. U.S.</u>, (Oct. 1989) Ct. Cl. No. 675-83L, slip opinion at 106.)

Although this case addresses seepage from canals, it indicates a willingness to hold Reclamation responsible for drainage problems in general.

[16]Performance standards may be expressed in a variety of ways. In the water quality field, the standards are usually expressed in terms of protection of existing or potential beneficial uses.

[17]The California Supreme Court made this clear in its decision of <u>National Audubon Society v. Superior Court</u>, (1933) 33 Cal. 3d 419, 441.

[18]The basin planning process constitutes implementation of the nonpoint source regulation requirements of the Clean Water Act (33 U.S.C. § 1329), and the Porter-Cologne Act (Calif. Water Code §§ 13240 - 263), to protect water quality.

[19]Central Valley RWQCB, Res. No. 88-195. The CVRWQCB requires water and drainage districts to prepare DOP's which are to contain voluntary strategies which, taken together, are intended to reduce drainage discharges to the level necessary to meet receiving water quality standards in the San Joaquin River. Districts are not, however, required to undertake any of these planned actions, and the regime does not impose specific standards on individual districts for drainage source control, collection, management, or disposal. EPA has determined that several of the water quality objectives do not satisfy the requirements of the Clean Water Act, and that the water quality standards for selenium and boron will not fully protect beneficial uses.

[20]United States v. SWRCB.

[21]Coal mining on private as well as public land is regulated by the Federal Government under the Surface Mining Control and Reclamation Act (SMCRA). When submitting permit applications, SMCRA requires operators to develop plans which specify the soil management practices that will be followed to ensure that mined land will be restored to support premining uses. These practices must meet performance standards as well as detailed criteria established by the Department of the Interior or by an approved State program.

The Federal Government regulates oil and gas leasing similarly under the Outer Continental Shelf Lands Act amendments. Exploration and leasing permits require preapproved development, production and management plans, which must specify the environmental safeguards to be employed. Exploration permits require the demonstration that the exploration will not unduly harm environmental resources, such as aquatic life, or result in pollution, among other requirements.

[22]Most treatment options aimed at addressing these contaminants are still in the laboratory test stage. Promising technologies are economically unattractive, thus there is little immediate prospect that the disposal problem will be solved by treatment. SJVDP, June 1990, at 3-17 through 3-24.

[23]Telephone conversation with Anthony Toto, CVRWQCB Staff, July 5, 1990.

[24]16 U.S.C. §§ 703 - 707; See U.S. Department of the Interior, May 31, 1985, at 5 - 6: "a party may be held criminally liable for incidental bird deaths if it fails to take precautions, reasonable under the circumstances, to prevent foreseeable bird mortality."

[25]Personal communication, Anthony Toto, RWQCB Staff, July 5, 1990.

[26]Peabody v. Vallejo, (1935) 2 Cal. 2d 351, 372 (interpreting Calif. Const. Art. X Sec. 2 as applicable to ground water).

[27]City of Los Angeles v. City of San Fernando, (1975) 14 Cal. 3d 199, 277-78, (surplus ground water); Stevens v. Oakdale Irrigation District, (1939) 13 Cal. 2d 343, 350 (imported water).

[28]California Water Code § 12922.

[29]California Water Code §§ 13700 et. seq., and § 2100.

[30]Anne Thomas, (Best, Best, and Krieger), January 1990, "Application of California Groundwater Law to Foreign Groundwater," Prepared for the Natural Heritage Institute, p. 1.

[31]Arizona Revised Statutes §§ 45-401 to -818.

[32]See Sawyer, 1988, "State Regulation of Groundwater Pollution Caused by Changes in Groundwater Quantity or Flow" *Pacific Law Journal*, V. 19, 1267.

[33]It must be acknowledged, however, that growers and districts would be allowed to profit from the sale of subsidized water. This revenue may help defray the costs of drainage management. It may have the appearance and

effect of a subsidy in that respect. Also, land retirement strategies will compensate landowners for foregoing irrigation. That compensation will likely reflect the value of subsidized water and, in this respect, operate as a subsidy.

# REFERENCES

*Publications*

Central Valley Regional Board, 1988. Resolution No. 88-195 Adopting Draft Amendments to the Water Quality Control Plan for the San Joaquin Basin (5C) for the Control of Agricultural Surface Discharges.
Frederick, K. D., 1989. *Qualitative Assessment of Policy Tools to Alter Water Use in the San Joaquin Valley*. Revised Draft, Natural Heritage Institute.
Gordon, M. F., 1990. *The Joint Exercise of Powers Authority: A Vehicle for Cooperation in Implementing Solutions to the San Joaquin Valley Drainage Problem*. Natural Hertage Institute.
Gray, B., 1989. A Primer on California Water Transfer Law, *Arizona Law Review*, 31, pp. 745.
McKevitt, J. J., 1989. U.S. Fish and Wildlife Service, Field Supervisor, Sacramento Office. Testimony presented to the Central Valley Regional Water Quality Control Board Workshop, *Waterfowl Impacts Occurring from Agricultural Evaporation Basins*.
Meyer, C. H., 1989. *Environmental Water Rights in California*. Preliminary Draft, Natural Heritage Institute.
National Research Council, 1989. *Irrigation Induced Water Quality Problems: What Can Be Learned from the San Joaquin Valley Experience*. National Academy Press, Washington, DC.
Ohlendorf, H. M. and Skorupa, J. P., 1989. Selenium in Relation to Wildlife and Agricultural Drainage Water. Paper presented at Fourth International Symposium on Uses of Selenium and Tellurium, Banff Alberta.
Roos-Collins, R., 1990. *Impact of Reclamation Law on Water Management and Environmental Mitigation in Federal Projects*. Natural Heritage Institute.
Sawyer, 1988. State Regulation of Ground-Water Pollution Caused by Changes in Ground-Water Quantity or Flow, *Pacific Law Journal*, 19, pp. 1267.
San Joaquin Valley Drainage Program, April 1990. *Final Report of the San Joaquin Valley Drainage Program*. Preliminary Draft.
San Joaquin Valley Drainage Program, June 1990. *Final Report of the San Joaquin Valley Drainage Program*.

San Joaquin Valley Drainage Program, 1989. *Preliminary Planning Alternatives for Solving Agricultural Drainage and Drainage-Related Problems in the San Joaquin Valley.*

Thomas, A., 1990. *Application of California Groundwater Law to Foreign Groundwater.* Best, Best & Krieger.

Toto, A., 1990. Personal Communication, Central Valley Regional Board Staff.

U.S. Department of the Interior, Office of Inspector General, 1985. *Report of Investigation 19.*

U.S. Department of the Interior, Office of the Solicitor, 1985. Memo from Frank K. Richardson, Solicitor, to the Secretary of the Interior.

U.S. Fish and Wildlife Service, 1988. *Deformed Waterbird Embryos Found Near Agricultural Drainage Ponds in the Tulare Basin.* Information Bulletin No. 88-49.

U.S. Fish and Wildlife Service, 1988. Applicability of MBTA to Operation of Evaporation Ponds in the San Joaquin Valley. Memorandum from Drainwater Studies Coordinator, Sacramento, California, to Deputy Regional Director and Regional Director - Region 1, Portland, Oregon,

## Cases

Baker v. United States, 1989. Ct. Cl. No. 675-83L, Slip Opinion.

Barcellos & Wolfsen, Inc. v. Westlands Water District, (E.D. Cal. 1986) No. CV-79-106-EDP.

California State Water Resources Control Board, 1984. Water Rights Decision 1600.

California v. United States, 1978. 438 U.S. 645.

City of Los Angeles v. City of San Fernando, 1975. 14 Cal. 3d 199.

Fox v. Ickes, 1937. 300 U.S. 82.

National Audubon Society v. Superior Court, 1933. 33 Cal. 3d. 419.

Peabody v. Vallejo, 1935. 2 Cal. 2d 351.

Stevens v. Oakdale Irrigation District, 1939. 13 Cal. 2d 343.

United States v. SWRCB, (1986) 182 Cal. App. 3d 82.

Westlands Water District v. United States of America, (E.D. Cal. 1986) CV-F-81-245-EDP.

## Statutes

### Federal

Clean Water Act, 33 U.S.C. 1251-1387.

Migratory Bird Treaty Act, 16 U.S.C. 703 et. seq.

Outer Continental Shelf Lands Act, 43 U.S.C. § 1331 et. seq.
Resource Conservation and Reclamation Act, 42 U.S.C. 6901-6992k.
Reclamation Project Act, 43 U.S.C. 371 et. seq.
Safe Drinking Water Act, 42 U.S.C. 300 et. seq.
Surface Mining Control and Reclamation Act, 30 U.S.C. 1201-1328.
Warren Act, 43 U.S.C. 523 et. seq.

State

California Health and Safety Code § 25208 et. seq.
California Toxic Injection well Control Act, Calif. Health & Safety Code 25159.10 et. seq.
California Toxic Pits Control Act, Calif. Health & Safety Code 25208 et. seq.
California Water Code.
Arizona Ground-Water Management Act, Ariz. Revised Stats. 45-401 to -818.

# 43 LEGAL ISSUES RAISED BY ALTERNATIVE PROPOSED SOLUTIONS TO KESTERSON WATER QUALITY PROBLEMS

Charles T. DuMars,
University of New Mexico, Albuquerque

## ABSTRACT

Legal entanglements in the San Joaquin Valley are complex. Holders of water rights under California law and Federal contracts have substantial judicial support. Environmentalists also hold a great deal of legal authority. Solution to the problems of agricultural drainage disposal at Kesterson Reservoir involve resolution of these conflicts.

## INTRODUCTION

This chapter does not provide a detailed analysis of all possible legal claims that might arise out of potential water quality control actions taken by institutions operating within the San Joaquin Valley (Valley). Rather, it delineates the various stakeholders and the legal options available to them in the context of implementing alternative solutions. A brief sketch of the problem area is essential to understanding the discussion that follows. Kesterson Reservoir (Kesterson), located on the west side of the Valley, was originally designed as a storage area to control the amount of water flowing down the proposed San Luis Drain into the western part of the Sacramento/San Joaquin Delta, on its way to the San Francisco Bay, and subsequently to the Pacific Ocean. For political reasons, the drain was never constructed past Kesterson, which consequently evolved into a large evaporation pond to dispose of agricultural drainage. While much of Kesterson is now covered with fill dirt, it originally encompassed 12 individual ponds with an average depth of 3 to 4 feet. The ponds, collectively referred to as Kesterson, incidentally provided a significant wildlife habitat.

Congress authorized construction of the San Luis Unit of the Central Valley Project (CVP) in 1960, in response to requests by the Westside Landowners Association (which later became the Westlands Water District) for a more reliable water supply for the west side of the Valley. The building of the drain was debatable even in 1965, when the public vented a concern for the environment and the potential pollution of the San Francisco Bay. In response, the California legislature requested a comprehensive study of the estuary. In 1967, apparently due to budget restraints, California declined to participate in the drain.

The U.S. Bureau of Reclamation (Reclamation) began construction of Kesterson in 1968 and completed it in 1971 at a cost of $7 million. During the first phase of construction, Reclamation completed the southern half of the drain. Kesterson was to function as a temporary holding pond to dispose of the collected agricultural drainage by evaporation and percolation until the second phase of construction was finished. However, the drainage system was not completed.

Some of the events which were factors in the Kesterson pollution problem stemmed from the release of the Environmental Impact Statement (EIS) required by the National Environmental Policy Act (NEPA) of 1969. Although the EIS supported the second phase of the construction, suits were filed, and a task force was formed to review the San Luis Unit. During this time, Reclamation continued to install the Westlands Water District subsurface drainage system, which began releasing substantial amounts of water into the San Luis Drain in 1979. By 1981, approximately 7,000 acre-feet per year, the upper limit that could be evaporated or disposed of at Kesterson, was draining into the reservoir ponds.

In 1980, Reclamation discontinued additional connection of onfarm drains to the main collector system until Kesterson could be enlarged and/or the second construction phase completed. The resulting drainage collected in water tables above shallow clay layers, causing them to rise and approach the surface in certain areas. By 1983, 25,000 acres in the Westlands Water District had perched saline water within 5 feet of the surface; 156,000 acres had perched water tables within 5 to 10 feet of the surface; and 91,000 acres had saline water between 10 and 20 feet of the surface. Until agricultural practices in the Valley change or other remedial strategies are implemented, the situation can only degenerate.

Significantly, the mineral content in the reservoir ponds elevated as water entered and evaporated. In 1982, the Fish and Wildlife Service (FWS) first observed the problems resulting from the escalating concentrations of trace elements when it was noted that fish species, including large-mouthed and striped bass, catfish and carp, had disappeared. In the spring of 1983, eggs from waterbirds exhibited decreased hatchability and deformities of the embryos.

The U.S. Geological Survey (USGS) subsequently linked high concentrations of selenium amassed in the ponds to the problem.

Awareness of the Kesterson problem has ignited the concerns of local, State, and Federal special interest groups, agencies, and legislators who represent different and often conflicting stakes in the drainage issue. At the Federal level, three agencies of the Department of the Interior are involved in resolution of the problem: USGS, FWS, and Reclamation. At the State level, California's Department of Water Resources (DWR), Department of Fish and Game (DFG), State Water Resources Control Board (SWRCB), and regional water quality control boards of the areas affected are involved.

Drainage problems in the Valley are the result of numerous and various hydrogeologic complications, but when reduced to their essence, they reflect the following basic truths. First, every activity which utilizes water degrades its quality to some degree. Second, if one brings new water into an area and the water is not totally consumed, provisions must be made for its drainage. Third, if water brought into an area contains salt, the salt remains even if the water is totally consumed. Fourth, for reasons no one really understands, boron, selenium, and other heavy metals often occur naturally in areas where shallow ground-water movement will leach them out and concentrate them because of their soluble nature. These substances can have serious deleterious health impacts on birds and other wildlife species living in the areas of peak concentration. Fifth, while members of the general public show little interest in the role of farm subsidies created through virtually interest-free construction of water projects, they express shock and concern when a byproduct of subsidized irrigation results in the loss of life--even if only to birds and other forms of wildlife. Sixth, once a coalition of environmental groups and entities, such as Ducks Unlimited, combines with others who have an overall concern for the environment, and with representatives of the local press, such as the *Sacramento Bee*, an otherwise unseen issue becomes important to the Federal bureaucracy in Washington, to irrigation farmers, and to residents of an area where the problem previously went largely unnoticed.

Specific solutions to this problem are expressed elsewhere in this volume. Thomas et al., for example, have reviewed, at considerable time and expense, the legal consequences of particular solutions. This chapter will not revisit that task. Rather, the solutions are lumped into four separate categories and analyzed in terms that should be applicable to any comparable problem arising elsewhere. These four categories are: (1) Imposition of regulatory standards, (2) economic incentives, (3) legislation mandating an end to the problem, and (4) cleanup and disposal. The first solution, imposition of regulatory standards, simply provides that, as of some specified date, there will be an internalized solution adopted by those producing the pollutant, ensuring that poor quality water does not leave the area. The second solution involves adopting economic incentives that will make it financially more attractive to the polluters to take measures to eliminate the pollution than to continue polluting. The third solution is to have the legislature mandate that in the future the problems will cease or the polluter will pay a fine or be subject to legal injunction. The last solution involves implementing a program of cleanup and disposal either by treating and removing salts, sending the unwanted substance to the ocean, or injecting the substance into the ground.

The ensuing discussion presumes that the above solutions are to be completed without an extensive infusion of new Federal money. If, as is often the case, the Federal Government opens its financial purse to alleviate the problem, there is very little need for legal discussion. Indeed, if farmers were paid in full for all of the lands in the Valley, and if water ceased to flow into the Valley altogether, the immediate problem of Kesterson would be solved. This is not to suggest that all of the problems would be solved. To the contrary, there likely would be wildlife relocation problems, population relocation problems, and problems of economic redistribution that would dwarf the existing circumstances. Such issues go to the wisdom of such a solution, not to the capacity of money to solve the problem, and are not discussed here.

A second assumption is that no overwhelming political sense of urgency about this problem will make the courts the standard bearer for a well-accepted principle of morality lying latent within the country's conscience. For example, the unanimous decision of the Supreme Court in Brown v. Board of Education of Topeka Kansas reflected an element of political consciousness regarding the fundamental immorality of racism. It spurred inevitable political change reflective of this view. Up to this point, although there has been an extensive amount of outrage among environmental groups and others regarding Kesterson, this concern in no way reflects the kind of moral imperative that can allow the judicial branch on its own strength to bring about social change. Rather, Kesterson reflects a host of competing values and institutions with sufficient stake to engage each other in the legal arena. No legal position is so weak as to be comparable with the slim legal reed ("separate schools for blacks are in fact equal") advanced in Brown.

## THE LEGAL STAKEHOLDERS

While variously stated in law, it is unanimously held that one can only seek redress under common law or statutes conferring rights or prohibiting infringement of rights, if one has a stake in the outcome of the controversy. The quantum of stake necessary to place one in the fray varies dramatically depending upon the kind of dispute and the proximity of the potential litigant to the controversy. For example, a person in New York may feel badly when he sees pictures of the unhatched birds at Kesterson on the educational channel of his television set. While he may express concern, it is unlikely that his feelings are enough to give him standing to sue. This is not to say that he is unimportant, as his feelings may cause him to send money to an entity in California that does have standing. Nevertheless, he has not suffered the kind of personal injury that will allow him to go to court to try to change the outcome of events in California. Conversely, a farmer who is told that he must purchase extensive infrastructure to improve his farm practices, or use less water, or grow different crops, is directly affected and can take legal action to challenge the proposed responsibility because his very livelihood is at stake. This is not to suggest that his feelings of being threatened by economic loss are more important than those of the individual living in New York. Rather, the principle of law developed suggests that an issue will only be fully and fairly litigated if the parties have some concrete stake in the outcome.

Moving away from the individual in New York and into California, suppose a person or group of persons in California have, as their principal collective goal, protecting wildlife, particularly the birds, of the Valley. Suppose further that some of their members actually go out into the area and photograph the birds or simply enjoy them. These individuals may have no direct financial interest in seeing the birds flourish; indeed, they may lose money through their support of bird interests, but the law would grant them standing to sue because they are affected by the activities in the area. In the Valley, persons in this category may also include organizations such as Ducks Unlimited who wish to promote the propagation of certain species of ducks, which ducks provide sport and revenue for those who hunt them. While these two groups make strange bedfellows, their interests are in many ways the same and are sufficient to assert legal rights. Other entities with standing to sue would include the rural domestic water users who might have their potable water supplies affected by the infiltration of heavy metals, and the irrigation districts themselves, whose members might be directly affected economically by a particular solution. Finally, the State of California, through the office of the Attorney General, would have standing to pursue what it perceives as the public good, assuming that the irrigation activities were considered to be a public nuisance. Further, the United States itself, through the Migratory Bird Treaty Act or the Clean Water Act, might be implicated if the death of the birds were considered a treaty violation or if water quality standards were violated.

From the above discussion, it can be readily understood that the range of potential actors and the variety of their interests is legion. For purposes of this chapter, and without suggesting in any way that irrigation farmers are less interested in the environment than others, we will assume: (1) That the interests of the farmers are mainly economic and (2) that the interests of the environmental groups are largely noneconomic and are more associated with the groups' perceptions of the quality of life that should be maintained in the area for the various forms of marsh and aquatic wildlife.

## THE PARTIES ACTED UPON

In every lawsuit there must be a defendant. And, in the same sense that the law requires the plaintiff to have a sufficient stake in the outcome to sue, the law likewise requires the defendant to have the power to implement some solution that is sought in the litigation. Suffice it to say that the range of potential defendants who have capacity to remediate the problem could include the irrigation districts, the farmers themselves, the Secretary of the Interior through the Bureau of Reclamation, and California agencies regulating water quality and quantity. While variously stated, the causes of action would all involve the defendants acting in some way inconsistently, or failing to act consistently, with the plaintiffs' interpretation of a statutory or constitutional right or some right found in the common law of the State.

While the distinction between substantive and procedural law is often ridiculed, academics persist in distinguishing between what they call substantive causes of action and procedural causes of action. The substantive cause of action does not attack the process through which a particular postulate or rule became law. Rather, the complaint is that the rule is not being applied correctly or that it is wrong. The procedural attack is based not on the premise that the rule itself is wrong, but on the premise that all arguments that could have been made were not made because the process was not fair, i.e., the decisionmaker did not have the opportunity to hear and consider all of the relevant issues because not all with an interest had the opportunity to participate. The relative merits of the different kinds of inquiries continue to be debated. Those who are "procedurephiles" are often more interested in the process of how a case gets decided than its outcome. Indeed, they may be critical of what they call judicial legislation as reflected in the "public trust" doctrine of water rights. Those who are fans of judicial independence and expect the courts to carry out their independent view of society's needs, are not impressed with the importance of procedural due process. Those in the latter camp often state that the key to a successful society is the substantive fairness of its laws, not necessarily the process of enacting them. In hell, they say, all laws would be unfair and because of this, all would be convicted, but the devil would see that procedural due process was meticulously observed. It is not the author's intent to become involved in this debate. The distinction between substantive and procedural remedies is introduced solely because it adds clarity to the discussion in this chapter.

## SUBSTANTIVE CAUSES OF ACTION

### Environmentalists' Position

The environmental parties will have at their disposal a host of legal challenges to whatever solution is posed. These challenges can be grouped roughly into Federal laws and State laws requiring nondegradation of the environment. The former laws include the Federal Clean Water Act, which expressly prohibits point-source pollution and requires, at a minimum, that nonpoint source pollution be regulated by the states pursuant to some reasonable plan. There also exists section 404 of that Act, which prohibits dredging and filling of wetlands without a permit. Likewise, there is the Resource Conservation and Recovery Act prohibiting storage of hazardous wastes, the Endangered Species Act, the Safe Drinking Water Act, and a host of other directly or tangentially applicable State laws. The latter laws include the Hazardous Waste Control Act, the Toxic Pits Control Act, and the Porter-Cologne Act, to name a few. Other substantive claims can be made under the water law of California, including the Public Trust Doctrine, which requires that California State Courts evaluate the implications to the public welfare of water quality decisions. There is also the doctrine of beneficial use of water, which can be construed to preclude "waste" of water by farmers and to require conservation in certain circumstances.

### Farmers' Position

Without assuming that in every case the farmer would be resistant to change or that his organization necessarily would be resistant to change, and assuming that a challenge were raised, farmers are not without a substantial arsenal with regard to water rights. First, they are the parties acted upon, so they hold some sort of vested property right in the water, either through contract or under California law. Any regulation that completely destroys that property interest will be labeled an unconstitutional "taking" in violation of the Fifth Amendment to the United States Constitution. Likewise, any action taken by one water user to reduce the water supply in the stream and transfer it elsewhere, or to reduce the water volume, will involve the rule of nonimpairment to existing rights. That is to say, under State law, a transfer will only be allowed if it does not impair the rights of others who are not a party to the transfer. With respect to affecting all water users of the district, it may be argued that all parties affected be joined in the action because their rights may be affected and thus are indispensable.

The United States Constitution requires equality of treatment in the application of regulatory sanctions or standards. Most of the Kesterson studies show that selenium is concentrated in pockets. Consequently, a regulation treating all farmers equally would punish some for a problem they are not causing, and would let others off easily when they are the predominate source of the injury. Furthermore, Federal Reclamation Law is replete with land mines for jurists. It is an unending, vague combination of Federal funding mechanisms and statements of public policy. Its principal purpose was to promote irrigated farming on small farms through subsidized water projects. The Reclamation Reform Act did little to clarify this issue other than to make clear that under certain circumstances, subsidies to large-scale agriculture were to be terminated. The ownership of water interests in such projects is still unclear. Is the contract right the property of the United States, formed by some form of Federal trust to carry out the interests of the Reclamation Act? Is the right the irrigation district's? Or is the right held in trust for the farmer? These are not insignificant legal questions, since relief can only be obtained from the individual with the capacity to bring about relief. Sorting out these questions will involve long and intractable litigation. Indeed, the Environmental Protection Agency (EPA) and Reclamation may find themselves at odds on this and other Federal law issues. Finally, sorting out the relationships under State law and among the irrigation districts will be no easy task. The districts are complex and varied and their jurisdictions may overlap, with power over the same land shared between governmental entities of equal legal status.

## PROCEDURAL CAUSES OF ACTION: FARMERS AND ENVIRONMENTALISTS

Both the farmers and the environmental groups have available a cadre of procedural legal arguments. Assuming one were to challenge a State law affecting some, but not all of the right holders in a district, the argument that all affected parties must be joined would immediately surface. This is true because water rights in such a district are interdependent. One person's return flow is another's irrigation water and so on. Adjudications of this kind taking in excess of 10 years are not unheard of in the West. Providing service of process to all affected parties can be extraordinarily expensive. Also, the adoption of any regulatory standard under California or Federal law requires adequate notice to all concerned. It is rare that such procedures can be carried out without at least some procedural error. These errors are reviewable in the courts of first instance and ultimately in the appellate courts.

In instances where damages or some other type of expenditure of funds is sought, and the Federal Government is a defendant, there is the possibility of sovereign immunity on behalf of the United States at a minimum, and conceivably on behalf of political subdivisions of the State. Even worse, if the United States is immune from suit and is indispensable to the action, then some courts have ruled that the suits cannot go forward at all. Assuming the relief is in the form of an injunction, then separate from the problem of immunity, the willingness of Federal judges to implement creative relief is being brought into question. In the context of Federal orders with respect to prisons, many are questioning the wisdom of such action. And of course if there is no Federal or State money available to implement the requested relief, it is unlikely that the court is going to send individual, named State or Federal officials to jail for failure to take action.

Finally, there is the NEPA. This Act would likely require the preparation of an environmental impact statement prior to the implementation of any Federal solution. While the Federal Government has not required such a document prior to Reclamation contract renewal, impact statements are being prepared after the fact as a result of such contact renewals. Certainly, the issuance of any permits by the Corps of Engineers under 404 would trigger this requirement, as would any Federally mandated retirement of acreage or change in water supply for wetlands or instream flows. This requirement can be cumbersome and the method of preparation is virtually always subject to judicial review with regard to the adequacy of the issues addressed, the sufficiency of the "scoping" process identifying the issues, the opportunity for comment, or whether or not the responsible agency actually took a "hard look" at the change being proposed. The economic and social consequences of these actions must be evaluated fully and debate on these questions is strong. These issues can be taken through the full range of judicial processes as high as the United States Supreme Court. With respect to virtually any major Federal action, an impact statement would have to consider, the effects on agricultural production, farm drainage, water quantity and quality, water allocation, wildlife habitat, recreation, and regional communities and economies, including the loss of tax bases. Thus, the possibility for error in the preparation of such a document is obvious.

Application of the above discussion of legal standing, potential defendants, and substantive and procedural legal claims is revealing. Assume, for example, that the following regulatory solution was adopted: The decision is made at the State level by statute to control absolutely the quantity of salts and heavy metals contained in the irrigation return flows from the irrigation works by establishing standards for each works. Assume further that there is a public administrative process for adopting these standards.

This scenario, while apparently straightforward, and currently available in California, contains numerous possibilities for legal involvement. Without suggesting anything regarding the merits of potential arguments, the following ones could be made. There inevitably would be an appeal by those who contend that the standards are so weak and generous to the farming community that the standards fail to meet the requirements established by the legislature. Likewise, there would be an appeal from the regulated parties claiming that the standards are severe to the point of being arbitrary and capricious, and that the adoption of such standards was never intended by the legislature. This appeal would come after an extensive war of experts at the administrative level, each challenging the credibility of the opposition's testimony.

On appeal to the judicial branch, there would be the argument by the farming community that these regulations are so severe as to constitute an illegal "taking" of their property without due process of law and without compensation. Likewise, there might be a United States constitutional challenge to these regulations based upon the unequal application of the law--that is, some parties contribute very little to the problem, but may be forced to curtail water use in the same proportion as those who contribute a great deal to the problem. There also could be litigation between the irrigation district and individual farmers, each claiming that the other shares a greater degree of responsibility. Further, there would be the argument that Federal law preempts the field in this area and that the sum of the Federal laws, including the Clean Water Act, the Resource Conservation and Recovery Act, and the Reclamation Act of 1902, did not anticipate states passing laws that virtually foreclose irrigation by such draconian, regulation of irrigation return flows. Additionally, if the issue of the regulation's enforceability in turn raises issues of scientific concern, then the battle of factual experts could become intense. The cost of implementing the regulation would be subject to great debate, as would be the levels of each of the substances that could be considered toxic or economically destructive; millions of dollars could be paid out for scientific breast beating.

On the environmental side, the argument would be made that the breadth of the state police power, as reflected in the Tenth Amendment to the United States Constitution, authorizes states to go as far as they choose in regulating water quality. Indeed, the water quality standards are necessary to protect the state of California's interest in its fisheries and wildlife in a manner akin to the protection of the freshwater shrimp in Mono Lake under the "public trust doctrine." As to the unconstitutional taking issue, the environmentalists would argue that there is no such taking because there is a valid regulatory purpose and the legislation is a good faith attempt to effectuate that purpose. Furthermore, the Clean Water Act plainly anticipates, and indeed mandates, cleanup of nonpoint source pollution by the states. With reference to the equal protection argument, environmentalists would declare that since this issue does not involve a special or suspect classification, such as discrimination based on race or in violation of some individual constitutional right, the legislation need not be drawn perfectly. If it is rational in purpose, it is not invalid solely because its effects are at times disparate.

On the procedural front, the arguments would be familiar. Obviously, if any Federal ratification was necessary to implement the regulations, then their ratification by a Federal officer would raise the inevitable NEPA argument, and however the impact statement was prepared, some opponent would find room for a colorable legal challenge. In addition, the impacts of such a regulation, to the degree they involved irrigating less land and returning less water (albeit water containing salts) to the river, would implicate NEPA, as would the acts of dredging and filling in "wetlands," because a dredge and fill permit would have to be obtained by the United States Army Corps of Engineers. Any challenge to such a regulation, or any attempt by court order to force a State agent to enforce the regulation, would raise questions of proper parties--who can and should be involved? If removal of some portion of water from the system and elimination of farm land from the area affects neighboring farmers, or the district itself, or the security interest that the Federal Government may have in any of the district works, then they are all conceivably necessary parties. These individuals must be served with process and given the opportunity to present testimony, to conduct discovery and to appear at trial.

An alternative, direct solution would be to: First, locate those lands that contribute to the problem and take such land out of production; require that those who are currently using water and contributing in part to the problem use less water; and ensure that this happens by rewriting the contracts between the Federal Government and the districts such that less water is permitted to enter. One could also change the Reclamation Law to emphasize fish and wildlife protection on an equal basis with agriculture and reallocate water for instream flows, dilution of drainage water, and wetland management.

The above proposition sounds simple enough. Implementation is not legally quite so simple. Any solution changing the basic purposes of the Federal Reclamation Law as applied to this region would involve not only a change in Federal law, but also would undoubtedly require implementing Federal regulations. The adoption of such regulations would have to be conducted according to Federal administrative procedures. Public hearings would be held and, finally, legal challenges would arise regarding whether the regulations go too far or do not go far enough in interpreting Congressional intent. The decision of which lands to retire would involve factual justification in the selection process. Those who did not wish to have their lands retired would fight the condemnation on the basis that it is not justified by any valid public purpose. Likewise, extensive litigation would take place over the values in the condemnation process, and the districts themselves would seek compensation based upon their injury because economics of scale in such projects are relevant in deciding the quantity of infrastructure to develop. Those districts that had incurred heavy debt and put infrastructure in place would argue that the compensation price must include the loss to the district of the tax base which previously had been available to pay for the cost of repaying loans. If water previously used for irrigation were now used to dilute salts, other arguments would be made. The argument would no doubt be made that the use of freshwater for dilution of salty water is not a beneficial use, and that such allocations would be subject to forfeiture or, at a minimum, are not favored under current water allocation law. Furthermore, arguments would be made that wholesale reallocation of water to wetlands is contrary to conservation practices in that it increases evaporation and may aggravate the problem, because the areas where the wetlands are to be created might themselves contain heavy metals which would leach back into the underground flows that ultimately reach the river. Certainly, those who held valid contract rights to water would demand compensation not only for the loss of the lands they could no longer irrigate, but also for their contractual right to water. The question would be raised, of course, as to whether the district or the individual holds the property right in the contract water, thus suggesting more litigation.

The environmentalists would argue, no doubt, that there is no loss in property as a function of the water loss. They would argue that any use which creates more damage through irrigation runoff than it does benefit through agricultural production should not be compensable. Rather, the irrigator should be forced to show that prior to the cutback in water availability, he was not wasting water, and that he cannot economically produce the same crops with lesser water using good conservation practices. Environmentalists would further argue that the use of water on wetlands and instream is beneficial, particularly given the history of riparian water law in California, the revenue generated by the duck hunting operations, and the fact that many of the species protected by the wetlands may be on the endangered species list.

On the procedural front, the NEPA arguments again would be invoked. There also would be questions of proper parties and issues of exhaustion of administrative remedies. The latter argument, not mentioned before, suggests that an individual cannot unilaterally challenge a Federal or State law in court until he or she has exhausted the remedies available before the agency itself. This doctrine is particularly relevant where administrative agencies are implementing new laws or regulations and are applying them for the first time. Under this doctrine, the agency's own interpretation of its laws is given great weight in the courts, and it is for this reason that the courts would stay their hands until the agency's remedies were exhausted. This process can be extremely time consuming and costly for all concerned.

A third approach could be to: Develop an effective program of marketing water outside of the district by allowing individual farmers to sell the rights they previously used on their land to nonpolluting sources, or by allowing farmers to sell some portion of their rights which they have saved through good conservation practices.

This solution seems almost ideal in its simplicity. As was said so often by the late Frank Trelease, "water runs downhill to gravity and uphill to money." However, economic solutions on paper are not necessarily so easily accomplished in practice. Indeed, some no doubt misinformed individuals have referred to economists as persons who spend a great deal of time proving that what works in practice, would not work in theory.

The water marketing solution on its face, in fact, does trigger less legal machinery. However, it assumes that water itself, or contractual rights to water, can be parceled easily and sold off in chunks like coal. Unfortunately, this is not the case. In order to sell a water right and transfer it to a new location, one must consider the impact on others. If in fact all of the districts in the area: (1) Shared equal priority dates or in most cases identical contract rights, (2) contained equal rights to the same duty of water, and (3) could move the water from district to district and from farm to farm without restriction or market it outside the district without hydrologic impacts, then this solution would be ideal. There are, unfortunately, a few problems here. First, to have a market, one must know who owns the product so that one knows whom to pay. Second, one must be able to move the product from place to place via a common carrier of some form. Third, one must know that his purchase of the product is not burdened by someone else's interest in that product. Unfortunately, in the case of water, it is rare that these circumstance exist. First, one would have to ask the legal question, "Who owns the product?" The district? The farmer? The United States? If the product is to be sold and consideration is paid, who gets the return on the investment? Assuming that the nonfarming purchase price is higher than the subsidized irrigation farmer's price, under Reclamation Law, who gets the "economic rent?" Furthermore, assuming one was to get a clear answer from the courts on this point as to ownership and profits, downstream water users, as mentioned earlier, may rely upon the irrigation return flow of the upstream user as their source of water. Do they have any property right to this water? Many states would answer that they do have such an expectation where private water rights in a public stream are involved. Furthermore, any such sale would be subject to some scrutiny on public trust or welfare grounds. If one creates a water market, it does not necessarily follow that rights would only be sold from lands that are producing selenium or other heavy metals. Agricultural productivity, not potential for pollution, would likely determine which land is sold. Under laws establishing a market, should an individual be given economic incentives to sell his water and be allowed to capture economic rents from government subsidies if his sale does not help to remedy the problem that the market system was developed to solve? Laws setting up water markets would, of course, trigger many of the same procedural problems discussed above.

Another solution would be to: Simply discharge the drainwater into the ocean by a canal system or treat the irrigation return flows through some sort of reverse osmosis process or selective ion exchange process, thereby removing the salts and the heavy metals, and then return the water to the hydrologic system.

From the above discussion, it does not require much imagination to conceive of the myriad attacks under Federal law that might be launched on a system intended to run the drainwater to some discharge point in a river or in the ocean. The Resource Conservation Recovery Act and the Clean Water Act come immediately to mind, not to mention the State laws that suggest conservation of water and elimination of waste is an underlying postulate of basic water allocation. Certainly, the EIS associated with such a solution would be immense and the "not in my back yard" syndrome would spawn a host of local planning and zoning laws that would attempt to prevent such a channel from passing through certain areas. While it is no doubt true that running the salts to the sea is simply speeding up the process nature already started, it was man and not nature that brought the salts into the Valley. The selenium may have been in place, but it was and is irrigation that has concentrated it. Thus, this solution, though straightforward from an engineering standpoint and logical from a simple input-output analysis, faces virtually all of the legal challenges raised above regarding other possible solutions.

As to treatment, putting aside for a moment the fact that when one removes the salt, a salt disposal problem will exist, this solution involves less legal entanglements than many of the others. There will be, of course, the fact that no one wishes to have the treatment facility near them, and the related problems that inevitably occur with respect to siting, but there will be fewer legal implications because the status quo will be preserved. However, there is one simple question before this simple solution can be implemented: Who will pay for it? The Federal Government was generous in paying for a desalinization plant on the Colorado River to ensure that the water quality of the Colorado River as it entered Mexico lived up to the requirements of minute 242 of the Mexican Water Treaty of 1948. However, no such treaty exists here. And, furthermore, it was posited at the beginning that none of these solutions assumes unlimited Federal dollars. Therefore, the costs would have to be paid regionally. It is indeed an axiom so clear as to be considered what the late Chester Smith called a "legal gem of the first water," that when costs are allocated to individuals, someone goes to court. The probability of suit is reduced if the costs of such a program are allocated to all the persons who benefit from treatment, including municipal, environmental, and other water users.

While it is true that challenges to taxes are difficult if they do not infringe upon a specific constitutional right, such an action nevertheless could be framed if the result of the tax were to put farmers out of business, eliminate the tax base of entire counties or destroy the collateral the United States may use to secure its farm loans. Realistically, however, the problem would be one of politics. If the political will and capacity to pay existed, treatment would be the solution least likely to engender legal debate. This would not appear, however, necessarily to be solving a problem. It would simply be a decision that any other solution aimed at modifying behavior simply was not worth the political and legal turmoil that would accompany it.

It is hoped that the above discussion indicates that the often-stated proposed solutions to the problems in the Valley lead to a maze of legal and institutional complexities. The discussion above is far from complete and a good legal scholar may conclude that some of the legal challenges might ultimately be unsuccessful. However, for each legal argument that may prove unsuccessful, there is likely to be another not mentioned to take its place.

There is a clear reason for this complexity: There is a great deal at stake from the standpoint of private property rights and from the standpoint of the public welfare. It is virtually axiomatic that when there is a substantial capital investment in a property right, there will be a host of laws on the books to protect that right. This is true in the Valley. The environmentalists in the Kesterson controversy, however, have substantial legal authority on their side. In modern society, environmental groups, fishermen and hunting groups, and indeed, the Congress of the United States have concluded that the public welfare values in water supplies that can sustain fish and wildlife for future generations are entitled to legal protection. The "final" solutions to the drainage problems will test the strength of these competing legal positions.

## REFERENCES

Publications

Thomas, G. A. and Leighton, M. T., 1989. *Options For Disposal of San Joaquin Valley Drain Water. Institutional and Regulatory Constraints and Opportunities.* Prepared for the U.S. Bureau of Reclamation.

Cases

Brown v. Board of Education, 1954. 347 U.S. 483.

## Statutes

### Federal

Clean Water Act. Public Law 92-500 et seq., as Amended 1977, Public Law 95-217, as Amended 1987, Public Law 100-4.
Federal Endangered Species Act. 16 U.S.C. § 1538 et seq.
Fish and Wildlife Act. 16 U.S.C. § 742a-j
Migratory Bird Treaty Act. 16 U.S.C. § 703 et seq.
National Environmental Policy Act. 42 U.S.C. §§ 4331-4335
Reclamation Act of 1902. 32 Stat. 388 et seq.
Reclamation Reform Act. 96 Stat. 1263
Resource Conservation and Recovery Act. 42 U.S.C. § 6901, et seq.
Rivers and Harbors Act, 1982. 33 U.S.C. § 401-13
Safe Drinking Water Act. 16 U.S.C. § 300 et seq.
U.S. Constitution. Amendment 5
U.S. Constitution. Amendment 14

### State

Agricultural Water Management Planning Act. California Water Code, §§ 10800-10855.
The Porter-Cologne Water Quality Act. California Water Code, § 13,000 et seq.
Toxic Pits Control Act. California Health and Safety Code, § 25208.1

# Ten: OUTLOOK AND DISCUSSION

# 44 MANAGEMENT OF THE SAN JOAQUIN VALLEY DRAINAGE PROGRAM: THE DICHOTOMY BETWEEN PRACTICE AND THEORY

Edgar A. Imhoff, San Joaquin Valley Drainage Program

## ABSTRACT

Unusual circumstances and needs have arisen in the management of a cooperative attempt to solve the agricultural drainage problem in the San Joaquin Valley (Valley), and have led to the development and application of strategies and methods that may be useful in other attempts to solve complex natural resources problems in an interagency, interdisciplinary setting. This chapter discusses the efforts to manage a major research and investigative program and to build upon that program a rational planning effort. Particular management problems and strategies discussed involve: funding, organization, interpretation, and use of research results, advisory committees, use of consultants, development of planning methods, analytical models, policy constraints, role of the media, and measuring success.

## BACKGROUND

The discovery in 1983 of selenium contamination at Kesterson Reservoir (Kesterson) raised two practical questions of immediate importance: (1) In the Western United States, are there other geographic associations between irrigation projects and fish and wildlife areas that have resulted in or may result in another "Kesterson"; and (2) within the Valley, what has caused the problems at Kesterson and what can be done to prevent other "Kestersons." A National irrigation drainage program was formed to answer the first question by investigating conditions throughout the Western United States. The San Joaquin Valley Drainage Program (SJVDP), the focus of this chapter, was formed to address the second question.

The basic organizational structure of the SJVDP is shown in figure 1. The roles and responsibilities of each of the organizational elements and the SJVDP's relationships to other major programs, termed "related programs" on figure 1, are described in the following:

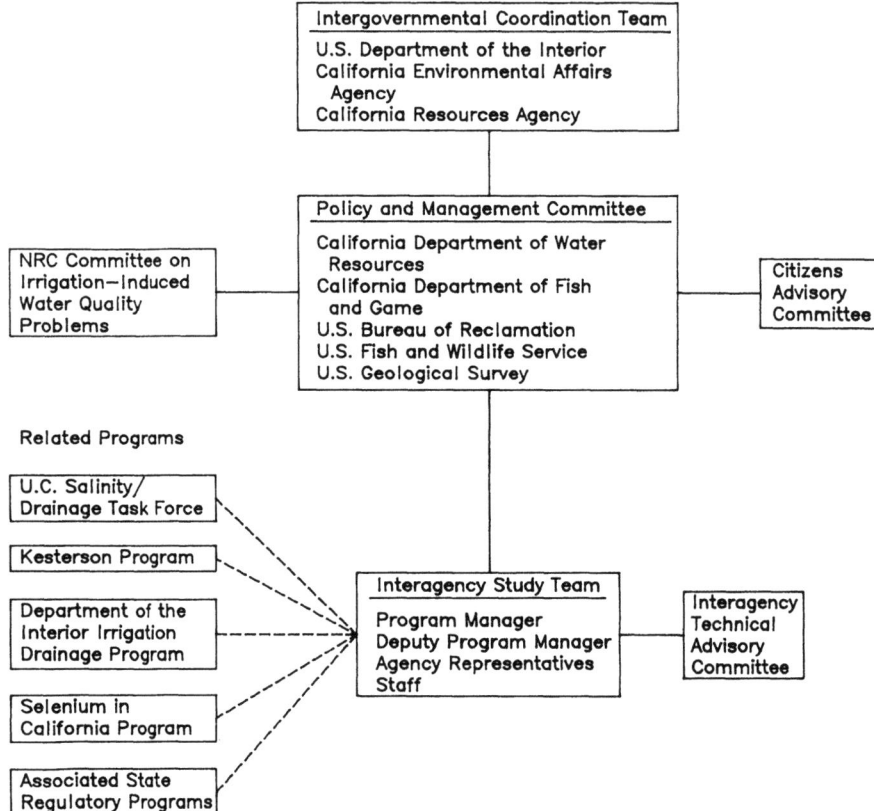

Figure 1. San Joaquin Valley Drainage Program organization chart.

An Intergovernmental Coordination Team was formed of policy-level appointees of the Secretary of the Interior and the Governor of California. The Intergovernmental Team provided broad guidance on SJVDP objectives and reviewed overall progress.

The Intergovernmental Coordination Team appointed a Policy and Management Committee (PMC), made up of the three Federal agency regional or district directors and the directors of two California State departments. (Federal and State regulatory agencies declined representation on the PMC, asserting their needs to preclude any potential conflicts of interest.) The PMC was responsible for providing specific guidance on SJVDP direction and priorities, allocating funds and personnel as needed to accomplish work, and acting on recommendations of the Interagency Study Team and advisory groups. The PMC functioned as the Board of Directors of the SJVDP and met several times each year.

A full-time Interagency Study Team was assembled as an interdisciplinary task force responsible for developing or obtaining information and analyses necessary for the formulation and evaluation of plans to solve drainage-related problems. The Program Manager and Deputy Program Manager, career managers and professionals selected by the PMC, reported directly to the Chair of that committee. The Program Manager served as Executive Secretary and ex officio member of the PMC. Each of the five-member agencies appointed a principal staff person to serve as primary liaison to their respective agencies (as Agency Representative), and also to work on the interagency team in their areas of professional expertise.

The National Research Council (NRC) was contracted to provide scientific oversight of the SJVDP through its Committee on Irrigation-Induced Water Quality Problems (CIIWQP). Much of the work of this committee was performed by subcommittees on: Economics and Policy, Systems Analysis, Public Health, Quality Assurance/Quality Control, and Treatment Technologies. The NRC-CIIWQP submitted occasional reports to the PMC and prepared a separate major summary report.

A Citizens Advisory Committee (CAC) was established to provide the SJVDP with information and viewpoints from the broad spectrum of organizations and individuals interested in and affected by drainage-related problems. The 14-member committee represented various affected interests and geographical areas.

An Interagency Technical Advisory Committee (ITAC) was formed of personnel from several Federal and State agencies and the California University System. This committee, unlike the PMC, included representatives from agencies with regulatory responsibilities. The ITAC provided technical advice on various aspects of the Program. Eight technical subcommittees provided direct support in the subject areas of: Agricultural water management, aquatic and fisheries biology, data management, geohydrology, ground water, public health, treatment and disposal, and valley biology. Several ad hoc working groups were formed, as needed.

The SJVDP, although a substantial program, was not the only major effort launched to address the selenium problem and the encompassing agricultural drainage problem. The programs and organizations listed on figure 1 as "related" investigations resulted in useful information. In addition, private individuals, local water organizations, and associations of local water districts launched their own inquiries and action programs toward solving the problems associated with agricultural drainage. These included the Grassland Task Force, Land Preservation Association, and Central Valley Agricultural Pond Operators.

## VIEWPOINT

In the sequence of the chapters of this book, I am probably the first author who is not commenting from a research viewpoint. My viewpoint is as manager of an investigative and public planning program which commissioned, conducted, or funded studies (including research) during the period 1985 to 1990. These studies, in the following fields of knowledge or professional disciplines served as the base for a major intergovernmental effort to solve agricultural drainage problems in the Valley: Agricultural engineering, bacteriology, biology, California history, chemical engineering, civil engineering, data management, demography, economics, fish and wildlife, geography, geology, hydrology, planning, political science, public health, sociology, soils, toxicology, and water law.

Certainly, I am unique among those authors in this book who are affiliated with the SJVDP as being the one person who will be asked to answer the inevitable audit questions that follow expenditures of tens of millions of Federal dollars for research, investigations, and planning. Routinely, the Federal offices of the Inspector General and the Congressional General Accounting inquire of managers like me: "Why did you spend $X for Y results, and what does it have to do with your objective Z?" Some small privilege, at least, should accrue to the person held most responsible for such decisions. The privilege I plan on exercising is to discuss, from a personal standpoint, the political environment in which the SJVDP study decisions were made; the analytical basis of SJVDP study decisions; and how, with the advantage of hindsight and using some of the ideas presented in this book, we might have more easily advanced the broad public interest in our attempt to solve agricultural drainage problems.

### *The Decisionmaking Environment Surrounding the SJVDP*

It has been my experience over some 30 years of managing studies that, whenever personal and organizational risks are low, it is relatively easy to systematically develop and adopt a study agenda. Unfortunately, that kind of environment for decisionmaking seldom exists and surely has not during the life of the SJVDP, 1984 to 1990.

In 1984, when the Kesterson debacle erupted, I was working in the office of the Department of the Interior's Assistant Secretary for Water and Science. A senior scientist of the U.S. Geological Survey (USGS) flew in from California to brief a roomful of us on the "mysterious" substance, selenium, that had just been assigned as the culprit in the deformity of waterbirds and in the reproductive failures of tens of thousands of tricolored blackbirds. The audience of that day and later audiences, including many water resources professionals, were caught offguard by the sudden potential importance of selenium to wildlife, to the health of people, and to the whole economy related to irrigated agriculture in the Western United States.

The first question from the Interior Department audience was raised by a chemical engineer who asked, "Is it (selenium) getting into foodstuffs?" The answer was that we didn't know. In fact, even some of the best of scientists knew very little about selenium--its source, occurrence, movement, fate, and effect. To remedy this, the first major work commissioned in July 1984 was a 5-year, comprehensive study by the USGS of the geohydrology and geochemistry of the Valley--with particular focus on the lands in the watershed that contributed to Kesterson. Soon, a second major thrust of studies was begun by the U.S. Fish and Wildlife Service (USFWS) to determine the occurrence, fate, and effect of selenium and other toxicants in wildlife and habitat.

From the standpoint of applying systems theory to project management, the above approach may be expeditious, but it isn't considered good management practice (Roman, 1986). Theoretically, the program effort should first have determined what decisions had to be made (at all levels of problem-solving potential) in order to set the study agenda of the SJVDP. Of course, such theory doesn't take into account the practical needs and political pressures of that time (1984-85), the spate of regulatory actions that had resulted in plugging of farm drains, and the intensive coverage by National media that treated Kesterson as the "Three-Mile Island of Western agriculture."

The massive research efforts in geohydrology and biology, while making SJVDP investigations somewhat lopsided, soon produced some answers and provided some very important clues to improving the design and focus of other studies.[1]

In addition to geohydrology and biology, there were also initial investments of hope and money in treatment technology, in the belief that removal of selenium from drainage water might soon prove technically--and perhaps even economically--feasible. In 1984-86, when the news media and political pressures were peaking, economists generally lagged in offering the kind of hope offered by the disciplines of geohydrology, biology, and engineering. While the scientists were promising to reveal cause and effects of the problem and engineers were hinting at a possible quick fix, economists were talking about determining how much growers could afford to pay. This was hardly a call to rally public support during 1984-85, although, as discussed later, the SJVDP did begin to invest in economic analyses, including modeling studies.

Initially, there was also an administrative difficulty in separating the cleanup of Kesterson from the prevention of other Kestersons, because both programs were funded largely by the U.S. Bureau of Reclamation (Reclamation). Prevention eventually became the study province of the five-agency SJVDP, leaving cleanup of Kesterson as primarily the responsibility of Reclamation.[2] It was not until late 1985 that there was a formal separation of the curative effort from the preventative effort, and many observers still refer to drainage investigations as "Kesterson studies." The separation of the programs helped the SJVDP move away from the emphasis on selenium, to give equal emphasis to boron and salt problems, and to begin to consider a long-term approach to solving drainage problems, as compared to the immediate cleanup that was being demanded by the publics and aggrieved parties at Kesterson.

Selenium, the focus of millions of dollars of scientific research (as a toxicant), has been even more instrumental as a political symbol. In early 1986, before accepting the job of SJVDP, Program Manager, I was forewarned that the mistrust and hostility that had been simmering for decades between National environmental organizations and Reclamation, the major water supplier to the Valley,[3] had boiled over with the Kesterson event. "They aren't going to let the Bureau up this time!", an authoritative political observer cautioned me. "They feel this is a great opportunity to bury the Bureau; wasting water was one thing, but poisoning a wildlife refuge is quite another. It is their cause celebre."

The realities of managing the SJVDP almost fulfilled this warning. Other than the normal differences one would find among members of a multidisciplinary group assembled from five agencies, divisiveness of motive and viewpoint created a difficult initial working environment within the SJVDP staff. Eventually, several measures were taken by managers, policymakers, and staff members themselves to improve the situation so that we could work together.

One of the most helpful of all administrative innovations to improve the climate for better professional work and study management was the appointment by the Chair of the PMC (California Department of Water Resources Director David Kennedy) of an advisory committee that, in intent and affiliation, spanned the spectrum of interested parties in the drainage problems. This advisory committee developed a facility for working well in addressing difficult issues. The performance of the CAC tended to significantly narrow the range of differences between agricultural and environmental interests. The general spirit of cooperation emanating from the CAC had a significant effect in improving internal working relations among the staff who, in expressing advocacy, were responding to many of the same pressures and viewpoints expressed by the CAC membership.[4]

The point of the preceding discussion is that intellectual and professional tasks are not the most difficult hurdles to be surmounted in initiating and managing a study program. Causism and advocacy have to be managed in any investigative body focusing on a complex and controversial subject.

Besides the initial obsession with selenium and the advocacy atmosphere, there was a third condition in the study environment of the SJVDP that frustrated SJVDP management. This condition concerned practical imperatives faced by the involved agencies and the extent to which the SJVDP would (or would not) contribute to meeting the needs raised by the imperatives. Here, for example, is a partial list:

- A judicial agreement (Barcellos settlement) that the Department of the Interior will develop a drainage facility plan for Westlands Water District by December 31, 1991.
- Federal "commitments" for drainage service to other water districts in the San Luis Unit.
- Similar State "commitments" for drainage service in State Water Project areas.
- An official (State) reevaluation of the water quality plans and objectives of the Central Valley and the Bay/Delta.
- The development of water quality criteria by the U.S. Environmental Protection Agency.
- Several specific regulatory orders affecting agricultural drainage.
- Litigation alleging upslope contributions to downslope drainage problems.

It was SJVDP policy to keep track of these imperatives, to exchange information needs, but to keep from getting entangled with discrete objectives and constraints. Of course it is impossible to stay entirely detached from events designed to change the very problem you are studying, as you study it.

## Developing and Carrying Out a Plan of Work

The purpose of the SJVDP, as outlined in a July 24, 1984, statement by then-Interior Secretary Clark was to conduct "... comprehensive studies to identify the magnitude and source of the (drainage) problem, the toxic effects of selenium on wildlife, and what actions need to be taken to resolve these issues." The Secretarial statement further noted: "Early action on the selenium problem is critical and is directly related to the ultimate question regarding the disposal of agricultural drainage water in the San Joaquin Valley from over 500,000 acres (200,000 hectares) of irrigated lands."

This initial policy proclamation was useful in launching the interagency program and in starting the research programs led by the USGS and USFWS. The mandate had to be converted, however, into a more practical guide for the Program staff. The goal statement developed by the SJVDP was: To identify measures to help solve immediate drainage and drainage-related problems and to develop comprehensive plans for long-term management of those problems. The first part of the goal recognizes the practical imperatives and serves the needs for immediate action (both physical and political); the second part of the practical goal statement recognizes the need for a long-term management framework. At the time policy agreement was reached on this general goal, preliminary results of the scientific studies were indicating the complexity and persistence of the drainage problem; it began to be apparent that there were no easy solutions. At that point, the SJVDP got very serious about designing a long-range plan for management of the problem. The planning method developed by the SJVDP is described in another chapter in this volume. From a manager's viewpoint, however, I wish to address the basis of the decisions made on analytical processes, study limits, and the conceptual model adopted for design of the plan.

The results of many of the key decisions about how the SJVDP would do its work are presented in the SJVDP "Prospectus" (February 1987), but the basis of choice of SJVDP content is not discussed. We presented major concerns (goals), established study boundaries, declared a planning process, and began the discussion of analytical methods. Subsequent policy reports by the SJVDP have added detail on those topics (SJVDP, 1987b, 1989, and 1990).

Four "concerns" were recognized by the staff and the PMC: Public health; water quality; agricultural lands and productivity; and fish and wildlife resources. The latter two concerns became primary goals to be balanced or equalized; the first two became constraints to be met as basic conditions in any balancing of the two primary goals.

By the fall of 1986, enough was known about the uptake of selenium in foodstuffs to surmise (later borne out by Klasing, August 1988) that selenium is not likely to be a problem in normal human diets and that foraging and possibly occupational hazards would be the major exposure route of concern. Hence, although public health's importance was not minimized, the subject was termed a constraint that could be met (in planning), and it was relegated to a study and monitoring category.

On a different basis, we chose to term water quality a constraint. Water quality agencies had declined to be members of the SJVDP's PMC, citing a need to maintain a separation between water supply agencies and water regulatory agencies. Also, very active water objective setting has continued throughout the duration of the Program. There has been a great deal of uncertainty about the actual criteria and numbers that would apply. It has been uncertain how much and what kind of agricultural waste could be accepted into receiving public waters. Generally, SJVDP planning has coped with this uncertainty by applying a rule that: You set the targets; We'll meet them. Because water quality regulations tend to be an invisible or moving target, we have had to rely on scientific study results (including some of our own work) to forecast what levels might prove protective and, in the long term, would probably be implemented.

The management decisions that we made about agricultural and fish and wildlife goals are, a priori, based on policymakers' selections of goals: that it is a good thing to sustain the productivity of existing lands on the west side of the Valley, but not at the expense of fish and wildlife, the habitat conditions for which must be restored to conditions preexisting prior to impact by selenium-contaminated drainage water.

To the disappointment of some observers and participants, we did not "stretch our wings" in the goal-setting process and soar into the rarefied air of reallocations of the public waters of the Central Valley or the State of California--although this was a common suggestion. Instead, as pedestrian as it may sound, our basic performance standard was simply stated: If we can develop a plan--which will be accepted publicly--to protect fish and wildlife and control the waterlogging and salinization problems at the lowest net social cost, we will have been successful.

In addition to the limits we imposed on goals, study boundaries also had to be limited to a reasonable size. Some decisions were made on the basis of science, others as a matter of preference (policy or analytical). Geographic boundaries of the SJVDP study area were declared on the basis of natural phenomena and civil division boundaries. For example, in the subsurface, the west side of the Valley is geochemically dissimilar from the east side (Gilliom et al., 1988).

Certain topics were excluded generally from study. For example, we did little work on pesticides because pesticides are not commonly found in west-side subsurface drainage waters. Our studies suggested that commingling of tailwater and subsurface drainage water, in any amount, should be avoided as an onfarm management practice.

Some of the study limits were not ideal, but they promised to be practicable. For example, the Valley is an international flyway, but we did not have the means or time to determine the fate of birds along the whole flyway.

Other limits were given to us, not by analysis, but by policy decisions made during the course of the study. Originally, one of our contract study activities was a model for screening out (eliminating) potential disposal sites for agricultural drainage water. Early in the planning effort, a policy decision limited any further out-of-valley disposal studies. Once again, in theory, investigators should not constrain themselves from a possible option. However, when you receive 60 formal individual complaints and protests from a mixed cadre of elected officials--from mayorships to the United States Senate--when you become a major media event and when there are legislative initiatives against your study (State and National), sometimes you suboptimize--settle for second best. Especially, you make such decisions if you need the assistance of aggravated parties in continuing your work and in the eventual implementation of your plan.

In the process of deciding on goals and setting limitations, we also determined a planning process. Essentially, it is based on the rational planning model (see Friedman, 1987) of the type used by regional planners throughout the United States. The particular model developed for the program (see Swain, this volume) holds public health and water quality as constraints, while first focusing on agricultural enhancement, at the possible expense of fish and wildlife (to an allowable minimum); then, it repeats the process, focusing on fish and wildlife enhancement, at the possible expense of agriculture. The precise recommended path is selected from between the two poles of emphasis, on the basis of evaluation and preference for certain levels of performance--i.e., effectiveness, net cost, risk, etc.

Initially, we managers harbored great hopes of developing the analytical tools necessary to ask a lot of "What-if" questions. The analytical power that we sought was to link site-specific water levels and salinity in crop root zones (sure signs of a drainage problem) to the extra costs/benefits that would accrue to a grower who took remedial action X or Y, or did not take such action. We saw sufficient promise in the power of a crop production/hydrology model to support its development--support, but not "bet the farm" on it, anticipating that design, operation, or data problems might cut short model development or negate its effectiveness. Our experience-based skepticism with models was well founded. The Westside Agricultural Drainage Economics (WADE) Model remains an excellent concept; it also remains, alas, in a stage of development.

At this point, I'll offer a conclusion prematurely in the context of this chapter. And I quote from the Draft Final Report (June 1990) of the SJVDP:

"Because of the complexities of the interactive factors involved in solving the drainage problems and the many unknowns, only limited success has been achieved in modeling the natural and cultural features of the problem area. This has prevented asking 'What-if' questions that could generate an infinite number of alternatives. Professional judgment, local experience, and public review will evidently continue to be the most important resources in developing a successful plan."

The point of the above discussion is that managers have to take risks, and experienced managers tend to hedge their decisions, preferring second best to failure. The investments that the SJVDP made in the idealistic WADE Model paid off immediately in organizing information and in several discrete evaluations, but the dream of the plan designers was not realized. The results of these limited uses of the WADE Model and of peer review commend it as a tool having enormous potential, but to paraphrase an ancient Roman poet, Martial:

> "He is, at 40 and 10, a man of promise still;
> methinks he needs eternity, that promise to fulfill."

## Improving Problem-Solving through Applications of Better Methods

In this section, I refer to some of the findings and recommendations presented in other chapters of this book and react to them as a manager seeking to improve application of analytical methods to problem-solving conducted within a political arena. To a commendable extent, the National Academy of Science/National Research Council (NAS/NRC) Committee (1989) has already undertaken this task, but their collective viewpoint had both the advantage and disadvantage of a detachment from day-to-day encounters with the drainage problem study environment. The NAS/NRC Committee, acting as an oversight committee, was enormously important in aiding the SJVDP to venture into subject areas not traditionally studied in California interagency activities, such as examination of water resources institutions. The committee members themselves also provided a great depth and ability in various disciplines, a reservoir of talent that could be tapped to improve Study Team performance. For example, the quality-assurance and quality-control advice and oversight was instrumental in improving the credibility of SJVDP results.

The atmosphere was generally one of beneficial tension between the notables on the NAS/NRC committee and SJVDP management. The persistent tug from the NAS/NRC was to attempt to pull the SJVDP out of "reductionist" thinking; the responsive tug from the SJVDP was in the direction of the technically accomplishable and the politically pragmatic--given the constraints of the study environment. Altogether, not a bad creative tension.

When I read many of the chapters in this book, I felt a similar tension arising between myself, as a manager/pragmatist, and some researchers. I will draw some conclusions and prescriptions that I hope are as constructive as those that prevailed between NAS/NRC and the SJVDP.

## Language and Audience

Few of us disciplinarians, including economists, make decisions in water resources. Appointed and elected officials, private entrepreneurs, and civil servants make decisions. If we want to tell them something, we need--at least in some part of our statements--to speak their language. We all use jargon within our practices and disciplines, but at some point in a paper (beginning or end) we need to talk about implications, about extension of the knowledge to others, and about application of the methods to real problems. Otherwise, we are informing our colleagues, but probably not influencing any deciders.

I found that several of the authors in this volume had managed to meet both needs rather well. Gilliom's last section is termed "Implications;" Morgan and Mercer point their conclusions specifically to decisionmakers; the word choice of authors such as Caswell, Letey, and Wichelns evidences their awareness of the invalley audiences, the deciders; and occasionally authors have gotten right to the point--witness, Moore et al.: "The era of 'get out and give us more money' has ended." Generally, however, we disciplinarians still tend to mask our messages with jargon.

## Improving the Application of Economic Models to Decisionmaking

I note several suggestions for the application of existing models to agricultural drainage problems. I think that, in general, it is a very good idea to look further at real applications of these models--to look for opportunities to apply them to some of the decisionmaking events that will occur in the future. It may be more academically rewarding to develop a new model than to adapt and apply an existing model, but what does the decisionmaker really need? In general, economic models have been little used in making decisions on water resources in California, suggesting that we must be doing something wrong in research and academia. Professor Just does a good job of stating the need: "...A general analytical framework has not been available that combines location-specific relationships between production practices and environmental characteristics of farmland in a way that can be aggregated consistently with the regional or National level for purposes of welfare and policy analyses." What has to be done to get around this? There are a number of existing models (see, for examples, Dinar and Zilberman) that might be used beneficially for decision-making on drainage problems.

## Available Data

Organizations and individuals who use models soon learn to adjust their appetite for sophisticated tools to the availability of data. A few authors of this book (e.g., Carson, Knapp, Just) emphasize the problem of data shortages. Carson's words are generally applicable to the situation we found with all the models we analyzed in the SJVDP: "...The data requirements to successfully implement the ... techniques are much more extensive and expensive than first believed."

There has been, I believe, a great deal of high-quality data developed during the SJVDP; one of the policy aims of the SJVDP is to make such data generally available. Much more good data are needed. Good data can make even a poor model seem worthwhile. The opposite cannot be said.

## Just How Involved?

Although it sounds prosaic, I will tell an anecdote to illustrate a conviction that I hold about the differences between theorists and deciders. Once when I was a topographic surveyor traversing lines across a large tropical island, I returned monthly to an ocean base to receive my new orders, which came in the form of aerial photographs with circles drawn on them in crayon by unknown person(s) in the engineering base at Fort Belvoir, Virginia. Oh! How my squad learned to hate that unknown person (whom I pictured as sitting idly bored in an air-conditioned office) as we waded through crocodile-infested waters and cut through some of the densest jungle in the world! On occasion, just a little more care with the crayon and a little empirical information in the hands of "control" might have put us on a grassy ridge or kept us out of a swamp. But orders were orders, and the crayon circle controlled our very movements.

That is how some real growers, wildlife managers, and water managers visualize researchers, planners, and especially regulators. They think we know too little about how water is actually managed. Many times they are right.

That is why I am encouraged by the fact that most of the authors in this book seem to have a familiarity with the west side of the Valley. We need to cultivate that more: and to know more about the considerable intricacies of moving water around and to be willing to discard and modify our ideas when they don't work--regardless of whether or not they are published in white literature.

*Timeliness*

No matter how relevant the theory, how brilliant the work, or how well-intentioned the effort, it means little unless it is timely. Politicians and managers act alike in looking for "windows of opportunity." There is a time when you can sell a study; and, if you deliver, there is a time when you can influence deciders with the results. But, if you let the analyses get hung up in peer review (which, of course, is vital), then you will be a researcher, but never an influencer. A second-best answer, given on time, far outweighs the brilliant insight delivered after the funds have been spent, after the course is set. Then, all we can do is profit a bit from the mutual assignment of blame and vow to do better next time.

## NOTES

[1]For example, the discovery by R. Gilliom et al. (1989) of the U.S. Geological Survey showed that, contrary to a commonly held belief, most of the waterlogging problems that were causing growers to drain their lands represented the top of a regional ground-water system that for hundreds of feet in vertical depth had become saturated with water over some decades as water in excess of plant needs, percolated past crop root zones and filled the subsurface. This discovery gave quite a different twist to a problem that has been viewed traditionally as largely due to a shallow perched water table.

[2]It is interesting to compare the costs of Kesterson cleanup (for about 1,000 hectares) to the preventative studies of the SJVDP (for about 1 million hectares): $30 million and an additional $2.7 million annually for Kesterson cleanup, as compared to $50 million for the SJVDP studies. Obviously, when possible, preventative steps (including study) are a better investment than cure of a Kesterson.

[3]National studies have since confirmed that, in many locations other than in the San Joaquin Valley, water supplies--Federal, State, private, municipal--contain selenium in elevated levels.

[4]Citizens Advisory Committee members: Chester McCorkle (Chair), UC Davis; Jean Auer (Vice-Chair), The Commonwealth Club of San Francisco; Jerald Butchert, Westlands Water District; James Crenshaw, California Sport Fishing Protection Alliance; Michael DiBartolomeis, California Department of Health Services; Thomas Graff, Environmental Defense Fund; Donald Anthrop, California State University, San Jose; Stephen Hall, Land Preservation Association; Clifford Koster, Farmer; Daniel Nelson, San Luis Water District; Polly Smith, League of Women Voters; Michael Stearns, Farmer/San Luis Canal Company; Joseph Summers, Summers Engineering, Inc.; Ronald Stork, Friends of the River.

# REFERENCES

Friedman, J., 1987. *Planning in the Public Domain.* UCLA Press.

Gilliom, R. J. et al., 1989. *Preliminary Assessment of Sources, Distribution, and Mobility of Selenium in the San Joaquin Valley, California.* U.S. Geological Survey Water-Resources Investigation Report 88-4186.

Klasing, S. A. and Pilch, S. M., 1988. *Agricultural Drainage Water Contamination in the San Joaquin Valley: A Public Health Perspective for Selenium, Boron, and Molybdenum.* Prepared for the SJVDP under U.S. Bureau of Reclamation contract.

National Research Council, 1989. *Irrigation-Induced Water Quality Problems.* National Academy Press, Washington, DC.

Rowan, D. D., 1986. *Managing Projects: A Systems Approach.* Elsevier Science Publishing Co., New York, NY.

San Joaquin Valley Drainage Program, 1987a. *Prospectus and Appendixes to Prospectus.*

San Joaquin Valley Drainage Program, 1987. *Developing Options: An Overview of Efforts to Solve Agricultural Drainage and Drainage-Related Problems in the San Joaquin Valley.*

San Joaquin Valley Drainage Program, 1988. *Formulating and Evaluating Drainage Management Plans for the San Joaquin Valley.*

San Joaquin Valley Drainage Program, 1990. *A Management Plan for Agricultural Subsurface Drainage and Related Problems on the Westside San Joaquin Valley.* Draft Final Report.

# 45 FUTURE RESEARCH ON SALINITY AND DRAINAGE

Henry J. Vaux, Jr., University of California, Riverside and
Kenneth K. Tanji, University of California, Davis and
UC Salinity/Drainage Task Force

## ABSTRACT

Future research on salinity and drainage should focus on achieving an acceptable balance between agricultural productivity and environmental quality. Future research agendas will place more emphasis on interdisciplinary research and institutional studies than did agendas of the past. This chapter identifies priority topics for future research. The most important of these is the need to find environmentally acceptable methods of maintaining salt balances over the long term. In the absence of such methods, irrigated agriculture and environmental quality prove incompatible.

## INTRODUCTION

The standard prescription of leaching and drainage to manage salt balances was formulated to ensure that crop yields would not be adversely affected by soil salinization. The research findings which led to this prescription of root zone salinity control were supported by a desire to sustain and enhance agricultural productivity in the face of threats posed by salinization. As such, most traditional agricultural research has involved efforts to maintain and increase the productivity of agricultural lands rather than focusing on offsite environmental impacts (van Schilfgaarde et al., 1974).

Historically, research agendas have also tended to reflect the view that the solutions to agricultural problems were largely technical in nature. Scientific advances were frequently adopted by irrigated agriculture with little or no regard for institutional and social settings. Additionally, research agendas tended to be formulated along disciplinary lines with little recognition of the fact that successful long-term solutions frequently require the collaborative efforts of investigators from several disciplines.

Despite the many advances in the management of soil and water salinity that have flowed from previous research, difficulties associated with salinity and drainage continue to plague irrigated agriculture. Although the phenomena that cause salinization are now better understood than they were in the past and the techniques of salinity management have improved over time, the problems of managing salinity have become more complicated. Significant impacts from the drainage of irrigated lands extend well beyond the farm gate. Percolating drainage waters may threaten ground-water quality. Lateral subsurface drainage flows may impose costs on growers and others who are remote from the area of drainage generation. Surface drainage flows are often discharged into waterways where they may compromise the quality of water supplies available to downstream users and threaten fish and other aquatic biota.

The widespread use of agricultural chemicals leaves residues in drainage water. Nitrates from fertilizers, pesticides, and other synthetic chemicals are found in drainage water, often in higher than acceptable concentrations. The quality of drainage water is also degraded by trace elements that occur naturally in the geological substrata and sediments. Some of these trace elements can bioaccumulate and biomagnify to create hazards for both aquatic and terrestrial biota. It is now widely recognized that virtually any water source can be degraded more readily than was previously thought. This is especially serious since contamination of existing water supply intensifies the problems of water scarcity which all water users face. California, for instance, currently faces significant difficulties in balancing essentially static water supplies with sharply increasing demands fueled by population growth. These difficulties are likely to increase substantially when the absolute size of the water supply begins to shrink because of degraded water quality (Vaux and Woods, 1990).

As a consequence, irrigation drainage flows are increasingly the subject of governmental regulation. New and more extensive regulation can be expected as additional information becomes available on drainage water constituents and as public attention is drawn to the water quality consequences of irrigation and drainage. The advent of new and more stringent regulations means that managing salt balances by traditional methods will become increasingly difficult. Regulations tend to constrain the ability of growers to manage salt balances by restricting both the extent and the circumstances of salt export from farms and from hydrologic basins. Thus new methods will have to be found to manage salt balances and drainage effluent.

Additionally there is an urgent need for new techniques to measure the impacts of drainage regimes (existing and new) on agriculture and on the natural environment. The research agenda necessary to develop new management strategies and assess their impacts must differ significantly from past research agendas if it is to reflect modern realities (National Research Council,

1989). In the remainder of this chapter the priority areas for research in salinity and drainage are identified and characterized. These priorities embody the need to manage irrigation waters in ways that minimize adverse environmental consequences while preserving agricultural productivity, to the extent possible. These priorities also incorporate the need for better water management institutions as well as the need for research that is integrated across disciplines.

## RESEARCH FOCUSED ON FARM-LEVEL WATER MANAGEMENT

Although much has been learned about how to manage water at the farm level to avert productivity losses attributable to salinity, far less is known about management of irrigation water to minimize the environmental consequences of drainage (Letey et al., 1986). Currently, leaching and drainage are essential to manage salt balances at the farm level. Ultimately, efficient management of drainage waters may require that drainage volumes be reduced in order to avoid adverse environmental impacts. This will probably require both improved operating methods for existing technologies as well as the use of new or emerging technologies. It will also require the development of better water management institutions (National Research Council, 1989).

New measurement techniques will be required both to facilitate research and to generate the information needed to develop more precise irrigation applications and improved methods for managing drainage water in the field, both of which will contribute to a decrease in the mass of salts discharge (Tanji, 1989). Research will be required to develop: Improved techniques for the measurement and identification of problems associated with slow rates of water infiltration; techniques for making real time assessments of soil salinity; and improved methods for characterization of shallow ground water bodies. The development of inexpensive tools to measure seepage and salinity would be particularly useful.

Although much information is now available about the behavior of selenium in soil and water environments, there are other potentially toxic trace elements which need further study (Deason, 1989). These include arsenic, boron, molybdenum, uranium, and vanadium. Studies are also needed to characterize the reactivity and mobility of toxic trace elements as well as the interactions among these trace elements and with other compounds. The conditions under which the simultaneous presence of two or more such trace elements increases their toxicity to plants and animals also need to be characterized. Additional investigations will be needed to describe and understand assimilation of trace elements by different crops. This work should include studies on plant uptake of trace elements as a well as studies of the potential crop toxicities in regions where one or more toxic trace elements can be mobilized in the soil.

The management of existing irrigation technology, including both water application and drainage removal methods, is conditioned by the type of crop, the characteristics of the soil, and various economic and legal incentives which confront the grower. The consequences of reducing drainage by minimizing the quantity of applied water require further study. Additionally, the impacts of moisture stressing on different crops at different times throughout the growing season have not been fully investigated (Vaux and Pruitt, 1983), nor are the effects of moisture stressing under a range of saline conditions over time completely understood. These gaps in knowledge point to the need to develop time dependent or dynamic crop water production functions under a wide variety of saline conditions. For the short term, dynamic crop water production functions need to be developed for a variety of annual crops under saline conditions. This is particularly important inasmuch as the crops grown in the areas most severely affected by salinity in California tend to be annual crops.

For the longer term, dynamic crop water production relationships need to be developed for permanent crops as well. Little is known about the longer-term consequences of moisture stressing on the longevity and productivity of permanent crops, particularly under saline conditions. Such studies will be both time consuming and expensive but will be required by regulations controlling non-point source emissions which will likely affect lands on which permanent, as well as annual, crops are grown. The results of intensive research on dynamic crop-water relationships could have additional benefits. More efficient management of agricultural water supplies, even where drainage is not a problem, could enable growers to economize their use of water, thereby permitting scare water supplies to be stretched further.

In devising methods to reduce drainage water volumes, it will be particularly important to account for the effects of spatial variability of soils. Irrigation practices cannot be adapted to variations in soil water intake rates without knowledge of what those variations are. Spatial variation in water intake rates probably cannot be measured deterministically and therefore optimal sampling strategies will need to be developed to allow growers to characterize field variability with reasonable accuracy (Letey et al., 1986). In many fields, water and solute flow preferentially through macropores such as root channels and earthworm holes. Preferential flows also occur in cracking soils. Virtually nothing in known about how such flows affect the uniformity of soil moisture and salt balances or about the consequences for both drainage and crop productivity of different irrigation regimes on such soils.

All of this work should facilitate the research needed to develop both improved water management technologies and better management of existing technologies. Improved methods of managing irrigation water in situations where soil water intake rates are variable will be particularly important. Here the issue to be addressed is not how to optimize crop productivity but rather how to characterize the joint effects on productivity and environmental quality of alternative irrigation strategies with waters of differing salt concentrations and soils of various uniformities.

New management strategies will also need to be developed for alternative irrigation technologies since closed conduit systems such as drip and sprinkler are managed differently than conventional surface irrigation systems under conditions of nonuniformity. The investigations of strategies should not be limited to research demonstration projects but should be conceived so as to account for the conditions which are actually faced by growers. Economic studies should be an integral part of these investigations so that the costs of different management regimes can be assessed. Costs should include losses in productivity and environmental impacts as well as the costs of actually managing the irrigation system being evaluated.

Research on different methods of drainage management is no less urgent (U.C. Committee of Consultants, 1988 and Letey et al., 1986). Design criteria for tile drainage systems are needed which account for intercepted shallow ground waters of varying qualities. The development of these criteria should be coordinated with studies of methods to manage shallow ground-water tables to satisfy a portion of crop-water demand. Initially, such studies should focus on use of shallow ground water as a source of supply in the short term, particularly during times of low flow in receiving waters and in periods of drought. For the long term it will be necessary to examine the extent to which and the conditions under which the use of shallow ground water to meet a portion of evapotranspiration demand is sustainable. Knowledge developed by this research should lead to characterization of the circumstances under which shallow ground water can be managed in situ and the circumstances under which it must be disposed of and drained.

Research aimed at developing salt-tolerant and drought-tolerant crops should continue (Shannon and Qualset, 1984 and Valentine, 1984). While the advent of genetic engineering increases the likelihood of successfully developing such crops, this effort should be conceived as a longer-term project since the probability of high payoff in the short-run is remote. However, even the development of viable salt- and drought-tolerant crops is unlikely to obviate completely the need to manage irrigation and drainage water with greater care in the future.

In addition to economic studies on the feasibility of alternative irrigation and drainage management schemes, institutional studies involving economists, political scientists, and legal scholars are needed to identify the array of incentives available to induce growers to manage irrigation and drainage waters for a variety of combinations of crop productivity and drainage water quality. Currently, many of the incentives faced by growers lead to practices which may exacerbate the drainage problem. For example, in some irrigation districts growers are obliged to pay for an annual water entitlement irrespective of whether they actually use the total entitlement. This practice, which is designed to ensure an even flow of revenues to defray the costs of impounding and delivering water, encourages growers to make substantial preirrigations (Natural Heritage Institute, 1990). These preirrigations have been identified as a major cause of excess drainage (U.C. Committee of Consultants, 1988). This perverse incentive, and others like it, arise because different agents in the water delivery and application process are attempting to achieve different objectives. In this case, the water purveyor is attempting to minimize the uncertainties surrounding the financing of a water delivery system and the grower is attempting to maximize profit. Studies are particularly needed, then, to design incentives which will allow different combinations of inconsistent objectives to be attained.

Further attention needs to be given to the role of price and pricing schedules in inducing careful management of irrigation and drainage. Estimates of the elasticity of demand for irrigation water under differing circumstances are still inadequate. The potential of increasing block rates as a means of inducing effective irrigation and drainage management has not been fully assessed. Studies of incentives should extend to the use of agricultural chemicals that contribute to the contamination of drainage waters. Incentives directed solely at water use may not be as effective as combinations of incentives directed at both chemical and water use. Institutional studies should not be limited to price and market incentives. Alternative systems of legal liability need further investigation as do other methods of centralized regulation. The technology standards and best management practices traditionally employed to improve water quality in the United States do not always lead to efficient levels of pollution control. There may be alternative methods that are both workable and more efficient and they should be investigated (Natural Heritage Institute, 1990). All of these studies should focus on incentives faced by individual growers.

## RESEARCH FOCUSED ON REGIONAL WATER MANAGEMENT

To be effective, drainage management strategies will have to be integrated at a regional scale in order to account for subsurface lateral flows of drainage water, the presence of aquifers which underlie more than one farm, and possible economies of scale in treatment and disposal technologies. Inasmuch as the impacts of drainage almost always transcend the individual farm, comprehensive assessments of environmental impacts and effective drainage management strategies should be undertaken at the regional scale. Research at the regional scale should focus on river basins or, where basins are very large, irrigation projects.

At this scale the development of effective management strategies will also require improved and more comprehensive data bases (National Research Council, 1986). These data bases will be needed both to devise management strategies and to assess the impacts of alternative management strategies. Improved techniques of measurement will be required to develop better data on the volume, quality, and direction of lateral subsurface flows. Instrumentation which will permit real time assessments of such flows would also be helpful. Data will also be needed to characterize the quality of ground water over time. Models will be required which link conditions on the land surface with ground-water conditions. Since significant degradation of ground-water quality often occurs over long time periods, such models will be needed to determine, in advance, the long term effects of water management practices on the quality of underlying ground water.

It will be particularly important to assess the long term sustainability of different management regimes. Data obtained at the farm level should be helpful in developing models which couple the vadose zone hydrology with ground-water hydrology and capture major water quality effects, including the migration of salts, nutrients, chemical residues and toxic trace elements. These regional models should have the capacity to predict the long-term effects of different water management regimes.

These modeling studies should also be directed at developing information that will be useful in the regulatory process. For example, the capacity to predict load/flow relationships for key non-point source pollutants such as salts, nutrients, and toxic trace elements will be particularly important. Such research could help in resolving the dilemma created by attempts to regulate pollutants by levels of concentration and help resolve the issue of whether regulation can be most efficiently accomplished through best management practices as opposed to other forms of regulation (Baumol and Oates, 1979).

Further research on the structural features of water delivery and drainage systems will also be necessary. In some irrigation districts, additional physical facilities will be required to permit water delivery on demand. Similarly, automated delivery systems gate will enhance the ability of growers to manage water more effectively at the field level. Indeed, should regulation take the form of on-farm best management practices, modern physical delivery facilities may be needed to permit growers to respond.

The economic and social consequences of both drainage regulations and new drainage management strategies will need to be evaluated. Studies which assess the economic and social outlook if the status quo is maintained will be particularly important since they will define a base line against which the potential impacts of new policies, regulations and management strategies can be assessed. Such studies will help identify which agricultural lands are likely to be urbanized thereby obviating further efforts to improve drainage water management in those areas.

Models of regional economies will also be necessary in assessing the implications of strategies which will not be sustainable over the long-run. The availability of such models will be particularly helpful in the process of selecting public policies to govern the discharge of non-point source pollutants. Definitive and credible information on the likely economic impacts of different pollution control policies will be just as necessary for the establishment of non-point source regulatory policy as it is for the establishment of other water pollution control policy.

Research efforts focused on the development of new techniques for treating and disposing of unusable drainage water are also needed. Studies should include investigations of the circumstances under which it is cost-effective to treat unusable drainage waters on a regional, as opposed to a farm-by-farm, basis. Attention should also be devoted to systems of rights and incentives needed to ensure that optimal treatment and disposal techniques are installed and operated efficiently.

A careful examination of disposal options for the long-run will be a critical component of the research agenda. Virtually all of the recent efforts to manage salt balances in California's Central Valley are stop-gap, short-term efforts which buy time. A series of such short-term efforts will not solve the problem over the longer-term, however. In the long-term, salt balances must be maintained if irrigated agriculture is to remain viable. It makes little sense to develop new technology and disposal methods if the undesirable end products produced by those methods cannot be disposed of in ways that are relatively benign to both agriculture and environmental quality (National Research Council, 1989).

If salts cannot be exported from irrigated regions to acceptable disposal sites or if they cannot be effectively concentrated and isolated for the long-term within these regions, attention will be required to identify the time spans over which it makes sense to maintain irrigated agriculture. An assessment of the environmental consequences of maintaining irrigated agriculture, even in the short-term, should be an integral part of these studies.

## CONCLUSIONS

Future research agendas must differ from those of the past in several ways if they are to reflect modern realities. First, the objective of enhancing agricultural productivity while ignoring the environmental consequences is no longer acceptable. The emergence of increasingly stringent drainage discharge regulations is but one manifestation of the intensifying social preferences for environmental quality. The new research agenda, then, will need to be premised on the assumption that future drainage and salt management will require striking a balance between sustaining agricultural productivity and maintaining environmental quality.

Second, the problems of irrigation and drainage management are not exclusively technical problems. There must be some recognition of the fact that water management institutions are not well suited for effective management of the residuals of irrigation. For example, existing water institutions have provided growers with little incentive to account for the off-farm impacts of drainage flows. Single purpose programs sometimes create conflicting or perverse incentives. In the future, effective management regimes will depend on the development of water management institutions that encompass and are responsive to the full array of impacts associated with drainage management efforts.

Third, to be fully responsive to the array of problems associated with drainage, future research will need to transcend disciplinary boundaries. The fact that environmental problems have sometimes been induced by advances in agricultural science suggests that narrowly focused research oriented to specific intellectual disciplines sometimes yields shortsighted or incomplete solutions. The adverse impacts of agricultural drainage waters occur well beyond the field and the farm gate. Research that focuses solely on physical and biological phenomena at the field level cannot be expected to identify consequences that occur beyond the field. An understanding of the chemistry and physics that govern the quality of irrigation drainage waters does not automatically confer an understanding of the economic, political, and legal consequences of drainage water disposal. Yet, it is important to note that these latter institutional consequences cannot always be clearly understood in the absence of knowledge about the underlying physical, chemical, and biological phenomena associated with the disposal of drainage water.

## REFERENCES

Baumol, W. and Oates, W., 1979. *Economics, Environmental Policy and the Quality of Life*. Prentice-Hall, Inc., Engelwood Cliffs, NJ.

Deason, J. P., 1989. Irrigation-Induced Contamination: How Real a Problem? *ASCE Journal of Irrigation and Drainage Engineering*, 115, pp. 9-20.

Letey, J.; Roberts, C.; Penberth, M.; and Vasek, C., 1986. *An Agricultural Dilemma: Drainage Water and Toxics Disposal in the San Joaquin Valley*. Special Publication 3319, Division of Agriculture and Natural Resources, University of California, Riverside.

National Heritage Institute, 1990. Workshop on Institutional Investigations on the San Joaquin Valley Drainage Problem: A Briefing Document. Unpublished Manuscript.

National Research Council, 1986. *Ecological Knowledge and Environmental Problem Solving: Concepts and Case Studies.* National Academy Press, Washington, DC.

National Research Council, 1989. *Irrigation-Induced Water Quality Problems: What Can Be Learned from the San Joaquin Valley Experience?* National Academy Press, Washington, DC.

Shannon, M. C. and Qualset, C. O., 1984. Benefits and Limitations in Breeding Salt-Tolerant Crops, *California Agriculture*, 30(10), pp. 33-35.

Tanji, K. K., 1990 (Ed.). *Agricultural Salinity Assessment and Management*, Manuals and Reports in Engineering Practices, No. 71, ASCE, New York, NY. In Press.

University of California Committee of Consultants on Drainage Water Reduction, 1988. *Opportunities for Drainage Water Reduction.* UC Salinity/Drainage Task Force and UC Water Resources Center, Riverside.

Valentine, R. C., 1984. Genetic Engineering of Salinity-Tolerant Plants, *California Agriculture*, 30(10), pp. 36-37

van Schilfgaarde, J.; Bernstein, L.; Rhoades, J. D.; and Rawlins, S. L., 1974. Irrigation Management for Salt Control, *ASCE Journal of Irrigation and Drainage Engineering*, 100 (IR3), pp. 321-338.

Vaux, H. J., Jr. and Pruitt, W. O. 1983. Crop-Water Production Functions. In: Hillel, D. (Ed.), *Advances in Irrigation.* Academic Press, New York, NY.

Vaux, H. J., Jr. and Woods, R. J., 1990. California's Water Resource: The Current Status and Future Challenges. In: Carter, H. O. and Nuckton, C. F. (Eds.), *Agriculture in California: On the Brink of a New Millennium, 1990-2010.* Agricultural Issues Center, University of California, Davis.

# 46 IRRIGATION TECHNOLOGY, INSTITUTIONAL INNOVATION, AND SUSTAINABLE AGRICULTURE

Vernon W. Ruttan,
University of Minnesota, St. Paul

### ABSTRACT

We are now in the third wave of social concern about the relationship between natural resources and sustainability of improvements in human well being. If we are successful in putting into place the policies and institutions needed to resolve current efficiency and equity concerns in Western water resource use we will be in a better position to respond to the more uncertain changes that will emerge as a result of future global climate change.

### INTRODUCTION

The belief that the application of science to the solution of practical problems represents a sure foundation for human progress has been a persistent theme in American intellectual and economic history. During the two decades following World War II, this belief was seemingly confirmed by the dramatic association between the progress of science and technology and rapid economic growth. Since the late 1960's, however, a view has emerged that the potential consequences of the power created by modern science and technology are obviously dangerous to the modern world and the future of humankind. The result has been to seriously question the significance for human welfare of scientific progress, technical change, and economic growth. The application of science and technology to the solution of problems is increasingly regarded as the source of even more intractable problems (Crouch, 1990).

One response has been to call for a new relationship between humankind and the natural environment. "Sustainable development--even in its less extreme versions--resurrects the classical concept of absolute scarcity" (Batie, 1990, p. 1084). The problems associated with irrigation development in areas such as the San Joaquin Valley (Valley) in California and the Indus Valley in Pakistan have often been cited as worst case examples.

## CONCERN ABOUT RESOURCES AND THE ENVIRONMENT

It is useful to provide some historical perspective on the issues of technological change, resource management, and the sustainability of agriculture systems as we move toward the first decade of the 21st century. The early 1990's is in the midst of the third wave of social concern since World War II--and the fifth since Malthus--about the relationship between natural resources and the sustainability of improvements in human well being (Ruttan, 1971 and 1990).

The first postwar wave of concern, in the late 1940's and early 1950's, focused primarily on the quantitative relations between resource availability and economic growth--on the adequacy of land, water, energy, and other natural resources to sustain growth. The reports of the President's Water Resource Policy Commission (1950) and the President's Materials Policy Commission (1952) were the landmarks of the postwar resource assessment studies generated by this wave of concern. The response to this first wave of concern was technical change. A stretch of scarcity, accompanied by higher prices, has not yet failed to induce the new knowledge and new technologies needed to locate new deposits, promote substitution, and enhance productivity. If the Materials Policy Commission were writing today, it would have to conclude that there has been abundant evidence "of the nonevident becoming evident; the expensive becoming cheap; and the inaccessible becoming accessible."

The second wave of concern occurred in the late 1960's and early 1970's. In this second wave, the earlier concern with the limits to growth imposed by natural resource scarcity was supplemented with concern about the capacity of the environment to assimilate the multiple form of pollution generated by growth. An intense conflict was perceived between the two major sources of demand for resource and environmental services. One source was the rising demand for environmental assimilation of residuals derived from growth in commodity production and consumption--asbestos in our ventilation systems, pesticides in our food, organic and chemical wastes in our water, and smog in our air.

The second source of concern was the rapid growth in consumer demand for environmental amenities--for direct consumption of environmental services--arising out of rapid growth in per capita income and high income elasticity of demand for environmental services. These included access to natural environments and freedom from pollution and congestion--for clean water, clean air, clean streets, and for safe food. The primary response to these concerns has been an attempt to design local incentive-compatible institutions designed to force individual firms and other organizations to bear the costs from externalities generated by commodity production and consumption--to make the polluters pay.

Since the mid-1980's, these two earlier concerns have been supplemented by a third. The more recent concerns center around the implications for environmental quality, food production, and human health of a series of environmental changes that are occurring on a transnational scale--issues such as global warming, ozone depletion, acid rain, and others. The institutional innovations needed to respond to these concerns will be even more difficult to design. They will, like the sources of environmental change, need to be transnational. Our experience with attempts to design incentive compatible transnational regimes such as the Law of the Sea Convention, or even the somewhat more successful protocols on reduction of chlorofluorocarbon (CFC) emissions, suggests that the difficulty of resolving free-rider and distributional equity issues will impose severe constraints on how rapidly effective transnational regimes to resolve these new environmental concerns can be put in place.

It is important to note that with each new wave of earlier concern, the concerns that had briefly faded have reemerged as part of the resource and environmental policy agenda. Thus the concern with material resources reemerged along with concern about the spillovers from agricultural and industrial intensification in the early 1970's. And both earlier concerns emerged again, along with the transnational resource and environmental issues, as a part of the broader resource and environmental policy agenda of the late 1980's and early 1990's.

The chapters in this volume deal primarily with the issues that were dominant in the first two waves of concern about resource and environmental policy. The global climate change issues that have dominated the third wave of concern may seem remote from issues such as selenium, ground-water contamination, salinization, or even the more mundane issue of water use efficiency in the Valley. But our capacity to resolve these "second wave" concerns will test our ability to deal with the broader effects of global climate change.

## CURRENT AND FUTURE CONCERNS

There can no longer be any question that the accumulation of carbon dioxide ($CO^2$) and other greenhouse gases--principally methane ($CH_4$), nitrous oxide ($N_2O$), and chlorofluorocarbons (CFC's)--has set in motion a process that will result in some rise in global average surface temperatures over the next 30 to 60 years (Ruttan, 1990). But there is substantial disagreement about whether warming due to greenhouse gases has already been detected. And there continues to be great uncertainty about the increase in temperature and changes in precipitation that can be expected to occur on any particular date or location in the future.

The bulk of the carbon dioxide omissions comes from fossil fuel consumption. Carbon dioxide accounts for roughly half of radiative forcing. Biomass burning, cultivated soils, natural soils, and fertilizers account for close to half of nitrous oxide omissions. Most of the known sources on methane are a product of agricultural activities--primarily enteric fermentation in ruminant animals, release of methane from rice production and other irrigated crops, and biomass burning. Estimates of nitrous oxide and methane sources have a very fragile empirical base. Nevertheless, it appears that agriculture and related land use and land use transformations could account for somewhere in the neighborhood of 25 percent of radiative forcing.

The alternative policy approaches to global warming have been characterized as <u>preventionist</u> and <u>adaptionist</u>. It seems clear that preventionist approaches would involve about five policy options. They include:

- Reduction in fossil fuel use, or capture of $CO^2$ emissions at the point of fossil fuel emissions combustion,
- Reduction in the intensity of agricultural production,
- Reduction in biomass burning,
- Expansion of biomass production, and
- Energy conservation.

The simple enumeration of these policy options should be enough to induce considerable caution in assuming that radiative forcing will be limited to anywhere near the present levels. Let me be more specific. Fossil fuel use will be driven, on the demand side, largely by the rate of economic growth in the third world and by improvements in energy efficiency in the developed and former centrally planned economies. On the supply side, it will be constrained by the rate at which alternative energy sources are substituted for fossil fuels. Of these, only energy efficiency and conservation will be likely to make any significant contribution over the next generation. Any hope of significant reversal of agricultural intensification, reduction in biomass burning, or increase in biomass absorption is unlikely to be realized within the immediate future. The institutional infrastructures that will be required do not exist and cannot be put in place rapidly.

An adaptionist strategy for agriculture implies moving as rapidly as possible to design and put into place the institutions needed to resolve the constraints that agricultural intensification is currently imposing on sustainable increases in yield. A particularly important example in the United States involves the policies needed to rationalize surface and ground water use in the West. If we are successful in putting into place the policies and institutions needed to resolve the current efficiency and equity concerns in resource use, we will be in a better position to respond to the more uncertain changes that will emerge as a result of future global climate change.

## INSTITUTIONAL INNOVATION

Technical effort can be redirected toward reducing environmental stress. Recent theoretical and empirical investigations have resulted in a perspective that views technical and institutional change as a dynamic response to changes or differences in resource endowments and to the social and economic environment (Hayami and Ruttan, 1971 and 1985; Binswanger and Ruttan, 1978; and Ruttan and Hayami, 1984). The value of social science knowledge is that it offers, or should offer if it is productive, the prospect of shifting the supply curve for international innovation to the right of reducing the costs of institutional innovation.

As long as the capacity to absorb residuals--as in the case of the Kesterson Reservoir--remained undervalued, both technical and institutional innovation were biased toward the excess population of residuals (in this case, the drainage water constituent selenium). The use of a resource that is priced below its social cost will grow even more rapidly than in a situation where substitution possibilities can occur only along a "given" production surface. The chapters in this volume clearly indicate that this process has been going on in a highly intensive manner in the study area. Our major agricultural commodity programs are operated by renting the land needed to balance supply and demand at target price levels. One effect has been to raise agricultural land prices above equilibrium levels. At the same time, capacity of the environment to absorb the residuals from crop and livestock production has been treated as a free good. As a result, technical change has been overly biased toward the development and use of land substitutes--plant nutrients and plant protection chemicals and crop varieties and management systems that reflect the overvaluation of agricultural land and the undervaluation of the social costs of containing or disposing of the residuals from agricultural production. The same biases have led to underinvestment in technological effort directed toward pest, soil, and water management systems consistent with the social value of environmental services.

An issue that has been the focus of a number of chapters in this volume is the perceived failure of Western water resource institutions to respond to changes in relative resource values and in the values of resource products relative to water resource supplies (Coontz, Dixon, Randall, DuMars, Weinberg and Willey, and Mercer and Morgan). It is surprising that they are surprised! It seems quite clear from historical experience that institutional change can be viewed as resulting from the efforts of economic units (households, farms, bureaus) to internalize the gains and externalize the costs of economic activity (Rausser and Zusman). These economic units operate in a society that is engaged in an effort to force them to internalize the costs and externalize the gains from this economic activity. The institutions that are put in place to achieve these distributional transfers generate resources that can be used to protect the historical patterns of income transfers. These economic resources are traded for the political resources needed to maintain existing distributional benefits. The failure to develop regulatory regimes that can interpret to agricultural producers and irrigation agencies the high social costs of treating an increasingly scarce resource--the capacity to absorb residuals--as a free good is now threatening the economic sustainability, eroding the environmental amenities of the region, and composing unacceptable health risks on the Valley residents. The institutional innnovations that would have been induced by the shift to the right in the demand for environmental quality have remained latent as a result of bias in the behavior of political markets.

## RESEARCH PRIORITIES

Several research priorities can be identified that would reduce the costs of realizing the technical and institutional changes needed toward more efficient use of land and water resources.

Design of Technologies and Institutions to Achieve More Efficient Resource Management. During the 21st century, water resources will become an increasingly serious constraint to agricultural production not only in traditional irrigated areas such as California but in more humid areas. Residuals from agricultural production continue to be a major cause of the decline in quality of both ground and surface water. Limited access to clean and uncontaminated water supplies is a major contributor to disease and poor health conditions in many parts of the developing world and in centrally planned economies. Global climate change can be expected to have a major differential impact on water availability, water demand, erosion, salinization, and flooding. The development and introduction of technologies and management systems that enhance water use efficiency represents a high priority for both natural and social science research because of both the short and intermediate term constraints on water availability and the longer run possibility of seasonal and geographic shifts in water availability. The identification, development, and introduction of more water-efficient crops and farming systems for dry land and saline environments is an important aspect of achieving greater water efficiency.

**Refinement of Technologies to Improve Agricultural Productivity.** Almost all increases in agricultural productivity over the next several decades will result from intensified production on land that is already used for crops and livestock. Increases will not come from opening up more land or developing large new sources of irrigation water. The necessary gains in crop and animal productivity will be generated by genetic improvements in plants and animals and from more intensive and efficient use of inputs such as water, chemical fertilizers, pest control chemicals, and more efficient feed conversion by animals. This implies high plant populations per unit area, new tillage procedures, new methods of delivering and managing water supplies, improved pest and disease control, more precise application of plant nutrients, and advances in soil and water management. Gains from these sources will be crop, animal, and location specific. They will require close articulation between the developers, suppliers, and users of new knowledge and new technology. One implication is that research on environmentally compatible farming systems should be intensified. There are a number of technical and institutional innovations in agriculture that could have both economic and environmental benefits. Among the possibilities is the design of a "third" or "fourth" generation of chemical biorational and biological pest management technologies. Another is the design of land use technologies and institutions that will contribute to reduced erosion, salinization, and ground-water pollution. These newer systems will be increasingly intensive in terms of the levels of knowledge, information, and management required.

**Research on Incentive Compatible Institutions.** Technical advantages, such as those suggested above, will make only incremental contributions to resolving the resource use problems facing such areas as the San Joaquin Valley in the absence of incentive-compatible institutional design. By incentive-compatible institutions I have in mind institutions capable of achieving compatibility among individual, organizational, and social objectives. For the issues being discussed in this book, this means compatibility among the objectives of individual water users, irrigation associations, State and Federal resource management and policy agencies, and the general public.

A clear agenda for the design of incentive-compatible institutions capable of resolving the technical and distributional issues of water and drainage management in the Valley unfortunately does not emerge from the chapters presented in this volume. One reason for this difficulty is that there has been no solution, even at the theoretical level, on which to base the design of incentive-compatible resource management policies and institutions. In spite of the very important contributions that have been reviewed here, it seems quite clear that our capacity to design institutions capable of achieving compatibility among individual, organizational, and social objectives is exceedingly limited. For the immediate future, we will be dependent on an evolutionary approach based primarily on trial and error, recognizing that error is expensive. The chapters in this volume suggest that at present we have very limited capacity to substitute social science knowledge for trial and error in the design of more effective resource management institutions. The prospect then is for the burden of the transaction costs involved in ad hoc approaches to policy and institutional design to remain very large.

This leads me to a final question. Why have social scientists devoted so little effort to the issue of institutional design? While dogma about the efficiency effects of lump-sum resource transfers keeps being repeated, a politically acceptable lump-sum transfer institutional design has never been developed. Research in the biological sciences, in contrast, extends from basic research to technology development--from basic biochemistry and genetics to plant breeding. But social science research rarely goes beyond analysis to give explicit attention to design. Until we do so, it is unlikely that society will place as much value on our contributions as we might prefer. If there is a demand for social science knowledge, it is derived from the demand for institutional innovation and improvements in institutional performance.

## PERSPECTIVE

In concluding, I want to address the problem of whether the public agricultural research system will respond to new challenges and opportunities such as: Releasing the biological and technical constraints on crop and animal productivity; meliorating the contribution of the agricultural sector to environmental degradation; and enabling the agricultural sector to adapt to the environmental changes emerging in response to the intensification of industrial production. Issues of both scientific and political capacity are involved.

Two decades of erosion in research capacity, particularly at the Federal level, have left the research system in a weakened position to respond to either- -let alone both--scientific or political concerns. The significance of this erosion is reinforced by the even more rapid decline in research support and capacity within Federal resource agencies and in the very limited support and capacity for mission-oriented research in the academic biological and environmental sciences.

The capacity of the agricultural research system to respond is also weakened by the political constraints within which it functions. The traditional agricultural research clientele--the organized commodity groups, elements of the agribusiness community, and members of Congress and State legislators with significant agricultural constituencies--are capable of bringing considerable pressure to limit the transfer of resources necessary to respond to the environmental research agenda. They doubt, correctly in my view, the capacity of the private sector to replace the traditional production-oriented research conducted by the public sector. Yet they have not demonstrated, in recent years, the political resources necessary to secure expanded funding, or even the funding necessary to prevent erosion of the capacity needed to respond adequately to constraints on agricultural production.

## REFERENCES

Batie, S., 1989. Sustainable Development: Challenges to the Profession of Agricultural Economics, *American Journal of Agricultural Economics*, pp. 1083-1101.

Binswanger, H. P. and Ruttan, V. W., 1978. *Induced Innovation Technology, Resources, and Development*. Johns Hopkins University Press, Baltimore, MD.

Coontz, N., 1991. Water Market Reforms for Water Resource Problems: Invisible Hand or Domination in Disguise. This Volume.

Crouch, M. L., 1990. Debating the Responsibilities of Plant Scientists in the Decade of the Environment, *The Plant Cell*, 2, pp. 275-277.

Hayami, Y. and Ruttan, V. W., 1971. *Agricultural Development: An International Perspective*. Johns Hopkins University Press, Baltimore, MD.

Mercer, L. J. and Morgan, W.D., 1991. Irrigation, Drainage, and Agricultural Development in the San Joaquin Valley. This Volume.

President's Material Policy Commission, 1952. *Resources for Freedom*. A Report to the President's Materials Policy Commission, U.S. Government Printing Office, Washington, DC.

President's Water Resources Policy Commission, 1950. *A Water Policy for the American People*. The Report of the President's Water Policy Commission U.S. Government Printing Office, Washington, DC.

Randall, A., 1991. Alternative Institutional Arrangements for Controlling Drainage Pollution. This Volume.

Rausser, G. C. and Zusman, P., 1991. Organizational Failure and the Political Economy of Water Resources Management. This Volume.

Runge, C. F., 1981. Common Property Externalities: Isolation, Assurance, and Resource Depletion in a Traditional Grazing, *American Journal of Agricultural Economics*, 63, pp. 598-506.

Ruttan, V. W., 1990. Challenges to Agricultural Research in the 21st Century. Presented at *Symposium on Technology and Economics*, National Academy of Engineering, Washington, DC.

Ruttan, V. W., (Ed.), 1990. *Resource and Environmental Constraints on Sustainable Growth in Agricultural Production*, Department of Agricultural and Applied Economics, University of Minnesota, St. Paul.

Ruttan, V. W., 1971. Technology and the Environment, *American Journal of Agricultural Economics*, 53, pp. 707-717.

Ruttan, V. W. and Yujiro H., 1984. Toward a Theory of Induced Institutional Innovation, *Journal of Development Studies*, 20, pp. 203-223.

Weinberg M. and Willey, Z. Creating Economic Solutions to the Environmental Problems of Irrigation and Drainage. This Volume.

# 47 IS THE DRAINAGE PROBLEM IN AGRICULTURE MAINSTREAM RESOURCE ECONOMICS?

V. Kerry Smith, North Carolina State University, Raleigh
and Resources for the Future, Washington, D.C.

## ABSTRACT

This chapter assesses the findings of the book in the context of environmental and resource economics. In particular, it addresses the following issues: (1) The usefulness of research directions and methodologies in addressing drainage and water quality problems and (2) how issues of drainage and water management relate to and affect the research agenda in environmental and resource economics.

## INTRODUCTION

After four dozen chapters explaining the complexities in evaluating water policies for agriculture, readers do not need further admonitions to convince them that it is difficult to analyze these programs. Instead, this chapter will offer some perspective on the various institutional, scientific, technological, and economic issues developed by earlier authors. I will focus on how these issues relate to: (1) The general descriptions usually offered to motivate public intervention; (2) the role of the natural and social sciences in evaluating the need for such action and measuring what might be gained if it is undertaken; and (3) the lessons that can be transferred from policy analyses of the drainage problem to other resource and environmental problems.

This evaluation and summary is developed in the next four sections. The first and second sections begin from a historical perspective. The first uses insights from Olson's *Logic of Collective Action* to organize several authors' (notably Weinberg and Wiley; Rausser and Zusman; Randall; Thomas; and Colby) descriptions of the evolution of Western water allocations and their effect on policies used to address the drainage problem. The second section considers how Eckstein (1958) might have evaluated the policy analyses developed in this volume for the San Joaquin drainage problem. Because many important dimensions of the drainage problem involve nonmarket outputs, models describing how they are "produced" and "valued" must replace the market prices that Eckstein advocated for evaluating the benefits and costs produced by water projects in the 1950's. Current models must blend the constraints imposed by science with the complex behavioral responses of economic agents.

Drainage problems involve private and impure public goods that are jointly affected by private actions and public policies. To evaluate the economic effects of these alternatives requires keeping score within a convenient metric. This means measuring peoples' values for outputs delivered within markets as well as those available outside markets. A debate is emerging over the appropriate methods for estimating these values. The San Joaquin studies include examples that apply primarily Contingent Valuation Methods (CVM) to estimate these values (see Cooper and Loomis and Loomis, Hanemann, Kanninen, and Wegge). Some of the chapters in this volume describe alternative perspectives on the credibility of benefit estimates developed with CVM. The third section summarizes whether sufficient information exists to resolve this question. Finally, the last section considers the role of economic information in policy responses to drainage problems, in particular, and environmental programs in general.

## COLLECTIVE ACTION AND THE DRAINAGE PROBLEM

Rausser and Zusman begin their discussion of the political economy of water resource management by identifying the pervasive tendencies for collective action in influencing the evolution of Western water policy. Several other chapters reinforce this explanation of how the present system developed into one that is characterized by several existing (or emerging) conflicts in three important areas.[1]

(1) Subsidized water prices assure the economic viability of many irrigation districts, but these subsidies are partially paid by taxpayers Nationwide and conflict with the interests of growing urban populations in the West.
(2) The drainage of water with high salinity and other pollutants (e.g., selenium) is required to manage salt balances and maintain agricultural productivity. However, the resources previously used for such drainage now create conflicts. Using them for disposal of polluted irrigation waters reduces the instream water quality for recreation and contaminates wetlands and ground water.

(3) The open access character of ground-water supplies, together with the prior appropriation system of water rights for surface waters, has created within-district externalities between farmers from polluted drainage and excessive rates of ground-water extraction.

These conflicts are examples of the types of behavior hypothesized in Olson's (1965) early description of the incompatibility of rational individual actions with collective rationality (for all individuals) when the decisions involve particular types of goods.[2] In a recent retrospective evaluation of Olson's book and the research it stimulated, Sandler (1990) describes the reasons for the failure of markets as arising from "quantity-constrained behavior." That is, the amount selected for a public good is not a choice variable from the individual's perspective. Rather it serves as a constraint to his behavior. Individuals will adjust their own private decisions on personal contributions to produce the public good incompletely. This connection motivated much of the research in this book.

Other connections have not been as clearly identified. They involve recent findings on the theoretical properties of production or preferences in formulating policies for resolving the free-riding problems Olson identified. Consider some examples. The production technology connecting individual economic agents' (i.e., farmers') actions to the level of the public good affects the incentives for free-riding. Most conventional treatments assume that the output of a public good is a linear function (usually a simple sum) of the amounts "produced" by each individual. However, if the technology for the public good is what Hirshleifer (1983) describes as the "weakest link" production process (i.e., public output corresponds to the minimum of the outputs selected by participating agents), then the process creates incentives that work against free-riding.

While it seems unlikely that the processes contaminating ground water or otherwise imposing drainage related externalities on farmers within the same irrigation district would adhere to this simple production principle, this does not preclude using some variation on this theoretical insight in policy design. One goal for intervention could well be altering the process through which each agent experiences the effects of others so that it approximately corresponds to a "weakest link" technology in its effects on each participant. Whatever the level of control for drainage wastewater, regional authorities could manage ground-water cleanup so it corresponded to the level implied by the farmer exercising the least control over his pollution. This should induce the farmers' private control actions to approximate efficient behavior. Of course, it also assumes the pollution control responses of all agents can be monitored.

An analogous result is found in Dixon's chapter. He uses game theory to evaluate the effects of how we characterize the strategic and technical features of interactions between farmers that lead to the externalities and the properties of efficient solutions. In his two-agent model, he found that differences in how the analyst describes each agent's knowledge and ability to respond to others' actions can dramatically affect the welfare differences between the uncontrolled and the efficient outcomes. Welfare losses that are estimated to arise in uncontrolled situations with myopia as compared with strategic responses that change over time (with information) differ by nearly 300 percent. Equally important, the influence of incomplete property rights depends on how much of the open access resource is available. If the available amount is large, interactions arising from attempts to capture the services of the resource will not be important until each agent's decisions begin to have implications for others. Discounting implies that recognition of pending depletion of an open access resource decreases in influence as the depletion is projected further into the future. Of course, the importance depends on both the availability of the resource's services and its role in contributing to what matters to the agent (in Dixon's terms, the marginal benefit functions).

A second example uses insights from the literature on altruism to illustrate how the specification of the production technology and/or preferences for the public good resulting from drainage control can influence the incentives for free-riding. Suppose the technology for control of drainage is a simple sum, so each farmer's control contributes additively to basin-wide quality. However, if the level of control is also assumed to provide "private" benefits (i.e., reducing salinity in the farmer's fields or yielding outputs from reallocation of salt loads), then the farmer's utility function resembles that used in Andreoni's (1990) analysis of impure altruism. Provided the private and public components of drainage control are imperfect substitutes, government policies designed to decrease taxes for those with incentives to undertake private control and to increase taxes for those with little private incentive (using tax revenues to support public control) will not crowd out the private initiatives. Specifications using the simple summation technology are analogous to pure altruism. Without recognizing this alternative specification for private control, they could be used to suggest that these policies would imply crowding out.

A third example can be found in Bergstrom and Cornes' (1981 and 1983) condition for the optimal amount of a pure public good to be independent of the distribution of private goods. They describe such preferences as conditional transferable utility. If Q designates the pure public good, $Y_i$, the income of the ith individual, and $U_i(Y_i, Q)$, a general expression for this person's utility, then their conditional transferrable functions will be $U_i(Y_i, Q) = A(Q)Y_i + B_i(Q)$. Note the $A(Q)$ is constrained to be equal across individuals, so changes in the distribution of $Y_i$ will not alter the contribution all individuals make to the sum of the marginal rates of substitution defining a Pareto efficient level of Q.

This analytical result has several implications for drainage policy. First, it specifies the only set of preferences where "history" (i.e., the prior assignment of entitlements and the associated income) does not affect the definition of efficient levels of the public good. Several authors in this volume note that policy design in controlling salinity must consider the income effects arising from any changes in the advantages bestowed on farmers from subsidized water rights. In practice, decisions on drainage control designed to improve Q would be expected to affect the distribution of income and hence the efficient levels of Q. Unless there is reason to believe that preferences conform to this specification, efficiency and equity will not be separable. Large changes in property rights can be expected to affect what is considered to be the new efficient level of Q. Second, a specialized version of transferable utility, quasi-linearity, can be used to establish that inefficiency increases with the number of agents involved in collective decisions about a pure public good.[3] Finally, these preferences also assure that honest revelation of private values for Q will be a dominant strategy (see Bergstrom and Cornes, 1983). Of course, this finding should not be surprising, given the fact that transferrable utility will not affect the level of the public good provided.

Overall, these examples illustrate how we can blend theory and practical experience to understand the features of applied problems and enhance policy design. Several chapters in this volume start this process.

## THE DRAINAGE PROBLEM AND BENEFIT COST ANALYSIS

Historical precedents for using benefit cost analysis to evaluate policy decisions have been identified as early as a report by President Jefferson's Treasury Secretary (see Kneese and Schulze, 1985). Nonetheless, systematic attention to assembling this information for policy purposes didn't begin until the mandates of the Flood Control Act of 1936. This volume's issues return to the same type of projects that motivated the development of the benefit cost methodology. Today, however, instead of looking afresh at the criteria for investment decisions, the analysis must face the legacy of management failures. Krutilla (1966) warned that analyses of these projects should not separate the economic evaluation of public investments from the criteria used in managing, particularly the use of correct beneficiary charges.[4] Many problems resulting from irrigation practices can be traced to inefficient pricing. Any form of ex post adjustment, once preferential charges have influenced the structure of production and entitlements to income generating assets (i.e., lands made productive by irrigation), they can be expected to affect efficiency and income redistribution. These impacts must become an integral part of the evaluation. Recognition of their potential effects on project evaluation (whether as public investments or regulatory decisions) is not new. It returns us to the writings of Krutilla and Eckstein (1958), Eckstein (1958), and others.

Because of these connections in the types of resources and the generic issues to be considered, the discussion here of applied welfare economics relevant to a new round of management decisions is related to Eckstein's (1958) discussion of the assumptions required to implement benefit cost analyses with early water projects. Five issues raised in his outline of the theoretical basis for benefit cost analysis are especially relevant:

- The role of competitive markets;
- The lack of physical independence;
- The influence of the income distribution;
- The effect of large policies on prices; and
- The valuation of "collective" goods.

While each of these (as well as seven other issues he identified) continue to be part of the evolving conceptual basis for benefit cost analysis, the focus of these discussions has diverged into two traditions. Applied welfare economists (or at least those interested in resource and environmental issues) in the United States have tended to focus on lack of physical independence and valuation of collective goods, while those in Western Europe have been more concerned with competitive markets and price changes arising from large policy impacts. Because the "U.S. school" considered problems in developed market economies (primarily the U.S.), they assumed market distortions were not important and project scale was limited in relation to the economy. By contrast, the applications considered by those in the European tradition were for developing countries where the outputs were usually marketed commodities but the price distortions (from government programs) and the scale of projects (in relationship to the economy) could not be assumed to be "small."

The analyses in this volume reflect a growing recognition that all five aspects can be important to modern policy issues. For example, as Berck, Robinson, and Goldman demonstrate, curtailment of agricultural water use in a regional economy will affect most prices and incomes. Scale matters! Moreover, price distortions and the effects of other types of agricultural programs can be important to the way the regional economy achieves equilibrium. Berck et al. note that Computable General Equilibrium (CGE) models can be used in developing consistent measures of before and after policy prices as well as measures for the appropriate shadow values of nontraded goods. The framework also allows evaluation of the influence of the policy perspective taken in judging proposed changes. For example, in evaluating a curtailment of water, how are price supports treated for some agricultural commodities?

While the Berck et al. analysis primarily focuses on regional impact analysis under alternative elasticity assumptions, their model offers a rich framework for evaluating both policies and the methods used for their evaluation. Consider three examples:

(1) Evaluating Benefit Approximations

Eckstein proposed simple rules of thumb for dealing with large projects and their price effects. He suggested that:

> ... crude measures for the limits of benefits (with large projects) can be found by using both sets of prices with, and without, the project. The case of outputs, the 'without' price, represents an upper limit to benefit, while the 'with' price is a lower limit, ... <u>the actual benefit is somewhere between the limits; if the change is not extreme, national benefit can be approximated by applying a price which is an average for the two to the increase in output</u> (Eckstein, 1958, p. 37, parenthetical phrase and emphasis added).

The Berck et al. model could easily evaluate the conditions when these and other rules provide reasonable approximations. Equally important, their model could be used to evaluate the importance of other partial equilibrium approximations inherent in many of the benefit and cost estimates used in policy analyses. How much of a general equilibrium perspective is required in defining consumer surplus estimates for policy changes?[5] Are approximations similar to those suggested by Eckstein or others by Harberger (1971) reasonable approaches for dealing with large-scale problems when CGE models are <u>not</u> available and there is no time or resources to develop them?[6] Guidance on these questions would extend the impact of the San Joaquin research beyond the specific policy questions involved in irrigation and drainage issues in the Valley to many large-scale policy issues in resource and environmental economics.

(2) Integrating Market and Nonmarket Effects

Economic policy analyses increasingly recognize the importance of a general equilibrium perspective for evaluating actions that are likely to have National effects, such as those associated with environmental programs (see Hazilla and Kopp, 1990 and Jorgenson and Wilcoxen, 1990). For the most part, these studies have taken the environmental quality improvements realized from such programs as given. Welfare costs measure the implications of the price increases required to meet quality standards for households, but they do <u>not</u> measure what might be described as the "net" effect for each household. Doing so would require incorporating the environmental quality (or pollution) measures in the expenditure relationships used to characterize households' preferences.[7] With such models, it would be possible to consider simultaneous changes in the "full prices" for activities that reduce pollution and the environmental quality improvements they provide.

Current analyses of these issues rely on separate "differentially partial" equilibrium analyses of how policies affect the prices of marketed goods and survey analyses of how people would value the "services" delivered by the policies (in the form of improved environmental quality). While the estimates can be compared as part of an overall benefit cost evaluation, assumptions about what is held constant in each type of evaluation are not likely to be consistent. Thus, the net gain or loss we would ideally like to measure is not being consistently measured. For small projects with limited effects, these could be ignored. For today's policy evaluations, they cannot. In principle, it is possible to take this next step within a CGE framework. This is especially true if we continue to tolerate the somewhat ad hoc approach that has been used to distill diverse estimates of the relevant economic parameters in parametizing the model. Crude indexes of environmental quality could be incorporated into the specifications for utility functions as a first step in computing these "net" effect solutions.

(3) Linking Scientific and Technical Constraints to Economic Models

Another aspect of the chapters in this volume also helps to identify the next round of research with regional CGE models for use in environmental policy analysis. It involves the treatment of natural resources in the production technologies of these models. The Berck et al. analysis simplified its description of production by using a mixed strategy to incorporate input substitution. In their high-elasticity framework, real value added is a Cobb-Douglas function of land, labor, and capital, but domestic and imported intermediate inputs are linked with a fixed coefficient (i.e., zero elasticity of substitution technology). Most significantly, for the objectives of the project, land and water are used in fixed proportions that are allowed to differ across agricultural sectors. In the low elasticity variant, capital is moved from the Cobb-Douglas master level to the Leontief technology as another fixed factor linked to land.

These models offer a first step. Just's chapter outlines some requirements in production models that should be considered in the next stage. His advise is to:

- Avoid imposing input to output separability often used for analytical convenience in past neoclassical models;
- Select models that allow varied levels of input substitution, especially when considering the effects of limitations on water availability;

- Distinguish production models at the farm from the sectoral level (fixed coefficient technologies might be plausible for the former but not the latter);
- Recognize the interactions between natural systems and agricultural production practices;
- Incorporate environmentally relevant heterogeneity across farms. Because land types are different, agricultural practices on one type can have very different impacts on environmental quality measures.

Examples of where CGE models have incorporated aspects of the first three types of complexity (primarily for nonagricultural problems) can be found in varying degrees of detail. To this author's knowledge, no existing models incorporate sufficient recognition of the interaction among production decisions, environmental systems, and microheterogeneity. This is important because these extensions would allow us to judge how reorganization at the farm level together with the policy targeting comparable to the progressive refinements in the conservation reserve program would affect our perceptions of the aggregate costs of realizing environmental goals. Berck et al. outlined a framework that can be extended to incorporate these refinements. Just described how conventional production models can be adapted to include environmental impacts. To take the next step requires the right data.

Eckstein was not sanguine about the prospects for valuing public goods. Dramatic progress has been made in the three decades since he offered his appraisal. But have we progressed far enough to genuinely inform the policy process? This is the next question to be confronted.

## VALUING ENVIRONMENTAL RESOURCES IMPACTED BY THE DRAINAGE PROBLEM

Several chapters in this volume mirror a growing professional debate over the ability of economic models, especially the different survey techniques loosely grouped under the general heading of CVM, to estimate the values for nonmarketed environmental resources. On the one hand, we find Loomis, Hanemann, Kanninen, and Wegge suggest that: "the Contingent Valuation Method (CVM) is the technique best able to measure California residents' willingness to pay for different levels of wildlife management in the San Joaquin Valley." CVM is described as a "widely accepted method" for these types of valuation tasks and the only one that can measure nonuse values. In contrast, Burness, Cummings, Ganderston, and Harrison's chapter concludes that "Research to date appears to make clear ('clear' in the sense we find no compelling contradictory evidence) that CVM surveys for one specific commodity yield values which cannot be defensibly attributed to that specific commodity."

Some commentaries (notably Phillips and Zeckhauser, 1989) writing outside traditional economics journals have taken stronger positions arguing, among other points, that:

- CVM embodies an approach to estimate people's values that has been rejected by economists and is therefore unlikely to realize acceptable levels of accuracy.
- CVM has been maintained in the professional literature because a small number of practitioners have pushed it, and despite "quiet doubts among most economists," it is self-sustaining. "Opponents are not sufficiently impressed to study the issue extensively, and relatively little academic credit is available for articulating skepticism, no matter how valuable it is" p. 522).

By calling attention to this debate, the chapters in this volume expose one of the most important questions facing applied welfare economics for decisions involving nonmarketed environmental resources. Namely, can estimates of how people value nonmarketed environmental resources be as accurate as those inferring people's values for the goods involved in market transactions? The discussion here of this issue addresses three questions:

(1) What is known, and how does it compare with valuing market-based commodities?
(2) Are the Burness et al. concerns over eliciting values for specific commodities as damaging as they imply?
(3) Is the Which-What-Which problem unique to CVM?

## What Do We Know?

The accuracy of values estimated by economists is impossible to judge because people's "true" values can never be known. This is true for both marketed and nonmarketed commodities. Because of this constraint, accuracy is usually judged by construct validity: Gauging the correspondence between CVM and other methods' estimates of the value people place on the same commodity; evaluating whether CVM estimates conform to some theoretically predictable behavior either alone or in relation to some other variables; or testing the estimates by their adherence to specific theoretical hypotheses.

Overall, comparative evaluations between methods (i.e., indirect v. survey) have found CVM comparable to the other methods.[8] For the most part, studies of purchase behavior (hypothetical and actual) have also found close correlations. Bishop and Heberlein (1979); Bishop, Heberlein, and Kealy (1983); and Dickie, Fisher, and Gerking (1987) all indicate reasonably strong support for CVM estimates and actual choices using constructed markets for hunting permits and strawberries. The only notable exception involves a recent study by Seip and Strand (undated) using consistency in the responses of a small sample of Norwegians who were asked if they would join the largest private environmental organization in Norway. The study involved three stages-- personal interviews asking in hypothetical terms whether respondents would join at the 1990 current annual membership fee; two mailed followups soliciting membership from those who said "yes" to the first hypothetical question about membership with no mention of the earlier interviews; and telephone interviews of a subset of those who said they would join and did not. Of 101 initially interviewed, 64 indicated they would join the group. However, only six actually joined in the second-stage mailed membership solicitation.

Because of the nature of the commodity (i.e., membership in the organization, defined in vague terms during the first stage), the change in format across the hypothetical and the actual solicitation (especially given the widespread use of mailed advertisements), and the possibility for changes in personal circumstances, this study cannot be considered a major contradiction to the earlier evidence. It may simply indicate that concerns over having clear and specific commodity definitions in CVM questions raised by Fischhoff and Furby (1988) and Mitchell and Carson (1989) are very important to performance of the method.

Nonetheless, two aspects of these results should be considered. First, reasonably close correspondence usually means that the two methods' estimates are significantly correlated, but often that CVM is a positive multiple (greater than unity) of the other "source" used to estimate the value of the nonmarketed commodity. Cummings, Brookshire and Schulze's (1986) earlier state-of-the-art summary reported estimates within ± 50 percent for use related values. The Monongahela comparison reported in Smith and Desvousges (1986) indicates that the relationship can be influenced by the choice-based method used for comparison. Nonetheless, using the preferred indirect method, CVM estimates were three to seven times larger than travel cost results, with smaller discrepancies associated with large (and presumably more easily discernable) changes in the water quality of the river.

While some critics have argued that this type of discrepancy is unacceptably large for many uses (see Phillips and Zeckhauser, 1989), it should also be placed in the context of price variability. Pratt, Wise, and Zeckhauser (1979), for example, reported comparable variation in the market prices for the same commodities across different stores in Boston. The range of the ratio of maximum to minimum prices for the 50 sets of identified commodities and services within Boston was from 1.11 to 6.67. When compared with this standard, the CVM estimates would be judged as quite comparable to the experience observed for marketed goods.

Second, comparisons between methods will be sensitive to the metric or standard used for evaluation. Estimates of people's values is only one. Dickie, Fisher, and Gerking (1987) used demand functions for strawberries derived from the two types of data. Smith et al. (1990) compare estimates of consumer surplus versus predicted purchase decisions across methods. While both standards indicate consistency, the strength of perceived agreement can be different using alternative standards.

In summary, more is known about the performance of CVM than the Burness et al. chapter might lead a "new entrant" to this field to conclude. However, the correspondence between CVM and choice-based methods is closest for commodities that most closely correspond to private goods. As the commodity to be valued is increasingly defined as a public or impure public good, there is less direct evidence. What there is suggests that it may be necessary to use a higher multiple in linking CVM to the choice-based indirect method. However, this does not imply that CVM is the culprit. Most of the successes in using choice-based indirect methods to value nonmarket goods arise because a model cleverly exploits some privately capturable component of the good or service that is revealed by observed choices. Something appears to have been missed in interpreting these results: the greater the ability to define the commodity for these models as a privately captured component of some impure public good, the less accurate the model is "revealing" the individual's value for the public good component of the impure public good.

## Are the Burness et al. Concerns Damaging?

The Burness et al. judgments on whether responses to CVM surveys can provide estimates for individual environmental resources reflect concerns over the so-called embedding hypothesis. To support their conclusions, they cited survey results reported by Kahneman and Knetsch (1991) indicating: (1) Estimates of individual willingness to pay are "... approximately constant for public goods that differ greatly in inclusiveness" (p. 10), and (2) the value assigned to more specific definitions for public goods "... varied by an order of magnitude depending on the depth of its embedding in the category for which willingness to pay (WTP) was initially assessed" (p. 20, parenthetical terms inserted). While there are significant problems with the design and implementation of the Kahneman and Knetsch survey (see Smith, 1990), they need not be addressed here to discuss Burness et al.'s reservations about CVM.

Burness et al. raise important questions, but they are not unique to CVM's estimates. Differences in the inclusiveness used to describe a public good arise because the definition of most public goods and descriptions of changes in the amounts available to people have never been systematically addressed in operational terms. Most policy relevant examples of public goods are impure in the sense that they have private and public components. For example, a substantial body of research confirms that people do value improvements in air quality. The economic research underlying this conclusion relies on hedonic property value models. Ideally, these models would use the air pollution relevant to each person's home location and perhaps exposures realized in the course of daily activities to measure the "amount" of air quality consumed. But does this aggregate completely reflect the way air quality (or avoiding exposure to air pollution) contributes to a person's well-being? To reflect other equally plausible contributions, the degree of inclusiveness could easily be expanded to include:

- Air pollution at the recreation sites a homeowner visits; or
- Air pollution in other cities where an individual travels.

Of course, not all of these air pollution measures would be expected to influence the hedonic price function specifying the prices required for an equilibrium in a market with heterogeneous houses. However, it is certainly plausible to hypothesize that these other air pollution measures would influence the individual buyer's marginal rate of substitution between air pollution and a numeraire good. Ignoring them would cause us to misinterpret the marginal rate of substitutions measured from a hedonic model and bias the willingness to pay estimates we derive from second-stage models estimated from using them with only the housing site's air pollution level.

Alternatively, air quality may be valued throughout the city because it is uniformly available, as in the altruism or impure public goods models. These issues are really what underlie the Kahneman-Knetsch inclusiveness questions. They are not unique to CVM. What is unique about CVM is that it forces the analyst to explicitly raise these issues in describing the "commodity" in a CVM question. For the indirect methods, the same issues are "buried" in the selection of proxy variables used to designate the specific environmental services of interest.

Of course, this same type of question arises in the practical implementation of microeconomics with any applied problem. How does the analyst define the commodity in practical terms when implementing a model? The use of applied microeconomics in evaluating when firms have market power (i.e., the ability to price with consistently positive economic profits) must address this question as a routine matter in defining the "commodity extent" of the relevant market.

The embedding question also reflects the need for consistency in partial equilibrium welfare measurement. Defining a sequence of changes in public goods as aggregates of subcomponents so that they are comparable requires that the progressive modifications in these nested components of each aggregate define the equivalent of consistent paths of integration (see Bockstael and Kling, 1988). In other words, both the changes and the levels of all components must be comparable if the values are expected to be comparable.

Again, this is not a problem unique to estimates from CVM surveys. Considered in the context of welfare measures for price changes, it is completely analogous to the differences derived from general versus partial equilibrium welfare measures (see Kokoski and Smith, 1987).

### Is Which-What-Which Unique to CVM?

Estimates of the total value of some change in a nonmarketed environmental resource routinely multiply the "average" estimate of household willingness to pay times the total number of households in some politically defined area (i.e., a county, state, or region). Burness et al. correctly observe that defining "which" households have these values is a part of the valuation task. This delineates the "geographic extent" of the market. Similarly, as observed above, defining how the environmental commodity in a valuation question relates to other "components" of environmental services delineates the "commodity extent" of the market. Resolution of these questions is not unique to CVM. It is required in implementing all applied microeconomic models with microlevel data, for both marketed and nonmarketed commodities alike. There is little doubt that resolving these questions will be different in each case, but they are fundamental to meaningful applications of economic analysis.

## ECONOMIC INFORMATION AND ENVIRONMENTAL POLICY

Clearly, I have concluded that agricultural water policies, especially as they have been described and analyzed for the Valley, raise nearly all of the important research questions that are now relevant to resource and environmental policy issues. While this volume contributes to an improved understanding of the economic dimensions (especially the efficiency tradeoffs associated with alternative water polices in the Valley), it also helps to identify the generic elements likely to be important in analyzing other environmental policies. Having said this, I recognize that the nature of the Valley's problems prevents us from transferring specific details to other policy issues, indeed even to those that might involve agricultural water policies in another region.[9] If this is so, is economic analysis really "worth the effort?"

For better or worse,[10] economic analyses of public investment, resource management, and regulatory decisions consider specific policies by developing two types of information: (1) That intended to gauge whether the proposed action improves efficiency; and (2) that directed at measuring its economic impacts (i.e., effects on prices, incomes, patterns of resource use, etc.). This chapter focused on developing the first type of information, which is most often interpreted as providing the basis for an "economic decision rule." So my discussion of the "worth" of economic analysis will be confined to it.

Economists are fond of answering questions about the worth of their "analysis" by using the very tools that they seek to evaluate. That is, they redefine the question in a value of information framework: can mistakes be avoided by using economic analysis to measure the net benefits of an action in advance? In this respecification, "mistakes" are defined as movements away from (i.e., negative net benefits) rather than toward (i.e., positive net benefits) an efficient allocation of resources.

A number of social scientists and policymakers consistently object to this characterization because, they argue, efficiency is not the goal of the action. Economic analysts are interpreted as advocates for efficiency because, the economists suggest, no one else involved in the decision process represents this perspective. From here the dialogue often deteriorates.

The problem can be traced to the redefinition of the question: can we measure the worth of applied efficiency indexes by using the realizations produced by them? The answer should be that direct economic analysis (using the efficiency indexes) changes the range of questions that should be asked and answered. The efficiency metric (if properly done) tells the decisionmaker whether the action has an "opportunity cost" from the perspective that real resources are available for something else. Actions leading to positive net benefits generate more gain than they cost. Those with negative net benefits may nonetheless be desirable, <u>but someone is forced to describe what they are intended to provide.</u>

By using these applied efficiency indexes, do we improve society's chances of getting what it wants? My answer is a decisive "maybe." It depends upon whether we know enough about any particular problem (or are given sufficient time to learn) so that economic measures do inform the process. In short, if we cannot rely on measures of net benefits when they indicate more gain than cost, then they fail to persuade others to consider seriously the questions about what else is involved.

By attempting to systematically assemble information for specific problems such as the San Joaquin drainage problem, economic analysis can inform the process. Equally important, by exposing the limitations in our methods, it should also serve to stimulate the research necessary to reduce the areas where it cannot inform decisions.

## NOTES

¹Collective action is not necessarily responsible for all the current conflicts. Some are externalities--unintended byproducts of the mix and level of activities induced by collective action. As such, they represent incomplete understanding of the natural and physical constraints on agricultural activities and the increasing concern of people over the quality of environmental amenities.

²Sandler (1990) described the three basic themes in Olson's book as:

(1) Group size is, in part, behind collective irrationality.

   (a) Large groups may not provide themselves with a collective good; hence, no individual or coalition within the group may satisfy the sufficient condition of a privileged group.
   (b) The larger the group, ceteris paribus, the greater the departures of Nash equilibrium from Pareto optimality.

(2) Group asymmetry, in terms of membership tastes and/or endowments, is related to collective irrationality.

   (a) Larger members (those with the greater endowments) will bear a disproportionate burden of collective provision . . . .
   (b) Asymmetric groups are more likely to be privileged.

(3) Group irrationality may be overcome through selective incentives (giving private benefit inducements) and institutional rules and design (pp. 3-4).

³See Cornes and Sandler (1986), pp. 83-84.

⁴Howe (1988) discusses reimbursement policy and the relationship between beneficiary charges for outputs of public water projects and their cost.

⁵The distinction drawn between general and partial equilibrium measures of the welfare change households experience as a result of some policy arises with how the analyst "connects" the policy to a set of price changes. The more complete the set of price changes actually associated with the policy allowed in the welfare computation, the more "general equilibrium" it is.

⁶Kokoski and Smith (1987) and Smith (1987) illustrate this type of analysis using much simpler models.

⁷For further discussion of how this might work, see Smith (1991) and for examples Math-Tech (1982) and Gilbert (1984).

⁸Smith and Desvousges (1986), Chapter 10, summarize most of the comparative evaluations available in the published literature up to about 1985.

[9] Ideally, the overall analysis of the San Joaquin problem would have been sufficiently advanced to allow an integrated treatment of all the economic and scientific information. This would take the form of specific policy evaluations. Because such an integrated evaluation was not presented in this volume, it is clearly the next step in the research program.

[10] Bromley's (1989) recent appraisal of the role of economics in policy analysis criticizes the economic information assembled. He has argued that efficiency indexes cannot be developed as the "objective information" envisioned here and in conventional discussions of the place of applied welfare economics.

His critique relates to the broad forces shaping societal goals and the evolution of property rights and other institutions over time. It does not answer whether using efficiency measures at "snapshot intervals" within the overall process can improve the performance of the process. And this seems to be one of the important implications of evaluating the importance of his criticism.

## ACKNOWLEDGEMENTS

Thanks are due Barbara Scott for improving this paper's exposition. Partial support for my research was provided by the University of North Carolina Sea Grant Program, Project Number R/MRD-18.

## REFERENCES

Andreoni, J., 1990. Impure Altruism and Donations to Public Goods: A Theory of Warm-Glow Giving, *Economic Journal*, 100, pp. 464-77.

Bergstrom, T.C. and Cornes, R.C., 1981. Gorman and Musgrave Are Dual: An Antipodean Theorem on Public Goods, *Economic Letters*, 7(4), pp. 371-78.

Bergstrom, T.C. and Cornes, R.C., 1983. Independence of Allocative Efficiency from Distribution in the Theory of Public Goods, *Econometrica*, 51, pp. 1753-67.

Bishop, R.C. and Heberlein, T.A., 1979. Measuring Values of Extra Market Goods: Are Indirect Measures Biased? *American Journal of Agricultural Economics*, 61, pp. 926-30.

Bishop, R.C.; Heberlein, T.A.; and Kealy, M.J., 1983. Hypothetical Bias in Contingent Valuation: Results from a Simulated Market, *Natural Resources Journal*, 23(3), pp. 619-33.

Bockstael, N.E. and Kling, C.L., 1988. Valuing Environmental Quality: Weak Complementarity with Sets of Goods, *American Journal of Agricultural Economics*, 70, pp. 645-62.

Bromley, D.W., 1989. *Economic Interests and Institutions*. Basil Blackwell, New York, NY.

Cornes, R. and Sandler, T., 1986. *The Theory of Externalities, Public Goods and Club Goods.* Cambridge University Press, New York, NY.

Cummings, R.G.; Brookshire, D.S.; and Schulze, W.D., 1986. *Valuing Environmental Goods: An Assessment of the Contingent Valuation Method.* Roman and Allanheld, Totowa, NJ.

Dickie, M.; Fisher, A.; and Gerking, S., 1987. Market Transactions and Hypothetical Demand Data: A Comparative Study, *Journal of the American Statistical Association,* 82, pp. 69-75.

Eckstein, O., 1958. *Water Resource Development: The Economics of Project Evaluation.* Harvard University Press, Cambridge, MA.

Fischhoff, B, and Furby, L., 1988. Measuring Values: A Conceptual Framework for Interpreting Transactions with Special Reference to Contingent Valuation of Visibility, *Journal of Risk and Uncertainty,* 1, pp.147-84.

Gilbert, C.C.S., 1985. *Household Adjustment and the Measurement of Benefits from Environmental Quality.* Unpublished Ph.D. Thesis, University of North Carolina, Chapel Hill.

Harberger, A.C., 1971. Three Basic Postulates for Applied Welfare Economics: An Interpretive Essay, *Journal of Economic Literature,* 9, pp. 785-797.

Hazilla, M. and Kopp, R.J., 1990. Social Cost of Environmental Quality Regulations: A General Equilibrium Analysis, *Journal of Political Economy,* 98, pp. 853-873.

Hirshleifer, J., 1983. From Weakest-Link to Best Shot: The Voluntary Provision of Public Goods, *Public Choice,* 41(3), pp. 371-86.

Howe, C., 1988. Public Intervention Revisited: Is Venerability Vulnerable? In: Smith, V.K. (Ed.), *Environmental Resources and Applied Welfare Economics.* Resources for the Future, Washington, D.C.

Jorgenson, D. W. and Wilcoxen, P.J., 1990. Environmental Regulation and U. S. Economic Growth, *The Rand Journal of Economics,* 21, pp. 314-340.

Kahneman, D. and Knetsch, J.L., 1991. Valuing Public Good: The Purchase of Moral Satisfaction, *Journal of Environmental Economics and Management.* In Press.

Mitchell, R.C. and Carson, R.T., 1989. *Using Surveys to Value Public Goods-- the Contingent Valuation Method.* Resources for the Future, Washington, DC.

Kneese, A.V. and Schulze, W.D., 1985. Ethics and Environmental Economics. In: Kneese, A.V. and Sweeney, J. (Eds.), *Handbook of Natural Resource and Energy Economics.* North Holland, Amsterdam.

Kokoski, M.F. and Smith, V.K., 1987. A General Equilibrium Analysis of Partial Equilibrium Welfare Measures: The Case of a Climate Change, *American Economic Review,* 77, pp. 333-341.

Krutilla, J.V., 1966. Is Public Intervention in Water Resources Development Conducive to Economic Efficiency? *Natural Resources Journal,* 6, pp. 60-75.

Krutilla, J.V. and Eckstein, O., 1958. *Multiple Purpose River Development: Studies in Applied Economic Analysis.* John Hopkins University Press, Baltimore, MD.

Math-Tech Inc., 1982. *Benefits Analysis of Alternative Secondary National Ambient Air Quality Standards for Sulfur Dioxide and Total Suspended Particulates.* Vol. II, Report to U.S. Environmental Protection Agency, Research Triangle Park, NC.

Olson, M., 1965. *The Logic of Collective Action.* Harvard University Press, Cambridge, MA.

Phillips, C.V. and Zeckhauser, R.J., 1989. Contingent Valuation of Damage to Natural Resources: How Accurate? How Appropriate? *Toxics Law Reporter*, pp. 520-529.

Pratt, J.W.; Wise, D.A.; and Zeckhauser, R.J., 1979. Price Differences in Almost Competitive Markets, *The Quarterly Journal of Economics*, 93, pp.189-212.

Sandler, T., 1990. The Logic of Collective Action: A Retrospective View. Preliminary Draft, Department of Economics Working Paper, Iowa State University, Ames.

Seip, K. and Strand, J. Undated. Willingness to Pay for Environmental Goods in Norway: A Contingent Valuation Study with Real Payment. Unpublished Paper, Energy and Society Research Program, SAF Center for Applied Research, Department of Economics, University of Oslo.

Smith, V.K., 1987. Harberger Versus Marshall: Approximating General Equilibrium Welfare Changes, *Economic Letters*, 25, pp. 123-26.

Smith, V.K., 1990. Arbitrary Values, Good Causes, and Premature Verdicts: A Reaction to Kahneman and Knetsch. Working Draft, Department of Economics, North Carolina State University, Raleigh.

Smith, V.K., 1991. Household Production Functions and Environmental Benefit Measurement. In: Braden, J. and Kolstad, C. (Eds.), *Measuring the Demand for Environmental Improvement.* North Holland, Amsterdam.

Smith, V.K. and Desvousges, W.H., 1986. *Measuring Water Quality Benefits.* Kluwer Nijhoff, Boston, MA.

Smith, V.K.; Liu, J.L.; Altaf, M.A.; Jamal, H.; and Whittington, D., 1990. How Reliable Are Contingent Valuation Surveys for Policies in Developing Economies. Working Draft, Department of Economics, North Carolina State University, Raleigh.

# 48 SUMMARY OF FINDINGS AND CONCLUSIONS

David Zilberman, University of California, Berkeley and
Ariel Dinar, University of California, Riverside and
San Joaquin Valley Drainage Program

Water is essential to human life. The living patterns and institutional structures of many societies have been affected by the accessibility of their freshwater supplies. The United States and many other areas of the world are now in a transitional period in terms of water. Gone are the days when water resources were considered to be unlimited and each increasing demand for water was met by a new water project. The availability of untapped water sources has declined as demand for water has increased over time. Changing water conditions require new approaches in policies and institutions. The emphasis may shift from development of new water sources towards better management and utilization of existing water sources. Similarly, growing concern for environmental quality problems and the limited capacity of the environment to absorb drainage residuals have substantially increased the need for more efficient management of agricultural drainage and runoff.

Economists play a crucial role in shaping new institutions and policies for management of water and drainage. To be effective in this role, economic research and analysis need to combine sound economic principles with knowledge of technological and institutional detail. This book has reviewed the economic approaches to address different issues associated with water and drainage management as well as relevant knowledge from other disciplines. Much of this knowledge is applied to address drainage problems in the San Joaquin Valley (Valley). Several major lines of economic research on water and drainage issues have been reviewed. Some of these are established lines of research and others are in early stages of development.

Much research was done on water management at field and farm level which provides information and knowledge on the relationship between water use and yield. There is, unfortunately, insufficient knowledge on quantitative interaction between water and other inputs, and impacts of different irrigation technology on productivity in different environments. Econometric structures of microlevel behavior have not provided much insight regarding parameters of production technology. They have been more successful in identifying behavioral patterns, providing substantial evidence that water use and irrigation and drainage technology choices by farmers are responsive to economic and environmental conditions in manners consistent with economic theory.

In recent years, there has been a proliferation of methodologies for analyzing regional impacts of resources management policies. These methodologies (CGE, Mathematical Programming, and Econometric Techniques) provide reasonable and consistent answers to the Valley water and drainage problems being considered. Although they vary in degree of detail and complexity, all provide limited capability to deal with dynamic processes and incorporate process of adoption and diffusion.

Value assessment of nonmarket environmental amenities is a growing area of research interest. This is a new field, going through the process of establishing and refining quantitative procedures, and some of the methodological argumentation have been presented in this book. Results of existing applications reassure and suggest that in the foreseeable future nonmarket valuation can be incorporated into the calculus associated with policy assessment and design.

Concepts and constructs of economic theory have been useful in analyzing water policies and institutions. The misallocation of resources associated with ground-water use and drainage generation were identified using game theory. Game theoretical arguments were also used to demonstrate that political considerations lead to exploitation of water resources and excessive generation of drainage and pollution. As water quality considerations become more prominent, conceptual analysis suggests that institutions (appropriative rights) and policies (conjunctive use, stabilization rules) should be multidimensional--treating and managing water of different quality separately but in concert.

Recent theoretical developments emphasize the study of problems in policy implementation. Implementation of "first best" policies (drainage taxes, for example) may be infeasible due to monitoring or enforcement problems. Transformation from property right to water market systems may encounter many obstacles and needs to be gradual. The definitions of products traded in water markets is not trivial and must incorporate consideration of quality and location.

The chapters in the book suggest room for interdisciplinary cooperation and such cooperation is evident in the development of water and drainage practices.

# FINDINGS AND CONCLUSIONS

Methodologies and knowledge of public health research can be incorporated into economic assessment of environmental, health-related regulations. The economic and legal studies on water institutions address similar problems and raise similar concerns.

The book also provides several interesting lessons about management of water and drainage in the Valley, the management of water systems in general, and directions for future research. General conclusions about the nature of the Valley's water and drainage problems can be drawn from the book's chapters:

(1) Any solution to the problem must consider the competing uses for water in the region. Traditionally, most of the emphasis has been on agricultural use. Even with all the disagreement on the CVM method, it is clear that recreation is a significant user of water resources in the Valley and any regional planning should allow recreation to be among the competing activities for water. We foresee that in the future California's Central Valley will become an even more important area for fishing and wildlife recreation in addition to maintaining its importance in agricultural production.

(2) Water availability does not pose any major physical constraint on agricultural production in the Valley. From the different studies reported in the book it is likely that water quantities for irrigation can be reduced by 25 to 30 percent without significantly affecting agricultural production. Furthermore, drainage generation can be drastically reduced without affecting agricultural production. Reduction in water use and drainage are costly and require capital investment as well as institutional reforms and may result in redistribution of resources.

(3) Heterogeneity with respect to locational characteristics and water quality dimensions should be recognized in analyzing and assessing production levels and drainage generation. Water of different qualities should be allocated to different locations for different products.

(4) Policy formulation should recognize dynamic processes of drainage generation and salinity accumulation as well as variability of temperature and rainfall over time. Ignoring the future impacts of present policies may cause irreversible damage, reducing the economic viability of the Valley in the long run.

(5) The existing water rights system is one of the major causes for excess drainage generation as well as low water efficiencies. Market movement will result in distributional change and require further investment in conveyances systems.

(6) The policymaking process is affected by many considerations including efficiency in resource allocation. Viable economic analysis must recognize the political economic forces behind the existing institutional

arrangements and consider them in any policy reform. In particular, the design of any policy reform should explicitly recognize distributional impacts, incorporate transfer mechanisms between groups to allow political acceptability, and institute monitoring and enforcement procedures.
(7) When it comes to the operation of a common property (ground water) or externality (drainage), the market forces and political systems tend to exploit resources, which may lead to crises. Only a crisis situation may induce drastic reform and changes in institutions and technologies that will allow more efficient use of the resources.
(8) Understanding the strategic behavior and interaction of different organizations in a water-drainage system is important to introducing incentive schemes and policies that lead to a stable solution (equilibrium).
(9) Many optimal technical solutions challenge the sovereignty of existing organizations (farms, water districts). For example, several solutions require production and land allocation plans to be coordinated between farms or water districts. Implementation of such plans may require new institutions for regional cooperation. Such plans will combine solutions at regional levels, where farm-level solutions can be induced by relatively simple measures.

The nature of a solution to Valley water and drainage problems is shown in the following:

The traditional approach to drainage problems is construction of a drainage canal to dispose of the drainage, but the cost of disposal has increased tremendously. Disposal to large scale drainage areas such as Kesterson or to the San Francisco Bay are no longer feasible physically or politically and disposal to the Pacific Ocean is too expensive, so other alteratives have to be considered. Some of the physical solutions include source reduction such as water conservation, land retirement, abatement (evaporation ponds, dilution and disposal to the river, treatment and disposal to the river), reuse and biological filtering (using, for example, Eucalyptus and halophyte), regional water quality stabilization, and conjunctive use of surface and ground water. Some of these solutions can be implemented at the farm level and some at the regional level.

The heterogeneity of the Valley suggests that no one solution is preferable everywhere, and the optimal solution may result in a quilt of locally adapted solutions. The challenge is to construct the institutional means to allow and induce the solutions.

For example, possible institutions can include penalties on drainage or on excessive water use, subsidies for drainage reduction, establishment of research

# FINDINGS AND CONCLUSIONS 937

and extension activities, regional drainage management authorities responsible for execution of regional solutions to coordinate and plan agricultural-related activities. Effective water markets and water transfers within the State should also be considered to help solve drainage problems. While it may be that all these measures are not sufficient to totally address the drainage problem, they may delay the need to construct a drainage canal and substantially reduce disposal costs.

The studies presented in this book and our understanding of the nature of drainage and water management problems and solutions in the Valley (as a general model) have provided us with some general conclusions about future research needs in economics and natural resources. A basic requirement of economic research on water management is to provide some mechanism to value water and drainage; these valuations are not simple. They may require adjustment for location and timeframe. One may also distinguish between different valuations under different institutional arrangements. Some of the chapters in this book present attempts to provide valuation of water, but the valuations are partial, addressing certain aspects and ignoring others. Even the general equilibrium models are partial in their analysis since they model the agricultural sector and its demand in varying degrees of detail but ignore: (1) Recreation demand for water, (2) dynamic opportunity cost associated with exhausting the region's environmental capacity to absorb drainage, and (3) excess demand and/or supply of water from outside the region.

Existing regional planning models and the general equilibrium model have had some success in predicting agricultural production patterns and resource allocation under alternative scenarios and assessing intersectoral impacts. Yet, an empirical framework is missing that combines all aspects of water use to provide reasonable valuation. A particularly challenging aspect in the design of such a framework is the incorporation of information derived from contingent valuation and other studies on demand for environmental amenities to achieve a multiuse demand framework.

Another area where the current research seems insufficient is the incorporation and representation of dynamic phenomena into policy analysis. Water quality, drainage, and salinity problems have many time-dependent aspects. Solutions to these problems require research, learning, and development of activities related to adoption and diffusion processes and inventories management--all dynamic activities. Use of static models may miss important aspects or even be irrelevant.

Development of empirical dynamic frameworks is quite a scientific and technical challenge. One particular problem is that dynamic modeling increases the dimension of the analysis many times over. One solution is to have several models with varying degrees of detail that will be considered in regional policy analysis. Dynamic models will operate with variables defined at much higher degree of aggregation than some static regional resource allocation models.

A third area of research that needs to be emphasized is institutional design. The existing political economic framework lacks the checks and balances to prevent overutilization of exhaustible resources. New policies cannot be implemented unless effective mechanisms are established for monitoring and enforcement. Interdisciplinary cooperation among economists, lawyers, engineers, etc. is needed to achieve applicable and useful results. That does not mean having scholars of different disciplines working closely together on a day-to-day basis, but rather more mutual awareness and exchange of ideas that will lead to a synergistic effort.

In assessing the economic research efforts addressing the Valley's water and drainage problems we find that much of the effort was allocated to regional and general equilibrium economic impact, mostly within a static framework. There was some effort to study the value of nonagricultural use of the region's water and much effort was allocated to assess the economics of farm-level water conservation and the role of water conservation policies in solving drainage problems. Very little economic research effort was allotted to studying the economics of technological solutions (other than farm level conservation) to the drainage problem. It is difficult to obtain optimal solutions without assessing biological approaches, conjunctive use, and regional transfer of water and drainage residuals. Study of the economics of these solutions has to be emphasized in the future in order to develop an overall plan. While this type of engineering economic analysis may seem mundane, it presents modeling challenges; furthermore, better knowledge of physical and biological relationships may lead to new insights in terms of economic incentives and policies.

Economic modeling should incorporate features of scientific systems and technological solutions into policy analysis. Generic modeling has limited value in generating specific solutions. For example, effective policy assessment requires comparison of the cost of different technologies and identification of locations where they are most desirable. Furthermore, the knowledge of scientific and technological features is essential to designing incentive schemes that will facilitate adoption of appropriate policies. Although the book exhibits state-of-the-art techniques and approaches in water management, many of the economic parameters are still not available for comparing alternatives because much is still unknown.

There is a need for scientific research to be policy relevant. Specifically, scientific research should identify phenomena and estimate parameters that are essential for policy development even though they may be of limited interest from the perspective of the scientific discipline per se. For example, in developing production functions for water and drainage management, one needs to understand how crop water relationships are related to land quality

# FINDINGS AND CONCLUSIONS

and weather. Policy-relevant research may require interaction between scientists of different disciplines addressing the same phenomena from different angles. For example, the construction of simulated crop production function for policy purposes may require the cooperation of soil scientists, entomologists, pomologists, agronomists, engineers, and economists.

Design of methodologies to obtain policy-relevant answers may become a challenge to top scientists and become part of the mainstream research agenda of certain disciplines. This can be done with appropriate incentives. Government agencies, agricultural experiment stations, as well as independent organizations, should establish research teams to conduct applied policy-relevant research projects. It is clear that agricultural experimental stations as well as projects such as the San Joaquin Valley Drainage Program (SJVDP) should be engaged in economics as well as natural science research that is policy relevant to fill in gaps in knowledge that will serve as the basis for policy.

This book is a hybrid of overview and general studies related to the Valley's water and drainage problem, assembled at the end of the SJVDP. In retrospect, it seems that the survey and overview efforts should have been taken in the early stages of the SJVDP so findings could have more directly affected its direction and evolution.

# Index

## A
Aquifer, semiconfined
   ground-water pumping strategies, 75-95
   in the San Joaquin Valley, 32-34
   selenium concentrations, 35-39

## B
Bureau of Reclamation, 578-581, 769, 772-774

## C
Commodity programs, 577-578
Computable General Equilibrium (CGE) model, 490-492, 496-501
   and policy evaluation, 492-496
   Social Accounting Matrix (SAM) in, 499-501
   water cutbacks in the San Joaquin Valley, 501-507
Consumer surplus (*see also* Contingent Valuation Method)
   Hicksian
      willingness to accept (WTA), 390-391, 394, 398
      willingness to pay (WTP), 390-391, 394, 398-402, 413-425, 448-450, 452-458
   Marshallian, 391
Contingent Valuation Method (CVM), 399-403, 413-425, 431-432, 452-455
   closed-ended (dichotomous-choice) questions, 414, 449, 452-453
   empirical evidence of problems, 437-441
   future research needs on, 441-444
   the "good cause" issue, 435-436
   logit model, 453-454
   open-ended questions, 449
   representation problem, 432-435
   validity, 921-926
   voter referendum format, 413-414, 421-423
Crop-water production function (*see* Production functions)

## D
Drainage-water contaminants (*see also* Selenium)
   and avian embryotoxicity
      deriving contamination thresholds, 358-360
      estimating risk thresholds, 360-363
      mathematical model for linking, 357-358
      predicting egg selenium levels, 354-360
      selenium bioaccumulation, 350-354
   bioaccumulation of, 330, 336-337, 340-341
   public-health risks
      exposure routes, 329-330
      populations at risk, 330-331
      selenium, 328, 332-338
      substances of concern, 328-329
Drainage-water management strategies (*see also names of individual strategies*)
   combinations, 132-133
   dilution (blending), 124-125, 129-131
   estimating regional net benefits, 63-68
   evaporation, 123-124, 128-129, 198, 848
   ground-water pumping, 72-81, 84-86, 89-95, 125-126, 131-132, 196-197
   ground-water storage, 849
   incentives
      a priori contractual liability, 543-545
      tradeable water pollution discharge permit (TDP), 541-543
      water markets, 538-541, 545-551, 804-812
      water pricing structures, 277-293, 844
      water transfers, 812-817, 844-845
   institutional
      legal remedies, 846-847
      water districts' role, 845-846
   irrigation efficiencies and, 61-64
   land retirement, 118-119, 126-127, 133-140, 195-196, 845
   master drain, 847-848
   regional analysis and, 52, 60, 68-69
   reuse, 197
   reuse versus dilution (blending), 99-113
   source control, 196, 536-538, 541-543, 843-845
   treatment, 122-123, 127-128
   uncertainty in planning, 189, 201-203
Drainage-water management strategies, evaluating
   dynamic optimization models, 599-613, 639-641, 662-675

economic model, 119-140
regional models, 60-68
  integrating, 563-571
  Westside Agricultural Drainage Economics (WADE), 465-484
by social costs, 126-138
systems approach, 187-203
Drainage-water reuse
versus dilution (blending), 99-100, 112-113
for drainage-water management, 197
economic evaluation, 102, 106-110
field tests, 111-112
on salt-tolerant crops, 110-113
simulation model, 100-106
and water quality, 100, 101
Dynamic optimization models (*see also* Economic models)
conjunctive ground and surface water systems, 618-634
crop rotations, 601-605
extension of Hotelling, 638-659
irrigation investment, 611-612
game theory, for ground water pumping
  analytical framework, 678-680
  assuming myopic behavior, 680-681
  closed-loop Nash equilibrium, 682-687
  open-loop Nash equilibrium, 681-682
  parallels with drainwater generation, 694-695
  trigger-strategy equilibria, 687-694
optimal control, for drainage projects, 675-676
  canal decisions, 665-666
  decision rules, 667-671
  drainage-source region, 662-665
  optimal outcomes, 671-673
  welfare maximization, 666-667
spatial variability, 599-600, 605-611

E

Economic model(s) (*see also model names, types, and applications*)
benefit-cost, for valuation of water use, 781-791
benefits of water-quality changes, 390-404
biomethylation feasibility, 180-181
consumer willingness to pay (WTP), 413-419
critique of Contingent Valuation Method (CVM), 432-444
drainage management strategies, 119-140
drainage-reduction price incentives, 277-285, 290-293
ground-water pumping, 87-95
irrigation technology adoption, 230-248, 297-303
modified production functions, 209-225, 252-271
penalty function, 726-727
reuse and dilution (blending), 100-106
simulation of hydrologic systems, 60-68
simulation versus mass-balance, 52-53
Social Accounting Matrix (SAM), 499-501
uncertainty, 701-704
  in benefit-cost analysis, 704-705
  in expected utility analysis, 705-706
  safety rule approach, 706-715
water marketing in the San Joaquin Valley, 545-551
water quality and appropriative rights, 826-830
wildlife hunting and viewing benefits, 448-460
Economic model(s), dynamic optimization
conjunctive ground and surface water systems, 618-634
drainage canal projects, 662-675
extension of Hotelling, 638-659
game theory, 678-694
spatial variability, 599-613
Economic model(s), regional, 512-514
California Agricultural Resource (CARM), 523, 559-560
Computable General Equilibrium (CGE), 490-507
econometric, 518-519, 522-527
input-output, 515-522
integrating WADE, CARM, and USMP, 563-571
multicrop land allocation, 581-585
United States Mathematical Programming (USMP), 560-563
Westside Agricultural Drainage (WADE), 465-484, 558-559
Environmental policy
and benefit-cost analysis, 917-921
and collective action theory, 914-917
economists' role in formulating, 926-928
evaluating
  drainage problems, 708-711
  pesticide residues in drinking water, 711-713
  shellfish contamination, 713-714,
preventionist and adaptionist, 905-906
research priorities
  farm-level water management, 895-898
  and policymakers needs, 887-890
  reducing costs of technical and institutional changes, 908-910
  and reductions in public research funds, 910-911
  regional water management, 898-900
  role of economics, 937-939
  traditional, 893-895
uncertainty, 701-702
Environmental regulation(s)
Ambient Water Quality Criterion, 328, 376

# INDEX

California Environmental Quality Act (CEQA), 832
Clean Water Act, 536-538
Federal Migratory Bird Treaty Act, 346
Hazardous Waste Management Act, 848
imperfect enforcement of, 717-718
    assumptions of farmers' behavior, 724-726
    fairness as a regulatory goal, 721-722
    penalties for violations, 726-730
    regulatory costs, 722-724
    regulatory goals, 718-721
    water allocation case study, 730-731
Migratory Bird Treaty Act, 843, 848
Toxic Pits Control Act, 848
Environmental resources
as a free good, 907-908
measuring nonuse values for, 433-444
option value for, 412, 433-434, 448
social concern for, 904-905
subsurface drainage reservoirs as nonrenewable, 637-641
typology of values for, 392-393
Epidemiology (*see* Public health risks, assessment techniques)

## G

Geographic information system (GIS), 190-191, 195-196
Government (*see* Environmental regulations, *names of agencies, and policy topics*)
Ground-water pumping
for drainage-water management, 75-79, 89-95, 125-126, 131-132, 196-197
existing strategies and conditions, 72-75, 82-83
future strategies and conditions, 76-79, 84-86
and geological formations, 72-75
and ground-water quality, 79-80
simulation models, 87-95, 682-694
Ground-water quality (*see also* Water quality)
and geological formations, 80-81
pumping strategies, 79-80

## H

Hotelling economic model
adding drainage disposal costs, 649-653
public policy implications, 657-659
San Joaquin Valley example, 653-657
socially optimal versus competitive outcomes, 645-649
socially optimal outcomes, 641-645
subsurface drainage reservoirs as a constraint, 637-641
Hydrologic systems
mathematical description, 58-59
processes, 54-58

in the San Joaquin Valley, 31-34, 143-145
simulation model, 60-68

## I

Irrigation
environmental impacts, 532-534
and hydrologic processes, 54-58
Irrigation management
conjunctive ground and surface water systems
    and drainage management, 627-634
    drainage problems, 625-626
    dynamic model, 618-621
    policy options, 622-624
and crop yield, 209, 210-212, 215
deep percolation, 209-210, 222-223
and drainage disposal costs, 224
fallow season, 221-222
models for assessing strategies, 209-225
salinity, 219-220
and uniformity, 210, 216, 217-218, 223-224

## K

Kesterson Reservoir (*see also* Drainage-water contaminants *and* Selenium), 173-174
avian embryotoxicity, 348, 356-357
biomethylation field tests, 175-176, 181
drainage problems, 859-862
public-health risks, 327-328, 330, 336-337
selenium concentrations in fish, 369-370
Kesterson Reservoir, potential litigation over, 862-864
procedural causes of action, 866-867
in drainage-water treatment, 872-873
in land retirement, 869-870
in regulatory standards, 867-869
in water marketing, 870-872
substantive causes of action, 865-866

## L

Land allocation, behavioral model for multicrop, 581-584
application for California, 585-591
and commodity programs, 584-585
and policy changes, 591-593
Land retirement, 845
compared with options, 118-119, 133-140
in drainage management, 118-119, 195-196
economic model results, 126-127
social costs criteria, 120-121, 126-127

## M

Mathematical models (*see also* Economic models)
estimating salt balance, 152-153
hydrological-political-economic, 735-756
Positive Mathematical Programming (PMP), 467-468
predicting avian egg selenium levels, 357-358

## P

Production functions (*see also* Economic models), 251-252
  Cobb-Douglas, 252
  comparing transient state and seasonal, 215-217
  crop-water, 217-225, 613-614
  data estimation efficiency, 258-261
  dual approach in multicrop (multioutput), 253-254, 260-261, 582-585
  modifications of Cobb-Douglas
    discrete (dichotomous) decisions, 261-262
    environmental pollution, 265-268
    excessive information costs, 264-265
    fixed allocatable inputs, 257
    generic versus specific inputs, 255
    heterogeneity in environmental analysis, 268-270
    multioutput production, 254-256
    nonjointness in inputs and outputs, 256-257
    risk aversion, 262-263
    separability, 255-256
  seasonal, 210-212
  transient state, 212-215
Public health risks
  assessment techniques
    biochemical, 322-325
    dose-response concept, 317-322
    and drainage management, 316-317, 324
  drainage contaminants, 327-341
Public policies (*see* Environmental regulations *and policy topics*)

## Q

Quasi-rent (*see* Technology adoption, irrigation)

## R

Regional analysis (*see* Economic models *and topics of analysis*)

## S

Sacramento River, 370-372, 375-377
Salinity
  and appropriative water use, 821-830
  in the Colorado River, 824-826
  and hydrologic systems, 54-58
  and irrigation management, 219-220
  shallow ground-water, 74-83, 89-95
Salt load
  balance equation, 152-153
  calculating, 153-155
  in the San Joaquin River Basin
    management, 160-165
    net accumulated, 155-157
    trends in accumulation, 158-160
San Francisco Bay, 371-372, 375-376
San Joaquin River, 41-43, 124-125, 129-133, 138-140, 188, 189, 370-372, 374-377
San Joaquin River Basin (*see also* Salt load, in the San Joaquin River Basin)
  salt imbalance, 153-160
  water quality, 149-152
San Joaquin Valley (Valley), 143-144
  agricultural development
    commercial period (1870-1920), 14-22
    expansion period (1820-1870), 10-14
    modern period (since 1920), 22-26
    and population growth, 11, 18
  water use history, 146-152
  drainage problems described, 71-72, 117-118, 187-188, 275-277, 411-412, 841-843
  drainage management options, 843-850
  evolution of Kings River water rights, 761-774
  geological formations, 30-34, 72-75
  ground-water pumping, 72-75
  hydrologic system, 31-34
  simulation models
    drainage-water policies, 708-711
    drinking-water policies, 711-713
    policy impacts, 564-571
    water cutbacks, 501-507, 520-522
    water marketing, 545-551
    water price increases, 522-527
  water allocation and pricing, 534-536
  wildlife hunting and viewing, 450-459
San Joaquin Valley Drainage Program (SJVDP)
  decisionmaking environment, 880-883
  drainage management strategies, 195-198
  Grasslands Subarea, 76-77, 80-81, 84-85
  land retirement analysis, 117-140
  organizational structure, 878-880
  planning process, 188-190
  policies, goals, and work plans, 883-887
  study area, 190
  systems planning approach, 190-203
    conceptual model, 191-194
    geographic information system, 190-191, 195-196
    illustration, 198-201
    problems, 201-203
  Westlands Subarea, 76, 78, 80-81, 85, 191-196, 198-201
Selenium
  concentration
    and saline shallow ground water, 35-39
    spatial variability, 39-40
  concentrations in fish
    estimating risks for fish, 377-379
    estimating risks for humans, 379-380
    future research needs on, 380-382
    in the San Joaquin Valley, 374-377

# INDEX

and salinity, 71-75
  in the San Joaquin River, 41-43
  in the San Joaquin Valley, 117-118, 169-1 170
    and avian embryotoxicity, 346-363
    as a criteria for land retirement, 118, 138-140
    in drainage water, 39-40
    and geological formations, 30
    and ground-water salinity, 35-39
    mobility from soil, 41
    public-health risks, 332-338
    and water management, 44-45
Selenium removal, via biomethylation
  biological processes, 171-173
  environmental parameters, 173-175
  environmental regulations, 176-178
  estimating the feasibility of, 180-181
  field tests, 175-176
  implementation, 178-180
State Water Resources Control Board, 831-834
Statistical models (*see also* Economic models)
  log-likelihood function, 415-419
  logit, 414-419, 421-423, 453-454

## T

Technology adoption
  and land use, 12, 19-21
  machinery, 11-12
Technology adoption, irrigation
  determinants, 300-301
  and drainage disposal costs, 237-243
  economic models, 297-303
  ground-water pumping, 22
  and input and output prices, 232, 236-244, 300, 302-309
  and land quality, 301-308
  normative studies, 304-305
  positive studies, 306-308
  public policy options and, 229-230, 245-248
  quasi-rent as a criteria, 232-234, 238-239
  simulation model, 230-248
  and the social cost of applied water, 637-649, 653-657
  and soil quality, 235-245
  for water conservation, 296-297
  and water prices, 304-306
  and water quality, 237, 244-245
  and weather, 233-235, 237, 240-244

## W

Water and irrigation districts
  Broadview Water District, 285-290
  drainage-reduction pricing policies, 277-293
  economic model, 740-741
  formation, 17-18

  Grassland Water District, 370-372, 374-377, 380
  Kings River Water Association, 761, 771-774
  model of political power structure, 744-745
  role in drainage-water management, 845-846
Water management
  alternative institutional approaches
    applied to California irrigation districts, 797-799
    game theoretic approaches, 794-797
    "market failure, government fix" approaches, 791-793
    modeling valuation of water use, 781-791
    property rights libertarian approach, 793-794
  Computable General Equilibrium (CGE) model, 490-507
  evaluating policies
    econometric models, 518-519, 522-527
    input-output models, 515-522
    regional models, 512-514
    in the San Joaquin Valley, 519-527
  lessons from the San Joaquin Valley, 935-937
  and selenium concentration, 44-45
  theoretical model of political economy, 735-736
    conjunctive water use, 749-752
    economic structure, 739-742
    hydrological-political-economic equilibrium, 746-749
    political power structure, 743-746, 753-756
    regulatory and physical systems, 736-739
  water marketing, 538-541
    in the San Joaquin Valley, 545-551
Water-pricing policies
  California, 535-536
  as incentives for drainage reduction, 277-280
  increasing block-rate structure, 279-280, 284-293
    Broadview Water District, 286-287
    estimating price responses, 290-293
  inverted block (tiered) rates, 477-482, 844
  irrigation district-level optimization, 280-285
Water projects
  Central Valley Project, 22-26, 148-149, 166
    San Luis Drain, 23-24
    San Luis Unit, 23-25, 149
  as a commercial enterprise, 14-19
  and land use, 24
  master drain in the San Joaquin Valley, 23, 847-848

Pine Flat Dam, 761-762, 768-774
State Water Project, 24-26, 166
Water quality
  measuring benefits of changes
    benefits defined, 390-393
    Contingent Valuation Method (CVM), 399-403
    Hedonic Pricing Method, 397-399
    public policy implications, 403-404
    Travel Cost Method, 394-397
  in Western rivers, 823-826
Water rights, 761-763
  appropriative
    State Water Resources Control Board's authority, 831-834
    economic model of water quality and, 826-830
  and water quality, 821-822, 835-836
  evolution of Kings River
    the Federal challenge (1927-1963), 772-774
    institutionalization period (1870-1927), 763-772
  innovative transfer arrangements, 812-817
  markets, 804-806
    environmentally based transfers, 810-811
    legal and political arrangements, 806-808
    transfers of Federal project water, 809-810
    underdevelopment, 811-812
  and neoclassical economic theory, 760-761, 774-775
  in San Joaquin Valley irrigation development, 13-14, 17-18
Water use(s), transfers among competing, 803-817, 844-845
Westside Agricultural Drainage Economics (WADE) model (*see also* Economic models), 465-472, 558-559
  agricultural production model, 472-474
  economic and hydrologic relationships, 482-484
  hydrology model, 474-475
  integrated with other regional models, 563-571
  salinity model, 475-476
  water-pricing application, 477-482
Wetlands and wildlife management, measuring benefits of improvements in
  California CVM survey, 419-424
  comparing benefits to costs, 424-425
  consumer willingness to pay (WTP), 413-414
  Contingent Valuation Method (CVM), 414-419
  log-likelihood function, 415-419
  logit model, 414-419, 421-423
Wildlife hunting and viewing
  effects of drainage water, 458-460
  estimating bird-viewing and hunting benefits, 452-458
  in the San Joaquin Valley, 450-459
  valuing, 448-450

MIX
Papier aus verantwortungsvollen Quellen
Paper from responsible sources
FSC® C105338

If you have any concerns about our products,
you can contact us on
**ProductSafety@springernature.com**

In case Publisher is established outside the EU,
the EU authorized representative is:
**Springer Nature Customer Service Center GmbH
Europaplatz 3, 69115 Heidelberg, Germany**

Printed by Libri Plureos GmbH
in Hamburg, Germany

# THE ECONOMICS AND MANAGEMENT OF WATER AND DRAINAGE IN AGRICULTURE

# THE ECONOMICS AND MANAGEMENT OF WATER AND DRAINAGE IN AGRICULTURE

Edited by

**Ariel Dinar**
University of California, Davis and USDA-ERS
(Formerly University of California, Riverside and
San Joaquin Valley Drainage Program)

**David Zilberman**
University of California, Berkeley

Springer Science+Business Media, LLC

**Library of Congress Cataloging-in-Publication Data**

The Economics and management of water and drainage / edited by Ariel Dinar, David Zilberman.
    p. cm.
    Includes bibliographical references and index.
    ISBN 978-1-4613-6801-4    ISBN 978-1-4615-4028-1 (eBook)
    DOI 10.1007/978-1-4615-4028-1
    1. Irrigation—California—San Joaquin River Valley. 2. Drainage--California—San Joaquin River Valley. 3. Water-supply, Agricultural—California—San Joaquin River Valley. I. Dinar, Ariel. II. Zilberman, David, 1947-
TC824.C2E28   1991
363.73 '94—dc20                                      91-16083
                                                                        CIP

---

**Copyright** © 1991 Springer Science+Business Media New York
Originally published by Kluwer Academic Publishers in 1991
Softcover reprint of the hardcover 1st edition 1991

All rights reserved. No part of this publication may be reproduced, stored in a retrieval system or transmitted in any form or by any means, mechanical, photo-copying, recording, or otherwise, without the prior written permission of the publisher, Springer Science+Business Media, LLC.

*Printed on acid-free paper.*

To our wives, Mati and Leorah, and children:
Roee, Shlomi, and Shira
and Shie, Eyal, and Aytan.

## Acknowledgements

*The editors would like to thank the following:*

The San Joaquin Valley Drainage Program and the U.S. Bureau of Reclamation for funding the research, writing, editing, and design that lead to the publication of this book;

The members of the book advisory committee, Margriet Caswell, Wilford Gardner, Charles DuMars, Herbert Snyder, and Henry Vaux, Jr. for providing helpful comments at the early stages of the book;

The numerous reviewers who helped us with their constructive comments in the review process;

Carmell Edwards of the Bureau of Reclamation's planning unit who edited and coordinated the book production;

The Bureau of Reclamation's word processing unit, directed by Sammie Cervantes and Diana Savignano, for their typing efforts;

Marilyn Murtos of the Bureau of Reclamation's graphics unit for the desktop publishing work and her dedication, without which this book would not have been completed;

And above all, many thanks to the authors for their full cooperation. This book is theirs.

*The editors would also like to acknowledge support provided by:*

Giannini Foundation of Agricultural Economics

School of Natural Resources, University of California, Berkeley

University of California Division of Agriculture and Natural Resources

University of California Water Resources Center

U.S. Department of Agriculture, Economic Research Service, Resources and Technology Division.

## Editorial Staff

Carmell L. Edwards, *Technical Publications Writer*
Marilyn D. Murtos, *Graphic Illustrator*
Sammie C. Cervantes, *Word Processing Supervisor*
Diana L. Savignano, *Editorial Assistant*

# Contents

| | | |
|---|---|---|
| **DEDICATION** | | v |
| **ACKNOWLEDGEMENTS** | | vii |
| **PREFACE**<br>*Jan van Schilfgaarde* | | xv |

### One: BACKGROUND AND SETTING     1

1. Introduction and Overview     3
   *Ariel Dinar and David Zilberman*

2. Irrigation, Drainage, and Agricultural Development in the San Joaquin Valley     9
   *Lloyd J. Mercer and W. Douglas Morgan*

3. Overview of Sources, Distribution, and Mobility of Selenium in the San Joaquin Valley, California     29
   *Robert J. Gilliom*

### Two: ENGINEERING, PHYSICAL, AND BIOLOGICAL APPROACHES TO DRAINAGE PROBLEMS     49

4. Hydrologic Aspects of Saline Water Table Management in Regional Shallow Aquifers     51
   *Mark E. Grismer and Timothy K. Gates*

5. Ground-water Pumping for Water Table Management and Drainage Control in the Western San Joaquin Valley     71
   *Nigel W. T. Quinn*

6. Reuse of Agricultural Drainage Water to Maximize the Beneficial Use of Multiple Water Supplies for Irrigation     99
   *James D. Rhoades and Ariel Dinar*

7. Land Retirement as a Strategy for Long-term Management of Agricultural Drainage and Related Problems     117
   *Craig M. Stroh*

| | | |
|---|---|---|
| 8 | San Joaquin Salt Balance: Future Prospects and Possible Solutions<br>*Gerald T. Orlob* | 143 |
| 9 | Removal of Selenium from Agricultural Drainage Water through Soil Microbial Transformations<br>*Elisabeth T. Thompson-Eagle, William T. Frankenberger, Jr., and Karl E. Longley* | 169 |
| 10 | A Conceptual Planning Process for Management of Subsurface Drainage<br>*Donald G. Swain* | 187 |

**Three: ECONOMICS OF FARM LEVEL IRRIGATION AND DRAINAGE**     207

| | | |
|---|---|---|
| 11 | Crop-Water Production Functions and the Problems of Drainage and Salinity<br>*John Letey* | 209 |
| 12 | Effects of Input Quality and Environmental Conditions on Selection of Irrigation Technologies<br>*Ariel Dinar and David Zilberman* | 229 |
| 13 | Estimation of Production Systems with Emphasis on Water Productivity<br>*Richard E. Just* | 251 |
| 14 | Increasing Block-rate Prices for Irrigation Water Motivate Drain Water Reduction<br>*Dennis Wichelns* | 275 |
| 15 | Irrigation Technology Adoption Decisions: Empirical Evidence<br>*Margriet F. Caswell* | 295 |

**Four: ENVIRONMENTAL AND PUBLIC HEALTH IMPACTS OF DRAINAGE**     313

| | | |
|---|---|---|
| 16 | Assessing Health Risks in the Presence of Variable Exposure and Uncertain Biological Effects<br>*Robert C. Spear* | 315 |
| 17 | Consideration of the Public Health Impacts of Agricultural Drainage Water Contamination<br>*Susan A. Klasing* | 327 |

18  Contaminants in Drainage Water and Avian Risk
    Thresholds
    *Joseph P. Skorupa and Harry M. Ohlendorf* 345

19  Preliminary Assessment of the Effects of Selenium in
    Agricultural Drainage on Fish in the San Joaquin Valley 369
    *Michael K. Saiki, Mark R. Jennings, and Steven J. Hamilton*

## Five: VALUING NON-AGRICULTURAL BENEFITS FROM WATER USE 387

20  Measuring the Benefits of Freshwater Quality Changes:
    Techniques and Empirical Findings 389
    *Richard T. Carson and Kerry M. Martin*

21  Willingness to Pay to Protect Wetlands and Reduce
    Wildlife Contamination from Agricultural Drainage 411
    *John Loomis, Michael Hanemann, Barbara Kanninen,
    and Thomas Wegge*

22  Valuing Environmental Goods: A Critical Appraisal of the
    State of the Art 431
    *H. S. Burness, Ronald G. Cummings, Philip T. Ganderton,
    and Glenn W. Harrison*

23  Economic Value of Wildlife Resources in the
    San Joaquin Valley: Hunting and Viewing Values 447
    *Joseph Cooper and John Loomis*

## Six: REGIONAL ECONOMIC ANALYSIS 463

24  A Regional Mathematical Programming Model to Assess
    Drainage Control Policies 465
    *Stephen A. Hatchett, Gerald L. Horner, and Richard E. Howitt*

25  The Use of Computable General Equilibrium Models to
    Assess Water Policies 489
    *Peter Berck, Sherman Robinson, and George Goldman*

26  Analyses of Irrigation and Drainage Problems: Input-
    Output and Econometric Models 511
    *W. Douglas Morgan and Lloyd J. Mercer*

27  Creating Economic Solutions to the Environmental
    Problems of Irrigation and Drainage 531
    *Marca Weinberg and Zach Willey*

| 28 | Impacts of San Joaquin Valley Drainage-Related Policies on State and National Agricultural Production<br>*Gerald L. Horner, Stephen A. Hatchett, Robert M. House, and Richard E. Howitt* | 557 |
|---|---|---|
| 29 | Cropland Allocation Decisions: The Role of Agricultural Commodity Programs and the Reclamation Program<br>*Michael R. Moore, Donald H. Negri, and John A. Miranowski* | 575 |

| Seven: | DYNAMIC ASPECTS OF IRRIGATION AND DRAINAGE MANAGEMENT | 597 |
|---|---|---|
| 30 | Optimal Intertemporal Irrigation Management Under Saline, Limited Drainage Conditions<br>*Keith C. Knapp* | 599 |
| 31 | Managing Drainage Problems in a Conjunctive Ground and Surface Water System<br>*Yacov Tsur* | 617 |
| 32 | Government Policies to Improve Intertemporal Allocation of Water Use in Regions with Drainage Problems<br>*Farhed Shah and David Zilberman* | 637 |
| 33 | Dynamic Considerations in the Design of Drainage Canals<br>*Ujjayant Chakravorty, Eithan Hochman, and David Zilberman* | 661 |
| 34 | Common Property Aspects of Ground-Water Use and Drainage Generation<br>*Lloyd S. Dixon* | 677 |

| Eight: | UNCERTAINTY, ENFORCEMENT, AND POLITICAL ECONOMY OF WATER SYSTEMS | 699 |
|---|---|---|
| 35 | Determination of Regional Environmental Policy Under Uncertainty: Theory and Case Studies<br>*Erik Lichtenberg* | 701 |
| 36 | Economic Aspects of Enforcing Agricultural Water Policy<br>*Andrew G. Keeler* | 717 |
| 37 | Organizational Failure and the Political Economy of Water Resources Management<br>*Gordon C. Rausser and Pinhas Zusman* | 735 |
| 38 | Water Market Reforms for Water Resource Problems: Invisible Hands or Domination in Disguise?<br>*Norm Coontz* | 759 |

## Nine: INSTITUTIONS, REGULATIONS, AND LEGAL ASPECTS OF WATER AND DRAINAGE PROBLEMS — 779

39 Alternative Institutional Arrangements for Controlling Drainage Pollution — 781
*Alan Randall*

40 Economic Incentives and Agricultural Drainage Problems: The Role of Water Transfers — 803
*Bonnie G. Colby*

41 Water Quality and the Economic Efficiency of Appropriative Water Rights — 821
*Mark Kanazawa*

42 Institutional and Legal Dimensions of Drainage Management — 841
*Gregory A. Thomas*

43 Legal Issues Raised by Alternative Proposed Solutions to Kesterson Water Quality Problems — 859
*Charles T. DuMars*

## Ten: OUTLOOK AND DISCUSSION — 875

44 Management of the San Joaquin Valley Drainage Program: The Dichotomy Between Practice and Theory — 877
*Edgar A. Imhoff*

45 Future Research on Salinity and Drainage — 893
*Henry J. Vaux, Jr. and Kenneth K. Tanji*

46 Irrigation Technology, Institutional Innovation, and Sustainable Agriculture — 903
*Vernon W. Ruttan*

47 Is the Drainage Problem in Agriculture Mainstream Resource Economics? — 913
*V. Kerry Smith*

48 Summary of Findings and Conclusions — 933
*David Zilberman and Ariel Dinar*

INDEX — 941

# PREFACE

Jan van Schilfgaarde, USDA Agricultural Research Service and National Research Council Committee on Irrigation-Induced Water Quality Problems

In 1982, a startling discovery was made. Many waterbirds in Kesterson National Wildlife Refuge were dying or suffering reproductive failure. Located in the San Joaquin Valley (Valley) of California, the Kesterson Reservoir (Kesterson) was used to store agricultural drainage water and it was soon determined that the probable cause of the damage to wildlife was high concentrations of selenium, derived from the water and water organisms in the reservoir. This discovery drastically changed numerous aspects of water management in California, and especially affected irrigated agriculture. In fact, the repercussions spilled over to much of the Western United States.

For a century, water development for irrigation has been a religiously pursued means for economic development of the West. The primary objective of the Reclamation Act of 1902 was, purportedly, the development of irrigation water to support family farms which, in turn, would enhance the regional economy (Worster, 1985).

Early on, it was abundantly clear that irrigation in arid regions would bring with it salinity problems that needed special management. It was 1886 when Hilgard warned that "the evils now besetting [California's Irrigation Districts] are already becoming painfully apparent; and to expect them not to increase unless the proper remedies are applied is to hope that natural laws will be waived in favor of California" (Hilgard, 1886). Over the years, principles of soil and water management to deal with salinity were developed and applied, so that agricultural production could be maintained or enhanced. Far more recently, it became recognized institutionally that such management required drainage and that the disposal of saline drainage water had an adverse effect on the receiving water body. The passage of the Colorado River Salinity Control Act (P. L. 93-320) in 1974 illustrates the point. The discovery of selenium at Kesterson in 1982 added another dimension; to the adverse effects of salinity in drainage water was now added the potentially toxic effect of a number of minor elements--selenium, molybdenum, uranium, as well as boron. These findings led to a growing recognition of contamination problems related to irrigation drainage at other waterfowl refuges. Intensive studies were initiated by the Department of the Interior in Nevada, Utah, Wyoming, Arizona, Montana, Texas, and other areas of California in 1986-87 involving contamination of wetland habitat by irrigation return flows.

During the last two decades, the presumed and actual dominance of irrigation as the preferred (certainly, favored) user of water at the expense of other uses, was beginning to be questioned. Concern with instream uses, the public trust doctrine, competition from municipalities--these and other issues gained in importance in academic discussions as well as in the real world, in public awareness and in court decisions. The loss of birds at Kesterson in 1982 by itself might not have been of great significance had it not been related, in a broader sense, to the reduction of wetlands in California's Central Valley from over 4 million acres to less than 300,000 acres; to the concurrent degeneration of Stillwater Refuge in Nevada; to the stress on migrating birds along the Pacific Flyway; and to the increasing influence of the nonfarming population interested in fishing, hunting, and rafting.

Be this as it may, the alarm set off by Kesterson caused the State and Federal governments to react strongly and, at first, in a manner approaching panic. Ultimately, the San Joaquin Valley Drainage Program (SJVDP) was organized, a group charged with investigating all aspects of the irrigation/drainage problem in the Valley and developing a set of recommendations for resolution of the dilemma. SJVDP had participation from three Federal agencies (U.S. Bureau of Reclamation, U.S. Fish and Wildlife Service, and U.S Geological Survey) and two state groups (California Department of Water Resources and California Department of Fish and Game). SJVDP, in turn, requested advice from the National Academy of Sciences in designing its studies in ways that could be defended in a scientific forum; this led to the Committee on Irrigation-Induced Water Quality Problems of the Water Science and Technology Board, which is part of the National Research Council. This committee argued, as one of the recurring themes during its 5-year life, that SJVDP needed to look beyond the "technological fix" at the many, complex institutional constraints and possibilities for finding rational and equitable solutions, and also at economic reality.

Thus, as chair of this committee, I welcomed the initiative by SJVDP leaders to organize this volume of technical papers and essays that deal primarily with economic and social aspects of the general problem of conflict resolution in water resources, and specifically with the Valley.

As of this writing, no specific solution, or plan of action, has formally emerged as the preferred course for the Valley. This writer doesn't expect any such miracle soon, but he does anticipate the presentation of a set of options that will lead to implementation of a series of steps to help mitigate the problem, together with the realization on the part of the public and of policymakers that complex, multifaceted problems in a dynamic society, at best, lead to gradual corrective steps and incremental improvements. He hopes, and fully expects, that these steps will include adjustments in institutions, including subsidy programs, water quality regulations, water contracts, and many others.

Because of the "Kesterson scare", a great deal of activity was generated both within or through the SJVDP and within academia. New physical and biological processes were analyzed and developed and old ones reevaluated. Effects of foreign substances on wildlife were assessed, and economic models were generated. New impetus was given to evaluating nonmarket values of natural resources and questions were raised about mechanisms for water transfers.

The book that lies before you makes no claim to being the exhaustive and definitive text on institutional or economic management of water quality problems. Rather, it contains a series of chapters that illustrate the diversity and complexity of such problems, while also demonstrating the substantial body of knowledge that is available for their analysis. The book is witness to the significant amount of work done in response to the problems discovered at Kesterson in 1982, often with support of the SJVDP. It also makes clear that the current level of understanding is far from complete. It covers some old material and some new. Some of the conclusions will come as no surprise to knowledgeable readers; others may be unexpected. It certainly surprised this writer to note an assessment that enhancing clean water supplies to the Valley refuges would reap a benefit of $3 billion per year.

The title of this book stresses economics; most of the authors are economists. One might conclude that the primary issues in the Valley are economic. They are not. Yet the tools of economics, when properly blended with the knowledge of geologists, biologists, soil scientists, public health specialists, and engineers, can be used effectively to clarify the issues and to illustrate some of the consequences of choices that must be made. As mentioned, the NAS Committee expressed concern that, without in-depth economic and policy analysis, a viable plan of action was not likely to emerge. This book's emphasis on economic analysis, in conjunction with input from other disciplines, demonstrates that the leaders of the SJVDP shared the committee's concern and responded to it.

As was the purpose of the NRC Committee's report (NRC, 1989), the hope with this book is that the policymakers confronted with similar problems elsewhere in the future will be able to benefit from the experience of the 1980's in the San Joaquin Valley of California. It provided a partial, yet coherent, record of analytical techniques and methods of evaluation applied to, and often developed for, the water management problems there encountered.

## REFERENCES

Hilgard, E. W., 1886. *Irrigation and Alkali in India*. College of Agriculture, University of California, Report to the President of the University, Bulletin No. 86, California State Printing Office, Sacramento, CA, pp. 34-35.

National Research Council, 1989. *Irrigation-Induced Water Quality Problems: What Can Be Learned from the San Joaquin Valley Experience*. National Academy Press, Washington, D.C.

Worster, D., 1985. *Rivers of Empire: Water, Aridity, and the Growth of the American West*. Pantheon Books, New York.

# THE ECONOMICS AND MANAGEMENT
OF WATER AND DRAINAGE
IN AGRICULTURE

# One: BACKGROUND AND SETTING

# 1 INTRODUCTION AND OVERVIEW

Ariel Dinar, University of California,
Riverside and San Joaquin Valley Drainage Program and
David Zilberman, University of California, Berkeley

In his statement, "Man is a complex being: he makes deserts bloom and lakes die," Gil Stern demonstrated two extremes associated with water management. On the one hand, water can be used as a vital source for living. Mismanagement of water, however, can jeopardize the environment along with its animal and human inhabitants.

Water scarcity and quality problems occur around the globe. These problems seem to become even more severe and acute over time to the extent that in many locales both the quality and quantity of water are exhausted. Thus, development of policies and design of institutions to sustain and enhance water resources utilization is critical at farm level as well as at regional and National levels.

Irrigation and drainage issues in the San Joaquin Valley (Valley) and similar sites around the world focus not only on agriculture, but also on environmental, water quality, and public health concerns. Various aspects of water and drainage problems are inextricably related to other resource questions currently on the public agenda. As a result of intensive efforts in the last decade, a substantial amount of new scientific information and analysis related to water resources has become available. Unfortunately, this knowledge is fragmented along disciplinary lines, with very little exchange of ideas among disciplines. This lack of communication increases the difficulty of formulating effective water policy.

Familiarity with all the aspects of the problem is essential to develop comprehensive water policy. Therefore, in this book, we attempt to take stock of the knowledge on water and drainage management from many disciplines and provide a base for systematic policy development.

Our objectives in putting together this book were severalfold: (1) To assemble state-of-the-art knowledge on policy and management of agricultural water and drainage, its methodologies and analytical tools, as well as empirical knowledge related to specific conditions in California and the Western United

States to serve as examples; (2) to use that knowledge in addressing various aspects of the problem including agricultural, environmental, public health, resources use, financial, social, and legal agendas; and (3) to provide insights and identify additional research necessary to fully assess alternative solutions for the Valley's drainage problems and to suggest possible solutions for regions with similar problems. While the book includes input from many scientific disciplines, the economic perspective is emphasized in the analysis and design of policy instruments and institutions.

The book comprises 10 parts that include both conceptual and empirical chapters based primarily on the Valley's experience. The first part provides general information on the Valley's historical and physical setting. Mercer and Morgan introduce the historical perspective of irrigated agriculture and associated drainage problems in the Valley. They show how competing demands for water by agricultural, municipal and industrial, and environmental interests add to the complications of drainage and salinity management. Gilliom provides the basis to understand the nature of selenium accumulation and distribution in Valley soils and drainage water. Based on extensive investigations conducted by the U.S. Geological Survey, he builds the case for maintaining the massive volume of high-selenium ground waters within the soil profile rather than removing it to the surface.

The second part of the book focuses on the various physical approaches to solve water and drainage problems that are currently being considered in the Valley. The Grismer and Gates chapter introduces a comprehensive long-term simulation model that accounts for major processes (including their stochastic nature) governing shallow ground water and the effects of these processes on water table depth, salinity, crop yield, and net farm income. Quinn provides information on the effect of ground-water pumping on depth and quality of the aquifer and the potential of water table management for drainage control. The chapter by Rhoades and Dinar introduces the concept and economic merits of agricultural drainage reuse. Their chapter demonstrates that the benefits to be derived from reuse of relatively low quality drainage water can best be realized by keeping this water separate from freshwater supplies and applying it on salt-tolerant crops. Stroh introduces the concept of land retirement as a drainage control device and analyzes a series of management alternatives such as treatment, dilution, pumping, and evaporation, evaluated against land retirement, to minimize social cost. Orlob presents the concept of restoring and reallocating natural salt loads to control salt balance and water quality within a region. Thompson-Eagle, Frankenberger, and Longley introduce the principles and application of a process to remove selenium from agricultural drainage water through soil microbial transformations. Swain explains and demonstrates the framework used in the planning process by the Federal-State interagency San Joaquin Valley Drainage Program.

Part three of the book presents farm-level studies on selection of irrigation and drainage reduction technologies and management practices under limiting conditions of salinity, water scarcity, and drainage. Letey provides the most recent developments in soil science modeling of multiseasonal crop-water relationships including soil salinity and drainage, and irrigation uniformity. Dinar and Zilberman present a comprehensive irrigation technology choice model that includes effects of input and output prices as well as weather conditions and the dual effects of soil and water quality. Just describes the current state of agricultural production analysis, emphasizing water productivity, and discusses estimation difficulties that arise when externalities (such as drainage and pollution) are the result of the production process. The Wichelns chapter provides the conceptual basis and case study evidence for a water pricing strategy to conserve water and reduce drainage generation. Caswell reviews the literature on adoption of irrigation technologies and its relevance for solving agricultural drainage problems. Her chapter explains factors affecting the adoption of new irrigation technologies and evaluates recent empirical findings.

Part four of the book focuses on environmental and public health impacts of drainage water. Spear introduces principles of modeling health risk. He demonstrates the difficulties of establishing a method to determine contamination standards and discusses the risk and reliability aspects involved. The chapter by Klasing summarizes the research conducted to evaluate the toxicity, metabolism, and environmental fate of drainage water contaminants, as well as potential routes of human exposure. Skorupa and Ohlendorf provide the latest findings on relationships between waterborne selenium and waterfowl egg selenium. They suggest that drainage water with selenium levels at or above 20 p/b should be considered widely hazardous to aquatic birds, and that a reasonable goal for chemical or biological decontamination technologies to prevent most avian toxicity would be waterborne selenium concentrations of <10p/b. To minimize avian contamination, a reasonable goal of purity would be <2.3p/b waterborne selenium. Saiki, Jennings, and Hamilton provide new findings on selenium content in fish from the Sacramento and San Joaquin Rivers systems and the San Francisco Bay complex. They note that high concentrations of environmental selenium can adversely affect the reproduction, growth, or survival of fish, and require public health advisories for people who consume contaminated fish.

Part five of this volume deals with valuing nonagricultural demands for water supplies. Carson and Martin review the techniques and empirical studies used to value nonmarket benefits of water-related public goods and the effects of water quality and quantity changes. Loomis, Hanemann, Kanninen, and Wegge provide the basis for estimating willingness to pay for alternative programs to protect and expand wetlands and reduce contamination of wildlife by agricultural drainage. The chapter by Burness, Cummings, Ganderton, and Harrison provides constructive criticism of the Contingent Valuation Method used to evaluate environmental goods and willingness of the society to pay for protecting the environment. They also focus on future research needed to derive meaningful value estimates for recreation, environmental, and other nonmarket goods. Cooper and Loomis quantify effects of agricultural drainage on the recreational demand for wildlife resources in the Valley.

Part six addresses water and drainage problems from regional and National economic perspectives. Hatchett, Horner, and Howitt present the framework for a mathematical programming model that links agricultural production to ground-water hydrology and drainage, demonstrating use of the model to evaluate the effect of water pricing on drainage reduction. Berck, Robinson, and Goldman explain the principles of a Computable General Equilibrium model and its use to estimate regional responses to water supply reduction. Morgan and Mercer present an input-output model, its principles, and use for regional calculation of changes in water supply. They also provide extensions based on regional econometric estimates. Weinberg and Willey analyze and discuss several policy options that include incentives, water markets, and water allocation as avenues to create economic solutions to environmental problems associated with irrigation and drainage in California. Horner, Hatchett, House, and Howitt present a model in which effects of different policies on the local, regional, and National economy are evaluated in conjunction with water and drainage problems. Moore, Negri, and Miranowski analyze the role of the Federal commodity and Reclamation programs in California. Their chapter also demonstrates the use of a behavioral model to evaluate effects of policy changes on cropland allocation and market prices.

Part seven presents studies on the dynamic aspects of irrigation and drainage management. Knapp introduces a field level dynamic model that considers crop rotations, spacial variability, and investment in irrigation technologies over time, in order to derive optimal production and investment decisions under saline, limited drainage conditions. Tsur surveys the economic literature on conjunctive water use and develops a framework for conjunctive use of ground and surface water when drainage is restricted. Shah and Zilberman develop a regional dynamic framework to evaluate the effect of government policy on the competitive allocation of exhaustible resources, using data from the Valley. The chapter by Charkovorty, Hochman, and Zilberman proposes an analytic framework to examine policy options such as pollution control at source, abatement, and disposal in a dynamic context. Dixon presents a game theory framework to the problem of managing ground water and drainage as a common property resource over time. An application to conditions in Kern County, California, is provided and analyzed.

Part eight emphasizes uncertainty, problems of enforcement, and the political economy of water and drainage systems. Lichtenberg introduces methodology to incorporate uncertainty about public health risk in analyzing regional environmental policies. Several examples of applications to pesticides and water quality regulations are provided to demonstrate tradeoff between agricultural production and public health risks. Keeler reviews the literature dealing with the economics of compliance with and enforcement of environmental regulations. Rausser and Zusman develop a political-economy model in which they demonstrate and analyze how regulatory problems due to unequal distribution of power can lead to failures of the market economy. Coontz presents a case study that demonstrates how growers in the Valley have been separated from control over most water resources through the emergence of powerful organizations that represent specific economic interests. These chapters indicate the need to recognize interdependent interests in developing plans for solving water and drainage problems.

Part nine of the book introduces institutional and legal aspects of regulating water quality and drainage management. Randall reviews alternative institutional arrangements for controlling drainage pollution. Regulations, price administration, property rights, and cooperative approaches are examined at the conceptual level. Colby reviews the mechanisms and benefits associated with water transfers to address water scarcity and agricultural drainage problems. Kanazawa develops a model for economic efficiency in water allocation based on the appropriative right doctrine. The feasibility of incorporating water quality considerations into the administration of appropriative rights in California is discussed in light of the model results. The chapter by Thomas provides a comprehensive discussion of the drainage- and water-related legal aspects of institutional arrangements that must be considered in attempting to solve Valley water and drainage problems. DuMars analyzes the legal consequences of various methods for allocating costs among users of irrigation water and return flows.

The last section of the book is devoted to outlook and discussion. Imhoff shares with us the difficulties in managing and directing a multimillion-dollar drainage study program subject to political and policy constraints from diverse interests. Special attention is given to developing practical management methods from theoretical studies. Vaux and Tanji focus on future research needs, suggesting a greater emphasis on interdisciplinary research to achieve an acceptable balance between agricultural productivity and environmental quality. Ruttan evaluates the information presented in the book and identifies some of the limitations and inadequacies of efforts to date from a policy and institutional perspective. He suggests some of the limitations that need to be overcome as well as identifying areas for future research. Smith addresses similar issues from the point of view of an environmental and resource economist. Finally, we, Zilberman and Dinar, summarize several common conclusions presented by various authors in the book, discuss the nature research efforts to solve water and drainage problems, and propose issues and subjects to be emphasized in future research in agricultural and resource economics.

# 2 IRRIGATION, DRAINAGE, AND AGRICULTURAL DEVELOPMENT IN THE SAN JOAQUIN VALLEY

Lloyd J. Mercer and W. Douglas Morgan,
University of California, Santa Barbara

## ABSTRACT

The irrigation history of the San Joaquin Valley (Valley) is composed of several phases starting with individuals digging ditches and continuing to the present State Water Plan. Various institutions have developed to assist in the expansion of irrigation. These include the water companies of the last century and the water districts created by the Act of 1887. Irrigation has produced an enormous rise in the number of farms, population, and harvested acreage in the Valley. Agriculture in the Valley has changed over time. Initially it was cattle ranching with the prime product being the hides. Around the middle of the last century wheat production began and by the 1870's, following the arrival of the railroad, the Valley was a vast wheat land. Drought and reduced yields resulting from continual cropping began to reduce wheat acreage by the 1890's. This was accelerated by the expansion of irrigation and the increase of intensive agriculture. The major agricultural activities which have developed in the present century are: (1) Dairying, (2) vegetables, (3) orchard crops, and (4) cotton. These four activities accounted for about 82 percent of harvested land in 1987. The great expansion of irrigation was associated with the large increase in ground-water pumpage and the development of the Central Valley Project pumping northern California water to the Valley. The State Water Project now provides additional northern California water to the Valley. The expansion of irrigation has been accompanied by the rise of severe drainage problems which pose a significant threat to the future of the Valley. Continued ground-water pumpage is seriously depleting the region's ground-water basin and has produced land subsidence. The Valley's agricultural future is at risk due to: (1) The demand for water for higher valued uses such as municipal and industrial supplies; (2) the environmental need for additional freshwater to maintain the Sacramento/San Joaquin River Delta and San Francisco Bay Estuary; and (3) the necessity to solve the drainage problem.

## INTRODUCTION

The purpose of this chapter is to examine the growth of the Valley with a focus on the expansion of irrigation and associated problems such as drainage and land subsidence. The development of irrigation and drainage problems will be reviewed in the framework of the Valley's agricultural development.

The irrigation history of the Valley includes seven overlapping phases. In the beginning, individuals dug ditches, or used mining ditches for irrigation. Such efforts initiated irrigation, but were minor in terms of acreage. Second, groups of farmers built ditches and canals. These efforts lacked the formal organization of the commercial enterprises to come next, but produced a significant increase in acreage irrigated. The irrigation systems of Miller and Lux, Carr and Hagin, and Tevis were typical of this form. Third, commercial enterprise entered the picture with the formation of ditch and canal companies. This represented a substantial leap forward, but still left much undone. Fourth, the passage of the Wright Act (1887) allowed water districts to be formed. Although litigation and other problems delayed the start of irrigation for even the best of the water districts for more than a decade, formation of water districts represented another substantial surge in acreage irrigated. The fifth stage was the growth of ground-water pumping after World War I which generated a substantial expansion in acreage irrigated. These first five phases all relied on the water from the Valley's rivers and ground-water aquifer. In the sixth phase, the development of the Central Valley Project provided a substantial infusion of northern California water to the Valley. This imported water, in conjunction with expanding ground-water irrigation, resulted in irrigated acreage almost doubling between 1940 and 1950 and more than doubling between 1940 and 1959. The last phase of irrigation development was the completion of major facilities of the State Water Project which provided another dose of northern California water for the Valley, financed with State funds rather than Federal funding. The following sections will trace the Valley's development through these seven phases, beginning with a brief look at the early days and the first settlers.

## THE EARLY PERIOD (FROM 1820 TO 1870)

Fur trappers ventured into the Valley by 1820. They found the area populated by two native Indian tribes: the Yokut Indians who held the valley floor from what is now Bakersfield to Stockton; and the Miwok Indians who controlled the foothill area from the Fresno River north into the Sacramento Valley.

The first settler to establish a home in the Valley was Jose Noriega, a Spaniard. Based on a Mexican land grant, Noriega built a home and stocked a ranch near Brentwood in 1836. Dr. John Marsh was the first American to become a permanent resident of the valley when he purchased Noriega's ranch, Los Meganos (sand dunes), late in 1837 and lived there until his death in 1857. Marsh began the era of the small, private irrigation projects. Captain John A. Sutter was another early irrigator. In 1839 he hired Spanish speaking Indians to dig irrigation ditches for him.

Mexican land grants were made in the Valley between 1836 and 1846. Under Mexican law an individual grant was limited to 11 square leagues (48,713 acres) with 1 irrigable, 4 dependent on rainfall, and 6 suitable for grazing. After the United States took over California, a total of 813 claims (for 12 million acres) were entered for land grants from the Spanish and Mexican period. The United States Land Commission and the Federal courts finally approved 604 of these claims for a total of about 9 million acres.

Early Valley agriculture by the Americans focused on the production of cattle. Hides were the primary product. Meat was considered to be of little value. Population and demand growth changed this over time. Melons, wheat, and corn were grown on the ranches for private use.

The discovery of gold at Sutter's Mill in January 1848 set off a surge of population growth which created the first large agricultural market. The demand for flour during the gold rush prompted the first attempts at wheat farming in the Valley. Some settlers discovered a better living could be made producing food than scrabbling for the elusive gold.

The initial experiment of John Wheeler Jones who planted 160 acres of wheat near the modern townsite of Manteca in 1855 showed that a bountiful crop of the grain could be grown in the Valley. A second wave of Americans in the 1860's began the wheat era with settlement in the region between the present cities of Stockton, Manteca, Modesto, and Escalon. In the fall of 1870 the McCapes planted the first wheat in Fresno County. The land south of Fresno proved ideal for wheat production. The original shipping point was Wheatville later renamed Kingsburg.

During a cycle of wet years, wheat production expanded enormously. Even without irrigation, wheat yields in the early years were high with 1,088 kilograms to 1,360 kilograms (a bushel is 27.2 kilograms) per acre reported; however, continued production of wheat depleted soil fertility and yields by 1890 were down to 272 kilograms per acre or less. Shallow plowing and lack of fallow by early producers contributed to the deterioration of yield. Drought ruined many promising crops.

The expansion of wheat production was associated with rapid advancement of machinery. A combination seeder and harrow was invented in 1860 as was the gang plow. In the beginning, harvesting primarily utilized the McCormick reaper and some stationary threshing was also done. The combine was developed between 1858 and 1868 with the manufacture of the standard combine starting in 1876. By the 1890's the combine was predominant. After 1900, steam and then gasoline tractors began to replace horses and mules. The catapillar tractor was invented in the San Joaquin Valley and the Catapillar Tractor Company formed.

The San Joaquin Valley was a huge wheat field by 1874 and California produced the most wheat of any state in the Nation in 1874 and 1875. Wheat production continued to grow with census figures on wheat acreage showing a growth of wheat acreage to 1890. In that year the total for the six counties was a little over 1.5 million acres compared to just under 200,000 acres in 1987. The decline of wheat started with the growth of intensive agriculture fueled by irrigation.

As settlement developed following the gold rush, county organization occurred. Initially California had only 27 counties, with San Joaquin as one of those original counties. Mariposa County dominated the entire Valley floor as well as the mountains and deserts to the Nevada border. New counties were created by dividing Mariposa. Tulare was organized in 1852 followed by: Stanislaus (previously part of Tuolumne County) in 1854; Merced in 1855; and Fresno in 1856. At the southern end of the Valley, Kern County, one of the State's largest counties with 8,160 square miles, was formed in 1866. Kings County was detached from Tulare County in 1893 and in 1909 a portion of Fresno County was transferred to Kings County.

Population in the Valley grew 21-fold between 1860 and 1920 compared with population growth of 9-fold Statewide during that time period. Except for the decade 1870 to 1880, rapid growth was continuous from 1860 to 1920. During this same period, expansion of irrigation produced growth in the number of farms. Between 1880 and 1950 the number of farms in the Valley grew from 4,511 to 38,593--an eight and a half fold expansion. Local urban populations and activities to service the agricultural sector grew rapidly. This commercial-industrial expansion fed by agricultural growth is indicated by the fact that the urban population represented 35.2 percent of Valley population in 1920. See table 1.

Two conditions are necessary for intensive agriculture: (1) Irrigation and (2) adequate transportation to move crops to market. The coming of the railroad in 1870 provided transportation, which fueled the rise of population and wheat output in the next decade and a half. The Central Pacific Railroad pushed by the Contract and Finance Company reached the Stanislaus River by the end of 1870 and was extended from Lathrop to Goshen by 1972. Construction south of Goshen was taken over by the Southern Pacific Railroad (owned by the same people) and reached Bakersfield in 1874. The Southern Pacific received land grant aid of 10 alternate sections per mile on each side of the rail for its construction.

Table 1. San Joaquin Valley population growth.

|      | *Population* | *Percent of California* |
|------|-----------|----------------------|
| 1860 | 22,064    | 5.8                  |
| 1870 | 44,150    | 7.9                  |
| 1880 | 65,096    | 7.5                  |
| 1890 | 113,162   | 8.8                  |
| 1900 | 136,805   | 9.2                  |
| 1910 | 253,373   | 10.7                 |
| 1920 | 468,725   | 13.5                 |

In 1896, a second railroad, the San Francisco and San Joaquin Valley Railroad, was built from Stockton to Bakersfield by Claus and John D. Sprechels. Branches connected Fresno with Tulare and Visalia. This rail was sold to the Santa Fe in 1898 and the Valley was then served by two major transcontinentals.

The railroads created the main towns and cities of the Valley. Merced, founded by the Contract and Construction Company, became the largest town in the county and the county seat in 1875. The Central Pacific laid out a town named Ralston in honor of one of the railroad's directors. Ralston refused the honor and in light of his modesty the town was named Modesto. The first load of Valley wheat was shipped by rail from this town on October 27, 1870, to Oakland. Riding horseback through the Valley looking for townsites, Stanford of the Southern Pacific, was impressed with a green wheat field in an arid waste and chose the site that became Fresno for the Southern Pacific's railway yards.

Although American settlers in the Valley had little or no experience with irrigation, it was soon obvious that irrigation would be required for long-term successful agriculture. The natural disaster of drought in this arid land highlighted the need for irrigation.

Water rights was an important issue in irrigation. The first California legislature in 1850 provided that the common law of England should govern wherever applicable if no statute had been specifically enacted. This action was to muddle the question of water rights in irrigation. Common law recognized the supremacy of riparian rights. Owners of lands lying along flowing streams were entitled to a full and undivided flow of that stream irrespective of the needs of nonriparian owners.

Miners in the gold rush were the first to make use of dams and flowing streams. They ignored riparian rights and gave precedence to the doctrine of appropriation. The taking of water was considered to be fine so long as it was for a beneficial purpose. Land riparian to the streams worked by miners belonged to the United States. The violation of riparian rights was ignored. The miners use of streams by appropriation established a precedent which made possible the first irrigation attempts.

In 1872, the California legislature adopted by statute the doctrine of appropriation. The California Supreme Court in Lux v. Haggin ruled in 1886 that riparian rights were dominant. The legislature responded by amending Civil Code 1410 to read that the use of water may be acquired by appropriation. This establishment of appropriation rights by the legislature was a challenge to the judicial doctrine of riparian rights. Dams were built and blown up. Litigation between irrigationists and cattlemen proceeded apace. In time, litigation led to appointment of a watermaster to dole out the water of an area.

In the early days, abandoned mining ditches were used for diversion of water to irrigate crops. Small scale ditches dug by individuals and later by groups of farmers provided the early irrigation in the Valley. Early attempts at irrigation were small and sporadic. A major expansion of irrigation began about 1870.

## THE MIDDLE PERIOD (FROM 1870 TO 1920)

Irrigation after 1870 took the form of commercial enterprise. Three distinct types operated in the Valley, involving:

1. Payment by irrigators to a canal or irrigation company on a quantitative basis with a specific charge per second-foot ($ft^3/s$) or acre-foot.
2. The sale of water rights to irrigators by a canal or irrigation company plus an annual fee for services rendered. Companies which at one time came under this classification included the Fresno Canal and Irrigation Company, the Consolidated Irrigation Company (and later District), and the Crocker-Hoffman Land & Water Company.
3. Irrigation enterprises which had as their primary purpose the sale of their landholdings. The purchaser of land secured a share in the water system. Examples were the 76 Land & Water Company with its 30,000 acres around the town of Traver and the Patterson Land Company with 19,000 acres in Stanislaus County.

South of the San Joaquin River Basin in the Tulare Lake Basin the closed river systems of the Kern, Tulare, and Kings Rivers created several inland lakes and a vast swamp. Reclamation of this swamp became an important goal of the local and Federal governments. The Kern Island Irrigating Canal Company began work on a 15-mile canal in 1870. This canal was completed in 1874 when the railroad reached Bakersfield. The construction of the canal led the Kern County Board of Supervisors to declare that the swampland in District 111 was reclaimed. By the late 1870's, the Kern Island Irrigating Canal possessed the capacity of irrigating 80,000 acres of land, making it one of the largest irrigation projects in the Valley at that time.

Kern County was the location of the first experiments in growing what would become one of the major crops of the Valley: cotton. Chester and Jewett of the Livermore and Chester Firm planted 133 acres of cotton in 1865. Their success demonstrated that cotton could be grown on lands adjacent to the Kern River. A second cotton planting occurred in 1872 when the California Cotton Growers and Manufacturers Association planted 1,200 acres. This group was dissolved the following year because costs exceeded returns.

In Merced County an even bigger irrigation project than the Kern Island Irrigating Canal Company was attempted when William Collier formed the Robla Canal Company in 1872. In 1873, the Farmer's Canal Company took over from Collier. This company built 50 miles of canals, but could not complete the project because of inadequate funding, a common plight of the canal companies in the 1870's and 1880's.

In 1868, Moses J. Church began construction of the Centerville Ditch, strongly opposed by the area's cattlemen. Church organized the Fresno Canal & Irrigation Company in 1870. This system delivered water to what is now the city of Fresno by way of Fancher Creek. A. Y. Esterby used that water to grow the wheat which resulted in the founding of Fresno.

San Francisco capitalists organized the San Joaquin & Kings River Canal Company in 1871 to irrigate land between Firebaugh and Newman. A dam was erected at the junction of the San Joaquin River and the Fresno Slough. This was the largest canal project in the Valley at that time. Prior to completion, the company sold the project to Henry Miller and Charles Lux, large landholders who had acquired substantial acreage in overflow land in the swamp where the lower Kern River emptied into Tulare Lake. Miller and Lux employed almost 400 men to complete the canal which was extended to Newman in Stanislaus County in 1878. The completed project was one of the most extensive in the Nation and irrigated 153,000 acres for the first time.

In 1875, O. P. Calloway set out to irrigate 35,000 acres of land north of the Kern River. When Calloway failed due to a lack of capital, the purchasers of his land gained a vested right to water along with the land. J. B. Hagin and W. B. Carr acquired his lands and the uncompleted project which they incorporated as the Kern River Land & Canal Company. The project carried water for 30 miles and irrigated 13,000 acres.

The Emigrants' Ditch was organized in 1876 on land north of the Kings River. Its success led to creation of the Centerville & Kingsburg Irrigation and Ditch Company. In the same year, the Herndon Ditch was begun in the area north of Fresno.

The year 1876 was a busy one on the irrigation scene. The Pioneer Ditch Company, a cooperative irrigation effort by small farmers, began that year. Hagin and Carr became stockholders in the Pioneer Ditch Company and advanced the funds necessary for its completion and also helped bring the Stine and Buena Vista Canals into operation. These canals plus the Pioneer irrigated about one-quarter of Kern Island and a little land north of the river.

Cooperative efforts again bore fruit when the Farmer's Canal Company built a canal to irrigate land in the vicinity of Panama. The Beardsley and McCord ditches were also constructed by small settlers and were capable of irrigating 10,000 acres.

C. H. Hoffman organized the Merced Canal and Irrigation Company in 1882 and bought out the Farmer's Canal Company. He brought in a San Francisco financier, William H. Crocker, and the Crocker-Hoffman Land and Water Company was formed. This company continued canal construction from 1882-88. The present Main Canal from the Merced River through the two tunnels to Lake Yosemite was completed. Gradual expansion of the Crocker-Hoffman system continued until 1922 when the Merced Irrigation District took over.

In October 1882, the 76 Land & Water Company began construction of what would be the largest irrigation canal in California. The plan was to irrigate 30,000 acres near Traver. Irrigation which began in 1884 made the Traver region supreme as wheat growing country. However, future drainage problems of the Valley were foreshadowed when irrigation ruined the land by forcing alkali to the surface. The 76 Land & Water Company eventually sold its canal system to the Alta Irrigation District.

The spread of irrigation systems is indicated by data from the 1890 census. In Kern County, there were 16 canals north of the Kern River and 15 canals south of the river. In Fresno County, there were 16 important irrigation systems taking water from the Kings, San Joaquin, and Fresno Rivers. These systems comprised 750 miles of canals and over a thousand miles of distributing ditches.

Laws of 1850 and 1855 encouraged the range cattle industry by declaring all unenclosed land to be legally free commons. Farmers had to fence their property to keep cattle out until the California legislature passed a no-fence law in 1874 which forced the cattlemen off the alluvial plains and stopped large hog drives. This law replaced one passed in March 1870 entitled, "An Act to Protect Agriculture and to Prevent the Trespassing of Animals Upon Private Property" which had applied only to a few counties. The 1874 law required the rancher to keep his livestock off the farmer's land and meant that fences were not required by farmers to protect their lands. This act and the growth of irrigation fostered the drive toward intensive agriculture.

Commercial projects in the 1870's and 1880's added significantly to acreage under irrigation in the Valley. The 1890 census reports 466,602 acres irrigated in the seven counties. By 1900, this acreage had grown to 723,219 although the totals shown for both Kern and Tulare Counties declined between 1890 and 1900. In 1910, a little over 1.3 million acres were irrigated. In the 20 years between 1890 and 1910, a total of 877,923 irrigated acres were added. The census reports 173,793 acres irrigated by irrigation districts in 1910. The almost threefold increase in acreage in the two decades received a boost from the irrigation district movement, although the bulk of the increase is apparently due to private efforts. After 1910, irrigation districts provided a vehicle for expansion of irrigation beyond the horizons possible with commercial enterprise.

In the 1870's and 1880's, persons knowledgeable about water studied European irrigation systems. In Spain, they found a majority of proprietors could develop an irrigation enterprise and compel the minority to share the expense. This idea appeared to have potential for California. On April 1, 1872, the California legislature passed "An Act to Promote Irrigation by the Formation of Irrigation Districts." Despite the best of intentions, however, this law remained practically inoperative.

The question of irrigation districts kept recurring in the legislature. In 1885, another strong effort was made to get an irrigation district law. The difficulty with these legislative attempts was that the districts to be created would rely on appropriative rights for water. Holders of riparian rights opposed extension of appropriative rights. The bill of 1885 was defeated in the legislature by strong opposition from riparian interests.

Democrats in Stanislaus County nominated C. C. Wright for the California Assembly in 1886. Wright campaigned solely on the promise to put through a law for irrigation and was elected by a large majority. Wright's bill in the legislature finally succeeded and was signed into law by the governor on March 7, 1887. The Wright Act provided for the formation of irrigation districts by a majority of the freeholders with the costs to be borne in equal proportion by those who benefitted. The districts were to be quasi-public corporations with the power to issue bonds and to tax. The 1890 census reports 26 irrigation districts. Various weaknesses in the Act and in the districts led to the failure of all but four in short order. Of 57 districts formed after 1887, only 9 were still in operation in 1915. In the next 40 years, about 100 new irrigation districts were formed.

Riparian interests took the Wright Act to court and challenged its constitutionality. Litigation continued for 9 years despite the fact the judiciary rebuffed the efforts of the anti-irrigationists until the United States Supreme Court affirmed the constitutionality of the Wright Act in 1896. The Wright Act was amended by the legislature the next year to correct some of its weaknesses and was further improved by the legislature in 1910.

As one of the most successful early water districts, Modesto provides an example of the changes in Valley irrigation and agriculture wrought by the Wright Act. Passage of the Act did not mean immediate irrigation; progress in the Modesto District was slow. A bond issue of $800,000 was approved December 19, 1887. On July 14, 1888, the board authorized issuance of only $500,000 in bonds to fund the District's works and on July 20, 1889 the board responded to requests to exclude the lands of 25 owners (about 28,000 acres) from the district. From the start, the Stanislaus River was the favored source of water. However, in 1890, the board voted to take water from the Tuolumne River as a cooperative effort between the Modesto District and another successful district, the Turlock District.

On June 18, 1890, the board purchased the Wheaton Dam and its associated water rights. The dam, built in 1852 and used by gold miners, was found to be inadequate for the purposes of the district. The Modesto and Turlock Districts jointly constructed La Grange Dam on the Tuolumne River a short distance from the town of La Grange. On its completion in 1893, La Grange Dam was the highest overflow dam in the world.

Meanwhile, anti-irrigationists, primarily cattlemen and riparian interests, fought on two fronts: the courts and the polls. The Modesto Irrigation District was continuously involved in litigation during its early years. Anti-irrigationists were elected to the board of directors of the irrigation district producing a board which did nothing. Work on the canals did not continue. During the next 4 years (1897-1900), the new board majority refused any tax levy for support of the District. The board failed to issue a call for the required general election in 1901, at which time pro-irrigationists took the matter to court and won. With the election of February 5, 1901, three pro-irrigationists were elected to the five member board. A notice was sent to the contractor to complete work on the District's canals on September 21, 1901 and both the lower and upper canals in the District were completed in 1903. In 1904, irrigation water began to flow in the Modesto District. This took 17 years after passage of the Wright Act in one of the most successful water districts.

After 1900, the irrigation districts provided the base for a new expansion. The period 1900-20 was one of rapid land subdivision, population growth, and growth of intensive agriculture in Stanislaus County as well as the other counties of the Valley. In Stanislaus County, this growth was propelled by the increased irrigation made possible by the Modesto Irrigation District. Many of the early intensive agriculturalists were European immigrants who had experience with irrigation and changed cropping patterns to vegetables and other intensive agriculture.

By 1920 the Crocker-Hoffman system irrigated 50,723 acres, but available water was insufficient compared to the potential land available for irrigation. The Merced Irrigation District with 171,700 irrigable acres was formed in 1919 to solve this problem. The District purchased the Crocker-Hoffman system for $2.5 million in 1922. A new, larger dam, Exchequer Dam, was built several miles further up the Merced River. A total of $16.5 million was voted to build the dam, related structures, and irrigation ditches. Construction was completed in 1927.

The census numbers clearly show the development of intensive agriculture (Table 2). In 1890, Stanislaus County had 2,699 dairy cows; by 1920, the County had 55,292. Similar expansion of dairy operations occurred in the other six counties with the smallest increase (a mere doubling) in Kern. Table 3 demonstrates the rising importance of the Valley as a percentage of total California agriculture. For milk cows, the Valley had a mere 7.6 percent of the State's total in 1880, but almost a third (32.9 percent) in 1910. Except for a decline in the 1940's, Valley dairy figures have remained around one-third of the State's total to the present day.

In 1890, Stanislaus County had 306,248 acres of wheat. This fell to 51,397 acres in 1920. With the exception of Kings and Kern Counties a similar pattern of declining wheat production is shown in the other counties. Wheat in the Valley relative to the State has fluctuated widely, but declined from over one-half in 1890 to just over one-third in 1987.

Again with the exception of Kings and Kern, hay production about doubled or more than doubled in all the counties. The acreage of irrigated alfalfa increased particularly fast, reflecting the growth of the dairy industry. The Valley hay production represented about 20 percent of the State total between 1890 and 1920, but this fraction doubled by 1987.

Over time Valley vegetable production increased and now represents about 40 percent of the State. The census reports vegetable acreage increased rapidly in San Joaquin, Stanislaus, Fresno, and Merced Counties.

Prior to 1920, rice and cotton acreage was minimal. Today, rice is still minor in the Valley compared to the State, but Valley cotton dominates with over 90 percent of the State total for 1987. Orchards, the other major modern crop in the Valley, was almost nonexistent up to 1920. However, by the 1930's and 1940's, the Valley represented about one-third of the State's orchards, expanding to almost 60 percent by 1987. Apparently cotton and orchards, the two big crops of the present day (56.9 percent of harvested acreage in 1987), require more water than was previously available.

Table 2. Agricultural statistics San Joaquin Valley[a].

| | 1880 | 1890 | 1900 | 1910 | 1920 | 1930 | 1940 | 1950 | 1959 | 1969 | 1978 | 1987 |
|---|---|---|---|---|---|---|---|---|---|---|---|---|
| Farms, number | 4,511 | 8,164 | 11,448 | 21,099 | 31,392 | 36,370 | 35,328 | 38,593 | 32,947 | 28,451 | 26,635 | 29,004 |
| Average size[c] | 817.8 | 637.3 | 662.3 | 308.1 | 219.9 | 199.5 | 210.5 | 272.4 | 302.0 | 378.1 | 378.1 | 335.0 |
| Average value, dollars | 9,063 | 18,282 | 11,613 | 16,429 | 28,873 | 20,420 | 15,321 | 42,463 | 104,622 | 186,551 | 397,808 | 624,542 |
| Harvested Cropland[c] | -- | -- | -- | -- | -- | 2,023,203 | 2,229,309 | 3,006,089 | 3,766,091 | 3,524,336 | 4,407,982 | 3,966,672 |
| Irrigated land[c] | -- | 466,602 | 723,219 | 1,344,525 | 1,951,733 | 2,236,360 | 1,706,137 | 3,199,574 | 3,766,091 | 3,736,014 | 4,494,834 | 4,015,257 |
| Cattle other than milk cows[d] | 120,088 | 254,867 | 289,457 | 529,433 | 338,991 | 331,360 | 427,255 | 505,538 | 942,571 | 1,237,584 | 1,168,516 | 1,246,420 |
| Milk cows[d] | 15,978 | 31,003 | 81,689 | 126,776 | 262,837 | 193,703 | 218,791 | 276,242 | 304,986 | 291,937 | 432,981 | 610,798 |
| Wheat[c] | 461,933 | 1,515,362 | 1,177,596 | 153,257 | 286,071 | 164,490 | 184,637 | 89,004 | 81,349 | 118,396 | 110,399 | 199,517 |
| Barley[c] | 117,165 | 225,397 | 295,928 | 381,716 | 272,232 | 312,328 | 487,675 | 561,896 | 662,891 | 471,335 | 444,747 | 105,165 |
| Rice[c] | -- | -- | -- | -- | 3,026 | 12,911 | 21,986 | 30,106 | 50,734 | 38,700 | 39,755 | 22,252 |
| Cotton[c] | -- | -- | -- | -- | 7,786 | 229,148 | 262,808 | 560,181 | 514,018 | 562,152 | 1,267,563 | 1,006,224 |
| Hay[c] | 78,049 | 258,809 | 433,594 | 469,973 | 432,482 | 474,977 | 457,622 | 523,589 | 592,383 | 552,955 | 571,024 | 645,426 |
| Vegetables[c] | -- | -- | 15,152 | 41,060 | 44,819 | 61,362 | 118,959 | 228,255 | 264,332 | 228,630 | 305,580 | 340,592 |
| Orchards[c] | -- | -- | -- | 781[b] | 442[b] | 569,387 | 506,571 | 586,228 | 674,109 | 852,685 | 1,089,273 | 1,252,345 |

[a]Includes San Joaquin, Stanislaus, Fresno, Merced, Kings, Tulare, and Kern Counties
[b]Small fruits only
[c]Acres
[d]Number

Table 3. San Joaquin Valley agriculture as percent of California[a].

| | 1880 | 1890 | 1900 | 1910 | 1920 | 1930 | 1940 | 1950 | 1959 | 1969 | 1978 | 1987 |
|---|---|---|---|---|---|---|---|---|---|---|---|---|
| Farms, number | 12.6 | 15.4 | 15.8 | 23.9 | 26.7 | 26.8 | 26.6 | 28.1 | 33.2 | 36.5 | 36.4 | 34.8 |
| Average size[a] | 177.0 | 157.0 | 167.0 | 97.2 | 88.0 | 89.1 | 91.5 | 102.0 | 81.4 | 82.4 | 84.6 | 91.0 |
| Average value, dollars | 124.0 | 138.7 | 119.0 | 99.9 | 109.4 | 81.0 | 94.1 | 105.0 | 101.7 | 85.7 | 76.3 | 107.0 |
| Harvested cropland[a] | -- | -- | -- | -- | -- | 35.1 | 34.1 | 37.8 | 46.9 | 46.1 | 50.1 | 52.3 |
| Irrigated land[a] | -- | 46.5 | 50.0 | 50.5 | 46.3 | 47.1 | 39.9 | 60.0 | 60.6 | 51.6 | 52.8 | 52.9 |
| Cattle other than milk cows[d] | 26.6 | 24.3 | 25.4 | 32.9 | 27.6 | 27.1 | 30.0 | 24.7 | 32.0 | 32.4 | 32.0 | 35.6 |
| Milk cows[d] | 7.6 | 9.8 | 26.6 | 27.1 | 33.7 | 32.8 | 35.8 | 38.7 | 40.5 | 45.0 | 51.5 | 57.1 |
| Wheat[c] | 25.2 | 53.3 | 43.9 | 35.8 | 26.3 | 26.4 | 31.3 | 14.3 | 24.1 | 29.3 | 18.6 | 35.5 |
| Barley[c] | 20.0 | 27.6 | 28.7 | 31.9 | 25.1 | 29.3 | 39.2 | 37.4 | 44.0 | 47.7 | 50.8 | 38.9 |
| Rice[c] | -- | -- | -- | -- | 2.3 | 15.7 | 21.0 | 9.9 | 16.8 | 9.9 | 8.2 | 5.6 |
| Cotton[a] | -- | -- | -- | -- | 8.9 | 76.4 | 83.1 | 65.0 | 62.6 | 84.7 | 83.3 | 92.8 |
| Hay[c] | 10.3 | 18.1 | 21.0 | 18.6 | 19.6 | 26.0 | 29.5 | 31.3 | 36.3 | 36.0 | 37.1 | 42.1 |
| Vegetables[a] | -- | -- | 58.6 | 27.0 | 20.6 | 15.5 | 25.2 | 42.5 | 40.2 | 33.8 | 33.7 | 39.5 |
| Orchards[c] | -- | -- | -- | 8.1[b] | 5.6[b] | 35.5 | 35.8 | 39.9 | 47.0 | 53.7 | 56.7 | 58.2 |

[a]Includes San Joaquin, Stanislaus, Fresno, Merced, Kings, Tulare, and Kern Counties
[b]Small fruits only
[c]Acres
[d]Number

## THE MODERN PERIOD (SINCE 1920)

Between 1910 and 1920 irrigated acreage in California increased by 1.555 million acres (including 607,208 in the Valley). The number of irrigated farms in the Valley rose from 15,140 in 1909 to 25,506 in 1919, an increase of over 68 percent. Even with a decline in Valley irrigated land after 1920, the total growth between 1910 and 1940 (from 1,344,525 to 1,706,137 irrigated acres) represented about 27 percent of the State.

A major factor in the growth of irrigation was improvements to the centrifugal pump developed shortly after World War I. Ground-water irrigation began in the Valley in the 1880's, but was limited to a few hundred wells before the invention of the centrifugal pump. The average depth of well in Kern County in 1890 was 551 feet with each well irrigating 125 acres. The 1900 census reports 152,566 acres irrigated in California with wells and tunnels. Well irrigation grew to 868,000 acres in 1920 and 1,463,272 acres in 1930. The Valley had 10,296 wells in 1920 (40.5 percent of the State) and 23,939 in 1930 (51.2 percent of the State). With the advent of the centrifugal pump, many thousands of deep wells were drilled. By the mid-1920's, California surpassed Iowa as the richest agricultural State in the Nation. When pumping began, the Valley underground aquifer may have held three quarters of a billion acre-feet of water, collected over thousands of years. Annual ground-water pumping increased from perhaps 100,000 acre-feet to about one million acre-feet by the 1950's. It has almost doubled since then. In the drought year of 1976, 6 million acre-feet were pumped in lieu of surface water. Overdraft in the valley is now about 1.5 million acre-feet per year. The water table has dropped sharply in the last 70 years and land subsidence has become a problem especially in Fresno and Merced Counties.

The significant growth of irrigation still left much land dry. Thought turned to ways to increase water for the Valley. Millions of acre-feet of water from northern California rivers ran to the sea every year presenting an inviting source for those desiring additional water for the Valley. Increased water supply for the Valley would come from importing Sacramento Valley surface water and from increased ground-water pumpage.

The first State Water Plan went to the California legislature in 1931. The Central Valley Project (CVP) was initially authorized as a State undertaking in 1933 with $170 million in revenue bonds approved by the voters. The project was to develop the agricultural potential of the Central Valley. Because of the condition of the bond market, the State was unable to proceed. Thus, the Central Valley Project Act was passed by Congress in 1935. The State plan became a Federal undertaking to be constructed and operated by the Bureau of Reclamation.

The initial facilities of the CVP included three dams (Shasta, Keswick, and Friant) and five canals (Madera, Friant-Kern, Delta-Mendota, Contra Costa, and Delta Cross Channel). In 1949, Congress authorized the construction and operation of the American River Division including Folsom Dam and Powerplant, Nimbus Dam and Powerplant, and the Sky Park Unit. The Sacramento Valley Canals Unit (Corning Canal, Tehama-Colusa Canal, and Red Bluff Diversion Dam) was authorized in 1950. The Trinity River Division (Trinity Dam, Lewiston Dam, Clear Creek, and Spring Creek Tunnels) was authorized in 1955. The San Luis Unit was authorized in 1960 as a joint Federal-State venture. This included San Luis Dam, Forebay Dam, 123 miles of canal, a pumping generating plant, and three pumping plants. Additional planned facilities included the Auburn Dam and Reservoir on the north fork of the American River. Construction was begun in 1967 on Auburn Dam, but was stopped in 1976. Completion is still under study.

In the late 1940's and early 1950's, the Army Corps of Engineers won a long battle with the Bureau of Reclamation and built Pine Flat Dam on the Kings River and Isabella Dam on the Kern River. This was followed with the Corps' construction of Terminus Dam on the Kaweah River and Success Dam on the Tule River. These projects made possible the irrigation of tens of thousands of acres of new land.

The current problem of drainage in the Valley first appeared in the 1880's. Irrigation caused the water table to rise and as early as 1886 there were reports of salinization and waterlogging in low-lying areas. The problem is a shallow and impermeable clay layer underlying much of the Valley. A few thousand acres are already out of production as a result of the drainage problems. Lost acreage could rise exponentially in coming decades without a drainage solution. By 1910 and 1911, salt and drainage problems had become acute in the Modesto Irrigation District. The U.S. Department of Agriculture demonstrated the use of subsurface tile drainage lines in 1909. Drains were built in the period 1908-14, followed by a more complete drainage system built in 1917. After this system was completed (1918-19) the situation improved.

In 1920, drainage enterprises in California comprised 3,010 miles of open ditches, 85.6 miles of tile drains and 1,131 of levees with an additional 204 miles of open ditches, 23 miles of tile drains and 120 miles of levees under construction. Of the 1,108,319 acres included in these drainage projects statewide, 270,626 were in the Valley.

Since completion of the San Luis Canal, subsurface irrigation return has caused a rise in the shallow water table and accelerated migration and concentration of soluble salts near the Valley trough. A major facility, referred to in the authorization of both Federal and State Water Projects as the "master drain," was to collect subsurface drainage water for disposal outside the Valley. This facility was never built. The San Luis Drain (to run north 110 miles from near Kettleman City to Kesterson Reservoir and 78 miles from Kesterson to the Delta) was required by the 1960 Federal law authorizing construction of the San Luis Unit. By 1975, only 85 miles of the San Luis Drain terminating in Kesterson Reservoir and 120 miles of collector drains were built. The drainage problem has been compounded since 1982 with the discovery of migratory bird deaths and deformities at Kesterson Reservoir. Evaporation and solar ponds are widely used by individual farmers in Kings and Kern Counties to handle drainwater. In mid-1984 five State and Federal agencies formed the San Joaquin Valley Drainage Program to investigate drainage problems on the west side of the Valley and to identify possible solutions.

The CVP greatly increased the irrigation water flow to the Valley. (Some State Water Project water began to flow to the Valley in 1968, but the amount is trivial by comparison.) The results of the CVP are demonstrated by the census figures. In 1940, there were 1,706,137 acres irrigated in the seven counties. By 1969, this expanded to 3,736,014 acres, more than doubling of irrigated acreage as the result of the CVP water and increased ground-water pumpage.

The changed mix of agricultural output also illustrates the substantial increase in irrigation water. Dairying which had become a major industry continued its upward trend with the number of dairy cows in the seven counties growing from 218,791 in 1940 to 291,937 in 1969. Not too surprisingly, hay output followed a similar pattern growing from 457,622 acres to 552,955 acres. Wheat continued its downward slide from 184,637 acres to 118,396 acres while barley held its own, falling only from 487,675 acres to 471,335 acres.

The other big winners were cotton, vegetables, and orchards. Cotton increased more than 100 percent from 262,808 acres in 1940 to 562,152 in 1969. Vegetables almost doubled from 118,959 acres to 228,630 acres. Orchards expanded substantially from 506,571 acres to 852,685 acres. Dairying, cotton, vegetables, and orchards accounted for about 82 percent of harvested acreage in the Valley in 1987.

Completion of the major Central Valley Project facilities left several tributaries of the Sacramento River flowing as wild rivers. Eager eyes turned northward in search of more water. The Feather River Project (State Water Project) was approved by the California legislature in 1951. The State Water Project includes 23 dams and reservoirs, 4 main canals and aqueducts, 6 powerplants and 22 pumping plants. The key to this project was construction of Oroville Dam on the Feather River. Dams and reservoirs would also be constructed on the upper tributaries of the Feather River at Lake Davis, Frenchman Lake, Antelope Lake, Dixie Refuge, and Abbey Bridge. The State Water Project was approved by the voters on November 8, 1960.

The State Water Project has entitlements for 4.23 million acre-feet of water annually, but at present can only deliver about 2.5 million acre-feet annually.

Full production by the State Water Project would require substantial new investment. About 59 percent of the project water is to be delivered south of the Tehachapi Mountains into Southern California with the Metropolitan Water District of Southern California receiving just over 2 million acre-feet or almost 50 percent of the project water. The San Joaquin Valley has entitlements for 1.355 million acre-feet annually (32 percent of the project water) of which Kern County Water Agency (the second largest contractor in the project) has the largest entitlement of 1.153 million acre-feet annually. This is enough water to irrigate 600 square miles.

Increases in irrigated acreage and crops since 1969 primarily reflect: (1) Additional CVP water produced by the Trinity River Division and the San Luis Unit, and (2) State Water Project water. Between 1969 and 1978, 758,820 additional acres were irrigated in the Valley with increased acreage of 72,225 in San Joaquin County; 33,884 in Stanislaus County; 106,529 in Merced County; 107,879 in Fresno County; 189,067 in Kings County; 64,354 in Tulare County; and 184,882 in Kern County.

Both harvested acreage and irrigated land declined significantly in the Valley between the 1978 and 1987 census. Cattle other than milk cows continue to expand reaching a peak number (1,246,420) in 1987. The number of cattle in the Valley now far exceeds the number in the ranch days of the last century. The dairy industry continues its rapid expansion with milk cows more than doubling from 291,937 in 1969 to 610,798 in 1987. Since the dairy industry suffers from excess capabilities in the United States, this continued rapid growth of the industry in the Valley is both troubling and surprising. Hay acreage continues to grow, but only by 13 percent (in 1987). Wheat holds its own and even rises slightly in the period. Barley acreage declines to less than one-quarter of its 1969 level by 1987. Rice reaches a peak in 1978 then declines sharply to 1987. Vegetables and orchards production continues to climb with vegetables increasing almost 50 percent between 1969 and 1987. Orchards also expand by almost half the period. Vegetables and orchards are not subsidized crops and their rapid growth in the Valley is a benefit to society. Cotton more than doubles between 1969 and 1978 then declines to about 80 percent above the 1969 level in 1987.

The history of the Valley shows the huge gain of irrigation to an arid, but fertile area. From the beginning it was clear that the agricultural potential of the Valley required water. At the start individuals and then groups worked to provide water. Their efforts were expanded by commercial enterprises in the 1870's and 1880's. The Wright Act (1887) created a mechanism to further expand irrigation. Ground-water pumpage increased starting in the 1880's. These first five phases of the valley's history all relied on the water of the Valley.

The next two phases (the Central Valley Project and the State Water Project) have relied on redirecting northern California water to the Valley. The result has been creation of an agricultural wonderland.

In the coming years it will be difficult to maintain the present agricultural level of the Valley because of: (1) The demand for water by higher valued uses, such as municipal and industrial supply; (2) the environmental need for additional water in the San Joaquin Delta and San Francisco Bay; (3) the necessity to find a solution to the drainage problem--a solution which may significantly affect agriculture in the Valley. Because of these factors, the Valley's history is likely to enter into a new phase in the not too distant future.

## REFERENCES

Bain, J. et al., 1966. *Northern California's Water Industry*. John Hopkins University Press, Baltimore, MD.

Beltrami, L. P., 1957. *The Modesto Irrigation District: A Study in Local Resource Administration*. M.A. Thesis, University of California, Berkeley.

California State Water Resources Control Board, 1987. *Regulation of Agricultural Drainage to the San Joaquin River*. Technical Committee Report.

Cooper, E., 1968. *Aqueduct Empire*. The Arthur H. Clark Company, Glendale, CA.

Cooper, M. A., 1954. *Land, Water, and Settlement in Kern County, California*. M.A. Thesis, University of California, Berkeley.

De Roos, R., 1948. *The Thirsty Land*. The Stanford University Press, Stanford, CA.

Fredericks, A., 1928. *Development of Irrigation in California*. M.A. Thesis, University of Southern California.

Gaffney, M., 1969. Economic Aspects of Water Resource Policy. *American Journal of Economics and Sociology*, XXVIII, pp. 131-144.

Graham, C. J., 1957. *The Settlement of Merced County, California*. M.S. Thesis, University of California, Los Angeles.

Graham, R., 1946. *An Epic of Water and Power: A History of the Modesto Irrigation District*. M.A. Thesis, University of the Pacific, Stockton, CA.

Harding, S. T. et al., 1950. The Central Valley Project and Related Problems, *California Law Review*, XXXVIII, pp. 547-793.

Henley, A. T., 1957. The Evolution of Forms of Water User Organizations in California, *California Law Review*, XLV, pp. 665-675.

Hutchison, C.B., 1946. *California Agriculture*. University of California Press, Berkeley.

Israelson, O. W., 1914. *A Discussion of the Irrigation District Movement*. M.S. Thesis, University of California, Berkeley.

Jewell, M. N., 1950. *Agricultural Development in Tulare County, 1870-1900.* M.A. Thesis, University of Southern California.

Kratka, G., 1937. *Upper San Joaquin Valley, 1771-1880.* M.A. Thesis, University of Southern California.

Meier, E., 1955. *Irrigation in West Fresno County, California.* M.A. Thesis, University of North Dakota, University, ND.

Reisner, M., 1987. *Cadillac Desert.* Penquin Books, New York, NY.

Rhodes, B., 1943. *Thirsty Land: The Modesto Irrigation District: A Case Study of Irrigation Under the Wright Act.* Ph.D. Dissertation, University of California, Berkeley.

Seckler, D., 1971. *California Water.* University of California Press, Berkeley.

Shaw, J. A., 1973. Railroads, Irrigation, and Economic Growth: The San Joaquin Valley of California, *Explorations in Economic History,* X, pp. 211-227.

Smith, W., 1939. *Garden in the Sun.* Powell Publishing Company, Los Angeles, CA.

Smith, W., 1932. *The Development of the San Joaquin Valley of California, 1772-1882.* Ph.D. Dissertation, University of California, Berkeley.

Smythe, William E., 1969. *The Conquest of Arid America.* University of Washington Press, Seattle.

Teilmann, H., 1963. The Role of Irrigation Districts in California's Water Development, *American Journal of Economics and Sociology,* XXII, pp. 409-415.

*U.S. Census of Agriculture.* Decenniel volumes for 1880-1987.

Wood, H. J., 1957. The Evolution of Forms of Water User Organizations in California, *California Law Review,* XLV, pp. 665-675.

# 3 OVERVIEW OF SOURCES, DISTRIBUTION, AND MOBILITY OF SELENIUM IN THE SAN JOAQUIN VALLEY, CALIFORNIA

Robert J. Gilliom, U.S. Geological Survey

## ABSTRACT

Soils in the tile-drained areas of the western San Joaquin Valley that yield high-selenium drainwater were derived from Coast Range marine sedimentary formations, were naturally saline, and probably contained abundant soluble selenium. Decades of irrigation have redistributed the most soluble forms of selenium from the soil into ground water and have caused the water table to rise 1 to 4 feet per year. The rising water table has caused a large area of farmland to require artificial drainage of ground water that contains high concentrations of selenium. The present areal distribution of selenium in shallow ground water reflects the natural distribution of saline soils and the depth distribution of selenium in ground water reflects the history of irrigation. The large volume of high-selenium ground water makes it desirable to leave this water where it is, rather than bring it to the land surface or allow it to move into parts of the aquifer that might be used for water supply.

## INTRODUCTION

Agricultural drainage problems in the San Joaquin Valley (Valley) have attracted National attention since 1983, when selenium in water from subsurface tile drainage systems was found to have toxic effects on waterfowl at Kesterson Reservoir (Kesterson). Other constituents of drainwater, particularly arsenic, chromium, boron, molybdenum, and dissolved solids, also cause water quality problems in some areas of the Valley, but selenium is the most widespread problem and the most limiting constraint on management alternatives.

The purpose of this chapter is to summarize the results of U.S. Geological Survey studies and related research by others on (1) the sources, distribution, and mobility of selenium in soils and ground water of the Valley, particularly the central part of the western Valley, and (2) the sources and concentrations of selenium in the San Joaquin River, where drainwater from about 77,000 acres of farmland is discharged. This broad overview of hydrologic and geochemical aspects of selenium contamination associated with subsurface agricultural drainage in the western Valley, which is summarized from Gilliom et al. (1989), provides the underpinnings for understanding the specific economic and management issues that are evaluated in more detail in the following chapters.

## GEOLOGIC SOURCE OF SELENIUM AND ITS DISTRIBUTION IN SOILS

Water quality problems associated with selenium are most likely in areas of the Valley where soils are formed of sediments from marine sedimentary rocks of the Coast Range (Barnes, 1985). The occurrence of Coast Range sediments and the highest soil selenium concentrations are clearly linked throughout the Valley (Tidball et al., 1986). Three areas of the western Valley--the alluvial fans near Panoche and Cantua Creeks in the central western Valley, an area west of the town of Lost Hills, and the Buena Vista Lake Bed area--have the highest soil selenium concentrations (figure 1). High concentrations of selenium occur in subsurface drainwater from some agricultural lands near, but not necessarily within, all three areas.

The central part of the western Valley is the largest area of the valley that has concentrations of total selenium in soil that exceed 0.36 mg/kg, the 90th percentile of concentrations in all Valley soils (Tidball et al., 1986). Within the central western Valley, concentrations of selenium are highest in soils between the alluvial fans of Cantua and Panoche Creeks, east of Monocline Ridge. Marine shales exposed in the area of Monocline Ridge probably are the primary sources of sediments that contribute to high total selenium concentrations in these soils.

Although high concentrations of selenium occur in shallow ground water and drainwater in parts of the central western Valley and in the other areas that have soils with high total selenium concentrations, a close spatial correlation between the locations of high-selenium soils and high-selenium ground water is not observed. Such a correlation does not occur because the distribution of soluble forms in soil can be different from the distribution of total selenium and only the soluble forms affect concentrations in ground water.

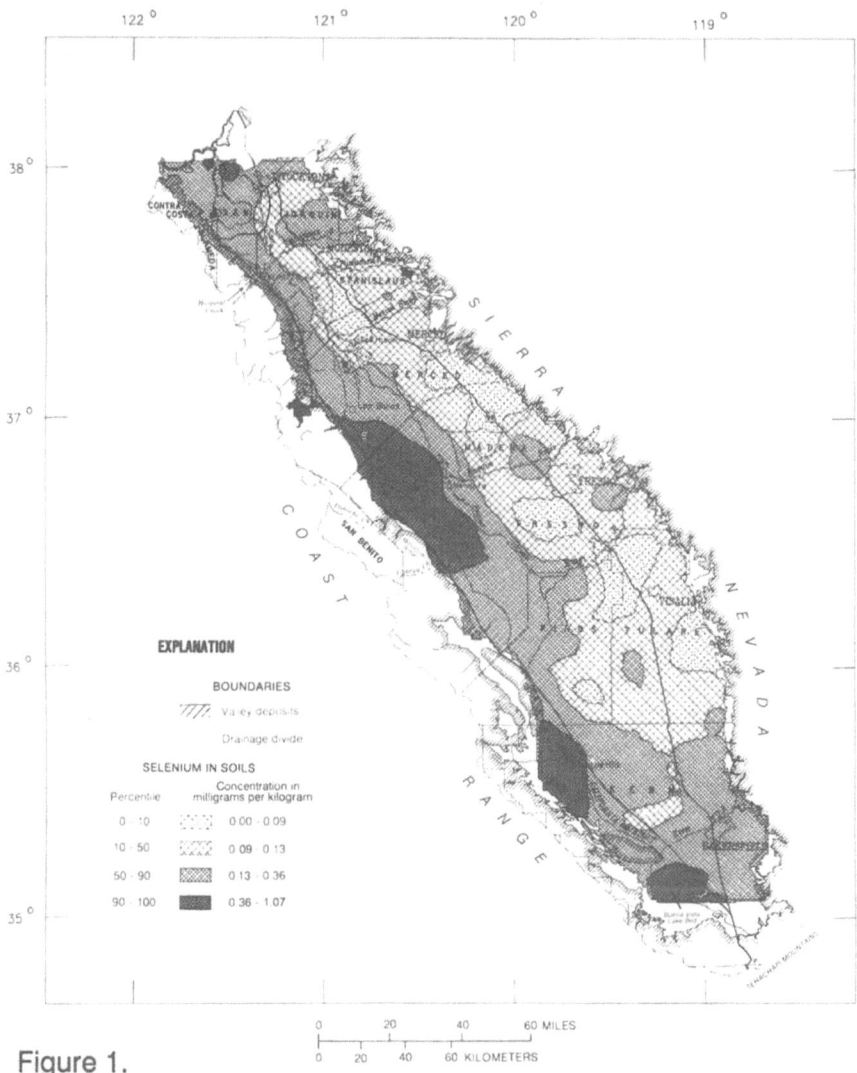

Figure 1.

## GROUND-WATER FLOW SYSTEM

One key to evaluating the origin and present-day distribution of agricultural drainage problems and high concentrations of selenium in ground water is to understand changes in the ground-water flow system that have occurred through time from natural to present-day conditions. These aspects of the flow system have been evaluated and described by Belitz and Heimes (1990). In the central western Valley, the Corcoran Clay Member of the Tulare Formation divides the ground-water flow system into a lower confined zone and an upper semiconfined zone (figure 2). The focus of this chapter is on ground water in the Coast Range alluvial sediments and Sierra Nevada sediments of the semiconfined zone. Under natural conditions, recharge to the semiconfined aquifer was primarily by infiltration of water from intermittent streams. Discharge of ground water under natural conditions was primarily by evapotranspiration and streamflow along the Valley trough.

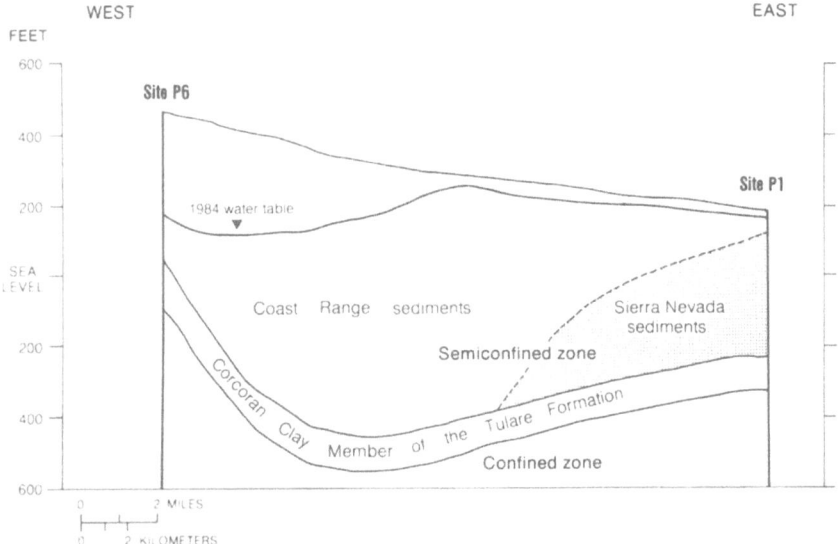

Figure 2.

The dominant influences on the ground-water flow system since the early 1900's, when irrigation began, have been increased recharge resulting from percolation of irrigation water past crop roots, and historic pumping of ground water from the confined aquifer beneath the Corcoran. The increased recharge and pumping between 1912 and 1967 simultaneously caused the water table to rise over a large part of the western Valley and resulted in a substantial component of downward ground-water flow. Importation of surface water beginning in 1967 led to further increases in application of irrigation water, and hence increased rates of recharge to the system, compared with earlier periods when ground water was the primary source of irrigation water. Concurrently, pumpage from below the Corcoran decreased since the mid-1960's, which has caused a decrease in the downward head gradient. The decreasing gradient causes a reduction in the downward flow of ground water through the Corcoran and causes the water table to rise more rapidly. The altitude of the water table has risen 40 feet or more over much of the central western Valley since 1952, causing more than 100,000 acres of farmland to have subsurface drainage problems (figure 3).

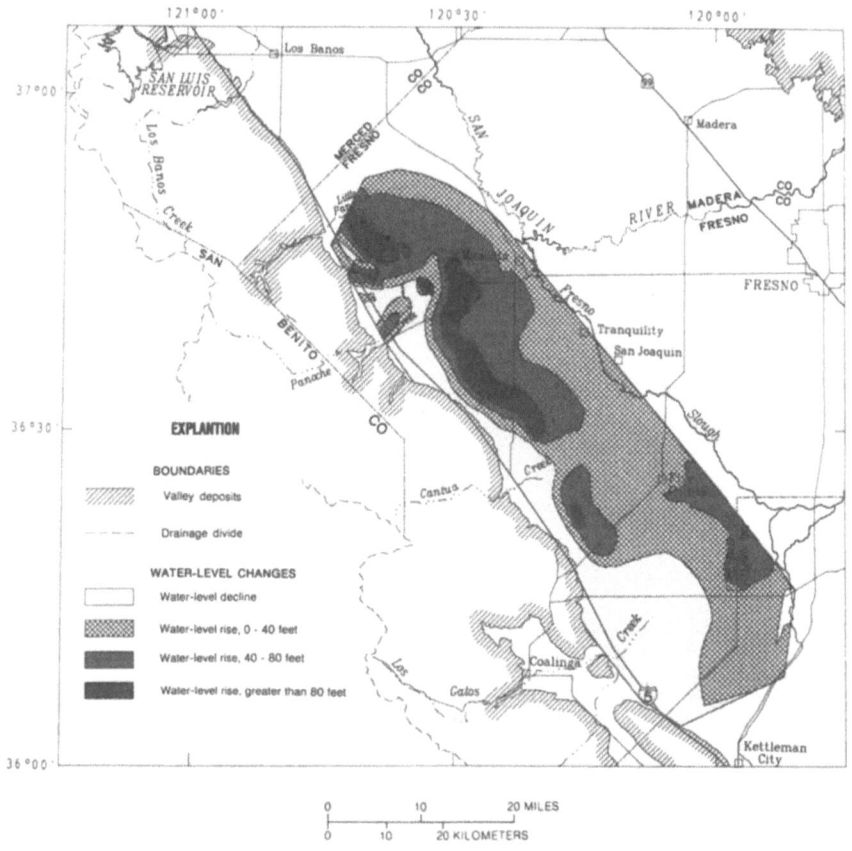

Figure 3.

A prominent feature of the present-day water table is the ground-water divide that parallels the Western boundary of the alluvial fans of the central western Valley (figure 4). The ground-water divide shifts westward between the major alluvial fans of Cantua, Little Panoche, Los Gatos, and Panoche Creeks, and shifts eastward near the upper parts of these fans. To the east of the ground-water divide, the water table lies at shallow depths, is a subdued replica of the topography, and the horizontal component of ground-water flow is eastward and northeastward. Ground-water flow occurs eastward across the

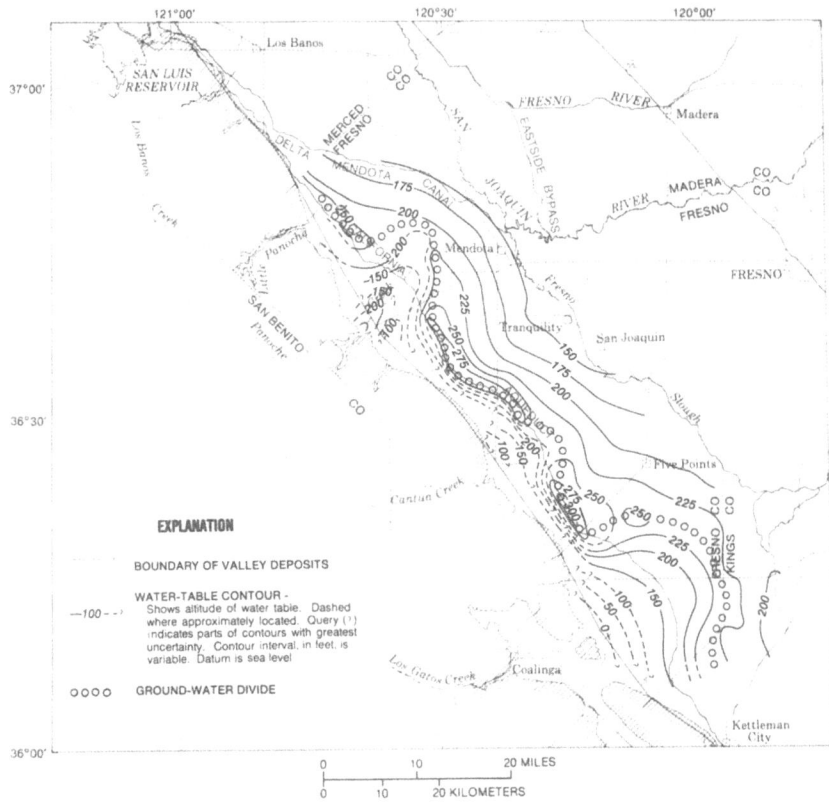

Figure 4.

Valley trough toward active production wells completed above the Corcoran in the Sierra Nevada sediments of the eastern Valley and downward toward the confined zone. West of the ground-water divide, the water table slopes steeply to the west, and flow is toward the west and downward toward the confined zone.

In the past, subsurface tile drainage systems have been installed in some areas east of the ground-water divide that have a shallow water table. Drainage systems have been effective in removing enough shallow ground water to maintain the water table at desired depths in the confines of drained areas, but their use is limited because of problems associated with disposing of the poor-quality shallow ground water that they collect.

## SELENIUM IN GROUND WATER

The feasibility of potential modifications or alternatives to conventional drainage systems for water-table control depends on the areal and depth distribution of selenium in ground water (Dubrovsky and Deverel, 1989). The highest concentrations of selenium in ground water occur in the upper part of the semiconfined zone in Coast Range alluvial sediments (figures 5 and 6). The upper part of the semiconfined zone contains water that originated from irrigation recharge. In the Coast Range alluvial sediments, irrigation water applied since 1952, which can be identified by the presence of tritium, has reached depths of at least 6 to 161 feet below the water table, with a median of 50 feet. Irrigation water applied before 1952 probably occurs in a 10- to 50-foot interval below these depths. Selenium concentrations in the upper part of the semiconfined zone, which ranged from less than $10\mu g/L$ to more than $1,000\mu g/L$, are correlated with ground-water salinity and the presence of oxic conditions (Deverel and Millard, 1988 and Deverel and Fujii, 1988).

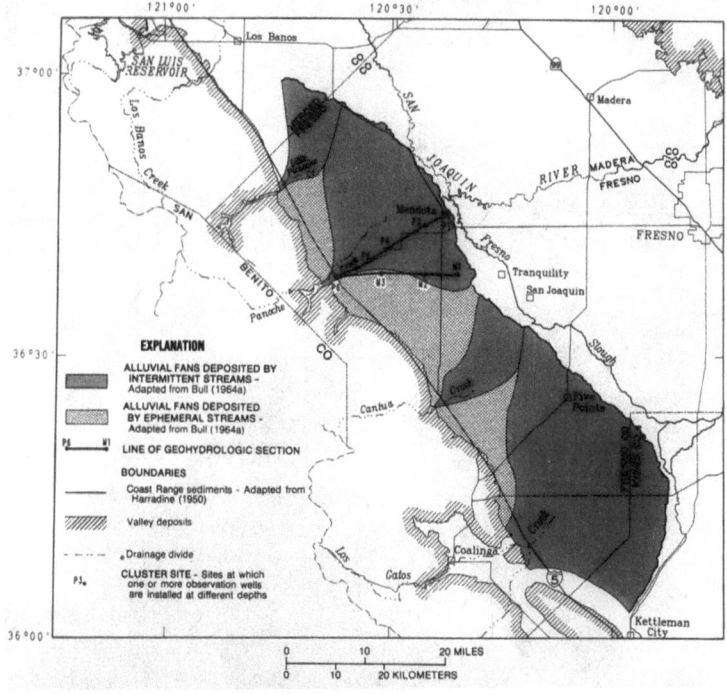

Figure 5.

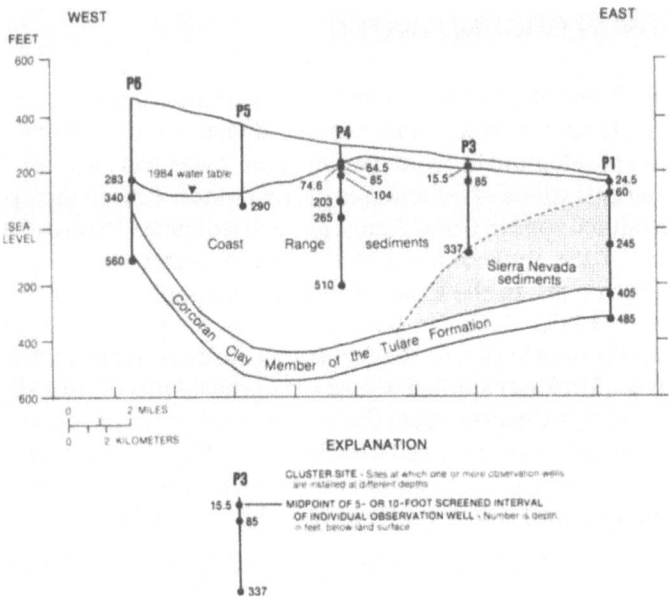
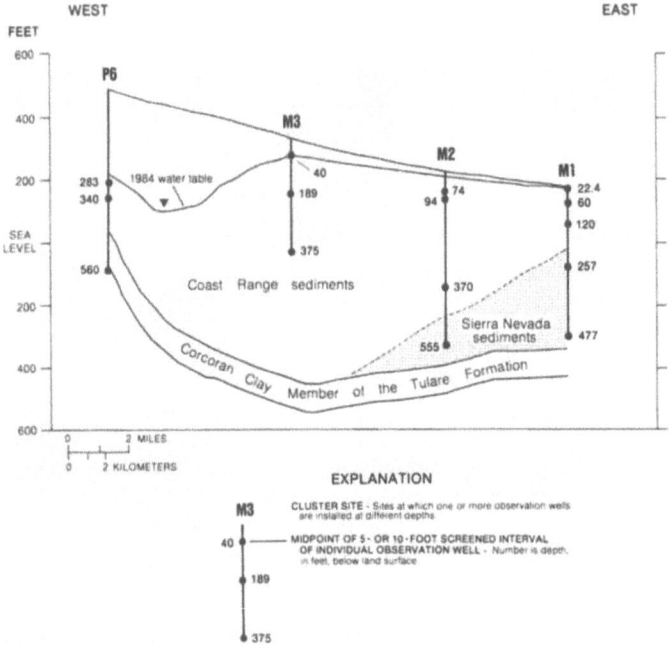

Figure 6.

Samples of native ground water that underlies the irrigation-derived water in the Coast Range alluvial sediments had selenium concentrations less than 10 $\mu$g/L. Selenium concentrations did not exceed 2 $\mu$g/L in ground water in the confined zone or in Sierra Nevada sediments of the semiconfined zone. The low concentrations in this ground water are attributable to reducing conditions and the low availability of selenium in these sediments. Selenium may be removed from solution if oxic ground water in Coast Range sediments moves into reduced Sierra Nevada sediments.

Two processes have had the greatest effects on the distributions of salinity and selenium concentrations in irrigation-derived ground water in the Coast Range alluvial sediments: leaching of soil salts and soluble selenium by infiltrating irrigation water and evaporative concentration. The areal distribution of selenium concentrations in shallow ground water (figure 7) generally correlates with soil salinity before agricultural development, reflecting the leaching

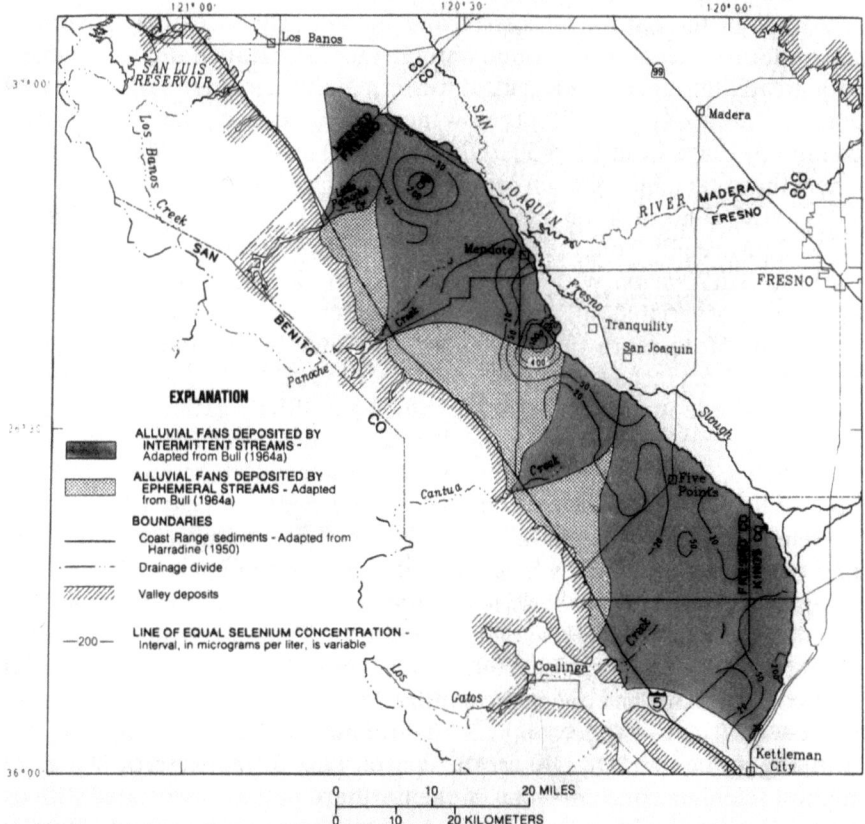

Figure 7.

of natural soil salts by irrigation. In areas where the water table has been near the land surface for extended periods, the highest concentrations of selenium have developed in shallow ground water as a result of evapotranspiration of ground water. Selenium concentrations of 20 $\mu$g/L to more than 1,000 $\mu$g/L occur in shallow ground water in the lower parts of Cantua, Little Panoche, Los Gatos, and Panoche Creek alluvial fans, and between the Panoche and Cantua Creek fans, where natural soil salinity was high and where the water table has been shallow for many years in some places (Deverel and Gallanthine, 1988).

In some irrigated areas where the water table is deep enough that evaporative concentration has not been substantive, recently recharged ground water near the water table has lower selenium concentrations than underlying ground water from early irrigation recharge. In these areas, most soluble forms of selenium already have been leached from the soil. This condition is common in the middle fan areas associated with the four largest streams, where selenium concentrations generally are less than 20 $\mu$g/L in shallow ground water (Deverel and Gallanthine, 1988).

Although the natural processes and human influences that govern the distribution of selenium in ground water in the Coast Range alluvial sediments vary greatly throughout the central western Valley, general patterns are evident. Within about 10 to 20 feet below the water table, selenium concentrations commonly range from 10 to 50 $\mu$g/L, but are 10 to 100 times higher than this where the water table has been near the land surface for an extended period and evaporative concentration has occurred. Water in this shallowest interval is derived principally from the most recent irrigation recharge, probably during the past 10 to 20 years. Within the range of 20 to 150 feet below the water table, an interval of variable thickness occurs in which selenium concentrations are commonly 50 to more than 1,000 $\mu$g/L. Water in this interval is derived principally from recharge of early irrigation water. Selenium concentrations in both of these depth intervals that are associated with irrigation recharge are in the highest part of the stated concentration ranges where natural soils were most saline. Native ground water, with selenium concentrations less than 10 $\mu$g/L, is below the ground water derived from irrigation recharge.

Selenium concentrations are low in ground water of most of the northern part of the western Valley, which is mainly in the western parts of Merced, San Joaquin, and Stanislaus Counties, and do not pose the same management problems as do the higher selenium concentrations that occur in the central western Valley (Dubrovsky, 1989). One existing production well in the semiconfined zone and one in the confined zone had selenium concentrations exceeding 10 $\mu$g/L--the present U.S. Environmental Protection Agency (EPA) drinking-water standard. As in the central part of the western Valley, the highest selenium concentrations in the northern part are associated with oxic Coast Range alluvial sediments and are correlated with salinity. With the exception of shallow ground water, the highest selenium concentrations in the semiconfined and confined zones were near Crow Creek, which contains higher selenium concentrations than most other Coast Range streams. A key finding from study of the northern part of the western Valley is evidence that recent irrigation water has moved into deep parts of the semiconfined zone and even into the confined zone. The deepest penetration probably is limited to the vicinity of wells, which have provided a pathway for rapid downward flow.

## SELENIUM IN TILE DRAINWATER

Effective management of existing tile drainage systems and assessments of the use of tile drains as a management option for the future require an understanding of how tile drains interact with the ground-water flow system and the levels and variability of selenium concentrations likely to occur in drainwater. Selenium concentrations vary greatly among drainage systems, from tens to thousands of micrograms per liter, but tend to be relatively consistent over time in drainwater from a particular system and are correlated with drainwater salinity (Deverel et al., 1989). The exception to their consistency over time is the first 1 to 5 years of drainage-system operation, when concentrations tend to be the highest and most variable. There are no clear seasonal patterns common to all systems in the area. The low temporal variability of selenium concentrations in water from existing mature drainage systems underscores the fact that drainage systems withdraw ground water, which tends to be of relatively constant chemical character over time in a particular place.

The high variability in selenium concentrations between existing drainage systems reflects the high spatial variability in shallow ground-water concentrations, the ages of the drainage systems, and variable hydrologic conditions at individual fields. Concentrations in drainwater from existing systems in Coast Range alluvial sediments are not predictable from a regionalized assessment of shallow ground-water concentrations derived from observation-well data. The age of the drainage systems, however, explained a significant part of the variance in median selenium concentrations.

Studies of individual drained fields show how local water-table history, geohydrologic conditions, irrigation history, and drainage-system design can markedly affect the type of water that is removed by the drains (Deverel et al., 1989). For example, figure 8 shows the flow system and selenium distribution for ground water beneath a field with a 15-year-old tile drain system. Most of the high-selenium ground water flowing into this drain system is from the sand zone located 30 to 60 feet below land surface. The highest selenium concentrations in ground water occur at varying depths below the water table in different

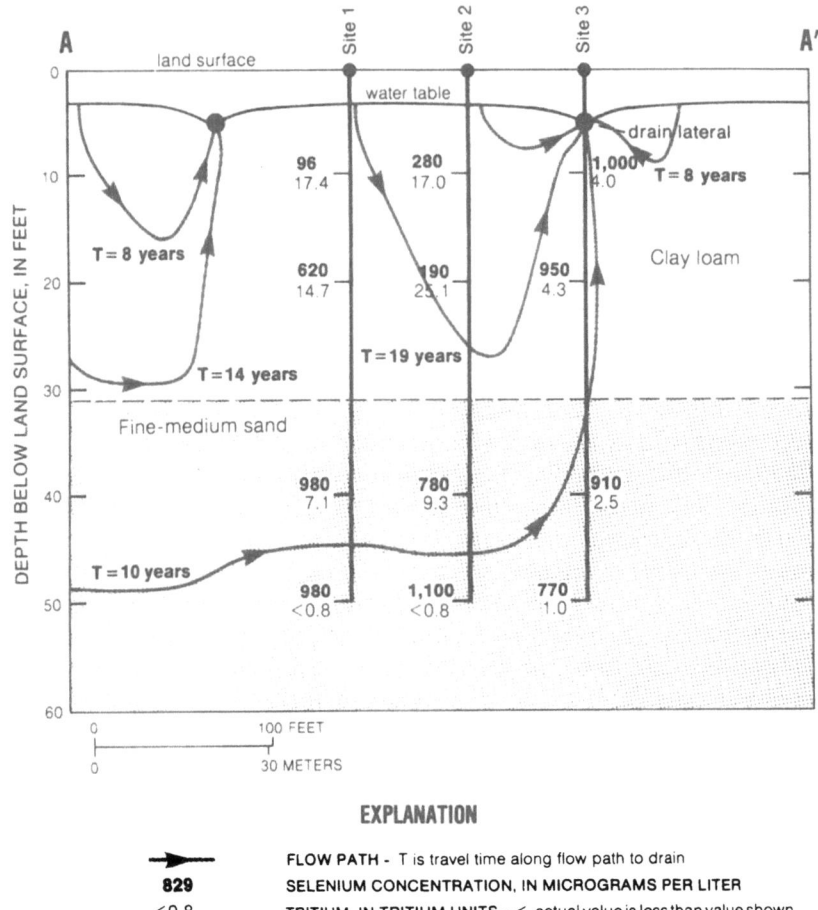

Figure 8.

areas. The design of a drainage system and local geohydrologic conditions determine the contributions of water from different depths to drainwater flow. These factors vary greatly between fields and are key reasons why the variability in selenium concentrations between drainage systems is so high.

## MOBILITY OF SOIL SELENIUM

Soils are a source of selenium to shallow ground water where irrigation occurs. Readily soluble forms of selenium in most present-day soils that have been irrigated are only a small fraction of the total selenium content, but the quantities of soluble selenium are substantially different among soils in different fields and at different depths (Fujii and Deverel, 1989 and Fujii et al., 1988). Forty-five years of irrigation, combined with 15 years of water-table control by a drainage system, resulted in highly leached soils throughout the unsaturated zone in a field drained for 15 years. Irrigation water infiltrating through these soils to the water table probably attains selenium concentrations in the range of 10 to 50 $\mu$g/L. In two fields irrigated for just as long, but drained for less than one-half the amount of time, saline soils with substantial quantities of soluble selenium were found at the 3-foot depth, even though the near-surface soils were highly leached. Water percolating through soils of these two fields still contains selenium concentrations greater than 100 $\mu$g/L.

Although dissolution of readily soluble soil salts is the primary mechanism of selenium release from saline soils, other mechanisms seem to be more important in highly leached soils. Selenium is correlated with salinity in soil-extract solutions for soils that have substantial quantities of soluble selenium, but such a correlation is not evident for highly leached soils with little remaining soluble selenium (Fujii, 1989). Saturation extracts of highly leached soils contained a substantial proportion of selenite, whereas selenate dominates in extracts of more saline soils. Processes such as sorption reactions involving selenite, dissolution of soil minerals, and release of organic selenium may control the movement of selenium from unsaturated soils after the most soluble salts have been leached.

## SOURCES AND CONCENTRATIONS OF SELENIUM IN THE SAN JOAQUIN RIVER

Drainwater from about 77,000 acres of tile-drained farmland eventually flows to the San Joaquin River. Flow of drainwater to the river occurs mainly through two tributaries: Salt and Mud Sloughs. Selenium concentrations in water from individual drainage systems that discharge to waterways that eventually reach these sloughs range from less than 10 to 4,000 $\mu$g/L, with water in larger collector drains ranging from 20 to 100 $\mu$g/L (Deverel et al., 1984 and Presser and Barnes, 1985).

During October 1985 through March 1987, two relatively distinct flow conditions occurred in the San Joaquin River: a combined low-flow period from October 1985 to mid-February 1986 and from mid-May 1986 through

March 1987; and a high-flow period from mid-February to mid-May 1986. Of total streamflow at the farthest downstream study site, the San Joaquin River near Vernalis, 49 percent occurred during the 15-month combined low-flow period and 51 percent during the 3-month high-flow period (Clifton and Gilliom, 1989).

Despite the greater quantity of streamflow during the 3-month high-flow period, 65 percent of the selenium load during the study period occurred during the 15-month low-flow period. During the low-flow period, Salt and Mud Sloughs contributed almost 80 percent of the Vernalis selenium load, despite contributing only 9 percent of the total streamflow, and within-reach gains or losses of selenium were not substantial (figure 9). The only major change in proportional sources of selenium loading to the river during the high-flow period was the increase from 3 to 20 percent of the total load from the upper San Joaquin River because of much greater streamflow.

During the low-flow period, median selenium concentrations were highest in Salt Slough (5.5 $\mu$g/L) and Mud Slough (8.8 $\mu$g/L). In Mud Slough, the maximum selenium concentration measured was 28 $\mu$g/L on July 15, 1986, and the maximum selenium concentration measured in Salt Slough was 22 $\mu$g/L on January 21, 1987. Median selenium concentrations in the San Joaquin River decreased from 5.2 $\mu$g/L downstream of Salt Slough to 1.0 $\mu$g/L near Vernalis, as water with low selenium concentrations entered the river from the eastside tributaries and other smaller inflows.

During the high-flow period, selenium concentrations in the San Joaquin River were lower than during the low-flow period because of greater dilution by Sierra Nevada runoff. In contrast, selenium concentrations increased with higher streamflow in Salt Slough. Median selenium concentrations were highest in Salt Slough (13 $\mu$g/L) and Mud Slough (3.9 $\mu$g/L). The maximum selenium concentration measured in Salt Slough was 20 $\mu$g/L on February 26, 1986, and the maximum selenium concentration measured in Mud Slough was 24 $\mu$g/L on April 16, 1986.

The EPA drinking-water standard of 10 $\mu$g/L was exceeded about 10 percent of the time in the San Joaquin River just upstream of the Merced River, and was not exceeded downstream of the Merced River. The proposed EPA aquatic-life criterion of 5 $\mu$g/L for selenium was exceeded in Salt and Mud Sloughs more than 60 percent of the time. In the San Joaquin River, just upstream of the Merced River, 5 $\mu$g/L was exceeded more than 40 percent of the time, and just downstream of the Merced River, it was exceeded more than 20 percent of the time. The 5 $\mu$g/L criterion was exceeded less frequently farther downstream in the San Joaquin River and never was exceeded at Vernalis.

The San Joaquin River contributes most of the riverine load of selenium to the Sacramento-San Joaquin Delta, even though the Sacramento River contributes most of the streamflow. Seldom do waters of the two rivers completely

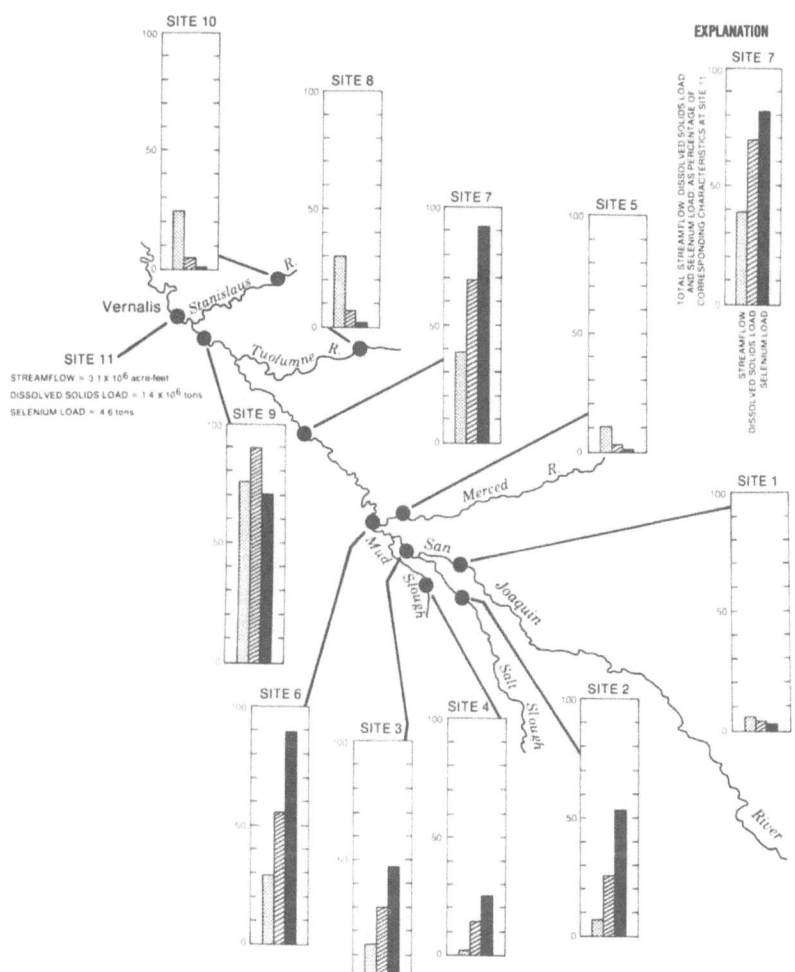

**Figure 9.**

mix before flowing out of the Delta to San Francisco Bay. At times, most of the San Joaquin River flow is withdrawn from the Delta, and flow through the Delta to the bay is almost entirely Sacramento River water. Selenium concentrations in tissues of the bivalve, corbicula, which inhabits the Delta, Suisun Bay, and the San Joaquin River, indicate that selenium from the San Joaquin River does not measurably affect the bioaccumulation of selenium in the southern Delta or Suisun Bay.

## IMPLICATIONS FOR WATER MANAGEMENT

Although this overview is based on preliminary results of studies in progress, these results have implications that could affect the development of water-management alternatives for the central part of the western Valley. Management alternatives are presently being developed by Federal, State, and local water-management agencies. In most instances, the implications are general in nature and cannot be quantified until the studies and associated simulation and empirical models are completed. These general implications may aid water managers and researchers in setting priorities for completing studies in progress and in guiding the development of study plans for unstudied areas of the Tulare Lake basin, which may have similar water quality problems.

The water table will continue to rise in the central western Valley if present irrigation practices continue in the absence of other changes in the ground-water flow system. The rising water table, which reflects net recharge to the ground-water flow system in excess of its capacity to discharge water, will enlarge the areas that have a shallow water table and associated drainage problems. The rise of the water table can be slowed or stopped by reducing ground-water recharge and increasing ground-water discharge.

Ground-water recharge can be reduced by increasing irrigation efficiency, reducing seepage losses from canals and water storage facilities, and changing or eliminating agricultural activities in some areas to reduce or eliminate irrigation. Ground-water discharge can be increased by tile drainage systems, pumping from wells, or increasing evapotranspiration in selected areas. Although tile drainage systems withdraw shallow ground water, which tends to have the highest selenium concentrations, dewatering wells may allow water-table control by withdrawal of deeper ground water that has low selenium concentrations. Evapotranspiration could be increased in some areas by eliminating artificial drainage and using salt-tolerant plants. Water-table management accomplished by removing water from Sierra Nevada sediments by dewatering wells could result in removal of selenium from ground water that moves from Coast Range sediments into Sierra Nevada sediments.

The large quantity of high-selenium ground water (50 to 1,000 $\mu$g/L) in the general range of 20 to 150 feet below the water table makes it desirable to use management practices that leave this water where it is, rather than bring it to the land surface or allow it to move into parts of the aquifer that may be used for water supply. Water-table control strategies based on increasing ground-water discharge need to be carefully evaluated with respect to their potential to affect the movement of water with high selenium concentrations.

The occurrence of ground water with 10 to 50 $\mu$g/L selenium in the upper 10 to 20 feet of the saturated zone indicates that, where evaporative concentration is controlled, continued irrigation with water that has low selenium concentration may result in a wider interval of this lower concentration water. Drainage strategies aimed at removing this ground water near the water table may be feasible in some areas. The highest concentrations of selenium in shallow ground water have developed and will continue to develop in irrigated areas where evaporative concentration is not controlled by water-table management.

Selenium concentrations vary greatly between drainage systems, but tend to be consistent over time in drainwater from a particular system after the first 1 to 5 years of operation. When selenium concentrations vary, they tend to be correlated with drainwater salinity. Periodic selenium measurements, on the order of two to four times per year, combined with frequent specific conductance monitoring, are an effective strategy for monitoring drainwater selenium concentrations.

If drainage flows from a particular drainage system are reduced, selenium concentrations in drainwater probably will remain relatively constant, resulting in a decreased selenium load. However, a reduction in dilution by irrigation water that contains low concentrations of selenium could lead to a gradual increase in drainwater selenium concentration if irrigation volume is decreased. In the long term, this may partially offset load reductions achieved by decreasing drain flows. Selenium concentrations in drainwater from existing systems in Coast Range alluvial sediments are not predictable from a regionalized assessment of shallow ground-water concentrations derived from observation-well data. Prediction of selenium concentrations for future drainage systems presents a particularly difficult problem because of highly variable geohydrologic and ground-water quality conditions at different sites. Accurate predictions of selenium concentrations for new drainage systems probably will require relatively detailed site-specific data on the depth distribution of selenium and the local geohydrology.

Even after the most soluble forms of selenium are leached from the soil by early irrigation, continued irrigation results in recharge to ground water that contains lower, but still undesirably high, selenium concentrations (10 to 50 $\mu g/L$). Therefore, long-term management alternatives will probably need to address the continued presence of such selenium concentrations in shallow ground water.

Continuation of present management practices will sometimes cause selenium concentrations in the San Joaquin River and Mud and Salt Sloughs to exceed Federal and State criteria for protection of aquatic life. Selenium concentrations in the sloughs and the river can be decreased by reducing selenium loading and adding dilution water from low selenium sources. Conjunctive management of the timing of selenium loading and addition of dilution water could be used to help meet water quality criteria.

# REFERENCES

Barnes, I., 1985. Sources of Selenium. In: *Selenium and Agricultural Drainage: Implications for San Francisco Bay and the California Environment*. Proceedings, 2d Selenium Symposium, March 23, 1985, Berkeley, CA., pp. 41-51.

Belitz, K. and Heimes F. J., 1990. *Character and Evolution of the Ground-Water Flow System in the Central Part of the Western San Joaquin Valley, California*. U.S. Geological Survey Water-Supply Paper 2348, 28p.

Clifton, D. G. and Gilliom, R. J., 1989. Sources and Concentrations of Selenium in the San Joaquin River. In: Gilliom, R. J. et al., 1989, *Preliminary Assessment of Sources, Distribution, and Mobility of Selenium in the San Joaquin Valley, California*. U.S. Geological Survey Water-Resources Investigations Report 88-4186, pp. 99-113.

Deverel, S. J.; Fio, J. L.; and Gilliom, R. J., 1989. Selenium in Tile Drain Water. In: Gilliom, R. J. et al., 1989, *Preliminary Assessment of Sources, Distribution, and Mobility of Selenium in the San Joaquin Valley, California*. U.S. Geological Survey Water-Resources Investigations Report 88-4186, pp. 77-91.

Deverel, S. J. and Fujii, R., 1988. Processes Affecting the Distribution of Selenium in Shallow Ground Water of Agricultural Areas, Western San Joaquin Valley, California, *Water Resources Research*, 24(4), pp. 516-524.

Deverel, S. J. and Gallanthine, S. K., 1989. Relation of Salinity and Selenium in Shallow Ground Water to Hydrologic and Geochemical Processes, Western San Joaquin Valley, California, *Journal of Hydrology*, 109, pp. 125-149.

Deverel, S. J.; Gilliom, R. J.; Fujii, R.; Izbicki, J. A.; and Fields, J. C., 1984. *Areal Distribution of Selenium and Other Inorganic Constituents in Shallow Ground Water of the San Luis Drain Service Area, San Joaquin Valley, California--A Preliminary Study*. U.S. Geological Survey Water-Resources Investigations Report 84-4319, p. 67.

Deverel, S. J. and Millard, S. P., 1988. Distribution and Mobility of Selenium and Other Trace Elements in Shallow Ground Water of the Western San Joaquin Valley, California, *Environmental Science and Technology*, 22(6), pp. 697-702.

Dubrovsky, N. M., 1989. Selenium in Ground Water of the Northern Part of the Western Valley. In: Gilliom, R. J. et al., 1989, *Preliminary Assessment of Sources, Distribution, and Mobility of Selenium in the San Joaquin Valley, California*. U.S. Geological Survey Water-Resources Investigations Report 88-4186, pp. 67-75.

Dubrovsky, N. M. and Deverel, S. J., 1989. Selenium in Ground Water of the Central Part of the Western Valley. In: Gilliom, R. J. et al., 1989, *Preliminary Assessment of Sources, Distribution, and Mobility of Selenium in the San Joaquin Valley, California.* U.S. Geological Survey Water-Resources Investigations Report 88-4186, pp. 35-66.

Fujii, R., 1989. Mobility of Soil Selenium. In: Gilliom, R. J. et al., 1989, *Preliminary Assessment of Sources, Distribution, and Mobility of Selenium in the San Joaquin Valley, California.* U.S. Geological Survey Water-Resources Investigations Report 88-4186, pp. 93-98.

Fujii, R. and Deverel, S. J., 1989. Mobility and Distribution of Selenium and Salinity in Ground Water and Soil of Drained Agricultural Fields, Western San Joaquin Valley, California. In: Jacobs, L. W. et al. (Eds.), *Selenium in Agriculture and the Environment.* Special Publication 23, Soil Science Society of America, Madison, WI., pp. 195-212.

Fujii, R.; Deverel, S. J.; and Hatfield, D. B., 1988. Distribution of Selenium in Soils in Agricultural Fields, Western San Joaquin Valley, California, *Soil Science Society of America Journal,* 52(5), pp. 1274-1283.

Gilliom, R. J. et al., 1989. *Preliminary Assessment of Sources, Distribution, and Mobility of Selenium in the San Joaquin Valley, California.* U.S. Geological Survey Water-Resources Investigations Report 88-4186, p. 129.

Presser, T. S. and Barnes, I., 1985. *Dissolved Constituents Including Selenium in Waters in the Vicinity of Kesterson National Wildlife Refuge and the West Grassland, Fresno and Merced Counties, California.* U.S. Geological Survey Water-Resources Investigations Report 85-4220, p. 73.

Tidball, R. R.; Severson, R. C.; Gent, C. A.; and Riddle, G. O., 1986. *Element Associations in Soils of the San Joaquin Valley, California.* U.S. Geological Survey Open-File Report 86-583, p. 15.

# Two: ENGINEERING, PHYSICAL, AND BIOLOGICAL APPROACHES TO DRAINAGE PROBLEMS

# 4 HYDROLOGIC ASPECTS OF SALINE WATER TABLE MANAGEMENT IN REGIONAL SHALLOW AQUIFERS

Mark E. Grismer, University of California, Davis and
Timothy K. Gates, Colorado State University, Ft. Collins

## ABSTRACT

This chapter considers hydrologic factors associated with irrigation and drainage in regions with saline shallow ground water. Soil water flow processes and the importance of regional analysis are described. Results of a simulation model which accounts for major processes governing shallow water table behavior in salinity-affected regions are discussed. The model is used to analyze the effects of irrigation-drainage management on water table depth, salinity, crop yield, and net economic returns to the grower over a 20-year planning period. Stochastic elements associated with soil hydraulic properties and irrigation applications have been incorporated into this model. These additions to the model indicate the importance of parameter uncertainty and enable results to be interpreted with the notions of stability and risk.

## INTRODUCTION

Irrigation projects in the basin areas of arid climatic zones through out the world are subject to problems associated with saline shallow ground water. These problems include waterlogging of crops and salinization of the root zone. These problems arise as a direct consequence of modifying the region's hydrologic system by importing irrigation water and its associated salt load. Salinity and shallow ground-water problems may be aggravated by the particular geologic setting of the region and lithologic interactions with the irrigation water. These problems have been well documented for several intensely irrigated regions in arid parts of the world (Al-Layla, 1978; IDP, 1979; Khan, 1980; and Singh, 1984).

One of the most studied irrigated regions with problems of saline shallow ground water is the western San Joaquin Valley (Valley) of California. Over 186,000 irrigated hectares in this region are presently adversely affected by saline shallow ground water, and it has been predicted that the affected area will grow by over 28,000 hectares in the next two decades. Problems in the impacted area stem directly from the water and salt load of the imported irrigation water and the slowly permeable soils and various substrata in the region. Altering the hydrologic system again so as to restore the predevelopment ground-water depth and quality conditions in the region would likely be both difficult and economically infeasible. Thus, suitable management practices must be developed. Effective management of the saline shallow ground water in such a way as to sustain agricultural productivity while minimizing adverse environmental impacts, however, requires thorough understanding of the region's hydrologic system.

Determining effective soil-water management practices at the regional scale is advantageous for project planning and evaluation purposes. Analysis at this scale enables incorporation of spatial variability aspects of water applications, soil properties, and crop yields, as well as the effects of lateral ground-water flows. Addressing saline ground-water problems of a large region may require decisions on large capital investments and benefit-cost analyses directed towards developing management practices which consider the overall region. Regional-scale analysis is necessary from a hydrologic perspective because saline shallow ground water at the local or farm scale is only a part of a larger regional ground-water system. Moreover, variability in soil hydraulic properties and water applications from farm to farm may need to be considered in terms of how they affect shallow ground-water depth, soil salinity, and thus, crop yield. A regional-scale approach is also useful in considering subsurface drainage requirements, including the economic benefits of designing regional drainage disposal facilities rather than numerous smaller facilities. Additionally, as a result of aquifer heterogeneities and interconnectedness, drainage flows and salt loads at the farm scale may have little to do with soil-water management practices at that particular farm.

Models addressing soil-water management practices at the regional scale include economic-policy based models described elsewhere in this book: Mass balance-models such as described by the San Joaquin Valley Drainage Program (SJVDP, 1989); deterministic ground-water models; and models based on optimization in the economic sphere of stochastic ground-water crop yield simulations. Mass-balance-type models typically simplify the ground-water system into a series of layered reservoirs into which water and salt are added, stored, or removed (depending on water applications); crop water use; drainage; and pumping. Details of the soil-water movement processes are generally not considered in this accounting scheme. Instead, spatial variability in practices and crop conditions is considered in terms of a number of individual "cells" within the region. Deterministic ground-water models at the regional scale involve averaging aquifer properties and water application and use factors over the entire region and then running the model for different combinations of management factors until the "best" solution is obtained. Due to the complexity of the ground-water simulations, the number of management alternatives may be limited as compared to the mass-balance-type models. In addition, no information is obtained regarding the probability of achieving the "best" solution, or what near-optimal solutions may require in terms of management strategies. Such deterministic models are often excellent for obtaining a first-approximation solution to complex regional problems. Stochastic simulation models incorporate the inherent variability in water applications and physical parameters affecting shallow ground-water response to irrigations. Such variability results in different crop yields which can then be linked to net economic returns to the grower. Thus, the optimal management problem is cast in a stochastic context which enables system responses to be evaluated with the concepts of stability and risk (Gates and Grismer, 1989). Obtaining solutions from this stochastic optimal management model is computationally intensive and presently requires "super computing power." Approximation methods which substantially reduce the computing requirements, however, have been introduced (Gates et al., 1989).

Presently, it appears that the more complex the various management alternatives may be (such as conjunctive surface and ground-water use), the less detailed the hydrologic analysis is in the simulation. Conversely, the detailed simulation of stochastic ground-water models includes relatively few management variables. Typically, a few management variables are considered in terms of regional planning target strategies. The value of sophisticated hydrologic models linked to economic returns is in providing economic analysis with a firm basis in actual physical processes rather than a hydrologic analysis driven by the economic system. For this reason, the focus here is on the key hydrologic factors involved in drainage of irrigated regions.

This chapter considers the geohydrologic factors affecting shallow ground water, and thereby, irrigation and drainage management in regions similar to those of the western San Joaquin and Imperial Valleys of California. Stochasticity in some of these factors will be described, as well as an optimization approach which considers the output of the stochastic shallow ground-water simulations. Finally, there will be a conceptual discussion of the impacts of drainage management practices on ground-water depth and salinity over time.

## GEOHYDROLOGIC SYSTEM DESCRIPTION

Irrigation and drainage management involves manipulation of the soil-water environment in the crop root zone to minimize salinity accumulation and adverse soil moisture conditions. Soil-water movement processes are also controlled by hydraulic parameters of the soil profile and water movement in and from the shallow aquifer. Figure 1 is a conceptual cross section of the lithologic profile of an irrigated region showing the primary water movement processes occurring as a result of irrigation and drainage activities. This section considers each of these processes at the local and regional scales, and how they are affected by irrigation and drainage.

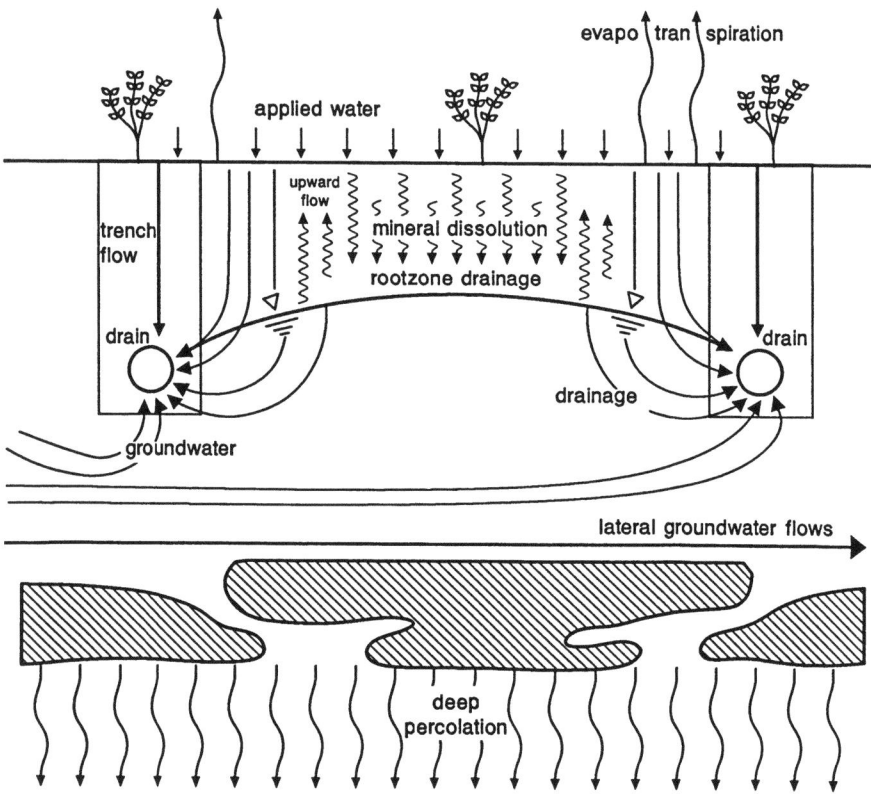

Figure 1. Schematic illustration of soil-water flow processes in the soil profile subject to irrigation and drainage.

The shallow ground water is recharged by the irrigation supply through seepage losses from conveyance systems and tailwater recovery ponds, spillage, and water applications to crops. In the Valley, there is also some water table recharge due to evaporation pond seepage. As with nearly all water imported to a region, a certain salt load is added to the ground-water system. Patterns of ground-water recharge within the region are variable due to both differences in soil hydraulic properties (infiltration and permeability) and water source location (canals and ponds, etc.). From a regional perspective, the key factor is the total ground-water recharge from all sources. Since most water management alternatives are related to the irrigation water supply, the scale of analysis should be such that the net effect of seepage control (e.g., canal lining) and water application management practices on ground-water recharge can be evaluated.

A large part of the applied water and seepage is used to satisfy crop water demands and is lost from the system by evaportranspiration (ET). Typically, the salt load of the applied water remains in the root zone or is leached progressively deeper into the soil profile. Naturally occurring salts in the root zone may also dissolve and become part of the soil water. Evaporation losses in water conveyance systems are typically minor compared to other losses. Local variability in evapotranspiration is generally associated with particular crops and available soil moisture. From a regional perspective, ET and evaporation represent a net water loss with no salt loss from the system. The scale of analysis should be sufficient to evaluate the affects of various crop management practices on system water losses.

The root and vadose zones include the most complex soil-water processes in the system. These processes are schematically illustrated in figure 2. The root zone forms the upper part of the vadose zone and is the extent of the soil depth below the ground surface containing most of the roots of a healthy crop. The vadose zone is bounded at the bottom by the free water surface of the water table. Irrigation and drainage management practices are designed to maintain soil salinity and moisture conditions in the root zone favorable to plant growth. A complete characterization of soil-water processes in this zone would require consideration of plant root water extraction patterns and unsaturated flow dynamics. From a regional perspective, however, the detail of these processes can be greatly simplified with mass-balance-type approximations, and the root zone is not distinguished from the rest of the vadose zone.

Soil-water processes in the vadose zone include downward flow of water and salt following irrigation (leaching), and upward flow of saline ground water at later times as part of the ET demand of the crop and drying of the soil profile. These processes do not occur simultaneously, though from a modeling perspective they often occur within the same time step. The upward flow process is relatively simple conceptually; shallow ground water and its salt load is taken

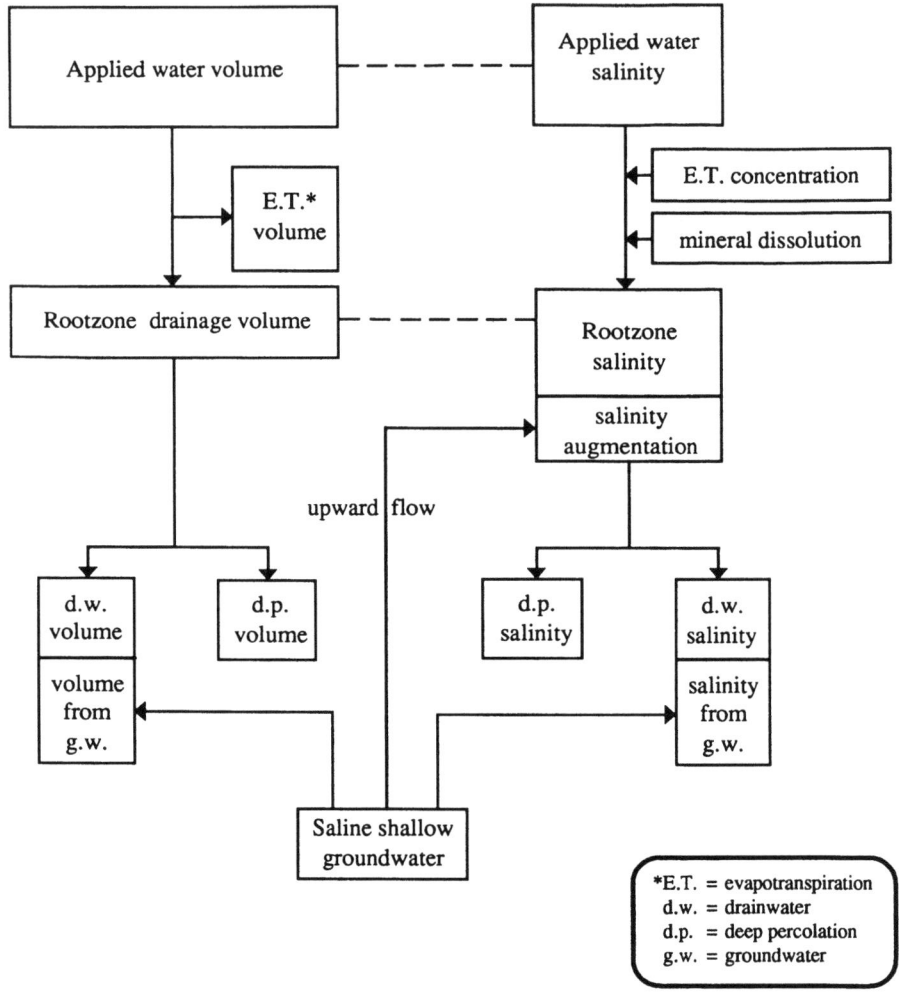

Figure 2. Schematic illustration of soil-water processes in the vadose zone of an irrigation/drainage system.

into the root zone where the water is transpired and evaporated, thereby resulting in increased root zone salinity. The next irrigation event reverses the upward direction of soil-water flow with downward soil-water flow leaching salinity from the root zone. Root-zone leaching involves the downward movement of only a fraction of the applied water, ideally all of the applied salt mass, and precipitation and dissolution of various minerals in the root zone until the leachate reaches the water table. At this point, root zone drainage mixes with the shallow ground water and is divided into drainage system flows, lateral shallow ground-water flows, and deep percolation flows into deeper ground-water zones (Grismer, 1990a).

Typically, the net effects of root-zone drainage and upward flow are to increase shallow ground-water salinity and piezometric surface. These adverse effects on the ground water are partially mitigated through water extraction by the subsurface drainage system. The presence of the drains or water "sinks" in the lithologic profile results in two-dimensional hydraulic gradients in the profile which converge on the drain. These gradients influence shallow ground-water flows over a range which is largely dependent on the magnitude of lateral and deep percolation flows removing water from the local system. Field data indicate that subsurface drainage system extract roughly 10-40 percent of the root-zone drainage volume in addition to flows through the backfill drainline trench and some shallow ground water. In some fields of the Imperial and San Joaquin Valleys, shallow ground-water accounts for 80-90 percent of the total drain flows. For these areas, drainwater reduction strategies based on irrigation management of deep percolation may have little effect on drainwater production and salt load.

Though not shown in figure 1, drainage wells in the shallow ground-water aquifer may also be used as "sinks" for root-zone drainage. Typically, these wells are more widely spaced than drainlines, and their practical use is usually limited to high permeability soils. These wells also tend to collect much deeper ground water than the lateral drainlines. Soil water flow processes to both lateral drainlines and drainage wells are similar.

The impact of drainage production and quality on saline shallow ground water, indicates that variability in the ground-water aquifer water quality and piezometric surface in local areas will control variability within the overall region. Local variability may be a small factor of concern if the shallow aquifer is of large areal extent.

Soil-water processes, as affected by irrigation-drainage management, determine the root-zone soil moisture and salinity conditions for the crop. Beyond the effects of neglecting fertilizer, herbicide and pesticide management, crop yield is limited by poor soil aeration and high soil salinity. Excessive soil moisture stunts root development and gaseous exchange in the rhizosphere for plant metabolism. Excessive soil salinity reduces available water extractable by plant roots and may cause ion toxicity to the crop. Differences in crop yield within a region, therefore, depend directly on the spatial variability in root-zone soil moisture and salinity, and indirectly on shallow ground-water depth and salinity.

## SYSTEM MODELING

In order to formulate a regional economic planning model, each of the soil-water and irrigation-drainage system processes considered in the previous section must be described mathematically. The complexity or detail of the mathematical description largely depends on the anticipated effect of each process on crop yield at the regional scale over relatively long time periods. For example, water infiltration into soil is a complex process involving unsaturated flow dynamics. However, in terms of irrigation system design, the infiltration rate and approximate wetting front depth can be adequately estimated from a simple power function and mass-balance equation, respectively. Complete analysis of the unsaturated flow dynamics would require solution of nonlinear partial differential equations. This solution would provide information about the distribution of soil moisture with depth and time which have little, if any, effect on irrigation system design. The computational effort required to achieve such a solution far exceeds that required for the simpler algebraic expressions, for little if any, greater benefit. Thus, when selecting a mathematical description of any of the processes discussed previously, the computational simplicity and level of accuracy required for regional analysis should be considered to provide for a tractable solution. This section briefly outlines some of the conceptual mathematical formulations used to describe soil-water flow processes at the regional scale following Gates and Grismer (1989).

Flow processes in the vadose zone at the regional scale are understood as fieldwide averages over time periods of weeks or months, during which considerable local and short-time processes are occurring. Consideration of the unsaturated flow dynamics and associated local variability would be an inappropriate level of detail for which little field data is available. Thus, for the root zone, simple mass balances of water and salt are sufficient. The root-zone mass balance for water would include inflows from applied water, precipitation, and upward flow from the water table; while outflows would include root-zone drainage and crop water use (ET). The capacity of the soil to store water is limited by the field capacity moisture content. The salt mass balance would include the salt masses associated with the flows described above in addition to mineral dissolution in the root zone. The quality of the root-zone drainage water may be adjusted to account for the leaching efficiency or the extent to which applied water displaces and mixes with saline root-zone water.

Root-zone flows associated with applied water, precipitation, and crop water use can be determined from water delivery records and climatologic data, respectively. Typically, root-zone drainage is then determined from the mass balance as the closure term. Root-zone drainage can also be estimated from the uniformity of water application to the field (Ben-Asher and Ayars, 1990). However, in the case of saline shallow ground water, it is important to estimate the upward flow component prior to calculating root-zone drainage. The primary purpose for including the upwater flow component in the root zone is to account for its salinization effects. The fact that root-zone drainage and upward flow cannot occur simultaneously in the time scale of hours, or a few days, is not important when considering these processes from the longer time scale of weeks. Upward flow can be quantified using empirical equations which depend on soil texture and water table depth (Grismer and Gates, 1988).

The principal mechanism of interaction between fields in the region is transport of water and salts in the saturated zone, that is, movement of the saline shallow ground water within the region. A number of numerical models are available to describe saturated flow within the ground-water system, all of which are based on solution of well-known governing flow equations (Bear, 1979). At the regional scale, salt transport in the ground water can also be described by advective transport equations since dispersive transport is of relatively minor significance (Anderson, 1979). Variability in the hydraulic parameters of the ground-water system is the primary source of stochasticity in the problem and its solution. Hydraulic conductivity is the key parameter of spatial variability in this case.

Additional processes that must be described when modeling the ground-water system include subsurface drainage flows and deep percolation flows out of the shallow system. Though usually designed to extract all of the root-zone drainage, subsurface drainage systems may extract only a fraction of the root-zone drainage in addition to other shallow ground water. The efficiency of the drainage system in extracting root-zone drainage depends on both the soil hydraulic properties and the depth and spacing of the drains. Soil hydraulic properties control the rates of recharge and ground-water movement and are adequately modeled by ground-water numerical models. Empirical equations may be developed to relate the efficiency of the drainage system in extracting root-zone drainage and the depth and spacing of the drains (Grismer, 1989 and 1990b). These equations can then be incorporated in mass balance equations considering flow between the root zone and shallow ground water. Deep percolation flows out of the shallow ground-water system can be modeled simply as a specified, or water table height-dependent flux out of the base of the system. Large deep percolation fluxes out of the system may have a large effect on the efficiency of the drainage system in removing root-zone drainage.

## REGIONAL PLANNING

Regional planning for optimal irrigation and drainage strategies must consider all of the economic factors under management control in the context of the geohydrologic processes affected by these strategies. Approaching development of management strategies in this context is important because complete control of the soil-water and ground-water processes is not practically, if at all, achievable. Costs associated with water supplies, delivery and application systems, drainage systems, and drainage disposal must be considered. Other nonwater-related production costs may be considered, but are not necessary for comparison or optimization of different water management strategies. Revenues obtained to offset nonwater costs reflect crop yields within the region. Geohydrologic phenomenon as influenced by a particular water management strategy determine the crop yield and its variability within the region. Finally, this section outlines the procedures used by Grismer and Gates (1989) to incorporate variability into a regional planning model.

Management of the water delivery and application system is the primary control variable for regional planning or optimization of management strategies. From a regional planning perspective, it is sufficient to determine the overall irrigation efficiency which must be achieved on the average to obtain maximum net benefits to growers in the region. Irrigation efficiency, defined as the ratio of regional ET rate to water application rate over the growing season, can be directly related to costs associated with the management or improvement of water delivery and application systems. Unfortunately, variable operational management of even the most uniform pressurized irrigation systems may result in poor irrigation efficiencies despite the capital investment. Costs associated with irrigation systems having different efficiencies have been cause for debate, though some analyses are available (Oster et al., 1988 and Letey et al., 1990). The debate involves assigning an efficiency level to a particular irrigation system rather than the cost of the system. Typically, less expensive surface irrigation systems have lower efficiencies than more expensive pressurized systems. Additional costs associated with the pressurized systems may be offset, however, by lower drainage costs due to reduced root-zone drainage resulting from the more efficient systems (Letey et al., 1990). Reduction in root-zone drainage alone may not be sufficient to substantially reduce drainwater production because actual drainwater volumes depend on the drainage system and behavior of the shallow ground water.

The second control variable is related to management of the subsurface drainage system. As with the water delivery and application systems, it is probably sufficient to determine an overall drainage efficiency for the region which leads to maximum net benefits. Drainage efficiency on a regional basis can be defined as the fraction of root-zone drainage water actually collected by the drainage system. At the local scale, an individual drainage system may be collecting some root-zone drainage waters from both the field where the system is installed as well as the root-zone drainage waters originating elsewhere and transported as part of flows in the shallow ground-water system. Costs associated with a particular drainage efficiency are difficult to establish. Direct costs include those associated with design and installation of the subsurface drainage. Indirect costs include those associated with crop yield losses from inadequate drainage (upward flow) and drainwater disposal. Depending on the soil type, drainage efficiencies generally increase as the depth and spacing of the drains decrease. Thus, increased intensity of drainage is associated with greater installation costs, but potentially lower disposal costs. Gates and Grismer (1989) defined drainage efficiency in terms of the ratio of the drainage design flow rate to the root-zone drainage flow rate to simplify analysis of costs associated with a particular drainage efficiency.

Both irrigation and drainage management are designed to influence the geohydrologic setting of the plants so as to promote a favorable growth environment. Soil hydraulic properties, water application, drainage intensities, and other factors affecting ground-water movement result in variable soil salinity, water-table depth, and crop yield, hence, net benefits. For example, low irrigation efficiencies in one particular area may result in locally low root-zone salinity, greater upward flow, a high water table (depending on the local drainage efficiency), and high crop yield while adversely affecting a neighboring area with augmented shallow ground-water salinity and drainage flows. Thus, this variability in water application, drainage intensity and shallow ground-water movement within the region must be accounted for computationally to develop realistic management strategies based on optimal net benefits to the growers throughout the region.

Gates and Grismer (1989) incorporated many of the factors described above into development of a stochastic model suitable for regional planning of optimal or near optimal irrigation and drainage management strategies in salinity-affected areas. A simplified flowchart of the computational procedure discussed below is shown in figure 3.

The computational procedure was based on stochastic simulation of groundwater transport and interaction with the root zone. The effects of the ground-water-root-zone interaction on water-table depth, root-zone salinity, and crop yield were obtained by solving the deterministic equations for 100 different realizations of the spatial random functions (Monte-Carlo simulation technique). The Monte-Carlo technique was selected over finite-order methods due to limitations in their applicability to nonlinear systems having a large coefficient of variation (CV > 0.2). Random fields in this study included normally distributed fields for the leaching efficiency, an upward flow parameter and water application efficiency, and a log-normally distributed depth-

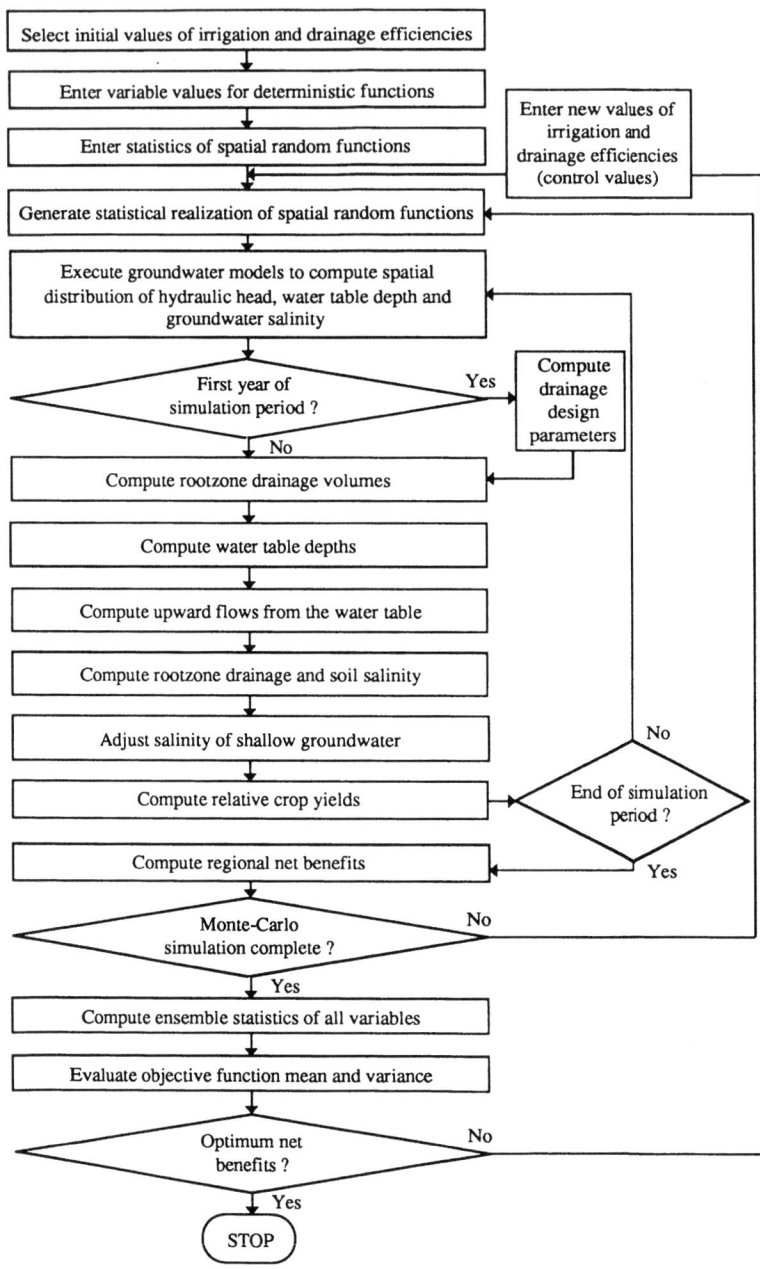

Figure 3. Simplified flow chart of computational procedure used by Gates and Grismer (1989).

averaged hydraulic conductivity field for the shallow ground water. Variability in porosity, though present, was found to have a negligible effect on the results. For each realization of the random field variables, the ground-water depths and salinity, soil salinity and crop yield were determined for every year over a 12-year period. Originally, a 20-year period was used, but it was found that steady-state conditions were reached after a 12-year simulation period. The mean and variance of regional net benefits within the region were then determined and another Monte-Carlo simulation conducted. This process continued until all Monte-Carlo simulations were complete. Finally the ensemble statistics were calculated, the overall regional net benefits (objective function) were evaluated and then compared with previous values until an optimal solution was obtained. Two different optimization methods were employed and these are described by Gates et al. (1989).

The stochastic optimization model was applied to a hypothetical region representative of conditions in the western San Joaquin Valley of California. A 20 km2 region was resolved into a grid of 60 square finite-difference blocks, each block being an independently managed cotton field approximately 32 ha (80 acres) in size. The shallow ground-water aquifer and perching clay layer were assumed to be of uniform thickness throughout the region. Lateral ground-water flows were specified along the boundaries parallel to the initial contours of hydraulic head, and no-flow boundaries were specified perpendicular to the contours. Values of deterministic physical parameters in the system were obtained from various studies in the region and included lateral flow rates and salinity, initial soil and aquifer conditions, aquifer porosity and specific yield, soil-water storage capacity, mineral dissolution and salinity, drainline slopes, and pre-irrigation application volumes. Deterministic economic parameters associated with potential crop yield, crop price, water application costs, drainage system costs, planning period, and interest rate were also included. A variable time step was used for each year which included a 28-day preirrigation season, four 23-day planting/emergence periods, a 3.7-day time step during the 3-month irrigation season, a 30-day maturity/harvest period, and three 41-day time steps during the fall/winter fallow period.

Illustrated in figure 4 is a response surface of expected regional net benefits as affected by the irrigation and drainage efficiency control variables. The optimal control strategy for this situation was an irrigation and drainage efficiency of 77 percent and 86 percent, respectively. The sharp decline in net benefits at irrigation efficiencies over 80 percent is a result of costs associated with conversion to pressurized irrigation systems and reductions in crop yield due to soil salinization from decreased leaching.

Also shown on the response surface of figure 4, are areas of near optimal solutions, within 5 percent and 10 percent optimum. These areas define the so-called "negotiation frontier" available to decisionmakers seeking near-optimal alternatives from which a final selection can be made based on institutional and sociopolitical issues. In this case, irrigation efficiencies of 75 percent to 80 percent fall within the 90-percent negotiation frontier. The analogous range for drainage efficiency is greater since the shape of the response surface appears to be largely controlled by the irrigation efficiency.

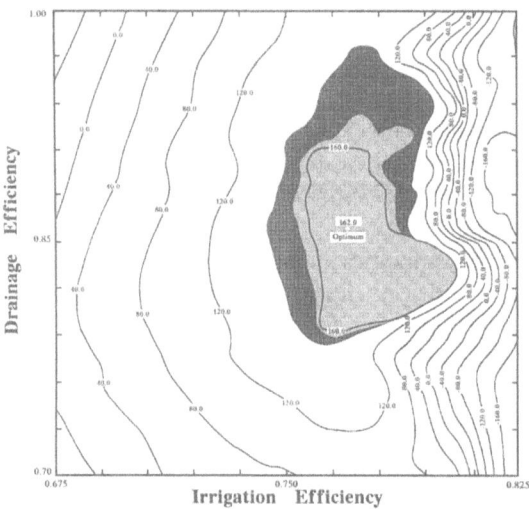

Figure 4. Response surface of expected annual regional net benefits($/ha). Dark and light shading represent the negotiation frontiers greater than or equal to 95% and 90% of the optimal, respectively.

Variability in net benefits at the optimal solution is also important and is indicated in the histogram of figure 5. The histogram of net benefits results from the 100 different realizations of the random variable parameters obtained for the particular management strategy. Although the expected value of annual net benefits is $162/ha, the range is nearly an order of magnitude--from approximatley $55/ha to $350/ha. The tenth percentile value is $106/ha. In other words, accounting for uncertainty due to spatial variablility in soil-water parameters and water applications, an analyst could expect a regional annual net benefit of $106/ha with 90 percent probability if the optimal management strategy is implemented.

## ANALYSES OF OPTIMAL MANAGEMENT STRATEGIES

The stochastic simulation model was used to determine the primary factors affecting the optimal solution, and to assess the effects of preirrigation depth, soil salinity and drainwater disposal costs on the optimal management strategy. These analyses provide valuable insight into development of various management alternatives for irrigated regions subject to saline shallow ground-water conditions. Sensitivity of the optimal solution to model variables is considered first followed by assessment of the effects of changes in initial conditions and costs.

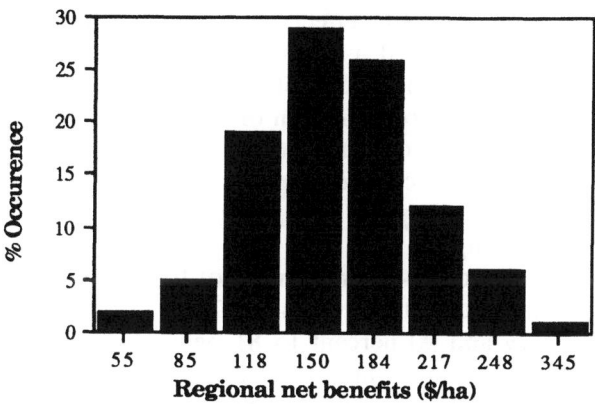

Figure 5. Frequency histogram of annual regional net benefits at the optimal management strategy.

Sensitivity can be evaluated by considering the effect of variance in the stochastic parameter on variance in the optimal solution. That is, the coefficient of variation is an indicator of the stability in regional net benefits as a function of the uncertainty accounted for in the model. Lower values of this coefficient indicate less variability, or a more certain solution. Variances in net benefits considered previously were most sensitive to variance in water applications, very sensitive to variance in hydraulic conductivity, and relatively insensitive to variance in the upward flow and leaching efficiency parameters. The variance in regional net benefits increased exponentially at approximately 1:1 with variance in water applications and hydraulic conductivity, respectively. These results suggest that field data collection should be directed primarily at characterizing variability in water applications and depth-averaged hydraulic conductivity.

The expected regional net benefit of the optimal management strategy is also sensitive to the mean value of hydraulic conductivity and the extent of mineral dissolution in the root zone, but relatively insensitive to drainage disposal costs and specific yield of the aquifer. For example, increasing the mean hydraulic conductivity from 0.02 m/day to 0.08 m/day increased the expected annual regional net benefits by $200/ha. This increase is due to reduced drainage costs resulting from increased drain spacings. Conversely, additional mineral dissolution of 0.5 kg/m3 reduced the expected value by over $300/ha due to salinization of the root zone and shallow ground water. The relationship between drainwater disposal costs and expected regional net benefits was linear where benefits declined at a rate of approximately $68/ha for every $0.10/m$^3$ (124 $/AF) increase in disposal costs. Doubling the soil specific yield resulted in an approximately 10-percent reduction in expected regional net benefits, probably due to the effect on depth to shallow ground water.

Gates et al. (1989) applied a stochastic optimization technique to the simulation model to examine effects of various parameters on the optimal management strategy (i.e., combination of irrigation and drainage efficiencies). From an implementation perspective, the optimal management strategy was relatively insensitive to changes in preirrigation depths, soil and groundwater salinity, drainwater disposal costs, and perching layer hydraulic conductivity over specified ranges.

At initial soil and ground-water salinities of 10 and 13 kg/m$^3$, respectively, the optimal management strategy ranged from 75 percent to 80 percent irrigation efficiency and 90 percent to 81 percent drainage efficiency for preirrigation depths ranging from 0.15 to 0.21 m (5.9 to 8.3 inches). Increasing preirrigation results in greater salt leaching during the preirrigation season, thereby requiring less leaching, that is, allowing for higher irrigation efficiencies during the growing season. Similarly, decreasing drainage efficiency by increasing preirrigation depth indicates that more upward flow of salts from the water table can be tolerated during the growing season because of improved preirrigation season soil salinity conditions. As preirrigation depth increased, expected regional annual net benefits also increased from $135/ha to $193/ha. Quantities of drainwater and salt load requiring disposal were similar for the optimal management strategies determined for each preirrigation depth. Had the management strategy been fixed at a particular combination of efficiencies, drainage quantities requiring disposal would have been greater. However, a decrease in preirrigation depth would also result in decreased net benefits because the particular strategy would no longer have been the optimal management strategy (Grismer et al., 1988).

Changing initial soil and ground-water salinities from 8 to 12 kg/m$^3$ and 11 to 15 kg/m$^3$, respectively, had little significant effect on optimal irrigation efficiency (78 percent to 77 percent), although optimal drainage efficiency increased from 79 percent to 93 percent. Expected annual regional net benefits decreased from $245/ha to $88/ha due to reduced crop yield caused by salinity. These results provide some insight into the costs, or benefits associated with salinity control in the region. In this case, additional preirrigation water may be useful in reducing the salinity hazard if the excess root-zone drainage can be successfully managed in the shallow ground water.

Increasing fixed costs of drainwater disposal from $0.10/m³ to $0.30/m³ ($124/AF to $370/AF) regardless of salt load had no effect on the optimal management strategy, although expected annual regional net benefits declined. For regional planning purposes, the lack of effect suggests that the optimal water management strategy is largely controlled by responses of the crop, root zone, and shallow ground water to irrigation and drainage rather than the economics associated with the externality of drainwater disposal.

Decreasing the hydraulic conductivity of the perching layer by an order of magnitude also had little effect on the optimal management strategy as irrigation and drainage efficiencies varied between 77 percent and 78 percent, and 85 percent and 86 percent, respectively. Expected annual regional net benefits decreased, however, from $166/ha to $137/ha. Decrease in net benefits occurs as a result of reducing deep percolation losses of water and salt from the system. From a practical perspective in the Valley, the information suggests that it may be economically advantageous for growers to increase leakage rates through the Corcoran Clay so as to dispose some of the deep percolation water and salt load in the deep ground water. However, this may not be an acceptable alterative in the sociopolitical arena (Oster, 1989).

Overall, results from these analyses indicate that for cotton farming on salinity-affected soils subject to saline shallow ground water, economically optimal regional irrigation and drainage efficiencies range from 75 percent to 80 percent and 79 percent to 93 percent, respectively. Implementation of this optimal water management strategy would require evaluation of irrigation systems having efficiencies of 75 percent to 80 percent which are consistent with the cost assumptions of the model. Conceivably, such efficiencies may be possible to achieve in well-managed surface irrigation systems. If so, it may not be economical to invest the capital necessary to upgrade surface irrigation systems to pressurized systems for areas subject to low-frequency irrigation due to water delivery schedules. In addition, subsurface drainage systems need only be designed to extract 79 percent to 93 percent of the root-zone drainage, allowing the remaining fraction to contribute to shallow ground-water storage, crop water use by upward flow, and deep percolation out of the system.

The stochastic optimization model was designed for regional planning which limits some of its applicability to farm-scale implementation practices. For example, the model considers only one annual irrigation schedule regardless of the type of irrigation system used. If the irrigation schedule is indeed more-or-less fixed by the water districts, it appears that surface irrigation system are the most economical. However, the model fails to consider a situation of continuous drip irrigation, or similar high-frequency irrigation systems. In addition, the costs associated with irrigation systems operated to achieve a particular irrigation efficiency are difficult to evaluate and apply. Clearly, the optimal management strategy is affected by these costs. Finally, drainage efficiencies on the order of 80 percent to 90 percent do not appear possible in the field, and it is not clear how such a management strategy may be economically implemented, especially in fields with existing systems. On the other hand, the model results indicate that complete recovery of root-zone drainage by the drainage system is not required for maximum net benefits.

The model also was unable to consider changing drainage practices over time. For example, as drainwater disposal costs increase and water applications are managed to reduce root-zone drainage, it may be desirable to close, or otherwise alter, the drainage system. On the other hand, the quantity and quality of drainwater collected by relatively deep drainage systems are largely controlled by the shallow ground water. Eventually, the drainwater quality may improve as saline shallow ground water is removed and then be usable for irrigation, or blending. A more comprehensive dynamic-based model may be necessary to incorporate the effects of changing management practices on shallow ground water, crop yield, and net benefits over time.

Physical conditions used in the model may not be applicable to every region of interest. The model presumes existence of a uniform clay layer at relatively shallow depth throughout the region. A more realistic description may include a fragmented clay layer covering a range of depths. Hydraulic conductivities used in the model were relatively low and salinities relatively high. In this respect, the model considered marginal fields having poor soil-water conditions. These conditions limit only the applicability of some of the results discussed previously, however, not the general applicability of the model to different regions when sufficient data is available.

## SUMMARY AND CONCLUSIONS

Crop cultivation in arid parts of the world often encounters problems of soil salinization and poor soil aeration resulting from saline shallow ground water. Saline shallow ground water is often present as a direct consequence of irrigation with imported water having a measurable salt load. Drainage systems are designed to alleviate problems associated with the shallow ground water and to maintain a root zone favorable to plant growth. Reductions in crop yield occur when the root zone is salinized and/or "waterlogged." This chapter outlines the key geohydrologic processes involved in this system; how these processes can be modeled; the need to approach development of irrigation and drainage management strategies from a regional perspective; and results of one such regional analysis utilizing a stochastic simulation model.

Many of the results indicate the importance of considering irrigation and drainage planning simultaneously, the need to consider planning from a regional perspective due to interaction of local soil-water flows with the regional shallow ground water, and the value of incorporating uncertainty in the analysis from both a planning and realistic ground-water simulation perspective. Physical parameters having the largest effect on regional net benefits for a particular management strategy included variability in water applications and hydraulic conductivity, and mineral dissolution in the root zone. Drainage disposal costs had a relatively moderate effect on regional net benefits.

Optimal management strategies were similar for all combinations of conditions considered. This similarity for a relatively broad range of conditions may have been due to failure to consider high-frequency irrigations and the large cost difference between surface and pressurized irrigation systems. Other limitations of the model were associated with computational problems which constrain the extent of realistic simulation. Nevertheless, the stochastic optimization of regional net benefits as controlled by the influence of saline shallow ground water on crop yield represents an approach towards economic analyses that is based on the hydrologic system of concern.

## REFERENCES

Al-Layla, M. A., 1978. Effect of Salinity on Agricultural in Iraq, *ASCE Journal of Irrigation and Drainage Engineering*, b4(IR2).

Anderson, M. P., 1979. Using Models to Simulate the Movement of Contaminants Through Groundwater Flow Systems, CRC Crit. Rev., *Environmental Contractor*, 9(2), pp. 97-156.

Bear, J., 1979. *Hydraulics of Groundwater*. McGraw-Hill Book Co., Inc., New York, NY.

Gates, T. K. and Grismer, M. E., 1989. Irrigation and Drainage Strategies in Salinity-Affected Regions, *ASCE Journal of Irrigation and Drainage Engineering*, 115(2), pp. 255-284.

Gates, T. K.; Wets, R. J. B.; and Grismer, M. E., 1989. Stochastic Approximation Applied to Optimal Irrigation and Drainage Planning, *ASCE Journal of Irrigation and Drainage Engineering*, 115(3), pp. 488-502.

Grismer, M. E. and Gates, T. K., 1988. Estimating Saline Water Table Contributions to Cotton Water Use, *California Agriculture*, 42(2), pp. 23-24.

Grismer, M. E.; Gates, T. K.; and Hanson, B. R., 1988. Irrigation and Drainage Strategies in Salinity Problem Areas, *California Agriculture*, 42(5), pp. 23-24.

Grismer, M. E., 1989. Drainage Efficiency and Drainwater Quality. *Proceedings of 11th International Congress on Agricultural Engineering*, Dublin, Ireland, 1, pp. 285-290.

Grismer, M. E., 1990a. Leaching Fraction, Soil Salinity and Drainage Efficiency, *California Agriculture*. In Press.

Grismer, M. E., 1990b. Deep Percolation, Drainage and Water Quality, Proceedings of ASCE National Conference on Irrigation and Drainage Engineering, July, Durango, CO., pp. 355-362.

Interagency Drainage Program (IDP), 1979. *Agricultural Drainage and Salt Management in the San Joaquin Valley.* Final Report.

Khan, I. A., 1980. Determining Impact of Irrigation on Ground Water, *ASCE Journal Irrigation and Drainage Engineering,* 106(IR4), pp. 331-344.

Letey, J.; Dinar, A.; Woodring, C.; and Oster, J., 1990. An Economic Analysis of Irrigation Systems, *Irrigation Science,* 11, pp. 37-43.

Oster, J. D., 1989. Alternative Irrigation Strategies in the San Joaquin Valley of California. Proceedings of the 1989 USCID Regional Meeting, Sacramento, CA.

Oster, J. D. et al, 1988. *Associated Costs of Drainage Water Reduction.* U.C. Committee of Consultants Report No. 2, U.C. Salinity/Drainage Task Force and U.C. Water Resource Center.

San Joaquin Valley Drainage Program, 1989. *Overview of the Use of the Westside Agricultural Drainage Economics Model (WADE) for Plan Evaluation.*

Singh, U. P., 1984. Agricultural Water Resources Management in India, *ASCE Journal of Water Resource Planning and Management,* 110(1), pp. 30-38.

# 5 GROUND-WATER PUMPING FOR WATER TABLE MANAGEMENT AND DRAINAGE CONTROL IN THE WESTERN SAN JOAQUIN VALLEY

Nigel W. T. Quinn, Cornell University and
San Joaquin Valley Drainage Program

## ABSTRACT

Drainage management strategies for control of salt and selenium contamination problems in the San Joaquin Valley (Valley) of California should account for subregional differences in ground-water chemistry and aquifer characteristics. Appropriate strategies include the reduction of surface applied irrigation water, the pumping of ground water to control shallow water tables, and the application of drainage reuse systems. The manner in which these techniques should be combined in a drainage management strategy depends on the quality of water supplied by each of these sources as well as economic considerations. In contrast to the other two strategies, ground-water pumping of the semiconfined aquifer of the western Valley has a finite life because of the limited volume of ground water available within the aquifer of acceptable quality for irrigation. Any reduction in the quality of applied irrigation water, due to blending with pumped ground water, and its consequent negative effect on crop yields, should be evaluated against the benefits realized by stabilizing saline water tables to levels below the active root zone. The rate at which the ground-water resource is depleted of water usable for direct application to crops is influenced by pumping rates, well field design, and existing chemical and textural stratification within the semiconfined aquifer.

## INTRODUCTION

Problems associated with drainage of irrigated agricultural lands on the west side of the Valley result from the necessary annual leaching of accumulated salts from the crop root zone. These salts, which threaten the future productivity of irrigated agriculture in the Valley, also include toxic and potentially toxic trace elements such as selenium, boron, arsenic, and molybdenum which are eliminated in subsurface drainage water. Impacts associated with the storage and disposal of this subsurface agricultural drainage include degradation of surface and ground-water quality, contamination of fish and wildlife habitats, and risks to public health. The scale and urgency of the problem call for creative approaches in formulating and evaluating alternatives for source control and drainage management and responsible stewardship of resources.

A wide range of strategies have been identified for managing drainage and drainage-related problems in the Valley (SJVDP, 1989). Of these strategies only source control and drainage recycling and reuse currently enjoy wide application. The use of ground-water pumping to deliberately lower water levels in areas affected by saline high water tables and to reduce or obviate the need for subsurface drainage facilities is not widely practiced, largely because of uncertainty about its potential. Except for two studies conducted by the San Joaquin Valley Drainage Program (SJVDP) (Schmitt, 1988 and 1989) there is no published literature on the technical and economic feasibility of such a strategy. This chapter describes the approach that was taken to evaluate ground-water pumping strategies in the context of a comprehensive and coordinated long-term plan for drainage reduction and management on the west side of the Valley.

## EXISTING GROUND-WATER PUMPING STRATEGIES

Pumped ground water is a supplemental water supply in those areas which periodically receive cutbacks in State or Federal project water during years of low precipitation or water shortage. The volume rate of pumpage of these wells and the qualities of the well discharge is largely dependent on the location with respect to the surficial geology of the western Valley, and the hydrogeologic characteristics of the deposits within which the wells are screened. The semiconfined aquifer contains four distinct hydrogeologic units (Coast Range Alluvium, Sierran Sands, Flood-basin deposits, and the Corcoran Clay), all of which differ in texture, hydraulic conductivity and oxidation state (figure 1). The stratigraphic juxtaposition of these geologic deposits influence the hydrochemistry of the semiconfined and confined ground-water aquifers on the west side of the Valley. An understanding of both the hydraulic properties and depth distribution of salts within these deposits is fundamental to the future design of pump well fields to control regionally high water tables.

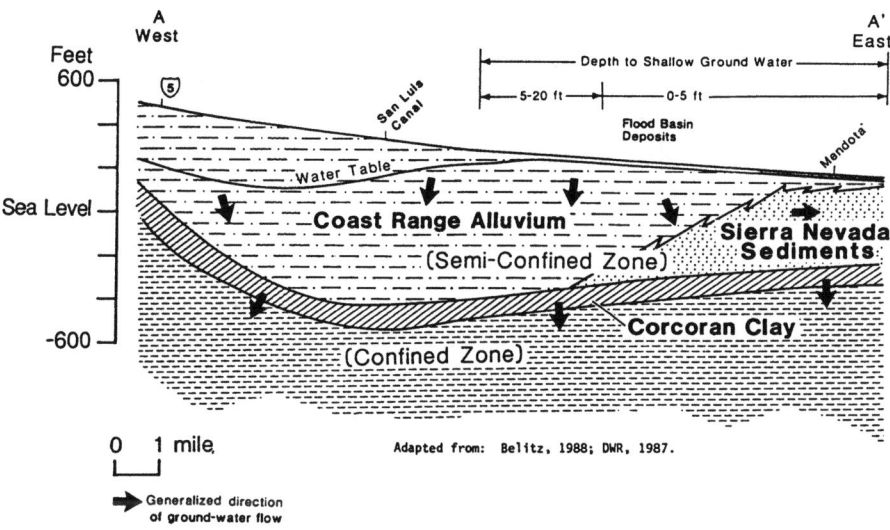

Figure 1. Generalized geohydrologic cross-section of the San Joaquin basin.

## Pumping from the Confined Aquifer Below the Corcoran Clay

The Corcoran Clay is a reduced, lacustrine confining layer ranging in thickness between 20 and 120 feet and is located at a depth of 400 feet at the Valley trough and 900 feet along the Coast Range (see chapter by Gilliom). This confining layer severely restricts vertical flow between the upper semi-confined and lower confined aquifers except where ground water is transmitted through the gravel-packs of wells, screened both in the confined and semiconfined aquifers. Many wells have been developed at depths of 450 to 1,000 feet below the land surface within the confined aquifer.

In the past 20 to 30 years, the volume of ground water extracted from the confined aquifer has declined with increased reliance on surface water supplies. The rate of vertical flow between the aquifers has similarly declined, to current values of between 0.05 and 0.4 acre-feet/acre-year in drainage problem areas, as the hydraulic gradient across the Corcoran Clay has been reduced (Belitz, 1988). Under current conditions, the net leakage of ground water across the Corcoran Clay helps to reduce the rate at which ground-water levels are rising, in the shallow ground-water affected areas on the west side of the Valley, and hence helps to reduce the volume of drainage requiring treatment and disposal. If current trends continue, however, this volume of leakage across the Corcoran Clay will decrease as the piezometric head within the confined aquifer rises. This process will lead to an increase in the land area affected by shallow ground-water tables. Continued pumping from below the Corcoran Clay can have a long-term beneficial effect in contributing to the control of rising shallow ground-water levels.

Although current salt and selenium levels in the confined aquifer do not preclude its use as an irrigation water supply, sustained pumping of the confined aquifer may change this situation over time. A number of wells exist in the study area which are screened in both semiconfined and confined aquifers, or which were poorly constructed, allowing hydraulic communication between the two aquifers. This can provide a pathway for selenium migration into the confined aquifer. Some large irrigation wells in the Panoche Water District are already pumping ground water with Total Dissolved Solids (TDS) of above 3,000 mg/L, well above the average TDS of sub-Corcoran ground water which is typically less than 700 mg/L TDS.

## Pumping from the Sierran Sands

The Sierran sands are reduced deposits which are derived from the Sierra Nevada range on the east side of the Valley. These deposits interfinger the Coast Range alluvium deposits along the Valley trough in the northern half of the Valley, referred to in figure 1 as the San Joaquin Basin. The Sierran sand deposits are 400 to 500 feet thick at the Valley trough and diminish in thickness to the east and to the west. Compared to the Coast Range sediment, the Sierran deposits are relatively inert, and contain few soluble salts.

Field studies have demonstrated good hydraulic connection between coarse sedimentary deposits and the overlying alluvial deposits in many of the areas currently affected by contaminated drainage problems. A short-term pump test during 1988-89 (Schmidt, 1988) indicated that the shallow water table could be lowered by approximately 0.5 foot over a radius of several hundred feet from the pumping wells. The wells were drilled to a depth of between 35 feet and 60 feet, perforated within the Sierran sands and pumped continuously at a rate of about 30 gal/min for 2 to 7 hours. A longer-term pumping test (Schmidt, 1989) conducted within the Sierran sands using a well, perforated from a depth of 112 feet to 244 feet, showed a similar response.

In the past, the quality of the ground water in the Sierran sands has been suitable for most uses. However, ground-water salinity has been reported higher than the 4,500 to 5,000 mg/L TDS values recorded in the long-term pumping test by Schmidt (1989). There appears to be a trend toward increasing salinity in the Sierran sands that underlie several of the drainage problem areas. Soluble selenium, however, is found only in trace amounts in the Sierran sand deposits (Deverel and Gallanthine, 1989). The current understanding is that the soluble selenium in ground water, which moves from the Coast Range alluvium into the reducing environment of the Sierran sands, is reduced to more strongly adsorbed species such as selenite or elemental selenium (figures 1 and 2, Gilliom et al., 1989). Although pumping from the Sierran sands seems to effect a natural removal of selenium in the reducing (redox) zone, there is a possibility that, over time, the boundary of this reducing zone will move eastward if pumping continues. This may affect the selenium adsorption potential of the Sierran sands (Gilliom et al., 1989).

## Pumping from the Coast Range Alluvium

The Coast Range alluvium contains both water-laid and mudflow deposits originating from the Coast Range. These deposits are normally well oxidized and high in soluble salts including selenium. The surficial deposits at the upper end of the fan are higher in sand and gravel than the deposits at the margin of the fan and have been more thoroughly leached over the course of time. Irrigation recharge in these zones has served to displace selenium and other soluble salts from the soil to the deep ground water. The lower or distal margins of the alluvial fans were natural discharge zones, prior to the development of irrigated agriculture, and were subject to a process of evaporative concentration, where water tables rose close to the land surface (Gilliom et al., 1989). Soluble salts precipitated in these areas, causing high concentrations of selenium and other trace elements to accumulate in the surface soils and in the shallow ground water. The alluvial deposits are approximately 850 feet deep along the Coast Range and pinch out at the Valley trough.

Water quality is a major concern in the semiconfined aquifer within the Coast Range alluvium deposits. Ground water below an average depth of 150 feet may prove to be moderately saline, but is likely to contain low (<5 p/b) selenium concentrations (Gilliom et al., 1989).

# GROUND-WATER PUMPING STRATEGIES FOR DRAINAGE MANAGEMENT

Ground-water pumping strategies to manage saline shallow water tables should be designed in accordance with the hydraulic properties and the distribution of salts and trace elements within the semiconfined aquifer. These strategies can take a number of forms as follows:

(1) Pumping by individual growers, primarily for water supply. This has an indirect effect on shallow water tables. Pumped water may be blended with

existing surface water supplies or used directly during the latter part of the season, when crops can tolerate higher salt concentrations in irrigation water, without affecting crop yields. Typically, the volume of water pumped would be limited by the cost of pumping and any seasonal shortfall in surface water deliveries. Existing pumps, screened in the deep semiconfined aquifer or in the confined aquifer, would typically be used in this instance.

(2) Pumping by growers deliberately to control local high saline water tables. Pumping may be practiced continuously or only during certain times of the year. Smaller capacity well pumps, spaced more closely together, would be appropriate to successfully apply this strategy. Adequate hydraulic communication between the layer in which the well screen is located and the water table is necessary to acheive a water table response using this pumping strategy.

(3) Pumping strategies adopted at the regional or the water district level. Wells could be operated by water districts or drainage districts to regionally lower water tables in drainage problem areas with the discharge either pumped directly into the surface water distribution system or treated by the district prior to discharge or reuse. A similar strategy was practiced by the Panoche Water District during 1989. Panoche Water District adopted a policy whereby growers received credit for ground water pumped which was applied against the cost of surface water deliveries. The pumped water was then blended with the district freshwater supply, thus diluting the salinity of the pumped water to a level suitable for direct application to field crops.

## FORMULATION OF INNOVATIVE GROUND-WATER PUMPING STRATEGIES

In formulating a plan which includes ground-water pumping for management of saline, shallow water tables the unique combinations of aquifer chemistry, hydrogeology, and institutional settings, known to exist within five planning subareas defined by the SJVDP (SJVDP, 1989), were considered. These subareas have been further divided into ground-water quality zones that cluster those areas which contain similar species and concentration levels of contaminants within the upper 20 feet of the semiconfined aquifer. Water quality zones (labeled A-D) for the Grasslands and Westlands subareas are shown in figures 2 and 3.

Although the contaminant species and the concentration of the contaminants in the shallow ground water bear some relation to the contaminants and their concentrations at lower depths, the general level of understanding of aquifer hydrochemistry is insufficient to allow confident predictions of the depth distribution and concentration of these contaminants. Hence, a reconnaissance study of the salt and contaminant distribution in the semiconfined

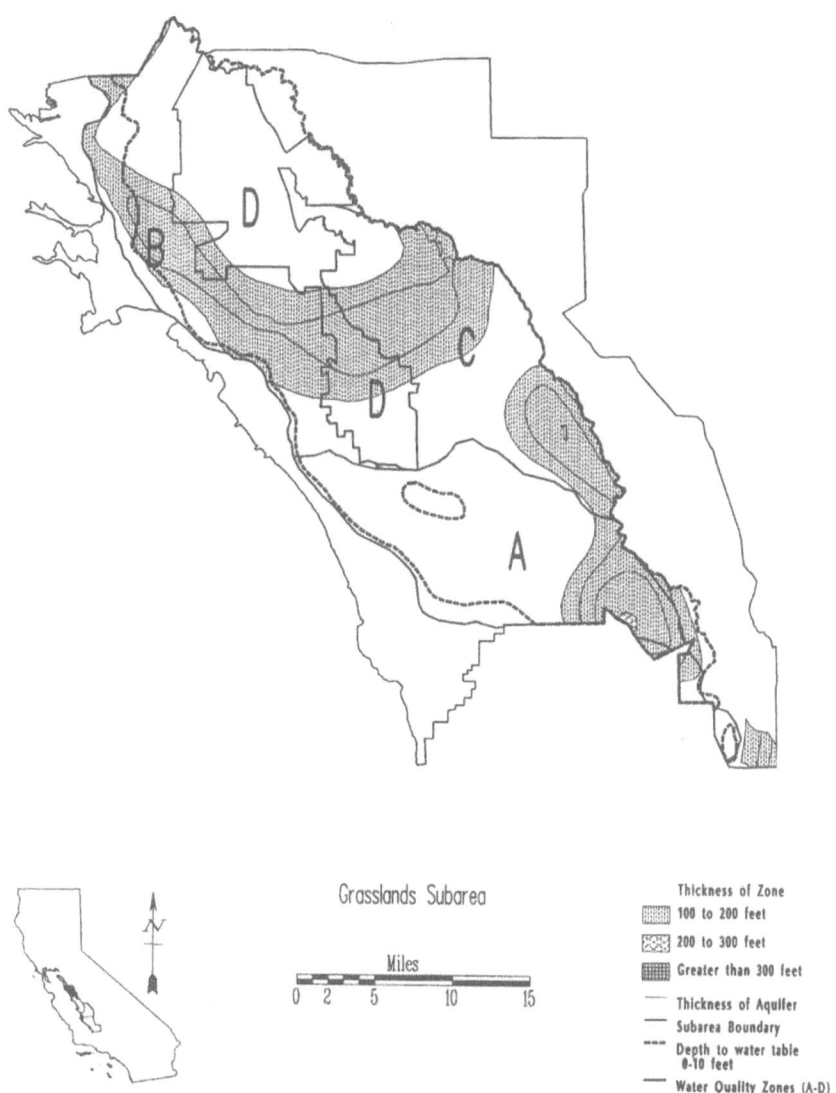

Figure 2. Zones of usable ground water in drainage problem areas with TDS < 1,250 mg/l in the semi-confined aquifer (Grasslands subarea). Water quality zones A - D denote different levels of TDS and Se in shallow ground water.

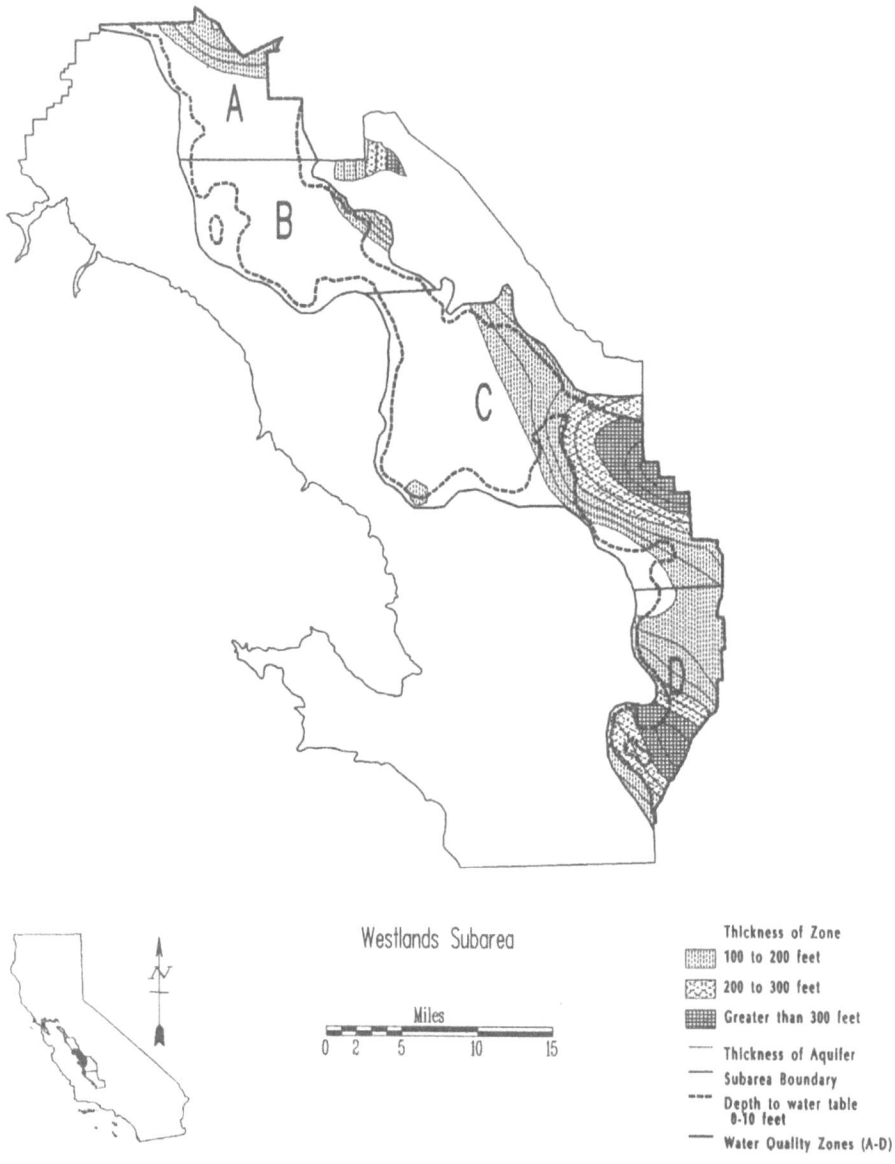

Figure 3. Zones of usable ground water in drainage problem areas with TDS < 1,250 mg/l in the semi-confined aquifer (Westlands subarea). Water quality zones A - D denote different levels of TDS and Se in shallow ground water.

aquifer and the confined aquifer was undertaken to help identify zones of ground water low enough in dissolved salts and boron to be usable as an irrigation water supply, or, sufficiently low in TDS, boron, arsenic, and selenium, to be usable as a supplemental water supply for wetlands. This involved an extensive review of well drilling logs, observation well records, and existing data base records for water quality data; hydraulic characteristics of the semiconfined aquifer; and records which describe the screened depth of existing wells throughout the western Valley. These data were subsequently interpreted to develop average characteristics of each water quality zone. Only the results for the Westlands and Grasslands subareas are discussed in this chapter.

## SURVEY OF GROUND-WATER QUALITY

Only two observation well cluster sites exist which allow direct observation of the distribution of salts with depth within the semiconfined aquifer. The only other sources of data are water quality samples drawn from active pumping wells. The initial search was conducted in the U.S. Geological Survey NWIS (WATSTORE) data bases for the years 1970 through mid-1989. This search was subsequently expanded to include the years 1958 through 1969, primarily to obtain data for parts of the Westlands subarea with no other available data.

Interpretation of the depth distribution of TDS and trace elements is difficult using these data, since the screened interval of these wells is extremely variable and in many cases is not known. The streamlines generated by each pumping well may draw water from a large vertical interval in the aquifer-- hence it is difficult to discern the nominal interface between zones of suitable and unsuitable ground-water quality in the semiconfined aquifer. Pumped ground water, usable for irrigation without blending, was defined as having a concentration of less than 1,250 mg/L TDS. The upper threshold of usability was chosen to 2,500 mg/L TDS, at which level it was assumed that some blending would be needed. These criteria were determined using salinity tolerance figures from FAO Publication 29.1 (Ayars and Westcot, 1985) and from discussions with water district managers.

Irrigation wells were assumed to be perforated over the lower 58 percent of their total depth. This figure was derived from a sampling of 75 irrigation wells with screened intervals reported in the U.S. Geological Survey WATSTORE data base. The assumed value of 58 percent was the median for the 75 wells, and was adopted to avoid unnecessarily complex interpretations in areas with numerous irrigation wells. The midpoint of a single or multiple screened interval(s) was used for the purpose of assigning an effective well depth correlated with water quality. In cases where no data on screened intervals were available, the assumption was made that the well was screened in the last 10 percent of its depth. A more complete description of this process is contained in Quinn et. al (1990).

Ground-water quality was evaluated using TDS as a measure of salinity. This was the most widely available parameter in the data bases queried. In order to develop an estimate of the volume of usable water resident in the semiconfined aquifer, assumptions were made concerning the overall patterns of salt, selenium, and boron distribution in the study area. These were:

(1) Water quality in the 0- to 50-foot-depth zone is generally unsuitable for direct agricultural use in a ground-water pumping alternative, especially in areas located downslope on the alluvial fans and in areas that still retain high levels of solublizable salts.

(2) In depth zones where ground-water quality estimates from existing wells indicate considerable variability, the deepest well meeting the criterion of 1,250 mg/L marks the top of the upper limit of the depth zone containing acceptable water.

(3) The mean concentration of ground water in the depth zone located between the Corcoran Clay and the upper limit of the depth zone (previously defined) is assigned to the ground-water quality zone.

(4) Where there are no wells that meet the water quality criterion, a conservative assumption is made that there is no potential for ground-water management in that ground-water quality zone.

The assumptions made above are conservative since most data are derived from existing pumping wells. Wells induce the greatest vertical movement of salts and other contaminants close to the well casing--this effect declines rapidly as distance increases from the well. Hence, water quality measurements made within the zone of influence of an agricultural pumping well are likely to reflect higher TDS and trace element concentrations than the average aquifer concentration, at equal depth, in areas not subjected to the same pumping stresses.

## INTERPRETATION OF GROUND-WATER QUALITY

Water quality in the 0- to 50-foot-depth zone of the Grasslands and Westlands subareas is highly variable, is generally poor to very poor, and is frequently underlain by water of lesser quality at depths ranging from 10 to 100's of feet. This pattern is attributable to the history of irrigation and leaching of salts and other soluble minerals from the crop root zone, and is related to the predevelopment distribution of soluble salts and the geohydrologic characteristics of the area (Gilliom et al., 1989 and Deverel et al., 1984). This finding has important implications for the operation of existing pumped well systems and the design of new ground-water pumping systems. The maximum usable life of the ground-water resource can be achieved by retarding the downward rate of the poor quality ground-water zone.

The thickness of ground-water zones meeting the salinity criterion of 1,250 mg/L are less in the eastern part of the study area than in the west. This reflects the regional pattern of higher soil and ground-water salinities at the margins of alluvial fans adjacent to the Valley trough, and the decreasing semiconfined aquifer thickness from the maximum which is located close to the midpoint of the alluvial fans (see the chapter by Gilliom). However, the Sierran sands which dominate the Valley trough typically contain better quality water than the Coast Range derived alluvium to the west. This influence is also seen on the eastern extremities of the Westlands subarea.

The semiconfined aquifer in the Grasslands subarea as a whole can be characterized as having less than 200 feet of ground water with salinities less than 1,250 mg/L. Boron concentrations are less limiting than salinity for agricultural applications. Shallow ground-water quality in the subarea, tends to be lowest in the south, adjacent to and on the Coast Range derived alluvium, and in the extreme north. In the southern third of the subarea salinity exceeds 7,000 mg/L, boron exceeds 8 mg/L, and selenium exceeds 50 $\mu$g/L. Much of the southwestern part of the subarea has not been characterized due to a lack of data at depths greater than 50 feet. In much of this uncharacterized area ground-water quality in the 0 to 50 foot depth is characterized by elevated salinity (1,600-14,000 mg/L), elevated boron (>8 mg/L) and elevated selenium (>50 $\mu$g/L).

In the Westlands subarea the Corcoran Clay lies 400 to 750 feet below ground level, increasing in depth from north to south. Ground-water quality is variable, with less than 50 to 200 feet of semiconfined aquifer having a ground-water quality less than 1,250 mg/L TDS. The map (figure 3) indicates increased storage of usable ground water within the semiconfined aquifer from west to east. This coincides with increasing depth to the Corcoran Clay in the Valley trough. There is an aquifer zone approaching 400 feet thick on the southern Los Gatos Creek Alluvial Fan. Shallow ground-water quality is generally poor, with much of the eastern part of the subarea characterized by salinity exceeding 7,000 mg/L, and boron exceeding 8 mg/L. Selenium exceeds 50 $\mu$g/L throughout much of the northern two-thirds of the subarea, and exceeds 20 g/L in about half of the southern third of the subarea.

## USABLE LIFE OF GROUND-WATER RESOURCES UNDER EXISTING CONDITIONS

In each of the subareas described above, where a usable volume of good quality ground water can be located below a saline and trace element contaminated zone, this usable volume is diminishing over time due to net outflow from the semiconfined aquifer across the Corcoran Clay. The magnitude of these outflows is related to the level of past and current pumping activity within the confined aquifer. The highest flows occur to the west of the drainage problem areas where sub-Corcoran pumping has been highest in the past. Using average outflow rates for the drainage problem areas, and taking into account the minimum, maximum, and mean aquifer resources of usable ground water with TDS less than 2,500 mg/L, an estimate can be made of the remaining usable life of these semiconfined aquifers (table 1). The remaining usable life is defined as the time it would take for the zone of poor quality ground water to completely displace the zone of presently pumpable ground water. Certain simplifying assumptions are made in this analysis:

- Ground-water pumping is continued at activity levels such that current piezometric levels in the confined aquifer remain unchanged.
- The heterogeneity of the aquifer and the high degree of textural stratification permit the assumption of plug flow conditions.
- Flow is uniform across the lower aquifer boundary of the semiconfined zone.

Table 1 shows that the mean time to displace pumpable ground water with saline ground water in the semiconfined aquifer ranges from 46 to 273 years. Because the depth of penetration of saline ground water is not constant over the water quality zones the minimum and maximum values should also be considered. The fact that most of the data obtained to make the estimation of the current depth of penetration of saline ground water was from pumped wells may bias the estimates to underreport the useable life of the ground-water resource.

The useable life of the ground-water resource within the semiconfined aquifer cannot be estimated with any degree of certainty from the current data. However, it may be concluded that in the absence of deliberate strategies to accelerate the downward movement of the saline zone within the semiconfined aquifer the resource should last beyond 2040, the planning period used by the SJVDP.

Table 1. Estimate of aquifer usable life under existing hydrologic conditions.

| Planning Subarea | WQ Zone | Area of Zone (acres) | Area with >50 ft of Usable GW (acres) | Percentage of Total Area (%) | Av. Corcoran Leakance (ft/year) | Aquifer Thickness w/Usable GW | | | Aquifer Life (TDS < 2,500 p/m)[b,c] | | | |
|---|---|---|---|---|---|---|---|---|---|---|---|---|
| | | | | | | Minimum (ft)[a] | Maximum (ft)[a] | Mean (ft)[a] | Minimum (years) | Maximum (years) | Mean (years) |
| Grasslands | A | 117,000 | 40,400 | 35 | 0.10 | 50 | 350 | 165 | 75 | 525 | 248 |
| | B | 94,000 | 74,000 | 79 | 0.30 | 50 | 200 | 134 | 25 | 100 | 67 |
| | C | 108,000 | 71,700 | 66 | 0.05 | 50 | 150 | 91 | 150 | 450 | 273 |
| | D | 89,000 | 67,800 | 76 | 0.30 | 50 | 150 | 91 | 25 | 75 | 46 |
| Westlands | A | 43,000 | 14,000 | 33 | 0.20 | 50 | 250 | 150 | 38 | 188 | 113 |
| | B | 67,000 | 43,000 | 64 | 0.25 | 50 | 350 | 183 | 30 | 210 | 110 |
| | C | 145,000 | 107,600 | 70 | 0.25 | 50 | 450 | 190 | 30 | 270 | 114 |
| | D | 46,000 | 36,000 | 78 | 0.30 | 50 | 400 | 222 | 25 | 200 | 111 |

[a] Ground-water quality of 2,500 p/m is assumed to be the threshold of acceptability for agricultural purposes.
[b] Assumes a drainable porosity of 0.15 for the semiconfined aquifer deposits.
[c] Aquifer life is the time interval for the zone of contaminated water to completely displace usable ground water (assuming plug flow).

## USABLE LIFE OF THE SEMICONFINED GROUND-WATER AQUIFER UNDER FUTURE PUMPING STRATEGIES

Deliberate pumping within the semiconfined aquifer to manage high saline water tables in the drainage problem area can significantly diminish the life of the ground-water resource. Additional ground-water pumping, managed to control water tables ($Q_{pump}$), has been assumed to remove, on average, a volume equivalent to the annual change in storage (recharge - leakage across the Corcoran Clay) and, in addition, a volume equivalent to a 0.1 foot/year change in applied water. Ground water recharge is currently assumed to be 0.7 acre-foot/acre-year in areas achieving salt balance within the crop root zone. This is equivalent to the average drainage yield for the Grasslands subarea calculated by the State Water Resources Control Board (SWRCB, 1988) and is 7 percent smaller than the average estimate for drainage-affected areas in the Westlands Water District (SJVDP, 1989). The average annual recharge is expected to decline by 0.2 acre-foot/acre with the adoption of conservation practices, as part of a comprehensive regional drainage management plan.

[1] $Q_{pump}$ = (0.7 acre-foot/acre - 0.2 acre-foot/acre - 0.1 acre-foot/year) + 0.1 acre-foot/acre

= 0.5 acre-foot/acre

For a well pumping within a well field and designed to control water tables over an area of 160 acres (1/4 section), the annual volume of pumpage can be calculated:

[2] $Q_{pump}$ = $(2640)^2$ (0.5 acre-foot/acre-year)

= 3.48 x $10^6$ ft³/year

This is equivalent to an annual average pumping rate of approximately 50 gal/min or 200 gal/min for a well pumped for only 13 weeks during the year. In a series of pump tests conducted at Broadview Water District and Firebaugh Canal Company sites in the Grasslands subarea (Schmitt, 1988), at depths ranging from 35 to 58 feet, maximum sustainable pumping rates of about 35 gal/min were measured. Water tables were observed to decline by more than 1 foot, in shallow piezometers several hundred feet from the pumped well, after several days. Subsequent tests (Schmitt, 1989) made in the same area but at greater depth (112 to 244 feet) and at at a pumping rate of 2,000 gal/min over a 14 day period, produced 0.5 foot of drawdown several hundred feet from the pumped well.

The effect of this rate of pumping on the life of the ground-water resource is illustrated in table 2. The table assumes the same mean and maximum aquifer thickness data reported in table 1 and includes the effect of pumping at a rate of 200 gal/min. The far-right columns in table 2 show the dramatic effect of pumping on aquifer life in those areas where the average flux across the Corcoran Clay is estimated to be 0.1 foot/year or less. The mean usable life (as previously defined) of the semiconfined aquifer within the Westlands subarea diminishes from an estimate of over 100 years to approximately 25 years in many of the aquifer zones. However, since the average economic life of a pumping well is between 20 and 30 years (SJVDP, 1990), this may still justify investment in this strategy.

Interpretation of these figures must be made with care, however, since they may overstate the length of time that a pumped well can continue to pump usable water. The strict plug flow assumption is not valid in the immediate vicinity of a pumping well; the drawdown cone into the well increases the hydraulic gradient between the well inlet and the water table, accelerating vertical flow along streamlines close to the pumped well. The well will generally produce a discharge higher in TDS than that contained in the aquifer several hundred feet from the well at the same depth from the surface. In addition, in areas where existing wells are used to implement the ground-water pumping strategy for drainage reduction, the length of the perforated well screen above the calculated mean screen elevation may considerably diminish the effective thickness of the pumpable ground-water zone. The estimated aquifer life figures may be more appropriately attributed to pumps located below the Corcoran Clay. However, as previously discussed, pumping within the confined aquifer is unlikely to have any pronounced effect on water tables in the vicinity of well except in the long term. This result has been verified by simulations performed using the ground-water model described in the next section and from preliminary simulations with an unpublished ground-water flow model (Belitz, 1990).

These models were used to more realistically evaluate the effect of semiconfined aquifer pumping on water tables and to better predict the usable life of individual wells located in a well field as part of a ground-water management plan. These models allow sensitivity analyses to be performed on the usable life of a single pumped well as affected by aquifer boundaries, pumping rates, depth of pumping, aquifer hydraulic properties, and aquifer chemistry.

Table 2. Estimate of aquifer usable life under future pumped conditions.

| Planning Subarea | WQ Zone | Area of Zone (acres) | Area with >50 ft of Usable GW (acres)[a] | Percentage of Total Area (%) | Av Corcoran Leakance (ft/year)[b] | Max Aquifer Thickness (ft) | Max Life Aquifer (years)[c] | Pumping Rate (ft/yr)[d] | Max Life w/Pumping (years)[e] | Mean Life w/Pumping (years)[e] |
|---|---|---|---|---|---|---|---|---|---|---|
| Grasslands | A | 117,000 | 40,400 | 35 | 0.10 | 350 | 525 | 0.4 | 105 | 74 |
|  | B | 94,000 | 74,000 | 79 | 0.30 | 200 | 100 | 0.4 | 43 | 14 |
|  | C | 108,000 | 71,700 | 66 | 0.05 | 150 | 450 | 0.4 | 50 | 91 |
|  | D | 89,000 | 67,800 | 76 | 0.30 | 150 | 75 | 0.4 | 32 | 10 |
| Westlands | A | 43,000 | 14,000 | 33 | 0.20 | 250 | 188 | 0.4 | 62 | 28 |
|  | B | 67,000 | 43,000 | 64 | 0.25 | 350 | 210 | 0.4 | 81 | 25 |
|  | C | 145,000 | 107,600 | 70 | 0.25 | 450 | 270 | 0.4 | 104 | 26 |
|  | D | 46,000 | 36,000 | 78 | 0.30 | 400 | 200 | 0.4 | 86 | 24 |

[a] Area within the water quality zone containing more than 50 ft. of aquifer with usable ground water.
[b] Estimated Corcoran leakance for the Grasslands subarea (zones A and C) and Westlands (zones A and B) are based on USGS regional flow model
[c] Maximum life of the aquifer prior to accounting for pumping above mean historic levels.
[d] Pumping rate based on 200 gals/min well on a 1/4 section operated for 13 weeks/year. No reduction in Corcoran leakance assumed.
[e] Aquifer life accounting for additional downward flux induced by pumping

## SIMULATION OF A SINGLE PUMPED WELL IN THE SEMICONFINED AQUIFER

A finite element model which simulates radial ground-water flow towards a pumping well, has been developed by Matanga (Frind and Mantanga, 1985 and Mantanga and Quinn, 1990) and is used here to develop a more detailed analysis of the long-term utility of a ground-water pumping strategy for drainage management. The model calculates steady-state heads in the semiconfined aquifer in response to pumping. Water table elevations of the ground-water flow system are based on the calculated heads. The model then generates a spatial distribution of stream functions, which allows the flowpath of salts originating from various levels in the aquifer to be traced into the pumped well. The arrival time of these salts from their origin, to the well, can be calculated. A mass balance of concentrations along these flow paths, at any point in time, is used to estimate the average concentration of the pumped water.

The model is applied to representative aquifer conditions in each water quality zone. These water quality zones divide the regional aquifer into subunits on the basis of shallow ground-water conditions, water table height, and depth distribution of contaminants. Boundary conditions include the lower boundary flux (leakance across the Corcoran Clay layer), upper boundary flux (rate of irrigation recharge), and lateral flows in to and out of the system.

Controlled pumping of ground water within the deep semiconfined aquifer is used to manage shallow water tables in those areas that meet the following criteria:

- A shallow water table is present. Shallow water tables are defined as being within 5 feet of the soil surface.
- There is a zone of pumpable ground water with a TDS of less than 1,250 mg/L and a thickness of at least 100 feet in the semiconfined aquifer.
- Ground water pumped can be used to supplement irrigation surface water supplies.

## THEORY

Ground-water flow can be described in terms of hydraulic head or stream functions. Hydraulic head is a flow parameter that can be measured in the field and represents energy per unit weight of ground water. Its derivative with respect to space in a specified direction represents a driving force of ground-water flow in that direction. The stream function is not measurable. It is a mathematical concept that is useful in generating flow paths (streamlines) of ground water. The streamlines allow visual evaluation of ground-water flow and are becoming popular in solution of problems related to salt migration through subsurface flow systems. A two-dimensional aquifer is considered which is heterogeneous and anisotropic with respect to hydraulic conductivity. A dual theory of hydraulic head and stream functions has been applied by Frind and Matanga (1985) to describe steady ground-water flow in two-dimensional systems.

## Flow Through a Stream Tube

A streamline is defined as the curve that is everywhere tangent to the specific discharge vector (Bear, 1979). In addition to this, the stream function is, by definition, constant along a streamline. A stream tube is a space between two adjacent streamlines, say $\Psi_1$, and $\Psi_2 = \Psi_1 + \Delta\Psi$. It can be shown that Darcy discharge Q through a stream tube is:

[3] $Q = \Delta\Psi$

Equation [3] implies that the Darcy discharge within a stream tube is equal to the numerical difference between the two boundary stream functions. If the width of the stream tube W(s) is known, Frind and Matanga (1985) have shown that the time t taken by a fluid particle to travel a distance s is:

[4] $t = (\theta/\Delta\Psi)\int W(s)ds$

where $\theta$ is the porosity; and the distance s is along a streamline. From equation [4], the travel time of a fluid particle is simply the area of the stream tube multiplied by a constant. Thus, the travel time can be determined directly from a streamline plot, without the need to differentiate heads.

For a stream tube with an average width w, equation (4) may be written as:

[5] $t = (\theta W s)/\Delta\Psi$

Each stream tube intersects the salinity profile of the aquifer at various points along the length of the stream tube. The travel times from each of these intersections can be independently calculated--hence at any time (t) the current salt concentration in the stream tube can be determined and the current concentration of the discharge from the well calculated as a combination of saline discharges along each stream tube. The salt concentration in the pumped ground water is given by:

[6] $C(t) = (C_1 n_1 + C_2 n_2 + ....C_z n_z)/(n_1 + n_2 + .....n_z)$

where C(t) is the salt concentration at time t since the start of pumping; $n_1$ is the number of stream tubes in which ground water of TDS equal to $C_1$ is being pumped by the well; $n_2$ is the number of stream tubes in which ground water of TDS equal to $C_2$ is being pumped; $n_z$ is the number of stream tubes in which the ground water of TDS equal to $C_z$ is being pumped.

## EVALUATION OF PUMPING STRATEGIES

The semiconfined aquifer contains interbedded layers of sand, silt, and clay. For modeling purposes a generalized aquifer texture profile was used, similar to the profile developed by Geomatrix Consultants for the Westlands Water District (Geomatrix, 1990). The aquifer is treated as an eight-layer flow system. An illustration of the textural profile and salt distribution with depth of the semiconfined aquifer is presented in figure 4. This figure also shows the approximate location of the pumping wells in the deep and shallow aquifer layers of the semiconfined aquifer.

The uppermost layers (0 to 30 feet) and (30 to 105 feet) in the semiconfined aquifer are assumed to be isotropic and are assigned equal hydraulic conductivities of 1 foot/day. The remainder of the semiconfined aquifer is divided into sand (aquifer) and clay (aquitard) layers. Sand deposits are assigned hydraulic conductivity values of 30 feet/day for shallow and deep aquifers in the Coast Range alluvium (30 to 105 feet) and (205 to 265 feet) and 90 feet/day for the Sierran sand deposits (205 to 265 feet) (Geomatrix, 1990). The sub-Corcoran aquifer (545 to 695 feet) is assumed to have the same hydraulic conductivity as the Sierran sand deposits. Although aquifer anisotropy has been reported in several field studies (Deverel and Fujii, 1989), these values are typically less than 10 for alluvial deposits. Anisotropy is defined as the ratio between the horizontal hydraulic conductivity ($K_{xx}$) and the vertical hydraulic conductivity ($K_{yy}$). Because of the general level of uncertainty regarding the hydraulic properties of aquifer materials in the western Valley and to avoid complexity, anisotropy was assigned a value of 1.0 for both aquifers and aquitards in this study. The clay layers (aquitards) are all assumed to have a hydraulic conductivity of 0.008 foot/day except for the Corcoran Clay which is assigned a conductivity of 0.004 foot/day (Geomatrix, 1990). These values compare favorably with the values of vertical conductivity used by Belitz (1990), in calibration of a steady-state ground-water flow model of the Panoche and Cantua Creek alluvial fans, on the west side of the Valley.

The ground-water flow system considered in this chapter is assumed to be at steady state and the water table is assumed to be horizontal under nonpumping conditions. This implies that horizontal flow is negligible at the water table. The operation of a pumping well induces symmetrical radial flow towards the

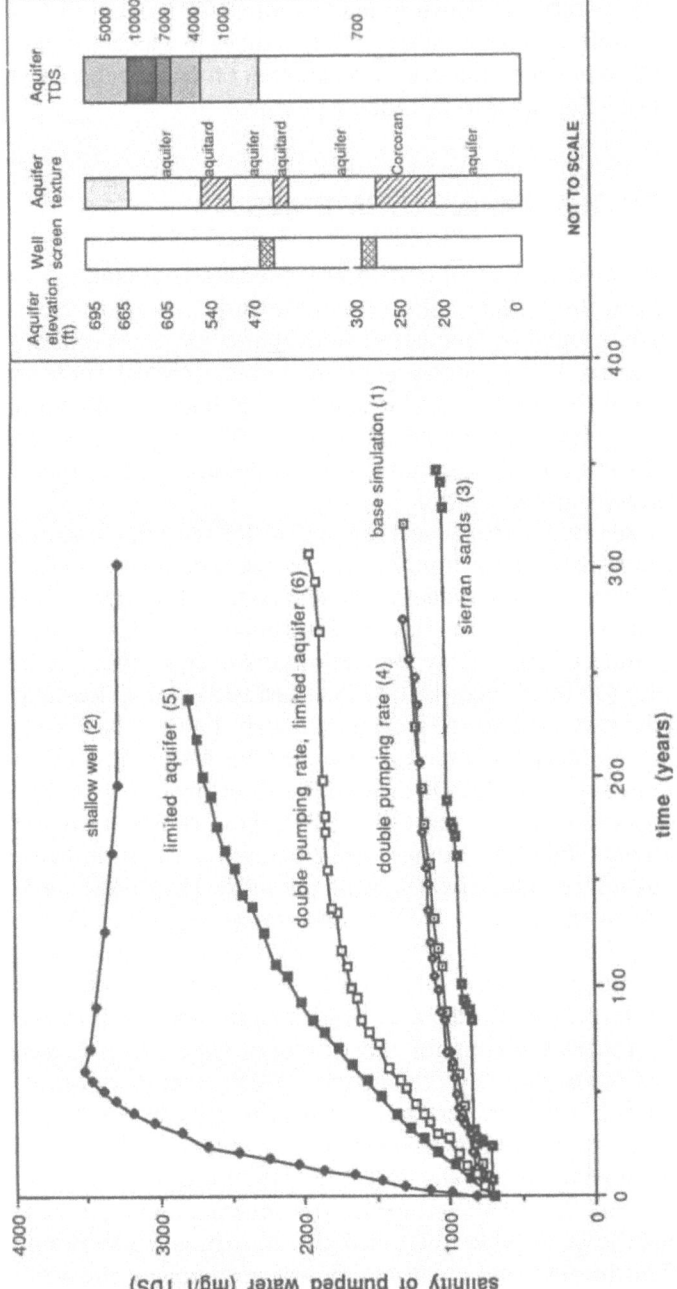

Figure 5. Comparison of the salinity of pumped well water over time for six configurations designed to lower shallow water tables in drainage problem areas.

pumping well, which is accommodated in the semiconfined aquifer by recharge across the water table boundary (vertical flow) and from flow across the radial boundary (horizontal flow). The vertical cross-section is 2,600 feet in width (well radius) and 700 feet deep (aquifer depth).

## Boundary Conditions

Solution of the flow equations presented in Matanga and Quinn (1990) requires specification of boundary conditions on three sides of the flow system. The well is treated as a specified-flow boundary with zero flow above the well screen. Flow at the well screen is assumed uniformly distributed and is based on the pumping rate. No-flow boundary conditions are assumed below the well screen. The water table boundary is designated a recharge boundary with specified flow ranging from 0.5 to 1.0 acre-foot/acre-year. The recharge rate is the volume rate of deep percolation below the crop root zone and is an inverse function of irrigation water use efficiency. A typical range of irrigation efficiency is from 50 percent to 85 percent over the study area. Higher efficiencies are associated with the finer textured soils in the basin rim that suffer shallow ground-water levels, and the lower values, with the upper alluvial fan deposits.

It is difficult to define the lower boundary of the flow system because the spatial flow distribution under pumping conditions is not known. Therefore, the bottom boundary was considered as a uniform specified-head boundary of 550 feet. This assumes a 150-foot difference in hydraulic head between the top of the aquifer and the base of the sub-Corcoran aquifer. This assumption is supported by the head contour map for a cross section of the San Joaquin Basin published in Belitz (1988).

The radial boundary of the semiconfined aquifer system was also simulated as a specified-head boundary. Boundary values of hydraulic head were calculated using the hydraulic conductivities previously described and an average flux of 0.315 foot/year across the Corcoran Clay layer (Belitz, 1990). To simulate an aquifer of limited extent the hydraulic conductivities of the radial boundary aquifer cells were set equal to 1 percent of their original values. This situation may occur where an aquifer, such as a relic sand channel, is truncated by a less permeable alluvial deposit. When simulating an aquifer of limited extent, the hydraulic properties of the clay layers were left unchanged.

## Salt Concentration

Salts are introduced into the flow system by recharge at the water table where they mix with resident salts in the shallow ground water. The salts then migrate through the flow system. Ground-water recharge is assumed to have a TDS of 5,000 mg/L after accounting for solubilization of native salts such as gypsum. At greater depths within the aquifer, several scenarios can be simulated with the TDS of the most concentrated aquifer zone ranging between 5,000 mg/L and 15,000 mg/L. In the examples which follow, the maximum TDS was assumed to be 10,000 mg/L TDS at a depth of between 45 feet and 90 feet below the ground surface. Ground-water quality immediately above the pumping well can range from 500 mg/L to 1,250 mg/L TDS in the zones designated as potential sites for pumping; 700 mg/L was used in this instance.

## Well Screens

Pumping wells are typically screened within deposits that yield the greatest quantities of good quality water. The length of the well screen used in this analysis was assumed to be 45 feet--designed to limit the mixing of good and poor quality water sources at the well. Shallow and deep pumping depths were 225 feet and 385 feet, respectively, to the top of the well screen.

## SIMULATION RESULTS

A number of steady-state simulation runs were made with the model for a series of well depths within the semiconfined aquifer; pumping rates; depth distribution profiles of TDS; and radial and lower boundary conditions (figure 4). In contrast to the analysis presented in tables 1 and 2, which assumed a strictly vertical ground-water flow system (plug flow), the model simulates flow in both vertical and horizontal directions.

The base run (1) in figure 4 is for a deep well (385 feet), pumped continuously at 200 gal/min in an aquifer of unlimited extent. The drawdown at steady state is 4.3 feet at the well and 2.7 feet at a radius of 2,400 feet from the well. The salt concentration of pumped ground-water increases slowly over time, reaching a TDS of 1,250 mg/L after 200 years. Flow to the well is primarily horizontal, with the most significant component of vertical flow closest to the well itself. A transient simulation conducted by Geomatrix Consultants for the same aquifer stratigraphy and a 250 mg/L pumping rate produced a maximum water table drawdown of 2.5 feet, immediately above the well, and a drawdown of between

0.5 and 1.0 foot at a radius of 2,400 feet from the well, after 130 days of pumping (Geomatrix, personal communication, June 1990).

Wells pumped at 200 gal/min on a 1/4-mile spacing and operated for 13 weeks per year were simulated, using an average annual pumping rate of 200 gal/min on a 1-mile grid, with the USGS regional flow model (Belitz, 1990). The effect of pumping in the aquifer layer, immediately above the Corcoran Clay, was to lower average water tables by 1 to 3 feet below predicted water tables for a base simulation by the year 2000. The base simulation assumed no increase in aquifer pumping.

Preliminary results show that, depending on the screened depth of the well pumps, up to 30 percent of the well discharge is contributed from below the well, and comprises water which would otherwise have moved into the confined aquifer and across the Corcoran Clay. The results also suggest that additional conservation measures are needed, in addition to ground-water pumping, if water tables are to be reduced below 1988 levels by the year 2000. Simulated water tables were greater than 1988 water tables, even with the addition of pumping at a rate of 200 gal/min (Belitz, 1990).

Pumping from a shallower aquifer depth (2) (figure 4) results in a greater depth to water table at steady state, but causes a more rapid rate of decline of water quality. The steady-state water table drawdown is 7.4 feet above the pumped well but water quality exceeded the 2,500 mg/L threshold, used to define the upper limit of usable ground water, after only 22 years. A greater volume of vertical flow was induced in this simulation than in the base run (figure 4).

Pumping from the deep Sierran sand aquifer (3) (figure 4) produces a drawdown similar to the base run of 4.1 feet; however, it is likely that this result is achieved much less rapidly. This concurs with Schmidt (1988). Also, Geomatrix Consultants have simulated a 2,000 gal/min pumping rate in the deep Sierran sand aquifer and achieved a drawdown of only 0.3 foot after 130 days of pumping.

Water quality deteriorates very slowly in (3), largely due to the fact that almost all flow to the well is horizontal. The graph in figure 4 shows a low initial TDS concentration at the pumping well and hence a long interval before the threshold TDS is exceeded. In the Sierran aquifer, the depth of penetration of ground water high in TDS is likely to be greater than is illustrated in the example profile in figure , even though there appears to be significant attenuation of selenium concentrations. This could significantly reduce the usable life of the well, though it would not likely affect the rate of water quality deterioration. Doubling the pumping rate (4) (figure 4) of the base simulation to 400 gal/min does not have a large effect on the rate of aquifer contamination. The concentration-time graph in simulation (4) is similar to that in (1) in figure 4. The final steady-state drawdown at the pumping well is identical for both simulations. This result is expected based on the fact that flow is primarily horizontal in the aquifer, which is assumed unlimited in areal extent.

If the aquifer is assumed to be limited in areal extent due, for example, to the physical presence of a fully penetrating aquitard at a radius of 2,600 feet from the pumped well, the effect of doubling the rate of pumping is more marked. In figure 4, the limited aquifer case (5) shows a much more rapid rate of deterioration of water quality than the base case. The TDS of pumped ground water exceeds the threshold for usable water for irrigation uses after 156 years. Steady-state drawdown at the pumping well is 5.2 feet, approximately 20 percent higher than the drawdown for the base run. When the rate of pumping is doubled (6) (figure 4), the steady-state drawdown increases marginally to 5.7 feet (30 percent higher than the base run); however, the rate of deterioration of water quality is much less. The threshold value of 2,500 mg/L is not exceeded, even after 300 years. This result must be interpreted with caution however, since it assumes that the piezometric head in the confined aquifer below the Corcoran Clay will remain constant. At higher pumping rates, a larger proportion of well pumpage is derived from the sub-Corcoran aquifer, which explains the lower TDS of well discharge in simulation (6). In reality, a higher volume of extraction from the sub-Corcoran aquifer will cause the piezometric head in the sub-Corcoran aquifer to decline more rapidly, making it increasingly expensive to extract ground water from this aquifer. Doubling the rate of extraction may in fact cause more rapid deterioration than the base case for an aquifer of limited areal extent.

The well pumping simulations summarized in figure 4 give a more optimistic appraisal of the merits of ground-water pumping to manage shallow water tables than the analysis presented in table 2. Comparing the two approaches shows the usable life of the aquifer under a ground-water pumping strategy to be very sensitive to the difference in elevation between the zone of poor quality ground water and the top of the well screen of the pumped well. Where this elevation difference is less than 100 feet, streamlines into the well screen capture a considerable volume of this poor water, even if flow into the well is primarily horizontal. In circumstances such as this, the usable life of the well can be as short as 20 years. The depth distribution of TDS also has a significant effect on the rate of degradation of the pumped well, and the time to reach the concentration threshold of 2,500 mg/L is considerably diminished if the initial TDS in the aquifer at the well screen is close to the criterion for usable ground water of 1,250 mg/L.

Which of the two analytical approaches best depicts the outcome of a decision to pump ground water to manage regional shallow water tables depends on the design of the well network and the manner in which the wells in this network are pumped. A large number of closely spaced wells, designed to provide drainage relief on a regional basis would likely produce a flow system in which vertical flow predominated. Hence, the rate at which the zone of poor water quality displaced the zone of usable ground water is mostly determined by the net rate of vertical flow in the semiconfined aquifer. This circumstance is also true of a heterogeneous aquifer of sand and clay materials where there is limited areal continuity of aquifer deposits and where flow to the pumping well has both lateral and vertical components. Conversely, where significant lateral continuity of aquifers is evident, lateral flow predominates and the second modeling approach may better describe the effects of various groundwater management strategies.

## CONCLUSIONS

This chapter has sought to highlight some of the hydrologic criteria to be considered in evaluating the use of ground-water pumping to address contaminated drainage problems on the west side of the Valley. Results have shown the importance of well field design and such factors as depth of pumping, pumping rate, and aquifer properties on achieving this purpose. The design of well fields will need to be based on a more thorough reconaissance of both the hydraulic properties of the aquifer and the distribution and depth of penetration of contaminants in the semiconfined aquifer than has been conducted to date. Lack of attention to this requirement for better survey data could lead to early abandonment of the well due to salinization of the crop root zone or a lack of water table response to pumping.

Design considerations aside, ground-water pumping appears to be a short- to medium-term option at best, and will likely hasten an ongoing process of aquifer degradation, shortening the usable life of the semiconfined aquifer in some cases to less than 25 years. Long-term economic considerations should also be evaluated as part of determining the feasibility of using water table management for drainage control.

## REFERENCES

Ayers, R. S. and Westcott, D. W., 1985. *Water Quality for Agriculture.* Irrigation and Drainage Paper 29.1, Food and Agriculture Organization of the United Nations.

Belitz, K., 1988. *Character and Evolution of the Ground-Water Flow System in the Central Part of the Western San Joaquin Valley.* U.S. Geological Survey Open-File Report 87-573.

Belitz, K., 1990. Unpublished Results From Scenario Runs Using the USGS Regional Aquifer Simulation Model. Report in Preparation.

California Department of Water Resources, 1987. *Map of Shallow Ground-Water Areas in the San Joaquin Valley.*

California State Water Resources Control Board, 1987. *Regulation of Agricultural Drainage to the San Joaquin River: Appendices A-H.*

Deverel, S. J.; Gilliom, R. J.; Fujii, R.; Izbicki, A. J.; and Fields, J.C., 1984. *Areal Distribution of Selenium and Other Inorganic Constituents in Shallow Ground Water of the San Luis Drain Service Area, San Joaquin Valley, California: A Preliminary Study.* U.S. Geological Survey Water-Resources Investigations Report 84-4319, 67 pp.

Deverel S. J. and Fujii, R. Processes Affecting the Distribution of Selenium in Shallow Ground Water of Agricultural Areas, Western San Joaquin Valley, California, *Water Resources Research*, 24(4), pp. 516-524.

Deverel S. J. and Gallanthine, S. K., 1989. Relation of Salinity and Selenium in Shallow Ground Water to Hydrologic and Geochemical Processes, Western San Joaquin Valley, California, *Journal of Hydrology*, 109, pp. 125-149.

Frind, E. O. and Matanga, G. B., 1985. The Dual Formulation of Flow for Contaminant Transport Modeling. 1. Review of Theory and Accuracy Aspects, *Water Resources Research*, 21(2), pp. 159-169.

Geomatrix Consultants, 1990. *Stratigraphy and Ground Water Quality of Selected Areas of the Lower Panoche Fan, Western San Joaquin Valley, California.*

Gilliom, R. J.; Deverel, S. J.; Fujii, R.; Belitz, K.; and Dubrovosky, N., 1989. *Preliminary Assessment of Sources, Distribution, and Mobility of Selenium in the San Joaquin Valley, California.* U.S. Geological Survey Water-Resources Investigations Report 88-4186, 129 pp.

Matanga, G. B. and Quinn, N. W. T., 1990. *Analysis of the Long-Term Sustainability of Water Quality for Irrigation from Pumping Wells Used to Manage High Saline Ground Water Tables.* Report to the San Joaquin Valley Drainage Program. In Preparation.

Nishimura, G., 1987. *Estimation of Ground-Water Pumping Costs.* Technical Information Record, San Joaquin Valley Drainage Program.

Nishimura, G. and Baughman, S., 1989. *Regulation and Timing of Salt Entry to the San Joaquin River.* Technical Information Record, San Joaquin Valley Drainage Program.

Page, R. W., 1986. *Geology of the Fresh Ground-Water Basin of the Central Valley, California, with Texture Maps and Sections.* U.S. Geological Survey Professional Paper, 1401-C, 53 pp.

Pinder, G. F. and Frind, E. O., 1972. Application of Galerkins's Procedure to Aquifer Analysis, *Water Resources Research*, 8(1), pp. 108-120.

Pinder, G. F. and Gray, W. G., 1977. *Finite Element Simulation in Surface and Subsurface Hydrology.* Academic Press, New York, NY.

Quinn, N. W. T.; Swain, W.C.; and Hansen, D. T., 1990. *Assessment of Ground-Water Pumping as a Management Option in Drainage Problem Areas of the Western San Joaquin Valley.* Technical Information Record, San Joaquin Valley Drainage Program. In Preparation.

San Joaquin Valley Drainage Program, 1989. *Preliminary Planning Alternatives for Solving Agricultural Drainage and Drainage-Related Problems in the San Joaquin Valley.* Final Report.

Schmidt, K., 1988. Report of Aquifer Tests for Shallow Wells in Firebaugh-Mendota Area. Unpublished Report Prepared for the San Joaquin Valley Drainage Program.

Schmidt, K., 1989. Results of 14-day Aquifer Tests Near Mendota. Unpublished Report Prepared for the San Joaquin Valley Drainage Program.

# 6 REUSE OF AGRICULTURAL DRAINAGE WATER TO MAXIMIZE THE BENEFICIAL USE OF MULTIPLE WATER SUPPLIES FOR IRRIGATION

James D. Rhoades, U.S. Salinity Laboratory and
Ariel Dinar, University of California, Riverside and
San Joaquin Valley Drainage Program

## ABSTRACT

This chapter provides conceptual arguments and empirical evidence to show that the blending approach typically used for water quality protection can result in economic losses to the agricultural community as a whole. A better strategy is suggested for dealing with the "disposal" of saline agricultural drainage waters which provides greater practical benefit to be derived from the total water supply than blending does. In this strategy the drainage water is intercepted, isolated from the good-quality water, and reused for the irrigation of suitably salt-tolerant crops in the same project. Ultimately, a greatly reduced volume of secondary drainage water is disposed of or treated in some manner other than blending.

## INTRODUCTION

The return of agricultural drainage to freshwater supplies is often claimed to be beneficial as a means to conserve water and increase water use efficiency in crop production (Davenport and Hagan, 1982). Water quality agencies commonly implement regulatory policies which permit agricultural drainage to be returned to a major water supply after it has been diluted with good quality water to a concentration level where the contaminate in the blend does not exceed a certain value (the so-called safe limit) deemed allowable. Such claims and regulatory policies do not recognize the potential deleterious effect that such blending can have on the usability of the total receiving water supply.

Broadly speaking, users of a water supply may be classified into two groups: (1) Those who concentrate the water in the process of use, and (2) those who do not. The first group of users suffer potential disbenefit under the "blending" philosophy of water quality protection. The blending process may limit the maximum practical benefit that can be derived from the total water supply. For example, the return of excessively saline waters to the water supply, even when sufficient dilution occurs to keep the salinity of the mixture within acceptable limits, reduces the quantity of the water supply that can be used productively to grow salt-sensitive crops (Rhoades, 1989). In this chapter, the factor limiting crop growth is assumed to be the presence of excessive total dissolved salts in the water (salinity), but an analogous case could also be made for boron or any other constituent that is toxic to plants.

The ultimate objective of water quality protection should be to permit the maximum practical benefit (use) from the available water supply. The purposes of this chapter are (1) to provide evidence--conceptual and empirical-- that the blending approach typically used for water quality protection can result in economic losses to the agricultural community as a whole and (2) to suggest a means to deal with the "disposal" of saline agricultural drainage which provides greater practical benefit from the total water supply than blending does. The theme of this chapter is that different water qualities may have relative advantages that can be realized by application to different crops at different periods in the growing season.

## CONCEPTUAL FRAMEWORK

In considering acceptable salinity levels for irrigation water, it is important to recognize that the higher its salinity[1], the lower the percentage of total volume that can be beneficially consumed for crop production.

Plants must have access to water of a quality that can be consumed without the concentration of salts (individually or totally) becoming excessive for adequate growth. In the process of transpiration, plants essentially perform reverse osmosis; the salts in the irrigation water are concentrated in the remaining unused soil water, which will ultimately become drainage. A plant will not grow properly when the salt concentration in the soil water exceeds its specific limit under the given conditions of climate and management (Bernstein, 1975). Thus, not all of the water can be consumed by a plant if a water supply contains salt.

A hypothetical case study was made to illustrate the effects of blending and of alternative drainage management practices on the usability of water supplies and the subsequent economic ramifications. The case study compares the losses in crop yield and in the volumes of consumable water resulting from salinity increase in a river system used for irrigation supply under different strategies of drainage management.

The salinity of the soil water resulting from irrigation was calculated from the irrigation water composition and the leaching fraction (L, the fraction of the infiltrated water not consumed in evapotranspiration and which passes beyond the root zone as drainage) after the method of Rhoades (1984c, 1986). In this method the water-uptake distribution within the irrigated root zone is assumed to be 40:30:20:10 by successive quarter-depth fractions; steady-state chemistry and "piston-displacement-type" water flow are also assumed. Relative crop yield was then calculated from the predicted average soil water salinity from knowledge of the plant tolerance to salinity and the assumption that crops respond to the average salinity within their root zone. Each of these assumptions is sufficiently true for the purposes of this chapter (Ingvalson et al., 1976 and Rhoades and Merrill, 1976).

More specifically, the average level of soil salinity (expressed as the electrical conductivity of the saturation-paste extract, $EC_e$) within the crop root zone resulting from the long-term irrigation with a water of $EC_{iw}$ was obtained from

[1] $EC_e = F_c \cdot EC_{iw}$,

where $EC_{iw}$ is the electrical conductivity of the irrigation water and $F_c$ is the related concentration factor appropriate for L. $F_c$ was obtained from the calculable relationship which exists between $F_c$ and L (Rhoades, 1982 and 1984c). The relative yields achievable as a function of average root zone $EC_e$ were obtained from the crop salt-tolerance tables of Maas (1986). Analogous values for boron and other specific solutes are also given in this same reference and could be used to evaluate yield losses for such specific solutes.

The fraction of the irrigation water that is consumed in evapotranspiration ($V_{et}/V_{iw}$) is related to L as

[2] $V_{et}/V_{iw} = (1-L)$.

For the purposes of this chapter, the volumes of $V_{iw}$ were normalized by expressing them relative to $V_{et}$, i.e., for the case where $V_{et}$ is taken to be equal to 1.

These relative results (yields and volume ratios) were used to evaluate whether blending drainage water with good quality water would be detrimental, or not, with respect to reducing the volume of the total water supply that could be used for crop growth without loss in yield.

## EMPIRICAL SPECIFICATIONS

The following case study illustrates the concept described above. A river of 500 units volume and an initial EC of 0.5 dS/m was assumed to pass through four successive identical agricultural projects. In each project, diversions from the river were used to irrigate four crops. Part of the applied water is consumed by the crops in evapotranspiration; the rest becomes drainage. Inefficiencies and losses of water in transit are not considered for our purposes. The following amounts were assumed to be consumed in each project for each crop: beans (15 units), alfalfa (40 units), wheat (25 units) and cotton (14.4 units). The water volumes applied to each crop were chosen sufficient to meet these ET levels and result in the following leaching fractions: beans ($L = 0.2$), alfalfa ($L = 0.1$), wheat ($L = 0.2$), and cotton ($L = 0.15$). These values of $L$ are not optimal under all conditions, but are representative of reasonably good management.

Three different irrigation/drainage management strategies were simulated. In Strategy I, each crop is irrigated with the river water (at its salinity level when entering each project) and all of the percolated drainage water from each crop is assumed to be collected by some drainage system and returned to the river below that project's irrigation diversion point. The resultant "blend" of river and drainage water is then assumed to flow on to the next downstream project where the process is repeated. This process is repeated through the succession of four projects. In Strategy II, the bean, alfalfa and wheat crops are also irrigated with the river water; however, their resulting drainage waters are not returned to the river. Instead they are combined and used to irrigate the cotton grown within the same project. The secondary drainage water resulting after cotton production is then returned to the river. This process is also repeated through the succession of four projects. In Strategy III, the crops are irrigated as in Strategy II, but the secondary drainage water from the cotton is not returned to the river. This relatively small volume of water is assumed to be disposed of by some other means (such as evaporation, desalting, deep-aquifer injection, etc.).

## ECONOMIC EVALUATIONS OF THE CASE STUDY RESULTS

The detailed physical results of drainage management Strategies (I, II, and III) are given in tables 1, 2, and 3, respectively. Summary results are given in table 4.

Compared to Strategy I, less irrigation water is diverted from the river and less drainage water is returned to it under the management practices of Strategy II. Whereas the volume and salinity of the river is the same under both strategies, less yield is obtained in Strategy II. All of this yield loss is accounted

Table 1. Results of simulation under conditions of Strategy I: No reuse made of drainage water in the projects; all drainage water discharged back to the river.[a]

| Crop | $EC_{iw}$ | $V_{iw}$ | $\overline{EC_e}$ | $V_{dw}$ | $EC_{dw}$ | % loss in yield |
|---|---|---|---|---|---|---|
| 1st project (Volume of river = 500; EC = 0.5 dS/m) | | | | | | |
| beans | 0.5 | 31.25 | .64 | 6.25 | 2.50 | 0 |
| alfalfa | 0.5 | 44.44 | .94 | 4.44 | 5.00 | 0 |
| wheat | 0.5 | 31.25 | .64 | 6.25 | 2.50 | 0 |
| cotton | 0.5 | 16.94 | .75 | 2.54 | 3.33 | 0 |
| 2nd project (Volume of river = 395.6; EC = 0.63 dS/m) | | | | | | |
| beans | 0.63 | 31.25 | .81 | 6.25 | 3.15 | 0 |
| alfalfa | 0.63 | 44.44 | 1.18 | 4.44 | 6.30 | 0 |
| wheat | 0.63 | 31.25 | .81 | 6.25 | 3.15 | 0 |
| cotton | 0.63 | 16.94 | .95 | 2.54 | 4.20 | 0 |
| 3rd project (Volume of river = 291.2; EC = 0.86 dS/m) | | | | | | |
| beans | .86 | 31.25 | 1.11 | 6.25 | 4.30 | 2 |
| alfalfa | .86 | 44.44 | 1.62 | 4.44 | 8.60 | 0 |
| wheat | .86 | 31.25 | 1.11 | 6.25 | 4.30 | 0 |
| cotton | .86 | 16.94 | 1.30 | 2.54 | 5.73 | 0 |
| 4th project (Volume of river = 186.80; EC = 1.34 dS/m) | | | | | | |
| beans | 1.34 | 31.25 | 1.73 | 6.25 | 6.70 | 15 |
| alfalfa | 1.34 | 44.44 | 2.52 | 4.44 | 13.40 | 5 |
| wheat | 1.34 | 31.25 | 1.73 | 6.25 | 6.70 | 0 |
| cotton | 1.34 | 16.94 | 2.02 | 2.54 | 8.93 | 0 |

[remaining volume of river = 82.4; EC = 3.03 dS/m]

[a] Volume diverted from river for irrigation in each project = 123.88 units; volume of drainage returned to river from each project = 19.48 units.

Table 2. Results of simulation under conditions of Strategy II: Drainage from bean, alfalfa and wheat crops used to irrigate cotton; drainage water from cotton discharged back to the river.[a]

| Crop | $EC_{iw}$ | $V_{iw}$ | $\overline{EC_e}$ | $V_{dw}$ | $EC_{dw}$ | % loss in yield |
|---|---|---|---|---|---|---|
| \multicolumn{7}{c}{1st project (Volume of river = 500; EC = 0.50 dS/m)} |
| beans | 0.50 | 31.25 | .64 | 6.25 | 2.50 | 0 |
| alfalfa | 0.50 | 44.44 | .94 | 4.44 | 5.00 | 0 |
| wheat | 0.50 | 31.25 | .64 | 6.25 | 2.50 | 0 |
| cotton | 3.15 [b] | (16.94)[c] | 4.76 | (2.54) | 21.0 | 0 |
| \multicolumn{7}{c}{2nd project (Volume of river = 395.6; EC = 0.632 dS/m)} |
| beans | 0.63 | 31.25 | .81 | 6.25 | 3.15 | 0 |
| alfalfa | 0.63 | 44.44 | 1.18 | 4.44 | 6.30 | 0 |
| wheat | 0.63 | 31.25 | .81 | 6.25 | 3.15 | 0 |
| cotton | 3.98[b] | (16.94)[c] | 5.99 | (2.54) | 26.5 | 0 |
| \multicolumn{7}{c}{3rd project (Volume of river = 291.2; EC = 0.858 dS/m)} |
| beans | .86 | 31.25 | 1.11 | 6.25 | 4.30 | 2 |
| alfalfa | .86 | 44.44 | 1.62 | 4.44 | 8.60 | 0 |
| wheat | .86 | 31.25 | 1.11 | 6.25 | 4.30 | 0 |
| cotton | 5.43[b] | (16.94)[c] | 8.19 | (2.54) | 36.2 | 2 |
| \multicolumn{7}{c}{4th project (Volume of river = 186.8; EC = 1.338 dS/m)} |
| beans | 1.34 | 31.25 | 1.73 | 6.25 | 6.70 | 15 |
| alfalfa | 1.34 | 44.44 | 2.52 | 4.44 | 13.40 | 5 |
| wheat | 1.34 | 31.25 | 1.73 | 6.25 | 6.70 | 0 |
| cotton | 8.46[b] | (16.94)[c] | 12.76 | (2.54) | 56.4 | 2 |

[remaining volume of river = 82.4; EC = 3.03 dS/m]

[a]Volume diverted from river for irrigation in each project = 106.94 units; volume of drainage returned to river from each project = 2.54 units.
[b]The weighted average EC of the drainage waters from the bean, alfalfa and wheat crops.
[c]The cumulative volume of the drainage waters from the bean, alfalfa and wheat crops.

Table 3. Results of simulation under conditions of Strategy III: Drainage from bean, alfalfa and wheat crops used to irrigate cotton; drainage water from cotton not discharged back to the river system.[a]

| Crop | $EC_{iw}$ | $V_{iw}$ | $\overline{EC_e}$ | $V_{dw}$ | $EC_{dw}$ | % loss in yield |
|---|---|---|---|---|---|---|
| \multicolumn{7}{c}{1st project (Volume of river = 500; EC = 0.5 dS/m)} |
| beans | 0.50 | 31.25 | .64 | 6.25 | 2.50 | 0 |
| alfalfa | 0.50 | 44.44 | .94 | 4.44 | 5.00 | 0 |
| wheat | 0.50 | 31.25 | .64 | 6.25 | 2.50 | 0 |
| cotton | 3.15[b] | (16.94)[c] | 4.76 | 2.54 | 21.0 | 0 |
| \multicolumn{7}{c}{2nd project (Volume of river = 393.06; EC = 0.5 dS/m)} |
| beans | 0.50 | 31.25 | .64 | 6.25 | 2.50 | 0 |
| alfalfa | 0.50 | 44.44 | .94 | 4.44 | 5.00 | 0 |
| wheat | 0.50 | 31.25 | .64 | 6.25 | 2.50 | 0 |
| cotton | 3.15[b] | (16.94)[c] | 4.76 | 2.54 | 21.0 | 0 |
| \multicolumn{7}{c}{3rd project (Volume of river = 286.12; EC = 0.5 dS/m)} |
| beans | 0.50 | 31.25 | .64 | 6.25 | 2.50 | 0 |
| alfalfa | 0.50 | 44.44 | .94 | 4.44 | 5.00 | 0 |
| wheat | 0.50 | 31.25 | .64 | 6.25 | 2.50 | 0 |
| cotton | 3.15[b] | (16.94)[c] | 4.76 | 2.54 | 21.0 | 0 |
| \multicolumn{7}{c}{4th project (Volume of river = 179.18; EC = 0.5 dS/m)} |
| beans | 0.50 | 31.25 | .64 | 6.25 | 2.50 | 0 |
| alfalfa | 0.50 | 44.44 | .94 | 4.44 | 5.00 | 0 |
| wheat | 0.50 | 31.25 | .64 | 6.25 | 2.50 | 0 |
| cotton | 3.15[b] | (16.94)[c] | 4.76 | 2.54 | 21.0 | 0 |

[remaining volume of river = 72.24 units; EC = 0.5 dS/m]
[drainage volume to be disposed of in each project is 2.54 units; EC = 21 dS/m]

[a] Volume diverted from river for irrigation in each project = 106.94 units;
[b] The weighted average EC of the drainage waters from the bean, alfalfa and wheat crops.
[c] The cumulative volume of the drainage waters from the bean, alfalfa and wheat crops.

Table 4. Comparison of simulation results for the three different strategies of irrigation and drainage management (I, II, III).

| Strategy | $\Sigma V_{et}^a$ | $\Sigma V_{iw}^b$ | $\Sigma V_{dw}^c$ | Yield Loss[d] Project 3 | Project 4 | $V_{rw}^e$ | $EC_{rw}^f$ |
|---|---|---|---|---|---|---|---|
| I | 417.6 | 495.5 | 77.9 | 2 | 20 | 82.4 | 3.0 |
| II | 417.6 | 427.8 | 10.2 | 4 | 45 | 82.4 | 3.0 |
| III | 417.6 | 427.8 | 0 | 0 | 0 | 72.2 | 0.5 |

[a]$RV_{et}$ = cumulative volume (units) of water used by crops in evapotranspiration in all four projects.
[b]$RV_{iw}$ = cumulative volume (units) of water diverted (pumped) from river for irrigation in all four projects.
[c]$RV_{dw}$ = cumulative volume (units) of drainage water discharged (pumped) back to river.
[d]Cumulative loss of crop yield within Projects 3 and 4.
[e]$V_{rw}$ = Volume (units) of river remaining in stream below Project 4.
[f]$EC_{rw}$ = EC of river below Project 4, dS/m.

for by the cotton grown in Projects 3 and 4 with drainage water. With Strategy III the volume of water diverted from the river is lower than with Strategy I and no drainage water is returned to it. As a result, the river salinity is low and constant throughout its entire length, the volume of the river is somewhat reduced compared to Strategy I, and no losses in crop yield occur anywhere in any project of the entire river basin. The percentage losses in crop yield predicted under the three strategies are 22, 49, and 0 for Strategies I, II, and III, respectively.

While the implementation of Strategy III is technically and agronomically feasible, economic considerations must also be favorable before such practices can be advocated. The analysis was therefore extended to include economic considerations and certain related externalities resulting from the diversion of irrigation water and the discharge and disposal of drainage.

Several general assumptions were made in the analysis. First it was assumed that the on-farm practices and behaviors of the individual farmers which influence profits are not appreciably changed as a result of implementing Strategy II or III. Since the agronomic analysis assumed steady state with regard to the soil salinity buildup, it was also assumed in the economic analysis. Third, it was implicitly assumed that the leaching fractions used for water application rates were achieved. No deficit irrigation was considered. Fourth, no technological change was considered in this analysis. The same agricultural and irrigation practices and technologies are assumed to be applied under all strategies presented.

For simplicity, all projects along the river were assumed to be identical in terms of resources, cropping patterns, technology, and cost. The differences resulting from Strategies I, II, and III implied in table 4 are assumed to be the only ones that distinguish the farms along the river.

The economic evaluation was based on the following functions which describe the long-term profit associated with the i-th farm activity under strategy k,

$$[3] \quad \pi^k_i = \sum_j^{J_i} \{[1-\beta^k_j] \cdot y_{ij} \cdot (P_j - H_j) - C_{ij} - F_{ij} - p^w \cdot w^k_{ij} - p^r \cdot r^k_{ij} - p^d \cdot d^k_{ij}\},$$

$$[4] \quad \Pi^k = \Sigma_i \pi^k_i - S(v,c),$$

where $\pi^k_i$ and $\Pi^k$ are the farm and regional profits, respectively; $J_i$ is the set of crops grown on farm i; $\beta^k_j$ is the yield loss (fraction) for crop j due to salt concentration in the applied irrigation water; $y_{ij}$ is the maximum yield level available when crop j is irrigated with fresh water; $P_j$ and $H_j$ are crop yield price and harvest cost associated with crop j, respectively; $C_{ij}$ and $F_{ij}$ are, respectively, nonwater variable cost and fixed cost for crop j; and $p^w$, $p^d$, and $p^r$ are, respectively, water price, drainage disposal cost, and water reuse cost. The latter two water costs represent, respectively, additional costs associated with the disposal of the drainage water (such as treatment, evaporation ponds, etc.), and associated with reuse (such as cost of pumping, ponds for intermediate storage, and monitoring). Note that all social costs that might result from drainage disposal are included in the variable $p^d$. The variables $w^k_{ij}$, $d^k_{ij}$, $r^k_{ij}$, are volume of applied freshwater, volume of disposed drainage water, and volume of drainage applied for reuse, respectively. Note that the total amount of applied irrigation water on crop j in farm i under strategy k is $w^k_{ij} + r^k_{ij}$. The variable $S(v,c)$ is the regional (social) cost associated with the volume of water remaining in the river (v) and its salinity (c). Obviously, S is a decreasing function of v and an increasing function of c. This topic falls beyond the scope of the current analysis and will not be considered here.

Potential yield and crop yield prices used in the economic analysis are presented in table 5 for the four crops. In order to determine effects of the different variables on single project and total regional profit, a sensitivity analysis was performed with respect to these variables. Ranges of values for $p^w$, $p^d$, and $p^r$ were used to account for effects of changes in water prices, drainage disposal costs, and water reuse costs on the profitability of the three strategies. For example, a zero value for $p^d$ was used where drainage disposal was not considered to cause a problem to society; and high levels of $p^d$ were used to account for the additional costs to make drainage disposal safe (pumping,

Table 5. Assumed crop yields, prices and production costs.

| Crop | Yield[b] (Ton/ac) | Crop price[a] ($/ton) | Harvest cost[b] ($/ton) | Non-water variable costs[b] ($/acre) | Fixed costs[b] ($/acre) |
|---|---|---|---|---|---|
| Alfalfa | 9.00 | 84.40 | 22.80 | 126.10 | 152.02 |
| Cotton | .36 (lint) | 1117.40 | 222.13 | 282.70 | 92.10 |
| Wheat | 2.70 | 124.20 | 20.05 | 161.00 | 87.85 |
| Beans | 8.80 | 450.00 | 252.90 | 427.70 | 149.92 |

Prices are 1987 constant dollars

[a]Long-term averages, based on Fresno County Agricultural Commissioner Annual Reports for the period 1977-1987.
[b]Based on University of California Cooperative Extension Cost Budgets for 1986-1987.

treatment, evaporation ponds, etc.). Changes in $P_j$ were used to account for the effect of crop profitability on the attractiveness of drainage water reuse. Following Dinar, Letey, and Knapp (1989), water costs of $15, $30, and $50 per acre-foot and disposal cost of $35 and $100 per acre-foot of drainage water disposed were assumed. A value of $150 per acre-foot was used, based on Stroh (1990) and adjusted to 1987 price level, for drainage treatment costs in the case of no river disposal (Strategy III). The additional cost associated with reuse was assumed to be $15 per acre foot after Knapp, Dinar, and Letey (1986), and adjusted for inflation since 1987. Where $p^d$ was taken to be $= 0$ for Strategies I and II where farmers are not restricted with the disposal of the drainage water. Where quality standards are imposed on the drainage disposal to the river, $p^d$ was taken to be $> 0$ to represent costs of different treatment levels needed to meet these standards. For Strategy III, nonriver disposal costs exist when $p^d \geq 0$. Such costs are associated with the use of evaporation ponds or other means of nonriver disposal of drainage water.

To normalize the results, the first project was used as a reference. The results for all other farms are compared to it. For Strategy I, the drainage water from each crop is disposed to the river, therefore, $p^d$ was charged to each crop. In Strategy II, drainage water from beans, alfalfa, and wheat are reused for cotton production. For this case, $p^d$ was taken as 0 for these crops but was taken as $\geq 0$ for cotton. When $p^d = 0$ for cotton, no cost is associated with the disposal of the drainage. For Strategy III all secondary drainage is assumed to be treated and disposed of by some means other than return to the river. Therefore, an analogous procedure was used to assign costs similar to the case of Strategy II, except that here $p^d$ was taken to be 150 to account for the additional treatment

Table 6. Relative farm and regional profitability associated with different reuse/disposal strategies.

| | Strategy 1 | | | Strategy 2 | | | Strategy 3 | | |
|---|---|---|---|---|---|---|---|---|---|
| | \-\-\-\-\-\-\-\-\-\-\-\-\-\-\-\-\-\-\-\-\-\-\-\-\-Disposal cost ($/unit)\-\-\-\-\-\-\-\-\-\-\-\-\-\-\-\-\-\-\-\-\-\-\- | | | | | | | | |
| | 0 | 35 | 100 | 0 | 35 | 100 | 0 | 35 | 100 |
| *Crop yield price index = 100* | | | | | | | | | |
| Project 1 | 100 | 100 | 100 | 100.1 | 100.5 | 101.5 | 100.1 | 100.6 | 101.5 |
| Project 2 | 100 | 100 | 100 | 100.1 | 100.5 | 101.5 | 100.1 | 100.6 | 101.5 |
| Project 3 | 99.7 | 99.7 | 99.7 | 98.2 | 98.6 | 99.1 | 100.1 | 100.6 | 101.5 |
| Project 4 | 97.3 | 97.3 | 97.3 | 79.1 | 79.5 | 80.2 | 100.1 | 100.6 | 101.5 |
| Region[a] | 100 | 100 | 100 | 95.4 | 95.5 | 96.3 | 100.8 | 101.3 | 102.2 |
| *Crop yield price index = 120* | | | | | | | | | |
| Region[a] | 100 | 100 | 100 | 96.2 | 98.4 | 99.3 | 104.3 | 105.9 | 107.3 |
| *Surface water price index = 150* | | | | | | | | | |
| Region[a] | 100 | 100 | 100 | 99.1 | 99.1 | 100.1 | 104.5 | 106.1 | 107.9 |

[a]Calculated as regional net income (net of treatment costs $150/unit of water) relative to the values in the case with no reuse.

and evaporation costs. No economies of scale as a result of regional cooperation were considered in this analysis.

Two crop prices were considered: a basic price and a price reflecting a 20 percent increase. The latter price level is used to represent situations where the cropping patterns include more profitable crops. A water price representing a 50-percent increase was also included for purposes of assessing the economics of the alternative drainage reuse and disposal strategies under different conditions of external costs.

Results of the economic analysis are given in table 6 for different disposal costs (0, 35, and 100). In the case of Strategy I, some yield loss occurs due to increased downstream salinity (see table 1) and due to the disposal costs associated with the volume of drainage water disposed of to the river. In Projects 1 and 2, no yield losses occur but there are disposal costs; farms in Projects 3 and 4 incur both yield losses and disposal costs. In Strategy II, no yield losses occur for Projects 1 and 2 and less drainage needs to be disposed of compared to Strategy I. Farms in Projects 3 and 4 suffer greater yield losses compared to Strategy I due to higher salinity in the irrigation water (both river and drainage waters), but disposal cost is lower since part of the drainage water is reused for irrigation. The ratio between disposal cost and yield price is such that farms in Projects 1 and 2 are more profitable with Strategy II compared to I, whereas the farms in Projects 3 and 4 are less profitable. Strategy II is also inferior on a regional basis compared to I. Because river quality remains the same and no yield losses occur in any project with Strategy III, all projects are more profitable than with Strategies I and II and the regional profit is higher.

Results where yield prices are higher (20 percent) and water prices are higher (50 percent) are presented only for the regional level. The regional profit is higher with increased crop price level and the gap between Strategy II and I decreases. For the case where irrigation water price is 50 percent higher, the profitability of reuse is improved. In all cases, reuse is more profitable as disposal costs increase. Note that an additional cost associated with the degradation of water quality in the river for other nonagricultural users was not included in the analysis. The economic appeal of Strategy III would increase when such costs exist.

## DISCUSSION

These results show that the normal loss of crop yield resulting when drainage water is returned to the good-quality water supply and the "blend" (the river in this case) is subsequently used to irrigate typical field crops can be avoided by using Strategy III. In this strategy the drainage water is intercepted, isolated from the good-quality water and reused for the irrigation of salt-tolerant crops in the same project. Not only is the loss in crop yield that results under conventional management (Strategy I) avoided but also the salinity of the river is maintained at a uniformly low level (0.5 dS/m in this case) throughout its entire length. All users have water of equal quality whether they irrigate from the upstream or downstream sections of the river. Reuse is made of the drainage water for irrigation without any loss of crop yield. The ultimate volume of drainage water needing disposal (or desalting) is minimized and distributed equally between all projects. In this strategy, all areas have water of equal quality and disposal problems are shared equally rather than progressively burdening downstream users as is typically the case (Strategy I).

The results of the case study clearly show that adding saline waters to good quality water supplies reduces the volume of the good-quality water supply that can eventually be consumed by salt-sensitive crops. The actual amount of such reduction will depend, of course, upon the relative volumes and concentrations of the receiving and waste-waters and upon the tolerances of the crops to be irrigated. When the growth-limiting factor is salinity, the ultimate fraction of water in a supply that can be used in crop growth is:

[5] $1-EC_{iw}/EC_m$

where $EC_{iw}$ is the electrical conductivity (concentration can be used alternatively) of the water supply and $EC_m$ is the maximum salinity (electrical conductivity, concentration, etc.) of the water in the root zone (on an $EC_w$ not $EC_e$ basis; essentially $EC_{dw}$) the plant can tolerate (i.e., draw water from and still yield about 85-100 percent). Values of $EC_m$ vary among the crop species, but typically they are (according to Bernstein, 1975) about 45 for such tolerant crops as cotton, sugar beets, and barley; 30 for intermediate crops like tomatoes, wheat, and alfalfa; and about 15 for sensitive crops like beans, clovers, and onions. Minimizing leaching and deep percolation always reduces the volume and salt load of the drainage water and usually minimizes pollution of the receiving water (van Schilfgaarde et al., 1974; Rhoades et al., 1974; and Rhoades and Suarez, 1977). For this reason, minimizing leaching and deep percolation should be the major goal of irrigation management. Except in situations where the waters cannot be, or have not been, fully utilized in their first "passage" through the root zone, the drainage water should be intercepted before it is discharged to water supplies of better quality and reused for irrigation (Rhoades, 1984d). While concentrations of salts in drainage waters are higher than those of the corresponding irrigation water supply, they are often within acceptable limits for growing suitably salt-tolerant crops (Rhoades, 1977 and 1986). The results of the case study here illustrate the merits of this management strategy. Under other circumstances, it might make economic sense to blend and to bear the consequences of the losses of water usability and of crop yield when the alternative costs of disposal are extreme.

In these simulations, conservation of salt was assumed in the calculations. In the real world, salt loading of the river from drainage return would probably be greater than that shown, and more so for Strategy I compared to II and III; hence, the benefits of the latter strategy is likely underpredicted in these simulations. More realistic calculations of the salt-loading processes could be made, as has been done by Rhoades and Suarez (1977).

A reuse strategy that avoids blending and is superior to that described in Strategy III has been proposed and demonstrated in field projects to be viable and advantageous in well managed irrigation projects (Rhoades, 1984a and b and 1987 and Rhoades et al., 1988a and b). In this reuse strategy, the two water supplies (good quality water and saline drainage water) are kept separate and used without blending. The saline drainage water is intercepted, isolated, and substituted for the conventional "good water" in suitable locations in the project when irrigating certain salt-tolerant crops grown in the rotation when they are in a suitably salt-tolerant growth stage (after seedling establishment); the "good water" is used at the other times. This successive use water with low and high salinity levels prevents the soil from becoming excessively saline while permitting the substitution, over the long period, of saline water for conventional water to meet a substantial portion (up to about 50 percent depending on the crop rotation, etc.) of the irrigation water needs for the area while also permitting the growth of salt-sensitive crops in the same fields. Results of extensive field experiments have demonstrated the credibility and feasibility of this strategy and these conclusions (Rhoades et al., 1988a and b and 1989).

Since continuous recycling, in the sense of a closed loop, is not possible, reuse efforts should ideally be designed so that the drainage waters intercepted and isolated from the major part of the project area are redistributed to a dedicated "reuse-area" within the project, or sequentially from areas where crops of lesser to greater salt tolerance are grown (often this occurs naturally from upslope to downslope lands); the ultimate minimized volume of drainage resulting in the reuse area must eventually be desalted or disposed of. This ultimate disposal should not be accomplished by discharging the drainage water into good quality water supplies, unless no other means is practical, for the reasons previously discussed.

## SUMMARY AND CONCLUSIONS

An example was given to show that irrigating salt-sensitive crops with blends of saline and good quality water supplies or diluting drainage with good quality water in order to meet discharge standards may be inappropriate under certain situations. Even though the concentration of the blend may appear to be low enough to be acceptable by conventional standards, the usability of the good quality water supply for growing salt-sensitive crops (or for other salt-sensitive water uses) is reduced through the process of blending. Each time the salt content of an agricultural water supply is increased, the degree to which it can be consumed before its concentration becomes excessive is decreased. More crop production can usually be achieved from the total water supply by sole use of the good quality water component. Serious consideration should be given to keeping saline drainage waters separate from the good quality water supplies, even when the saline waters are to be used for irrigation. They can be used more effectively as a substitute for the conventional water supply in the irrigation of certain crops grown in the rotation after seedling establishment. The feasibility of such reuse for irrigation has been demonstrated in field studies in California. Reuse of drainage water for irrigation of suitably salt-tolerant crops reduces the volume of drainage water needing ultimate disposal and the offsite pollution problems associated with the discharge of irrigation return flows. The practice of blending or diluting excessively saline waters with good quality water supplies should only be undertaken after consideration is given to how it affects the volume of consumable water in the total supply and overall beneficial use.

## NOTES

[1]Salinity, a term referring to the total content of soluble inorganic constituents in the water, is generally limiting in this regard, but certain individual plant-toxic constituents, such as boron, may be in special cases. The term salinity will be used herein in a general sense to mean the presence of total dissolved salts and/or individual toxic constituents, like boron.

## REFERENCES

Bernstein, L., 1975. Effects of Salinity and Sodicity on Plant Growth, *Annual Review of Phytopathology*, 13, pp. 295-312.

Davenport, D. C. and Hagan, R. M., 1982. *Agricultural Water Conservation in California with Emphasis on the San Joaquin Valley*. Technical Report 10010, Department of Land, Air, and Water Resources, University of California, Davis.

Dinar, A.; Knapp, K. C.; and Letey, J., 1989. Irrigation Water Pricing to Reduce and Finance Subsurface Drainage Disposal, *Agricultural Water Management*, 16, pp. 155-171.

Ingvalson, R. D.; Rhoades, J. D.; and Page, A. L., 1976. Correlation of Alfalfa Yield with Various Indices of Salinity, *Soil Science*, 122, pp. 145-153.

Knapp, K. C.; Dinar, A.; and Letey, J., 1986. Onfarm Management of Agricultural Drainage Problems: An Economic Analysis, *Hilgardia*, 54(4), pp. 1-31.

Maas, E. V., 1986. Salt Tolerance of Plants, *Applied Agricultural Research*, 1(1), pp. 12-26.

Rhoades, J. D., 1977. Potential for Using Saline Agricultural Drainage Waters for Irrigation. *Water Management for Irrigation and Drainage*, ASCE Proceedings, Reno, NV., pp. 85-116.

Rhoades, J. D., 1982. Reclamation and Management of Salt-Affected Soils After Drainage. *Rationalization of Water and Soil Resources and Management*, Proceedings of the First Annual Western Provincial Conference. Lethbridge, Alberta Canada, pp. 123-197.

Rhoades, J. D., 1984a. New Strategy for Using Saline Waters for Irrigation. *Water-Today and Tomorrow,* Proceedings of the ASCE Irrigation and Drainage Special Conference, July 24-26, Flagstaff, AZ., pp. 231-236.

Rhoades, J. D., 1984b. Salt Problems from Increased Irrigation Efficiency, *ASCE Journal of Irrigation and Drainage Engineering,* 111(3), pp. 218-229.

Rhoades, J. D., 1984c. Using Saline Waters for Irrigation. Proceedings of International Workshop on Salt-Affected Soils of Latin America, Maracay, Venezuela, October 23-30, 1983, pp. 22-52; *Scientific Review on Arid Zone Research,* 2, pp. 233-264.

Rhoades, J. D., 1984d. Reusing Saline Drainage Waters for Irrigation: A Strategy to Reduce Salt Loading of Rivers. In: French, R. H. (Ed.), *Salinity in Watercourses and Reservoirs,* Chapter 43, pp. 455-464.

Rhoades, J. D., 1985. Principles of Salinity Control on Food Production in North America. Proceedings of International Conference on Food and Water, May 26-30, 1985, Texas A&M University.

Rhoades, J. D., 1986. *Use of Saline Water for Irrigation: Water Quality.* Special Issue Bulletin, National Water Resource Institute, Burlington, Ontario Canada.

Rhoades, J. D., 1989. Intercepting, Isolating, and Reusing Drainage Waters for Irrigation to Conserve Water and Protect Water Quality, *Agricultural Water Management,* 16, pp. 37-52.

Rhoades, J. D. and Merrill, S. D., 1976. *Assessing the Suitability of Water for Irrigation: Theoretical and Empirical Approaches.* Soils Bulletin 31, Food and Agriculture Organization of the United Nations, pp. 69-109.

Rhoades, J. D.; Oster, J. D.; Ingvalson, R. D.; Tucker J. M.; and Clark M., 1974. Minimizing the Salt Burdens of Irrigation Drainage Waters, *Journal of Environmental Quality,* 3, pp. 311-316.

Rhoades, J. D. and Suarez, D.L., 1977. Reducing Water Quality Degradation through Minimized Leaching Management, *Agricultural Water Management,* 1-2, pp. 127-142.

Rhoades, J. D.; Bingham F. T.; Letey, J.; Dedrick, A. R.; Bean, M.; Hoffman, G. J.; Alves, W. J.; Swain, R. V.; Pacheco, P. G.; and LeMert, R. D.; 1988a. Reuse of Drainage Water for Irrigation: Results of Imperial Valley Study. Hypothesis, Experimental Procedures and Cropping Results, *Hilgardia,* 56 (5), pp. 1-16.

Rhoades, J. D.; Bingham, F. T.; Letey, J.; Pinter, P. J. Jr.; LeMert, R. D.; Alves, W. J.; Hoffman, G. J.; Replogle, J. A.; Swain, R. V.; and Pacheco, P. G., 1988b. Reuse of Drainage Water for Irrigation: Results of Imperial Valley Study II. Soil Salinity and Water Balance, *Hilgardia,* 56(5), pp. 17-44.

Rhoades, J. D.; Bingham, F. T.; Letey, J.; Hoffman, G. J.; Dedrick, A. R.; Pinter P. J.; and Replogle J. A., 1989. Use of Saline Drainage Water for Irrigation: Imperial Valley Study, *Agricultural Water Management,* 16, pp. 25-36.

Stroh, C. M., 1990. Land Retirement as a Strategy for Long Term Management of Agricultural Drainage and Related Problems. This Volume.

Suarez, D. L. and Rhoades J. D., 1977. Effect of Leaching Fraction on River Salinity, *ASCE Journal of Irrigation and Drainage Engineering*, 103(IR2), pp. 245-257.

van Schilfgaarde, J.; Bernstein, L.; Rhoades, J. D.; and Rawlins, S. L., 1974. Irrigation Management for Salt, *ASCE Journal of Irrigation and Drainage Engineering*, 100(IR3), pp. 321-338. Closure: 102(IR4), pp. 467-469.

van Schilfgaarde, J. and Rhoades, J. D., 1984. Coping with Salinity. Water Scarcity. In: Engelbert, E. A. (Ed.), *Impacts in Western Agriculture*, University of California Press, pp. 157-179.

# 7 LAND RETIREMENT AS A STRATEGY FOR LONG-TERM MANAGEMENT OF AGRICULTURAL DRAINAGE AND RELATED PROBLEMS

Craig M. Stroh, U.S. Bureau of Reclamation and San Joaquin Valley Drainage Program

## ABSTRACT

Land retirement--the policy of removing land from irrigation--is compared to other drainage management alternatives for areas where shallow ground-water tables contain elevated levels of selenium. These drainage management alternatives, which are being evaluated by the San Joaquin Valley Drainage Program (SJVDP), are: Drainage water treatment, evaporation, dilution, and ground-water pumping. Candidate lands are identified by the ground-water concentration of selenium. The criterion for determining whether to retire a candidate parcel is the minimization of social cost. The value of water made available through retirement for other uses is another important consideration. In several of the cases analyzed, land retirement is the least-cost management alternative.

## INTRODUCTION

This chapter presents an assessment of land retirement as a strategy for the long-term management of agricultural drainage and drainage-related problems being considered by the SJVDP in California. These problems can be viewed from both the short- and long-term perspectives. The short-term perspective focuses on contaminant problems associated with the production and management of drainage water while the long-term perspective emphasizes drainage problems related to salinity and boron buildup in the soil as it affects the sustainability of agriculture. Waterfowl and waterbird deformities and deaths at Kesterson Reservoir and at agricultural drainage evaporation ponds elsewhere in the San Joaquin Valley (Valley) have been attributed to elevated concentrations of selenium in drainage water. This has focused attention and research efforts on contaminant issues. Contaminant-related problems (lethal effects on wildlife and potentially harmful public health effects) must be solved as a part of, or prerequisite to, solving the problems of agricultural sustainability posed by salt and boron accumulations in the soil. The latter may be considered economic/social issues which will need to be addressed at some future date. The emphasis in this chapter is on contaminants which pose a pollution problem.

## CONCEPT OF LAND RETIREMENT

The term land retirement refers to the policy of removing land from commercial irrigated agricultural production. Retiring irrigated land in areas subject to seasonally high ground-water tables containing elevated concentrations of dissolved selenium (>5 p/b) (EPA, 1987) is a means of reducing both the present and projected future quantity of selenium-contaminated drainage requiring management. Selenium locked in the soil profile and present in ground water does not directly damage agricultural production, including the quantity and quality (for public health concerns) of crops or fish and wildlife resources. Selenium, however, may become a problem for the environment when concentrated in surface discharge of agricultural drainage, surfacing ground water, or when insufficiently diluted in streams, surface water deliveries, or evaporation ponds. Damages to environmental resources and public health are social costs that could be avoided through implementing a policy of land retirement. This makes land retirement a potentially important nonstructural alternative for drainage management--that is, one that requires little or no construction of facilities. The rationale for public intervention is the avoidance of environmental and potential public health damages that may result when excessive concentrations of selenium are released to the surface environment.

Although retiring land from irrigation would reduce drainage volume and salt loads as well as contaminants such as selenium, the focus here is solely on selenium in order to identify candidate lands for retirement and then to determine whether they should be retired. Identification of candidate lands is based on hydrologic considerations while the decision to retire candidate lands is an economic one. At present, selenium is the only drainage water constituent conclusively identified as having lethal effects on wildlife (Ohlendorf, 1989).

Land retirement does not refer to land going out of irrigation due to adverse economic conditions, such as the loss of markets which make irrigated agriculture unprofitable in the long run. Nor does it refer to the possible situation where worsening agricultural drainage problems, such as high and/or saline water tables, cannot be successfully managed while simultaneously permitting the farmer to earn an adequate profit. These can be thought of as examples of "land abandonment" resulting from the interplay of economic forces. Land retirement is used here to mean the conscious determination to cease irrigation as a preferred social policy.

## ECONOMIC ANALYSIS

The criterion used to determine whether candidate land would be retired from irrigated agricultural production is the social cost associated with alternative drainage management methods. The objective is to utilize land retirement as an in-lieu-of drainage management method in order to minimize social costs (environmental, public health, and other costs resulting from the production, management, and disposal of drainage).

### Externalities

No attempt has been made to estimate biologically and physically the full scope of possible effects--adverse as well as beneficial--of drainage management alternatives on public health and environmental resources. The full range of impacts caused by selenium may not even be known, and the known impacts are often difficult to quantify. Any estimate of both beneficial and adverse effects would likely be incomplete. Instead, it has been assumed that objectives and regulations designed to protect public health and environmental resources would be met by all alternatives and that damages to the environment would be avoided. In view of this approach, the minimization of social cost (rather than the maximization of benefits) is the appropriate planning objective.

Conceptually it would be useful to think in terms of drainage producers and receivers. Producer lands would discharge drainage through artificial drains (as a by-product of irrigating crops) while receivers would be the water bodies or streams (or lands on which the water body is located) to which drainage would be discharged. Land parcels which are both drainage producers and receivers could be characterized as being either "net producers" or "net receivers." Damages created by drainage producers and incurred by other receivers would be a form of externality.

From a regional hydrologic perspective, the volume of lateral subsurface drainage flow along the west side of the Valley is relatively minor, probably no more than 10 percent, compared to the vertical flow (Belitz, 1988). Therefore, regionally at least, lateral flows and the externalities they may cause appear to be a minor problem. At the other areal extreme--the individual field--lateral subsurface flows can be more significant and can contribute to the rise and extent of shallow water tables underlying downslope lands or add to the subsurface drainage discharge of downslope lands underlain by onfarm drainage systems. Contributing to the rise of shallow water tables or to increased drainage flows of downslope lands is an obvious externality to the downslope grower/landowner.

In reality, it is virtually impossible, given the current state of the hydrologic science, to identify conclusively what upslope parcel of irrigated land contributes to the subsurface drainage flow experienced by a downslope parcel. Subsurface flows may be intercepted from many upslope irrigated areas, and as the affected area increases in size, the relative importance of lateral flows as a contributor to shallow ground water and subsurface drainage flows diminishes. Given the difficulty in identifying or measuring the quantity of ground water contributed to subsurface drainage flows by the irrigation of land upslope, the emphasis of the analysis is on lands that actually produce drainage or are underlain by high water tables.

The downward movement through the soil profile of shallow ground water with high selenium concentrations may possibly degrade good quality ground water beneath it. The location, extent, and time rate of degradation depend on a number of factors, including selenium concentration in the ground-water layers, thickness of ground-water layers, irrigation and drainage management practices, and soil permeability. Hydrologic and soil conditions and irrigation and drainage practices vary significantly throughout the west side of the Valley, and soil conditions can vary significantly even within a field. Given the uncertainty as to the extent and degree of degradation of lower lying ground water (and whether it is even reliably measurable), this type of possible degradation (externality) is not considered. For similar reasons, possible lateral migration of ground water high in dissolved selenium into usable surface waters, such as rivers and streams, is not considered.

## Economic Model

An economic model has been developed to compare the social costs of land retirement to the social costs of other drainage management measures.

Minimizing the social costs of drainage management/disposal is the policy objective, and the costs of four management techniques are compared to the cost of land retirement to select the least-cost method. These techniques are: Treatment, evaporation, dilution, and ground-water pumping (SJVDP, 1989a). Costs of management combinations, such as treatment with discharge to evaporation ponds, are also estimated. Retirement of farmland is preferred when total social costs are less than the total social costs of other drainage management methods or combination of methods.

The net social cost of land retirement is represented by the agricultural income forgone when land is no longer irrigated, minus the incremental value of irrigation water for other uses. Water formerly used to irrigate retired land would be available for other uses, and its increased value, if any, for other uses would be subtracted as a cost offset from the loss of agricultural income.

## DATA AND EMPIRICAL SPECIFICATIONS

This section discusses the data and empirical specifications for drainage management alternatives and begins with those common to all alternatives.

### Common Assumptions

Assumptions concerning the values of parameters common to all management methods are discussed below. Parameters may take on a range of values to simulate a range of conditions that actually occur now or may occur in the future.

Land value = $1,500/AC and $2,500/AC
Rate of return in agriculture = 5% and 10%
Irrigation water use = 2.5 AF/AC and 3.0 AF/AC
Increase in water value = $0/AF and $100/AF
Drainage yield coefficient = 0.2 AF/AC to 0.6 AF/AC
Drainage contributing area = 5,000 AC and 20,000 AC

Land values and the rate of return on land in agriculture are used to estimate annual agricultural income. Another method frequently used to estimate annual farm income is the farm budget technique which models a farm operation and generates estimates of revenues, costs, and net income. Application of this technique is usually time consuming and requires a great deal of information specific to selected farming areas. The approach adopted instead focuses on land as an investment, and the annual income from land (and attached water) can be estimated by an expected rate of return. Net returns to land in agriculture can be expected to vary because of differences in farmland, including soil fertility, security and price of water supply, drainage conditions (e.g., no problems expected, problems likely in the future, problems existing and corrected, problems existing and uncorrected), and proximity to markets.

The range of land values is based on a survey of recent sales of agricultural land throughout the Westlands Water District. The rate of return range spans what is thought to be acceptable in agriculture and other sectors of the economy. The annual return to farmland is equal to land value multiplied by the rate of return.

Agricultural water use per acre of land is an estimate of the amount of water used in irrigation which would be available for alternative uses if land were retired from irrigation. The water use values selected are based on experience in the Valley (KCWA, 1985 and 1989 and WWD, 1989).

The additional value created by water in an alternative use depends on the use selected. Three levels of additional values per AF ($0, $50, $100) have been assumed and could be represented by agricultural use elsewhere ($0 increase) or municipal/industrial use ($50, $100), for example. It should be noted that the value of water is not necessarily equal to the price paid for it. For example, the price paid by a grower to a water district for water supplied by the Bureau of Reclamation or California State Water Project is generally less than the value created by that water in agricultural use (Frederick and Gibbons, 1986).

The drainage coefficient, measured as the volume of drainage water produced by an onfarm subsurface drainage system on a per-acre basis, is affected by irrigation and drainage management practices. It is assumed that growers will adopt practices to minimize drainage production. First, source control is the least-cost method to reduce drainage water production (SJVDP, 1989a); second, society would probably expect landowners to invest in source control before incurring costs to retire land. These costs are estimated to be in the range of $50-$60 per acre (SJVDP, 1989a).

It is assumed that drainage alternatives, including land retirement, will be implemented on a reasonably large scale and that total drainage from a contributing area would be treated as a unit. This would simplify administration and allow exploitation of any economies of scale.

## Drainage Management Alternatives

Four drainage management alternatives being considered by the SJVDP include treatment to remove selenium; evaporation of drainage effluent; dilution prior to discharge into the San Joaquin River (River), evaporation, or reuse; and ground-water pumping (SJVDP, 1989a).

## Treatment

The anaerobic-bacteria drainage water treatment process provides the basis for the analysis of treatment performance and costs. A treatment cost function was developed from performance/cost information (Nishimura, 1989).

Annual treatment plant capacity varies from 1,000 AF to 12,000 AF. Treatment costs vary with the volume and selenium concentration of drainage water being treated. The target selenium concentrations in the treated water are 10 p/b and 20 p/b. Costs for achieving the targets have been estimated for influent drainage with a selenium concentration of 300 p/b. Nishimura (1989) has indicated that the existing cost estimates for achieving the 10 p/b and 20 p/b targets are reasonable for selenium influent concentrations of approximately 200 p/b and perhaps less. Given that treatment costs are reasonably invariant to influent selenium concentrations (over at least a range of likely values), treatment costs are expressed in terms of drainage quantity and target selenium concentrations in the treated water.

An equation imposed on the Nishimura data was used to estimate the cost of treating drainage to the level of 10 p/b or 20 p/b selenium in the product water. The cost equation used for the estimate is the following:

$$C = \exp(-34.09)Q^{.924}S^{4.694}$$

where  C  = annual treatment cost to remove selenium, in $million
Q = quantity of drainage treated, in AF/YR and
S = reduction in selenium concentration achieved by treatment, in p/b

Treatment itself does not constitute a complete drainage management alternative since selenium concentrations could, at best, be lowered to 10 p/b. This would exceed the threshold selenium concentration of 5 p/b (EPA, 1987) and could result in damages if discharged to the environment. Treatment could be combined with other management methods, however, such as evaporation or River discharge.

## Evaporation

An evaporation pond system is assumed to consist of three cells among which drainage effluent is routed as the concentration of dissolved salts and other constituents, such as selenium, increases. They eventually precipitate out in the last cell. The system is expected to evaporate 4 AF of drainage per acre of surface water area annually. In cases where elevated concentrations of selenium in evaporation ponds could pose a threat to waterfowl safety, vegetative growth on levees would be eliminated, and hazing programs would be employed to frighten off waterfowl and waterbirds. In addition, alternate freshwater habitat near the evaporation ponds would be provided to attract the birds.

It is assumed that a regional entity constructing evaporation ponds would locate them on the least productive and most drainage problem-plagued agricultural land within its boundaries and that the value of this land would be one-half that of irrigated land. In addition, water formerly used to irrigate this land would be available for alternative uses.

The following summarizes data used to estimate evaporation costs:

Construction cost = $2,425/AF evaporated
Land cost = $750/AC (50% agricultural land value)
Operation and maintenance = $9/AF evaporated
Hazing and vegetation control = $10.25/AF evaporated
Alternative habitat = $137/AF and $194/AF evaporated
Useful life of ponds = 20 and 30 years

Evaporation costs are based on construction and operation and maintenance costs developed by SJVDP (1990) while costs for hazing, vegetative control, and alternate habitat are taken from Bradford et al. (1989).

*Dilution*

Direct discharge to the River is limited by its capacity to assimilate selenium as implied by selenium objectives proposed for the River. Drainage could also be diluted with freshwater prior to discharge to the River. In an attempt to avoid the cost of providing alternate habitat, drainage could be diluted with freshwater before discharge to evaporation ponds. The following summarizes the empirical data utilized in the dilution analysis:

River discharge:
  Cost of freshwater = $12/AF
  Selenium concentration of freshwater = 1 p/b
  San Joaquin River selenium objective = 3 p/b, 5 p/b, 8 p/b
  Selenium concentration in drainage = 150 p/b

Evaporation pond discharge:
  Cost of freshwater = $40/AF
  Pond selenium influent standard = 1.5 p/b, 5 p/b
  Selenium concentration of freshwater = 1 p/b

The costs of freshwater for River dilution are from SJVDP (1990) and for evaporation pond dilution from Bradford et al. (1989). The concentration range of dissolved selenium in drainage is characteristic of portions of the west side of the Valley (SJVDP, 1990), and the selenium concentration in freshwater is developed in SJVDP (1989b). The selenium standard for pond influent to avoid the need for alternate waterfowl habitat is from EPA (1987) and UCCC (1988), and the selenium concentrations in treated drainage water is developed from Nishimura (1989).

The 5 p/b selenium objective for the River and its tributaries (all but critically dry years) refers to the base quality of water flowing in these watercourses. Once the selenium concentration in a stream reaches 5 p/b, the concentration of selenium in discharges to the River cannot exceed this level or the objective itself will be exceeded. If the selenium concentration in the receiving water is less than 5 p/b, then some capacity to assimilate drainage with a selenium concentration in excess of 5 p/b exists. The rate of flow and selenium concentration in the receiving water will determine the allowable volume and selenium concentration of drainage that can be discharged. If the volume and selenium concentration of drainage exceed the levels consistent with River selenium objectives, dilution with freshwater or treatment prior to discharge would be required. The quantity of freshwater needed for dilution depends on the River's assimilative capacity, the volume and selenium concentration of drainage water, and selenium concentration in freshwater. Dilution cost depends on the price and required quantity of freshwater.

## Ground-Water Pumping

Ground-water pumping could be used to manage shallow ground-water tables and reduce the volume of subsurface drainage requiring management. The amount of ground-water pumping that could be incorporated into drainage management plans depends not only on the cost of pumping but also the quality of ground water and the safe yield of the aquifer. It is assumed that the pumped water would be used to supplement the irrigation supply so that surface water deliveries would be reduced. The following summarizes the data utilized in developing the costs of the ground-water pumping alternative:

Well depth = 250 FT and 600 FT
Pumping depth = 100 FT
Pumping rate = 200 gal/min and 340 gal/min
Annual pumpage volume = 0.4 AF/AC and 1.5 AF/AC
Energy cost = $0.085/kWh
Capital cost = $22,300 and $39,800
Annual OM&R cost = 5 percent of capital cost

This alternative includes four cases: (1) Pumping 0.4 AF/AC at 200 gal/min from a 250-ft-deep well; (2) pumping 0.4 AF/AC at 200 GPM from a 600-ft-deep well; (3) pumping 1.5 AF/AC at 340 gal/min from a 250-ft-deep well; and (4) pumping 1.5 AF/AC at 340 gal/min from a 600-ft-deep well. The capital costs correspond to the two well depths, and the actual pumping depth remains constant for all cases. The parameter values are developed in SJVDP (1990).

Deep percolation, which contributes the buildup of the ground-water table, averages approximately 11 inches in the Westlands area (Burt and Katen, 1988). Onfarm water conservation is assumed to eliminate 3 inches of deep percolation, and 3 inches are assumed to percolate through the Corcoran Clay. The other 5 inches would be eliminated through pumping.

Halophytes, a very salt-tolerant plant, could be grown as a forage crop and be irrigated with large volumes of drainage water. In this case, 1.5 AF/AC of ground water would need to be pumped to avoid a buildup of the water table.

In cases 1 and 2, 0.4 AF/AC of pumped ground water would replace surface water as part of the irrigation supply. The water thus freed would be available for alternative uses, and any increase in value in an alternative use would be a cost offset. In cases 3 and 4, 1.5 AF/AC of pumped ground water would be available as an irrigation supply and a like amount of surface water would be freed for other uses. Any increase in value of this water would act as a cost offset to this alternative.

## EMPIRICAL RESULTS

The social costs of drainage management alternatives are presented below. Costs have been calculated under a variety of assumptions regarding the values of key parameters to reveal their sensitivity to a range of possible future conditions. All economic values are expressed on an annual basis. Costs are first presented for the individual management alternatives followed by costs of combined management alternatives.

### Land Retirement

The social cost of land retirement is calculated for several assumptions concerning irrigated land values, irrigation water use, and the social value of irrigation water in alternative uses. Part of the annual net social cost is given by the return to irrigated land forgone because of retirement. Forgone annual agricultural income to land ranges from $75 per acre to $250 per acre.

The cost offset associated with releasing water from retired land for alternative uses ranges from $0/AC to $300/AC retired and is based on an increased value of water released from irrigation of $0/AF and $100/AF and a quantity of water released annually equal to 2.5 AF/AC to 3 AF/AC.

Annual net social costs of land retirement are shown in table 1. A negative value means that economic value actually increases when irrigated land is retired and the water is transferred to a higher valued use, such as the municipal/industrial sector. The values shown are expressed on a per-acre basis. It should not be inferred that continually increasing quantities of water could be transferred from agriculture to alternative uses without an effect on the relative value of water in these uses. Demand for water, including water for M&I uses, is not unlimited, and its marginal economic value will decline as greater quantities are made available and used.

Table 1. Net social cost per acre of land retired ($).

|  |  | Annual Return to Land Forgone ($) | | | |
|---|---|---|---|---|---|
|  |  | 75 | 125 | 150 | 250 |
| Increase in Water Value ($) | 0 | 75 | 125 | 150 | 250 |
|  | 125 | -50 | 0 | 25 | 125 |
|  | 150 | -75 | -25 | 0 | 100 |
|  | 250 | -175 | -125 | -100 | 0 |
|  | 300 | -225 | -175 | -150 | -50 |

## Treatment

Costs to reduce the selenium concentration in agricultural drainage have been estimated for several conditions. The treatment process is subject to economies of scale, and these were accounted for by selecting alternative acreages as drainage contributing areas and different drainage coefficients. Total annual treatment costs on a per acre-foot basis for treated are shown in table 2.

Table 2. Annual treatment cost ($ per AF).

|  |  | Drainage Quantity (AF) | | | | | |
|---|---|---|---|---|---|---|---|
|  |  | 1,000 | 2,000 | 4,000 | 8,000 | 10,000 | 12,000 |
| Target Se Concent (p/b) | 10 | 335 | 318 | 302 | 286 | 281 | 278 |
|  | 20 | 284 | 270 | 256 | 243 | 239 | 236 |

Table 3 summarizes total annual selenium treatment costs.

Table 3. Total annual drainage effluent treatment cost ($1,000).

|  |  | Drainage Quantity (AF) | | | | | |
|---|---|---|---|---|---|---|---|
|  |  | 1,000 | 2,000 | 4,000 | 8,000 | 10,000 | 12,000 |
| Target Se | 10 | 335 | 636 | 1,207 | 2,290 | 2,814 | 3,330 |
| Concent (p/b) | 20 | 284 | 540 | 1,024 | 1,942 | 2,387 | 2,825 |

## Evaporation

The net social cost of using evaporation ponds to manage drainage effluent could vary for four reasons. First, the drainage yield on irrigated land can vary. Second, since it is assumed that the ponds are constructed on formerly irrigated land, water used to irrigate the pond area would be released for alternative uses, amounting to 2.5 AF/AC to 3 AF/AC. Third, the increase in the social value of water released could vary. Finally, alternate waterfowl habitat, hazing, and vegetative control would be required if the selenium concentration of drainage influent posed a threat to waterfowl.

The annual costs shown in table 4 assume:

- Evaporation land cost is $750/AC.
- Cost of alternate waterfowl habitat is $0 (none required), $137/AF evaporated, and $194/AF evaporated.
- Quantity of water released from use on irrigated land converted to ponds is 2.5 AF/AC.
- Increase in the social value of water released from irrigation on land converted to ponds is $0 (no increase), $50/AF, and $100/AF.

Costs per AF evaporated do not vary with the drainage coefficient and total volume evaporated.

Table 4. Annual evaporation cost ($ per AF evaporated).

| Alternate Habitat Cost | Increase in Social Value of Water | Evaporation Cost |
|---|---|---|
| 0 | 0 | 265 |
| 137 | 0 | 412 |
| 194 | 0 | 469 |
| 0 | 50 | 234 |
| 137 | 50 | 381 |
| 194 | 50 | 438 |
| 0 | 100 | 203 |
| 137 | 100 | 350 |
| 194 | 100 | 407 |

These costs do not include any provision for periodic clearing of solid salts from the ponds, the expansion of ponds, or permanent disposal of salts in the Valley, adjacent foothills, or the Pacific Ocean.

## Dilution

Once the assimilative capacity of the River has been reached, additional drainage discharge would require dilution with freshwater. The capacity of the River to assimilate selenium dissolved in drainage water discharged from the Grasslands area has been estimated under typical wet year and dry year hydrologic conditions. During wet years, flow in the River measured near Newman is nearly 2.4 million AF and during dry years is nearly 490,000 AF. In wet years, when River selenium objectives for SJVDP planning purposes (SJVDP, 1990) assume values of 3 p/b and 5 p/b, drainage discharge would be constrained during months of relatively low flow (summer through early winter). In dry years, when the selenium objective assumes values of 5 p/b and 8 p/b for planning purposes, drainage discharge would be severely restricted by the capacity of the River to assimilate drainage throughout the entire year. Table 5 shows the monthly distribution of drainage discharge that could be assimilated by the River under selected selenium objectives and hydrologic conditions:

Table 5. Drainage discharge (1,000 AF).

| Month | Wet | | Critically dry | |
|---|---|---|---|---|
| | 3 p/b | 5 p/b | 3 p/b | 8 p/b |
| Oct | 0.5 | 0.8 | 0.6 | 2.1 |
| Nov | 0.3 | 0.6 | 0.3 | 1.0 |
| Dec | 0.6 | 1.0 | 0.3 | 1.1 |
| Jan | 0.1 | 0.5 | 0.4 | 1.3 |
| Feb | 3.6 | 7.1 | 0.3 | 1.2 |
| Mar | 7.0 | 7.0 | 0.5 | 1.6 |
| Apr | 5.5 | 5.5 | 0.7 | 2.6 |
| May | 3.0 | 5.7 | 0.6 | 1.9 |
| Jun | 1.6 | 2.9 | 0.2 | 0.8 |
| Jul | 0.7 | 1.2 | 0.2 | 0.7 |
| Aug | 0.5 | 0.9 | 0.2 | 0.6 |
| Sep | 0.5 | 1.0 | 0.2 | 0.6 |

Discharge of additional quantities of drainage with dissolved selenium concentrations exceeding the selenium objective of the River would require dilution. Tables 6 and 7 show the quantities of freshwater and the cost of freshwater required to dilute drainage generated from 20,000 drained acres before discharge to the River for selected river selenium drainage yields and drainage coefficients. The drainage coefficient of 0.7 AF/AC represents an existing areawide average and is assumed to represent yield from drained and on which improved efficiency measures have not been adopted (SJVDP, 1989a).

Table 6. Quantity of freshwater (1,000 AF).

| | | Drainage Coefficient (AF/AC) | | | |
|---|---|---|---|---|---|
| | | 0.2 | 0.4 | 0.6 | 0.7 |
| Selenium | 3 | 294 | 588 | 882 | 1,029 |
| Objective | 5 | 145 | 290 | 435 | 507.5 |
| (p/b) | 8 | 81.2 | 162.4 | 243.6 | 284.2 |

The cost of freshwater for dilution prior to discharge to the River is assumed to be $12/AF.

Table 7. Cost of freshwater ($).

|  |  | \multicolumn{4}{c}{Drainage Coefficient (AF/AC)} |
| --- | --- | --- | --- | --- | --- |
|  |  | 0.2 | 0.4 | 0.6 | 0.7 |
| Selenium | 3 | 3,528 | 7,056 | 10,584 | 12,349 |
| Objective | 5 | 1,740 | 3,480 | 5,220 | 6,090 |
| (p/b) | 8 | 974 | 1,949 | 2,923 | 3,410 |

## Ground-Water Pumping

Table 8 summarizes annual costs of the ground-water pumping alternatives on a per-acre served basis and for 20,000 acres served excluding the potential cost offset.

Table 8. Ground-water pumping cost without offset.

|  | ($/AC) | ($/20,000 AC) |
| --- | --- | --- |
| Case 1 | 26.40 | 528,000 |
| Case 2 | 43.50 | 870,000 |
| Case 3 | 46.15 | 923,000 |
| Case 4 | 63.20 | 1,264,000 |

Table 9 shows the costs of ground-water pumping over a 20,000-acre area with a released water cost offset of $50/AF and $100/AF.

Table 9. Ground-water pumping cost with offset.

|  | ($50/AF) | ($100/AF) |
| --- | --- | --- |
| Case 1 | 128,000 | -272,000 |
| Case 2 | 470,000 | 70,000 |
| Case 3 | -577,000 | -2,077,000 |
| Case 4 | 236,000 | -1,736,000 |

Negative values mean that the increase in the value of released water outweighs the cost of ground-water pumping so that total social value actually increases as a result of this alternative.

## Combination Alternatives

Individual drainage management alternatives might not have the capacity to control high-selenium drainage while avoiding potential risks to public health or damage to environmental resources. In such cases combined alternatives might be necessary to create management techniques that would avoid these adverse effects. Given the possible limitations on the quantity of freshwater available for dilution purposes in Grasslands, especially during critical years, drainage could be treated to reduce the volume of dilution water required. Dilution of drainage water with freshwater or treatment and dilution prior to discharge to evaporation ponds could eliminate the need for vegetative control, hazing, and alternate waterfowl habitat to prevent waterfowl losses at ponds. Treatment alone prior to evaporation would not eliminate the need for vegetative control, hazing, and alternate waterfowl habitat since the best that treatment can do is lower the selenium concentration in the product water to 10 p/b.

<u>Treatment and Dilution (River)</u>. Table 10 shows the costs of treating and then diluting drainage water prior to discharge to the River. The selenium concentration of the product water is assumed to be 10 p/b and the River selenium objective 8 p/b. Drainage is produced from 20,000 AC with selected drainage coefficients.

Table 10. Cost to Treat and Dilute Drainage ($ per AF).

| Drainage Coefficient (AF/AC) | | |
|---|---|---|
| 0.20 | 0.40 | 0.60 |
| 305 | 290 | 281 |

<u>Evaporation Combinations</u>. Dilution could be combined with evaporation in two ways. First, drainage water could be diluted with freshwater prior to discharge to evaporation ponds. Second, drainage water could be treated to lower its selenium concentration and then blended with freshwater prior to discharge to a pond.

For example, blending 1 AF of 50 p/b selenium drainage water with freshwater to a standard of 5 p/b selenium would require 11.25 AF of freshwater at a cost of $450 with freshwater at $40/AF. Blending 4,000 AF of drainage produced from 20,000 AC (0.2 AF/AC) would require 45,000 AF of fresh water at a cost of $1,800,000. The annual cost to evaporate the blend would be $12,985,000 for a total annual cost of $14,785,000. This translates to $740/AC drained. Blending the same volume of drainage containing greater concentrations of dissolved selenium or meeting more stringent selenium influent standards would be more expensive.

Treating 4,000 AF of drainage from 20,000 AC (0.2 AF/AC) and blending the product water (with a selenium concentration of 10 p/b) to an evaporation pond influent standard of 5 p/b selenium would require 5,000 AF of freshwater costing $200,000. Subsequent evaporation would increase the annual cost to $3,792,000. This amounts to $190/AC drained. Treating and blending drainage with greater concentrations of dissolved or meeting more stringent selenium influent standards would also raise the cost.

## Cost Comparison

Costs of alternative drainage management methods have been estimated and compared under selected assumptions concerning key variables that affect costs. Figure 1 displays the costs of seven drainage management alternatives (land retirement with the three levels of social value of released water; evaporate; evaporate with alternate habitat; treat, dilute, and evaporate; and dilute and evaporate). Development of these costs was based on the following assumptions:

- 20,000 AC are drained.
- Selenium concentration of treated drainage water is 10 p/b.
- Annual cost of alternate evaporation pond habitat is $137/AF of water evaporated.
- Annual return to irrigated land is $150/AC.

Annual social cost is measured vertically and three values for the drainage coefficient are plotted horizontally. The first three bars in each histogram represent the social cost of land retirement in the cases when the irrigation water released remains in agriculture ($3,000,000) or is transferred to another use with an increase in social value of $50/AF ($500,000) or $100/AF (-$2,000,000). The negative sign of the latter signifies an actual increase in social value when released water is utilized in a higher value use. Land retirement costs are invariant with drainage yield.

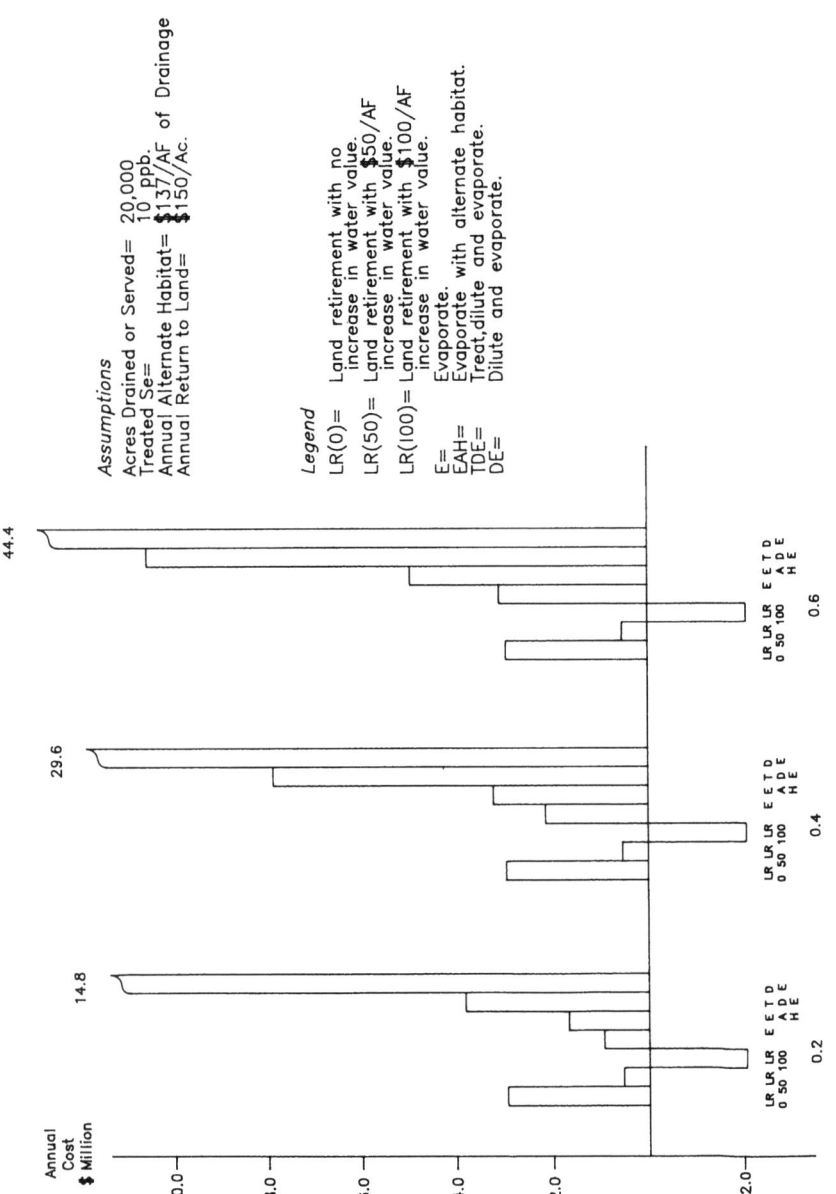

Figure 1. Cost of land retirement and selected drainage management alternatives.

"Evaporate" represents the costs of evaporating drainage water which does not contain elevated concentrations of selenium and is included for comparison purposes. "Evaporate (Alternate Habitat)" represents the costs of evaporating drainage water with selenium concentrations exceeding the 1.5 or 5 p/b standard, as appropriate, and requiring alternate habitat to protect waterfowl and waterbirds. "Treat, dilute, and evaporate" depicts the cost of treating drainage to 10 p/b selenium concentration, diluting with 1 p/b selenium concentration freshwater costing $40/AF, and evaporating the 5 p/b selenium blend. "Dilute and evaporate" shows the cost of diluting drainage of 50 p/b dissolved selenium with freshwater having 1 p/b selenium and costing $40/AF, and evaporating the 5 p/b selenium blend. Evaporation is the least costly drainage management method except for land retirement when the water released from irrigation would be used for other purposes with an increase in social value of at least $50/AF.

The costs of the four cases of ground-water pumping are compared to land retirement costs in figure 2. Where alternative water uses are not associated with increases in social value created by released water, the ground-water pumping alternative is less costly than land retirement. For a $50/AF increase in social value of water in alternative uses, ground-water pumping is still less costly than land retirement, but the difference in costs narrows, especially with regard to case 2 of the ground-water pumping alternative. For the $100/AF increase in water value, only case 3 of the ground-water pumping alternative is less costly than land retirement and this cost differential is less than 4 percent.

Ground-water pumping costs are compared to evaporation costs in figure 3. The conservative assumption is made that water freed by ground-water pumping is utilized in agriculture with no increase in value. For even the lowest drainage coefficient ground-water pumping is less costly than evaporation including the situation where alternate habitat is unnecessary.

Several factors serve to limit the usefulness of ground-water pumping a an effective drainage management alternative. The semiconfined aquifer lying above the Corcoran Clay can be interspersed with clay layers which can have the effect of lessening the hydrologic continuity of the aquifer. This could increase ground-water pumping energy requirements and costs above the levels reported herein which assume a reasonably continuous aquifer in the vertical direction. Water pumped from the semiconfined aquifer would be used in irrigation, and the SJVDP in its planning, limits the quality of this water to 1,250 p/m total dissolved solids (TDS). This pumping would draw shallow ground-water downward and degrade the quality of the semiconfined aquifer as mixing occurs. Accelerating the rate of ground-water pumping as a drainage management alternative would hasten this degradation, and some preliminary analyses by the SJVDP indicate that, depending on localized aquifer characteristics and pumping rates, the 1,250 p/m TDS limit could be reached in a little

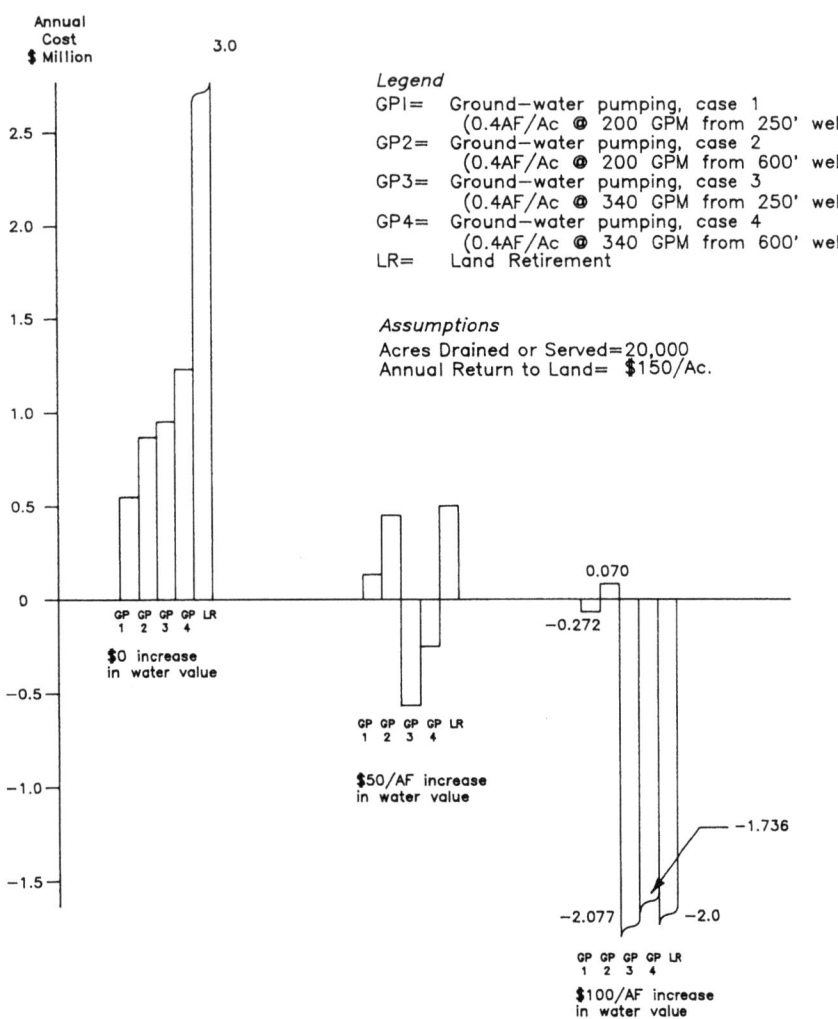

Figure 2. Ground-water pumping and land retirement cost.

# LAND RETIREMENT

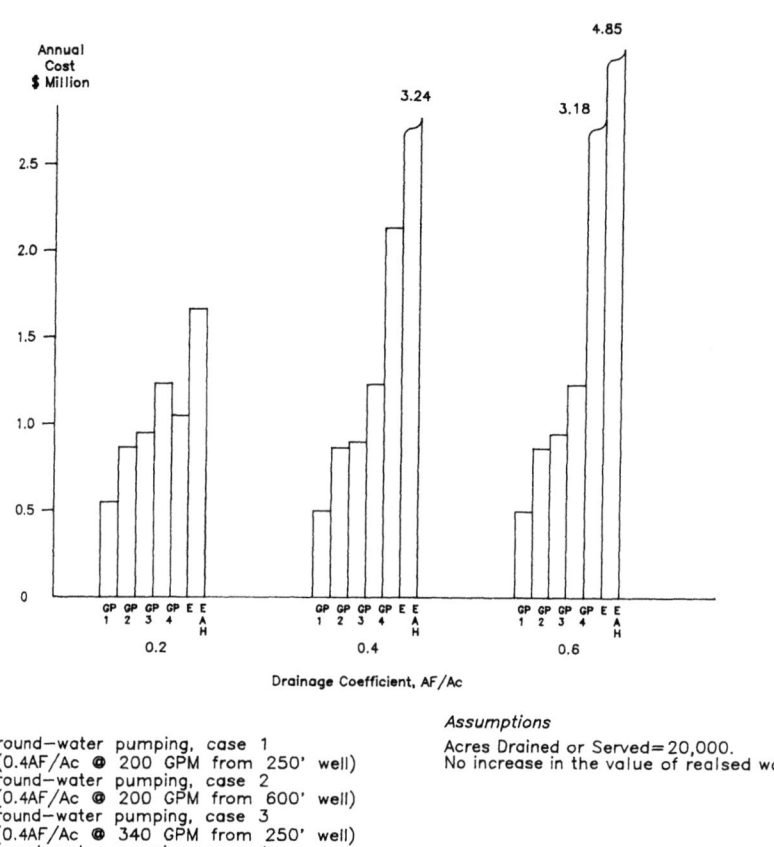

**Figure 3.** Cost of ground-water pumping and selected drainage management alternatives.

as 10-30 years. Furthermore, accelerating the rate of degradation of the semiconfined aquifer may not be consistent with the ground-water quality policy in the Central Valley Basin Water Quality Control Plan.

For the more restrictive River selenium objective of 3 p/b, the cost of freshwater required to dilute drainage from 20,000 drained acres ($3,528,000-$12,348,000 annually depending on the drainage coefficient) exceeds the annual cost of retiring land ($3,000,000) when water released is utilized in agriculture with no increase in economic value. For the 5 p/b River selenium objective, the annual cost of freshwater for dilution ranges from $1,740,000 to $6,090,000 annually and exceeds the annual cost of land retirement (with no increase in the value of water) except for the lowest drainage coefficient (0.2 AF/AC). In this case, the annual cost of freshwater would be $1,740,000, but the annual cost of lowering the drainage coefficient to 0.2 AF/AC has been estimated to be on the order of $60/AC (SJVDP, 1989a) increasing the overall cost of dilution by $1,200,000 annually. In this event, there is little difference between the cost of land retirement and dilution. For a River selenium objective of 8 p/b, the cost a freshwater for dilution alone would exceed the land retirement cost (no increase in value of water) only for the highest drainage coefficient, 0.7 AF/AC. When the drainage coefficient is 0.2 AF/AC, the annual cost of freshwater and additional irrigation/drainage management costs would exceed $2,000,000 for 20,000 drained acres.

Two important additional considerations have a bearing on the use freshwater for dilution. The first relates to the very use of freshwater for dilution. The Central Valley Basin Water Quality Control Plan (SWRCB, 1975) does not include dilution of contaminants, such as selenium, as a beneficial use of waters of the River. The second relates to the quantity of freshwater that would be required for dilution. The 8 p/b selenium objective applies only during critically dry years when freshwater for dilution simply might not be physically available. In addition, strong pressure would undoubtedly be brought to prevent the use of very limited freshwater supplies for dilution during a critical year.

## CONCLUSIONS

A key consideration for deciding to retire irrigated land underlain by shallow water tables containing elevated levels of selenium is the increase in social value created by water in alternative uses. In the case of the highest values for alternative uses for released water ($100/AF), land retirement is less costly than all alternatives considered except case 3 of ground-water pumping. Cases 3 and 4 of ground-water pumping both involve the release of 1.5 AF/AC for alternative uses. When water released by land retirement is used in irrigation with no increase in economic value, several other alternatives have lower costs than land retirement, and of these, ground-water pumping is the least cost. Questions related to aquifer continuity and ground-water quality degradation, however, cast doubts on this alternative as a viable drainage management alternative even for the short term. Further site-specific research in candidate pumping locales should be undertaken prior to implementation.

In comparing the costs of land retirement with the costs of alternatives that physically manage drainage water, the value of the drainage coefficient is critical. Where there is no increase in social value created by water in alternative uses, evaporation generally has a lower cost even when alternate freshwater waterfowl habitat must be provided in the vicinity of the ponds. The exception occurs when the annual cost of alternate waterfowl habitat ranges near its upper limit ($194/AF) and drainage yields are relatively high (0.6 AF/AC). In this case land retirement is less costly than evaporation even if the released water were used in agriculture with no increase in economic value.

Providing alternate freshwater habitat near high-selenium evaporation ponds, which are also subject to hazing and vegetative control, has not been proven to be thoroughly effective in attracting waterfowl away from the ponds. Wildlife biologists are hopeful that the technique will succeed in reducing, if not eliminating, bird deformities and deaths. If it proves unsuccessful, evaporation of high selenium concentration drainage water may not be permitted in the long term.

Management of the solid residue remaining after drainage water evaporates from ponds could pose problems. Where selenium is not present in concentrations which threaten wildlife, residue could be removed from the ponds and disposed elsewhere, as in a landfill or conveyed to the Pacific Ocean. If the residue is left in the ponds, the ponds would probably require covering with earth to keep the material dry. Additional ponds would need to be constructed as existing ones gradually fill up. Where selenium is a problem, the residue might have to be removed to a Class I hazardous wastesite. The storage capacity of these sites is limited, and the only one near the drainage problem area is located in the Kettleman Hills at the western edge of the Valley.

In areas with drainage access to the River, land retirement is less expensive than dilution under the most stringent River selenium objective (3 p/b). As the selenium objective is relaxed, the cost of dilution relative to that of land retirement improves. Costs for the dilution alternative shown above, however, include only the cost of freshwater and omit costs that might be incurred in conveying either the freshwater prior to dilution or the resulting blend.

At present, treatment does not offer a sole solution because the technology can lower the selenium concentration in the product water to no better than 10 p/b. Discharge of this water to evaporation ponds or the River would require dilution or alternate waterfowl habitat. Further treatment research should continue in anticipation of developing cost-effective methods to neutralize elevated selenium concentrations in drainage.

## REFERENCES

Belitz, K., 1988. *Character and Evolution of the Ground-Water Flow System in the Central Part of the Western San Joaquin Valley, California.* U.S. Geological Survey Open-File Report 87-573.

Bradford, D. F.; Drezner, D.; Shoemaker, J. D.; and Smith, L., June 1989. *Evaluation of Methods to Minimize Contamination Hazards to Wildlife Using Agricultural Evaporation Ponds in the San Joaquin Valley, California.*

Burt, C. M. and Katen, K., 1988. *Water Conservation and Drainage Reduction Program.* Technical Report to the Office of Water Conservation, California Department of Water Resources. Westside Resource Conservation District, 1986-87.

California State Water Resources Control Board, 1975. *Water Quality Control Plan for the San Joaquin Basin (5C).*

Environmental Protection Agency, 1987. *Ambient Water Quality Criteria for Selenium.*

Frederick, K., (Ed), with the assistance of Gibbons, D., 1986. *Scarce Water and Institutional Change.*

Kern County Water Agency, May 1985. *Water Supply Report, 1984.*

Kern County Water Agency, May 1989. *Water Supply Report, 1988.*

Nishimura, G., 1989. Selenium Treatment Cost Projections for Land Retirement Study. Memorandum, San Joaquin Valley Drainage Program.

Ohlendorf, H. M., 1989. Bioaccumulation and Effects of Selenium in Waterfowl. In: L. W. Jacobs (Ed.), *Selenium in Agriculture and the Environment,* Special Publication No. 23, Soil Science Society of America, Madison, WI., pp. 133-177

San Joaquin Valley Drainage Program, August 1989a. *Preliminary Planning Alternatives for Solving Agricultural Drainage and Drainage-Related Problems in the San Joaquin Valley.*

San Joaquin Valley Drainage Program, September 1989b. Technical Information Record, Supporting Document to *Preliminary Alternatives for Solving Agricultural Drainage and Drainage-Related Problems in the San Joaquin Valley.*

San Joaquin Valley Drainage Program, September 1990. Technical Information Record, Supporting Document to Final Report.

University of California Committee of Consultants on San Joaquin River Water Quality Objectives, February 1988. *The Evaluation of Water Quality Criteria for Selenium, Boron, and Molybdenum in the San Joaquin River Basin.*

Westlands Water District, November 1989. *Water Conservation and Drainage Reduction Programs 1987-88.*

# 8 SAN JOAQUIN SALT BALANCE: FUTURE PROSPECTS AND POSSIBLE SOLUTIONS

Gerald T. Orlob, University of California, Davis

### ABSTRACT

The San Joaquin River Basin in California is presently in a state of salt imbalance with salt loads derived by natural inflow, importations, and accretions within the basin exceeding the loads carried from the basin by hydrologic outflow and extrabasin transfers. Trends in the rates of accretion and excretion parallel the development of the basin's water resources for agricultural use. Accompanying this development there has been a progressive depletion of the natural outflow of the San Joaquin River, a principal tributary of the Sacramento-San Joaquin Delta estuarine system, and degradation of the quality of water available to users dependent upon the main stem of the river as a primary source of supply. In the lower river, this degradation has been mitigated in recent years by releases of high quality water impounded by New Melones Dam on the Stanislaus River. Supplies for water quality control from this source are limited and may not be sufficient to meet quality and flow targets in the future in the face of competing demands and continuing degradation of quality in the San Joaquin River. This chapter traces the historical development of salt loading in the basin from 1930 to 1989 by means of a basinwide salt balance accounting of principal accretions and excretions. Alternative scenarios of water quality control, and including reallocation of yield from east side reservoirs, seasonal storage of saline drainage in ground-water systems, and control of imported salt loads, are explored.

### INTRODUCTION

California's San Joaquin Valley (Valley) is situated between the crests of the Sierra Nevada and Coast Ranges, extending from the northern boundary of the virtually landlocked Tulare Basin to the Sacramento-San Joaquin Delta, the estuarine system to which it is tributary. As shown in figure 1, the Valley is drained by four major stream systems which discharge into the main stem above

Vernalis, the location of the U.S. Geological Survey (USGS) stream gauge and the principal water quality monitoring station. These streams and the Bureau of Reclamation (Reclamation) Delta-Mendota Canal comprise the water supply system for one of the most productive irrigated agricultural areas in the world.

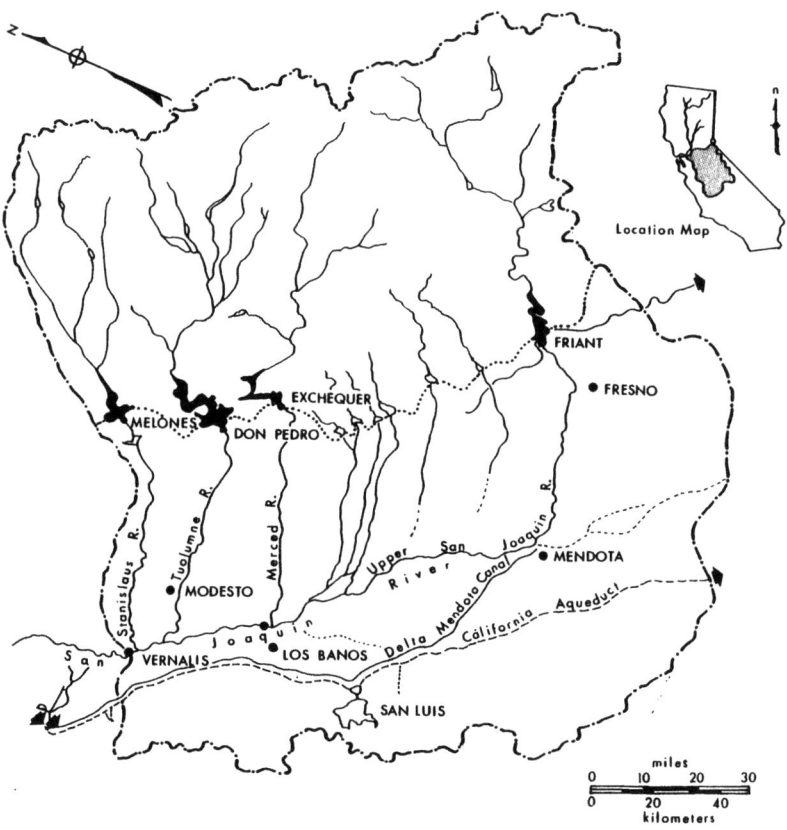

Figure 1. San Joaquin River Basin.

Since the turn of the century, the Valley has experienced a steady growth in irrigated agriculture, with a corresponding increase in consumptive utilization of the Valley's resources. The demand for water is out of phase with the natural hydrological cycle, characterized by peak snowmelt runoff in the late spring and early summer months, necessitating the impoundment of runoff in storage reservoirs for reallocation during the precipitation-deficient irrigation season. A result of this impoundment, which has steadily increased in capacity as irrigation water demand has grown, is a modification of the basin's hydrology to one of controlled flow, consisting primarily of agricultural tailwaters during the March to October irrigation season. Exportation of impounded runoff to extrabasin locations like the Tulare Basin to the south and to the city of San Francisco, and the importation of excess runoff from the Sacramento Valley in Northern California through the Delta-Mendota Canal, have also shifted the water balance of the Valley to one dictated almost entirely by water use demands.

The most apparent consequence of these changes in the Valley hydrologic system is a reduction in the long-term average annual runoff of the San Joaquin River near Vernalis as measured by the USGS. A comparison of the mean water year (October to September) runoff for the 15-year period 1930 to 1944, prior to the advent of the Central Valley Project (CVP), and the comparable period 1952 to 1966 indicates a reduction of about 1,300,000 acre-feet per year, after correction for natural hydrologic variation. Most of this decrease in runoff is evident during the irrigation season due to capture of snowmelt that prior to about 1949 produced peak discharges during April, May, and June of even the driest years.

Because most of the impounded water is diverted to irrigated agriculture with consumptive use ranging from 60 to 80 percent, the residual streamflow in the main stem of the San Joaquin River is comprised largely of drainage, tailwater, and ground-water accretions that carry the burden of dissolved mineral salts at increased concentrations. Often since 1949 water quality at downstream locations has been degraded to levels unsuitable for sustained productive agriculture. Also, expansion of irrigation into areas previously considered marginally productive due to high soil salinity has exacerbated the water quality problem for downstream users by adding additional burdens of salt leached from these soils carried out of the basin by its principal drainage course. A changing hydrologic regimen, progressive degradation of water quality, and expansion of irrigated agriculture have combined to shift the San Joaquin Basin into a state of salt imbalance, i.e., salt is being added to the basin at a rate exceeding the rate of removal. To find a solution to this problem it is necessary to establish the historic trends in water development that have contributed to the imbalance, to quantify the present status of salt load accretions to and excretions from the basin, and to examine critically alternative scenarios of future change that could achieve an equilibrium favorable to productive agriculture throughout the Valley.

## HISTORICAL PERSPECTIVE

### Agricultural Water Use

The earliest irrigated agriculture in the Valley dates from the late 1800's, the postgold rush era when unsuccessful placer miners turned to farming, adapting water diversion facilities to the needs of irrigation. Initially, diversions were made directly from streams with crudely constructed log or rubble barriers that directed the water into earthen ditches or wooden flumes. Often these were temporary structures, like the "sack dam" on the main stem of the San Joaquin River near Mendota, that were installed following the spring snowmelt period when river stages declined and were washed out during the next flood. The first permanent diversion dam to provide water for irrigation in the Valley was probably a wooden crib structure constructed on the Merced River by the Robla Canal Company, incorporated on March 30, 1870 (McSwain, 1978). The first artificial impoundment for irrigation supply, designed also for the Robla system, was achieved by constructing a 50-foot earthen levee at Lake Yosemite in 1887. Water was first diverted into the lake on February 1, 1888, creating an impoundment with a capacity of 7,425 acre-feet. From that time to the present, storage capacity dedicated primarily to agricultural use has steadily increased. The historical development of this capacity is depicted in figure 2.

Similar direct diversions were made from the Tuolumne River for the Turlock Irrigation District in 1899 and the Modesto Irrigation District in 1903 (Paterson, 1987 and USGS, 1989). The Stanislaus Water Company commenced diversion from the Stanislaus River in 1904. The first impoundments of these streams began with Modesto Reservoir in 1911, Turlock Lake in 1915, and Woodward Reservoir on the Stanislaus in 1918. By 1918 irrigation water storage capacity totaled 75,000 acre-feet on the three east side streams. Construction of Don Pedro Dam on the Tuolumne River for the Modesto and Turlock irrigation districts in 1922, Exchequer Dam (impounding Lake McClure) on the Merced River in 1926, and Melones Dam on the Stanislaus River in 1926 raised the total capacity of irrigation storage in the San Joaquin Basin to 646,000 acre-feet. The main stem of the San Joaquin River remained without storage for irrigation water supply until the construction of Friant Dam in the mid-1940's by Reclamation. This added 520,500 acre-feet, bringing the basin's total irrigation supply storage to 1,166,500 acre-feet by the end of the decade.

Major expansion of storage capacity occurred in 1968, 1971, and 1978 with the construction of large multipurpose projects on the Merced (New Exchequer), the Tuolumne (New Don Pedro), and the Stanislaus (New Melones). By 1978 the total installed storage capacity of impoundments dedicated primarily to irrigation supply reached 6,419,000 acre-feet, or 112 percent of the mean annual (1930-89) unimpaired runoff of the principal tributaries of the San

# SALT BALANCE

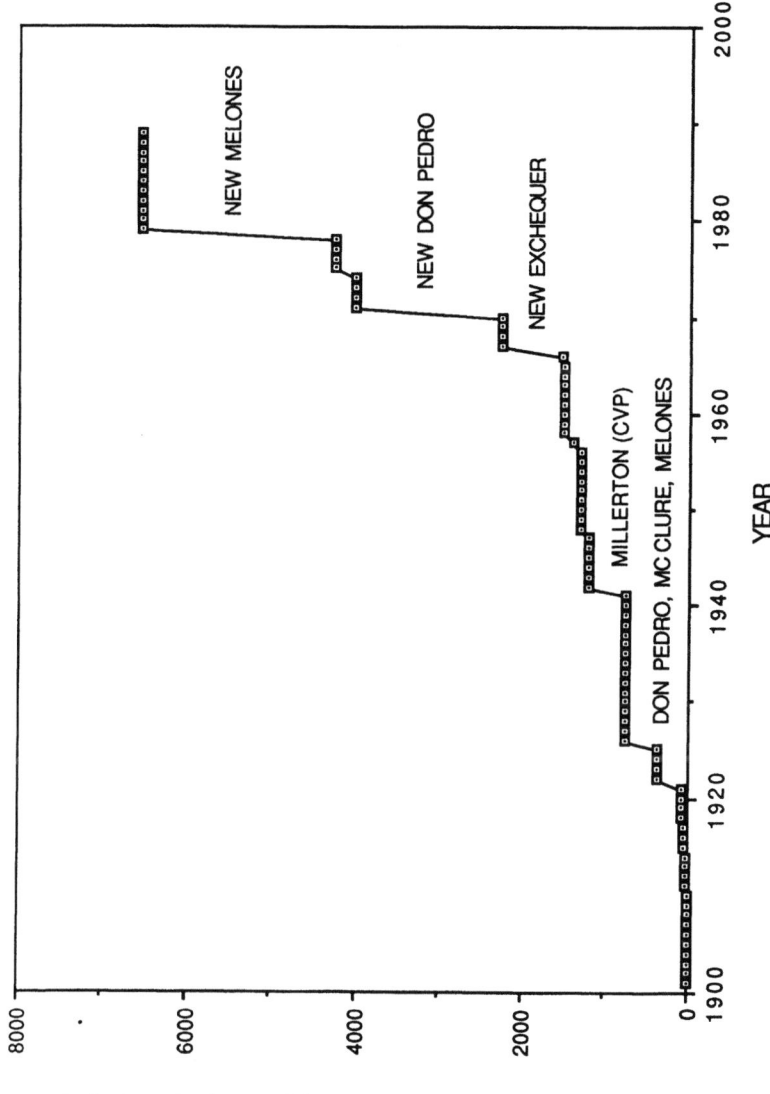

Figure 2. Installed storage capacity of irrigation projects - San Joaquin Basin (1900-1989).

Joaquin River system. (The unimpaired runoff is the estimated full natural flow of the river at the location of the principal reservoir.)

Historical growth of irrigated acreage in the Valley below the principal points of diversion roughly parallels the pattern of developed storage capacity. Figure 3 shows the historical annual diversions from the four major tributaries by five principal irrigation districts and the Central Valley Project. A steady growth of agricultural water use occurred during the first half century. About 1950 when major expansion of storage capacity was initiated, the diversion variability also increased significantly, apparently a consequence of developing available water resources to nearly the limits of the basin's natural supply. The aggregate diversion rate of seven major water users in the period after 1950 averaged approximately 3,360,000 acre-feet per year, roughly 55 percent of the average unimpaired flow available at the rim stations, i.e, at the location of the supply reservoirs. In dry years (that is, years in the lower quartile of unimpaired runoff), average diversions in the four subbasins have been about equal to the natural supply. In the Tuolumne subbasin, the most developed of the four principal subbasins, dry year diversions have exceeded unimpaired flows by about 24 percent. In four of the driest years, 1968, 1976, 1977, and 1981, the rates of diversion actually exceeded the calculated natural inflows to all of the reservoirs, a condition that could only have been met with the ability to carry over storage from prior years. This historic record indicates a present level of agricultural development on the east side of the Valley that is virtually at the limit of available supply.

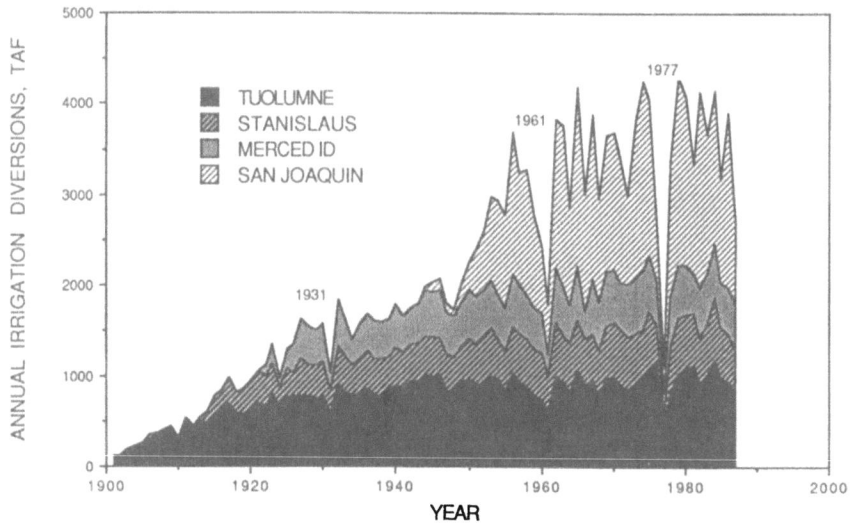

Figure 3. Irrigation diversions, 1900 - 1987 (four San Joaquin basins).

Historical water use on the west side of the Valley was governed primarily by the natural hydrology of the San Joaquin main stem through the 1940's, with virtually no regulation to assure dependable supplies throughout the irrigation season. Diversions were made by temporarily damming the river channel and "wild flooding" of pasture lands. Construction of Friant Dam and the Delta-Mendota Canal by Reclamation as integral components of the CVP allowed nearly complete regulation of the system after 1951. Water was delivered from the Delta through the Delta-Mendota Canal to the Mendota Pool in gradually increasing amounts until, by the mid-1960's, annual deliveries exceeded 1.5 million acre-feet per year. Addition of the San Luis Unit to the CVP, with deliveries beginning in 1965, increased importation of more than 2.5 million acre-feet per year, of which about 700,000 acre-feet were delivered to areas outside of the natural San Joaquin Basin. The pattern of deliveries through the Delta-Mendota Canal for the period 1951 to 1989 is shown in figure 4. It is estimated that about 60 to 80 percent of this water is used consumptively during the irrigation season.

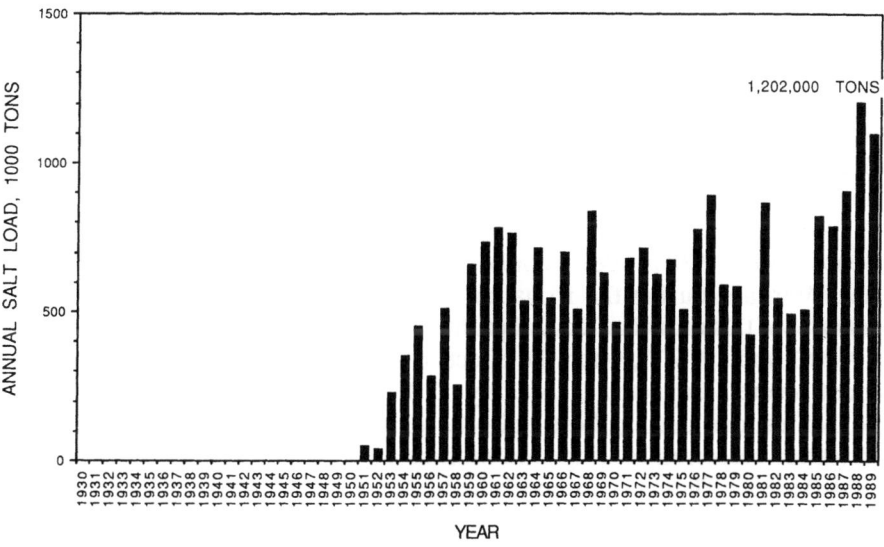

Figure 4. Annual accretion salt load - Delta Mendota Canal (1930-1989).

## Water Quality

Data on water quality in the San Joaquin Basin are very sparse for the early years of water resource development, although such data as do exist indicate severe deterioration of quality since about 1950. This is illustrated in figure 5, a comparison of three sets of water quality observations made at intervals of about 25 years in the vicinity of Mossdale, near the point where the river enters the Delta. The data depicted are for 3 dry years of natural flows within the lower quartile of record, 1908, 1933, and 1959. The 1908 data were derived from a special survey conducted in 1906 and 1908 by USGS (Van Winkle and Eaton, 1910); the 1933 data were developed by correlation of total dissolved solids (TDS) and chloride data from a program of grab sampling conducted by the California Department of Water Resources (DWR) from 1929 to 1971 (DWR, 1971); and the 1959 data were derived from continuous electrical conductivity recordings by Reclamation (1990).

In 1908, before any major water development in the basin, water quality was generally excellent and relatively unaffected throughout most of the year, except during the late summer period when some minor effects of evaporative concentration are evident. A maximum TDS concentration of 410 milligrams per liter (mg/L) was recorded, with an annual average for 1908 of 220 mg/L. In 1933, following the first stages of water resource development, the situation changed slightly; a maximum of 415 mg/L was observed, but the average annual TDS level was elevated to about 315 mg/L. Concentrations were sustained in the range of 300 to 400 mg/L through the spring and summer months. In 1959, after the CVP commenced operation, the influences of water development were very apparent. The peak TDS concentration reached 610 mg/L in July and the average for the year was more than double that of the dry year 1933 and about four times the average for 1908.

The progressive deterioration of San Joaquin River water quality is evidenced by the 1930 to 1989 record of mean TDS concentrations for the station near Vernalis during the irrigation season, as shown in figure 6. Several features of this record reflect unique hydrologic events, such as the droughts of 1959-61 and 1976-77, when the effects of upstream water resource development severely impacted the quality of the lower San Joaquin River. TDS concentrations during these episodes exceeded 800 mg/L throughout the irrigation season and during some months mean concentrations were in excess of 940 mg/L. These events may be contrasted with water quality experiences during comparable droughts in the period prior to the mid-1950's, notably during the period 1930-34 when TDS concentrations rarely exceeded 500 mg/L. It was on the basis of this experience that a target level of 500 mg/L was incorporated in the authorizing legislation for the New Melones Project and has subsequently become the objective for water quality control at Vernalis. The historical record for the Vernalis station after 1978, when the project began full operation, indicates successful regulation of water quality within the target limit, especially during the drought of 1986-90. During the irrigation seasons in this period up to 150,000 acre-feet of so-called "interim water" not

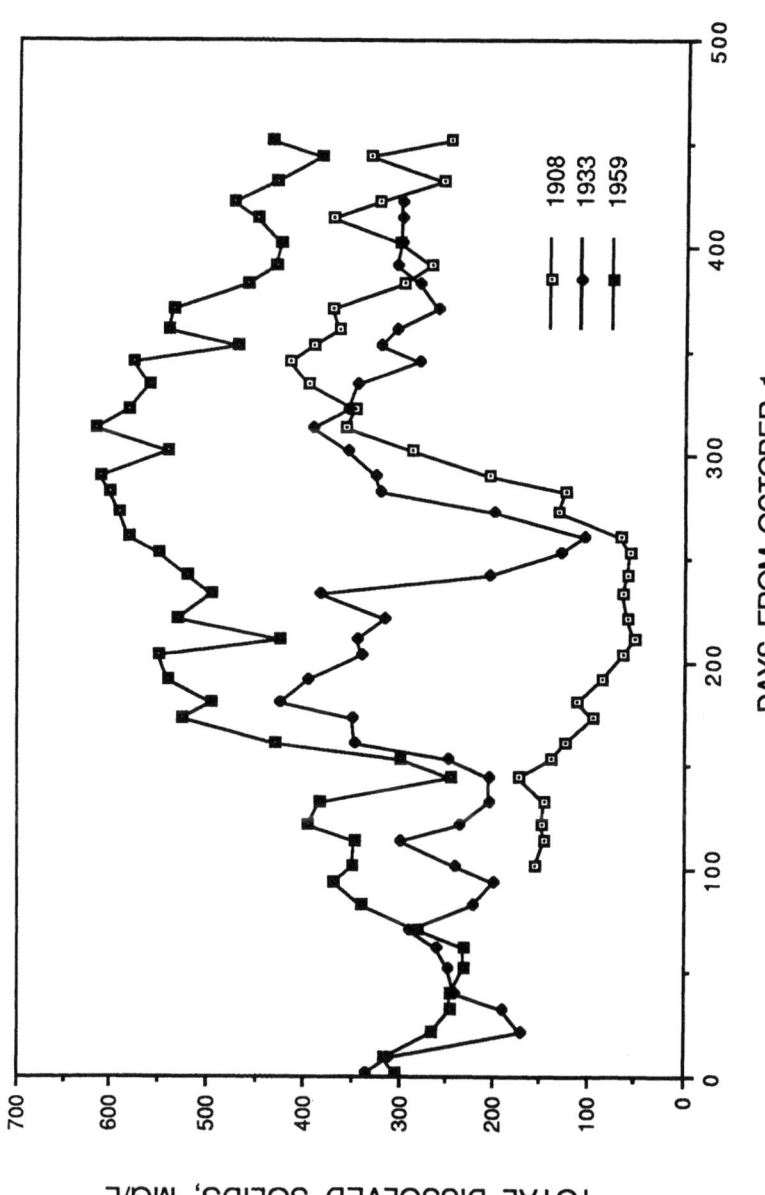

Figure 5. Total dissolved solids, San Joaquin River at Mossdale - three dry years (1908, 1933, and 1959).

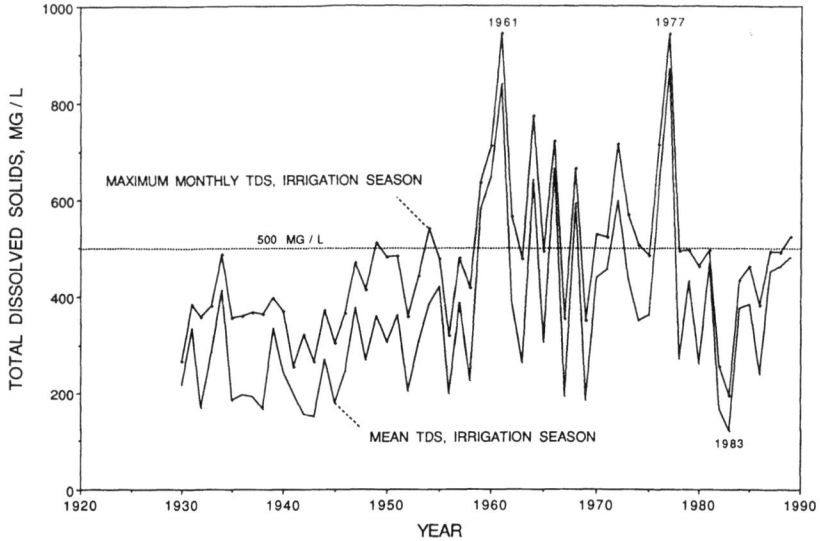

Figure 6. Mean and maximum monthly TDS during irrigation season, San Joaquin River near Vernalis (1930-1990).

committed to contract by Reclamation was allocated for water quality control to supplement the 70,000 acre feet normally reserved for this purpose. Without this additional allocation the experiences of the prior droughts would no doubt have been repeated.

## SALT BALANCE

### Balance Equation

Over an extended period of natural hydrology, say several decades without the intervention of man, the net accumulation rate of dissolved mineral salts in the surface waters of the San Joaquin Basin should approach a limit value of zero; that is, the rate of excretion from the basin through its natural outlets should be virtually equal to the rate of salt accretion from sources extraneous to or within the basin. This relationship is characterized generally by the salt balance equation

[1]  $dS/dt = dA/dt - dE/dt$

where dS/dt is the rate of salt load accumulation (tons/year), dA/dt is the rate of accretion, and dE/dt is the rate of excretion.

Natural accretion of salt for a defined surface water hydrologic unit includes conveyance through upstream boundaries by natural inflow, accretion from ground waters, importation with precipitation, and production by natural processes of weathering and leaching. Natural excretion includes salt load carried from the basin by runoff and losses through ground water. Anthropogenic processes of accretion and excretion include importation of water for irrigation or other consumptive uses, e.g., Delta-Mendota Canal, expansion of irrigated lands, extrabasin transfers like Friant-Kern Canal and Hetch Hetchy Aqueduct, application of soil amendments and fertilizers, harvesting and exportation of crops, and various water using domestic and industrial processes. A schematic representation of the principal salt balance processes in the San Joaquin Basin is shown in figure 7. The principal accretions and excretions included in the salt balance calculations are identified as follows.

| Location | Code |
|---|---|
| San Joaquin River near Vernalis | SJV |
| Stanislaus River at Melones Reservoir | ST |
| Tuolumne River at Don Pedro Reservoir | TU |
| Merced River at Exchequer Reservoir | ME |
| Chowchilla River at Buchanan Reservoir | CH |
| Fresno River near Daulton | FR |
| San Joaquin River at Millerton Reservoir | SJF |
| Madera Canal | MA |
| Friant-Kern Canal | FK |
| Hetch Hetchy Diversion | HH |
| Delta-Mendota Canal | DMC |
| Westlands Irrigation District | WL |

In this analysis the annual salt load excretion rate from the San Joaquin Basin (tons/year) is calculated as the water-year sum of monthly runoff at Vernalis times the mean monthly TDS concentration. Accretions to the basin are calculated to include salts introduced at the rim of the Valley, i.e., at the locations of the impounding reservoirs on the six principal tributary streams; importation through the Delta-Mendota Canal, less transfers to service areas outside the basin; and net production of salt from other processes required to achieve a balanced condition in the basin over the period 1930 to 1949. Net exchange of ground-water salts with surface waters of the basin is set to zero for the base period 1930 to 1949.

Salt load calculations were performed for the period 1930 to 1990, utilizing flow and water quality data derived from records of the USGS (1989), Recla-

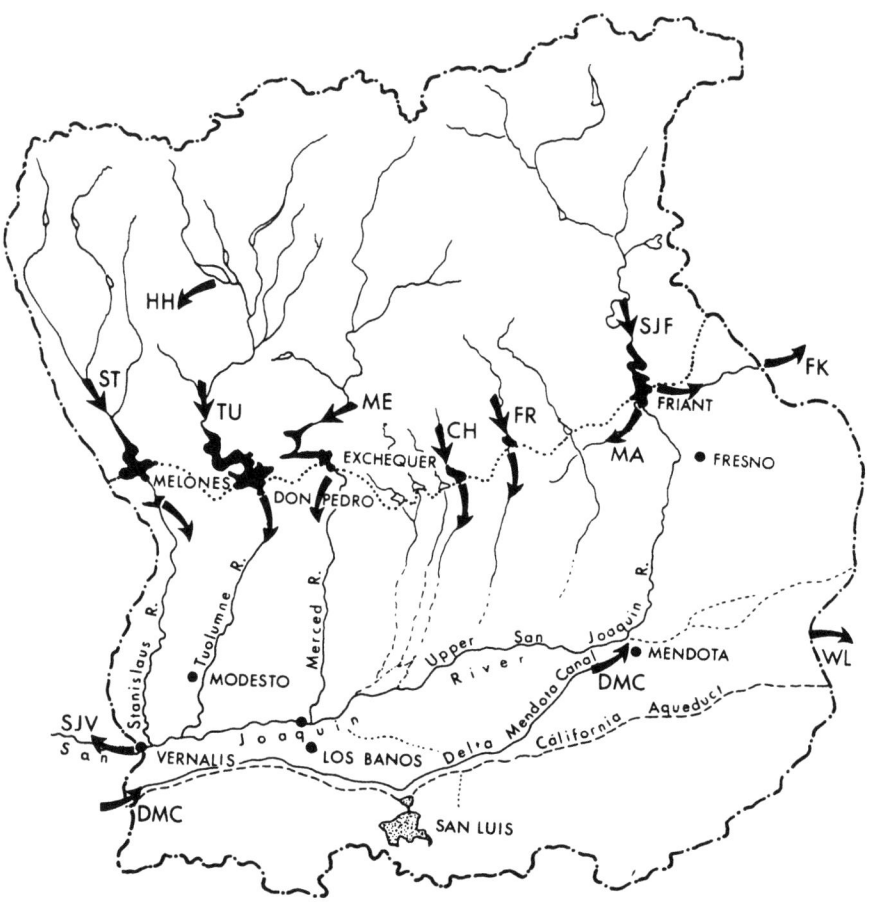

Figure 7. Sources of salt load accretion and excretion - San Joaquin Basin.

mation (1990), and DWR (1971, 1987, and 1990). Annual salt loads for the river at Vernalis and the Delta-Mendota Canal are based on the sum of loads computed at monthly intervals. The accretion load from the Delta-Mendota Canal to the San Joaquin Basin was adjusted for extra-basin transfer by taking the difference between the total entering the canal at the Tracy Pumping Plant and that portion of the delivery that was supplied to the Westlands Water District outside of the basin boundary. The contribution of the Westlands service area to the basin salt load was derived from the district area within the basin, estimated at 30 percent of the total, plus the contribution of the tile drained area south of the basin divide during the period 1979-86 when the San Luis Drain was operating.

Rim station loads were calculated from the annual unimpaired inflows at the individual stations, that is Tuolumne River at Don Pedro Reservoir, San Joaquin River at Millerton Lake, etc., and the estimated TDS concentration of the flow at these locations. Because water quality data were sparse at these stations, an average inflow TDS concentration of 50 mg/L was assumed. A comparison of salt loads calculated in this manner, with a more precise calculation using gauged flows and observed TDS concentrations, indicated that the estimated loads for rim sources were within 5 percent of actual loads.

Salt load contributions from ungauged or unmonitored sources, such as weathering of soils, accretions (or excretions) from ground water, etc., were estimated by comparing accretion and excretion rates during the period 1930-49 when the basin was assumed to be in a salt balanced condition. During this period annual excretion rates were invariably higher than accretion rates calculated from rim flows. The incremental accretion rate, attributed to ungauged and unmonitored sources, was found to be related to hydrologic conditions, that is, a function of annual unimpaired runoff for the basin. Analysis of the 20 years of record indicated average accretion rates for each of four quartiles of unimpaired runoff as shown in table 1.

Table 1. Average accretion rates.

| Year Class | Unimpaired Runoff, TAF/year | Annual Salt Load, 1000 tons |
|---|---|---|
| Dry | Less than 3,500 | 206 |
| Below Normal | 3,500 to 5,600 | 216 |
| Above Normal | 5,600 to 7,500 | 343 |
| Wet | Above 7,500 | 1,121 |

These incremental loads were applied to all years of the period 1930-89, assuming that accretions from these sources were unmodified by developments in the basin in succeeding years.

## Net Accumulated Salt Load

The differences between annual accretion and excretion for the San Joaquin Basin over the period 1930-89 are shown graphically in figure 8. It is apparent that the historic pattern of the estimated annual balance retains much of the variability associated with the hydrology of the basin, but that it also characterizes anthropogenic alteration of the basin's natural tendency toward equilibrium. While the first two decades show yearly fluctuations that may be mainly related to the randomness in the runoff record, the period from 1950 reflects a pronounced shift toward an annual accumulation of salt that is largely induced by increased importation through the Delta-Mendota Canal.

The pattern of annual salt imbalance exhibited in figure 8 suggests that after about 1950 the basin was accumulating salt at rates greatly exceeding the rates of excretion. Although some of the randomness associated with hydrologic processes and other uncertain phenomena are also exhibited in this part of the record, the general indication is one of a basin out of balance, gradually accumulating salt. This is illustrated more dramatically in figure 9 which shows the integrated sum of annual salt balances over the period 1930-89. The slope of the plot is indicative of the temporal rates of accumulation associated with hydrologic conditions, and to some extent project operation. The overall trend is positive, that is, the basin accumulated salt throughout the period 1950-89 at an average rate of approximately 466,000 tons per year. The corresponding accretion and excretion rates were 1,432,000 and 966,000 tons per year, respectively.

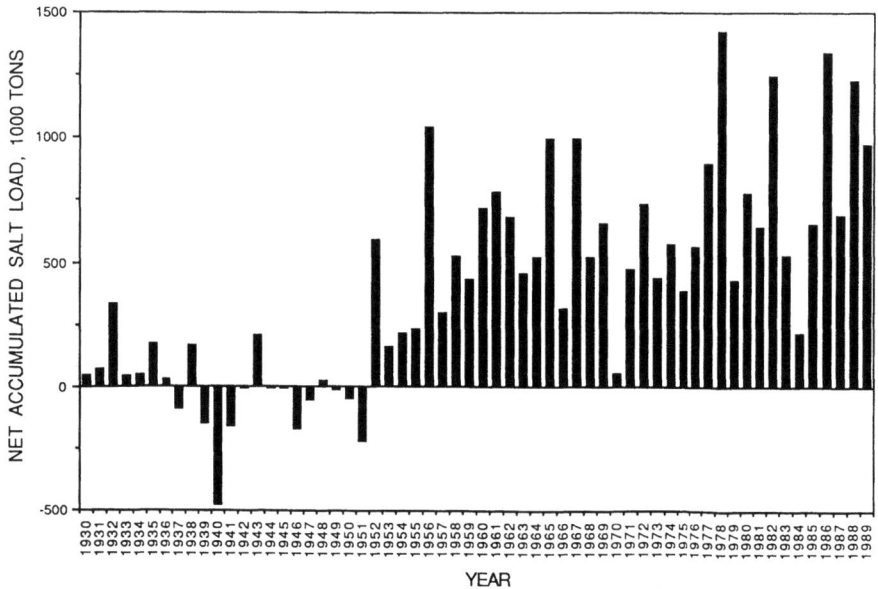

Figure 8. Net accumulated salt load - San Joaquin Basin (1930-1989).

# SALT BALANCE

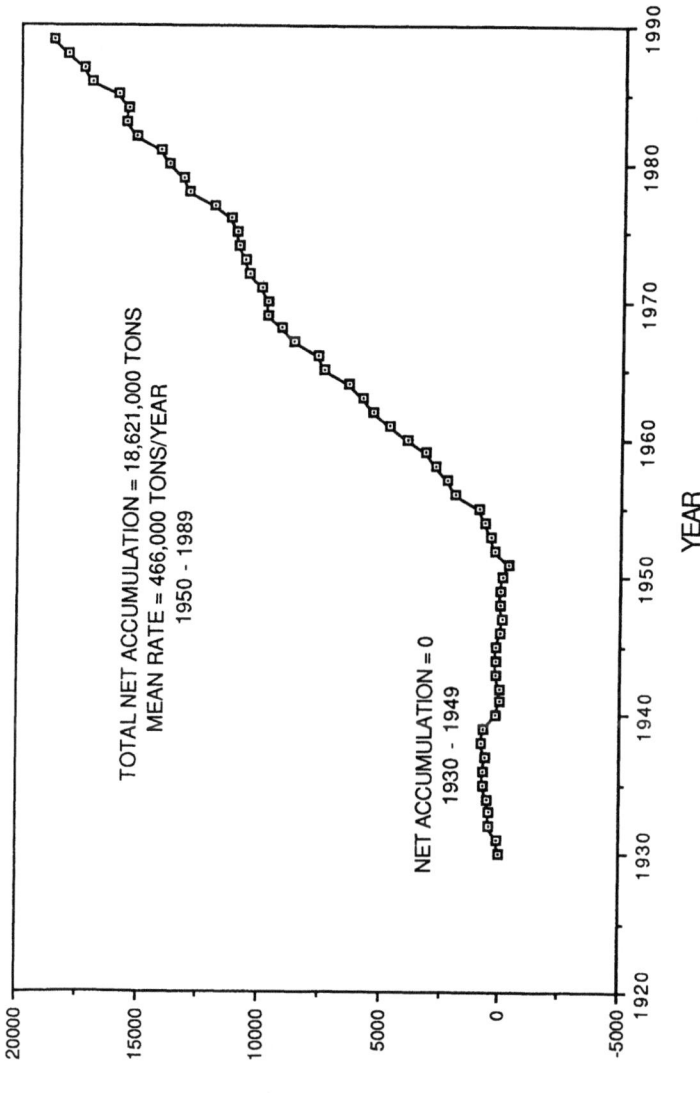

Figure 9. Net salt load accumulation - San Joaquin Basin (1930-1989).

## Trends in Salt Load Accumulation

To explore the future of salt loading in the basin it is instructive to examine the trends in the two general processes, accretion and excretion. These are integrated separately as depicted in figure 10, showing the divergence between the two integrals as the net basin accumulation.

An implication of the patterns shown is that both accretion and excretion rates (slopes of the curves) have been increasing in recent years. Neglecting hydrologic variability inherent in both data sets and fitting the two integral curves with third order polynomials, the general trends of accretion and excretion are:

Accretion:

[2] $A = 44.157 + 564.59 t + 10.740 t^2 + 4.7308e^{-4} t^3$ $\qquad R^2=0.999$

Excretion:

[3] $E = -1959.3 + 1004 t + 10.663 t^2 + 0.15983 t^3$ $\qquad R^2=0.998$

where A is cumulative accretion, E is cumulative excretion, and t is time in years from 1929.

It is of interest to estimate from the trends of these processes when the rates will be equal, that is, when the basin will have attained a new salt balance condition, assuming no remedial measures to change the present state of the system. The equilibrium time is estimated by equating the first derivatives of equations [2] and [3] and solving for t. Such a calculation suggests that equilibrium would be achieved by about the year 2007, at which time the rates of both accretion and excretion would be about 2,243,000 tons per year.

This is probably an unrealistic expectation in that it presumes a continued increase in accretion which, in light of the recent steps taken to reduce accretions from the San Luis Drain and Kesterson Reservoir, is more likely to remain steady or decrease. Alternatively, if it is assumed that the present (1990) average rate of accretion remains steady at about 1,860,000 tons/year, as indicated by the slope of the accretion curve (equation [2]), then the estimated equilibrium date will be about the year 1999.

It is reasonable to anticipate that a new equilibrium state will be at a condition of water quality in the lower reaches of the river system that will be considerably poorer than has been experienced historically. There is already evidence of adverse changes in quality in the increasing salt load emanating from the basin at Vernalis and in the extraordinary amounts of New Melones water that have been released in recent drought years to cope with quality

# SALT BALANCE

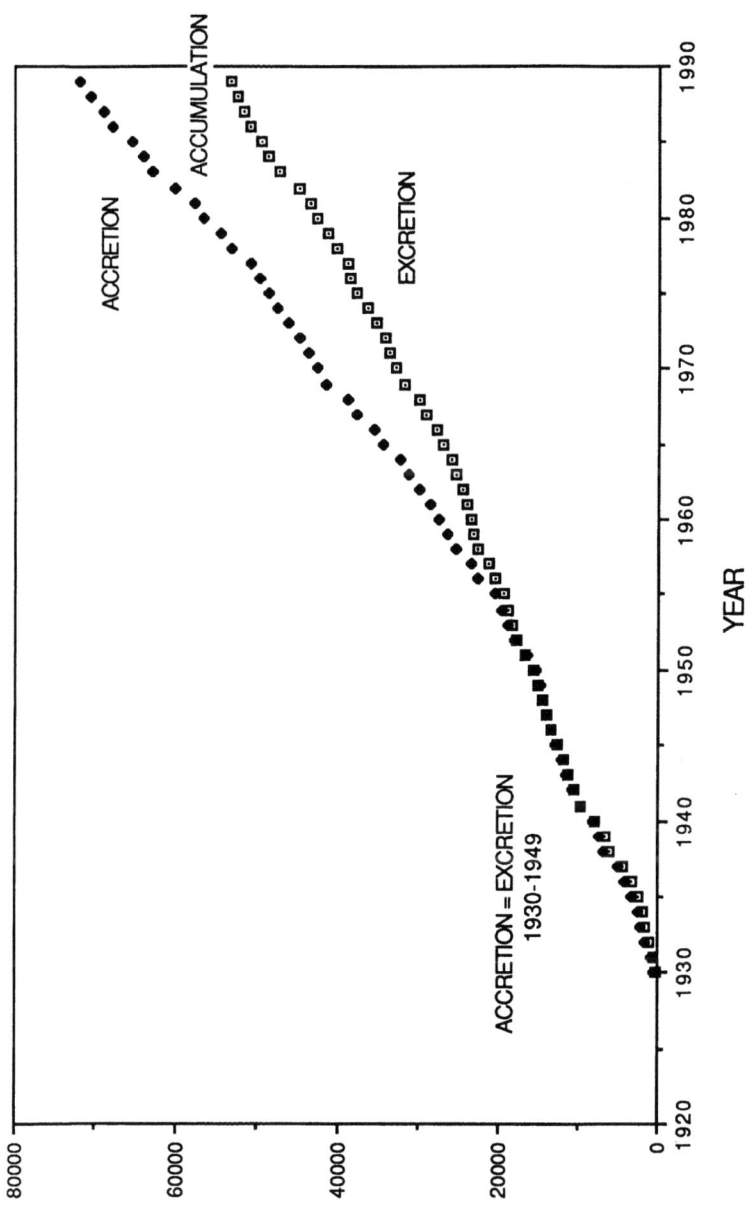

Figure 10. San Joaquin salt balance (1930-1989).

degradation in the San Joaquin River above its confluence with the Stanislaus River. It is appropriate then, with the historic trends in focus, to examine some alternatives for reducing accretions to the system and/or managing existing water resources to mitigate adverse quality effects.

## SALT LOAD MANAGEMENT

### Salt Load Accretion Reduction

There are not many promising solutions for the present condition of salt imbalance in the San Joaquin Basin, at least not in the short term. However, it may be possible to slow the rate of salt accretion to the basin by improving the quality of water in the vicinity of the Delta-Mendota Canal diversion point in the South Delta. Recent rates of pumpage by the CVP have been close to capacity, delivering about 3 million acre-feet per year, but unfortunately at inferior qualities in the range of 400 to 500 mg/L TDS. Allowing for about 700,000 tons delivered outside of the basin, the Delta-Mendota Canal accretion to the basin under these conditions is estimated to be in the range of 1,250,000 to 1,560,000 tons per year. A 100 mg/L TDS improvement in quality at the pumps would correspond to an accretion load reduction of about 312,000 tons per year. Based on the analysis presented above, and recent trends in basin accretion and excretion rates, it would require a quality improvement of about 130 mg/L at the Delta-Mendota Canal head to bring the net rate of salt load accumulation down to zero. There are at least two promising possibilities for achieving this goal, both of which involve controlling the influence of San Joaquin River quality and salt burden on the quality of inchannel water supplies in the South Delta.

The first possible control measure is to improve the quality of water entering the estuary at Vernalis by augmenting the supply with releases from New Melones Dam on the Stanislaus River. This strategy is built into the New Melones Project in the form of a provision for water quality control releases up to 70,000 acre-feet per year. It has been extended in recent years (1988-90) by augmenting water quality releases with up to 150,000 acre-feet of "interim" water not under contract by Reclamation. The benefit to South Delta water quality of these releases is evidenced in the experience during the irrigation season of 1988, as illustrated in figure 11. During the period from March through mid-August of this year, approximately 220,000 acre-feet of water, in addition to riparian and fish releases, were released from the Project for water quality control. The quality target for control at Vernalis, 500 mg/L TDS or less, was successfully met throughout the irrigation period, despite the comparatively poor quality of water from upstream sources.

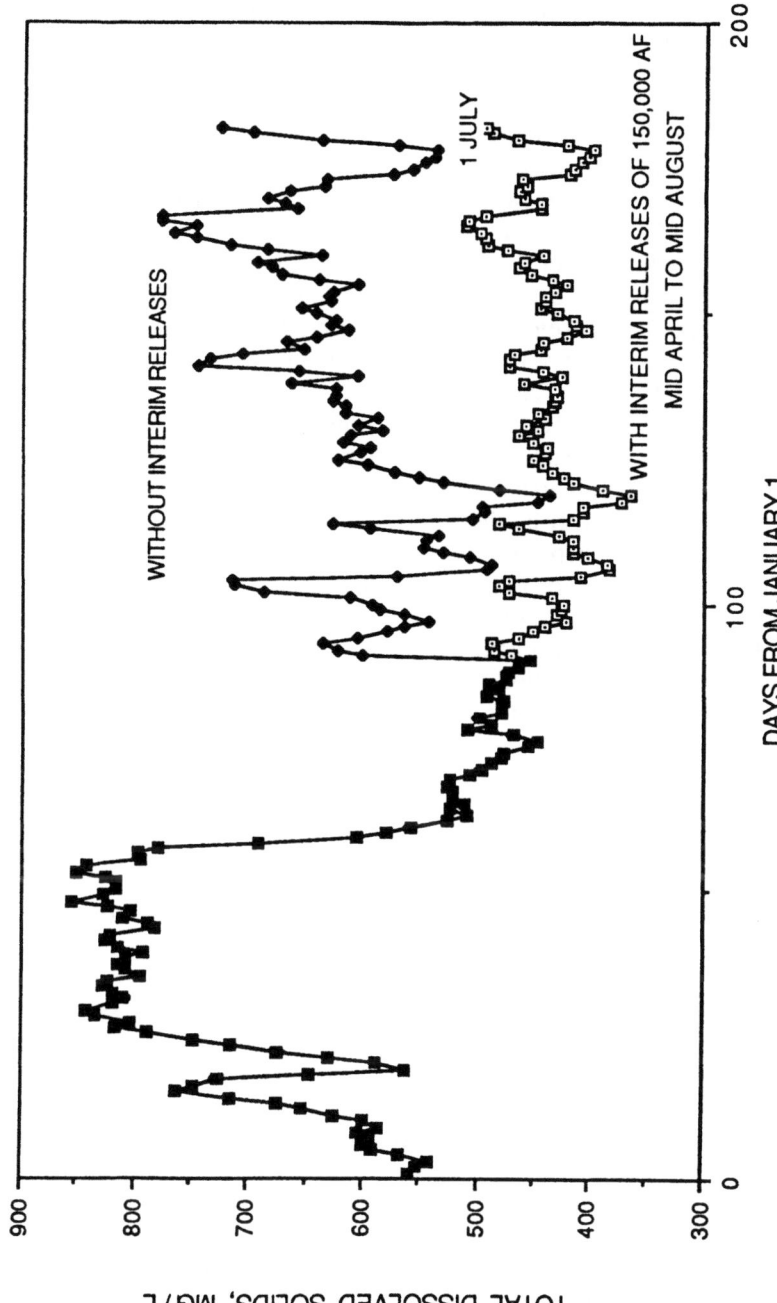

Figure 11. San Joaquin River quality (with and without 150,000 AF of interim releases from New Melones Reservoir, 1988 conditions).

However, despite the obvious benefit to water quality in the South Delta channels of New Melones releases during 1988, the quality of water available at the CVP pumps remained in the higher range of historic experience. The average quality of export through the Delta-Mendota Canal was only slightly better than the inflow at Vernalis, 400 mg/L v. 466 mg/L at Vernalis. This condition is an apparent consequence of "short circuiting" or "recycling" of San Joaquin water and its salts westward through the South Delta channels to the pumping plant forebay. Studies performed by the author with mathematical models indicate that under the conditions prevailing at present as much as half of the water exported originates with the San Joaquin River inflow at Vernalis. Clearly, steps taken to improve quality of this source will directly benefit the overall salt balance in the basin by reducing the salt load imported through the Delta-Mendota Canal.

A second possible control strategy, management of flow and water levels in South Delta channels by means of hydraulic barriers, addresses the problem of recycling San Joaquin salt loads back through the Delta-Mendota Canal. Under a plan devised jointly by the South Delta Water Agency, DWR, and Reclamation, control structures would be installed on selected channels to regulate flows under the action of tides in such a way as to preclude cross-Delta flows, assure unidirectional flows in most channels, maintain water levels, and prevent salt accumulation during the irrigation season. Under these conditions it is estimated that the proportion of San Joaquin flow that would arrive at the pumps likely would be 20 percent or less of the total drawn to the pumps from the central portion of the Delta.

## Inbasin Salt Load Management

The goal of salt load management is to distribute the burden of salts derived from the basin or imported to the basin so that the salt concentrations in basin waters do not impair specified beneficial uses. Two general approaches are available. One is to control the allocation of water that carries the salt, either in its native state or as modified by consumptive use. Another is to extract the salt from the water by an appropriate method so that it can be removed in dry form. A variant of both approaches is to concentrate it sufficiently to assure economic conveyance to a point of ultimate disposal where it will not reenter the system. The possible consequences of each can be examined in the framework of the San Joaquin Basin.

**Streamflow reallocation.** Users downstream of Vernalis on the San Joaquin River require a water supply sufficient in quantity and velocity to assure the flow in Delta channels and prevent stagnation and salt accumulation. This flow, estimated to be about 20 percent greater than normal consumptive use during the irrigation season, has been achieved in recent years incidental to the release of water from New Melones Reservoir for quality control. However, without the resource of interim water provided in addition to the project's reservation of 70,000 acre-feet, such flows could not be maintained.

An alternative water quality control strategy that would shift some of the burden from New Melones would be to apportion the deficiencies in requisite flow among the several major tributary streams of the San Joaquin River proportional to their relative contribution to the natural flow without upstream development. Analysis of the mean annual "full natural flows" at the rim stations for the four major tributaries leads to division in the following proportions: Stanislaus, 19 percent; Tuolumne, 33 percent; Merced, 16 percent; and Upper San Joaquin, 32 percent. It would be logical to allocate water from these sources to downstream users only to the extent that they are "surplus," that is, exceed the natural flow at the rim stations.

To test this strategy the historic experience for the calendar year 1988 was chosen. As noted previously, this was a year during which 220,000 acre-feet of water was released from New Melones Reservoir for water quality control. Figures 12(a) and 12(b) illustrate the result of this reallocation of the basin's water resources. Figure 12(a) shows the effect of allocating flow from the east side reservoirs to meet the targeted South Delta inflow and the net deficiency that would have to be made up by releases from New Melones Reservoir. Figure 12(b) indicates the corresponding quality that would result near Vernalis. A significant difference between this scenario and actual conditions during 1988 was that the New Melones allocation required to meet the targeted flows and water quality at Vernalis would be reduced by 138,000 acre-feet, a quantity that would be made up by releases from the other east side reservoirs. A related benefit of this management alternative would be improved water quality in most reaches of the main stem of the San Joaquin River between Vernalis and the mouth of the Merced River.

<u>Westside drainage control.</u> This alternative considers temporary retention of high salinity drainage waters within the areas of origin during the irrigation season with later release during the off-irrigation high runoff period. Possible retention schemes include temporary storage in shallow ground-water aquifers, surface ponds, and conveyance systems.

To test this strategy the same base hydrology and water quality conditions for 1988 were used. Also, the East Side streamflow allocation procedure described above was used to apportion deficiencies. Salinity and flow data developed by the USGS (1988) proved sufficient to determine the distribution of salt loads among the various contributing drainage areas along the main stem of the river between a gauge near Stevinson and Vernalis. The salt load assumed to be

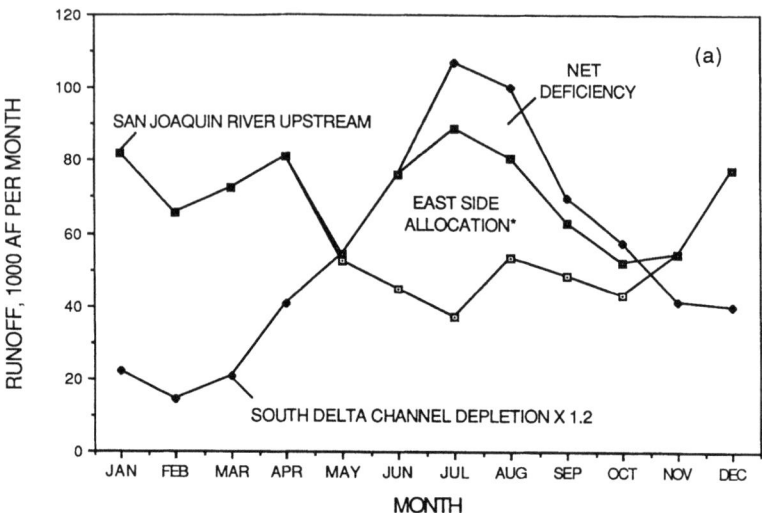

*Note*: Gross deficiency allocated in proportion to unimpaired runoff of tributaries, where Q(allocated) > Q(unimpaired).

Figure 12a. Effects of East Side allocation - 1988.

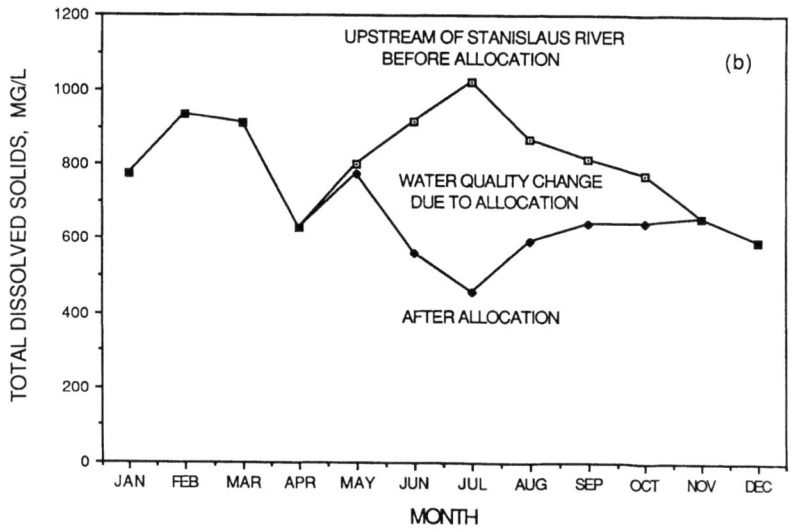

Figure 12b. Water quality near Vernalis after East Side allocation - 1988.

subject to control under this alternative originates in the west side drainage area between Stevinson and Newman, including flows from Mud Slough and Salt Slough. For the purposes of this study, the mean irrigation season salt load from these sources is determined to be 40,000 tons per month, half of which could be temporarily retained in storage for the 7-month period, March through September. This quantity would be returned to the river at a uniform rate during the other 5 months of the year. For purposes of water balance calculations it was assumed that the salt load would be carried at a mean concentration of 2,500 mg/L. The drainage flow that would otherwise have carried these salts to the river is subtracted from historic flow.

Figure 13 illustrates the effect of this drainage control scheme, including the east side streamflow allocation of the previous scenario, on the quality of the San Joaquin River near Vernalis. It is noted that quality at this location is markedly improved during the irrigation season as a combined result of the salt load reduction and east side flow allocations considered in this alternative. The result of this two-stage strategy to improve quality and flow at Vernalis is to further reduce the releases required from New Melones Lake by about 191,000 acre-feet.

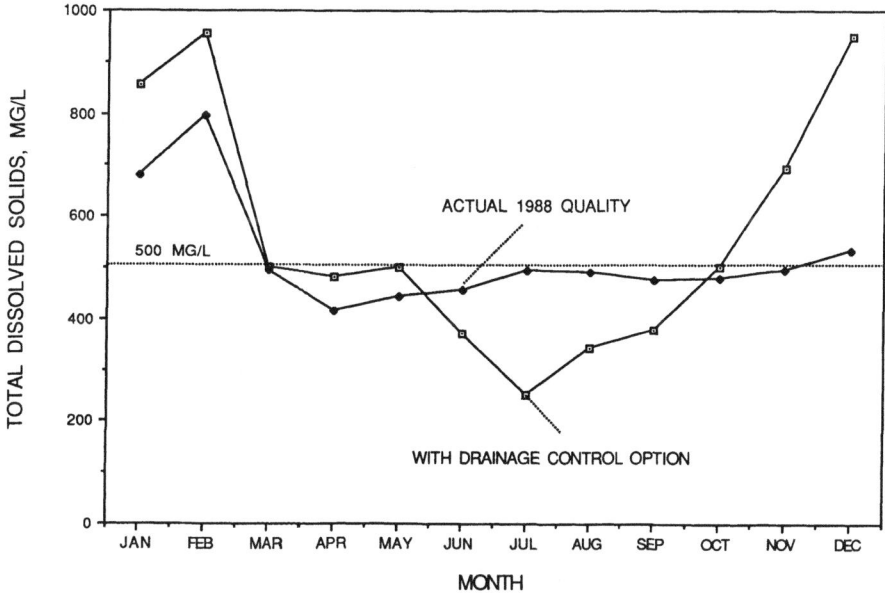

Figure 13. Effect of drainage control option - water quality near Vernalis (1988).

## CONCLUSIONS

The San Joaquin Basin is presently in a state of salt imbalance, with the rate of salt accretion exceeding the rate of excretion. The difference between these rates represents an accumulation of salt in storage within the basin that threatens the runoff quality of the main stem of the San Joaquin River. This, in turn, imposes a burden on downstream agricultural users in the Sacramento-San Joaquin Delta who must divert increased amounts of poorer quality water from Delta channels in order to assure sustained productivity.

Degradation of San Joaquin River water quality has been primarily a combined result of intensive water resource development throughout the basin which has reduced its outflow, importation of salt through the Delta-Mendota Canal which has substantially increased the rate of salt load accretion, and expansion of irrigated agriculture into areas of marginal productivity on saline soils. The present state of the system is such that even if salt balance is achieved in the near future it will be at a level of equilibrium between higher rates of both accretion and excretion, a condition that portends even poorer water quality in the main stem of the river. Solutions of the salt balance problems of the basin must incorporate measures to reduce salt load importation, to control entry of high salinity drainage into the main channel of the river, and to reallocate water resources equitably among all users in accordance with needs and entitlements.

Reducing salt load importation focuses on improving the quality of water-imported through the Delta-Mendota Canal, which is dependent to a high degree on controlling the quality of the San Joaquin at Vernalis and the influence of this source on the South Delta channels. Measures such as the proposed South Delta tidal barrier system should result in improving water quality at the Tracy Pumping Plant by forcing unidirectional flows in South Delta channels, thereby eliminating accumulation of salts in stagnant zones. Relocation of the CVP intake to a location farther north in the Delta, perhaps in conjunction with revisions in the State Water Project (SWP) export facilities, will also improve water quality and reduce imported salt load. Of course, any reductions in export rate as a result of increased irrigation efficiencies or reductions in irrigated acreage will also contribute to reduced accretion rates.

Steadily increasing salt loads carried by the river at Vernalis can only be mitigated by reducing saline drainage from upstream sources. The obvious choices for control measures are those sources in west side irrigated areas south of the confluence of the Merced and San Joaquin Rivers. Closure of the San Luis Drain has probably already resulted in some reduction in saline accretions to the river, although loads delivered through Salt and Mud Sloughs remain as major contributions to the total burden carried by the river below the mouth of the Merced. Temporary storage of saline drainage from these sources in ground-water reservoirs or surface water conveyance facilities during the irrigation season with later release at times of high runoff is an attractive short-term mitigative measure that could substantially improve water quality downstream.

The burden of water quality and flow control in the lower reaches of the river system needs to be shifted from the single source, New Melones Reservoir on the Stanislaus River, to a more equitable allocation among the other major tributary streams. It is reasonable to distribute the responsibility for maintenance of downstream quantity and quality in proportion to the unimpaired runoff of these streams. This may require adjustments in appropriative entitlements throughout the basin. However, in the interest of balanced utilization of a water supply that is virtually already exploited to the limit of natural supply such adjustments deserve serious consideration. The future of the San Joaquin Basin demands it.

## REFERENCES

California Department of Water Resources, 1971. *Four Day Chloride Sampling Program, Various Stations in the Sacramento-San Joaquin Delta, 1929-1971.*

California Department of Water Resources, 1987. *California Central Valley Unimpaired Flow Data,* Second Edition, p. 37.

California Department of Water Resources, 1990. Cedec System, Basic Hydrologic Data, 1900-Date. In: McSwain, K. R., *History of the Merced Irrigation District, 1919 - 1978.* SpectroChrome Graphics, Inc.

Paterson, A. M., 1987. *Land, Water, and Power - A History of the Turlock Irrigation District, 1887-1987.* The Arthur Clarke Co., Glendale, CA.

U. S. Bureau of Reclamation, 1990. *Monthly Summaries of Delta Water Balance and Water Quality at Selected Stations in the Sacramento-San Joaquin Delta, 1950-1990.*

U. S. Geological Survey, 1989. *Water Supply Papers,* Various Issues 1910-1989.

U. S. Geological Survey, 1985. *Water Budgets for Major Streams in the Central Valley, California, 1961-1977.* Open-File Report 85-401, p. 87.

Van Winkle, W. and Eaton, F. M., 1910. *The Quality of Surface Waters of California.* U. S. Geological Survey Water-Supply Paper 237, p. 139.

# 9 REMOVAL OF SELENIUM FROM AGRICULTURAL DRAINAGE WATER THROUGH SOIL MICROBIAL TRANSFORMATIONS

Elisabeth T. Thompson-Eagle and William T. Frankenberger, Jr., University of California, Riverside and Karl E. Longley, California State University, Fresno

## ABSTRACT

Recent developments in biological deselenification of contaminated soils and water are supporting a new approach in remediation of selenium-enriched matrices. Laboratory research has shown that soil micro-organisms can detoxify seleniferous sediments by transforming soluble selenium compounds into a volatile form, dimethylselenide, that is nonhazardous to rats. The authors have discovered that under optimum conditions this naturally occurring bioremediation technique can be accelerated to the point where there is a significant decline in the original soil selenium inventory. This chapter focuses on the optimization and characterization of this microbial transformation with emphasis on on-farm applications, process operation, environmental regulations pertaining to selenium bioremediation, and a feasibility analysis.

## INTRODUCTION

Selenium is a naturally occurring metalloid associated with specific geological formations on the earth's surface. The chemistry and, in particular, the geochemistry of selenium is similar to that of sulfur, but selenium is less abundant (0.09 mg kg$^{-1}$) in the earth's crust (Lakin, 1972). Most of the seleniferous soils in the world are found in arid and semiarid regions. Recent attention has focused on California soils because of a combination of geology, climate, salinity, high water tables, intensive irrigation practices, and death and deformities of wildlife attributed to selenium toxicosis. Selenium is thought to be the primary constituent responsible for gross toxicological problems in wildlife inhabiting drainage water disposal basins. The toxicological effects of selenium on birds in certain areas of the San Joaquin Valley include high incidences of embryonic mortality and multiple developmental abnormalities as well as reduced growth and mortality of adults (Ohlendorf, 1989). The threshold between selenium deficiency and selenium toxicity is extremely narrow. In 1980, the U.S. Environmental Protection Agency (EPA) established a water quality criterion for dissolved selenium at $10\mu g$ $L^{-1}$. The California State Water Resources Control Board (SWRCB, 1987) stipulates that the selenium content in the Central Valley drainage water must not exceed 5 p/b.

A concerted effort is being made to reduce elevated selenium concentrations in agricultural drainage waters and sediments, and to encourage growers to become more conservative in their use of irrigation water. Research at University of California, Riverside, has focused on accelerating a natural microbial process which results in selenium volatilization. This transformation permanently removes selenium from water, soils, and sediments reducing the selenium inventory in polluted areas of the western San Joaquin Valley.

## BACKGROUND INFORMATION

The San Joaquin Valley of California is one of the most productive agricultural regions in the world and relies heavily on irrigation water to supplement rainfall. The climate is semiarid with an average annual rainfall of about 7 inches (Westlands Water District, 1989). Over 35 major commercial crops are grown in this area including cantaloupe, tomato, lettuce, onions, beans, carrots, garlic, grapes, and cotton. Irrigation leads to the mobilization and buildup of salt in the soil profile. Water management problems in the Valley have existed since the 1880's. Continued production of agricultural commodities is dependent on controlling salinity and the accumulation of toxic trace elements such as selenium. Selenium is one of four constituents (boron, molybdenum, selenium, total dissolved solids) identified as being of primary concern in subsurface agricultural drainage water by the SWRCB (Johns and Watkins, 1989). Drainage water varies in selenium concentration from <1 to 943 p/b (median 13 p/b) (personal communication, Karl Longley, Department of Civil & Surveying Engineering, California State University, Fresno). Currently, the drainage water is disposed of into 21 active evaporation ponds (Wass, 1990) which cover a total area of 6,650 acres in the Central Valley (Westcot et al., 1988 and personal communication, Dennis Westcot, California Regional Water Quality Control Board, Sacramento, CA). Approximately 20,000 more acres of evaporation ponds are in the planning and construction stages (Westcot, 1988).

## MICROBES AND SELENIUM VOLATILIZATION

Micro-organisms play an important role in the cycling of many different elements including carbon, nitrogen, and sulfur. They are also considered to be an essential component in the recycling of selenium (Doran, 1982). Microbes are involved in the following selenium transformations: immobilization, mineralization, oxidation, reduction, and methylation. No selenium methylation will occur if the soil or water is sterilized (Thompson-Eagle and Frankenberger, 1990a; Frankenberger and Karlson, 1989a; and Doran, 1982). Methylation of selenium results in the release of gaseous selenium from polluted water, sediments, and soil. It is believed that all selenium species are subject to biomethylation but the highest dimethylselenide emissions have been recorded from selenite, selenate, and selenomethionine (Chau et al., 1976 and Doran, 1982). Micro-organisms with the capacity to methylate selenium are apparently widespread and have been isolated from diverse environments (table 1). Volatilization through methylation is thought to be a protective mechanism used by micro-organisms to detoxify their surrounding environment. The process permanently removes selenium from soil and/or water. The predominant group of selenium methylating organisms isolated from soils and sediments are bacteria and fungi (Doran, 1982 and Karlson and Frankenberger, 1988b), while in water, bacteria are thought to play a more dominant role (Thompson-Eagle and Frankenberger, 1990d). Although other organisms including protozoans (*Tetrahymena thermophila*), plants, and animals are known to volatilize selenium (Drotar et al., 1987 and Doran, 1982), the advantage of using micro-organisms in a bioremediation program is that large numbers of indigenous organisms can multiply within a relatively short period of time. Under optimal conditions, this microbial process can be accelerated to such an extent that there is a significant decline in the selenium inventory. When the process is complete, all soil optimization amendments cease and the micro-organisms naturally decline back to their indigenous levels.

## BIOMETHYLATION PRODUCTS - TOXICITY AND FATE

Dimethylselenide is the main product of biomethylation in seleniferous soils and water with smaller quantities of dimethyldiselenide also being produced (Doran, 1982; Frankenberger and Karlson, 1988; and Thompson-Eagle et al., 1989). Dimethylselenide is 500 to 700 times less toxic than aqueous selenite and selenate ions and is nonhazardous to rats (Franke and Moxon, 1936; McConnell and Portman, 1952; and Frankenberger and Karlson, 1988). Once selenium is methylated, it is released into the atmosphere, diluted and dispersed by air currents away from the contaminated source. Dimethylse-

Table 1. Microorganisms that methylate selenium.

| Microorganism | Source | Se Substrates | Se concn. of medium(p/m) | Aerobic | Anaerobic | Se product | Reference |
|---|---|---|---|---|---|---|---|
| **Bacteria** | | | | | | | |
| *Aeromonas sp.* | Lake sediment | SeO$_3$ | 5 | X | | (CH$_3$)$_2$Se | Chau et al., 1976 |
| *Flavobacterium sp.* | | | | | | (CH$_3$)$_2$Se | |
| *Pseudomonas sp.* | | | | | | Unknown volatile Se | |
| *Corynebacterium sp.* | Seleniferous soil | SeO$_3$, SeO$_4$, Se° | NA | X | | (CH$_3$)$_2$Se | Doran & Alexander, 1975 |
| *Pseudomonas fluorescens* | Evap. pond sediment | SeO$_4$ | 0.8 | | X | (CH$_3$)$_2$Se<br>(CH$_3$)$_2$Se$_2$<br>CH$_3$SeSCH$_3$ | Chasteen et al., 1990 |
| Unidentified sp. | Evap.pond water | Mainly SeO$_4$ | 1.2 | X | | (CH$_3$)$_2$Se | Unpublished data, Thompson-Eagle & Frankenberger |
| **Fungi** | | | | | | | |
| *Cephalosporium sp.* | Garden soil | SeO$_3$<br>SeO$_4$ | 457<br>418 | X | | (CH$_3$)$_2$Se | Barkes & Fleming, 1974 |
| *Fusarium sp.* | | | | | | | |
| *Penicillium sp.* | | | | | | | |
| *Scopulariopsis sp.* | | | | | | | |
| *Scopulariopsis brevicaulis* | Bread | SeO$_3$, SeO$_4$ | 15 | X | | (CH$_3$)$_2$Se | Challenger & North, 1934 |
| *Schiophyllum commune* | Wood | SeO$_4$ | 366 | X | | (CH$_3$)$_2$Se | Challenger & Charlton, 1947 |
| *Aspergillus niger* | Bread | SeO$_4$ | 1316 | X | | (CH$_3$)$_2$Se | Challenger et al., 1954 |
| *Candida humicola* | Sewage | SeO$_3$, SeO$_4$ | 46,418 | X | | (CH$_3$)$_2$Se | Cox & Alexander, 1974 |
| *Acremonium falciforme* | Evap. pond sediment | $^{75}$SeO$_3$* | 100 | X | | (CH$_3$)$_2$Se | Karlson & Frankenberger, 1989 |
| *Penicillium citrinum* | | | | | | | |
| *Ulocladium tuberclatum* | | | | | | | |
| *Acremonium falciforme* | Evap. pond sediment | SeO$_4$ | 0.79 | X | | (CH$_3$)$_2$Se<br>(CH$_3$)$_2$Se$_2$ | Chasteen et al., 1990 |
| *Penicillium citrinum* | | | | | | | |
| *Penicillium sp.* | Sewage | SeO$_3$ | 457 | X | | (CH$_3$)$_2$Se | Fleming & Alexander, 1972 |
| *Alternaria alternata* | Evap.pond water | SeO$_3$, SeO$_4$ | 1,100 | X | | (CH$_3$)$_2$Se | Thompson-Eagle et al., 1989 |

NA indicates that the data is not available.
*The radioisotope $^{75}$Se was utilized in this experiment.

lenide reacts with OH and $NO^3$ radicals and ozone ($O^3$) within a few hours to yield products which are as yet unknown (Atkinson et al., 1990); however, it is likely that these oxidized products may be scavenged onto aerosols or sorbed onto particulates which have a relatively long residence time (7-9 days) in the atmosphere (Mosher and Duce, 1989), and can travel considerable distances (Mayland et al., 1989).

## ENVIRONMENTAL PARAMETERS AFFECTING BIOMETHYLATION

### Soil and Water Quality

The quality of agricultural drainage water in evaporation ponds is highly dependent on the geologic setting of the eluted soil (Westcot, 1990) and may vary widely at different locations in the Valley. Total dissolved solids (TDS) in different pond waters range between 1,300 and 390,000 mg $L^{-1}$ (Westcot et al., 1988). High selenium concentrations in drainage water, evaporation pond water, and sediments are associated with sulfur and sodium (Fujii and Deveral, 1989; Westcot, 1990; and Tanji, 1990). High rates of selenium volatilization have been demonstrated in naturally saline sediments (22 dS $m^{-1}$) collected from Kesterson Reservoir (Merced County, CA) (Frankenberger and Karlson, 1989b). Apparently the indigenous microflora have a high tolerance to saline conditions. Evaporation pond water is alkaline ranging from pH 8.1 to 9.9 (Tanji and Grismer, 1988). The optimum pH for selenium biomethylation in seleniferous Kesterson sediments (pH 7.7) is 8.0 (Frankenberger and Karlson, 1989b). Despite the variation in physical and chemical properties of sediments, drainage water, and evaporation ponds, all samples tested so far contain indigenous populations of micro-organisms capable of accelerated selenium volatilization (Frankenberger, 1989; Frankenberger and Karlson, 1988, 1989b; Frankenberger and Thompson-Eagle, 1988; Karlson and Frankenberger, 1988a,b, 1989, 1990; Thompson-Eagle, Frankenberger, and Karlson, 1989; and Thompson-Eagle and Frankenberger, 1990a,b,c,d). Thus there is no need to inoculate seleniferous environments with selenium-methylating organisms to accelerate volatilization.

## SELENIUM CONCENTRATION AND SPECIATION

Selenium is unevenly distributed in the Central Valley and is mainly concentrated on the west side of the Valley (Deveral et al., 1984). The land surrounding Kesterson Reservoir in the northern Valley contains moderate to low amounts of selenium while the Panoche Fan has relatively high levels of selenium (Burau, 1989). Many of the drainage water disposal ponds are located in the former Tulare Lake Basin. Evaporation ponds contain selenium levels between 0.3 and 2000 $\mu$g selenium $L^{-1}$ (Thompson-Eagle et al., 1989; Tanji and Grismer, 1988; and Westcot et al., 1988) while sediments collected from Kesterson Reservoir range between 1 and >700 p/m (Frankenberger, 1989 and Frankenberger and Karlson, 1988). The rate of selenium volatilization is highly dependent on the selenium inventory and is intermediate between first-order (exponential) and zero-order (linear) kinetics. The higher the selenium levels, the greater the volatilization rates. The dominant forms of selenium in agricultural drainage water and evaporation pond water are $SeO_4^{2-}$ and $SeO_3^{2-}$ (Cutter, 1989; Fujii and Deveral, 1989; Izbecki, 1984; and Presser and Barnes, 1984). In sediments, $SeO_3^{2-}$, $SeO_4^{2-}$, $Se°$ and organoselenium compounds are frequently present. Without the addition of a carbon amendment, microbial selenium volatilization of $SeO_3^{2-}$ in sediments is an order of magnitude higher than $SeO_4^{2-}$ (Karlson and Frankenberger, 1989). However, in the presence of an available carbon source, this difference virtually disappears (Karlson and Frankenberger, 1989). It takes more energy for the microbe to reduce $SeO_4^{2-}$ to a selenide ($Se^{2-}$) than $SeO_3^{2-}$, but when carbon is nonlimiting, both selenium species are readily volatilized at near equal rates. The most readily methylated organoselenium compound is selenomethionine (Doran and Alexander, 1977 and Karlson and Frankenberger, 1989).

## CLIMATE

The average annual precipitation (1976-87) at the Tranquility Field Station in the Westlands Water District (275,000 ha of irrigated lands in the western San Joaquin Valley) is 14.4 cm per year (personal communication, Gerald Robb, Westlands Water District, Fresno, CA). Selenium volatilization in soil occurs at a maximum rate in field-moist soil (-33 kPa) (Frankenberger and Karlson, 1989b). Both air-drying and waterlogging results in low levels of biomethylation (Frankenberger and Karlson, 1989b). The average monthly solar radiation at the Tranquility Field Station ranges between 151 and 6,623 langleys/day and air temperatures between 2 °C (December) and 32 °C (July). Biomethylation is strongly affected by temperature and increases 2.6-fold with each 10 °C rise in temperature. The optimum temperature for selenium volatilization is 35-40 °C (Frankenberger and Karlson, 1989b). With higher temperatures, the vapor pressure of dimethylselenide increases. Raising the temperature from 10 to 25 °C doubles the vapor pressure of dimethylselenide and raising it from 25 to 40 °C doubles it again (Frankenberger and Karlson,

1988). A thermal differential permits gaseous exchange of dimethylselenide between the atmosphere and soil air at the intermediate soil surface. Average monthly windspeeds at the Tranquility Field Station range between 3 and 6 miles per hour. The pressure and suction effects of high winds replenish the soil atmosphere particularly within barren soils. The effect of air turbulence on the transfer of dimethylselenide vapor in soils suggests that mass air flow could be high. A wind speed of 5 miles per hour can penetrate several centimeters into soil and mulches (Farrell et al., 1966). Even without mass flow, fluctuations in air pressure at the soil surface result in considerable mixing, enhancing transport beyond that due to diffusion.

## OPTIMIZATION OF BIOMETHYLATION

### Micro-organisms

Selenium-methylating fungi isolated from seleniferous soils of the western San Joaquin Valley are aerobic, tolerant to highly saline conditions, and able to withstand extreme osmotic stress in soil (Karlson and Frankenberger, 1989). Without the presence of micro-organisms, no selenium methylation occurs in these soils, thus indicating that this is a biotic reaction. The micro-organisms are heterotrophic in nature requiring organic sources for carbon and energy. The growth of methylating organisms can be stimulated by specific organic amendments.

### Soil Nutrients

Short-term studies with seleniferous sediments demonstrate that selenium biomethylation is carbon-limited and is accelerated by providing organic amendments including saccharides, amino acids, and especially pectin and proteins (Frankenberger and Karlson, 1988). In some soils, nitrogen may also be a limiting factor. The optimum carbon/nitrogen ratio for selenium volatilization in Los Banos clay loam was found to be 20:1 (Karlson and Frankenberger, 1988b).

### Field Work

Microbial volatilization of selenium is now being considered as a cost-effective remediation technique to detoxify seleniferous sediments at Kesterson Reservoir and at other evaporation pond sites in the Central Valley (Frankenberger, 1989). Field experiments initiated in July and October, 1987 at Kesterson pond 4 (selenium inventory range, 10 to 209 mg selenium kg$^{-1}$), Kesterson pond 11 (selenium inventory range of 1 to 11 mg selenium kg$^{-1}$), Sumner Peck Ranch pond 6 (selenium inventory range of 1 to 9 mg selenium kg$^{-1}$) and in September 1988 with excavated, homogenized San Luis Drain (SLD) sediment of 86.5 to 100.4 mg selenium kg$^{-1}$, have identified the most effective sediment treatments to accelerate selenium volatilization (Frankenberger and Karlson, 1988). In all cases, simply irrigating and rototilling the soil promoted high rates of dimethylselenide release with a subsequent dramatic decrease in the residual selenium (figure 1). Within a 2-year period, the cleanup goal of 4 p/m was achieved in the less contaminated Pond 11 sediment (Frankenberger, 1989). One amendment, casein (a milk protein), was particularly effective in promoting biomethylation at Kesterson pond 4, with a 68-percent selenium inventory removal in 23 months (figure 1). Citrus peel was also an effective amendment in promoting selenium volatilization. Citrus peel contains approximately 30 to 35 percent pectin on a dry-weight basis. Selenium dissipation rates at Kesterson Reservoir decreased during the winter season, most likely as a result of decreased temperatures and waterlogged conditions. Soil temperatures in the field seasonally varied between 4 and 50 °C. There was a diurnal peak of dimethylselenide emission during the midafternoon and a seasonal one during the summer months (Frankenberger and Karlson, 1988).

Field studies demonstrated that selenium biomethylation is dependent on selenium concentration, soil type, organic matter content, temperature, moisture, nutrients, and cofactors. A first-order rate equation was used to estimate the time required for cleanup of the seleniferous sediments. Half-lives ranged from 1.5 to 5.5 years for the pond 4 sediment and 2.5 to 4.5 years for the SLD sediment. This bioremediation approach is projected to require between 3 and 7 years to reach the cleanup goal of 4 p/m for sediments containing relatively high levels of selenium.

## ON-FARM APPLICATIONS

### Environmental Regulations

There are several new laws which are applicable to agricultural drainage treatment facilities containing high levels of selenium. The Frankenberger-Karlson selenium volatilization process is designed to treat the sediments of evaporation ponds. Thus the first step in implementing this process is to completely evaporate the pond water. During evaporation, the selenium concentration in the water is likely to increase and possibly reach a hazardous level of 1,000 p/b. At this point, the water in the evaporation pond falls under the jurisdiction of the Toxic Pits Cleanup Act (TPCA), the Department of Health Services (DOHS), and Subchapter 15 of the SWRCB. The TPCA is enforced by the Regional Water Quality Control Boards. If the pond meets specific criteria of the 1988 amendments, AB 3843 and AB 2875, it may qualify for an exemption from remediation until January 1, 1993. In order to be exempt from the rulings of the DOHS, the applicant must demonstrate either that the material has mitigating physical or chemical characteristics or that the operation is managed so as to be nonhazardous to the health and safety of humans, livestock, and wildlife. If exempted from DOHS regulations, then an exemp-

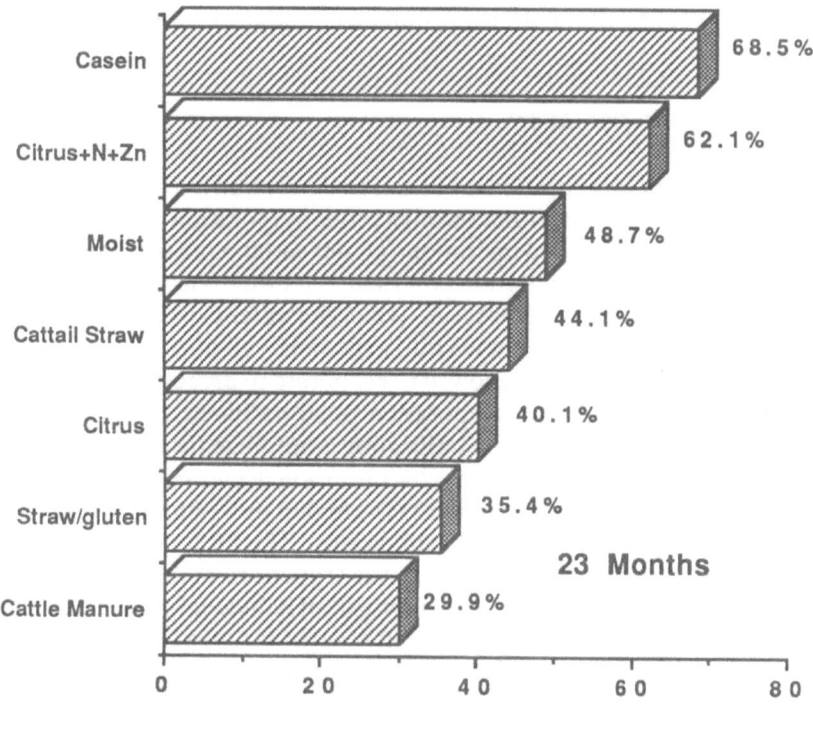

Figure 1. Soil selenium depletion in response to specific amendments added to Kesterson Pond 4 sediments.

tion from the SWRCB may be obtained. Neither the State of California nor EPA has established ambient air quality standards for selenium compounds. However, it is unlikely that field dimethylselenide emission will be considered as air pollution either by the EPA, the California Air Resources Board (ARB), or the Regional Air Pollution Control Districts given that the maximum modeled field ambient air selenium concentration of 2,409 ng m$^{-1}$ under a worst-case scenario of severe stagnating winds and optimum selenium volatilization at Kesterson Reservoir (U.S. Bureau of Reclamation, 1988) is well below the EPA-derived acceptable superfund site inhalation exposure of 3,500 ng m$^{-3}$ (U.S. Bureau of Reclamation, 1986). Obtaining approval from these regulatory agencies to implement volatilization as a treatment process may be time consuming especially if regulatory decisions are inconsistent. It is highly recommended that the legislation adopt guidelines specifically for the treatment of agricultural drainage water by biological treatment processes.

## Process Implementation

The selenium volatilization process can be operated in a continuous manner by two different systems, either through series evaporation pond management, or through a primary pond operation. Under serial pond management, water flows into a primary evaporation pond and as it evaporates, the salts including selenium increase in concentration. When the water closely approaches 1 mg selenium L$^{-1}$, the water is pumped into a secondary treatment unit, known as a flash evaporation pond and allowed to evaporate to dryness. This process can be repeated until the flash evaporation pond sediment approaches 100 mg kg$^{-1}$ (by weight). At this point, volatilization is optimized as a treatment process and continues until the selenium concentration decreases significantly to an acceptable concentration (i.e., 4 p/m). The higher the selenium inventory in the sediment, the greater the volatilization rates. The flash evaporation pond is then placed back into service and receives effluent from the primary evaporation pond. The primary pond system consists of individual units which undergo cycles of pond filling and evaporation to dryness until the sediment concentration approaches 100 mg kg$^{-1}$ (by weight). As with the series pond management system, volatilization is optimized until the sediment selenium concentration is reduced to an acceptable level and the pond becomes available for operation again. The evaporation and volatilization cycles can be repeated for many years in both systems before disposal of the accumulated salts is necessary.

## LAND MANAGEMENT

In order to calculate the land requirements for a series pond operation system with selenium volatilization as a remediation technique, selenium volatilization and evaporation rates need to be considered. The optimal, pessimistic, and average selenium volatilization rates at the Sumner Peck Ranch (Fresno Co., CA) are 4.86, 0.97, and 1.94 g/ac-d, respectively. Other assumptions for this calculation include an additional 20 percent land requirement for structures such as embankments, a projected average drainage flow rate of 0.5 acre-feet (ac-ft) per acre-year, maximum water depths in the flash evaporation ponds of 2 feet or less, and an evaporation rate of 4.7 ac-ft/ac-yr (Westlands Water District). The calculated land requirements assuming an average volatilization rate of 1.94 g/ac-d for various selenium concentrations in the drainage water are summarized in table 2.

Table 2. Land requirements for a Se volatilization series pond operation system with an irrigation flow rate of 0.5 ac-ft year (Frankenberger, Karlson and Longley, 1990).

| Se concn. | Volatilization area | Evaporation area | Total nonproducing area |
|---|---|---|---|
| $\mu g/L$ | % total ac | % total ac | % total ac |
| 10 | 1.9 | 11.4 | 13 |
| 30 | 5.5 | 11.1 | 17 |
| 50 | 8.9 | 10.8 | 20 |
| 70 | 12.0 | 10.4 | 22 |
| 100 | 16.5 | 10.0 | 27 |
| 300 | 38.8 | 7.9 | 47 |
| 500 | 53.2 | 6.5 | 60 |

*Example*: A drainage water system with a Se concentration of 50 $\mu$g L-1 and a flow rate of 0.5 ft/yr would have a land requirement of 20% of the total land being drained with 9% being bioremediated through Se volatilization and 11% used for evaporation.

Economic analyses show that this system is compatible with current irrigation practices (Frankenberger, Karlson, and Longley, 1990). If the irrigation flow rate is reduced to 0.25 ac-ft/yr, then the percentage of land required to treat 50 $\mu$g L$^{-1}$ drainage water would decrease from 20 percent to 11 percent of the total acreage (Frankenberger, Karlson, and Longley, 1990). The process could be further optimized by ensuring that the secondary or flash evaporation ponds are put into use no later than April or May so that evaporation can be completed by midsummer and volatilization initiated during the higher sediment temperatures conducive to high dimethylselenide production.

## Process Operation

The operation of the series evaporation pond system requires the pumping of water from primary to secondary ponds. No transfer of water is necessary when operating the single evaporation-treatment pond system. The volatilization process itself requires a carbon source, aeration, and water. Cattle manure is the most effective carbon source to promote high rates of volatilization at the Sumner Peck Ranch. Cattle manure can be applied at the rate of 10 to 15 tons per ac-yr (dry weight). The carbon source is disked into the soil to a depth of 6 inches since most of the selenium is concentrated within the upper 6 inches of evaporation pond sediments. Disking the soil not only fulfills the oxygen requirement of the microbes (particularly during the warm months of April to October), but also breaks up a soil crust that might form with irrigation. The soil is kept slightly moist using a sprinkler irrigation system. Flooding is avoided at all times to prevent oxygen depletion. Both water and soil selenium levels must be carefully monitored to ensure the success of this deselenification process. Wildlife need to be discouraged from using the primary ponds as sources of food by controlling aquatic vegetation and invertebrate populations. The disking operation will discourage their use of secondary ponds.

## FEASIBILITY ANALYSIS

Three major factors will affect the costs of implementing microbial volatilization to detoxify seleniferous sediments: (1) The amount of land removed from production, (2) the income generated from the land, and (3) the costs of the treatment process. For the purpose of the calculations here, a steady state analysis was used. The amount of selenium volatilized is assumed to equal the amount of selenium transported into the evaporation/treatment facilities by the drainwater annually and the amount of water evaporated is presumed to equal the volume of incoming drainwater. Pond construction is assumed to cost $1,500 per acre and it is to be expected that ponds will be built on land with the least profit potential. The net income loss associated with taking land out of production is calculated at two different levels of $50 and $100 per acre. Manure costs are $8 per ton with transportation and spreading costs of $1.25 per ton. Water is assumed to cost $17.50 per ac-ft with 0.5 ac-ft per acre being applied annually. Labor and equipment costs are budgeted at $16.65 per ac-yr. The annual costs per acre of productive land being drained have been calculated with respect to these variables (Frankenberger, Karlson, and Longley, 1990) and are summarized in table 3. Costs range between $92 and $151 per ac-yr based on varying water and land quality. If drainage flows were increased from 0.5 to 0.75 ac-ft/ac-yr for 70 lg Se $L^{-1}$ drainage water, the per ac-yr cost would increase by 33 percent but if the drainage flows were reduced to 0.25 ac-ft/ac-yr then the cost would be reduced by 49 percent (Frankenberger, Karlson, and Longley, 1990). Further analyses showed that reducing the useful life of the treatment system from 40 to 20 years increased treatment costs by 7 percent (Frankenberger, Karlson, and Longley, 1990). These costs do not include the mitigation costs to discourage wildlife and waterfowl use of these aquatic environments.

Table 3. The annual costs per acre of land with respect to the Frankenberger-Karlson Se volatilization process (Frankenberger, Karlson and Longley, 1990).

| Drainage water production | Se concn. | Income loss of land out of production | |
|---|---|---|---|
| | | 50 | 100 |
| ac-ft/ac-yr | µg/L | $/ac | |
| 0.5 | 50 | 92 | 104 |
| | 70 | 108 | 122 |
| | 100 | 133 | 151 |

## CONCLUSIONS

At the present time, field projects involving volatilization of selenium from sediments at Kesterson Reservoir, Sumner Peck Ranch, and the San Luis Drain Sediment are showing a rapid decline in the soil selenium inventory. The economic and feasibility analyses indicate that costs of selenium volatilization as a bioremediation technique could be kept well below $150 per acre of productive land being drained. Conservative use of irrigation water by growers together with the Karlson-Frankenberger selenium volatilization process could prove to be of extreme value in sustaining high productivity on the west side of the Central Valley. Bioremediation of seleniferous drainage water through soil microbial transformations looks quite promising and future research will focus on generating more information to successfully initiate a full-scale selenium volatilization operation.

## ACKNOWLEDGEMENTS

We wish to thank Dr. Dennis Nef (CSUF) for his technical assistance in the applied economics of this work.

## REFERENCES

Atkinson, R.; Aschmann, S. M.; Hasegawa, D.; Thompson-Eagle, E. T.; and Frankenberger, Jr. W. T., 1990. Kinetics of the Atmospherically Important Reactions of Dimethyl Selenide, *Environmental Science and Technology.* In Press.

Barkes, L. and Fleming, R. W., 1974. Production of Dimethylselenide Gas from Inorganic Selenium by 11 Soil Fungi, *Bulletin of Enviromental Contamination and Toxicology,* 12, pp. 308-311.

Burau, R. G., 1989. Selenium in Arid and Semiarid Soils, *ASCE Journal of Irrigation and Drainage Engineering,* 115, pp. 42-47.

California State Water Resources Control Board, 1987. *Regulation of Agricultural Drainage to the San Joaquin River.* Technical Committee Report.

Challenger, F. and Charlton, P.T., 1947. Studies on Biological Methylation. Part X. The Fission of the Mono- and Di-Sulphide Links by Moulds, *Journal of the Chemical Society,* pp. 424-429.

Challenger, F.; Lisle, D. B.; and Dransfield, P. B., 1954. Studies on Biological Methylation. Part XIV. The Formation of Trimethylarsine and Dimethylselenide in Mould Cultures from Methyl Sources Containing $^{14}C$, *Journal of the Chemical Society,* pp. 1760-1771.

Challenger, F. and North, H. E., 1934. The Production of Organo-Metalloid Compounds by Micro-Organisms. Part II. Dimethylselenide, *Journal of the Chemical Society,* pp. 68-71.

Chasteen, T. G.; Silver, G. M.; Birks J. W.; and Fall, R., 1990. Fluorine-Induced Chemiluminescence Detection of Biologically Methylated Tellurium, Selenium, and Sulfur Compounds, *Chromatographia.* In Press.

Chau, Y. K.; Wong, P. T. S.; Silverberg, B. A.; Luxon, P. L.; and Bengert, G. A., 1976. Methylation of Selenium in the Aquatic Environment, *Science,* 912, pp. 1130-1131.

Clark, P. J.; Zingarro, R. A.; Irgolic, K. J.; and McGinley, A. N., 1980. Arsenic and Selenium in Texas Lignite, *International Journal of Environmental and Analytical Chemistry,* 7, pp. 295-314.

Cox, D. P. and Alexander, M., 1974. Factors Affecting Dimethylarsine and Dimethylselenide Formation by Candida Humicola, *Microbial Ecology,* 1, pp. 136-144.

Cutter, G. A., 1989. Fresh Water Systems. In: Ihnat, M. (Ed.), *Occurrence and Distribution of Selenium*, CRC Press, Inc., Boca Raton, FL., pp. 243-262.

Deveral, S. J., 1984. *Areal Distribution of Selenium and other Inorganic Constituents in Shallow Ground Water of the San Luis Drain Service Area, San Joaquin Valley, California. A Preliminary Study.* U.S. Geological Survey Water-Resources Investigation Report 84-4319, p. 67.

Doran, J. W., and Alexander, M., 1975. Microbial Formation of Dimethylselenide, *Abstracts American Society for Microbiology*, 188.

Doran, J. W., 1982. Micro-Organisms and the Biological Cycling of Selenium, *Advances in Microbial Ecology*, 6, pp. 1-32.

Doran, J. and Alexander, M., 1977. Microbial Formation of Volatile Selenium Compounds in Soil, *Soil Science Society of America Journal*, 41, pp. 70-73.

Drotar, A.; Fall, L. R.; Mishalanie, E. A.; Tavernier, J. E.; and Fall, R., 1987. Enzymatic Methylation of Sulfide, Selenide and Organic Thiols by Tetrahymera Thermophila, *Applied and Environmental Microbiology*, 55, pp. 2111-2118.

Farrell, D.A.; Greacen, E. L.; and Curr, C. G., 1966. Vapor Transfer in Soil Due to Air Turbulence, *Soil Science*, 102, pp. 305-313.

Fleming, R. W. and Alexander, M., 1972. Dimethylselenide and Dimethyltelluride Formation by a Strain of Penicillium, *Applied Microbiology*, 24, pp. 424-429.

Franke, K. W. and Moxon, A. L., 1936. A Comparison of the Minimum Fatal Doses of Selenium, Tellurium, Arsenic and Vanadium, *Journal of Pharmacology and Experimental Therapy*, 58, pp. 454-459.

Frankenberger, W. T., Jr., 1989. *Dissipation of Soil Selenium by Microbial Volatilization at Kesterson Reservoir.* Prepared for the U.S. Department of Interior, Bureau of Reclamation, Contract No. 7-FC-20-05240.

Frankenberger, W. T., Jr. and Karlson, U., 1988. *Dissipation of Soil Selenium by Microbial Volatilization at Kesterson Reservoir.* Prepared for the U.S. Department of Interior, Bureau of Reclamation, Contract No. 7-FC-20-05240.

Frankenberger, W. T., Jr. and Karlson, U., 1989a. *Land Treatment to Detoxify Soil of Selenium.* U.S. Patent and Trademark Office, Patent No. 4,861,482.

Frankenberger, W. T., Jr. and Karlson, U., 1989b. Environmental Factors Affecting Microbial Production of Dimethylselenide in a Selenium-Contaminated Sediment. *Soil Science Society of America Journal*, 53, pp. 1435-1442.

Frankenberger, W. T.; Karlson, U. Jr.; and Longley, K. E., 1990. *Microbial Volatilization of Selenium from Sediments of Agricultural Evaporation Ponds January 1990.* Prepared for the State Water Resources Control Board, Interagency Agreement No. 7-125-250-1.

Frankenberger, W. T., Jr., and Thompson-Eagle, E. T., 1988. *In Situ Volatilization of Selenium, II. Evaporation Ponds.* Prepared for the San Joaquin Valley Drainage Program. U.S. Bureau of Reclamation Contract No. 7-FC-20-05110.

Fujii, R., and Deveral, S. J., 1989. Mobility and Distribution of Selenium and Salinity in Groundwater and Soil of Drained Agricultural Fields, Western San Joaquin Valley of California. In: Jacobs, L. W. (Ed.), *Selenium in Agriculture and the Environment,* Soil Science of Society America Special Publication No. 23, Madison, WI., pp. 195-212.

Izbecki, J. A., 1984. *Chemical Quality of Water at 14 Sites Near Kesterson National Wildlife Refuge, Fresno and Merced Counties, California.* U.S. Geological Survey Open File Report 84-582.

Johns, G.E., and Watkins, D. A., 1989. Regulation of Agricultural Drainage to San Joaquin River, *Journal of Irrigation Drainage Engineering,* 115, pp. 29-41.

Karlson, U., and Frankenberger, W. T. Jr., 1988a. Determination of Gaseous Selenium-75 Evolved from Soil, *Soil Science Society of America Journal,* 52, pp. 678-681.

Karlson, U., and Frankenberger, W. T. Jr., 1988b. Effects of Carbon and Trace Element Addition on Alkylselenide Production by Soil, *Soil Science Society of America Journal,* 52, pp. 1640-1644.

Karlson, U. and Frankenberger, W. T. Jr., 1989. Accelerated Rates of Selenium Volatilization from California Soils, *Soil Science Society of America Journal,* 53, pp. 749-753.

Karlson, U. and Frankenberger, W. T. Jr., 1990. Alkylselenide Production in Salinized Soils, *Soil Science,* pp. 56-61.

Lakin, H. W., 1972. Selenium Accumulation in Soils and its Absorption by Plants and Animals, *Geological Society of America Bulletin,* p. 83.

Mayland, H. F.; James, L. F.; Panter, K. E.; and Sondregger, K. E., 1989. Selenium in Seleniferous Environments. In: Jacobs, L. W. (Ed.), *Selenium in Agriculture and the Environment.* Soil Science Society of America Special Publication No. 23, Madison, WI. , pp. 15-50.

McConnell, K. P. and Portman, O. W., 1952. Toxicity of Dimethylselenide in the Rat and Mouse. *Proceedings of the Society of Experimental Biological Medicine,* 79, pp. 230-231.

Mosher, B. W. and Duce, R. A., 1989. The Atmosphere. In: Ihnat, M. (Ed.), *Occurrence and Distribution of Selenium,* CRC Press, Boca Raton, FL., pp. 295-325.

Ohlendorf, H. M., 1989. Bioaccumulation and Effects of Selenium in Wildlife. pp. 133-177. In Jacobs, L. W. (Ed.), *Selenium in Agriculture and the Environment.* Soil Science Society of America Special Publication No. 23, Madison, WI., p. 233.

Presser, T. S. and Barnes, I., 1984. *Selenium Concentrations in Waters Tributary to and in the Vicinity of the Kesterson National Wildlife Refuge, Fresno and Merced Counties, California.* U.S. Geological Survey Report 84-4122, pp. 1-25.

Tanji, K., 1990. Pond Bed Materials and Salt Crusts. Agricultural Evaporation Ponds. *Abstracts, March 1990 UC Salinity/Drainage Task Force Annual Research Conference, Sacramento, CA.*

Tanji, K. and Grismer M., 1988. *Evaporation Ponds for the Disposal of Agricultural Waste Water.* Submitted to California State Water Resources Control Board, Contract No. 5-190-150-0 and U.S. EPA, Contract No. C-060000-23-0. Department of Land, Air and Water Resources, University of California, Davis.

Thompson-Eagle, E. T.; W. T. Frankenberger, Jr.; and U. Karlson, 1989. Volatilization of Selenium by Alternaria Alternata, *Applied and Environmental Microbiology*, 55, pp. 1406-1413.

Thompson-Eagle, E. T. and Frankenberger, W. T. Jr., 1990a. Volatilization of Selenium from Agricultural Evaporation Pond Water, *Journal of Environmental Quality*, 19, pp. 125-131.

Thompson-Eagle, E. T. and Frankenberger, W. T. Jr., 1990b. Protein-Mediated Selenium Biomethylation in Evaporation Pond Water, *Environmental Toxicology and Chemistry.* In Press.

Thompson-Eagle, E. T. and Frankenberger, W. T. Jr., 1990c. Microbial Volatilization of Selenium from Seleniferous Sediments and Water. *Toxic Substances in Agricultural Drainage - Emerging Technologies and Research Needs,* Proceedings of the November 1989 US CID/Bureau of Reclamation Seminar, Sacramento, CA.

Thompson-Eagle, E. T. and Frankenberger, W. T. Jr., 1990d. Selenium Biomethylation in Alkaline, Saline Pond Water, *Water Resources Research.* Submitted.

U.S. Bureau of Reclamation, 1986. *Final Environmental Impact Statement, Kesterson Program.*

U.S. Bureau of Reclamation, 1988. *Air Quality Impacts of Enhanced Selenium Volatilization at Kesterson Reservoir.* Prepared by CH2M Hill.

Wass, L., 1990. Location and Physical Characteristics of Subsurface Agricultural Drainage Evaporation Basins. Agricultural Evaporation Ponds. *Abstracts, March 1990 UC Salinity/Drainage Task Force Annual Research Conference, Sacramento, CA.*

Westcot, D., 1988. Reuse and Disposal of Higher Salinity Subsurface Drainage Water. A Review, *Agricultural Water Management*, 14, pp. 483-511.

Westcot, D., 1990. Trace Element Buildup in Drainage Water Evaporation Basins, San Joaquin Valley. Agricultural Evaporation Ponds. *Abstracts,*

*March 1990 UC Salinity/Drainage Task Force Annual Research Conference, Sacramento, CA.*

Westcot, D.; Rosenbaum, S.; Grewell, B.; and Belden, K., 1988. *Water and Sediment Quality in Evaporation Basins Used for the Disposal of Agricultural Subsurface Drainage Water in the San Joaquin Valley, California.* Central Valley Regional Water Quality Control Board Report, p. 50.

Westlands Water District, 1989. *Water Conservation and Drainage Reduction Programs 1987-1988.*

# 10 A CONCEPTUAL PLANNING PROCESS FOR MANAGEMENT OF SUBSURFACE DRAINAGE

Donald G. Swain, San Joaquin Valley Drainage Program

## ABSTRACT

Management of subsurface agricultural drainage is a complex resources management issue. In California, for example, over one-half of the San Joaquin Valley (Valley) has no natural outlet for surface or subsurface drainage. As of 1987, an area of approximately 750,000 acres had ground-water levels within 5 feet of the surface and this shallow water area is continuing to expand. This affects agricultural production and introduces obstacles to conventional irrigated farming. In addition, most of the drainage contains relatively high levels of contaminants such as selenium which greatly complicate the drainage disposal effort. The variability and interrelationships among hydrology, soil characteristics, land use, institutions, and opportunities for drainage disposal within the drainage problem area required the San Joaquin Valley Drainage Program (SJVDP) to consider a systems approach to evaluate alternative solutions and select a recommended plan. This approach includes identification, quantification, and awareness of the interdependency of major variables.

## INTRODUCTION

Over the last four decades, Valley drainage problems have been the theme of countless technical studies, Congressional hearings, conferences, symposiums, legal analyses, and debate. Prior to the 1980's, proposed plans addressed drainage in a traditional manner, primarily as a salt-management problem. Most of these earlier planning efforts focused on collection, transport, and disposal of drainage out of the basin (IDP, 1979). Efforts were focused on maintenance of a salt balance at a level which would not restrict Valley crop selection and yield.

In the early 1980's, concentrations of some of the constituents of subsurface drainage water were found to be toxic to fish and wildlife. Along with this discovery, increasing costs and increasing restrictions placed on traditional means of drainage disposal required that a new philosophy be developed for the SJVDP investigations.

In comparison to previous planning efforts, for example, alternative disposal sites considered by SJVDP were limited to those located within the Valley in order to reduce environmental risk associated with concentrations of agricultural drainage contaminants. Specifically, the disposal sites for consideration were limited to (in Valley) shallow ground-water aquifers, evaporation and solar gradient ponds, and diluted discharge to the San Joaquin River. Increased water conservation (source control and reuse of agricultural drainage water) and modified land use were also developed as major components of all SJVDP plans. Source control options addressed adequate incentives, such as a tiered water price design and possibly water marketing, to encourage farmers (both those with and those contributing to drainage problems) to invest in conservation measures and thus reduce drainage volume and salt load.

Additionally, none of the previous planning processes (prior to SJVDP) had incorporated such components as: (1) Retirement of specific lands with shallow ground-water problems characterized by high levels of salinity and high selenium concentrations, (2) a significant increase in the irrigation application efficiency (distribution uniformity of applied water), (3) reuse of drainage water to grow very salt-tolerant plants including halophytes, (4) large scale pumping of ground water to manage shallow water levels, or (5) reallocations of available surface and ground water, which served as the foundation for SJVDP planning alternatives.

## SJVDP PLANNING PROCESS

The relationship between land classification and use, agricultural cropping patterns, irrigation practices, quantity and quality of applied irrigation water, location and quality of the shallow ground water, and quantity and quality of subsurface drainage is critical to accurately define drainage management problems (existing and anticipated), and to recommend alternative solutions.

For the purposes of SJVDP planning and this discussion, drainage problem areas are defined as those lands with depth to shallow ground water table of less than 5 feet of the land surface at some time during the irrigation period (March through September). Within these shallow water-table areas, salinity and trace-element concentrations are normally high enough to reduce irrigated agricultural productivity and/or complicate the process of leaching salts from the root zone.

Other planning parameters included restrictions on conventional disposal methods due to concentrations of selenium and other trace elements characteristic of drainage from most of the study area. In addition to the potential adverse affects to fish and wildlife exposed to high selenium levels, potential health risks associated with consumption of that wildlife, were important program concerns. Water-quality objectives being established by the California State Water Resources Control Board to protect beneficial uses of the San Joaquin River were also considered to be important constraining factors.

Planning objectives adopted by the SJVDP thus included methods to: (1) Minimize potential health risks associated with subsurface agricultural drainage water, (2) protect existing and future reasonable and beneficial uses of surface and ground water from impacts associated with drainage water, (3) sustain productivity of farmlands on the west side of the Valley, and (4) protect valley fish and wildlife resources. Planning criteria were adopted to achieve these objectives.

Two levels of planning criteria were considered representing an existing set of conditions and regulations and a more restrictive set. Level A represents the conditions as defined by existing water-quality and land-use objectives, irrigation districts and wildlife habitat managers, and available water supply (SJVDP, 1990). Level B includes more stringent water-quality and environmental regulations that would result in greater restrictions on drainage disposal methods and increased cost. The two levels were used to ensure that alternative plans reflect accomplishments for a reasonable range of future conditions and potential changes in policies and regulations.

The most likely set of planning criteria (a combination of level A and B) served as the basis for developing the performance standards used to formulate the recommended plan. Options were selected as plan components based on their effectiveness in achieving performance standards, cost, proven technology, ease of implementation, and environmental acceptability. Individual components were designed to be phased in such a way as to avoid major commitments of economic and environmental resources as long as possible. It also became evident that the reliability of predictions beyond a 10-year period is severely limited because of the many variables and conditions such as: rate of increase in the shallow water table area; onfarm drain installation; agricultural economic conditions; local, State, and Federal regulatory actions; and the collective results of individual decisions by water- and land-use managers over approximately 2.5 million acres of land. Thus, forecasted conditions which served as the basis for recommended plans were only considered to be of sufficient reliability to schedule implementation of plan components for a 10-year period. However, the planning process and range included in assumptions allow for reformulating plans to reflect changing conditions occurring throughout the 50-year planning horizon.

The basis for an effective economic and environmental evaluation of any plan is the projection of what would occur in the absence of the plan. The SJVDP projection of future conditions, defined as expected conditions without a coordinated and comprehensive plan over the 50-year timeframe, is based on a continuation of existing trends.

Proposed actions were compared against these "future-without" conditions as a basis for justifying the investment of public and/or private funds for facilities necessary for the management of subsurface drainage water. A reevaluation of the projections both with and without a comprehensive action should be made at a minimum of every 5 years to reassess the need for modification of the plan and/or policies as necessary to effectively resolve the problems described within the study area.

The SJVDP study area includes those lands which have, or contribute to, drainage or drainage-related problems. The area, which generally lies on the west side of the Valley, was divided into five subareas with unique hydrologic and/or institutional characteristics. That portion of each subarea with shallow ground water (0-20 feet in depth) was further divided into water-quality zones, with their boundaries defined by the similarity of the shallow ground-water quality. This enabled the formulation of plans to provide solutions for small homogeneous units of area.

In excess of 70 different drainage and drainage-related management options were identified by the SJVDP (1989a). Several of these options have been combined to form approximately 13 planning components used in formulating alternative drainage management plans for the 16 individual water-quality zones. The application of most of these planning components depended on the water-quality and physical characteristics of the individual subareas.

## CONCEPTUAL MODEL

Several conceptual models have been used by various agencies and organizations over the past 30 years to assist in the formulation of plans to provide drainage management to service lands in the Valley (IDP, 1979). These models, however, did not include some of the major externalities which affect tradeoffs between source control, reuse, and land-use changes with appropriate economic considerations (such as the value of conserved irrigation water supply). These models also lacked the level of detail to allow for constraining factors and evaluation of tradeoffs.

The conceptual planning model presented here relies on a combination of geographical information system (GIS) analysis and spreadsheet analysis to create data displays and assess the projected accomplishments of various plan components. A GIS analysis identifies areas with common characteristics and thus similar drainage management opportunities. The model links the various plan components into a system subject to physical, environmental, and agricultural constraints. It also helps identify and prioritize additional research, analysis, and monitoring needed to better assess the relative merits of options and/or plans.

The extent of shallow ground-water area at the end of each of the two timeframes was estimated. For the example referred to in this chapter, the problem area for the year 2040 was assumed to be equal to the 1987 lands (latest shallow water table survey (DWR, 1987)) with 0- to 10-foot water depth. These lands represent the maximum drainage problem area. Based on the 1987 data and the 2040 level of shallow ground water, an approximation was made of the year 2000 by interpolation between these two points using a simple polynomial.

The Westlands Subarea (SJVDP, 1990), shown in figure 1, is used here to illustrate the application of the planning model. In 1987, approximately 100,000 acres of this district had a water table less than 5 feet in depth at some time during the crop growing season. This problem area is expected to increase to 170,000 acres by the year 2000 and 227,000 acres by 2040.

The applicability of the individual plan options have been determined using a set of performance standards based on the water- and land-use planning criteria discussed earlier. Those standards which apply to the Westlands Subarea are included in table 1. The applicability of the individual drainage management options as plan components for the individual water-quality zone is presented in table 2.

Four water-quality zones are delineated in the Westlands Subarea. The water-quality characteristics of each zone, included in table 3, were determined by calculating the area weighted average concentration of salinity, and the trace elements boron, selenium, and molybdenum, within each zone using a GIS.

The volume of subsurface drainage water estimated for this subarea equals the assumed average applied water (2.5 feet) plus effective precipitation (0.3 feet), plus upflux contributions to crop evapotranspiration (ET) (minimal contribution with subsurface drains in place) (0.1 feet), minus crop ET (2.2 feet), minus vertical movement downward across the Corcoran Clay (0.1 feet), and minus pumping from the semiconfined aquifer (varies among the water-quality zones) (0.0 feet). The net amount of lateral movement contributes to the available upflux within the shallow ground-water area. The estimated average potential drainage volume yield for this subarea is 0.75 acre-foot/acre. This value approximates the 1982-85 average yield from drained lands within Westlands Water District which occupies most of the subarea. (This value represents the pre-1986 estimated hydrology and irrigation management practices.)

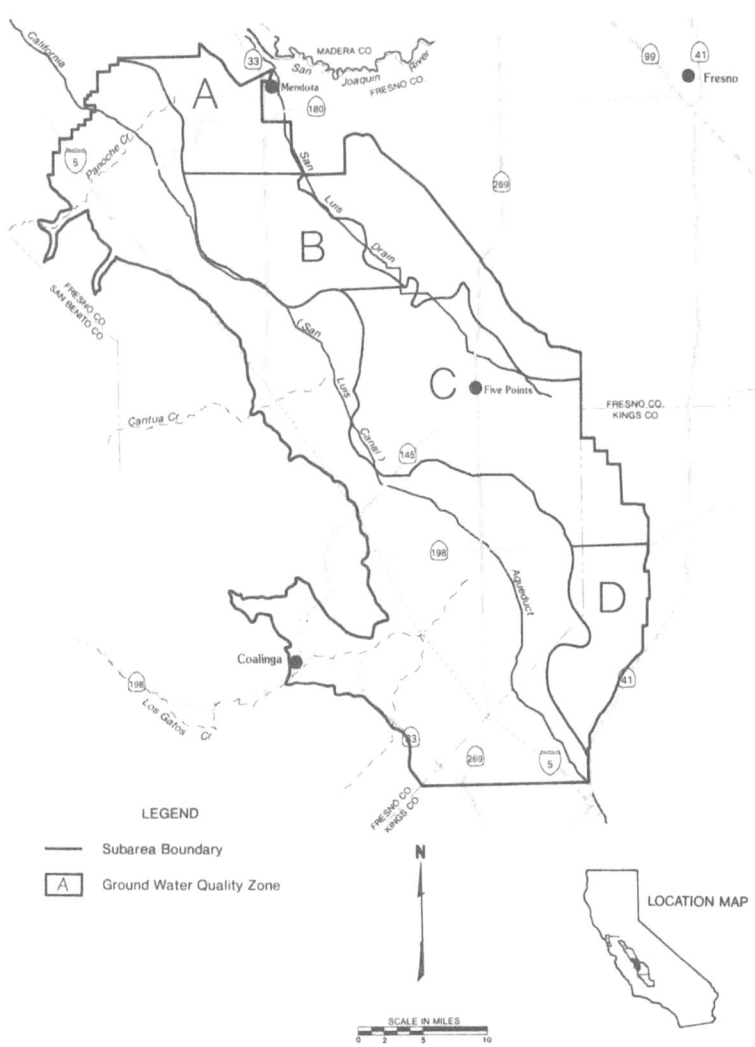

Figure 1. Westlands Subarea - ground-water quality zones.

Table 1. Performance standards used to formulate recommended plan for Westlands Subarea.

| Category | Feature | Planning Criteria |
|---|---|---|
| Water Quantity (Mean Monthly) | Agricultural Water Supply | Salinity $\leq$1,250 p/m TDS<br>Boron $\leq$0.75 p/m<br>or<br>1,250 p/m TDS $\leq$ salinity $\leq$2,500 p/m TDS<br>Boron $\leq$2 p/m<br>(with dilution or restricted use) |
|  | Wetland Water Supply | Salinity $\leq$1,250 p/m TDS<br>Boron $\leq$1 p/m<br>Selenium $\leq$2 p/b |
|  | Reuse of Subsurface Drainage on Salt-Tolerant Plants | Eucalyptus Trees $\leq$10,000 p/m TDS<br>Halophytes $\leq$25,000 p/m TDS |
|  | Evaporation Pond Influent Quality | Selenium $\leq$2 p/b - (No Alternate Habitat)<br>Selenium >2 and <50 p/b - (Alternate Habitat)<br>Selenium $\geq$50 p/b - (No Evaporation Ponds) |
| Water Quantity | Pumping Semiconfined Aquifer | Aquifer Thickness $\geq$200 feet[a]<br>with Salinity <1,250 p/m TDS[b] |
|  | Alternate Habitat Water Supply (Evaporation Pond Mitigation) | Supply - 10 Acre-feet/Acre/Year |
|  | Design Limit to Regional Deep Percolation | Supply - 0.4 Acre-foot/Acre/Year[c] |
| Land Use | Wildlife Habitat | Alternate Habitat<br>Equal in Size to Evaporation Pond<br>Area Where Se Influent >2 and <50 p/b |
|  | Retirement of Irrigated Agricultural Lands | Lands with >50 p/b Se Conc. in Shallow<br>Ground Water and Relatively Low<br>Productivity Due to High Salinity<br>and Poor Drainage Conditions |

[a]Where ground-water geology data is limited a minimum aquifer thickness of 300 feet was used to insure a well life of at least 20 years.
[b]As salinity of pumped water exceeds 1,250 p/m TDS, its use as an irrigation water supply becomes limited; however, it is considered usable up to 2,500 p/m for very salt tolerant crops.
[c]That portion of applied irrigation water passing the root zone which requires drainage management. An additional 0.1 to 0.3 acre-foot/acre of deep percolation is assumed to pass the Corcoran Clay layer.

Table 2. Applicability of drainage management options[a] for Westlands Subarea.

| Subarea and Water Quality Zone | Land Retirement[b] | Drainage Source Control | Ground Water Mgmt.[c] | Salt Tolerant Trees | Halo- phytes | Evaporation System | | |
|---|---|---|---|---|---|---|---|---|
| | | | | | | Ponds Without Alternate Habitat | Ponds with Alternate Habitat[d] | Enhanced Evap. System + Solar Ponds |
| A | Y(5k acres) | X | NA (<200 ft thick) | X | X | NA (>2 p/b Se) | NA (≥50 p/b Se) | X |
| B | Y(15k acres) | X | NA (<200 ft thick) | NA (>10k p/m TDS) | X | NA (>2 p/b Se) | NA (≥50 p/b Se) | X |
| C | Y(13k acres) | X | Y(38k acres) | X | X | NA (>2 p/b Se) | NA (≥50 p/b Se) | X |
| D | NA (<50 p/b Se) | X | Y(24k acres) | X | X | NA (>2 p/b Se) | X | NA (<50 p/b Se) |

[a]Application of options based on recommended plan performance standards in table 1.
[b]A combination of ≥50 p/b selenium concentration in the shallow ground water and relatively low land productivity due to high soil salinity and poor drainage conditions (USBR Class 4 or equivalent Storie soil classification) were used to select lands on which irrigated agriculture would be discontinued.
[c]Option limited by the aquifer thickness and quality of the ground water (less that 1,250 p/m TDS).
[d]New evaporation ponds can be used when drainage water selenium concentration >2 p/b and is <50 p/b; only if ponds can be made bird-safe or bird-free. Measures necessary to make ponds bird-free will include alternative habitat with an adequate firm water supply.
X Option is applicable without any limitation in its application.
Y Option is applicable but limited to the quantities within the parentheses.
NA Option not applicable due to its failure to meet the performance standard in parentheses (see table 1).

Table 3. Quantity and quality of problem water in Westlands Subarea.

| Subarea and Water Quality Zone | Problem Water Volume[a] | | Average Concentrations[b] Year 1990 | | | |
|---|---|---|---|---|---|---|
| | Year 2000[c] (AF) | Year 2040[d] (AF) | TDS (p/m) | B (p/m) | Se (p/b) | Moly (p/b) |
| A | 12,800 | 24,300 | 7.3 | 6 | 300 | 160 |
| B | 15,300 | 29,000 | 1.2 | 11 | 310 | 880 |
| C | 40,000 | 75,600 | 7.1 | 12 | 68 | 140 |
| D | 12,800 | 24,300 | 9.9 | 11 | 15 | 310 |
| Total[e] | 80,900 | 153,200 | 8.5 | 11 | 140 | 310 |

[a]Volume of water removed from the shallow ground-water aquifer annually if on-farm tile drains were installed (limited by the rate of drain installation).
[b]Area weighted mean values (drain and shallow well data collected during 1984-1988 period) for each zone.
[c]Limited by the rate of drain installation or implimentation of alternate management practices.
[d]Limited to 90 percent of the area with 0-5 foot shallow ground-water depth.
[e]Water quality values represent the area weight concentrations for the subarea.

## Geographical information system (GIS)

A GIS was used extensively in the presentation, analysis, and interpretation of the available geohydrological, water quality, land classification and use, and physical facility data associated with the agricultural drainage problem. The GIS graphically portrays information and, through a series of overlays, can facilitate the selection, location, and sizing of plan components. For example, the GIS was used to present and analyze available information on:

- Areas with shallow ground water within 5 feet of the land surface.
- Shallow ground-water areas where the selenium concentration in ground water is greater than 50 p/b.
- Land areas with high soil salinity and poor drainage characteristics.
- Areal locations of the semiconfined aquifer suitable for ground-water pumping to lower high water tables.
- Average concentrations of various water-quality parameters within the individual water-quality zones.

This information, summarized in table 4, was the basis for sizing the recommended plan components for the Westlands Subarea. For example, the GIS was used to analyze and present three-dimensional semiconfined aquifer ground-water quality data within a particular water-quality range. This information was used to select areas where pumping from the semiconfined aquifer could maintain shallow ground-water levels below 5 feet of the land surface and provide a water supply of an acceptable quality over a 20-year period.

There are essentially five components to any plan for the Westlands Subarea. They include: (1) Retirement of lands from irrigated agriculture, (2) source control (deep percolation reduction), (3) pumping from the semiconfined aquifer to lower shallow water levels, (4) reuse of drainage to irrigate salt-tolerant crops, and (5) evaporation and salt-disposal system. Each one of these components has the capacity to reduce the potential drainage volume. Pumping from the semiconfined aquifer to lower the shallow water levels and the evaporation system will also result in salt disposal (storage).

The following is a brief discussion of the five separate plan components utilized in this model and of their effectiveness in reducing drainage volume.

**Retirement of irrigated agricultural lands.** The lands selected for retirement included those with selenium concentrations within the shallow ground-water system that exceed 50 p/b and those with poor drainage characteristics. The lands with poor drainage characteristics are defined as those which fall into the class 4 land productivity designation used by the Bureau of Reclamation or classes 4, 5, and 6 of the Storie Index. Land classification indicates general

Table 4. Plan formulation GIS data for Westlands Subarea.

| | (Acres in 1000's) | | | | | |
|---|---|---|---|---|---|---|
| Water Quality Zones | Area with Shallow Ground-Water Depth <5 ft. Year 2040 | Area with Shallow Ground-Water Se Conc. ≥50 p/b Year 1990 | Class 4 & 6 Lands Year 1990 | Candidate Lands for Retirement Year 1990[a] | Aquifer Thickness with Salinity <1,250 p/m TDS Year 1990 | |
| | | | | | ≥200 ft | ≥300 ft |
| A | 23 | 20 | 5 | 5 | 0 | 0 |
| B | 45 | 38 | 15 | 15 | 0 | 0 |
| C | 123 | 57 | 16 | 13 | 38 | 20 |
| D | 36 | 0 | 24 | 0 | 25 | 13 |
| Total | 227 | 115 | 60 | 33 | 63 | 33 |

[a]Lands which are candidate for retirement include those that overlie shallow ground water are classed as 4 (due to soil salinity and texture), and have ground-water selenium levels ≥50 p/b.

agricultural productivity while the selenium criterion approximates the level above which it would concentrate in evaporation ponds to the hazardous waste level specified in the California Toxic Pits Act (Act). Construction of ponds designed to meet Act criteria would be extremely expensive. Land retirement can eliminate the potential drainage volume by 0.75 acre-foot/acre.

<u>Source control.</u> The drainage volumes could be reduced by adopting more water-conserving irrigation technologies, improving irrigation system management, and improving irrigation scheduling. (Actions are already being taken by farmers to reduce deep percolation by adopting these options.) A variety of alternative technologies are capable of improving existing systems, which for planning purposes are representative by 1/2-mile irrigation furrow runs without a return system which represent pre-1986 conditions. In formulating the recommended plan, it was assumed that 1/2-mile furrows would be replaced by furrow irrigation with 1/4-mile runs and a return system. The return system provides an opportunity to increase the uniformity of the distribution of applied water and thus reduce deep percolation. Improved technology, plus improved management and irrigation system methods, are assumed to reduce the net deep percolation (subsurface drainage volume) by an average of 0.35 acre-foot/acre within the shallow ground-water area.

<u>Pumping from the semiconfined aquifer.</u> Subsurface drainage can be managed by pumping from the semiconfined aquifer at a sufficient rate to maintain the water level below the plant root zone without onfarm drains. The pumping rate is assumed to average 0.4 acre-foot/acre, which is equal to the total drainage yield (0.75 acre-foot/acre) minus the effect of increased source

control (0.35 acre-foot/acre). The application of this option in the recommended plan was limited to those areas where the aquifer thickness is greater than 300 feet which have water quality of less than 1,250 p/m Total Dissolved Solids (TDS).

This option would have a limited life because water-quality in the semiconfined aquifer would deteriorate as more, shallow, saline ground water migrates downward. The selection of an aquifer thickness of 300 feet ensures that pumping would continue a minimum of 20 years and possibly as long as 50 years with ground-water quality adequate for limited agricultural use. This depth provides a safety factor needed because: (1) variation in TDS with depth is unknown and (2) it is difficult to gauge exact pump-screen depth. After the salinity exceeds the 2,500 p/m TDS objective, pumping would be discontinued and subsurface drains installed or pumping would be continued but water use would be restricted to very salt-tolerant crops, such as cotton, barley, and wheat. Irrigation with undiluted ground water of this quality will require freshwater for leaching of salts and seed germination. This water could also be used to irrigate eucalyptus trees.

Reuse by irrigation of salt-tolerant crops. Subsurface drainage from most areas in Westlands exceeds 3,000 p/m. This water is normally considered a waste, requiring disposal. One method of reducing the volume of drainage water requiring disposal includes applying it to very salt-tolerant plants, such as eucalyptus trees and halophytes (salt-loving plants). The maximum quality limits for irrigation water used in these crops are specified in table 2. They would be grown in a sequence with subsurface drainage from the tree plantations used to irrigate the halophyte crop. Crops such as cotton could be used as the first crop in the sequence, where drainage water quality is less than 3,500 p/m TDS and where freshwater is available for the first irrigation supply. ET rates for eucalyptus trees and halophytes were assumed to be 5.0 and 3.0 acre-feet/acre, respectively. For each crop, underlying drains and high leaching fractions will be used to maintain nonlimiting salt balance within the crop root zone.

The advantages of reusing drainage for irrigation of these plants are twofold. First, the overall cost is less than that of an evaporation pond because of the potential value of the crops grown and the high ET rate. Second, trees, when irrigated with water having a selenium concentration exceeding 2 p/b, is assumed to offer fewer adverse environmental effects than evaporation ponds. It is likely that there is an overall increase in wildlife habitat resulting from growing these crops.

Additional research and demonstration projects are needed to verify the effectiveness of the crops to utilize this saline drainage water with high levels of boron as an irrigation supply and their effect on the environment.

Evaporation systems. Evaporation systems include: (1) Evaporation ponds that have been traditionally constructed in the study area, (2) evaporation ponds with facilities and actions needed to protect wildlife, and (3) accelerated evaporation systems combined with solar energy generation ponds in areas where selenium is sufficiently high to cause invocation of the Act. The evaporation system selected depends on selenium concentrations in the influent water. Traditional evaporation ponds dispose of drainage at a rate of about 4.0 acre-feet/acre, and an accelerated evaporation system may dispose of much higher rates depending on the concentration of dissolved solids in the influent water. The effluent from the accelerated evaporation system will be the source of the brine layer in solar ponds. It was assumed that salts will remain in the ponds which serve as a final disposal site. Proper safeguards, including a comprehensive monitoring program would be included to ensure that the environment is not significantly affected adversely by these drainage disposal facilities. Additional research is also needed to perfect the accelerated evaporation system and the solar ponds.

## PLANNING CONCEPT ILLUSTRATION

The sequence of applying the plan components to reduce the volume of drainage water and dispose of salts is illustrated in figure 2. The applicability of the drainage management options as plan components is presented in table 2, and their effectiveness has been discussed. The plan developed for water-quality Zone C of the Westlands Subarea illustrates the application of this planning model. The geohydrology, water quality, and soil conditions in this portion of the study area offer an opportunity to apply all available drainage management options as plan components. It is projected that, in the year 2000, about 53,000 acres in this zone would have ground water within 5 feet of the surface. The associated potential drainage volume would be 40,000 acre-feet. (As mentioned earlier, some steps have been taken by individual farmers since 1985 to reduce drainage volume and manage drainage water in a manner similar to that described below.)

The first step would be retiring about 6,000 acres of class 4 land overlying shallow ground water with selenium concentrations of at least 50 p/b. (An additional reduction in irrigated land area results from changing land use from agriculture to drainage reuse and disposal facilities.) This would eliminate 4,500 acre-feet of drainage at 0.75 acre-foot/acre, leaving a drainage potential of 35,500 acre-feet.

Second, application of onfarm source control techniques would reduce deep percolation on 44,400 acres by 0.35 acre-foot/acre. This would reduce potential drainage by another 15,500 acre-feet. It is assumed that, since 1986 (the year

# CONCEPTUAL PLANNING PROCESS

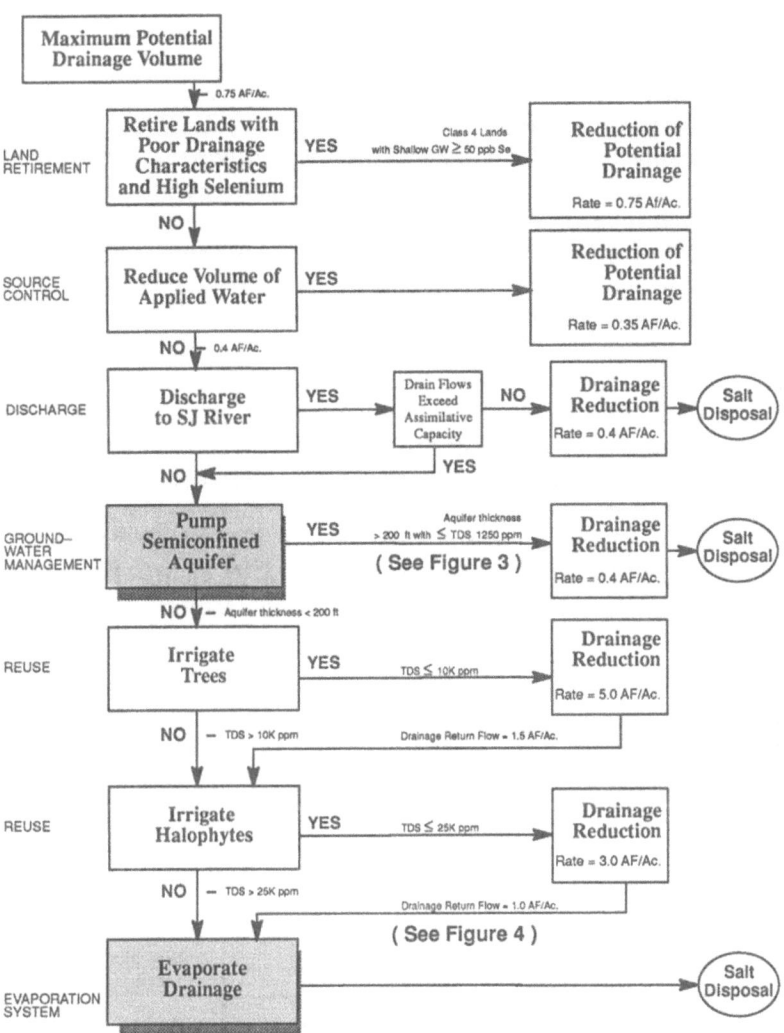

Figure 2. Plan formulation sequence.

when actions by State and Federal agencies began to restrict drainage disposal opportunities), the farming community has significantly reduced the amount of deep percolation in the shallow ground-water area. The 15,500-acre-foot target reduction would be achieved by 2000, leaving a drainage potential of 20,000 acre-feet.

Third, a regime of ground-water pumping would be implemented on 10,000 acres overlying the semiconfined aquifer where the thickness is at least 300 feet and water quality less than 1,250 p/m TDS. A pumping rate of 0.4 acre-foot/acre (equal to the rate of deep percolation after source control measures are applied) would reduce potential drainage and increase the available water supply by 4,000 acre-feet. Pumped ground water would be available for limited use until its quality deteriorated to a level above 2,500 p/m TDS. Figure 3 illustrates the criteria and assumptions used to determine the utility of this component. As a result of applying this drainage reduction measure, drainage requiring management would be reduced to 16,000 acre-feet.

Fourth, drainage water removed by onfarm drains would be used to irrigate salt-tolerant crops. The quality of drainage water is less than 10,000 p/m TDS and, therefore, could be used to irrigate both salt-tolerant trees and halophytes. The combined effort would reduce the drainage volume by about 13,000 acre-feet with the irrigation of about 2,100 acres of land planted to these salt-tolerant trees and 800 acres planted to halophytes. In addition, an additional 2,200 acre-foot reduction in drainage volume would result from using irrigated

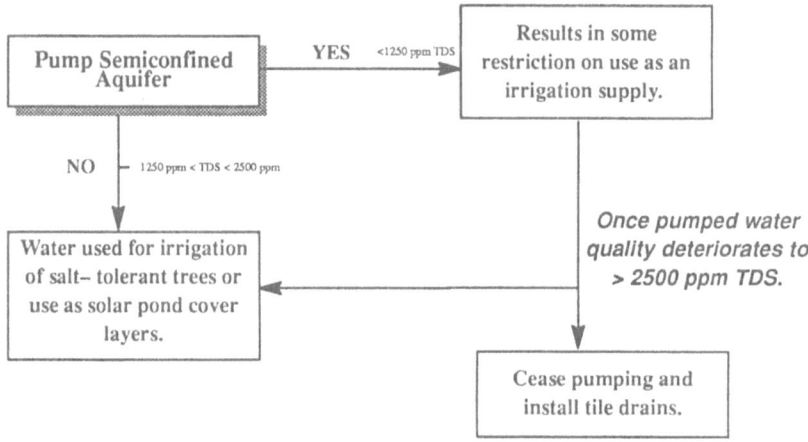

Figure 3. Plan formulation squence - pump semiconfined aquifer.

agricultural land to grow salt-tolerant plants. The remaining drainage volume is 800 acre-feet.

The final step includes the disposal of the remaining 800 acre-feet of drainage resulting from irrigation of the halophyte crop. An evaporation system of about 200 acres would be needed and would include an accelerated evaporation system and solar radiant ponds designed to generate electrical energy and store salts. Figure 4 diagrams the decisions related to the use of various methods of evaporating drainage.

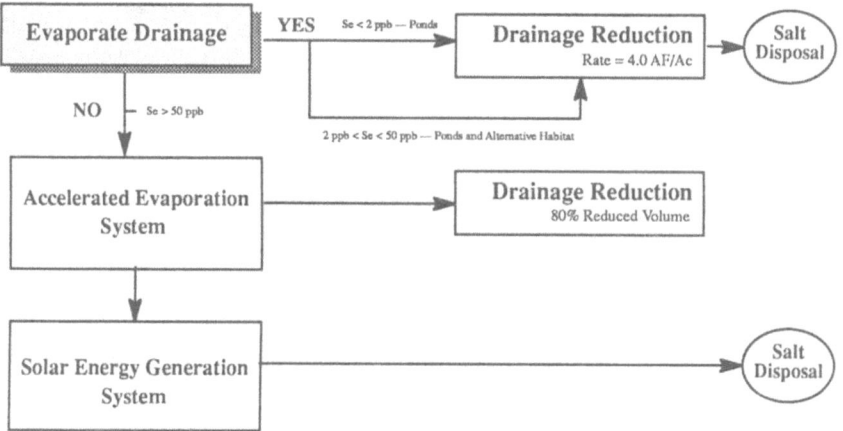

Figure 4. Plan formulation sequence - evaporate drainage.

Figure 5 illustrates the concept of drainage-water reuse and disposal. Salts are assumed to be conservative, and as consumptive use of drainage water increases, salt concentrations increase proportionately. The size of disposal facilities would be greatly reduced by reuse. Another advantage of reuse is the potential economic value of crops grown. Although additional research and demonstration projects are needed to accurately determine benefits, costs, and environmental effects associated with reuse, it seems likely that a net benefit would accrue, including increased wildlife habitat.

## PROBLEMS AND LIMITATIONS OF THIS APPROACH

This conceptual model is based on a series of anticipated drainage-related decisions made by farm owners/managers, water or drainage district managers,

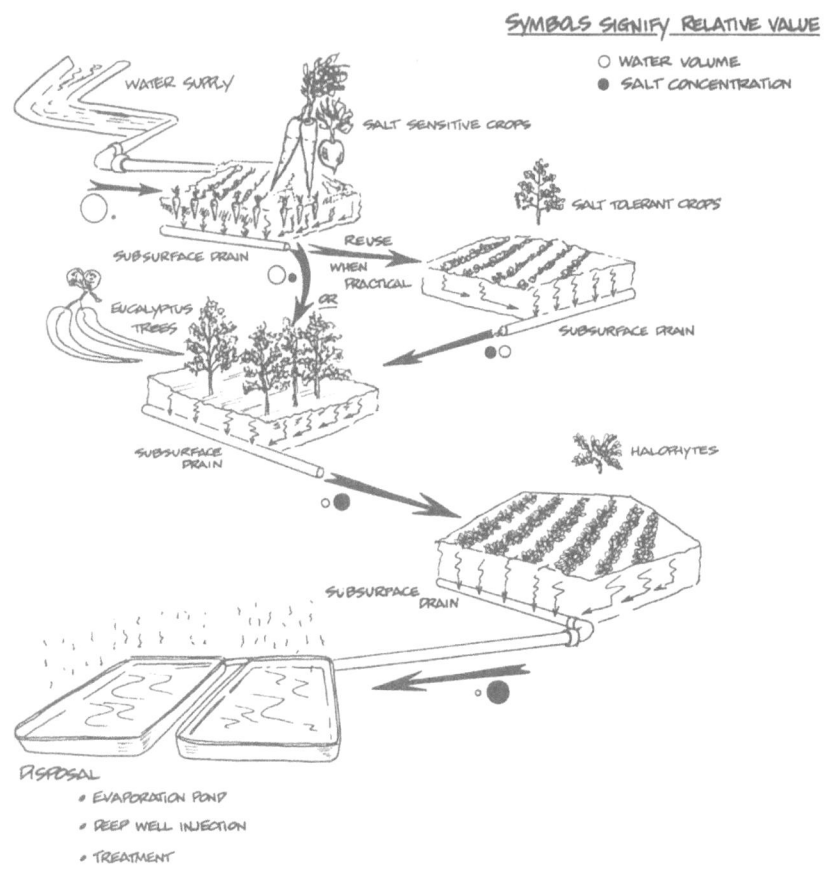

Figure 5. Drainage water reuse and disposal.

and Federal and State government bureaucracies responsible for providing water supplies, drainage management services, and environmental regulations. The model includes one sequence of decisions based on a given set of assumptions. It does not produce an economic optimal solution nor evaluate the relative importances of a full range of future decisions or the probability of their occurrence. It essentially illustrates how the problem can be solved.

It is highly likely that many of the future drainage management decisions will be made by individual farm managers based on short-term economics. Existing trends analysis may be as valid as use of an agricultural production optimization model. The rate of implementing a drainage management system is driven, in part, by high water table areas. These conditions can change by season; i.e., high ground-water pumping during a drought may lower water tables and delay a decision to install drains.

A district or farm manager could use the conceptual model to make short-term decisions regarding how to best invest their limited funds to resolve their drainage and drainage-related problems. The linkage of the conceptual model with an optimization model could also provide regional planners with a more effective tool to make decisions on a 5- to 10-year time horizon. The availability of data on semiconfined aquifer geohydrology, water-quality characteristics of the shallow ground water, agro-economics of growing salt-tolerant crops, and effective incentives for encouraging onfarm conservation, are important for effective decisionmaking, maintainance of agricultural productivity, and protection of the environment. This information needs to be readily available and understandable to the decisionmakers. A clearinghouse of this information and the interpretive tools would be vital if effective long-term decisions are to be made over the entire 2.5 million acres of irrigated agricultural lands on the west side of the San Joaquin Valley, regardless of available models.

## CONCLUSIONS

All planning models used in resource management decisions must include the ability to organize a large volume of data, often from multidisciplines, and evaluate alternative scenarios. The conceptual model discussed in this chapter serves these functions. The combination of using spreadsheets and a GIS system provides the decisionmaker at any level of management a better understanding of how a management system can be formulated to achieve its drainage management objectives. The simplicity of this model allows easy access to the important variables plus an understanding of the cause and effect relationships. Although not an optimization model, it does provide direction with regard to resources or drainage management. The model also provides a better understanding as to where additional research, analysis, and monitoring are needed to better assess the relative merits of the alternative options and/or plans.

## REFERENCES

Ayers, R. S. and Westcot, D. W., 1985. *Water Quality For Agriculture*. Irrigation and Drainage Paper 29, Rev. 1, Food and Agriculture Organization of the United Nations.

Bradford, D. F.; Drezner, D.; Shoemaker, J. D.; and Smith, L., 1989. *Evaluation of Methods to Minimize Contamination Hazards to Wildlife Using Agricultural Evaporation Ponds in the San Joaquin Valley, California.*

Burt, C. M. and Katen, K., 1988. *Technical Report to the Office of Water Conservation, California Department of Water Resources, on Westside Resource Conservation District--1986/1987 Water Conservation and Drainage Reduction Program.*

California Department of Food and Agriculture, Agricultural Research Branch, 1990. *A Farming System for the Management of Salt and Systems in Irrigated Land (Agroforestry).*

California Department of Water Resources, 1987. *Present and Potential Drainage Problem Areas, Data Map, San Joaquin Valley.*

California State Water Resources Control Board, 1975. *Water Quality Control Plan for the San Joaquin Basin (5C).*

Central Valley Regional Water Quality Control Board, 1988. *Amendments to the Water Quality Control Plan for The San Joaquin Basin (5C) For the Control of Agricultural Subsurface Drainage Discharges.* Draft Report.

Central Valley Regional Water Quality Control Board, 1989. *Staff Report on the Regulation of Evaporation Basins with Consideration of Impacts to Waterfowl.* Presented to Regional Board at its 337th Regular Meeting.

CH2M Hill, 1989. *Irrigation System Costs and Performance in the San Joaquin Valley.*

DeBruyn, D., 1987. Multiple Use of Agricultural Water Supplies, *Ambient Water Quality Criteria for Selenium,* U.S. Environmental Protection Agency.

Frederick, K. (Ed.), with the Assistance of Gibbons, D., 1986. *Scarce Water and Institutional Change.*

Interagency Drainage Program (IDP), 1979. *Agricultural Drainage and Salt Management in the San Joaquin Valley.*

Lee, E. W.; Nishimura, G. H.; and Hansen, H. L., 1988. *Agricultural Drainage Water Treatment, Reuse, and Disposal in the San Joaquin Valley of California, Part I: Treatment Technology.* Technical Report, San Joaquin Valley Drainage Program.

Nishimura, G., 1989. Selenium Treatment Cost Projections for Land Retirement Study. Memorandum, San Joaquin Valley Drainage Program.

Rhodes, J. and Dinar, A., 1990. Reuse of Agricultural Drainage Water to Maximize the Beneficial Use of Multiple Water Supplies for Irrigation. This Volume.

San Joaquin Valley Drainage Program, 1987. *Developing Alternative Future-Without-Project Scenarios for Agricultural Lands and Wetlands in the San Joaquin Valley.*

San Joaquin Valley Drainage Program, 1988. *Formulating and Evaluating Drainage Management Plans for the San Joaquin Valley.* Technical Report.

San Joaquin Valley Drainage Program, 1989a. *Preliminary Planning Alternatives for Solving Agricultural Drainage and Drainage-Related Problems in the San Joaquin Valley.*

San Joaquin Valley Drainage Program, 1989b. Technical Information Record (Draft), Supporting Document to *Preliminary Alternatives for Solving Agricultural Drainage and Drainage-Related Problems in the San Joaquin Valley.*

San Joaquin Valley Drainage Program, 1990. Draft Final Report.

San Joaquin Valley Drainage Program, 1990. Technical Information Record, Supporting Document to Final Report.

University of California Committee of Consultants on San Joaquin River Water Quality Objectives, 1988. *The Evaluation of Water Quality Criteria for Selenium, Boron, and Molybdenum in the San Joaquin River Basin.*

Westlands Water District, 1989. *Water Conservation and Drainage Reduction Programs, 1987-1988.*

# Three: ECONOMICS OF FARM LEVEL IRRIGATION AND DRAINAGE

# 11 CROP-WATER PRODUCTION FUNCTION AND THE PROBLEMS OF DRAINAGE AND SALINITY

John Letey, University of California, Riverside

## ABSTRACT

Crop yield and deep percolation functions for irrigation under saline conditions can be simulated using a seasonal model. A transient state model provides for multiseasonal simulation of yield, deep percolation, salt distribution and water distribution for any given set of irrigation management regimes. Applications of these models, combined with an irrigation uniformity model, for various irrigation-drainage situations are summarized.

## INTRODUCTION

A production function is commonly referred to as the relationship between inputs to and the output of a production process. Growing crops is the production process under consideration in this chapter. Water represents the input and crop yield and deep percolation (DP) of water beyond the root zone are considered outputs. Although DP is not commonly considered to be an output of a crop-water production function, it becomes important as a source for externalities in the economics of agricultural drainage water.

Water is but one of several inputs to crop production process. Plants typically grow to the level that is allowed by the component provided in the least amount. Substitution between inputs does not normally lead to higher yields. For example, if the nitrogen supply can produce a maximum yield of 50 units, a yield higher than 50 units cannot be achieved by additional levels of phosphorus, potassium, water, etc. This chapter assumes that other inputs have been applied at a level so that water is the limiting factor in crop production. Furthermore, yields are reported on a relative rather than absolute basis. In other words, a relative yield (RY) of 1.0 indicates that water was not a limiting factor in the production and that the absolute yields would be determined by the level of other input factors.

Level of water application can alter the nonlimiting levels of other inputs, particularly mobile plant nutrients. Water application leading to large amounts of DP leach mobile nutrients from the root zone and the leached nutrients must be compensated by applying higher levels. Because interactions between water and other crop production inputs are greatest at water applications leading to DP, these interactions must be considered in applying crop-water production functions at high levels of water application.

Irrigation water supplies have differing levels of dissolved salts (salinity), which can influence crop yields. Thus crop-water production functions must account for salinity.

In general, two approaches to estimate crop-water production functions are apparent in the literature. One approach synthesizes production functions from theoretical and empirical models of individual components of the crop-water process. The second approach estimates production functions by statistical inference from observations on alternative levels of crop yield, water applications, soil salinity, and other variables. This chapter will only consider the first approach. A more comprehensive review on production functions, including the second approach, is provided by Letey, Knapp, and Solomon (1990).

Models to compute production functions can be broadly classified as being either transient or seasonal. Transient models use basic waterflow and salt transport equations with initial soil conditions to compute salt and water distributions in the soil at various times. A water uptake (root extraction) term is added to the flow equation to account for water removal by transpiration. The root extraction term provides linkage between the soil-water-salinity status and crop yield. Seasonal models compute yield from total applied water of a given salinity during the season. Both types of model produce production functions which are only valid for uniform water application across the field. However, in most fields, depth of applied water may vary considerably with position. Thus, field-level production functions may differ from those estimated assuming uniform irrigation. A later section of this chapter covers the integration of crop-water production functions with spatial variability.

## SEASONAL PRODUCTION FUNCTION MODEL

Letey et al. (1985) and Letey and Dinar (1986) described the development of a seasonal production function model. The relationships of yield v. evapotranspiration (ET), yield v. average root zone salinity, and average root zone salinity v. leaching faction were combined to develop an equation that relates yield to the amount of a seasonal applied water of a given salinity. A linear relationship between yield and ET was used in the model. A linear relationship between total dry matter production and ET has commonly been reported in the literature. However, the marketable product, such as cotton lint, may not be linearly related to ET. For these cases, the model must be first used to compute the total dry matter production and an additional relationship between the total dry matter and the marketable product yield is required to develop the production function for the marketable product.

The piece-wise linear relationship proposed by Maas and Hoffman (1977) was used to relate yield to average root zone salinity in the seasonal model. Hoffman and van Genuchten (1983) presented a theoretical relationship between average root zone salinity and leaching fraction which was based on steady-state assumption. Combination of these three relationships provided a model that could be used to compute the yield, leaching volume, or electrical conductivity (EC) of the water percolating below the root zone for given quantities of seasonal applied water (AW) of given salinities. The seasonal AW includes quantities of water applied before planting the crop but excludes runoff.

Letey and Dinar (1986) used the model to compute production functions for several crops. Yields were recorded on a relative scale and AW was scaled to pan evaporation ($E_p$) to facilitate transfer of the production function to locales other than those from which the data were measured. The computed production functions for several crops were included in the report. The production functions for tomato are presented in figure 1 to illustrate the nature of the crop-water production functions.

Crop yields from experiments were compared to those computed by the seasonal model for several crops (Letey et al., 1985 and Letey and Dinar, 1986). When the salinity of the irrigation water varied over the season, the weight-averaged EC of irrigation waters was used. Measured and computed yields correlated well. Letey and Dinar (1986) discussed the usefulness and the limits of the model in detail.

A problem with the use of the seasonal model occurs when the soil salinity at the beginning of the crop season differs greatly from the salinity of irrigation water. Steady-state conditions are not achieved during one growing season so the model is less reliable under these conditions. This limits the utility of the model in accounting for the dynamics of the soil salinity, particularly for a multiseasonal analysis involving crop rotations. Knapp (1990) replaced the steady-state soil salinity relation by a dynamic equation for soil salinity which assumed piston-flow conditions. This modification made it useful for a dynamic analysis as reported by Knapp in another chapter in this book (Knapp, 1990).

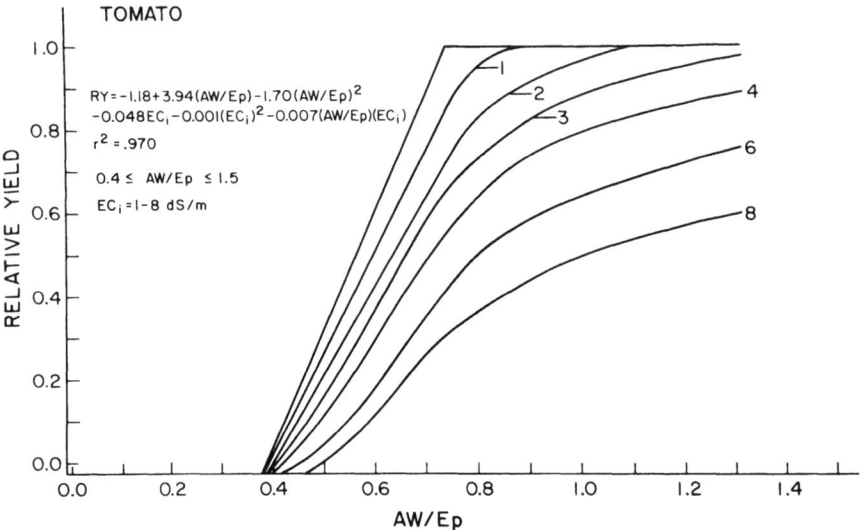

Figure 1. Computed relative yields of tomato for various quantities of applied water which are scaled by pan evaporation. Each curve is for given EC of irrigation (dS/m).

The seasonal model has utility only when there is free drainage below the root system. If the water table is sufficiently high such that water can be extracted from the water table, the seasonal model cannot be accurately used.

## TRANSIENT STATE MODELS

The transient state models consist of two major components. One component is a numerical method to compute water and salt flow in soils so that their distribution within the soil can be determined as a function of time. Bresler and Hanks (1969) provide an example of this component. The second component consists of a root extraction term which removes water from the soil as it is transpired by the plant, thus linking the soil-water-salinity state to plant response. Root extraction terms which have been proposed can be identified by two categories. The root extraction term, as presented by Nimah and Hanks (1973), is an example of the first category. The root extraction term is basically a microscale waterflow equation from the soil into the root. Cardon and Letey (1991) have reported that this type extraction term does not adequately account for soil salinity effects.

The other category of root extraction term is based on empirically observed plant response to the soil-water-salinity status. This type root extraction term (Feddes et al., 1978 and Belmans et al., 1983) can be expressed as

[1] $S = \alpha(H)S_m(z)$

where $\alpha(H)$ is a dimensionless water stress response function equivalent to the ratio between the actual (S) and the potential ($S_m$) extraction rates and

[2] $S_m = T_p/L$

where $T_p$ is potential transpiration and L is rooting depth. $S_m(z)$ can be adjusted at various depths to account for root distribution. The root extraction rate equals the potential transpiration rate when $\alpha(H)$ equals 1 and becomes less than potential transpiration when $\alpha(H)$ is less than 1. The water potential (H) can be separated into matric potential (h) and osmotic (salt) potential ($\pi$).

A linear relationship between yield and transpiration is assumed and maximum yield is associated with potential transpiration. Note that this assumption is common to the seasonal water production function model as well as both root extraction functions for the transient state model. Empirical relationships between h or $\pi$ and plant response can be used to establish the relationship between T and h or $\pi$. It was suggested by van Genuchten and Hoffman (1984) that measured salt tolerance response functions of various crops could be expressed by a smooth S-shaped curve between relative yield and the average salt concentration of the root zone. The expression after converting salt concentration to osmotic potential is

[3] $Y = Y_m/[1 + (\pi/\pi_{50})^p]$

where Y is yield, $Y_m$ is maximum yield, $\pi_{50}$ is the osmotic potential at which the yield is reduced by 50 percent and p is an empirical constant found to be approximately 3 for several crops (van Genuchten and Hoffman, 1984). Using the assumptions, $S/S_m = T/T_m = Y/Y_m$ from equations 1 and 3, then

[4] $\alpha(\pi) = 1/[1 + (\pi/\pi_{50})^p]$

If matric potential and osmotic potential have similar, but not necessarily linearly additive, effects on yield and thus on transpiration, then

[5] $\alpha(H) = 1/\{1 + [(a_1 h + \alpha_2 \pi)/\pi_{50}]^p\}$

where $h_{50}$, $\pi_{50}$, $a_1$, and $a_2$ are presumed to be parameters specific to the crop, the soil, and the climate.

Cardon and Letey (1991) compared the yields computed from a transient state model using the root extraction term as expressed in equation [5] to experimentally measured yields of corn in Israel (Shalhevet et al., 1986) and found good agreement (figure 2). The experimental variables were four irrigation intervals (3.5, 7, 14, and 21 days) and a range of irrigation water salinities ranging from 1.5 to 11.1 dS/m.

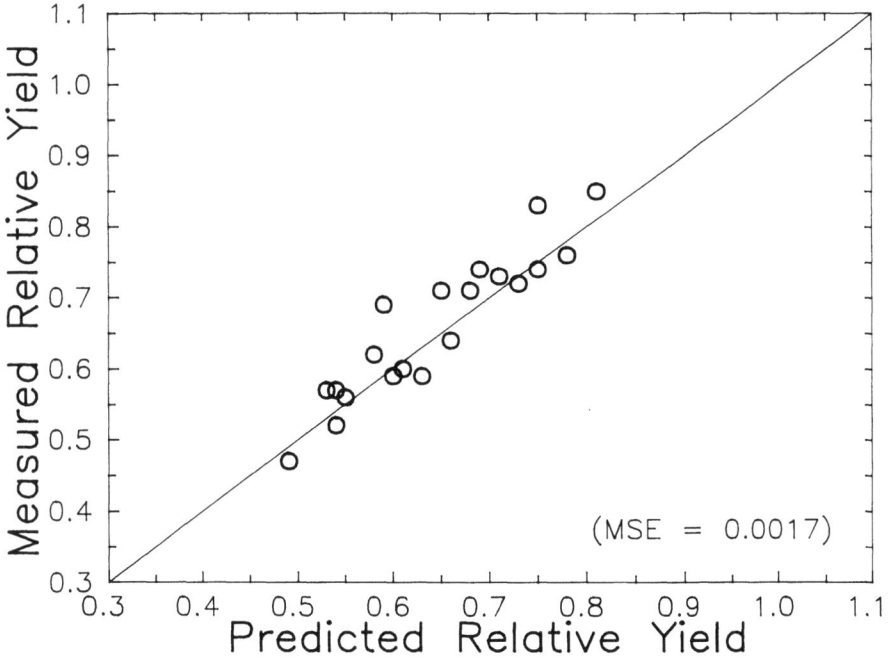

Figure 2. Comparison between measured and predicted relative yields of corn using a transient state model.

The following is a brief description of a simulation procedure using the transient state model. A depth is selected well below the root system to establish a lower boundary which can have one of three possible fixed conditions. A constant matric potential is selected for the case with free drainage without a water table above the boundary layer. When the water table is above the boundary layer, either a zero flow or constant flow boundary condition is imposed. An initial distribution of water and salt in the soil profile from the surface to the lower boundary layer is imposed. Thereafter, a water application representing either precipitation or irrigation of any specified interval and amount is programmed. Between water applications, water is extracted by the crop at a rate dependent upon the climate, crop, and soil-water-salt status. The program continuously monitors the water and salt distribution in the soil profile. At the end of the growing season, the relative yield is determined by taking the ratio of the computed ET to the maximum ET. The fallow period is simulated by eliminating the root extraction term but allowing evaporation to occur from the surface. Precipitation events are programmed as before. This sequence of events can be programmed over any number of years using any combination of crops and irrigation management strategies.

## COMPARISON OF SEASONAL PRODUCTION FUNCTION MODEL TO TRANSIENT STATE MODEL

The seasonal model has the advantage of having modest input requirements, and the annual yield for a given seasonal water application amount of given salinity is computed by solving one equation. A production function relating yield to seasonal applied water can be computed by simply solving the equation for a series of AW values. Use of the piston-flow assumption as proposed by Knapp (1991) allows the model to have utility for dynamic modeling over multiseasons. The model lends itself well to an optimization analysis; its shortcomings include an inability to cope with high water table situations and account for irrigation scheduling effects.

The transient model has the advantage of simulating production over multiseason, multicrop, and dynamic irrigation and precipitation events. Further, it has utility under conditions with high water table as well as free drainage conditions. The main shortcoming is that considerable input data and computer time are required to simulate a growing season, the time being dependent upon the complexity of irrigation events. The transient model is rather cumbersome in deriving a production function curve. The computed seasonal yield is related to the programmed irrigation events. A series of simulations with different irrigation applications is required to produce a production function curve. Thus, the transient model is extremely useful in simulating the effects of a proposed management scheme but is not convenient in an optimization analysis.

Both approaches assume one-dimensional flow. They compute relative yield to AW where AW is expressed as an equivalent depth of water. AW in the field can be ambiguous. It can represent water discharged to the field including runoff or the amount that infiltrated the soil which is the difference between discharge and runoff. Here, AW denotes infiltrated water because runoff does not contribute to crop production or deep percolation. Because water discharge and runoff are usually measured by volume, AW is computed by dividing the volume by the area of the land that was irrigated. Applied water computed in this way represents the average value ($\overline{AW}$) for the field. The actual AW at any given location in the field may be considerably different from $\overline{AW}$ because irrigation water is never applied uniformly. Variable AW values at different locations in the field lead to variable yields at the different locations. The production function for a field would relate the average yield, $\overline{Y}$ to $\overline{AW}$. As will be reported below, the relationship between $\overline{Y}$ and $\overline{AW}$ differs from the relationship between Y and AW as computed from the models.

Analytical procedures to combine Y v. AW functions with irrigation uniformity have been proposed. Warrick and Yates (1987) presented tables for response functions and average yields for three water distributions. Except for extremely low applications of water, yields under nonuniform application are lower than yields under uniform application for a given AW value. For a piecewise linear production function, applying more water overcomes the effects of uniformity, i.e., maximum yield can be achieved by applying high amounts of water. It is assumed that the soil can transmit the higher applications of water without becoming waterlogged. If the production function is quadratic, nonuniformity cannot be overcome by applying more water and yields under nonuniform conditions are lower than yields under uniform conditions.

Analytical solutions are usually restricted to specific crop-water production function and applied water distribution (Solomon, 1984). Letey et al. (1984) presented numerically computed production functions for nonuniform irrigation. With high-speed computers, this procedure can be used for any infiltrated water distribution or Y v. AW function.

Warrick (1989) expanded the seasonal model of Letey et al. (1985) to account for spatial variability of seasonally available water. Additionally the results were generalized such that the water amount, irrigation salinity level and yield were expressed in dimensionless forms. This simplified the inputs and also allowed a presentation of results in a single compact table for a wide range of conditions. The information presented by Warrick (1989) is considered to be applicable to many practical scenarios for which a lack of input parameters or need for expediency rules out more comprehensive modeling.

Feinerman et al. (1983) concluded that economically optimal values of AW increase with decreasing uniformity for piece-wise linear production functions but the optimal level of AW is lower for the nonuniform application than for the uniform application if the production function is quadratic. Whereas yields under nonuniform application are lower than yields under uniform application for a given $\overline{AW}$ value, the DP is higher under nonuniform application. The economic analyses of Feinerman et al. (1983) assumed the costs or benefits of

DP were negligible. Assigning a cost to DP alters the conclusions from an economic analysis as will be discussed later.

## APPLICATIONS

The next three sections illustrate the application of the crop-water production function models to salinity-drainage problems. The first of these sections will apply the transient state model to the case of a high water table with neither internal drainage nor installed drainage system. The second section analyzes the consequences of irrigation water management on water lost by evaporation during the winter fallow season in the San Joaquin Valley of California by using the transient state model. Finally, completed studies on optimal irrigation management which utilized the seasonal salinity model and/or the irrigation nonuniformity models will be reviewed.

### High Water Table Situation

Soil salinity is frequently associated with areas having high water tables such as the western San Joaquin Valley of California. If irrigation is less than the amount required to replace the ET losses, water can move from the water table into the root zone in response to the hydraulic gradient and contribute water to the crop. The seasonal model cannot handle this situation; however, the transient state model does.

The transient state model is for one-dimensional flow. The presence of the water table enhances the opportunity for lateral flow particularly beneath the water table. For example, under nonuniform irrigation, water could move up from the water table into the root zone in areas receiving low amounts of water. On the other hand, parts of the field receiving excess water would have a rise in water table. The difference in height of water table associated with the nonuniform irrigation would create a hydraulic gradient causing a lateral flow from high to low positions of the water table. This phenomenon tends to reduce the effects of nonuniform irrigation from a water distribution point of view. In other words, redistribution of water within the water table tends to smooth out the nonuniformities of surface application. Although lateral flow reduces the nonuniformity from a water content point of view, the consequences on the nonuniformity of salinity are intensified. Lateral flow not only transports water but also salt. Thus, if one part of the field is continuously underirrigated because of poor irrigation uniformity, water moving up to the root zone from the water table transports and accumulates salts within the root zone. In this regard, the nonuniformity of salt distribution across the field related to nonuniform irrigation is intensified by the presence of a water table.

The transient state model can be manipulated to partially account for lateral flow. As water moves up from the water table under deficit irrigation conditions, water and salt is programmatically added to the lower part of the profile to reestablish the water table prior to an irrigation event.

Inasmuch as the transient state model equipped to simulate multiyear production under high water table conditions has just recently been formulated, only fragmentary preliminary results can be reported. The following conditions were assumed in the simulations reported here. The water table was assumed to be initially at a depth of 150 cm, an impermeable layer existed at 250 cm depth, and no drainage was allowed. The water within the water table had an EC of 9 dS/m and the soil profile above the water table was initially nonsaline. Nonsaline (0.4 dS/m) irrigation waters were applied to cotton. Scheduling of irrigation was typical of practices in the Western Valley. Precipitation was assumed to be 14.4 cm per year. Under the simulated irrigation management program, a preirrigation of 18 or 33 cm was imposed. Irrigations during the growing season applied either 1.0 or 0.6 of the potential crop ET for the time period between irrigations. The simulation was conducted with no lateral flow or lateral flow imposed as described above. The simulation was conducted for four or six growing seasons depending upon the management practice.

Irrigation which applied 1.0 crop ET resulted in maximum yields throughout the simulation, so detailed results will not be reported. The results from the 0.6 ET irrigations are summarized in table 1, with data presented for the relative total dry matter as well as the relative cotton lint yields. Relatively large decreases in dry matter production can be imposed with relatively low cotton lint losses. The relationship between cotton lint yield ($Y^L$) and total dry matter ($Y^t$) was derived from the data of Davis (1983) which produced the following relationship:

[6] $\quad Y^L = -0.361 + 0.194Y^t - 0.00489(Y^t)^2$

Inseason irrigations equal to 0.6 ET were not adequate to provide full production even though water was supplied from the water table. Assuming no lateral flow, the 33-cm preirrigation (imposed during the second and succeeding years) provided considerably higher yields than the 18 cm preirrigation. The 33-cm preirrigation provided both leaching at the beginning of the season and a higher water supply for the crop. Even though the yields declined gradually with time, quite high cotton lint yields were maintained by this treatment.

Table 1. Cotton production under high water table with various pre-irrigation levels and later flow conditions and in-screen irrigations equal to 0.6 ET.

| Lateral flow | Preirrigation (cm) | Year | | | | | |
|---|---|---|---|---|---|---|---|
| | | 1 | 2 | 3 | 4 | 5 | 6 |
| | | \multicolumn{6}{c}{Relative Dry Matter, %} | | | | | |
| No | 18 | 79 | 65 | 61 | 60 | 59 | 58 |
| No | 33 | 79 | 77 | 76 | 75 | | |
| Yes | 18 | 84 | 82 | 72 | 64 | 61 | 60 |
| Yes | 33 | 84 | 92 | 89 | 87 | | |
| | | \multicolumn{6}{c}{Relative Cotton Lint, %} | | | | | |
| No | 18 | 93 | 83 | 79 | 78 | 77 | 76 |
| No | 33 | 93 | 92 | 91 | 90 | | |
| Yes | 18 | 96 | 94 | 88 | 82 | 80 | 78 |
| Yes | 33 | 96 | 98 | 97 | 97 | | |

The imposition of lateral flow provided an additional water and salt supply for the crop. Considering the 18-cm preirrigation, higher yields were achieved with lateral flow as compared to no lateral flow, particularly during the first few growing seasons. The difference in yield between lateral flow and no lateral flow diminished to small values after a few seasons. This result is the consequence of the buildup of salt from lateral flows that accumulated with time and had a negative effect on yields. With a 33-cm preirrigation, the lateral flow provided higher yields than no lateral flow over successive years because of the additional water for leaching at the beginning of the crop season and increased water supply during the growing season.

The distribution of salt immediately after preirrigation for the cases without lateral flow are depicted in figures 3 and 4 for preirrigation levels of 18 and 33 cm, respectively. The salt concentration is based upon the electrical conductivity of the solution at water content equivalent to saturation. The larger preirrigation caused the salts to be leached to a greater depth than the lower

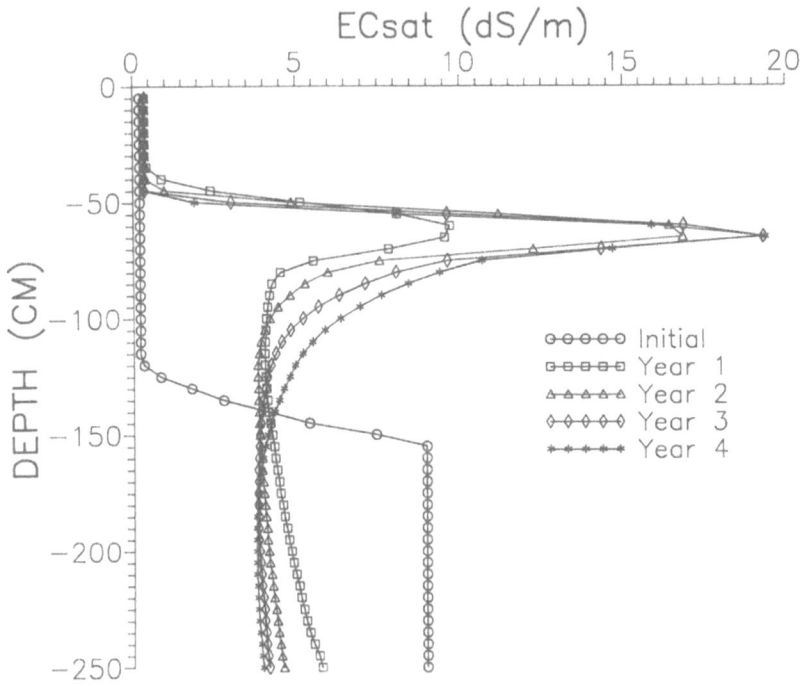

Figure 3. Distribution of soil salinity for various years after an 18-cm preirrigation and no lateral flow.

preirrigation treatment. The peak salt concentration is also lower under the 33-cm preirrigation than for the 18-cm preirrigation. Thus, the higher yields with the higher preirrigation levels are associated both with water being supplied on an annual basis and the salts being leached deeper into the profile with a lower peak concentration during the preirrigation.

These simulated results suggest that cotton production can be maintained for several years in the presence of a saline high water table if a nonsaline irrigation water supply is available and irrigation is properly managed. This information is useful for an economic analysis of drainage and irrigation management options in high water table areas. Additional scenerios must be simulated to provide adequate information on a full range of management options.

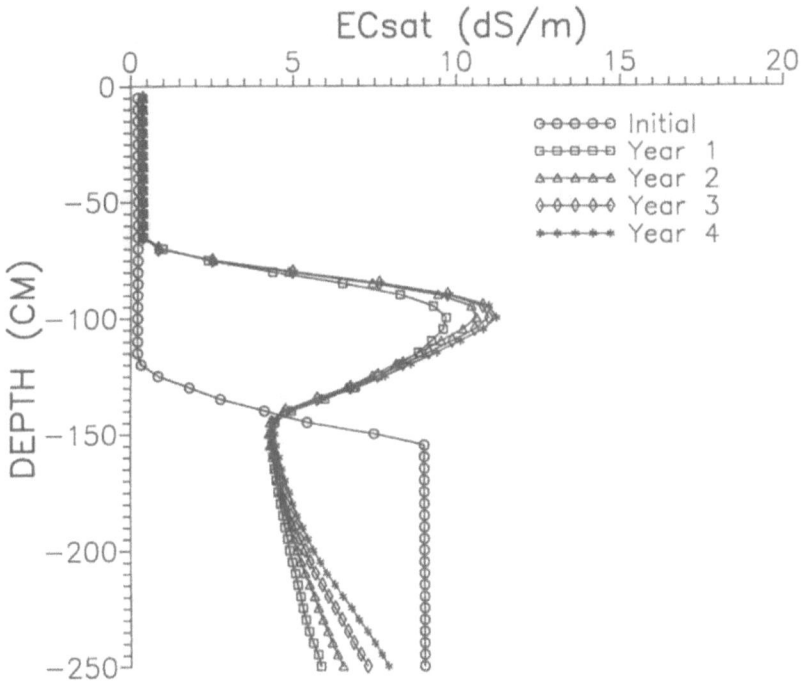

Figure 4. Distribution of soil salinity for various years after a 33-cm preirrigation and no lateral flow.

## Water Management During the Fallow Period

Greatest attention is given to irrigation management during the growing season as it should be. Nevertheless, the dynamics of water and salt flow during the fallow period can also be affected by water management. During the fallow period in California, water in the profile is lost by evaporation and added by precipitation. The evaporation rate is controlled by climatic factors as long as the soil surface is moist. If the soil surface becomes dry, the rate of evaporation is controlled by movement of water through the soil to the soil surface. The hydraulic conductivity of the soil (ease with which water flows through the soil) decreases with decreasing water content. Scheduling of the last irrigation may have significant consequences on the net evaporation from the soil during the fallow period.

The transient state model computes the amount of evaporation during the fallow period. The amount of evaporation during successive years of irrigating cotton with 1.0 and 0.6 ET is depicted in table 2. The evaporation during the fallow period is higher under the 1.0 ET than the 0.6 ET irrigation even though no irrigation is applied during the fallow period in either case. This result is the consequence of having larger water contents in the soil profile at the end of the growing season under the higher water application. The increase in evaporation in successive years under 1.0 ET is the result of relatively high preirrigation which resulted in higher soil-water profile contents in successive years.

Table 2. Amount of evaporation (cm) during successive annual fallow periods for irrigation of 1.0 and 0.6 ET and a pre-irrigation of 33 cm.

|        | Year |      |      |      |
|--------|------|------|------|------|
|        | 1    | 2    | 3    | 4    |
| 1.0 ET | 15.2 | 17.1 | 19.3 | 20.7 |
| 0.6 ET | 12.2 | 12.0 | 12.0 | 12.0 |

Results shown in table 2 identify the significance of preirrigation timing. Until recently, the irrigation water calendar year in western San Joaquin Valley was from January 1 to January 1. Under this arrangement, any water allotment to the farmer that was not used during the growing season was applied before the water year terminated. This practice had two consequences: (1) To increase the water content in the soil during the fallow period resulting in higher evaporation as depicted by the data presented in table 2, and (2) create a moist profile that could cause considerable deep percolation in the event of later rains. Delay of preirrigations until near the cropping season has water saving advantages. A dry soil profile during the fallow period also reduces the transport of salts to the soil surface.

## Management with Subsurface Drainage Systems

Subsurface drainage systems are typically installed when the water table is near the soil surface. The collected drainage effluent must be disposed by some means and if the drainage water is severely degraded by salts and/or potentially toxic elements, disposal of these waters imposes a cost. The cost may be in the form of environmental degradation or a monetary cost to some segment of society for their appropriate disposal. Under these conditions, deep percolation is an important, costly output of the crop production process. Letey and Dinar (1986) used the seasonal model to simulate and report drainage production functions for several crops irrigated with saline waters. The DP increased with increasing salinity of the irrigation water for a given AW value. Significantly, irrigation with saline waters produced DP even under low values of AW. Increasing the value of AW, however, resulted in increasing levels of DP.

Irrigation uniformity influences the DP function as well as yield. The yield and DP functions are depicted in figure 5 for cotton irrigated with nonsaline water under different degrees of irrigation uniformity. The irrigation uniformity is characterized by the Christiansen uniformity coefficient (CUC) where CUC = 100 represents perfectly uniform irrigation and uniformity decreases with decreasing values of CUC. Note that with increasing levels of applied water, both the yield and DP increased. For a given level of water application, decreasing uniformity results in decreasing yields but increasing levels of DP. Thus, there is a cost associated with decreasing uniformity of irrigation in the form of decreased yields and increased DP.

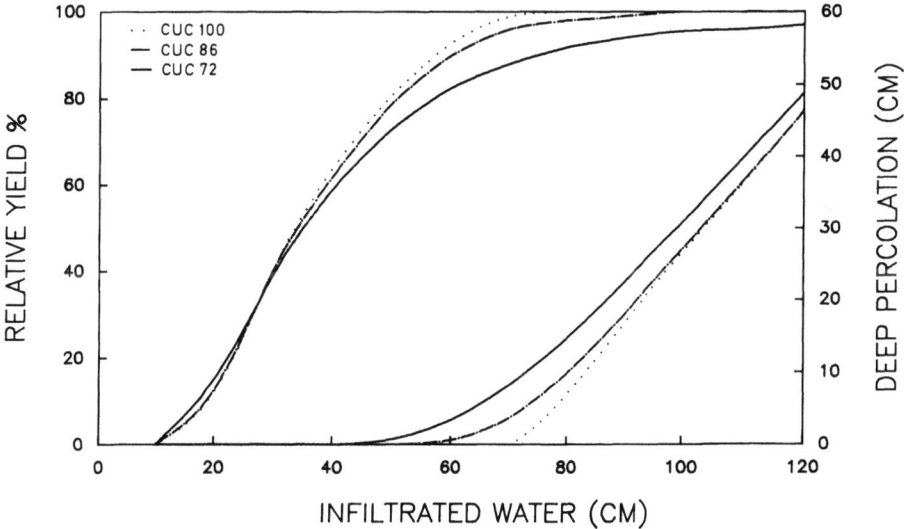

Figure 5. Relationship of relative cotton lint yield and the amount of deep percolation to seasonal amount of infiltrated water for various values of irrigation uniformity as characterized by Christiansen's uniformity coefficient (CUC).

The optimal (profit maximizing) level of irrigation is dependent on the cost for drainage water disposal. Knapp et al. (1986) conducted an economic analysis of onfarm management of agricultural drainage water. They evaluated three drainage conditions: (1) Unlimited natural drainage, (2) drainage water disposal in an onfarm evaporation pond, and (3) access to a free off-farm disposal facility. The optimal water application and associated profits were reduced when disposal was with an evaporation pond as compared to the other conditions and the reductions increased with decreasing irrigation uniformity. Under perfectly uniform irrigation, the optimal applied water and profits did not differ greatly for the various drainage options. Similar results were reported by Dinar et al. (1985) in a field-scale economic analysis of the combined effects of salinity, irrigation uniformity, and drainage requirements for a specific crop.

Higher levels of irrigation uniformity lead to higher profits particularly when drainage costs are imposed. Irrigation systems and/or management leading to higher uniformity usually impose a cost on the farmer. If the yield and drainage functions in crop production are known, the optimal irrigation management can be evaluated if irrigation uniformities can be assigned to an irrigation system. Irrigation uniformity for any irrigation system can be highly variable based upon design, maintenance, and management. Letey et al. (1990) assigned a typical uniformity for each irrigation system. The pressurized irrigation system such as linear move sprinkler and drip were assigned higher uniformity than furrow irrigation. The cost associated with the more uniform irrigation systems were higher than for the less uniform. For the conditions of their study, the increased costs to provide more uniform irrigation were not economically justified when no drainage disposal costs were entailed. However, as the cost for drainage disposal increased, the optimal irrigation system switched from the less uniform to the more uniform systems.

Deep percolation often results in nonpoint water pollution which cannot be quantitatively traced to a given farmer; thus neither can the cost associated with environmental degradation be assigned to the farmer. Under this condition environmental degradation is an externality and the farmer behaves as if the drainage cost was zero. This leads to AW and DP that exceed economically efficient levels. Dinar et al. (1989) evaluated irrigation water pricing policies to reduce and finance subsurface drainage disposal. A direct charge on drainage waters equal to their disposal costs induced economically efficient water application but required a water drainage volume monitoring system. A flat fee on irrigation water could be set to induce an economically efficient water application but, in this case, the revenues generated by the imposition of the extra charge greatly exceeded the cost of disposal and lead to low farmer profits. Tiered water pricing, whereby the unit water price is increased as volume increases, could be used to induce economically efficient applications.

However, the revenues generated by the tiered price was less than the cost for disposal.

A water marketing system, whereby individual farmers could market a portion of the water available for irrigation, was evaluated for its effects on water conservation in agriculture, drainage, and environmental pollution reduction, and improved economic efficiency using a microlevel production model (Dinar and Letey, 1990). The results suggested that under a variety of conditions, such water market enables the farmer to both invest in an improved irrigation technology and pay for safe disposal of drainage produced on his field. Societal benefits included the reduction in environmental pollution and benefits to the urban sector from additional water for its consumption.

## CONCLUSIONS

The seasonal model has advantages of rapidly producing complete crop-water production functions. Furthermore, output from the model can rather easily be combined with irrigation uniformity analysis to develop field level crop-water production functions which account for nonuniformity of irrigation. The model is not applicable to conditions of high water tables where significant amounts of the crop ET is derived from the water table. The transient state model as presently developed has the capacity to simulate crop yields, water distribution, salt distribution, evapotranspiration, and deep percolation on a multiseasonal basis for a wide range of water management practices. Since the model in its most complete form has only recently been developed, an adequate number of simulations which can be compared with experimental data have not been made. These models can simulate the interrelationships between crop production and irrigation-drainage management which are important in the economic analysis of management.

## REFERENCES

Belmans, C.; Wesseling, J. G.; and Feddes, R. A., 1983. Simulation Model of the Water Balance of a Cropped Soil. *SWATRE J. Hydrology,* 63, pp. 271-286.

Bresler, E. and Hanks, R. J. 1969. Numerical Method of Estimating Simultaneous Flow of Water and Salt in Unsaturated Soils, *Soil Science Society of America Journal,* 33, pp. 827-832.

Cardon, G. E. and Letey, J., 1991. Crop Production Simulation Using Models of Water and Solute Flow and Plant Water Uptake, *Soil Science Society of America Journal.* Manuscript Submitted.

Davis, K. D., 1983. Trickle Irrigation of Cotton in California. Proceedings, Western Cotton Production Conference 9-11, Las Cruces, NM., pp. 34-48.

Dinar, A.; Letey, J.; and Knapp, K. C., 1985. Economic Evaluation of Salinity, Drainage and Non-Uniformity of Infiltrated Irrigation Water, *Agricultural Water Management*, 10, pp. 221-233.

Dinar, A.; Knapp, K. C.; and Letey, J., 1989. Irrigation Water Pricing Policies to Reduce and Finance Subsurface Drainage Disposal, *Agricultural Water Management*, 16, pp. 155-171.

Dinar, A. and Letey, J., 1990. Agricultural Water Marketing, Allocative Efficiency and Drainage Reduction, *Journal of Environmental and Economic Management*. In Press.

Feddes, R. A.; Kowalik, E. J.; and Caradny, H., 1978. *Simulation of Field Water Use and Crop Yield*. Halstead Press, John Wiley & Sons, New York, NY., p. 188.

Feinerman, E.; Letey, J.; and Vaux, H. J., 1983. The Economics of Irrigation with Non-Uniform Infiltration, *Water Resources Research*, 19, pp. 1410-1414.

Hoffman, G. J. and van Genuchten, M. Th., 1983. Soil Properties and Efficient Water Use: Water Management for Salinity Control. In: Taylor, H. M. et al. (Eds.), *Limitations to Efficient Water Use and Crop Production*, American Society of Agronomics, Madison, WI., 53711, pp. 73-85.

Knapp, K. C., 1992. Optimal Intertemporal Irrigation Management under Saline Limited Drainage Conditions. This Volume.

Knapp, K.; Dinar, A.; and Letey, J., 1986. On-Farm Management of Agricultural Drainage Water: An Economic Analysis, *Hilgardia*, 54(4), pp. 1-31.

Letey, J.; Vaux, H. J, Jr.; and Feinerman, E., 1984. Optimum Crop Water Application as Affected by Uniformity of Water Infiltration, *Agronomy Journal*, 76, pp. 435-441.

Letey, J.; Dinar, A.; and Knapp, K. C., 1985. Crop-Water Production Function Model for a Saline Irrigation Waters, *Soil Science Society of America Journal*, 49, pp. 1005-1009.

Letey, J. and Dinar, A., 1986. Simulated Crop-Water Production Functions for Several Crops when Irrigated with Saline Waters, *Hilgardia*, 54(1), pp. 1-32.

Letey, J.; Dinar, A.; Woodring, C.; and Oster, J. O., 1990. An Economic Analysis of Irrigation Systems, *Irrigation Science*, 11, pp. 37-43.

Letey, J.; Knapp, K.C.; and Solomon, K., 1990. Crop-Water Production Functions under Saline Conditions. In: Tanji, K. K., (Ed.), *Agricultural Salinity Assessment and Management*, ASCE, New York, NY.

Maas, E. V. and Hoffman, G. J., 1977. Crop-Salt Tolerance - Current Assessment, *ASCE Journal of Irrigation and Drainage Engineering*, 103(IR-2), Proceeding Paper 12993, pp. 115-134.

Nimah, M. N. and Hanks, R. J., 1973. Model for Estimating Soil, Water, Plant and Atmospheric Interrelations: I. Description and Sensitivity, *Soil Science Society of America Journal*, 37, pp. 522-527.

Shalhevet, J.; Vinten, A.; and Meiri, A., 1986. Irrigation Interval as a Factor in Sweet Corn Response to Salinity, *Agronomy Journal*, 78, pp. 539-545.

Solomon, K. H., 1984. Yield-Related Interpretations of Irrigation: Uniformity and Efficiency Measures, *Irrigation Science*, 5, pp. 161-172.

van Genuchten, M. Th. and Hoffman, G. J., 1984. Analysis of Crop-Salt Tolerance Data. In: Shainberg, I. and Shalhevet, J. (Eds.), *Soil Salinity Under Irrigation - Ecological Studies*. Springer-Verlag, New York, NY., pp. 258-271.

Warrick, A. W., 1989. Generalized Results for Crop Yield Model with Saline Waters, *Soil Science Society of America Journal*, 53, pp. 1641-1645.

Warrick, A. W. and Yates, S. R., 1987. Crop Yield as Influenced by Irrigation Uniformity, *Advances in Irrigation*, 4, pp. 169-180.

# 12 EFFECTS OF INPUT QUALITY AND ENVIRONMENTAL CONDITIONS ON SELECTION OF IRRIGATION TECHNOLOGIES

Ariel Dinar, University of California, Riverside and San Joaquin Valley Drainage Program and David Zilberman, University of California, Berkeley

### ABSTRACT

The use of modern irrigation technologies has been proposed as one of several possible solutions to water scarcity, limited drainage, and associated problems in irrigated agriculture. These technologies should be assessed within economic decisionmaking frameworks which could be applied to guide farmers and water districts in irrigation and technology choices, to assist public policymakers in designing policy instruments to increase conservation and reduce drainage and runoff, and to aid developers of irrigation technologies in the design and marketing of new products. An economic model is developed in this chapter, which includes many of the aspects previously developed, and also takes into consideration new aspects such as weather conditions and the dual effects of soil and water quality. The results provide several general insights regarding the impacts of different irrigation technologies and input qualities on productivity and profitability. The results also illustrate differences in outcomes associated with crop selections as affected by weather, input quality, and technology selection.

### INTRODUCTION

Water scarcity and water quality deterioration in many irrigated agricultural areas (Messer, 1982) have caused irrigation and drainage management to be issues of major concern. Moreover, financial limitations and the scarcity of public funds for investment in large scale water and drainage projects make the need for efficient onfarm irrigation and drainage management even more urgent.

A wide range of onfarm solutions can be considered to address irrigation water scarcity and drainage problems. These include improved management practices; installation of subsurface drainage systems; changes in cropping patterns (to more salt-tolerant crops); modification of existing irrigation systems (such as reduced furrow length, increased flow rate, and installation of gated pipes); or adoption of new irrigation technologies such as drip, sprinkler, or LEPA systems.

These potential solutions should be assessed within economic decisionmaking frameworks which could be applied to guide farmers and water districts in their irrigation and technology choices, to assist public policymakers in designing policy instruments to increase conservation and reduce drainage and runoff, and to aid developers of irrigation technologies in the design and marketing of new products.

Two types of approaches have been advanced for normative analysis of irrigation technology choices. The first, which will be referred to as "conceptual models" (such as Caswell and Zilberman, 1985), attempt to qualitatively identify adoption patterns of irrigation and drainage technologies under alternative policies and environmental conditions, as well as suggesting directions and hypotheses for further empirical research. The second type of analysis which will be referred to as "empirical models" involve developing specific algorithms and solutions for irrigation and drainage problems, and incorporating physical relationships determined by soil and water scientists.

The empirical literature on irrigation choices consists on several lines of research: (1) Simulated or estimated production function studies which relate crop yields to irrigation choices under alternative physical conditions,[1] (2) procedures for estimating economically optimal allocations of irrigation water across a growing season using crop-growth simulation models (Boggess and Amerling, 1983 and Musser and Tew, 1984), (3) studies on optimal timing and allocation of water of different qualities (e.g., Yaron et al., 1980 and Feinerman et al., 1983), (4) analysis of optimal choice of irrigation technology and water use under various conditions (e.g., Hornbacker and Mapp 1988; Letey et al., 1990a; and Dinar et al., 1989).

This chapter attempts to generalize, combine, and assess findings of both approaches to normative analysis. A general conceptual framework is introduced for assessing irrigation choices which is followed by an empirical model analyzing physical (land and water quality and weather) and economic (output and water prices and drainage cost) impacts on water use and technology choice by San Joaquin Valley (Valley) growers.

The results of these two analyses will be used to evaluate existing water and drainage management practices, to suggest alternative policies to promote water conservation and drainage reduction, to assess the capabilities and limitations of existing models, and to suggest new avenues for research.

## CONCEPTUAL FRAMEWORK

The main components and results of a general model developed in Dinar and Zilberman (1990) are presented here as a conceptual framework for assessing irrigation choices. The main components of the model are as follows:

1. The production function equation.

Let y = f(e, i, c) where e is effective water, the amount of water not lost due to runoff, deep percolation or evaporation, and used by the crop for growth. Irrigation effectiveness is defined here as the ratio of effective irrigation to applied water. It is assumed that the higher the irrigation effectiveness, the higher the yields, and that this effect diminishes as effectiveness increases. The variable i is a technology index. Each technology i combines managerial effort with physical equipment. The technologies are enumerated according to their costs, and costlier technologies are assumed to have higher irrigation effectiveness. The old technology, say furrow irrigation, is denoted by i = 0. It is assumed that more capital intensive technologies tend to increase yield by making water application more responsive to specific crop and field characteristics and irrigation frequency. The variable c denotes weather. It can be measured in degree days or pan evaporation levels. Higher c represents higher temperature (and wind). It is assumed that higher values of c tend to increase yield but also to increase evaporation.

2. The effective irrigation equation.

Let e = h(a, i, q, c) where variable a denotes water application, and q denotes quality. It can be either a measure of water or soil salinity, or of soil water-retaining capacity. It is assumed that higher quality results in greater irrigation effectiveness, which diminishes as quality improves. It is also reasonable to assume that capital intensive technologies improve irrigation effectiveness; thus, in terms of water use effectiveness, these technologies augment quality. The gain in effectiveness associated with capital intensive technologies is likely to decline with quality and increase with specific weather conditions. Other reasonable assumptions are that effective water use tends to increase with the amount of applied water, and that higher temperature tends to reduce irrigation effectiveness (by increasing evaporation).

3. Drainage function.

Drainage per acre is denoted by z, with the drainage generation function being z = g(a, i, q, c). The term drainage can denote runoff or deep percolation.

It is assumed to decline with quality, and the rate of decline diminishes as quality increases. Obviously, drainage rates increase as more water is applied. Capital intensive technologies are assumed to generate less drainage.

4. Prices of output, water, and drainage.

Prices of output, water, and drainage are given by P, W, and V, respectively. The fixed cost per acre associated with each technology is $k_i$. This fixed cost includes annualized repayment of investment in training and equipment and annual setup and maintenance costs that are independent of the quantity of water applied. The technologies are ordered according to their capital intensity and then it is assumed that $k_{i+1} > k_i$.

## The Farmer's Choice Problem

Assuming profit maximizing behavior, the farmer's choices involve selecting an irrigation technology, i, and water application level, a. The problem of choice can be presented as

$$\max_{a, \delta_i} \sum_{i=0}^{I} \delta_i \{P \cdot f[h(a, i, q, c), i, c] - W \cdot a - V \cdot g(a, i, q, c) - k_i\}$$

where $\delta_i$ is a dichotomous variable that can assume a value 0 if technology i is not selected and 1 when it is selected. For simplicity it is also assumed that only one technology can be used on a given plot.

The farmer's problem involves a joint discrete-continuous choice. For analytic simplicity, let the decision analysis be conducted in two stages: First, optimal water use will be determined for each distinct technology and, second, profits will be compared across technologies. Obviously, the technology with the highest profit will be selected and its water use will prevail. If none of the technologies yield positive profit, the farm will not remain in operation.

## Optimal Water Use Within a Technology

Let $\pi_i$ denote maximum quasi-rent per acre obtained under technology i. It is determined by the choice of per acre applied water $a_i$ with technology i, that is by

[1] $\pi_i = \max_{a_i} \{P \cdot f[h(a, i, q, c), i, c] - W \cdot a - V \cdot g(a, i, q, c) - k_i\}$.

First order conditions for optimality imply that water will be applied at a level where the value of marginal productivity of applied water is equal to the marginal cost of water. The value of marginal productivity of applied water is the product of: (1) Output price, (2) the marginal productivity of effective water, and (3) the marginal effectiveness of applied water. The marginal cost of water is the sum of the market price of water and the marginal cost of drainage associated with water application.

Under reasonable assumptions, Dinar and Zilberman (1990) showed that an increase in quality affects water use in three ways:

- Marginal effectiveness effect--increase in quality increases irrigation effectiveness which results in an increase in applied water.
- Marginal productivity effect--increase in quality tends to reduce the marginal productivity of applied water which results in a reduction in applied water.
- Marginal drainage effect--increase in quality tends to reduce drainage cost which results in an increase in water use.

The overall effect depends on the relative importance of these three effects. Consider the case of a simple multiplicative relationship between effective and applied water (Caswell and Zilberman, 1986). In this case $h(a, i, q, c) = a \cdot h(i, q, c)$.

It is argued by Caswell and Zilberman (1986) that in most situations without drainage considerations, less water is applied as quality increases. The introduction of an additional drainage cost may reverse this tendency; lands of higher quality will use more water.

Considering the impact of climatic change on water use, an increase in average temperature has several distinct impacts on water use:

- Marginal effectiveness effect-- tendency to reduce water application rates as irrigation water efficiency declines with higher temperature.
- Marginal productivity effect of applied water -- tendency to increase water application rates as marginal productivity of effective water increases due to decline in effective water with higher temperature.
- Marginal productivity effect--tendency to increase water application rates as marginal productivity of effective water increases with temperature.
- Marginal drainage effect--the effect of higher temperatures to increase evaporation and reduce drainage.

The overall effect depends on the relative importance of these four effects. Consider again a case with no drainage concerns. The analysis of increased temperature impacts on applied water suggests a resulting increase in water application both due to high productivity associated with the higher temperature and the need to compensate for evaporation loss.

## Technology Choice

Once optimal water application is determined for each technology, it determines the quasi-rent ($\pi_i$) of the technology, and the technology with the highest quasi-rent per acre is selected given that the quasi-rent is neither negative nor less than land rent.

For simplicity, consider the case of two technologies, denoted by i=0 and 1 for old and new technologies, respectively. The new technology is selected when

$$\pi_1 > \pi_0, \pi_1 \geq r \geq 0$$

where r is the rental rate of land.

The quasi rent difference between the two technologies can be written as

$$\Delta\pi = P \cdot \Delta y - W \cdot \Delta a - V \cdot \Delta z - \Delta k$$

where $\Delta y$ is the yield difference, $\Delta a$ is the difference in applied water, $\Delta z$ is the difference in drainage volumes, and $\Delta k$ is the difference in fixed cost. Selection of the new technology is likely to increase yield and reduce applied water rates and drainage volumes. If these impacts overcome the extra cost that the new technology entails, the new technology is selected.

To relate technology choice patterns to quality and temperature, one needs to investigate how the quasi-rent difference changes as a result of variations in these factors.

The quasi-rent of each technology increases with quality, reflecting the values of the increase in effective water and the decline in drainage associated with a marginal increase in quality.

When drainage is not taxed Caswell and Zilberman (1986) identified conditions that lead to a decline in the quasi-rent differential between technologies as quality increases. This outcome is very likely since under most circumstances the improved quality associated with modern technologies tends to decline with quality, thus the differences in water use and yield decline with quality and enhance the decline of the quasi-rent differences.

Introduction of drainage considerations may operate in the same fashion, and increase the decline in the quasi-rent differential associated with increased quality. The reason is that the marginal contribution of quality to drainage declines with quality, but it declines faster using old technology.

With very high quality land there is not likely to be much difference between water use and output under different technologies. Thus, at these qualities the traditional technologies are likely to be more profitable because they require a smaller investment. As $d\Delta\pi/dq < 0$ it is likely that there will be a range of high qualities where the old technology will be applied, but there might be a range of qualities under a certain critical level, for which the new technology is more profitable.

Increase in temperature is likely to increase profitability associated with the modern technology due to water saving and yield effects. Considering the impact of changes in temperature on technology choice, one can use similar reasoning to show that if the new technology has a higher yield effect of temperature that dominates the water use effect, the new technology will be selected.

## DATA AND EMPIRICAL SPECIFICATIONS

Parameters available from several empirical studies were used to develop a simulation model to consider the irrigation technology choices of cotton and tomato growers in the Valley. This simulation incorporates the impacts of most factors considered in the empirical literature including water quality, soil quality, and temperature.

The empirical model is based on modifications to a production function model suggested by Letey et al. (1985) that was modified by Letey and Dinar (1986) to capture water quality effects and weather conditions. It was applied to a 1-ha field under conditions prevailing in the west side of the Valley, assuming constant returns to scale irrigation technologies. The empirical model included water quality and soil quality components in the production and drainage functions, and was very similar to the one used in Dinar et al. (1989a). In addition to the specifications in Dinar et al. (1989), a soil quality variable was introduced representing available soil moisture in the root zone (water retention capacity minus permanent wilting).[2] In the empirical analysis the grower's objective function was maximized with respect to applied water quantities and selected irrigation technologies.

Cotton and tomatoes were selected for the analysis as being major Valley crops using a variety of irrigation technologies. Pan evaporation values of 165 and 200 cm/season for cotton, and 139.6 and 157.1 cm/ season for tomatoes were used based on U.S. Department of Commerce data for 1977-88. Combinations of values for pumping drainage water and disposal costs were used to represent degrees of drainage problems. Several drainage disposal situations were simulated using three values for disposal costs ($3, $8 and $12/ha-cm) and pumping costs equal to $0.08/ha-cm.

The irrigation technologies considered for the analysis were furrow (1/2 mile) representing the old irrigation technology, and linear move sprinklers, Low Energy Precise Application (LEPA), and drip systems as the pressurized modern irrigation technologies. Costs for the irrigation technologies and related management expenses were taken from University of California Committee of Consultants (1988) and University of California Cooperative Extension budgets (1985). It was assumed implicitly that irrigation equipment is designed properly and that irrigation performances are adequate. (For more details, the reader is referred to Dinar and Zilberman, 1990.) Crop yield prices of $1,100; $1,665; and $1,998 per ton of lint for cotton ($0.50, $0.75, and $0.90 per pound, respectively) are based on U.S. Department of Agriculture (USDA) data (1989). Prices for tomato yield of $40, $50, and $60 per ton were taken from California Tomato Growers Association (1986). Maximum potential yields for the western Valley were estimated at 1.6 ton/ha and 85 ton/ha for cotton lint and for tomatoes, respectively (Westlands Water District annual reports for various years). Water prices ranged from $1 to $8/ha-cm. Salt concentration in the irrigation water was 0.7 EC (mmhos/cm) and 4 EC. Available soil moisture values used to represent land quality were .06, .14, and .19 for loamy sand soil, loam soil, and silty clay soil, respectively (Westlands Water District, 1984), with higher values representing higher land quality.

Applied water, as used in this chapter, represents the infiltrated water available for crop production or deep percolation. It was adjusted to the particular soil type using available soil moisture values. Runoff from furrow systems was considered to be recycled and made available for reuse in irrigating the crop. The computations were done assuming no effective precipitation.

In most of the analysis (following the U.C. Committee of Consultants, 1988) the annual cost of the different technologies were calculated using an interest rate of 5 percent.

## RESULTS

The empirical model was applied to a variety of environmental conditions and input and output prices. These were analyzed for their effect on the optimal (profit maximizing) values of applied irrigation water, drainage volumes, irrigation technology selected, yield, and resulting profit.

Results for the performances of the four irrigation technologies under different economic and environmental conditions are presented in table 1 for cotton and in table 2 for tomatoes. The results are based on one case where drainage is not a limiting factor and two cases where drainage disposal costs are moderate and high ($3 and $12/ha-cm). The tables show that:

(1) The range of yield reduction relative to the maximum potential yield is between 1 percent (for some situations with drip irrigation) to 33 percent (for low land quality and high drainage disposal costs in the case of furrow). This reflects both the effects of land quality and irrigation technology.

(2) There are substantial variations in drainage generation, mostly attributed to differences in technology. For example, furrow may produce several hundred times the amount of drainage that drip generates. Differences in water use under similar conditions can be up to 30 percent and differences in yield may be at most 33 percent.

(3) In all cases modern technologies reduce the impact of land quality differences on yield, applied water, and drainage production. For example, irrigated tomatoes with good quality water on low quality land use 15 percent more water, generate about 25 percent more drainage, and produce 5 percent less output than on high quality land; this results in a 40-percent reduction in profits.

(4) In most cases modern technologies (drip and LEPA) use less water and generate slightly more output than furrow irrigation; however, this result may not hold when drainage costs are high. In the case of cotton, high drainage costs may result in increased water use and much higher output under drip and LEPA technologies than under furrow and sprinklers. In the case of tomatoes, high drainage costs substantially reduce differences in water use between technologies, but modern technologies produce up to 33 percent more yield.

(5) As expected, reductions in water quality (up to EC=4) result in higher water use and tend to reduce yield especially with furrow irrigation. These effects are much more pronounced for tomatoes than for cotton. Moreover, with furrow irrigation, the losses due to reduced water quality become much greater as soil quality declines. For example, in the case of high quality land the profit difference between furrow irrigation with high and low quality water was about $50/ha, and on low quality land the difference was about $400/ha. When drainage cost was introduced, this gap increases.

(6) Warmer weather (pan evaporation) resulted in reduced profits in all cases. For cotton, the loss due to higher pan evaporation values across the board was around $80 to $100/ha. In the case of tomatoes, the losses were $20 to $80/ha. Higher temperatures generally resulted in more applied water.

The selected technologies are presented in table 3 for cotton and tomatoes. Results are presented for low and high crop prices; for lowest and highest

Table 1. Technology performances for cotton under various economic and environmental conditions.

| Technology | Low Land Quality | | | | Medium Land Quality | | | | High Land Quality | | | |
|---|---|---|---|---|---|---|---|---|---|---|---|---|
| | $x^a$ | f | a | g | $x$ | f | a | g | $x$ | f | a | g |
| Drainage disposal cost=$0/ha-cm; $s^b$=0.7; W=4.0; P=0.75; c=165 | | | | | | | | | | | | |
| Furrow | 908 | 1,504 | 106 | 28 | 962 | 1,535 | 102 | 22 | 1033 | 1,547 | 96 | 15 |
| Sprinkler | 855 | 1,558 | 93 | 13 | 918 | 1,552 | 87 | 7 | 951 | 1,568 | 86 | 4 |
| LEPA | 934 | 1,568 | 86 | 4 | 934 | 1,568 | 86 | 4 | 934 | 1,568 | 86 | 4 |
| Drip | 933 | 1,597 | 85 | .85 | 933 | 1,597 | 85 | .85 | 933 | 1,597 | 85 | .85 |
| Drainage disposal cost=$12/ha-cm | | | | | | | | | | | | |
| Furrow | 716 | 1,372 | 79 | 9 | 830 | 1,433 | 80 | 7 | 922 | 1,474 | 80 | 5 |
| Sprinkler | 760 | 1,468 | 79 | 5 | 859 | 1,511 | 79 | 3 | 910 | 1,539 | 80 | 2 |
| LEPA | 890 | 1,545 | 81 | 2 | 890 | 1,545 | 81 | 2 | 890 | 1,545 | 81 | 2 |
| Drip | 933 | 1,597 | 85 | .85 | 933 | 1,597 | 85 | .85 | 933 | 1,597 | 85 | .85 |
| Drainage disposal cost=$0/ha-cm; s=4.0, W=4.0, P=0.75; c=165 | | | | | | | | | | | | |
| Furrow | 821 | 1,493 | 119 | 42 | 947 | 1,527 | 106 | 27 | 1022 | 1,545 | 98 | 17 |
| Sprinkler | 852 | 1,539 | 94 | 14 | 917 | 1,552 | 87 | 7 | 950 | 1,568 | 86 | 4 |
| LEPA | 933 | 1,568 | 86 | 4 | 934 | 1,568 | 86 | 4 | 934 | 1,568 | 86 | 2 |
| Drip | 933 | 1,597 | 85 | .88 | 933 | 1,597 | 85 | .88 | 933 | 1,597 | 85 | .88 |
| Drainage disposal cost=$12/ha-cm | | | | | | | | | | | | |
| Furrow | 494 | 1,342 | 87 | 19 | 737 | 1,431 | 85 | 13 | 891 | 1,469 | 81 | 7 |
| Sprinkler | 752 | 1,464 | 79 | 5 | 856 | 1,510 | 79 | 3 | 910 | 1,539 | 80 | 2 |
| LEPA | 890 | 1,545 | 81 | 2 | 890 | 1,545 | 81 | 2 | 890 | 1,545 | 81 | 2 |
| Drip | 933 | 1,597 | 85 | .88 | 933 | 1,597 | 85 | .88 | 933 | 1,597 | 85 | .88 |
| Drainage disposal cost=$0/ha-cm; s=0.7; W=4.0; P=0.75; c=200 | | | | | | | | | | | | |
| Furrow | 821 | 1,488 | 121 | 27 | 895 | 1,516 | 116 | 21 | 950 | 1,539 | 113 | 16 |
| Sprinkler | 758 | 1,523 | 108 | 13 | 826 | 1,548 | 104 | 7 | 860 | 1,564 | 103 | 5 |
| LEPA | 855 | 1,567 | 104 | 5 | 855 | 1,567 | 104 | 5 | 855 | 1,567 | 104 | 5 |
| Drip | 854 | 1,595 | 102 | .81 | 854 | 1,595 | 102 | .81 | 854 | 1,595 | 102 | .81 |
| Drainage disposal cost=$12/ha-cm | | | | | | | | | | | | |
| Furrow | 640 | 1,362 | 93 | 9 | 749 | 1,417 | 94 | 7 | 844 | 1,464 | 95 | 6 |
| Sprinkler | 666 | 1,458 | 94 | 5 | 722 | 1,496 | 93 | 2 | 823 | 1,529 | 95 | 2 |
| LEPA | 812 | 1,529 | 95 | 2 | 812 | 1,529 | 95 | 2 | 812 | 1,529 | 95 | 2 |
| Drip | 854 | 1,595 | 102 | .81 | 854 | 1,595 | 102 | .81 | 854 | 1,595 | 102 | .81 |

$^a$ $\pi$=profit  a=applied water  
 f=cotton yield  g=drainage volume  

$^b$ s=salt concentration in the irrigation water  
 W=Price of irrigation water  

P=Cotton yield (lint) price  
c=Pan evaporation

# SELECTION OF IRRIGATION TECHNOLOGIES

Table 2. Technology performances for tomatoes under various economic and environmental conditions.

| Technology | Low Land Quality | | | | Medium Land Quality | | | | High Land Quality | | | |
|---|---|---|---|---|---|---|---|---|---|---|---|---|
| | $x^a$ | $f$ | $a$ | $g$ | $x$ | $f$ | $a$ | $g$ | $x$ | $f$ | $a$ | $g$ |
| **Drainage disposal cost=$0/ha-cm; $s^b$=0.7; W=4.0; P=50; c=139.6** | | | | | | | | | | | | |
| Furrow | 540 | 78.22 | 171 | 71 | 689 | 79.79 | 152 | 51 | 785 | 82.25 | 149 | 47 |
| Sprinkler | 525 | 79.01 | 131 | 31 | 645 | 80.43 | 119 | 18 | 701 | 81.59 | 116 | 14 |
| LEPA | 773 | 83.96 | 130 | 27 | 773 | 83.96 | 130 | 27 | 773 | 83.96 | 130 | 27 |
| Drip | 733 | 84.87 | 104 | .92 | 733 | 84.87 | 104 | .92 | 733 | 84.87 | 104 | .92 |
| **Drainage disposal cost=$12/ha-cm** | | | | | | | | | | | | |
| Furrow | 56 | 63.95 | 116 | 25 | 262 | 69.83 | 118 | 23 | 459 | 73.90 | 116 | 19 |
| Sprinkler | 229 | 74.00 | 116 | 19 | 431 | 80.43 | 119 | 18 | 541 | 76.78 | 105 | 6 |
| LEPA | 607 | 81.24 | 115 | 13 | 607 | 81.24 | 115 | 13 | 607 | 81.25 | 115 | 13 |
| Drip | 733 | 84.87 | 104 | .92 | 733 | 84.87 | 104 | .92 | 733 | 84.87 | 104 | .92 |
| **Drainage disposal cost=$0/ha-cm; W=4.0; P=50; c=139.6** | | | | | | | | | | | | |
| Furrow | 143 | 71.14 | 203 | 107 | 559 | 79.14 | 173 | 72 | 735 | 80.55 | 148 | 47 |
| Sprinkler | 505 | 79.34 | 136 | 35 | 640 | 80.46 | 120 | 19 | 700 | 81.57 | 116 | 14 |
| LEPA | 773 | 83.96 | 130 | 27 | 773 | 83.95 | 130 | 27 | 773 | 83.95 | 130 | 27 |
| Drip | 733 | 84.86 | 104 | .96 | 733 | 84.86 | 104 | .96 | 733 | 84.87 | 104 | .96 |
| **Drainage disposal cost=$12/ha-cm** | | | | | | | | | | | | |
| Furrow | -619 | 50.41 | 124 | 41 | -23 | 67.87 | 130 | 36 | 323 | 71.36 | 118 | 22 |
| Sprinkler | 173 | 74.41 | 120 | 22 | 420 | 80.14 | 119 | 18 | 539 | 76.74 | 105 | 6 |
| LEPA | 607 | 81.24 | 115 | 13 | 606 | 81.24 | 115 | 13 | 607 | 81.24 | 115 | 13 |
| Drip | 732 | 84.86 | 104 | .96 | 732 | 84.86 | 104 | .96 | 733 | 84.87 | 104 | .96 |
| **Drainage disposal cost=$0/ha-cm; s=0.7; W=4.0; P=50; c=157.1** | | | | | | | | | | | | |
| Furrow | 485 | 75.64 | 168 | 57 | 619 | 79.55 | 167 | 54 | 706 | 82.21 | 167 | 52 |
| Sprinkler | 440 | 78.95 | 147 | 34 | 568 | 80.44 | 134 | 20 | 627 | 81.48 | 130 | 15 |
| LEPA | 707 | 81.66 | 131 | 16 | 707 | 81.66 | 131 | 16 | 707 | 81.66 | 131 | 16 |
| Drip | 675 | 85.00 | 118 | .91 | 675 | 85.00 | 118 | .91 | 675 | 85.00 | 118 | .91 |
| **Drainage disposal cost=$12/ha-cm** | | | | | | | | | | | | |
| Furrow | 37 | 65.68 | 130 | 26 | 196 | 69.43 | 130 | 23 | 376 | 73.83 | 130 | 20 |
| Sprinkler | 130 | 73.83 | 130 | 20 | 330 | 73.39 | 118 | 9 | 466 | 76.72 | 118 | 6 |
| LEPA | 538 | 76.72 | 118 | 6 | 538 | 76.72 | 118 | 6 | 538 | 76.72 | 118 | 6 |
| Drip | 674 | 84.81 | 117 | .90 | 674 | 84.81 | 117 | .90 | 674 | 84.81 | 117 | .90 |

$^a$ $\pi$=profit    a=applied water    $^b$ s=salt concentration in the irrigation water    P=Cotton yield (lint) price  
f=cotton yield    g=drainage volume    W=Price of irrigation water    c=Pan evaporation

Table 3. Optimal irrigation technologies under various environmental conditions.

| Crop price | Water price | Drainage disposal cost | s=0.7 Land quality | | | | | | s=4 Land quality | | | | | |
|---|---|---|---|---|---|---|---|---|---|---|---|---|---|---|
| | | | H[a] | M | L | H | M | L | H | M | L | H | M | L |

Cotton

| ($/lb) | ($/ha-cm) | ($/ha-cm) | Pan evaporation (cm) | | | | | | Pan evaporation (cm) | | | | | |
|---|---|---|---|---|---|---|---|---|---|---|---|---|---|---|
| | | | -----165----- | | | -----200----- | | | -----165----- | | | -----200----- | | |
| .50 | 1 | 0  | F[b] | F | F | F | F | F | F | F | F | F | F | F |
| .50 | 1 | 3  | F | F | F | F | F | S | F | F | S | F | F | L |
| .50 | 1 | 12 | F | D | D | D | D | D | F | D | D | F | D | D |
| .50 | 4 | 0  | F | F | F | F | F | F | F | F | L | F | F | F |
| .50 | 4 | 3  | F | F | S | F | F | - | F | F | L | F | F | - |
| .50 | 4 | 12 | F | D | D | F | - | - | F | D | D | F | - | - |
| .90 | 1 | 0  | F | F | F | F | F | F | F | F | F | F | F | F |
| .90 | 1 | 3  | F | F | D | F | D | D | F | D | D | F | D | D |
| .90 | 1 | 12 | D | D | D | D | D | D | D | D | D | D | D | D |
| .90 | 8 | 0  | F | D | D | F | D | D | F | D | D | F | D | D |
| .90 | 8 | 3  | F | D | D | F | D | D | D | D | D | F | D | D |
| .90 | 8 | 12 | D | D | D | D | D | D | D | D | D | D | D | D |

Tomatoes

| ($/ton) | ($/ha-cm) | ($/ha-cm) | Pan evaporation (cm) | | | | | | Pan evaporation (cm) | | | | | |
|---|---|---|---|---|---|---|---|---|---|---|---|---|---|---|
| | | | ----139.6---- | | | ----157.1---- | | | ----139.6---- | | | ----157.1---- | | |
| 40 | 1 | 0  | F[b] | F | L | F | F | L | F | L | L | F | F | L |
| 40 | 1 | 3  | F | L | L | F | L | L | L | L | L | F | F | L |
| 40 | 1 | 12 | D | D | D | D | D | D | D | D | D | D | D | D |
| 40 | 4 | 0  | - | - | - | - | - | - | - | - | - | - | - | - |
| 40 | 4 | 3  | - | - | - | - | - | - | - | - | - | - | - | - |
| 40 | 4 | 12 | - | - | - | - | - | - | - | - | - | - | - | - |
| 60 | 1 | 0  | F | F | L | F | F | L | F | L | L | F | F | L |
| 60 | 1 | 3  | L | L | L | L | L | L | L | L | L | L | L | L |
| 60 | 1 | 12 | D | D | D | D | D | D | D | D | D | D | D | D |
| 60 | 8 | 0  | D | D | D | D | D | D | D | D | D | D | D | D |
| 60 | 8 | 3  | D | D | D | D | D | D | D | D | D | D | D | D |
| 60 | 8 | 12 | D | D | D | D | D | D | D | D | D | D | D | D |

s = Salt concentration in the irrigation water
[a] Land quality: L=Low quality  M=Medium quality  H=High quality
[b] F=Furrow  S=Sprinklers  L=LEPA  D=Drip

irrigation water prices, and for a range of drainage disposal costs ($0, $3, and $12/ha-cm). In cases with negative profits, no technology was selected.

Under conditions with low water prices and no drainage cost, furrow was the preferable technology. This was always the case for cotton. For tomatoes, furrow was preferred for conditions of high land quality, but LEPA was preferred for low land quality. For medium land quality, furrow was preferred when water quality was high and LEPA, when water quality was low. A combination of high water and drainage prices could result in abandoning farming when output prices were low, especially in the case of tomatoes, or adoption of drip irrigation. Even when water prices were low, high drainage costs may lead to adoption of drip, especially on low land quality. Low output prices and moderate drainage taxation may lead to adoption of LEPA for the case of tomatoes.

The impact of soil quality on technology choice is stronger than pan evaporation and water quality changes. Still, there is enough evidence that reductions in water quality and increases in pan evaporation tend to enhance adoption of modern technologies.

Profits associated with the optimal solutions are presented in tables 4 and 5 for cotton and tomatoes, respectively. Profits tend to increase with land and water quality, and decline with pan evaporation values. The rate of decline in profits as environmental conditions become more limiting, is dependent on the technology selected.

As assumed in the conceptual model, the farmer faces the problem of selecting a technology type and quantity of applied irrigation water for a given crop. However, the grower may chose simultaneously between crops. A comparison between cotton and tomatoes can provide additional insight into the selection problem.

To make that comparison one can use the current market prices for cotton and tomatoes ($0.75/lb and $50/ton, respectively) as the base case. At that price level and below, cotton is preferred to tomatoes for all combinations of environmental and cost values. Cotton will be selected by the farmer with the associated optimal decisions on water application rates and the optimal technologies. However, with a 20-percent increase in market prices for cotton ($0.90/lb) and tomatoes ($60/ton), tomatoes are preferred in most cases. Only when the water price is $8/ha-cm, drainage disposal cost is $0/ha-cm, and land quality is highest, would cotton be preferred to tomatoes or so equal in return to land and management that the grower may be indifferent.

The analysis so far is applicable for cases without irrigation equipment. For farms with well-functioning traditional irrigation technology,[3] the irrigation capital cost under the traditional technology can be considered as zero. Results for cotton and tomatoes (not presented) suggest that the range of conditions

Table 4. Returns to land and management in the optimal solution for the case of cotton.

| Cotton price ($/Lb) | Water price ($/ha-cm) | Drainage disposal cost ($/ha-cm) | s=0.7 Land quality | | | | | | s=4 Land quality | | | | | |
|---|---|---|---|---|---|---|---|---|---|---|---|---|---|---|
| | | | Pan evaporation (cm) | | | | | | Pan evaporation (cm) | | | | | |
| | | | 165 | | | 200 | | | 165 | | | 200 | | |
| | | | H[a] | M | L | H | M | L | H | M | L | H | M | L |
| .50 | 1 | 0 | 486 | 460 | 420 | 458 | 430 | 387 | 480 | 443 | 376 | 458 | 430 | 387 |
| .50 | 1 | 3 | 433 | 389 | 433 | 404 | 355 | 299 | 422 | 354 | 326 | 404 | 355 | 299 |
| .50 | 1 | 12 | 361 | 321 | 321 | 333 | 295 | 295 | 335 | 321 | 321 | 333 | 295 | 295 |
| .50 | 4 | 0 | 198 | 156 | 99 | 123 | 78 | 23 | 187 | 124 | 81 | 123 | 78 | 23 |
| .50 | 4 | 3 | 171 | 119 | 71 | 98 | 44 | - | 156 | 73 | 71 | 98 | 44 | - |
| .50 | 4 | 12 | 127 | 65 | 65 | 58 | - | - | 97 | 65 | 65 | 58 | - | - |
| .75 | 1 | 0 | 1342 | 1314 | 1265 | 1312 | 1280 | 1228 | 1336 | 1296 | 1217 | 1312 | 1280 | 1228 |
| .75 | 1 | 3 | 1274 | 1220 | 1191 | 1241 | 1180 | 1165 | 1263 | 1191 | 1191 | 1241 | 1180 | 1165 |
| .75 | 1 | 12 | 1191 | 1191 | 1191 | 1165 | 1165 | 1165 | 1191 | 1191 | 1191 | 1165 | 1165 | 1165 |
| .75 | 4 | 0 | 674 | 608 | 599 | 532 | 465 | 456 | 656 | 599 | 599 | 532 | 465 | 456 |
| .75 | 4 | 3 | 652 | 599 | 599 | 514 | 456 | 456 | 629 | 599 | 599 | 514 | 456 | 456 |
| .75 | 4 | 12 | 609 | 599 | 599 | 476 | 456 | 456 | 599 | 599 | 599 | 476 | 456 | 456 |
| .90 | 1 | 0 | 1857 | 1829 | 1776 | 1826 | 1794 | 1737 | 1851 | 1810 | 1726 | 1826 | 1794 | 1737 |
| .90 | 1 | 3 | 1781 | 1725 | 1712 | 1745 | 1687 | 1687 | 1769 | 1712 | 1712 | 1745 | 1687 | 1687 |
| .90 | 1 | 12 | 1712 | 1712 | 1712 | 1687 | 1687 | 1687 | 1712 | 1712 | 1712 | 1687 | 1687 | 1687 |
| .90 | 8 | 0 | 1166 | 1117 | 1117 | 1019 | 971 | 971 | 1148 | 1117 | 1117 | 1019 | 971 | 971 |
| .90 | 8 | 3 | 1139 | 1117 | 1117 | 993 | 971 | 971 | 1117 | 1117 | 1117 | 993 | 971 | 971 |
| .90 | 8 | 12 | 1117 | 1117 | 1117 | 971 | 971 | 971 | 1117 | 1117 | 1117 | 971 | 971 | 971 |

s=Salt concentration in the irrigation water
[a]Land quality: L=Low quality  M=Medium quality  H=High quality

**Table 5.** Returns to land and management in the optimal solution for the case of tomatoes.

| Tomato price ($/Ton) | Water price ($/ha-cm) | Drainage disposal cost ($/ha-cm) | s=0.7 Land quality | | | | | | s=4 Land quality | | | | | |
|---|---|---|---|---|---|---|---|---|---|---|---|---|---|---|
| | | | Pan evaporation (cm) 139.6 | | | Pan evaporation (cm) 157.1 | | | Pan evaporation (cm) 139.6 | | | Pan evaporation (cm) 157.1 | | |
| | | | H[a] | M | L | H | M | L | H | M | L | H | M | L |
| 40 | 1 | 0  | 410 | 368 | 323 | 385 | 342 | 298 | 386 | 323 | 323 | 385 | 342 | 298 |
| 40 | 1 | 3  | 280 | 261 | 261 | 249 | 235 | 235 | 261 | 261 | 261 | 249 | 235 | 235 |
| 40 | 1 | 12 | 196 | 196 | 196 | 177 | 177 | 177 | 196 | 196 | 196 | 177 | 177 | 177 |
| 40 | 4 | 0  | -   | -   | -   | -   | -   | -   | -   | -   | -   | -   | -   | -   |
| 40 | 4 | 3  | -   | -   | -   | -   | -   | -   | -   | -   | -   | -   | -   | -   |
| 40 | 4 | 12 | -   | -   | -   | -   | -   | -   | -   | -   | -   | -   | -   | -   |
| 50 | 1 | 0  | 1245 | 1199 | 1163 | 1214 | 1172 | 1138 | 1218 | 1163 | 1163 | 1214 | 1172 | 1138 |
| 50 | 1 | 3  | 1089 | 1081 | 1081 | 1051 | 1051 | 1051 | 1081 | 1081 | 1081 | 1051 | 1051 | 1051 |
| 50 | 1 | 12 | 1045 | 1045 | 1045 | 1025 | 1025 | 1025 | 1045 | 1045 | 1044 | 1025 | 1025 | 1025 |
| 50 | 8 | 0  | 317 | 317 | 317 | 206 | 206 | 206 | 317 | 317 | 317 | 206 | 206 | 206 |
| 50 | 8 | 3  | 317 | 317 | 317 | 206 | 206 | 206 | 317 | 317 | 316 | 206 | 206 | 206 |
| 50 | 8 | 12 | 317 | 317 | 317 | 206 | 206 | 206 | 317 | 317 | 316 | 206 | 206 | 206 |
| 60 | 1 | 0  | 2088 | 2036 | 2003 | 2057 | 2002 | 1978 | 2053 | 2003 | 2003 | 2057 | 2002 | 1978 |
| 60 | 1 | 3  | 1921 | 1921 | 1921 | 1885 | 1885 | 1885 | 1921 | 1921 | 1921 | 1885 | 1885 | 1885 |
| 60 | 1 | 12 | 1893 | 1893 | 1893 | 1873 | 1873 | 1873 | 1893 | 1893 | 1893 | 1873 | 1873 | 1873 |
| 60 | 8 | 0  | 1166 | 1166 | 1165 | 1054 | 1054 | 1054 | 1165 | 1165 | 1165 | 1054 | 1054 | 1054 |
| 60 | 8 | 3  | 1165 | 1165 | 1165 | 1054 | 1054 | 1054 | 1165 | 1165 | 1165 | 1054 | 1054 | 1054 |
| 60 | 8 | 12 | 1165 | 1165 | 1165 | 1054 | 1054 | 1054 | 1165 | 1165 | 1165 | 1054 | 1054 | 1054 |

s=Salt concentration in the irrigation water
[a]Land quality: L=Low quality  M=Medium quality  H=High quality

under which the traditional technologies are preferred increase substantially for a farm with a well-functioning furrow irrigation system.

For example, a comparison of the results (not presented) suggests that in cases with good water quality ($s = .7$ EC), when water price is \$4/ha-cm and with no drainage disposal cost, it is optimal to continue to use furrow irrigation in cotton production when a well-functioning furrow system exists, while it is optimal to use LEPA irrigation on new cotton operations. Comparing the results for tomatoes (not presented) shows fewer differences (than in the case of cotton) in optimal choices between production units with well-functioning furrow systems and new operations.

Results where the annual capital cost of the irrigation technology was calculated using the nominal interest rate of 10 percent (not presented) suggest a reduced range of conditions under which modern technologies are preferable than when a 5-percent interest rate is assumed.

## CONCLUSIONS

The results from the empirical model provide several general insights. In particular:

(1) Modern irrigation technologies serve to substantially reduce impacts of quality and weather differences on profitability. Drip, LEPA, and sprinkler irrigation technologies serve as "land quality augmenting" and "weather improving" since their impact on profitability tends to become stronger at lower qualities. Thus, the introduction of these technologies is likely to serve the reduction of some locational premiums, especially when water price and drainage cost are high.

(2) The impacts of changes in one input quality indicator are not independent of the level of other input-quality indicators. In particular, it seems that, for a given technology, an increase in land quality is likely to have a stronger impact on the optimal profits if water quality is lower.

(3) Since lower water qualities tend to amplify the losses associated with lowering the quality of land using furrow irrigation, the attractiveness of modern irrigation technologies for lower quality land tends to increase substantially when the water quality is poor. Thus, locations which have both low water and soil qualities are likely to be among the first to adopt modern irrigation technologies.

(4) Profitability of tomatoes is more responsive to changes in most parameters and tends to vary more than the profitability of cotton. Tomato profits under furrow irrigation are especially vulnerable to environmental and economic conditions; therefore, the set of circumstances under which adoption of

a modern irrigation technology is the preferred strategy is greater for tomatoes than for cotton. Reductions in water and land qualities strongly reduce profits for tomatoes irrigated with furrow.

(5) Categorization of technologies as "input saving" or "yield increasing" may be misleading. Changes in circumstances may lead to change in the nature of the gains associated with the use of a new technology. When drainage is free of cost, the main impact of drip (relative to furrow) is "water saving"; but it also has a relatively small "yield increasing" effect. With high drainage costs ($12/ha-cm), drip has strong "yield increasing" effects, but it is also "water use increasing." Thus, the introduction of drip (and other modern technologies) may not always guarantee water conservation.

(6) At current low water prices (between $1 to $3/ha-cm in much of the Valley) and without drainage fees, the use of furrow irrigation in both cotton and tomato production is consistent with profit maximization. Furrow irrigation results in substantial generation of drainage and water use.

(7) Taxation on drainage at about $12/ha-cm ($144/acre-foot) would cause drip irrigation to become, almost always, the most profitable technology, and the adoption of drip would nearly eliminate the generation of drainage. Higher tax levels on drainage can reduce it significantly by reducing water use with furrow irrigation and inducing adoption of modern irrigation technologies on lower and medium quality lands.

Profitability of production is likely to be strongly affected by a high drainage tax. When output price is low and water price is high, the tax may cause both cotton and tomatoes to be profitable. The impact of a tax on tomato profits is higher because of the higher volume of drainage generated by tomato production; thus, it may include a transition from tomato to cotton production.

## DISCUSSION

The analysis has shown that without intervention, given the low cost of water and absence of taxes for drainage, there is no reason why farmers would switch from traditional irrigation. This section discusses some of the practical issues associated with selection of policy intervention to induce conservation, and some of the problems associated with existing normative models for policy analysis.

The use of taxation to induce drainage reduction may make it politically unfeasible (Buchanan and Tullock, 1975). Thus, one may consider alternative policies to force transition from furrow to modern technologies, such as drip irrigation. One possible policy is a substantial subsidy for the adoption of drip systems. A subsidy of $250/ha for the use of drip may achieve the desired outcome, but will require very high public expenditures. A tax-subsidy scheme, where adopters of modern technology receive a subsidy (say, $100/ha annually for adoption of drip) that is combined with a drainage fee of, say, $4/ha-cm, would not be as costly to the Government, may lead to a desired reduction in drainage generation, and may be palatable to farmers.

In spite of the powerful effect of drainage fees, it is unrealistic to expect actual taxation on drainage because of its nonpoint pollution nature and monitoring difficulties. Instead, one can introduce an imputed tax on drainage added to the price of water resulting in differential water pricing which depends on crop, water quality, land quality, and irrigation technology. If, for example, the cost of drainage were $3/ha-cm, a furrow-using tomato grower with high quality land should have extra water costs of $0.75/ha-cm (12·5/80 using table 1). There should not be an added cost on drip. Obviously, this procedure is complex and may require use of a simple formula based perhaps on technology only.

Throughout this analysis, the authors encountered limitations associated with the use of normative models for prediction of irrigation technology choices. The models suggest the use of drip and other modern irrigation technologies in many situations where they are not currently being used. Some of the reasons for the disagreement between prescribed and observed behavior may include:

- Unaccounted costs. There are high training and adjustment costs associated with the transition to a new irrigation technology both at the farm and water district levels. For example, drip requires continuous delivery of water and may need more extensive supervision than furrow. Quantifying these costs is not simple, but they are a substantial deterrent to adoption.
- Reliability. The assumptions on technology performance ignored some problems of clogging and other difficulties that are more likely to occur with drip and other new technologies. These should be part of the adjustment costs that have not been considered before.
- Uncertainty. The farmer generally has many years of experience with furrow rather than with LEPA or drip systems. He is uncertain about the applicability of the model results and how they may apply to his acreage with its unique characteristics. Moreover these results are based on estimates which are subject to much randomness themselves. These types of "scientific" uncertainties in addition to the usual ones (such as price and yield) may make risk-averse farmers wary of adopting a technology even though some scientists' calculations suggest that it is in his best interest. This does not mean that farmers do not pay attention to profits or that profits calculated using simulation models cannot predict adoption. Caswell and Zilberman among others, have shown that increases in

simulated profit differences increase the likelihood of adoption. Thus, normative models should be combined with adoption models to predict adoption.
- Differences between existing and new operations. Most normative models constructed to select optimal technology implicitly assume a capital investment in technologies. This is not the case for many existing operations with well-functioning furrow systems. Thus, analysis of technology choices with normative models should be performed under several assumptions regarding initial irrigation technology and cost of its modification. Such analysis would allow divergence in predicting between optimal irrigation technologies for new and existing operations, and the impact of alternative policies on these choices.
- Discounting and credit considerations. The discount rates used to derive annual irrigation capital costs have a significant impact on technology choice. While applied normative models are likely to use the real discount rate, assuming that relative prices do not change over the relevant planning horizon, farmers may behave according to different discounting criteria. It is quite plausible that the discount rate used by farmers in their decisionmaking analysis (for investment) is higher than the real interest rate and may be a major influence in reducing the adoption rates of capital intensive irrigation technologies (drip and LEPA).

The last issue that should be brought up is the relation between conceptual and empirical models. Conceptual models serve to present hypotheses for positive models and to suggest developmental direction for the empirical models. Conceptual models argue that modern irrigation technologies are yield increasing and are more likely to be practiced on low quality land and with low quality water, without providing any detailed data. In essence, they save policymakers and research and marketing managers time and resources. They introduce considerations that are not quantifiable in empirical models which explain gaps between normative models and reality.

Finally, conceptual models suggest the direction of improvement in empirical models. For example, the empirical model used here is not capable of significantly capturing: (1) Yield effect of drip irrigation and (2) weather effect on yield. The second deficiency may produce even quantitatively false results because in many cases the yield effects of weather conditions overcome its water cost increasing effects. Yield effects are considered to play a major role in the introduction of drip irrigation in California. Thus, the effect has not been captured by empirical models. It is only natural that our ability to "exactly" model natural procedures will lag behind our ability to conceptualize and identify elements of a complex system. Therefore, the empirical models present a "partial" subsystem in detail, and the conceptual models identify the missing pieces.

## NOTES

[1] Hexem and Hedey (1978) and Vaux and Pruitt (1983) reviewed an extensive body of literature on water crop production functions mostly under nonsaline conditions. The survey by Letey et al. (1990) reviewed also production function studies under saline and limited drainage conditions.

[2] Depth of water remaining in a soil at a uniform soil moisture tension of about -15 bars of atmospheric pressure, which is the approximate tension at which plants irreversibly wilt due to moisture stress. It is expressed as a depth of water in inches per 1 foot of soil depth (Doorenbos and Pruitt, 1984).

[3] According to Dinar and Campbell (1990), 49 percent of the San Joaquin Valley Drainage Program study area is presently irrigated with furrow and 28 percent with border irrigation.

## REFERENCES

Boggess, W. G. and Amerling, C. B., 1983. A Bioeconomic Simulation Analysis of Irrigation Investments, *Southern Journal of Agricultural Economics*, 15(2), pp. 85-92.

Buchanan, J. M. and Tullock, G., 1975. Polluters' Profits and Political Response: Direct Controls Versus Taxes, *American Economic Review*, 65(1), pp. 139-147.

California Tomato Growers Association, Inc., 1986. *Negotiated Prices*.

Caswell, M. and Zilberman, D., 1985. The Choices of Irrigation Technologies in California, *American Journal of Agricultural Economics*, 67(2), pp. 224-234.

Caswell, M. and Zilberman, D., 1986. The Effects of Well Depth and Land Quality on the Choice of Irrigation Technology, *American Journal of Agricultural Economics*, 68, pp. 798-811.

Dinar, A.; Knapp, K. C.; and Letey, J., 1989. Irrigation Water Pricing to Reduce and Finance Subsurface Drainage Disposal, *Agricultural Water Management*, 16, pp. 155-171.

Dinar, A. and Campbell, M. B., 1990. *Adoption of Improved Irrigation and Drainage Reduction Technologies in the West Side of the San Joaquin Valley, Part I: Literature Review, Survey Methods and Descriptive Statistics*. San Joaquin Valley Drainage Program.

Dinar, A. and Zilberman, D., 1990. *The Economics of Irrigation and Drainage (Reduction) Technology Choice: The Role of Input Quality and Environmental Conditions.* San Joaquin Valley Drainage Program.

Doorenbos, J. and Pruitt, W., 1984. *Crop Water Requirements.* Irrigation and Drainage Paper 24, Food and Agriculture Organization of the United Nations.

Feinerman, E.; Letey, J.; and Vaux, H. J. Jr., 1983. The Economics of Irrigation With Nonuniform Infiltration, *Water Resources Research,* 19(6), pp. 1410-1414.

Hexem, R. and Heady, E. O., 1978. *Water Production Functions and Irrigated Agriculture.* Iowa State University Press, Ames.

Hornbaker, R. H. and Mapp, H. P., 1988. A Dynamic Analysis of Water Savings from Advanced Irrigation Technology, *Western Journal of Agricultural Economics,* 13(2), pp. 307-315.

Letey, J. and Dinar, A., 1986. Simulated Crop-Water Production Functions for Several Crops When Irrigated with Saline Waters, *Hilgardia,* 54(1), pp. 1-32.

Letey J.; Dinar, A.; and Knapp, K. C., 1985. Crop-Water Production Function Model for Saline Irrigation Waters, *Soil Science Society of America Journal,* 49, pp. 1005-09

Letey, J.; Dinar, A.; Woodring, C.; and Oster, J., 1990a. An Economic Analysis of Irrigation Systems, *Irrigation Science,* 11, pp. 37-43

Letey, J.; Knapp, K.; and Solomon, K., 1990b. Crop-Water Production Functions. In: Tanji, K. K. (Ed.), *Agricultural Salinity Assessment and Management.* ASCE, New York, NY. Forthcoming.

Messer, J., 1982. International Development and Trends in Water Reuse. In: Middlebrooks, E. J. (Ed.), *Water Reuse,* Ann Arbor Science Publishers, Ann Harbor, MI.

Musser, V. N. and Tew, B. V., 1984. Use of Biophysical Simulators in Production Economics, *Southern Journal of Agricultural Economics,* 16(1), pp. 74-86.

University of California Committee of Consultants, 1988. *Drainage Water Reduction, Associated Costs of Drainage Water Reduction,* No. 2. University of California Salinity/Drainage Taskforce and Water Resources Center.

University of California, 1985. *Various Crop Budgets for Fresno County.* UC Cooperative Extension, Davis, CA.

U.S. Department of Commerce, N.O.A.A. Climatological Data, California, Various Years.

U.S. Department of Agriculture, 1989. *Cotton and Wool, Situation and Outlook Yearbook,* Economic Research Service, CWS-57, p. 47.

Vaux, H. J. Jr. and Pruitt, W. O., 1983. Crop-Water Production Functions, *Advances in Irrigation,* 2, pp. 61-97.

Westland Water District, 1984. *Water Conservation and Management Handbook.*

Yaron, D.; Bresler, E.; Bielorai, H.; and Harpenist, B., 1980. A Model for Optimum Irrigation Scheduling with Saline Water, *Water Resources Research,* 16(2), pp. 251-67.

# 13 ESTIMATION OF PRODUCTION SYSTEMS WITH EMPHASIS ON WATER PRODUCTIVITY

Richard E. Just, University of Maryland, College Park

### ABSTRACT

This chapter discusses methodologies and data requirements for estimation of key technical and behavioral parameters related to water productivity, including available empirical results. The discussion emphasizes the use of microlevel production systems in assessing aggregate substitution possibilities between water-related externalities and production and the associated need for data with microlevel distributional detail.

## INTRODUCTION

Estimation of production functions has interested economists since statistical and econometric methods were first applied to economic data (Tolley et al., 1924). Optimal allocation of scarce inputs to production activities so as to maximize satisfaction of human wants depends crucially on understanding the relationships whereby production inputs are transformed into consumer goods. Accordingly, the historical objective of agricultural production economists has been to identify the technological relationships that determine the maximum combinations of agricultural outputs that can be produced from given combinations of productive inputs (Heady and Dillon, 1961).

As alternative behaviors have been recognized, economists have realized that optimal production conditions cannot be studied independent of behavioral criteria. In the simplest behavioral paradigm of profit maximization, optimal input combinations can be derived for given levels of existing or expected output based simply on input prices, or optimal production can be determined simply on the basis of input and output prices. In this context, a duality exists such that production can be represented either as a function of input quantities (production function) or as a function of input and output prices (obtained as the derivative of a profit function).

As potential environmental considerations have been realized in agricultural production, these problems have become more complex and the interest in agricultural production relationships has expanded to include environmental economists. The concern with the two-way relationship between inputs and outputs has been replaced with concern for the three-way relationship which also includes pollution such as agricultural drainage (Hochman and Zilberman, 1978). This adds to the complexity of identifying the technological relationships and also to the demands for data to estimate these relationships. In addition, the heterogeneous nature of agricultural production and its acute interaction with the effectiveness of environmental policies has begun to become painfully obvious (Just and Bockstael, 1990). The degree to which the pollutants of agricultural production (such as drainage) are an environmental problem depends heavily on local conditions. Thus, the way heterogeneity among farms is reflected in the technological relationships governing agricultural production is critical to the understanding of agricultural environmental problems in general and agricultural drainage in particular (Just and Antle, 1990).

The purpose of this chapter is to review the state of agricultural production analysis at this time, consider the extent to which agricultural production analysis has been expanded to include environmental considerations, review the empirical methodologies which are applicable to the study of agricultural drainage, and consider the feasibility of adequately accounting for the heterogeneity in agriculture using these approaches.

## AN OVERVIEW OF AGRICULTURAL PRODUCTION ANALYSIS

The historical objective of agricultural production economists has been to identify the technological relationships that determine the maximum quantities of agricultural outputs that can be produced from given combinations of productive inputs (Heady and Dillon, 1961). Much of this work has its roots in the estimation of production functions pioneered by Cobb and Douglas (1928) which represents output quantities in terms of single-product production functions, e.g.,

[1] $y_i = f_i(x_{1i}, x_{2i}, ..., x_{ni})$

where $y_i$ is the quantity of output i produced and $x_{ji}$ is the quantity of input j allocated the production of $y_i$.

As the study of agricultural production evolved, these estimated production functions were used to investigate profit maximizing approaches to agricultural production and to make recommendations for production plans and input use to farmers (e.g., Swanson et al., 1973). For example, if output i has price $p_i$ and input j has price $w_j$, then profit is maximized by choosing input levels $x_{ji}$ such that

[2]  $p_i \, \partial f_i/\partial x_{ji} = w_j$, i = 1,...,m, j = 1,...,n,

which are the first order conditions for profit maximization assuming all inputs are available in unlimited quantities and sales of outputs are unlimited at the associated prices. These profit maximizing input levels can be easily computed once the production functions in [1] are estimated.

While some of the early work on production function estimation relied on carefully generated experimental data (e.g., Heady et al., 1964), most production function estimation has relied on market generated data at either the micro or aggregate levels. The difficulty with data generated in this context is that the first order conditions in [2] tend to result in highly correlated input variables. For example, all input levels tend to be increased simultaneously when output price increases. The associated multicollinearity makes precise estimation of production functions from market data difficult.

## JOINT ESTIMATION OF BEHAVIORAL RELATIONSHIPS

Early on, Marschak and Andrews (1944), recognized that if the data were generated under profit maximization, then the first order equations in [2] would provide additional information that could be used to identify the production function coefficients. They proposed an approach for simultaneous estimation of both the production function in [1] and the first order conditions in [2] in the Cobb-Douglas case. This approach generally enables more precise estimation of the production function parameters assuming profit maximization and that the form of the production function is known.

### Duality and Estimation of Production Systems

Based on the earlier work of Shephard (1953), McFadden (1978) recognized econometric possibilities made possible by a duality between production functions and profit functions (or cost functions) that exists under profit maximization. With this approach, the researcher can begin from an arbitrary specification of the profit function, e.g.,

[3] $\pi_i = \pi_i(p_i, w_1, w_2, ..., w_n)$,

rather than the production function in [1]. In principle, the profit function is obtained by solving [1] and [2] for $y_i$ and $x_{ji}$ in terms of prices $p_i, w_1, ..., w_n$ and substituting into $\pi_i = p_i y_i - w_1 x_{1i} - ... - w_n x_{ni}$. A mutually consistent output supply and set of input demand specifications can be obtained as simple derivatives of an arbitrary profit function, e.g.,

[4] $y_i = \partial \pi_i / \partial p_i = y_i(p_i, w_1, w_2, ..., w_n)$,

[5] $X_{ji} = -\partial \pi_i / \partial w_j = X_{ji}(p_i, w_1, w_2, ... w_n)$,

This is tractable under a much more flexible representation of technology than when the production function is specified arbitrarily, in which case derivation of the output supply and input demands requires solution of a set of nonlinear first-order equations. This flexibility has led to a flood of popularity for the dual approach. Application of flexible forms has permitted the testing of many hypotheses about the structure of production in more general settings than were possible before. Notwithstanding this popularity, however, the dual approach has several difficulties and limitations as discussed below.

## Multiple Output Production Analysis

As the analysis of agricultural production has evolved, its complex nature has received increasing attention. First, most farms not only use a wide variety of inputs but also produce a wide variety of outputs. Early attempts to analyze multioutput production problems simply generalized the common single-output production function in a single equation format, i.e.,

[6] $g(y_1, y_2, ..., y_m) = f(x_1, x_2, ..., x_n)$

where $x_j$ is the total quantity of input j applied in joint production of all outputs. For example, Klein (1947) proposed a generalization of the Cobb-Douglas function with

[7] $y_1 y_2^{\delta 2} ... y_m^{\delta m} = \alpha_o x_1^{\alpha 1} x_2^{\alpha 2} ... x_n^{\alpha n}$.

# ESTIMATION OF PRODUCTION SYSTEMS

This simplistic approach is inadequate for several reasons as explained below.

## Generic Versus Specific Inputs

An important conceptual difficulty with such relationships as applied to agricultural production is that most agricultural inputs are allocated by farmers to specific production activities. For example, generic inputs such as tractor and labor hours, fertilizer, and pesticides are allocated to specific production activites such as wheat, corn, and soybean production.

From the standpoint of determining an optimal production plan, a farmer needs to know not only how much water to apply on his farm but how much water to use on a given cotton field.

The total (or generic) and the allocated (or specific) input quantities must follow the simple accounting relationships

$$[8] \quad x_j = x_{j1} + x_{j2} + \ldots + x_{jm}.$$

Thus, the Klein generalization of the Cobb-Douglas production function in [7], for example, becomes

$$[9] \quad y_1 y_2^{\delta 2} \ldots y_m^{\delta m} = \alpha_o (X_{11} + \ldots + X_{1m})^{\alpha 1} \ldots (X_{n1} + \ldots + X_{nm})^{\alpha n}.$$

The absurd implication of equation [9] is that increasing the amount of water applied to wheat production offers the farmer a choice of, say, either increasing wheat production or corn production. Just et al. (1983) point out that the same absurd implication follows from any single-equation multioutput production function in the general form of equation [8].

## Separability

In terms of the historical methods used in agricultural production analysis, the problem of multiple outputs was addressed more reasonably by the use of linear programming techniques. These techniques, which were introduced to agricultural production analysis by Waugh in the early 1950's, assume fixed proportions production functions for individual outputs in which case [1] becomes

$$[10] \quad y_i = \min_j \{X_{ji}/a_{ji}\}$$

and the joint production possibilities frontier is described by

[11] $Ay \leq x$

where $A = \{a_{ij}\}$, $y = (y_1, y_2, ..., y_m)'$, and $x = (x_1, x_2, ..., x_n)'$. Linear programming techniques have found wide use in agricultural production analysis and can represent a considerable degree of complexity in agricultural production. For example, the detailed choice of input characteristics can be represented by defining sufficiently large numbers of inputs.

The source of the absurd implication of [9] is evident by comparison to [11]. To make this point, consider a general multioutput production function given by $h(y,x) = 0$ where $x = \{X_{ij}\}$ is a matrix representing the allocation of total inputs, x, to specific production activities following [8]. Since this general form is not tractable for most purposes, the approach in [6] and [9] is to gain tractability by assuming separability of $h(y,x)$ with respect to inputs and outputs which implies that $h(y,X) = g(y) - f(X)$. Where $h(y,X)$ is a scalar function, this implies that the allocation of inputs does not determine the mix of outputs. For example, the allocation of land and water to cotton versus alfalfa does not limit the combination of cotton and alfalfa that is produced.

Just et al. (1983) argue that tractability can be attained by making alternative assumptions that better characterize agriculture than does separability with respect to inputs and outputs. They argue that output combinations are essentially determined by the allocation of inputs to various production activities (aside from stochastic conditions). These assumptions rather than separability, characterize the programming approach in [10] and [11] and account for its popularity and intuitive applicability to the problem of helping farmers determine farm production plans (including input allocations).

## Nonjointness

Although the programming approach is more useful than the single equation production function for some multioutput production problems in agriculture, its fixed proportion limitations impose nonjointness with respect to both inputs and outputs (Lau, 1972). A more general approach is to allow substitution of inputs while retaining the additive physical accounting relationships for allocatable inputs. For example, labor can substitute for pesticides in weed control.[1] Under plausible assumptions, $h(y,X) = 0$ can be solved for a vector function $y = f(X)$ in which case each output quantity is determined according to [1] and production activities are linked only by the physical accounting relationships in [8]. Thus, a production system is obtained where nonjointness is imposed only with respect to inputs and not outputs (Lau, 1972).[2]

The consideration of input substitution in production is crucial for the study of water issues. When water availability is altered substantially, for example, with the construction of a public water project or the implementation of water use restrictions, the amount of water available for production may change greatly, forcing a substantial change in the combinations with which water is combined with other inputs. These changes can only be understood by sufficiently capturing input substitution possibilities.

## Malleable Capital and Allocatable Fixed Inputs

In the theory of the firm, some inputs cannot be adjusted in the short run because of various kinds of fixities, lead times required for adjustment, etc. Such fixed inputs were traditionally tied to the production of a single output and were fixed not only in the sense that their quantities were fixed in the short run but also in the sense that they could not be shifted to the production of alternative outputs in the short run. Shumway et al. (1984) have pointed out, however, that many fixed inputs in agricultural production, while fixed at the farm level, can be freely allocated among production activities. For example, land and machinery can be freely allocated between production of wheat and barley. They suggest that allocatable inputs are an important source of jointness in agricultural production and argue that the preponderance of econometric evidence that rejects input nonjointness does not discriminate between true input nonjointness and apparent nonjointness due to fixed but allocatable inputs.

The recognition of this class of inputs is important for the production analysis of water because total water availability is fixed on many farms by irrigation project allocations. This water is typically priced below the marginal cost of delivery and use is determined by contractual restrictions. Nevertheless, in most cases the water and associated land can be allocated to production of alternative crops in the short run. Moore, Negri, and Miranoswki in this volume demonstrate the applicability and utility of modeling water as an allocatable fixed input under these conditions. Their results show that water policy can play a significant role in determining the pattern of agricultural production and the aggregate amounts of various crops that are produced.

## Maximal Use of Data and Information

Just et al. (1983), developed an empirical approach to model nonjoint multioutput technologies with fixed but allocatable inputs. Their approach recognizes that some of the data, particularly the allocation data, is missing in most agricultural production analyses. Nevertheless, efficiency in estimation demands use of all available information from both techological and behavioral assumptions in estimating the multioutput production problem. Thus, the estimated system must be tailored to data availability in each circumstance.

Generally, agricultural production problems are characterized by (a) production relationships which describe the outputs as functions of the input allocations, e.g., equations such as [1] for $i = 1,...,m$, (b) the behavioral criterion described by its associated conditions, e.g., the first order conditions in [2], (c) accounting relationships that tie total input use (purchases) to the allocations, e.g., equations such as [8] for $j = 1,...,n$, and (d) a determination of which of the allocatable inputs are fixed in total as opposed to freely available at market prices as well as the total levels at which they are fixed.

For example, the problem of profit maximization in the context of relationships such as [1] and [8] can be represented as

$$\max_{y, X, x_v} p'y - w'x_v$$

subject to $y = f(X)$ and $Xe = x$

where p is an m x 1 vector of output prices, y is an m x 1 vector of output quantities, w is an $n_v$ x 1 vector of variable input prices, $x_v$ is an $n_v$ x 1 vector of total variable input quantities, X is an m x n matrix of variable input allocations, $e = [1\ 1...1]'$, x is an n x 1 vector of total input quantities, $x' = [x_f' x_v']$, and $x_f$ is an $n_f$ x 1 vector of fixed allocatable input quantities. By developing a Lagrangian and deriving first order conditions, one finds that the solution to this problem is characterized by conditions

[12] $\quad p = \lambda;\ \phi_v = w;\ \lambda_i\ \partial f_i/\partial x_{ji} = \phi_j,\ j=1,...,n,\ i=1,...,m;\ y = f(X);\ Xe = x$
$\quad\quad\ \ (m)\quad\ \ (n_v)\quad\quad (m\cdot n)\quad\quad\quad\quad\quad\quad\quad\quad\quad\quad (m)\quad\quad (n)$

where $\lambda$ is a vector of shadow prices associated with the production function constraints, $\phi = [\phi_f'\ \phi_v'] = [\phi_1 ... \phi_n]$ is a vector of shadow prices associated with the accounting relationships, and numbers in parentheses represent the number of scalar equations implicit in each relationship. This gives $2m + n + n_v + m\cdot n$ nonredundant equations in $3m + 2n + n_v + m\cdot n - 1$ variables (see Just et al., 1983).

The number of observable equations in [12] depends on how many variables are observed. The maximum number of nonredundant equations that can be expressed solely in terms of observable data is the number of observable variables less the number of exogenous variables. The number of exogenous variables is $m + n - 1$ which includes the $m + n_v - 1$ prices in p and w (considering one price as the numeraire under homogeneity) and the $n_f$ quantities of fixed inputs in $x_f$. Several likely cases with associated maximum numbers of nonredundant equations are as follows. For observed data (y,p,x,w,X), i.e., where all data other than shadow prices are observed, one can solve $\lambda$ and $\phi$ out of [12] obtaining

[13] $p_i \partial f_i/\partial X_{ji} = w_j$, $j=n_{f+1},...,n$, $i=1,...,m$; $y=f(X)$; $Xe=x$;

$p_i \partial f_i/\partial X_{ji} = p_1 \partial f_1/\partial X_{j1}$, $j=1,...,n_f$ $i=2,...,m$.

Next, let the input allocation matrix be partitioned as $X = (X_f'\ X_v')'$ where the first $n_f$ rows included in $X_f$ represent the fixed input allocations and the last $n_v$ rows included in $X_v$ represent the variable input allocations. Then where the variable input allocations are not recorded and observed data is (y,p,x,w,xf), one can solve the system in [13] for $X_v = X_v(p,w,x_f)$ and aggregate obtaining $x_v = X_v e = x_v(p,w,x_f)$ to replace the first $n_v \cdot m$ equations. Then $x_v$ can be replaced accordingly in the last $n_f \cdot (m-1)$ relationships obtaining the system

[14] $x_v = x_v(p,w,x_f)$; $y = f(X)$; $X_f e = x_f$;

$p_i \partial f^*_i/\partial X_{ji} = p_1 \partial f^*_1/\partial X_{j1}$   $j=1,...,n_f$ $i=2,...,m$.

where $f^*_i$ denotes evaluation of the derivatives at $x_v = x_v(p,w,x_f)$.

If no input allocations are recorded and observed data consists of (y,p,x,w), one can further solve $x_f$ out of [14] obtaining

[15] $p = p(y,x)$, $w = w(y,x)$, $h(y,x) = 0$

which are the inverse supply and demand functions and a single equation relationship of outputs and aggregate inputs similar in spirit to equation [6]. Note, however, that the latter single equation relationship here embodies behavioral information and is not simply a technical relationship. Rather, it is a reduced form equation summarizing the interaction of behavioral and technical information in the larger underlying but unobservable system.

A comparison of the systems in [13]-[15] demonstrate how information is often thrown away and efficiency in estimation is lost. If the system in [15] is estimated when data are available to estimate [14] or if [14] is estimated when data are available to estimate [13] then the precision of estimates is reduced and the detail and range of uses is limited. Since water is an allocated input and in many instances the total quantity available to a farm is fixed, these considerations are particularly important for production analyses of water and drainage.

## Generalization of Duality for Multioutput Production

While the bulk of the work of Just et al. (1983), including the empirical example was done in a primal setting, the same principles apply to dual frameworks. In most applications of duality to multioutput production problems, the allocated nature of some inputs has not been considered. Instead, the estimated equations typically have consisted of

[16] $y = y(p,w,x_f), x_v = x_v(p,w,x_f)$.

Comparing to [15], the system in [16] is a full information system only if the observed data consists solely of (y,p,w,x).

Some studies have criticized this use of duality for multioutput production problems because it becomes inferior as data availability improves (Shumway et al., 1986). This is true, for example, if any of the input allocations are observed. Of course, the allocation of land among crops is almost always observed.

Chambers and Just (1989) have recently demonstrated, however, that specifications under duality can be generalized to handle the multioutput problem when some allocations are observed. This is done by structuring the profit function in terms of the allocatable fixed inputs so, for example, the short-run profit for a single output follows a function $\pi_i = \pi_i(p_i,w,X_{fi})$ where $X_{fi}$ is the ith column of $X_f$. In words, the short-run profit for the ith output is a function of the price of ith output, the prices of the variable inputs, and the quantities of the allocatable fixed factors allocated to the ith output. The overall structured profit function then follows

[17] $\pi(p,w,x_f) = \max_{X_{f1},...,X_{fm}} [\pi_1(p_1,w,X_{f1}) + ... + \pi_m(p_m,w,X_{fm}) \mid X_{f1} + ... + X_{fm} = x_f]$.

In the framework of (17), the full information system corresponding to [13] under duality is

[18] $X_{ji} = -\partial \pi_i(p_i,w,X_{fi})/\partial w_j, j=n_{f+1},...,n,\ i=1,...,m; \pi_i = \pi_i(p_i,w,X_{fi}), i=1,...,m;$

$Xe = x; \partial \pi_i(p_i,w,X_{fi})/\partial X_{ji} = \partial \pi_1(p_1,w,X_{f1})/\partial X_{j1}, j=1,...,n_f, i=2,...,m.$

and the full information system corresponding to [14] is

[19] $x_v = x_v(p,w,x_f)$; $\pi_i = \pi_i(p_i,w,X_{fi})$, $i=1,...,m$; $X_f e = x_f$;

$\partial \pi_i(p_i,w,X_{fi})/\partial X_{ji} = \partial \pi_1(p_1,w,X_{f1})/\partial X_{j1}$, $j=1,...,n_f$, $i=2,...,m$.

where $x_v(p,w,x_f) = - \partial \pi_1(p_1,w,X_{f1})/\partial w - ... - \partial \pi_m(p_m,w,X_{fm})/\partial w$.

The advantage of the systems in [16], [18], and [19] over those in [13]-[15] is flexibility. By arbitrarily specifying the output-specific profit functions, all estimated relations can be derived by simple differentiation rather than solution of systems of nonlinear equations. As demonstrated by Chambers and Just (1989), this approach can be used to test and distinguish true nonjointness in production from the apparent nonjointness caused by allocated fixed inputs.

Using dual methods in multioutput production problems with allocatable inputs adds to functional flexibility. With added flexibility, the same data set must identify more parameters and the estimated relationships may fit the data better. However, the indirect implications of estimated relationships may become more absurd.

The results of Chambers and Just (1989) serve to illustrate this point. They estimate a system similar to [19] and then use the estimated profit functions to estimate the unobserved allocations of variable inputs.

Examining the estimates more closely, however, reveals that the primal Cobb-Douglas estimates are more plausible that the dual Cobb-Douglas results. This is evident upon comparing to the production norms used in the region of Israel where the data were generated. Direct interviews with farms advisers in the region reveal that most farmers are observed to produce within the production norms. While both the primal and dual approaches produce estimated allocations for many farms outside of this range, the dual estimates are entirely outside of the range while the distribution of primal estimates is centered on the range with a large share of the estimates within the range of production norms. Clearly, the primal estimates are superior to the dual estimates by this standard.

## Discrete Versus Continous Decision Variables

Not all decisions in agricultural production are continuous decisions. For example, decisions to install irrigation equipment on a given field are dichotomous. Similarly, a decision to switch from sprinkler to drip irrigation is dichotomous. When dichotomous decision variables are present, production analysis is further complicated. Essentially, one must consider a separate production or profit function for each technology or government program choice. For example, profit without irrigation might be represented by $\pi_i(p_i,w,X_{fi})$ and profit with irrigation by $\pi^*_i(p_i,w,X_{fi}) - K_i$ where $K_i$ is the necessary annualized investment. Then irrigation is chosen if $\pi^*_i(p_i,w,X_{fi}) - K_i > \pi_i(p_i,w,X_{fi})$. Where characteristics and, thus, the benefits to irrigation vary over farms according to some well-defined probability distribution, this induces a logit or probit estimation problem with respect to the dichotomous choice with the rest of the estimation problem structured as above. If the choice is not simply dichotomous but involves a number of distinct alternatives, then the problem can be represented by a multinomial logit or probit structure.

## Risk Aversion

As the study of agricultural production has been refined, researchers have come to realize that not all farmers seek to maximize profits. In reality, agricultural production is a risky venture and, in general, farmers are willing to trade off some expected profit in order to reduce risk. A number of alternative criteria have been suggested to represent this phenomenon including safety first, maximin (maximizing the minimum profit outcome), minimax regret (minimizing the maximum regret), etc. As research on the role of risk in agricultural production has evolved, however, this phenomenon has been represented most commonly by the maximization of expected utility (Anderson, 1977).

The introduction of risk aversion in the analysis of agricultural production raises a host of difficult issues. First, the first-order conditions for expected utility maximization are generally far more complex than for expected profit maximization. An analytical solution is possible only under a handful of specifications of the utility function and stochastic distributions of variables affecting profits (Collendar and Zilberman, 1985). Second, the standard methods of duality are not applicable; one cannot recover the production function from the profit function or vice versa because behavioral parameters are also involved. While useful combinations of input demands and output supplies can be recovered from an indirect expected utility function, the necessary properties to impose in specifying such a function are not well understood (Pope, 1982). Third, risk averse behavior raises the possibility that some inputs will be used to reduce risk instead of or in addition to increasing production. This obviates the need to model the risk as well as the production effects of agricultural inputs (Just and Pope, 1978). Furthermore, the introduction of risk effects of inputs means that cost functions can no longer be determined solely by expected production but must be characterized by the entire stochastic distribution of output (Pope, 1982).

The consideration of the risk effects of inputs is important for water issues because irrigation tends to have a risk effect contrary to standard representations. Lee (1987) has found that irrigation has a risk-reducing effect in a semiarid environment. In the traditional case of multiplicative disturbances in production functions, all inputs increase risk in proportion to expected output. Thus, consideration of the distinct risk effects of irrigation is crucial for the case of risk averse decisionmakers.

The first primal approach to full information estimation of behavioral criteria and production technology under risk aversion with distinct risk effects of inputs was proposed by Just and Pope (1977). Basically, their approach was to use a production function of the form $y_i = f_i(X_{\cdot i}) + g_i(X_{\cdot i})\varepsilon_i$ where $X_{\cdot i} = (X_{1i},...,X_{ni})'$ and $\varepsilon_i$ is a random disturbance reflecting stochastic influences on production of output i. For risk increasing inputs $\partial g(X_{\cdot i})/\partial X_{ji} > 0$ while $\partial g(X_{\cdot i})/\partial X_{ji} < 0$ for risk reducing inputs. To apply their approach in a framework parallel to those above, one would

$$\max_{y, X, x_v} E[U(p'y - w'x_v)]$$

subject to

$$y_i = f_i(X_{\cdot i}) + g_i(X_{\cdot i})\varepsilon_i, \; i = 1,...,m$$

$$Xe = x$$

where $U(\cdot)$ is a utility function conditioned on current wealth and $E(\cdot)$ is the expectation operator. The solution to this problem is characterized by

[20] $\phi_v = E[U'w]$; $y_i = f_i(X_{\cdot i}) + g_i(X_{\cdot i})\varepsilon_i$, $i = 1,...,m$; $Xe = x$;

$$E[U'\{p_i \partial f_i(X_{\cdot i})/\partial X_{ji} + p_i \varepsilon_i \partial g_i(X_{\cdot i})/\partial X_{ji}\}] = \phi_j, \; j=1,...,n, \; i=1,...,m$$

where $U'$ is the first derivative of utility with respect to profit. The system in [20] corresponds to [12] in the case of profit maximization. Full information application requires reducing these equations to the maximum number of observable equations in each circumstance of data availability. For practical purposes, however, this approach is difficult to apply except under the assumption of constant absolute risk aversion and normality.

## Alternative Behavioral Criteria

One of the major advances in microeconomic research over the past two decades has been recognition of the problems of information acquisition and processing. When information is difficult to gather and analyze, farmers may not utilize inputs strictly according to profit maximization or expected utility maximization conditions. Alternatively, farmers may find that following accepted practices and extension recommendations or infrequent determination of profit maximization input use sufficiently approximates maximization of profit in their individual circumstances.[3]

The applicability of this behavior has been analyzed recently by Just et al. (1990) on the same data set used in the primal profit-maximization analysis of Just et al. (1983) and the dual profit-maximization analysis of Chambers and Just (1989). Noting that agricultural extension specialists as well as farmers tend to characterize agricultural production by enterprise budgets, a production framework is developed where farmers are supposed to behave as if their production functions follow constant returns to scale with fixed input/land ratios. These ratios are assumed to be based on regional averages with modifications for seasonal and farm specific conditions.

This framework corresponds to the case where accepted practices tend to be followed due to the excessive information cost. Enterprise budgets produced by extension activities either play a role in guiding accepted practices or in summarizing the practices that are followed.

Specifically, where the only fixed allocatable input is land ($n_f = 1$), the variable input/land ratios are assumed to follow

$$X_{ji}/X_{fi} = \alpha_{ji} + \beta_{jk} + \gamma_{jt}$$

where k indexes the farmer and t indexes the season. Thus, $\alpha_{ji}$ represents the accepted practice, $\beta_{jk}$ represents the farm specific modification, and $\gamma_{jt}$ represents the season specific modification. In this context, the full information system corresponding, for example, to [14] and [19] is

[21] $x_v = (\alpha_1 + \beta_k + \gamma_t)X_{f1} + ... + (\alpha_m + \beta_k + \gamma_t)X_{fm};\ y = f(X);\ X_f e = x_f;$

$p_i \partial f^{**}{}_i / \partial X_{fi} = p_1 \partial f^{**}{}_1 / \partial X_{f1},\ i=2,...,m$

where $f^{**}{}_i$ denotes evaluation of the derivatives at $X_{ji} = (\alpha_{ji} + \beta_{jk} + \gamma_{jt})X_{fi}$ and $\alpha_i, \beta_k,$ and $\gamma_t$ are vectors with respective elements $\alpha_{ji}, \beta_{jk},$ and $\gamma_{jt}$.

Just et al. (1990) find that the system in [21] not only fits the data better but produces much more plausible estimates of the unobserved variable input allocations for a case of Israeli desert agriculture as compared to the profit maximization estimates.

These results suggest that, while estimation of agricultural production functions, behavior, and input allocations can be improved by joint estimation of the behavioral criterion, the appropriate criterion may not be clear.

Furthermore, the choice of criterion can greatly affect the indirect estimates of behavior obtained.

## APPLICATION OF AGRICULTURAL PRODUCTION ECONOMICS TO ENVIRONMENTAL PROBLEMS

Subject to the various caveats indicated above, agricultural production analysis can be extended rather readily to consideration of environmental issues. To do this, the concern with the two-way relationship between inputs and outputs must be augmented to consider the three-way relationship between inputs, outputs, and the pollution generated by agricultural production. Runoff and ground-water contamination are examples of such pollutants.

Agricultural production relationships can be augmented to include pollution in two ways. One way is to regard pollution as an input in the production process. In this case, the production function might be represented as

$$y_i = f_i(X_{\cdot i}, Z_i)$$

where $Z_i$ is the quantity pollution generated by production of output i. With this approach, the decisionmaker can exercise a tradeoff between production and pollution for a given mix of input usage following $\partial f_i(X_{\cdot i}, Z_i)/\partial Z_i$. This approach is similar to assuming pollution is a joint output with production and that joint output technologies can be represented completely in a single equation relationship such as [6].

An alternative way to include pollution is to regard pollution as a joint output of a production relationship which is uniquely determined by the choice of inputs. For example, for each output one might have

[22] $y_i = f_i(X_{\cdot i}), Z_i = h_i(X_{\cdot i})$

where $Z_i$ is the amount of pollution resulting from production of the ith output and the function $h_i$ is a function describing how pollution depends on the input mix used. In this case, a tradeoff between output and pollution cannot be exercised for a given combination of inputs.

For most agricultural pollution problems, the representation in [22] appears to be appropriate. Many forms of agricultural pollution are closely related to input use. For example, in the case of pesticide or chemical contamination of ground water, pollution is essentially equal to the amount of pesticide or chemical use times a leaching coefficient which measures the rate of leaching into ground water given the characteristics of the soil on which application is made. On irrigated lands, the rate of leaching or other runoff problems will depend also on irrigation practices. In all of these cases, the amount of pollution is determined by input decisions. To the extent that a tradeoff exists between output and pollution for a given input use, it exists as a reduced form embodying, for example, the possibilities of changing irrigation timing with a given total water quantity. By describing the input decisions in sufficient detail, these tradeoffs tend to be eliminated.

Once the technology of pollution generation is represented as in [22], remaining issues in representing the agricultural production system depend on the type of policy instruments that regulate pollution. If the generation of a pollutant is regulated directly by a policy instrument, then an additional constraint must be added to the problem. In the case of nonpoint source pollution, which characterizes most agricultural pollution problems, the pollutant cannot be regulated directly because it is not measured. Policy instruments are thus limited to price or quantity controls on the production inputs and outputs or constraints on the choices of technology (production functions) available.

Price controls such as taxes on polluting inputs are easily considered by simply substituting [22] for the production function relationships in [12] and modifying the appropriate input prices. If total use of an input (e.g., water) on a farm is regulated then that input becomes an allocatable fixed input in the framework of [12]. If input use per acre is regulated, then one of the input decisions with its accompanying first order condition is essentially eliminated from the problem. If the choice of technology is altered, e.g., by requiring pollution abatement practices, then the production functions must be modified accordingly.

In all of these cases, however, the solution of the production problem under profit maximization is characterized generally by

[23]  $p = \lambda; \phi_v = w; \lambda_i \partial f_i / \partial X_{ji} = \phi_j, j=1,...,n, i=1,...,m; Xe = x;$

$y = f(X); Z = h(X); z = Z'e$

where $Z$ is vector of output specific pollution quantities and $z$ is total pollution over all production activities. In the system in [23], the prices of some inputs or outputs may include a tax, some of the total allocatable input quantities may represent constrained levels of total farm use of inputs associated with pollution, and the production functions may reflect the technology associated with required pollution abatement practices.

If risk aversion or other behavioral criteria are applicable, then similar modifications in systems such as [20] or [21] are appropriate. Similar comments apply to dual frameworks although the advantages of the dual framework for these problems may be somewhat reduced in the case of certain policy instruments and certain cases of data availability.

## EMPIRICAL METHODOLOGIES FOR ENVIRONMENTAL ANALYSIS OF AGRICULTURAL PRODUCTION

Incorporating environmental considerations in agricultural production analysis adds to the complexity of identifying the technological relationships and also to the demands for data to estimate these relationships. Levels of pollution generated in the production of individual crops and on the farms as a whole represent additional dimensions of observability that are needed. If pollution levels are observable, then the dimensions of full information estimation are increased. In the more common case of nonpoint source agricultural pollution problems, however, neither the levels of pollution generated by individual production activities nor the total level of pollution are observable. For example, the total level of pesticides reaching an aquifer or the total amount of runoff from a given farm cannot be observed. At best, these quantities can be approximated based on the production practices and characteristics of the farm using physical models produced by other disciplines. Thus, in effect, several equations of [23] must be eliminated because of lack of data and the estimated system reduces essentially to the production systems in [12]-[21] that did not include environmental variables.

As a result, feasible production analysis for environmental purposes differs little from other production analyses. The possibility of developing meaningful results for environmental purposes depends crucially on the availability of physical models that can be used to approximate useful measures of pollution from information on production input quantities and farm characteristics.

There is a diverse literature that addresses potential interactions between agriculture and the environment through these linkages. Since the 1970's, a variety of models have been developed and are continuing to be developed to quantify soil erosion, chemical runoff into surface water, and chemical transport through soils to ground water (Lorber and Mulkey, 1982). These models are composed of systems of differential equations that express changes in environmental quality as functions of management actions and farm characteristics and require detailed information regarding the timing of input decisions and location-specific characteristics of farmland (Donigain and Dean, 1985).

Similarly, models for evaluation of the effects of chemicals on humans and other species utilized dose response relationships which range from simple linear models to more complex models that take into account repeated exposures (Rowe, 1983).

The literature on soil erosion externalities and their management is the most extensively researched (e.g., Loehr et al., 1978). In this literature, soil transport models are used to link agricultural production practices to the pollution measure of interest. Anderson et al. (1985) address location-specific pesticide contamination of ground water using a physical model to translate the results of an economic model of production into pollution measures of interest.

While many such models have been developed, there are two important constraints in their use. First, although research has begun to link the agricultural production process to environmental quality, a general analytical framework has not been available that combines location-specific relationships between production practices and environmental characteristics of farmland in a way that can be aggregated consistently to the regional or National level for purposes of welfare and policy analyses. Such a framework has been developed recently by Antle and Just (1990) as a generalization of earlier work by Hochman and Zilberman (1978) and by Just and Zilberman (1988).

Second, even given an appropriate analytical framework, the data needed to apply these models at a meaningful level for policy analysis are not available. Statistically, valid samples that combine on a location-specific basis both management practices and environmental variables do not exist. Production and environmental data have tended to be collected independently so that their joint use in translating production analyses into useful environmental implications is not facilitated. That is, because of the great heterogeneity that exists in agriculture with respect to many environmental problems, production practices must be linked to environmental implications at a microlevel before aggregation to a meaningful level for regional policy analysis is supported. Antle and Just (1990) argue that the generation of joint production-environmental data bases with microlevel detail is crucial for extending production analyses to useful environmental implications.

## ACCOUNTING FOR HETEROGENEITY IN ENVIRONMENTAL ANALYSIS OF AGRICULTURAL PRODUCTION

The importance of heterogeneity is receiving increased attention in agricultural production analysis as environmental interest rises. Farmland characteristics vary widely. Some farmland is much more environmentally sensitive than other land and the characteristics that make farmland productive are not necessarily highly correlated with the characteristics that cause environmental sensitivity. Wide differences in farmland characteristics has prompted legislation directed at specific types of farmland, e.g., wetlands and sodbusting legislation. Other targeted legislation is needed. Chemical use, for example, may present much more significant problems on sandy soils where leaching into ground water occurs quickly and directly. Soil erosion may be a significant problem on steep or loose sandy soils but not on level or heavy clay soils. Runoff may be a much bigger problem when a lucrative fishery is located nearby. If environmental policies are administered universally without regard to location-specific environmental characteristics, various pollution problems may be improved only marginally in environmentally sensitive areas while needlessly imposing a heavy toll on production efficiency elsewhere.

Market prices and policies affect farmers' incentives at both the extensive and intensive margins. Besides determining agricultural production, these decisions have environmental impacts through two distinct but interrelated mechanisms. Decisions at the extensive margin determine which particular acres of cropland are put into production, and thus determine the environmental characteristics of the land in production. Production decisions at the intensive margin determine the application rates of chemicals, water use, tillage practices, etc., given the acres in production. Once these production decisions are determined for each unit of land in production, physical relationships translate the production decisions and the environmental characteristics into the quantity of pollution associated with each unit of land in production.

To develop a useful framework for policy analysis in this context requires developing a farm production model that shows how production decisions on both the extensive and intensive margin for each unit of land depend on prices, policies, and farm-specific characteristics. The distribution of farm and environmental characteristics in a region then induces a distribution of production behavior and environmental characteristics with respect to the land in production. This joint distribution provides the basis for aggregation of outputs, inputs, and pollution to a meaningful regional level for environmental policy analysis. Once these relationships are developed, one can proceed to analyze the tradeoffs between production and pollution that are associated with various policy alternatives.

Unfortunately, the current prospects are limited for taking these considerations into account in extending agricultural production analysis to the analysis of drainage policy issues. First, knowledge of production functions at the microlevel is limited. While some of the smooth production relationships used by agricultural economists have fit well at aggregate levels, there is some evidence that similar results do not apply in small homogeneous environments. Some studies show, for example, that crop yields at this level follow Von Liebig functions that explain yield as a function of evapotranspiration (Stewart et al., 1974) or applied water (Grimm et al., 1987). Berck and Helfend (1989) demonstrate that heterogeneity of characteristics such as soil fertility and water holding capacity can explain how Von Liebig production functions at the microlevel aggregate up to the smooth production function forms used effectively by agricultural economists at the aggregate level. If these arguments are correct, then accepted specifications that have worked well in aggregate level production analyses may be of limited use in extending production analysis to environmental considerations where the important interaction is at a location-specific level.

Second, the data base available to support production analysis at the location-specific level is inadequate. Data that represent the distribution of soil fertility and water holding capacity across all farmland, for example, are limited and few efforts have been made to incorporate the data that are available into location-specific models of production. As policies are adopted that impose a heavier burden on producers that use water and generate drainage excessively, technologies will be altered. Irrigation may cease on some lands while water conserving technologies such as drip and trickle irrigation may be adopted on others. The advantages of irrigation are generally less on high quality land than on low quality land. Thus, a sufficiently detailed inventory of land quality is necessary to determine the effects of water related policies on either production or pollution. In particular, the joint effect of production and pollution will depend critically on the extent to which characteristics associated with high productivity are correlated with the characteristics associated with high pollution (Antle and Just, 1990).

Third, the rather independent disciplinary research that has evolved on modeling agricultural production versus the physical process by which production decisions and location-specific land characteristics are translated into pollution has resulted in models that are poorly suited for linkage. Production models tend to incorporate market variables of interest to economists such as prices and quantities of purchased inputs. The physical models produced by agricultural production scientists, however, tend to focus on scientific laboratory measurements of evapotranspiration, dew fall, capillary rise, etc. Research is needed to design and link these models more effectively for joint use.

## SUMMARY AND CONCLUSIONS

This chapter assesses the current state of agricultural production analysis. The techniques of agricultural production analysis have been generalized in recent years to consider (1) Improved efficiency in estimating multioutput production relationships through joint estimation of behavioral criteria, (2) the allocated nature of many agricultural inputs and the unobserved nature of some of these input allocations, (3) the applicability of duality and other methods in adding flexibility and relaxing separability and nonjointness in representations of technology, and (4) generalization of behavioral criteria to include risk aversion and rules of thumb associated with costly information acquisition and processing. This methodology could be extended readily to incorporate environmental considerations directly in agricultural production analysis. However, the nonpoint source nature of pollution in agriculture prevents direct observability. As a result, environmental analysis with agricultural production models depends crucially on the development and availability of physical models that can translate agricultural production decisions and location-specific farm characteristics into estimates of pollution measures of interest.

Environmental analysis, and drainage analysis in particular, appears to depend crucially on location-specific characteristics. Moreover, the feasible tradeoffs between environmental quality and production depend crucially on how local characteristics which influence production are correlated with characteristics that influence pollution and on how this joint distribution is altered as various policy instruments are imposed and adjusted. Existing research on agricultural production tends to be at too aggregate a level for these purposes because agriculture is characterized by considerable heterogeneity even at a subfarm level. Available data are inadequate for making this determination because of the independent way in which environmental and production data bases have been developed. Economic models of production and physical models of pollution have also been developed independently resulting in structures and specifications that are incompatible.

While this chapter abstracts from explicit dynamics, quantitative applications will require addressing the full range of issues that arise in applied production economics research including the dynamic aspects of the economic and physical models. For example, the physical models of soil erosion and chemical transport and fate generally involve dynamic processes that relate the farmers' intraseasonal and interseasonal management decisions to environmental impacts. Thus, much remains to be done in extending agricultural production analysis to environmental analysis in general and to drainage analysis in particular.

## NOTES

[1]This substitution can be represented crudely in, say, a linear programming framework by including a sufficient number of activities to approximate the choice of tradeoffs between labor and pesticides that is available. However, this approach does not lend itself well to econometric methodologies for estimation of technical relationships or to the joint empirical consideration of technical and behavioral relationships discussed below.

[2] A further relaxation of input nonjointness as well has been proposed by Mittlehammer, Matulich, and Bushaw. In their case, the equation $h(y,X) = 0$ becomes a multivariate equation. This results in a production system with generality but they stop short of providing a practical and tractable empirical approach with this generality.

[3] In some modern frameworks that account for costs of information acquisition and processing, these generalizations can be incorporated into the analysis of agricultural production in the broad context of profit maximization. However, the optimal combinations of inputs and outputs that result in these cases differ from the characterization of profit maximizing levels that have been derived in traditional frameworks. No empirical studies to date have derived optimal levels of productive inefficiency due to costs of information acquisition and processing in this broader context of profit maximization.

[4] A Von Liebig function assumes that output increases linearly in the input up to some maximum. The maximum for each input may be determined by application of other inputs in which case the Von Liebig function corresponds closely to the fixed proportions production functions of programming models.

## REFERENCES

Anderson, G. D.; Opaluch, J. J.; and Sullivan, W. M., 1985. Nonpoint Agricultural Pollution: Pesticide Contamination of Ground-Water Supplies, *American Journal of Agricultural Economics*, 67, pp. 1238-1243.

Anderson, J. R., 1977. Perspectives on Models of Uncertain Decisions. In: Roumasset, J. A.; Boussard, J.; and Singh, I. (Eds.), *Risk, Uncertainty and Agricultural Development*, Agricultural Development Council, New York, NY.

Antle, J. M. and Just, R. E., 1990. Effects of Commodity Program Structure on Resource Use and the Environment. In: Bockstael, N. E. and Just, R. E. (Eds.), *Commodity and Resource Policy in Agricultural Systems*, Springer-Verlag, Heidleberg.

Berck, P. and Helfend, G., 1989. *Reconciling the Von Liebig and Differentiable Crop Production Functions*. Department of Agricultural and Resource Economics, University of California, Berkeley.

Caswell, M. and Zilberman, D., 1985. The Choices of Irrigation Technologies in California, *American Journal of Agricultural Economics*, 67, pp. 224-234.

Chambers, R. G. and Just, R. E., 1989. Estimating Multioutput Technologies, *American Journal of Agricultural Economics*, 71, pp. 980-995.

Cobb, C. W. and Douglas, P. H., 1928. A Theory of Production, *American Economic Review*, 18, pp. 139-165.

Collendar, R. N. and Zilberman, D., 1985. Land Allocation under Uncertainty for Alternative Specifications of Return Distributions, *American Journal of Agricultural Economics*, 67, pp. 779-786.

Donigian, A. S. and Dean, J. D., 1985. Nonpoint Source Pollution Models for Chemicals. In: Neely, W. B. and Blau, G. E. (Eds.), *Environmental Exposure from Chemicals*, CRC Press, Inc., Boca Raton, FL.

Grimm, S.; Paris, Q.; and Williams, A. W., 1987. A Von Liebig Model for Water and Nitrogen Crop Response, *Western Journal of Agricultural Economics*, 12, pp. 182-192.

Heady, E. O. and Dillon, J., 1961. *Agricultural Production Functions*. Iowa State University Press, Ames.

Heady, E. O.; Madden, J. P.; Jacobsen, N. L.; and Freeman, A. E., 1964. Milk Production Functions Incorporating Variables for Cow Characteristics and Environment, *Journal of Farm Economics*, 46, pp. 1-19.

Hochman, E. and Zilberman, D., 1978. Examination of Environmental Policies Using Production and Pollution Microparameter Distributions, *Econometrica*, 46, pp. 739-760.

Just, R. E. and Antle, J. M., 1990. Interactions Between Agricultural and Environmental Policies: A Conceptual Framework, *American Economic Review*, 80, pp. 197-202.

Just, R. E. and Bockstael, N. E., 1990. *Commodity and Resource Policy in Agricultural Systems*. Springer-Verlag, Heidelberg.

Just, R. E. and Pope, R. D., 1977. On the Relationship of Input Decisions and Risk. In: Roumasset, J. A.; Boussard, J.; and Singh, I., *Risk, Uncertainty and Agricultural Development*, Agricultural Development Council, New York, NY.

Just, R. E. and Pope, R. D., 1978. Stochastic Specification of Production Functions and Economic Implications, *Journal of Econometrics*, 7, pp. 67-86.

Just, R. E. and Zilberman, D., 1988. A Methodology for Evaluating Equity Implications of Environmental Policy Decisions in Agriculture, *Land Economics*, 64, pp. 37-52.

Just, R. E.; Zilberman, D.; and Hochman, E., 1983. Estimation of Multicrop Production Functions, *American Journal of Agricultural Economics*, 65, pp. 770-780.

Just, R. E.; Zilberman, D.; Hochman, E.; and Bar-Shira, Z., 1990. Input Allocation in Multicrop Systems, *American Journal of Agricultural Economics*, 72, pp. 200-209.

Klein, L. R., 1947. *The Use of Cross-Section Data in Econometrics with Application to a Study of Production of Railroad Services in the United States*. National Bureau of Economic Research, Washington, DC.

Lau, L. J., 1972. Profit Functions of Technologies with Multiple Inputs and Outputs, *Review of Economics and Statistics*, 54, pp. 281-289.

Lee, J. G., 1987. *Risk Implications of the Transition to Dryland Agricultural Production on the Texas High Plains*. Unpublished Ph.D. Dissertation, Department of Agricultural Economics, Texas A&M University.

Loehr, R. C.; Haith, D. A.; Walter, M. F.; and Martin, C. S., 1978. Best Management Practice for Agriculture and Silviculture, Proceedings of the 1978 Cornell Agricultural Waste Management Council, Ann Arbor, MI., Ann Arbor Science Publishers, Inc.

Lorber, M. and Mulkey, L. A., 1982. An Evaluation of Three Pesticide Runoff Loading Models, *Journal of Environmental Quality*, 11, pp. 519-529.

Marschak, J. and Andrews, W. H., 1944. Random Simultaneous Equations and the Theory of Production, *Econometrica*, 12, pp. 143-206.

McFadden, D. 1978. Cost, Revenue and Profit Functions. In: Fuss, M. and McFadden, D. (Eds.), *Production Economics: A Dual Approach to Theory and Applications*, North-Holland, New York, NY.

Pope, R. D., 1982. To Dual or Not to Dual? *Western Journal of Agricultural Economics*, 7, pp. 337-351.

Rowe, W. D., 1983. *Evaluation Methods for Environmental Standards*. CRC Press, Boca Raton, FL.

Shephard, R. W., 1953. *Cost and Production Functions*. Princeton University Press, Princeton, NJ.

Shumway, C. R.; Pope, R. D.; and Nash, E. K., 1984. Allocatable Fixed Inputs and Jointness in Agricultural Production: Implications for Economic Modeling, *American Journal of Agricultural Economics*, 66, pp. 72-78.

Stewart, J. I.; Hagan, R. M.; and Pruitt, W. O., 1974. Functions to Predict Optimal Irrigation Programs, *ASCE Journal of Irrigation and Drainage Engineering*, 100, pp. 179-199.

Swanson, G. R.; Taylor, C. R.; Welch, L. F., 1973. Economically Optimal Levels of Nitrogen Fertilizer for Corn: An Analysis Based on Experimental Data, 1966-71, *Illinois Agricultural Economics*, 13, pp. 16-25.

Tolley, H. R.; Black, J. D.; and Ezekiel, M. J. B., 1924. *Inputs as Related to Output*. U.S. Department of Agriculture Bulletin 1277.

Waugh, F., 1953. Applicability of Recent Developments in Methodology to Agricultural Economics, *Journal of Farm Economics*, 35, pp. 692-706.

# 14 INCREASING BLOCK-RATE PRICES FOR IRRIGATION WATER MOTIVATE DRAIN WATER REDUCTION

Dennis Wichelns, University of Rhode Island, Kingston

## ABSTRACT

This chapter reviews the potential role of price incentives in modifying farm-level irrigation decisions to coincide with socially optimal choices when drainwater contains toxic elements. An optimization framework for irrigation districts facing drainwater discharge constraints is presented and implications for water pricing structures are discussed. Details of a block-rate pricing program that was implemented in a California irrigation district during 1989 are described and farm-level responses to the program are discussed. Results of the program include reductions in applied water for some crops in 1989 and reductions in the total volume of drainwater collected in the district.

## INTRODUCTION

Much of the subsurface drainwater collected from farmland on the west side of the San Joaquin Valley (Valley) contains selenium, boron, molybdenum, and other elements occurring naturally in local soils. California has recently implemented water quality guidelines that require a reduction in the amount of salts and selenium entering the San Joaquin River in drainage water from local farms. The State Water Resources Control Board (SWRCB) has suggested that the regional guidelines can be achieved by reducing the volume of drainwater discharged from the region by 30 percent throughout a drainage problem area of 94,000 acres.

Subsurface drainage water is generated when applied water exceeds crop water requirements (measured as evapotranspiration (ET) rates for a given crop) either through inefficient irrigation, or nonuniform infiltration. Water that percolates below the crop root zone contributes to high water tables that are sometimes saline and can restrict plant growth and limit crop yields. Subsurface drainage systems are installed to collect and remove the accumulated water to maintain productivity when natural drainage is inadequate.

Collection and disposal of drainwater in the Valley is described by both point source and nonpoint source characteristics. Deep percolation on farm fields without drainage systems can contribute to high water tables that are drained by systems installed on neighboring farms. Drainage collected in individual systems may be generated by irrigation on overlying fields or may result partially from nonpoint source contributions to the water table from other farms in the area. When this occurs, it may be difficult and very costly to determine the amount of drainwater produced by individual farmers.

Point source outlets of drainage water are created when farmers install subsurface drainage systems and pump drainwater into ditches and other waterways. It is relatively easy to monitor the volume discharged from individual drainage systems using flow meters placed on outlet pipes. Efficient drainwater reduction policies must consider nonpoint contributions to the point source outlets, however, to achieve regional objectives at minimum cost. Standard point source control policies implemented at drainage outlet pipes will result in high marginal costs to reduce drainwater for some farmers and very low marginal costs for others.

The mixture of point and nonpoint source contributions that describe drainwater generation suggests that point source policies alone are inappropriate to motivate reductions in drainwater volumes. Nonpoint source pollution policies will be required to achieve efficient solutions to drainage externalities. Candidate policies include incentives and regulations that motivate adoption of best management practices, taxes or subsidies on inputs that generate or reduce nonpoint source pollutants, and restrictions on the use of selected inputs.

Several studies have examined the use of nonpoint policies in agricultural settings. Griffin and Bromley (1982) describe the conceptual foundations of nonpoint taxes, standards, and incentives for reducing surface runoff. Shortle and Dunn (1986) examine the implications of selected nonpoint policies in a stochastic setting when farmers and regulators have access to different information regarding farm-level management. Knapp, Dinar, and Nash (1990) describe the efficiency and distributional impacts of alternative nonpoint policies in an empirical analysis of cotton production in the Valley.

Gardner and Young (1988) examine combinations of nonpoint pollution policies to control irrigation-induced water quality problems in the Colorado River. Dinar, Knapp, and Letey (1989) examine the effect on farm-level decisions of drainage fees, higher water costs, and increasing block-rate prices for irrigation water. Results of these studies promote the use of economic incentives that modify the price of inputs which generate or reduce effluent from nonpoint sources. One example of such a program is the use of increasing block-rate prices for irrigation water.

This chapter examines the use of increasing block-rate prices for irrigation water to motivate reductions in drainage water. A conceptual framework describing the potential role of price incentives in modifying farm-level decisions to coincide with socially optimal choices is briefly reviewed. An optimization model for irrigation districts facing drainage discharge constraints is presented and implications for water pricing structures are discussed. A block-rate pricing plan that was implemented in a California irrigation district during 1989 is described in detail. Results of the program describe farm-level response to the pricing program, observed reductions in applied water for some crops, and reductions in the total volume of drainwater.

## PRICE INCENTIVES AND PRIVATE V. SOCIAL OPTIMALITY

Irrigations in excess of crop water requirements generate deep percolation as an increasing function of applied water. Deep percolation contributes to the volume of water collected in subsurface drains either directly by entering drain lines installed beneath an irrigated field or indirectly by adding to a regional high water table. Empirical findings (Wichelns and Nelson, 1989) suggest that the marginal drainwater product (MDWP) of applied water increases as the amount of irrigation water applied to a field increases.

The off-farm costs associated with disposal of drainwater containing toxic elements include effects on aquatic wildlife, fisheries, and instream uses of receiving waterways. These costs are likely an increasing function of the load of toxic elements in drainwater discharged from agricultural sources. A positive relationship between the load of toxic elements and the volume of drainwater discharged implies that the damage function associated with irrigation is also an increasing function of applied water. This suggests that the marginal off-farm costs of irrigation increase with applied water.

In the absence of regional or district-level drainage policies, the off-farm effects of drainwater disposal are external to farm-level decisions regarding optimal water use. Farmers choose profit-maximizing irrigation depths by setting the marginal value product (MVP) of applied water equal to its price, or the marginal cost of water delivery. Optimal farm-level irrigation depths, $x^*_f$ in figure 1, will exceed the socially optimal levels, $x^*_s$. Socially optimal irrigation depths are those where the MVP of applied water is equal to the marginal cost of water delivery plus the marginal external cost of drainwater disposal.

Policy tools available to motivate farm-level selection of $x^*_s$ include restrictions on applied water, promotion of best management practices, and pricing schemes that raise the price of irrigation water to reflect the external costs associated with drainwater disposal. Three classes of pricing alternatives are conceptually relevant to the drainwater externality arising from irrigation: (1) An increase in the flat-rate price of irrigation water, (2) a continuously variable water price, and (3) a set of increasing block-rate prices. Each of these alternatives can motivate the socially optimal amount of irrigation water, but management costs and the sum of payments extracted from farmers vary among them.

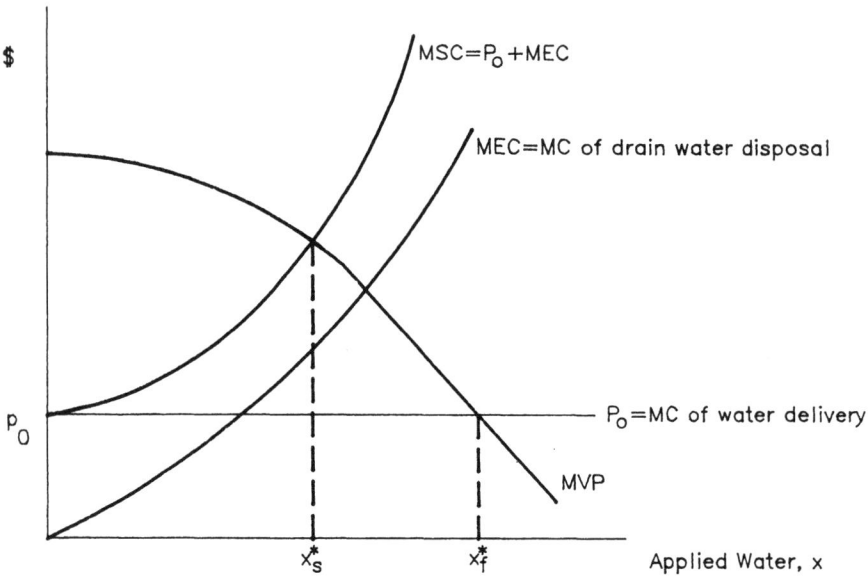

Figure 1. Private and social optima when drainwater containing toxic elements is generated as a result of irrigation.

An optimal increase in the flat-rate price of irrigation water is achieved by raising the water price by an amount equal to the marginal external cost of irrigation, evaluated at the optimal amount of water use, $x^*_s$. This effectively shifts the supply curve of water upward from $p_0$ to $p^*$, where $p^* = p_0 + \tau^*$, or the sum of marginal water delivery cost and the marginal external cost of applying $x^*_s$ (figure 2). This program would motivate farmers to select $x^*_s$, but would extract payments that exceed the total external costs of drainwater disposal. Program payments received under this scheme would equal area abcd, while the total external costs of drainwater disposal are represented by area acd.

An alternative plan would adjust the price of water at each quantity supplied, so that the supply curve becomes the marginal social cost curve. Water price would be described by $p^* = p_0 + \tau(x)$, where $\tau(x)$ represents the marginal external cost of drainwater disposal (figure 2). This plan would motivate farmers to select $x^*_s$ and the revenue raised from water sales would just equal the private and social costs of water delivery and drainwater disposal. Both the program payments received and the total external costs would equal area acd with this plan. Potential problems with imposing this price structure include farm-level acceptance and district-level management of a water price that increases continuously as irrigation depths increase.

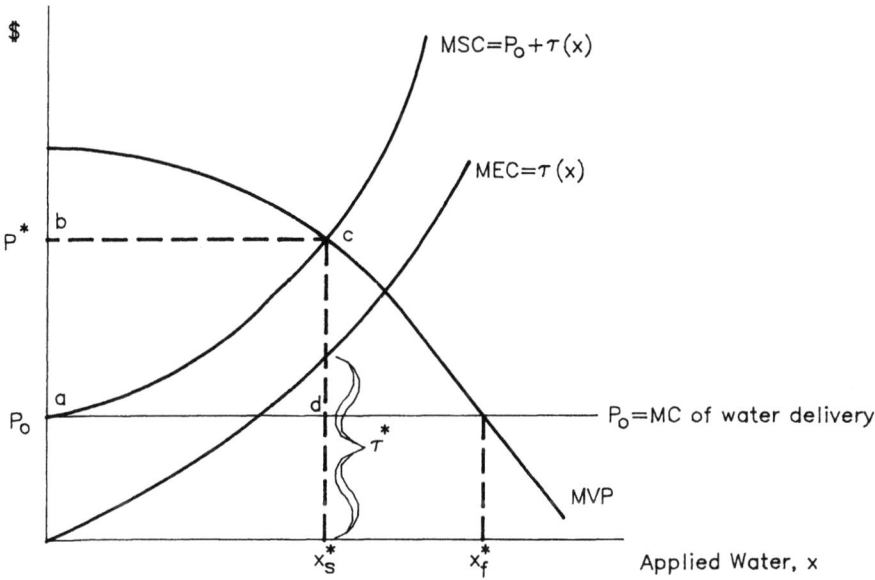

Figure 2. Alternative price incentives to motivate farm-level selection of socially optimal irrigation depths.

A third option that incorporates features of both the single price plan and the continuously adjusting price plan is increasing block-rate pricing. This program adjusts the price of water in discrete steps, or blocks, as the volume of applied water increases. Water price would be described by $p^*_b = p_0 + p_b[\tau(x)]$, where $p_b$ represents the price block associated with the marginal external cost of drainwater disposal (figure 3). This program can also motivate selection of

$x^*_s$, while minimizing excess payments collected from farmers. The sum of shaded areas above and below the marginal social cost curve can be made approximately equal to zero in a well-designed block-rate pricing program.

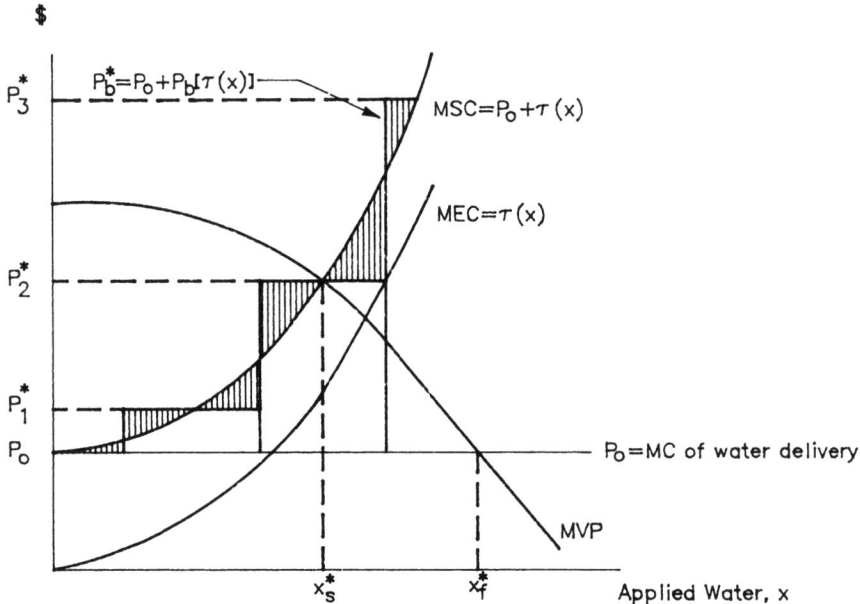

Figure 3. An increasing block-rate price structure to motivate farm-level selection of socially optimal irrigation depths.

## DISTRICT-LEVEL OPTIMIZATION

Current water quality guidelines for the west side of the Valley require that irrigation districts reduce the load of toxic elements in drainwater entering the San Joaquin River. Initial attempts to achieve this goal will include reducing the volume of drainwater discharged from individual irrigation districts. The load of toxic elements is expected to decrease when drainwater volumes are reduced. In the absence of a charge on drainwater disposal, the marginal external costs of farm-level irrigations are described by the impact of marginal additions to the volume of drainwater requiring disposal when discharges are constrained at the district level.

Irrigation districts must motivate farmers to choose irrigation depths that will result in attainment of district-level drainwater reduction goals. Pricing policies to motivate drainwater reduction are consistent with the charter and purpose of irrigation districts that were formed to provide farm-level delivery of water obtained through contracts with Federal and State agencies. The districts are nonprofit organizations and must recover the cost of water delivery and other district functions including drainwater management. Price incentives to motivate drainwater reduction are appropriate within this framework.

A district-level objective function can be described as minimizing the costs of water delivery and drainwater management, subject to meeting the derived demands for irrigation water expressed by member farmers. An alternative description includes maximization of the sum of net returns to members, subject to water delivery and drainwater management costs. In either case, the price of irrigation water is endogenous at the district level. Its value is chosen to raise funds sufficient to recover district costs, while minimizing water costs that are passed along to member farmers.

A district-level drainwater reduction goal becomes an additional constraint on either of the objective function formulations. A Lagrangian describing net revenue maximization subject to a drainwater discharge restriction is given by:

$$[1] \quad \max L = \sum\sum p_k y_{ik} a_{ik} - \sum\sum w x_{ik} a_{ik} - \sum\sum (PC + HC)_{ik} a_{ik} + \lambda [Z - \sum\sum z_{ik} a_{ik}]$$

where: $y_{ik}$ = yield of crop k for farmer i (units of yield/acre),

$p_k$ = price of the kth crop ($/unit of yield),

$a_{ik}$ = area planted to crop k by farmer i (acres),

$x_{ik}$ = irrigation depth (acre-feet/acre),

w = price of water ($/acre-foot),

PC = nonwater production costs ($/acre),

HC = harvest costs ($/acre),

Z = drainwater volume discharge constraint (acre-feet),

$z_{ik}$ = collected drainwater, equivalent depth (acre-feet/acre),

$\lambda$ = Lagrange multiplier describing the marginal value of the drainwater discharge constraint.

Necessary conditions for an interior optimum require that the marginal value product of irrigation water applied by the ith farmer on the kth crop is equal to the price of irrigation water, plus a term that describes the marginal impact of an additional unit of applied water on the value of the objective function, given that drainwater is generated as a result of irrigation:

$$[2] \quad p \frac{\partial y_{ik}}{\partial x_{ik}} = w + \lambda \frac{\partial z_{ik}}{\partial x_{ik}}, \text{ or } MVP_{ik} = w + \lambda \cdot MDWP_{ik}$$

where $MVP_{ik}$ is the marginal value product of irrigation water in crop production and $MDWP_{ik}$ is the marginal drainwater product. A set of these conditions exists for all farmers i and all crops k.

The second term on the right hand side of equation [2], $\lambda \cdot MDWP_{ik}$, is similar to the marginal external cost of applied water, $\tau(x)$, described in the general framework above. A key distinction in the present framework is that the MDWP of applied water may vary among farmers and crops. A set of marginal external cost functions, rather than a single curve, will be relevant in formulating policy alternatives.

Farmers will select optimal amounts of applied water by setting the marginal value product equal to price. This will result in over-irrigation from the district's perspective, when a drainwater discharge restriction is in place and when marginal drainwater products are positive. A district-level policy restricting farm-level irrigation depths or raising the farm-level price of water will be needed to achieve the desired drainwater reduction goal.

One method of inducing farmers to select irrigation depths that are optimal from the district's perspective is to modify the price of water using one of the three programs described above. A higher single price of water equal to $w + \rho_{ik}^*$, where $\rho_{ik}^* = MDWP_{ik}$ evaluated at $x^*_{ik,d}$ (figure 4), is one alternative, where f denotes farm level and d denotes district level. A second plan is to implement a continuously increasing water price equal to $w + \rho_{ik}(x_{ik})$, so that the supply curve becomes the district-level marginal cost curve. A third alternative is to implement increasing block-rate pricing where the water price is described by $p^* = w + p_b[\rho_{ik}(x_{ik})]$. Any one of these plans can be designed to motivate farm-level selection of optimal irrigation depths.

Several issues arise in developing a water pricing policy that incorporates the district-level impacts of drainwater generated by individual farmers. Marginal drainwater products likely vary among farms and crops according to variation in soils and drainage system characteristics (Wichelns and Nelson, 1989). A full range of individual surcharges for all farmers and all crops, $\rho_{ik}$, would be required to achieve the optimal pattern of irrigations throughout the district.

Such a program may be very costly to design and may be inconsistent with an established policy regarding equality of water prices for all farmers in the district.

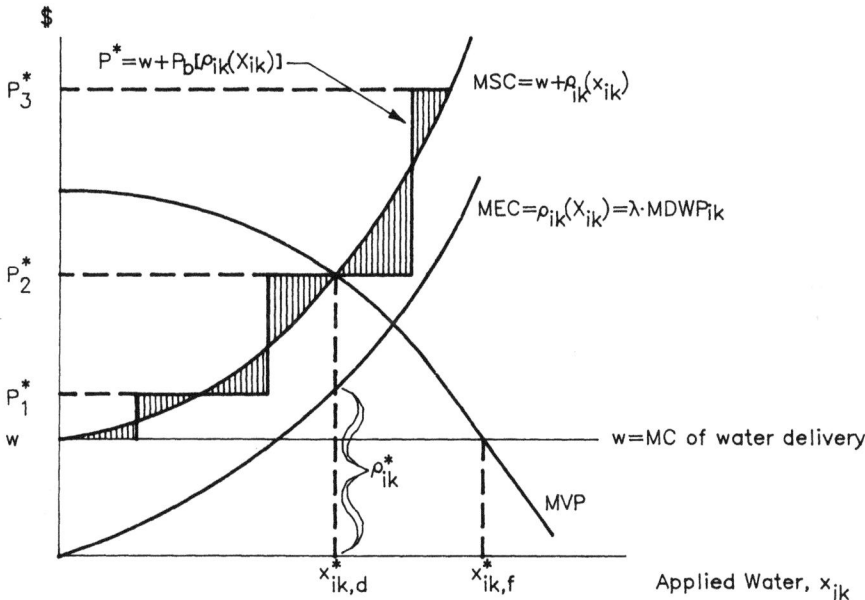

Figure 4. Alternative price structures to motivate farm-level selection of district-level optimal irrigation depths.

An additional concern is that the value of $\lambda$ may not be known with certainty, as short-term and long-term components are difficult to measure. The value of an incremental increase in the volume of drainwater released from a district in any one year includes the short-term reduction in farm-level costs of water conservation measures or recirculating drainwater on the farm. Long-term components of the value include the impacts on crop production from releasing additional salts that would otherwise accumulate in district soils. An approximation of the value of $\lambda$ can be derived from current expenditures to maintain drainwater discharge options in districts where drainage outlets are purchased or leased. These costs provide a lower-bound estimate of the average value of drainwater disposal.

The marginal drainwater products described in the necessary conditions for an optimum represent direct farm-level effects of applied water. In cases where contribution to individual drainage systems from subsurface lateral flows is significant, additional terms are required to fully describe the marginal impacts of irrigation water on all fields. Equation [2] requires modification to include cross-partial derivatives that describe subsurface flow effects:

$$[3] \quad p \frac{\partial y_{ik}}{\partial x_{ik}} = w + \lambda \frac{\partial z_{ik}}{\partial x_{ik}} + \lambda \frac{\partial z_{i'k'}}{\partial x_{ik}}, \text{ where } i \neq i' \text{ and } k \neq k'$$

Equation [3] describes the case where irrigation water applied on one farm field contributes to drainwater collected by a drainage system installed beneath another farmer's field. The marginal impact of applied water is described through partial derivatives involving drainwater collected from more than one field.

A theoretically complete and accurate pricing structure designed to motivate drainwater reduction would include unique surcharges for all farms and fields with significantly different marginal drainwater products. A large amount of the hydrologic information required to determine correct values for the surcharges is not available, however, given the nonpoint nature of drainwater generation. A complex pricing structure, if feasible, would represent a major departure from historical water prices and may not be acceptable to farmers. A reasonable alternative to a completely specified pricing program is one that includes some variability in surcharges among farms or fields, while remaining feasible and being acceptable to district farmers.

Increasing block-rate pricing programs are appropriate for achieving irrigation and drainage policy goals in the Valley. Programs can be tailored to achieve very general water conservation objectives or specific drainwater reduction goals. Price blocks can be defined to vary among farm fields according to soil conditions, drainage system characteristics, or the crops being grown. The degree of complexity in the pricing program can be selected according to farm-level comprehension and acceptance, and the ability of district personnel to monitor water deliveries and manage the block-rate pricing scheme.

Increasing block-rate prices motivate reductions in the amount of water applied to farm fields by raising the per-unit price of water in discrete steps, or blocks, as the cumulative amount of water delivered to a field increases throughout the season. A complete pricing program includes a set of water prices and corresponding irrigation depths at which the prices become effective. Significantly higher prices for water applied in excess of specified levels motivate reductions in the total amount of applied water, while raising the average cost of water by a relatively small amount. This feature of block-rate pricing makes the program attractive in situations where a significant incentive for water conservation is desired, while having minimal impact on the total farm-level expenditure for water.

## BROADVIEW PROGRAM

The 10,000-acre Broadview Water District, located near Firebaugh, California, implemented an increasing block-rate pricing structure for irrigation water in October 1988. The goal of the program is to motivate adoption of water conservation practices that reduce the volume of drainwater collected in a set of 25 subsurface drainage systems installed beneath 6,500 acres in the district. Motivation for the new price structure is provided by regional water quality guidelines that require a reduction in the amount of salts and selenium entering the San Joaquin River in drainwater released from farms in the region.

The appropriate block-rate pricing structure for achieving the drainwater reduction goal includes crop-specific tiering levels (irrigation depths at which the price of water rises) and field-level accounting of water deliveries. Crop-specific tiering levels allow larger amounts of water to be applied on crops with higher water requirements, before higher prices become effective. A pricing program that includes only a single tiering level that is imposed on all crops would distort farm-level cropping pattern decisions in favor of crops with relatively low water requirements. This design is sensible when the ultimate goal of the program is water conservation, but will impose unnecessary farm-level costs when the goal is drainwater reduction.

Field-level accounting of water deliveries motivates careful water application on all farm fields, regardless of soil or drainage system characteristics. Broadview soils range in texture from relatively coarse (sandy loam) to relatively fine (clay and clay loam). Surface irrigations on the coarser soils generate more drainwater than those on the finer soils, as longer set times are required to move water across the field. A program that allows farmers to purchase irrigation water at prices based on the average depth of water applied on several fields provides less incentive to reduce applied water on coarse-textured soils when a farmer has both coarse- and fine-textured fields. The potential to significantly reduce drainwater is greater when farmers are motivated to reduce applied water on the coarse-textured soils.

## Program Design

Prior to implementing the block-rate pricing program at Broadview, the single price of irrigation water was $16 per acre-foot. This price is sufficient to recover the variable costs of delivering irrigation water, including the price paid to the U.S. Bureau of Reclamation and the cost of energy required to lift the water through a series of ponds and laterals in the district. Fixed costs of district operations are recovered through an annual assessment of $42 per acre.

The average annual cost to operate and maintain Broadview's drainwater discharge outlet is about $21 per acre-foot of collected drainwater, or $3.15 per acre-foot of water delivered to all fields in the district. One goal of the block-rate pricing program is to recover district drainage costs through higher prices for units of water with the highest marginal drainwater products. The base price of $16 per acre-foot is retained in the new price structure for water delivered up to crop-specific tiering levels. Additional water is priced at $40 per acre-foot, the price considered sufficient to recover total drainage disposal costs through payments for water purchased in the higher price block. Payments for 0.4 acre-foot of water in the $40 price block generate the same additional revenue as 3.0 acre-feet of water priced at a flat-rate surcharge of $3.15 per acre-foot.

A two-price block-rate design was selected for Broadview to motivate drainwater reduction, while maintaining an easily understandable price structure. Pricing schemes with several blocks were considered too complex for use in the first year of a pilot program. In future years a more detailed block-rate pricing program that follows the marginal social cost curve of applied water more closely will be implemented. At this time, the two-price program provides the desired motivation for drainwater reduction. Farmers and irrigators can easily monitor cumulative irrigation depths, relative to the two-price blocks, throughout the season.

Tiering levels for the 1989 crop year were determined by reducing district wide average irrigation depths for 1986 through 1988 by 10 percent. This approach incorporates locally relevant crop water requirements, soil characteristics, and irrigation practices. Tiering levels based on district average irrigation depths motivate farmers to reduce or eliminate excessive irrigations, while not requiring changes in cultural practices where irrigation depths are already less than the district average. A 10-percent reduction was chosen to motivate a 15-percent reduction in drainwater volume throughout the district (Wichelns and Cone, 1989).

Average irrigation depths for the six major crops grown during 1986 through 1988 range from 2.06 feet on alfalfa seed to 4.58 feet on sugarbeets (table 1). The 3-year average for cotton and tomatoes in Broadview is 3.2 feet, while the average irrigation depth for rice in an adjacent water district is 5.65 feet. All of the data presented in table 1 describe farm-level deliveries of irrigation water and include surface runoff and deep percolation. Deliveries are measured using propeller meters placed in field turnout gates. Onfarm recirculation is not practiced in Broadview because the district collects all of the surface and subsurface drainage water. Some portion of the drainage water is recirculated and blended with freshwater delivered to farm fields.

Table 1. Crop-specific average irrigation depths and selected tiering levels for the increasing block-rate pricing program in the Broadview Water District.

| Crop | Applied Water | | | 86-88 Average | 1989 Block-Rate Pricing Program Tiering Level |
|---|---|---|---|---|---|
| | 1986 | 1987 | 1988 | | |
| | | *(Feet)* | | | |
| Cotton | 3.21 | 3.13 | 3.27 | 3.20 | 2.9 |
| Tomatoes | 3.21 | 3.29 | 3.15 | 3.22 | 2.9 |
| Melons | 2.15 | 1.99 | 2.20 | 2.11 | 1.9 |
| Wheat | 2.01 | 2.55 | 2.35 | 2.30 | 2.1 |
| Sugarbeets | 5.01 | 3.81 | 4.92 | 4.58 | 3.9 |
| Alfalfa Seed | 2.13 | 2.24 | 1.80 | 2.06 | 1.9 |
| Rice[a] | 5.72 | 5.24 | 5.99 | 5.65 | 5.1 |

[a] Rice irrigation data for 1986 through 1988 are from Firebaugh Canal Water District.

Ten-percent reductions in the average irrigation depths result in tiering levels that range from 1.9 feet for alfalfa seed and melons to 5.1 feet for rice (table 1). Cotton and tomato fields are allowed 2.9 feet at the lower water price, while sugarbeet fields are allowed 3.9 feet. Tiering levels for barley and wheat are 1.7 feet and 2.1 feet.

## *Irrigation Summary*

Irrigation practices observed during the 1989 crop year indicate that farmers responded positively to both the spirit and technical details of the water pricing program. Farmers displayed increased interest in irrigation depths throughout the season and shared information regarding irrigation practices and crop management. Several farmers worked with consultants to schedule irrigations.

These and other farmers implemented improved irrigation practices including shortened furrow lengths, reduced set times, and alternate-furrow application during some irrigations.

Applied water was reduced on tomato, melon, and sugarbeet fields in Broadview in 1989, while average water applications increased on cotton and wheat fields. The average cotton and wheat irrigation depths of 3.34 feet and 3.02 feet are higher than those observed during any of the previous 3 years. The average depths applied on tomatoes (2.72 feet), melons (1.93 feet), and sugarbeets (3.73 feet) are lower than those observed during the same time period. The average depth applied on alfalfa seed (1.84 feet) is lower than that observed during 2 of the previous 3 years.

Much of the increase in the average irrigation depths on cotton and wheat in 1989 can be explained by ET rates that exceeded average levels in critical months. Reference ET was higher than average from April through August and cotton growth degree days accumulated rapidly, early in the season. The first irrigation was applied during late May and early June on most fields. Farmers report that the variety of cotton grown in Broadview (GC-510) is less tolerant of water stress and must be irrigated more regularly than varieties grown in the past. Fields of this variety must be irrigated more frequently, after irrigations have been started, to prevent cracking of the predominantly clay soils in Broadview.

Higher than average reference ET values were sustained in July and August, resulting in a continuation of cotton irrigations beyond the normal period. In past years, the majority of cotton fields have received three summer irrigations in addition to the preirrigation applied in winter. In 1989, only 4 cotton fields (12 percent) were irrigated three times in summer, while 14 fields (44 percent) were irrigated four times and 13 fields (41 percent) received a fifth irrigation.

Irrigations on 7 of 32 cotton fields were completed without exceeding the 2.9 feet of water available at $16 per acre-foot. Water costs on remaining fields rose from $2.64 per acre on a field that received 3.01 feet of water to $36.24 per acre for a field receiving 4.41 feet.

Rainfall in winter usually contributes a portion of the crop water requirement for grain crops that are planted in November and December in Broadview. In December 1988, rainfall was higher than the 3-year average for this month and stimulated early germination. Higher than average reference ET during December and January promoted rapid growth of the grain crops. Rainfall during January through April was below average and did not contribute to crop water requirements. Irrigations began in January and continued through May, resulting in an average irrigation depth of 3.02 feet. This exceeds the average for the previous 3 years by 0.72 feet.

None of the grain fields completed the season within the lower price block. Water costs rose from $5.52 per acre on a field that received 2.13 feet of water to $65.04 per acre on a field receiving 4.61 feet.

Sprinklers were used for the first irrigation on the single sugarbeet field in Broadview and quarter-mile furrows were used during remaining irrigations. The total water applied to the sugarbeet field was 3.73 feet, which is less than the average for the previous 3 years. The farmer reported that irrigations were managed more closely in 1989 and that additional time was allowed to elapse between irrigation events. The sugarbeet field was irrigated completely within the lower price block. Sugarbeet tonnage was equal to that achieved in previous years and the sugar content was higher than expected.

The average irrigation depth for alfalfa seed was 1.84 feet in 1989, which is similar to the 3-year average for this crop. Two of the five alfalfa seed fields received water from the $40 price block, with increased water costs ranging from $4.08 to $7.92 per acre.

Irrigations on tomato and melon fields in 1989 provide additional examples of successful reductions in applied water, as a result of improvements in water management in Broadview. Farmers report that these crops are particularly susceptible to overwatering and that investments in irrigation management are rewarded with yield improvements. Nonuniform application of irrigation water has been known to damage tomato and melon plants in some portions of a field, while satisfying crop water requirements in others. Soil characteristics and water requirements often vary widely within the typical 160-acre field in Broadview.

Tomato and melon farmers shared information on irrigation practices and observed yield effects in discussions that were stimulated, in part, by the water pricing program. Several farmers responded by increasing their efforts to manage irrigations more closely on these crops. In past years, most irrigations were applied to an entire field over a period of many days, using 12-hour and 24-hour sets. Farmers now apply water to selected portions of fields, according to detailed measurements of actual water requirements. Irrigation sets of 6 and 8 hours are becoming commonplace on tomatoes and melons, especially during the final irrigations, when the potential for yield improvements and drainwater reduction may be greatest.

Four of seven tomato fields and five of nine melon fields completed the season within the lower price block. Increases in water costs on fields receiving water in the $40 price block ranged from $6.00 to $16.08 per acre for tomatoes and from $1.20 to $6.48 per acre for melons.

## Drainage Summary

The volume of drainwater collected in district sumps during the 1989 crop year suggests that the ultimate goal of the pricing program has largely been achieved. Drainwater was reduced below the average volume collected during 1986 through 1988 at 15 of 25 drainage system sumps. Twelve drainage systems produced less water than the volume collected during 1988. Reductions at individual drainage systems between 1988 and 1989 range from 5.3 acre-feet (4.1 percent) at one system to 181.0 acre-feet (23.4 percent) at another.

The total volume of drainwater collected in Broadview sumps in crop year 1989 is 249.5 acre-feet (6.3 percent) less than the 3-year average annual volume, but 108.7 acre-feet (3.0 percent) greater than the volume collected in 1988. A large part of the increase in volume between 1988 and 1989 is attributed to a small set of drainage systems in the district. Four drainage systems were expanded in size between 1988 and 1989, and a fifth system displayed an unexplained 55 percent increase in drainwater, following a 128-percent increase between 1987 and 1988.

In 1989, Broadview farmers planted more acres of crops that produce large volumes of drainwater and fewer acres of crops that produce small volumes of drainwater. The area planted in cotton, tomato, and rice increased by 300, 170, and 178 acres between 1988 and 1989. Sugarbeet and melon plantings declined by 304 and 256 acres. The total volume of drainwater collected in Broadview sumps increased by only 108.7 acre-feet (3.0 percent) in 1989, despite this change in the district's cropping pattern and the increased size of four drainage systems.

Further evidence of reduced drainwater generation is provided by examining drainwater volumes collected by a subset of Broadview sumps. The subset is formed by eliminating from calculations the four drainage systems that were expanded between 1988 and 1989 and the one that has shown exceptional increases in volumes for the past 3 years. The remaining 20 drainage systems collected a total volume of 3,057.9 acre-feet in 1988 and 2,706.8 acre-feet in 1989. The 3-year average annual volume for this subset is 3,520.7 acre-feet. The volume of drainwater collected by these 20 sumps in 1989 was 813.9 acre-feet (23.1 percent) less than the average annual volume and 351.1 acre-feet (11.5 percent) less than the volume collected in 1988.

## An Estimated Price Response Function

Data generated in the first year of the Broadview pricing program provide an opportunity to estimate farm-level responsiveness to increasing block-rate prices. The ratio of applied water to crop water requirement is examined for

1988, when the price of water was a flat-rate $16 per acre-foot, and for 1989, when the block-rate program was in effect. The average ratio of applied water to crop water requirement declined for cotton, melons, and tomatoes between 1988 and 1989, while increasing for wheat (table 2). The decrease in the melon ratio is statistically significant.

Table 2. Average ratio of applied water to crop water requirement for selected crops, Broadview Water District, 1988 and 1989.

|  | Year | Number of Fields | Ratio of Applied Water to Crop Water Requirement | |
|---|---|---|---|---|
|  |  |  | Mean | Std. Dev. |
| Cotton | 1988 | 22 | 1.63 | 0.17 |
|  | 1989 | 25 | 1.57 | 0.16 |
| Wheat | 1988 | 3 | 2.18 | 0.49 |
|  | 1989 | 8 | 2.27 | 0.69 |
| Melons | 1988 | 5 | 2.67 | 0.47 |
|  | 1989 | 4 | 1.94 | 0.08 |
| Tomatoes | 1988 | 3 | 2.15 | 0.53 |
|  | 1989 | 3 | 1.86 | 0.14 |

The responsiveness of irrigation depths to the higher marginal price of water in 1989 is examined using a linear regression model that describes the ratio of applied water to crop water requirement as a function of the crop grown, soil characteristics, field size, and the marginal price of water:

$$[4] \quad AWCWR_{ikt} = (\beta_0 + \beta_1 d_1 + \beta_2 d_2 + \beta_3 d_3) + \beta_4 SOIL_{ikt} + \beta_5 ACRES_{ikt}$$
$$+ (\beta_6 + \beta_7 d_1 + \beta_8 d_2 + \beta_9 d_3) \cdot PRICE_{ikt} + \epsilon_{ikt}$$

where AWCWR is the ratio of applied water to crop water requirement for the crop grown by farmer i on field k in year t, SOIL is an index of soil texture measured as percent of clay in unit of volume for field k (1=clay, ..., 12=sand where clay soil is associated with the highest percent clay and sand with the lowest), ACRES is the size of field k (acres), PRICE is the marginal block price of applied water ($/acre-foot), $d_1$ through $d_3$ are crop dummy variables that allow testing for significant differences in response function intercepts and slopes ($d_1$=wheat, $d_2$=melons, $d_3$=tomatoes, and cotton is the bench mark crop), $b_0$ through $b_9$ are regression coefficients to be estimated, and e is a random error term.

The estimated equation explains 65 percent of the observed variation in the ratio of applied water to crop water requirement. The cotton response function intercept and shift coefficients for wheat and melons are statistically significant (table 3). The soil index coefficient is significantly positive, indicating that estimated response functions for coarse-textured soils lie to the right of curves for fine-textured soils. The field size variable is negative, suggesting that higher ratios of applied water to crop water requirement are observed on smaller fields, but the estimate is not statistically significant.

Table 3. Ordinary least squares estimates of a price response function for applied water, Broadview Water District, 1988 and 1989.[a]

| Variable | Coefficient | Estimate | t-Statistic |
|---|---|---|---|
| *Cotton Intercept* | $\beta_0$ | 1.19 | 4.12 |
| Wheat Shift | $\beta_1$ | 0.67 | 2.09 |
| Melon Shift | $\beta_2$ | 1.65 | 6.02 |
| Tomato Shift | $\beta_3$ | 0.49 | 1.51 |
| Soil Shift | $\beta_4$ | 0.09 | 3.20 |
| Size Shift | $\beta_5$ | -0.001 | -1.48 |
| *Cotton Slope* | $\beta_6$ | -0.002 | -0.56 |
| Wheat Shift | $\beta_7$ | -0.002 | -0.21 |
| Melon Shift | $\beta_8$ | -0.040 | -3.87 |
| Tomato Shift | $\beta_9$ | -0.005 | -0.45 |

[a] Number of observations = 73, $R^2$ = 0.65.

Slope coefficient estimates allow calculation of response elasticities that describe the relative change in the applied water ratio as the price of water changes. Only the estimated melon slope coefficient (-0.040) is significantly different than that for cotton. The estimated response elasticity for melons is -0.82 when evaluated at the $40 marginal price of water and the mean of applied water ratios observed in 1989. The ratio of applied water to crop water requirement for melons is expected to decline by 8.2 percent in response to a 10-percent increase in the marginal price of water. Although the program was conducted for only 2 years and the data used for this analysis does not allow much variation, it is clear that melons are more sensitive to water application changes than the other crops analyzed, and therefore, the coefficient for melons was found to be significant.

The coefficient for melons is significant because melons are particularly sensitive to changes in irrigation practices. Additional data collected during 1990 and 1991 may include greater variation in water prices and may describe significant price responsiveness for crops other than melons.

## SUMMARY

Increasing block-rate pricing for irrigation water provides an innovative method for motivating changes in irrigation practices to reduce the volume of drainwater collected beneath farm fields. Higher prices for water applied in excess of selected tiering levels provide significant incentive for reducing applied water, while having minimal impact on the total farm-level expenditure for water. The framework is general in its approach to motivating careful use of water resources and in reducing the off-farm effects of irrigation. Details of individual pricing structures will vary with program objectives and the level of water delivery accounting that can be undertaken by individual water districts.

## REFERENCES

Dinar, A.; K. C. Knapp; and J. Letey, 1989. Irrigation Water Pricing Policies to Reduce and Finance Subsurface Drainage Disposal, *Agricultural Water Management*, 16, pp. 155-171.

Gardner, R. L. and Young, R. A., 1988. Assessing Strategies for Control of Irrigation-Induced Salinity in the Upper Colorado River Basin, *American Journal of Agricultural Economics*, 70(1), pp. 37-49.

Griffin, R. C. and Bromley, D. W., 1982. Agricultural Runoff as a Nonpoint Externality: A Theoretical Development, *American Journal of Agricultural Economics*, 64(3), pp. 547-552.

Knapp, K. C.; Dinar, A.; and Nash, P., 1990. Economic Policies for Regulating Agricultural Drainage Water. *Water Resources Bulletin*, 26(2), pp. 289-298.

Shortle, J. S. and Dunn, J. W., 1986. The Relative Efficiency of Agricultural-Source Water Pollution Control Policies, *American Journal of Agricultural Economics*, 68(3), pp. 668-677.

Wichelns, D. and Cone, D., 1989. An Increasing Block-Rate Pricing Program to Motivate Water Conservation and Drainwater Reduction. *Toxic Substances in Agricultural Water Supply and Drainage*, Proceedings of the Second Pan-American Regional Conference on Irrigation and Drainage, USCID, June 8-9, Ottawa, Canada, pp. 137-147.

Wichelns, D. and Nelson, D., 1989. An Empirical Model of the Relationship between Irrigation and the Volume of Water Collected in Subsurface Drains, *Agricultural Water Management*, 16, pp. 293-308.

# 15 IRRIGATION TECHNOLOGY ADOPTION DECISIONS: EMPIRICAL EVIDENCE

Margriet F. Caswell, University of California, Santa Barbara

## ABSTRACT

The widespread adoption of low-volume irrigation technologies in arid regions may reduce the agricultural demand for water and lessen the production of contaminated drainage water. Research on the adoption and diffusion of irrigation technologies is relatively new, but it is based on a well-established adoption literature in both development economics and industrial organization. Many of the irrigation adoption studies are derived from a threshold model that shows how adopters and nonadopters can be differentiated by a critical level of a heterogeneous characteristic such as land quality. Several empirical normative studies demonstrated the effects of price changes and environmental policies on the profitability of each irrigation system. In these normative models, it was assumed that the most profitable technology would be adopted, so the resulting output supply, water demand, and drainage production could also be computed. The positive models used data on actual adoption decisions obtained primarily in the United States and Israel to estimate the probability of adopting modern irrigation technology. The most significant factors influencing adoption appear to be land quality and water cost savings. The theoretical models and empirical evidence support what has been observed--that microirrigation equipment is used most often with high value crops, low quality land, saline waters, or where water costs are high. These promising results will form the basis of further work that is needed to predict the consequences of pricing and environmental policies on the adoption of low-volume irrigation technologies and, hence, on the agricultural sector and the resource base.

## INTRODUCTION

Throughout history, the management of water for agriculture has allowed civilizations to develop and prosper in arid areas. Currently, agriculture accounts for about 70 percent of global water use and although only 17 percent of the world's cropland is irrigated, that land produces one-third of the global harvest (Postel, 1990). The Central Valley of California is one of the most productive areas in the world due largely to the development of water supplies for irrigation. Often, however, irrigation brings with it the seeds of its own destruction in the form of waterlogging and high levels of salinity which directly affect crop productivity.

Recently, the external effects of irrigated agriculture have been recognized, adding a new urgency for discovering ways to sustain agriculture in arid regions. Drainage water containing potentially toxic naturally occurring and manufactured constituents has contaminated ecosystems and ground-water supplies. In addition, the development of new freshwater supplies has been slowed primarily due to high costs and environmental concerns. There is also an increasing demand for the relatively fixed supply of water caused by population growth and the awareness that instream flows and freshwater habitats have social value. Irrigated agriculture must now compete with other uses for increasingly scarce water resources. Since irrigation in California's Central Valley uses more than 80 percent of the developed supplies, a reduction in agricultural water use could have a significant impact on water demand.

Water conservation in agriculture is being looked to as a new "source" of water. Although some of the conserved water will be consumed in growing urban areas, some will be diverted for environmental uses such as wetland habitat and increasing instream flows. Water conservation (reducing the amount of water actually applied to the land) may also reduce drainage flows. The concentrations of contaminants of concern in the drainage water may increase, but the volume needing to be disposed of will be reduced.

One of the ways through which water might be conserved is the wide-scale adoption of low-volume irrigation systems. Modern systems (e.g., drip, microsprinklers, and low energy precision application (LEPA)) use energy and equipment to produce the necessary pressure to deliver water directly to the root area of the crop on almost a continuous basis. Little water is lost to evaporation, runoff, weed competition, or deep percolation. Traditional surface irrigation systems (e.g., flood, border, and furrow) use gravity to deliver large amounts of water to a field at relatively infrequent intervals. The relative advantage of low-volume technologies will depend on the quality of the land on which it is used. On land having soils with good water-holding capacity that has been laser leveled the irrigation efficiency[1] of traditional systems approaches that of the modern pressurized equipment. The efficiency level drops quickly, however, as land quality deteriorates. Crop productivity per acre depends on water application uniformity to a great extent (Letey et al., 1990), and uniform conditions cannot be achieved with gravity systems on slopes or if soil composition varies. The average irrigation efficiency for furrow irrigation systems with half mile furrows in California's San Joaquin Valley is 64 percent while a well-managed drip irrigation system has an efficiency of 90 percent. Also, the annual cost of these systems is quite disparate: $19/year for furrow and $248/year for drip (CH2M Hill, 1989).

The adoption of low-volume irrigation systems in the Central Valley could slow the rate of waterlogging and reduce the volume of contaminated drainwater. Unfortunately, new technologies are never instantly adopted by all potential users, and the innovation may never completely saturate the market. Decisionmakers need to assess the time pattern of adoption as well as its eventual extent in response to alternative policies designed to encourage adoption. The impacts of adoption on agricultural output, farm profitability, and the demand for water must be estimated in order to judge the success of any policy.

The purpose of this chapter is to survey various studies that have attempted to empirically explain observed adoption and diffusion patterns. The first section of the chapter will review the key factors of adoption in general. A theoretical model applied to the adoption of irrigation technologies will be discussed and a survey of the empirical studies will be presented. Possible future research directions will be offered in the concluding section.

## ADOPTION OF AGRICULTURAL TECHNOLOGIES

### General Agricultural Technology Adoption Models

When a new technology is developed and introduced in an agricultural region, some farmers will begin using the technology immediately, some will adopt the technology at a later date, and some may never use it. Economic growth often depends on technological advances, and the success of an innovation depends on the extent of adoption and the pattern of adoption over time (diffusion). Early adoption studies (Mansfield, 1961 and Rogers, 1962) did not have a firm theoretical basis although the importance of economic incentives for adoption was recognized. It has been observed for years that aggregate diffusion over time follows an S-shaped pattern and can be described by several sigmoid mathematical formulas. Rogers (1962) characterized the S-shaped pattern as reflecting the adoption of the innovation by three groups of farmers ("early adopters," "followers," and "laggards"), but did not identify the source of heterogeneity that would explain the difference. Mansfield (1961) explained that adoption was merely a process of imitation, and the diffusion of the technology could be modeled as a spread of rumors or disease. Griliches (1957) conducted the first econometric study of technology diffusion. He estimated the share of land using the new technology at any time, $P(t)$, as the following logistic function of time:

$$P(t) = K[1 - e^{-(a+bt)}]^{-1}$$

where K is the maximum adoption level (saturation), a is the initial share using the technology, and b is the rate of adoption. Griliches explained the variation in K and b as due to locational differences in profitability. Unfortunately, observing the shape of the diffusion curve is not sufficient to determine the underlying causes of the dynamic process. The theoretical framework to explain the observed patterns of the adoption of irrigation technologies should be based on a model of individual decisionmaking by farmers which includes the characteristics of the farm, the crop, and the technologies from which the farmer can choose. Changes over time should be generated by equations of motion that describe how farmer decisions are affected.

Much of the work on the determinants of technical change has concentrated on the adoption and diffusion of agricultural technologies in less developed countries (LDC's) or of industrial technologies in developed countries. The majority of studies related to the adoption of low-volume irrigation technologies are derived from a combination of theories from both bodies of literature. Many irrigation studies also confirm the general results found in work on other agricultural technologies.

There are two primary types of adoption research. The first is normative and the second is positive. Normative models use economic theory to indicate what "should" be obtained. Conceptual normative models are developed to formulate hypotheses that can be tested empirically and to determine important components in the analysis. Normative empirical work is usually based on an engineering approach by computing profits, water use, etc., based on assumed parameters for production functions, costs, and efficiencies. Positive models, on the other hand, try to analyze what people are actually doing rather than what they should do. Empirical data is used to test hypotheses developed by conceptual normative analyses or to be analyzed using econometric techniques. Positive models identify factors that actually affect adoption and assess the importance of those factors on the adoption decision. The parameters from a positive model can then be used in simulation studies to formulate policy scenarios and to predict future conditions.

A new technology or innovation will change the marginal rate of substitution between inputs in a production process, therefore, the past production relationship has changed. An individual, once aware of the new technology, can choose to adopt (purchase, install, and use) the innovation. For "lumpy" technologies (e.g., tube wells or tractors) this would be a dichotomous decision. The data used for empirical analysis would be the discrete choices made by agents (adopt or not adopt). With divisible technologies such as new crop varieties or fertilizers, the choice may be the extent to which the innovations are used by each firm. The data obtained for analysis would be a continuous measure of the extent of adoption such as the share of capacity on which the new technology is used (e.g., adoption on 80 percent of the acreage). The aggregate extent of adoption (the number of firms or acres) will determine overall output and input use.

The econometric model that is used for analysis will be determined by the available data. Data are not as readily available for the adoption decisions made for individual farm plots (a dichotomous choice). Either logit or probit techniques can be used with discrete data (Amemiya, 1981) to estimate the probability of adoption for the new technology. Most irrigation studies use data on the share of capacity utilizing the new technology at the local, regional, or National level. Least-square statistical techniques can be used to analyze continuous (share) data.

Some studies use cross-section data gathered during a single time period. One of the most common econometric techniques to use with cross-section data is to estimate a logit relationship of the form

$$\log [P_{1n}/P_{on}] = \sum_{j=1}^{I} \alpha_{ij} X_{ijn}$$

where $P_i$ is the probability of using technology i (i = 0 for the traditional technology and i = 1 for the modern system), n indicates the observation, and j represents the explanatory variables. For the simultaneous estimation of discrete and continuous adoption decisions, the Heckman procedure can be used.

Although the following discussion will deal primarily with the introduction of a single and distinct technology, there are many agricultural innovations that are either improvements on an old technology or parts of a technology "package." For instance, sprinkler irrigation techniques were not widely used until after aluminum pipes became readily available. Sprinkler systems with iron pipes may be thought of as a different technology than aluminum systems, but data may be unavailable that differentiate between the two. High yield variety crops are often adopted as part of a technology package that includes fertilizer and irrigation. Sometimes individuals adopt all of the components at the optimal level while others only use a subset of the package. The measurement of the intensity of adoption will depend on the definition of the multidimensional technology.

Interest in the determinants of adoption for agricultural technologies has spawned a large body of literature, and many studies have been conducted that have identified variables that may be important in explaining adoption behavior. Feder et al. (1985) and Thirtle and Ruttan (1988) give a full discussion of these models, and only a brief review of key adoption factors for general agricultural technologies will be presented below. Economic variables may be the most important determinants of adoption. It has been shown that price increases for an output have a positive effect on adoption, and increases in the costs of inputs have a positive effect on the adoption of input-augmenting technologies. The prices of complementary and substitute inputs will affect adoption as well. The price of the technology has the expected negative effect on adoption, and the declining investment cost over time that is common with many new technologies will generate the diffusion process. Farm size is usually shown to be positively related to the speed and intensity of adoption of a new agricultural technology (Feder and O'Mara, 1981), but the results may be related to other factors such as credit constraints, land quality, or access to information rather than economies of scale.

Risk may be another important consideration. If the new technology reduces yield risk, the probability of adoption will increase (Just and Zilberman, 1983). Irrigation equipment reduces dependence on variable rainfall, hence reduces risk. On the other hand, some high yield variety crops are productive under only the best conditions which may increase risk and lower the probability of adoption. Price risks will affect expected profitability, and there are risks inherent in government policies as well. Learning will increase the probability of adoption. Technology producers often improve methods and lower the capital cost of the equipment over time. Farmers also learn to use the new technology more efficiently (Hiebert, 1974 and Lindner, 1980). Often the time since the technology's introduction is used as a proxy for learning. Learning itself may be a form of risk reduction as the uncertainties inherent in a new technology are dissipated. Tenurial arrangement effects on adoption have been of particular concern in studies of developing countries. There are conflicting empirical results as to whether land ownership or tenancy increases the tendency to adopt new technologies (Schutjer and Van der Veen, 1977). Technologies that are tied to the land (e.g., irrigation equipment) tend to be adopted first when the land is owner operated. There is apparently no difference in the adoption pattern between owner and tenant for technologies such as high yield variety crops.

Education, agricultural extension activities, and other measures of human capital play a positive role in adoption because skill is needed to address change and adapt to the more sophisticated technology (Welch, 1978). The introduction of drip irrigation to California was initiated by a Cooperative Extension agent in San Diego County (Caswell et al., 1984), and the pattern of adoption suggests that the information disseminating role of the extension agent was a very important factor in the rapid diffusion of the sophisticated technology in that area (Caswell, 1982). Capital markets also play a role in adoption because most new technologies require an investment outlay. Restrictions on credit and financing based on farm size, income, or land value will slow the adoption process (Weil, 1970). Land quality has a positive (negative) effect on adoption if there is a positive (negative) correlation between the resource-augmenting characteristic of the technology and asset quality. For instance, the higher the quality of land, the more likely a high-yield variety crop will be planted (Hiebert, 1974 and Gladwin, 1979), but it is less likely that low-volume irrigation equipment will be installed (Caswell and Zilberman, 1986). Environmental conditions will play an increasingly important role in technology efficiency, hence, adoption. The relative advantage of remote sensing techniques was found to depend on cloud cover and terrain (Caswell, 1989), and saline conditions will determine irrigation effectiveness for each technology (Dinar and Knapp, 1986; Feinerman, 1983; and Feinerman and Vaux, 1984).

Many of the factors above are intuitively appealing reasons for the adoption of an agricultural innovation by an individual. They do not, however, explain the causes of the dynamic diffusion process. Most economists would prefer that adoption was determined by differences in economic parameters among individuals rather than psychological differences between potential adopters and that the change in the level of adoption over time would be generated by dynamic changes in an economic parameter. The normative threshold model introduced by David (1975) showed how adopters and nonadopters could be separated by a critical (threshold) level of a heterogeneous characteristic (e.g., firm size, income, resource quality), and that the critical level is determined by prices and costs. Schultz (1975) had suggested that there was a period of disequilibrium after the introduction of a new technology, but the threshold model shows that each adoption level represents a new equilibrium. Changes in the level of adoption over time will be generated by shifts in the economic parameters. Some changes in the economic factors can be exogenous (e.g., lower capital costs for the innovation over time), and some changes will result from the adoption process itself. Cochrane (1958) introduced the concept of the "technological treadmill" to describe the reduction in gains from adoption if either output or input prices were endogenously determined. If, for instance, a new technology enabled farmers to substantially increase output, the price of the crop might fall, thus lowering profits.

Since many theory-based empirical studies of irrigation technology adoption are derived from a normative threshold model, that model of irrigation system choice will be briefly sketched in its general form.

## Threshold Model of Irrigation Technology Adoption

It is assumed that farmers in a competitive industry will try to maximize profits and that they have the choice of more than one irrigation technology, at least one of which is an innovation. A manager will choose the technology that yields the highest profit given the quality of the land, q, the market price for the crop, P, and the price of water[2], W. Land quality can represent the water-holding capacity of the soil, slope, or other characteristics of importance for irrigation. A more complete description of this model can be found in Caswell and Zilberman (1986). Extensions to this model can be found in Dinar and Zilberman in this volume.

Per acre yield, y, using technology i is a function of the amount of water beneficially used by the plant, e, $[y_i = f(e_i)]$ where $f(\cdot)$ is a standard production function with $f(0) = 0$, $f'(\cdot) > 0$, $f''(\cdot) < 0$ (primes denote derivatives). For simplicity, only two irrigation technologies will be discussed[3]: the modern technology (drip or LEPA when $i = 1$) and the traditional one (flood or furrow when $i = 0$). It is assumed that each irrigation technology transforms the amount of water actually applied to a field, $a_i$, into effective water, $e_i$, according to a concave function $h_i(q)$ that depends on land quality. This is equivalent to the American Society of Civil Engineers (1978) definition of irrigation efficiency as the ratio of water beneficially used by the crop to the total applied water $[h_i(q) = e_i/a_i]$. Since the U.S. Soil Conservation Service defines land-use capability classes with respect to traditional irrigation methods, the irrigation efficiency measure can be normalized on the traditional system so that $h_0 = q$, $h_1(q) \geq q$, and $h_1'(q) \geq 0$. Therefore, a technology switch is equivalent to a change in asset quality from q to $h_1(q)$. The relative advantage of the modern irrigation technology is negatively correlated with asset quality, so $h'' < 0$.

It is also assumed that the annualized investment cost, I, for the new technology is greater than for the traditional one $(I_1 > I_0)$, and that investment costs do not depend on the land quality[4]. Therefore, the operational profit that can be earned for each land quality can be written as:

$$\Pi_i(q) = P \cdot f(h_i(q) \cdot a_i) - W \cdot a_i - I_i.$$

The first-order condition that must hold at the optimum is that the value of the marginal product of applied water must be equal to the price of water $[P \cdot f' \cdot h_i = W]$, and the solution to this equation will yield the optimal variable input use for each technology.

Changes in the underlying economic variables (P and W) will also affect the optimal values of output supply and input demand. An increase in output price or a reduction in water price will increase output, applied water demand, and the optimal amount of water that will be used beneficially by the crop. In addition, an increase in land quality will increase output and increase the optimal level of effective water use. Profits for each technology are a positive function of land quality. This suggests that there will be a land quality, $q_i^m$ where $\Pi_i(q_i^m) = 0$. Farms with a quality greater than this marginal quality would earn the operator positive profits using technology i, and below that level use of technology i would not be profitable.

It is assumed that a farmer would compare the profitability of the two technologies and would choose the one which gave the highest return. The modern system has a higher water-use efficiency and will produce higher yields than the traditional system, but the investment cost is higher as well. When both technologies are used, there will be at least one asset quality, $q^s$, for which profits will be equal for both modern and traditional technologies $[\Pi_1(q^s) = \Pi_0(q^s)]$. The farmers with qualities lower than $q^s$ but greater than $q_1^m$ will adopt the modern technology while those with higher qualities will retain traditional methods. $q^s$ represents a single switching land quality—the threshold of adoption. After the introduction of the modern technology, the irrigated land base will expand since the range of land qualities that can be used profitably is increased from $q_1^m$ to $q_0^m$. Farms with a land quality that is between $q_0^m$ to $q^s$ will switch from the traditional to modern technology while those farms with high quality assets ($q > q^s$) will not find it profitable to adopt the new technology. The actual number of acres on which the modern technology is used will depend on the distribution of land qualities.

The values of $q_1^m$ and $q^s$ are derived from the profit function, and as such are themselves functions of output and input prices and investment costs. The marginal land quality will decrease (more land will be brought into production) if output price increases or if water or investment costs for the new technology decline. The adoption threshold will increase (more converted acreage) if crop price or the costs of the traditional system increase or if the cost of the modern system declines. The effect of an increase in water costs on the switching quality will depend on the water-savings characteristics of the modern technology.

Caswell and Zilberman (1990b) included endogenous price changes in an equilibrium analysis of technical change. They showed how Cochrane's 1958 "treadmill" hypothesis could be explained using a threshold model. The secondary impacts of technology adoption may determine the perceived successfulness of the innovation.

## REVIEW OF IRRIGATION STUDIES

The results of a worldwide irrigation survey (Abbott, 1984) showed that the United States and Israel were the largest users of microirrigation technologies with respect to acreage. From 1975 to 1982, the use of low-volume irrigation equipment increased over 200 percent in the United States and over 700 percent in Israel. It should therefore not be surprising that the majority of economic research on the adoption of these technologies has been in these two countries. According to Abbott (1984), the modern technologies were used primarily on high value crops, on sandy soils, in greenhouses, on sloped land, where water quality is low, and where water and labor are expensive or in short supply. Although yield increases were reported to be an important determinant of adoption in Europe, water savings were the primary reason for adoption in most of the world.

Mapp (1988) describes changing irrigation practices in the High Plains area. He suggests that rising water costs due to ground-water depletion appear to have increased the incentive for farmers to adopt LEPA systems despite the substantial fixed costs associated with purchasing the modern system, but neither a theoretical nor an empirical model of adoption were presented.

## Normative Models

Several normative models of adoption were developed to show the potential profitability of modern irrigation systems under various policy or environmental regimes. These models could be used to indicate the tendency to adopt but not used for a prediction of the level of adoption because they only estimate profit levels, not behavior.

Letey et al. (1990) use an engineering approach to analyze the profitability of five irrigation systems (furrow, hand-move sprinkler, linear-move sprinkler, subsurface drip, and LEPA) used on cotton fields in the San Joaquin Valley of California. Uniformity (as measured by the Christiansen uniformity coefficient) is the primary measure of land quality, and two drainage policies (a drainage disposal charge and an upper limit on drainage volume) are assessed with respect to their affect on profitability. Profits are computed using a crop-water production function, representative costs in the area, and average system uniformities. They find that the relative profitability of the modern systems was positively related to the size of the yield advantage and to the reduction in drainage volumes. Feinerman et al. (1983) showed how the distribution of land quality affected profitability and demonstrated how the agricultural land base would be extended to lower quality lands with the introduction of an input-augmenting irrigation technology.

The study by Hornbaker and Mapp (1988) compares center pivot and LEPA systems used to irrigate grain sorghum. They use a "real time" dynamic risk-neutral model to simulate yields, water use, and profits. The simulation is generated by a plant growth model, generated costs, and historical weather data. The results show that LEPA will be more profitable in areas with high water costs (deep wells or high delivery prices) although water savings are significant in all cases.

Caswell and Zilberman (1986) use two hypothetical production functions (Cobb-Douglas and quadratic) to estimate water use, output, energy demand, and quasi-rents as functions of well depth and pressurization requirements for traditional, sprinkler, and drip irrigation technologies. The results of the analysis show that the quadratic production function produces results that are consistent with observed behavior. With that function, yield effects for the modern technologies will depend on land quality and well depth (water cost). The new technologies will not be adopted in locations where they do not increase yield. Water savings were also found to be significant, and land rent will be expected to increase for land using modern technologies. In Caswell and Zilberman (1990a), a numerical simulation was used to generate S-shaped diffusion curves in an area of heterogeneous land quality and changing economic conditions. They also show that the diffusion of the modern irrigation technology may also cause output supply and input demand to follow a sigmoid pattern over time.

Caswell et al. (1990) expanded the normative threshold model to include external effects of input use and the role of environmental policies (e.g., drainage charges) in conserving water and reducing pollution. The theoretical model shows that adoption would tend to increase when a pollution tax was high and the modern technology effectively reduces pollution. Using a quadratic production function, a numerical simulation based on the general characteristics of cotton production in California is used to estimate crop yields, water use, and drainage levels that could be expected under several crop price/drainage tax regimes. Caswell et al (1990) found that although farmers can reduce water use while retaining their current irrigation technology, the largest conservation gains will come from switching technologies. The simulation confirms the theoretical results that adoption is more likely with growers having lower land quality, higher value crops, higher water costs, and more severe drainage problems. Dinar et al. (1989) also included drainage production into their analysis. They conclude that taxing or regulating drainage production will increase the probability of adoption for technologies that improve irrigation efficiency.

Casterline et al. (1989) used a numerical example based on conditions found in the Central Valley of California for cotton growers to compare profits, yields, water use, and drainage per acre for each irrigation technology under several policy scenarios. They show that the relative gains of adopting modern irrigation equipment increase as water prices rise and drainage controls become more onerous.

## Positive Models

Much of the work on the adoption of modern irrigation technologies is very recent and conducted primarily in the United States and Israel. Empirical work has been limited by a lack of data, particularly with respect to time series data needed to assess diffusion. Despite these limitations, several empirical studies of diffusion have been completed, and they are consistent with the results of the general agricultural adoption literature and they confirm the hypotheses generated by the normative models.

Nieswiadomy (1988) studied the irrigation practices on the High Plains of Texas. He analyzed the change in the usage of all irrigation inputs for cotton and grain sorghum growers in the area using farm level data for the years 1970 to 1980. Expenditures on irrigation equipment were used as the measure of the intensity of adoption. The elasticities of substitution between water and three irrigation technologies were estimated to assess the substitutability (or complementarity) of water saving capital for water. Nieswiadomy found that the likelihood of adoption for the water-saving irrigation technology increased as the prices of water and output increase or the quality of land declines.

The majority of the remaining studies reviewed here were derived to test hypotheses generated by a threshold model similar to the one presented above. That threshold model was tested by Caswell and Zilberman (1985) using cross-section data from the Central Valley of California obtained from county level questionnaires on the share of each perennial crop grown with each technology. Multinomial logit techniques were used to estimate the probability of adopting either sprinkler or drip irrigation systems instead of surface systems as a function of water cost, crop, water source (ground or surface source) and location. Explicit land quality data were not available, so location was used as a proxy. The results of the analysis show only small regional effects which is not surprising since the location variable would also capture climate, marketing, and other effects as well as land quality. Crop effects are significant, confirming that some crops are more responsive to irrigation method than others. There is a significant water cost effect, and users of ground water (an expensive but constant source) are more likely to adopt the modern system than those relying on surface deliveries.

An empirical analysis of national data using the threshold framework was conducted by Negri and Brooks (1990). They used cross-section data from 5,145 ground-water-using farms to estimate a binomial logit model of the adoption of sprinkler rather than gravity irrigation systems as a function of ground-water pumping costs, labor costs, climate, slope, soil/land quality, and location. Land quality appeared to be the most important determinant of adoption although water costs were also significant. Farms with soils of low water-holding capacity were more likely to adopt the land-quality augmenting technology.

Casterline et al. (1989) also used National data (by state) to assess the adoption of modern irrigation technologies versus traditional ones over a 30-year period. They show major regional differences in the pattern of irrigation technology adoption over time and that the diffusion of center-pivot irrigation in Nebraska behaves as an S-shaped function of time. Casterline et al. (1989) also present evidence that government support programs encouraged the adoption of center-pivot irrigation in the Plains states beyond what would have been optimal.

Lichtenberg (1989) looked at the adoption of center pivot irrigation equipment and the changing cropping patterns for field crops in Nebraska. He captured the simultaneous nature of the crop/technology choices by using multinomial logit techniques on county level shares of adoption for the years 1966 to 1980. The log odds of adoption were regressed on a quadratic function of constructed expected own crop price, expected hay price (baseline case), cost of the center pivot system, average land quality, the squares of each term, and interaction terms. Lichtenberg's results show that land quality exerts a marked influence on cropping patterns and that the introduction of center pivot (a land-quality augmenting technology) has induced significant changes in cropping patterns. He confirmed that technologies which have input augmenting characteristics that are negatively correlated with quality will tend to be adopted rapidly on lower qualities of land.

Recent work by Cason et al. (1990) also deals with the simultaneous choice of crop and irrigation technology. Using cross-section, time-series data on the share of acreage using each technology/crop combination from three states over 14 years, they show that costs are a stronger determinant of adoption (two to three times) than revenue. The coefficients were then used to estimate the acreage and input use effects that would result from changes in electricity costs. Since drip irrigation was not included as a technology choice in the original study, the introduction of a low-volume technology was simulated in the second stage.

The studies by Shrestha and Gopalakrishnan (1988 and 1989) dealt with the adoption of drip irrigation for sugar cane production in Hawaii. At the time of the study, 60 percent of the sugar cane in the state was irrigated, and 82 percent of that acreage was irrigated by drip rather than furrow methods. Although the theoretical framework described in the papers was based on the Caswell and Zilberman (1986) threshold model of adoption, the empirical work and the discussion of the diffusion process were only loosely tied to the theory. Quadratic yield equations were estimated using cross-section time-series data as functions of water use, crop and land quality characteristics, and irrigation method. The yield equations show that the determinants of adoption include more than economic variability alone. Heterogeneous land characteristics and water-use efficiency of the modern technology appear to be significant determinants of adoption. A probit equation was also estimated to determine the probability of adoption as a function of yield, location, time, and agricultural practices. In interpreting their results, Shrestha and Gopalakrishnan imply that the positive coefficient on year shows that learning-by-doing is a strong force in adoption over time.

Empirical studies by Feinerman and Yaron (1990) and Fishelson and Rymon (1988) use cross-section time-series data gathered from a survey of cotton farmers in Israel from 1976 to 1983. Feinerman and Yaron (1990) use the annual growth rate of the share of drip irrigation (compared to sprinkler) as the dependent variable, and they hypothesize that the growth is a function of the relative profitability of drip, the increase in cotton area, the cumulative area of drip, and the year. They conclude that profitability is a major motive for adoption with the understanding that profitability varies with respect to soil, slope, weather, etc. Fishelson and Rymon (1988) use a linear logistic function to estimate the saturation level of adoption and the coefficients of a time trend. Their results do generate a logistic pattern for the adoption process.

Dinar and Yaron (1990) empirically test the theoretical hypotheses that the adoption of modern irrigation technologies in Israel on citrus is positively related to increases in input and output prices, farm size and organizational structure, and negatively related to input quality and age of the trees. They use a linear regression with the share of adoption for low-volume irrigation as the dependent variable. Their results confirm the hypotheses. Dinar and Yaron also represent the speed of adoption (the time lag between technology introduction and adoption) as functions of the same variables, and their results show that diffusion will be most rapid when the conditions above hold.

## CONCLUSIONS AND FUTURE RESEARCH

Studies of irrigation technology adoption are relatively new. They are derived from earlier work on the adoption of both agricultural technologies in developing countries and industrial innovations. Those empirical studies based on a theoretical economic framework are derived primarily from the threshold model of adoption (David, 1975) that shows that there may be a critical level of a heterogeneous characteristic which will separate adopters from nonadopters. The simple threshold model presented here developed a theoretical framework of individual decisionmaking by tying the agronomic and engineering relationships of the alternative technologies to the prevailing economic conditions when land quality is heterogeneous. Most of the studies that were reviewed confirmed the hypotheses derived from the theoretical model. The adoption of modern irrigation technologies is more likely on lower quality land, when crop and input prices are high, and when the cost of switching technologies is low.

Several important research lines remain to be explored. Institutional impediments to adoption (legal, political, social, economic, and bureaucratic) should be considered. Credit constraints based on income or land value have been found to slow the adoption process in some LDC's, and such constraints may become important in the United States if environmental regulations and drainage problems reduce land values. Water conveyance systems may also be a constraint. Farmers who receive water from delivery agencies that are only able to deliver water on an intermittent basis would be unable to adopt drip irrigation without investing in water storage facilities. Current water law may also affect a farmer's choice of irrigation technology. The incentive to conserve water would be diminished if farmers feared that their future rights to deliveries would be negatively affected by reducing their water demand. The benefits to be gained by removing institutional constraints to adoption need to be assessed.

Uncertainty issues should be included in future studies and results obtained from mathematical simulations need to be confirmed using field data. Also, adoption-induced endogenous price changes identified by Caswell and Zilberman (1990b) need to be assessed, and the magnitude of the secondary effects on output supply and input demand estimated. As data become available, the initial models reviewed here will be expanded and tested. Hopefully, the results can be used to improve the management of water supplies and drainage flows in California and in other arid regions.

## NOTES

[1] Irrigation efficiency is defined here as the ratio of water effectively used for crop growth to the amount of water actually applied to the field.

[2] For simplicity, only a single variable input is used for the presentation. W can also represent a vector of input prices in a more general model.

[3] Although most of the discussion compares two technologies, the farmer has a suite of alternatives from which to choose at any time. In addition to irrigation equipment, furrow diking, shortened runs, irrigation scheduling, and other management strategies can be employed.

[4] Although this assumption is not strictly true, including a more realistic function adds complexity but no further insights into the theory.

# REFERENCES

Abbott, J. S., 1984. Micro Irrigation--World Wide Usage, *ICID Bulletin*, 33(1), pp. 4-9.

Amemiya, T., 1981. Qualitative Response Models: A Survey, *Journal of Economic Literature*, 19, pp. 1483-1536.

American Society of Civil Engineers, 1978. On-Farm Irrigation Committee.

Cason, T. N.; Casterline, G. L.; and Uhlaner, R. T., 1990. *A Discrete Choice Analysis of Agricultural Production Decisions.* University of California.

Casterline, G.; Dinar, A.; and Zilberman, D., 1989. The Adoption of Modern Irrigation Technologies in the United States. In: Schmitz, E. (Ed.), *Free Trade and Agricultural Diversification: Canada and the United States.* Westview Press, Boulder, CO.

Caswell, M. F., 1982. *Diffusion of Low-Volume Irrigation Technology in California Agriculture.* Unpublished Ph.D. Dissertation, University of California, Berkeley.

Caswell, M. F., 1989. Better Resource Management Through the Adoption of New Technologies. In: Botkin, D.; Caswell, M.; Estes, J.; and Orio, A. (Eds.), *Changing the Global Environment: Perspectives on Human Involvement.* Academic Press, New York, NY.

Caswell, M. F.; Lichtenberg, E.; and Zilberman, D., 1990. The Effects of Pricing Policies on Water Conservation and Drainage, *American Journal of Agricultural Economics.* Forthcoming.

Caswell, M. F. and Zilberman, D., 1985. The Choices of Irrigation Technologies in California, *American Journal of Agricultural Economics*, 67(2), pp. 224-234.

Caswell, M. F. and Zilberman, D., 1986. The Effects of Well Depth and Land Quality on the Choice of Irrigation Technology, *American Journal of Agricultural Economics*, 68(4), pp. 798-811.

Caswell, M. F. and Zilberman, D., 1990a. *The Diffusion of Resource-Quality-Augmenting Technologies: Output Supply and Input Demand Effects.* University of California Working Paper.

Caswell, M. F. and Zilberman, D., 1990b. *Supply and Demand Effects When Output Changes in Response to the Adoption of a New Technology.* University of California Working Paper.

Caswell, M. F.; Zilberman, D.; and Goldman, G. E., 1984. Economic Implications of Drip Irrigation, *California Agriculture*, 38, pp. 4-5.

CH2M Hill, 1989. *Irrigation System Costs and Performance in the San Joaquin Valley.* Report Prepared for the San Joaquin Valley Drainage Program.

Cochrane, W. W., 1958. *Farm Prices: Myth and Reality.* University of Minnesota Press, Minneapolis.

David, P. A., 1975. *Technical Choice, Innovation and Economic Growth*. Cambridge University Press.

Dinar, A. and Knapp, K. C., 1986. A Dynamic Analysis of Optimal Water Use under Saline Conditions, *Western Journal of Agricultural Economics*, 11(1), pp. 58-66.

Dinar, A.; Knapp, K. C.; and Letey, J., 1989. Irrigation Water Pricing Policies to Reduce and Finance Subsurface Drainage Disposal, *Agricultural Water Management*.

Dinar, A. and Yaron, D., 1990. Influence of Quality and Scarcity of Inputs on the Adoption of Modern Irrigation Technologies, *Western Journal of Agricultural Economics*, 15(2), pp. 224-233.

Feder, G.; Just, R. E.; and Zilberman, D., 1985. Adoption of Agricultural Innovations in Developing Countries: A Survey, *Economic Development and Cultural Change*, 33, pp. 255-298.

Feder, G. and O'Mara, G. T., 1981. Farm Size and the Adoption of Green Revolution Technology, *Economic Development and Cultural Change*, 30, pp. 59-76.

Feinerman, E., 1983. Crop Density and Irrigation with Saline Water, *Western Journal of Agricultural Economics*, 8(2), pp. 134-140.

Feinerman, E.; Letey, J.; and Vaux, H. J., Jr., 1983. The Economics of Irrigation with Nonuniform Infiltration, *Water Resources Research*, 19, pp. 1410-1414.

Feinerman, E. and Vaux, H. J., Jr., 1984. Uncertainty and the Management of Salinity with Irrigation Water, *Western Journal of Agricultural Economics*, 9(2), pp. 259-270.

Feinerman, E. and Yaron, D., 1990. Adoption of Drip Irrigation in Cotton: The Case of Kibbutz Cotton Growers in Israel, *Oxford Agrarian Papers*. Forthcoming.

Fishelson, G. and Rymon, D., 1989. Adoption of Agricultural Innovations: The Case of Drip Irrigation of Cotton in Israel, *Technological Forecasting and Social Change*, 35, pp. 375-382.

Gladwin, C. H., 1979. Cognitive Strategies and Adoption Decisions: A Case Study of Nonadoption of an Agronomic Recommendation, *Economic Development and Cultural Change*, 28, pp. 155-174.

Griliches, Z., 1957. Hybrid Corn: An Exploration in the Economics of Technological Change, *Econometrica*, 25, pp. 501-522.

Hiebert, D., 1974. Risk, Learning, and the Adoption of Fertilizer Responsive Seed Varieties, *American Journal of Agricultural Economics*, 56, pp. 764-768.

Hornbaker, R. H. and Mapp, H. P., 1988. A Dynamic Analysis of Water Savings from Advanced Irrigation Technology, *Western Journal of Agricultural Economics*, 13(2), pp. 307-315.

Just, R. E. and Zilberman, D., 1983. Stochastic Structure, Farm Size, and Technology Adoption in Developing Agriculture, *Oxford Economic Papers*, 35, pp. 307-328.

Letey, J.; Dinar, A.; Woodring, C.; and Oster, J. D., 1990. An Economic Analysis of Irrigation Systems, *Irrigation Science*, 11, pp. 37-43.

Lichtenberg, E., 1989. Land Quality, Irrigation Development, and Cropping Patterns in the Northern High Plains, *American Journal of Agricultural Economics*, 71(1), pp, 187-194.

Lindner, R. K., 1980. *Farm Size and the Time Lag to Adoption of a Scale Neutral Innovation*. University of Adelaide, Australia Working paper.

Mansfield, E., 1961. Technical Change and the Rate of Imitation, *Econometrica*, 29, pp. 741-765.

Mapp, H. P., 1988. Irrigated Agriculture on the High Plains: An Uncertain Future, *Western Journal of Agricultural Economics*, 13(2), pp. 339-347.

Negri, D. H. and Brooks, D. H., 1990. Determinants of Irrigation Technology Choice, *Western Journal of Agricultural Economics*. Forthcoming.

Nieswiadomy, M. L., 1988. Input Substitution in Irrigated Agriculture in the High Plains of Texas, 1970-80, *Western Journal of Agricultural Economics*, 13(1), pp. 63-70.

Postel, S., 1990. Saving Water for Agriculture, *State of the World, 1990*. W. W. Norton & Co.

Rogers, E., 1962. *Diffusion of Innovations*. Free Press of Glencoe, New York, NY.

Schultz, T. W., 1975. The Value of the Ability to Deal with Disequilibrium, *Journal of Economic Literature*, 13, pp. 827-846.

Schutjer, W. and Van der Veen, M., 1977. *Economic Constraints on Agricultural Technology Adoption in Developing Countries*. U.S. Agency for International Development, Occasional Paper No. 5, Washington, DC.

Shrestha, R. B. and Gopalakrishnan, C., 1988. *Water Use Efficiency Under Drip Irrigation: An Econometric Analysis*. University of Hawaii Working Paper.

Shrestha, R. B. and Gopalakrishnan, C., 1989. *Adoption and Diffusion of Drip Irrigation Technology: An Analysis Using a Joint Discrete and Continuous Choice Framework*. University of Hawaii Working Paper.

Thirtle, C. G. and Ruttan, V. W., 1988. *The Role of Demand and Supply in the Generation and Diffusion of Technical Change*. Harwood Academic Publishers, New York, NY.

Weil, P. M., 1970. The Introduction of the Ox Plow in Central Gambia. In: McLaughlin, P. F. (Ed.), *African Food Production Systems: Cases and Theory*. Johns Hopkins University Press, Baltimore, MD.

Welch, F., 1978. The Role of Investment in Human Capital in Agriculture. In: Schultz, T. W. (Ed.), *Distortion of Agricultural Incentives*. Indiana University Press, Bloomington.

# Four: ENVIRONMENTAL AND PUBLIC HEALTH IMPACTS OF DRAINAGE

# 16 ASSESSING HEALTH RISKS IN THE PRESENCE OF VARIABLE EXPOSURE AND UNCERTAIN BIOLOGICAL EFFECTS

Robert C. Spear, University of California, Berkeley

## ABSTRACT

The impact of water quality on human health is a longstanding concern. In the developing world pathogenic agents remain the principal issue, but in developed countries chemical contamination and the threat of chronic or delayed toxic response is the focus of public concern and regulatory activity. The nature of environmental exposures to chemicals and the population variability in response make it very difficult to determine population risks from traditional epidemiological studies. This has given rise to quantitative risk assessment, an activity which attempts to predict risk from secondary evidence concerning environmental transformation and transport, exposure mechanisms, and biological response probabilities. This process contains inherently uncertain and ambiguous elements which lead to conflict between scientists who become involved in legal or regulatory proceedings. The public, responding to the scientists' and regulators' inability to assure safety, has tended to lose confidence in governmental regulation of environmental health threats. New methods of measuring human exposure and preclinical response to chemicals may provide a renewed impetus to base risk assessments on epidemiological data rather than the unverifiable assumptions inherent in quantitative risk assessment as presently practiced. However, the political and legal aspects of risk acceptability are likely to continue to outweigh the technical dimensions of the problem.

## INTRODUCTION

Water quality and its impact on the health of human populations has been a central concern of the public health establishment for hundreds of years and remains so today. In the developing world, both the quantity and the quality of water available to the population have a strong influence on infant mortality as well as on morbidity associated with enteric disease in the adult population. In general the engineering solutions that lead to acceptable water quality are well developed and it is often mainly the lack of economic resources that prevents their application in the developing world.

In contrast, the water quality issues that concern developed countries have changed from a focus on infectious agents, which are largely under control, to a concern with chemical contaminants. Moreover, this concern is generally associated with low levels of chemical exposure that do not lead to acute disease, but may be expressed by chronic or delayed toxic responses which will be almost impossible to detect by traditional epidemiological methods. An exception may be the concern with reproductive effects associated with groundwater contamination by organic solvents, but a more common concern is for the carcinogenic potential of the trihalomethanes, for example, that are found in some domestic water supplies as a result of chlorination for the control of infectious agents.

The importance of this transition from a focus on infectious agents, whose effect is expressed in the human population very shortly after the exposure takes place, to potentially cancer causing chemicals whose effects are likely to be detected clinically perhaps several decades after exposure, has caused a major change in the way that the health effects of environmental chemicals are evaluated. First, there is an understandable reluctance to wait 10 or 20 years to determine if elevations in cancer rates may be due to some environmental factor and then to implement control. Secondarily, it is almost impossible to link environmental exposure to cancer incidence given the small elevations in cancer rates that might generally be expected combined with the high background level of some cancers and the high level of mobility that characterizes many American populations.

The potential health effects associated with agricultural drainage water fall in this class of environmental health issues in which traditional epidemiological surveillance is likely to be of little use. If there are any public health impacts of drainwater usage or disposal, they are not likely to lead to acute effects that are easily detected by the usual means of health surveillance. On the other hand, if there are exposures sufficient to cause chronic or delayed toxic effects, they are likely to effect small numbers of people who are temporally and geographically distant from one another by the time such effects might be expressed; that is, they are likely to be epidemiologically invisible.

The current approach to dealing with such problems is via chemical-by-chemical analyses which attempt to infer, from secondary evidence, the potential magnitude of the human health risk. This process has been most highly developed in the context of carcinogenic risks and falls within the general category of activities called quantitative risk assessment (National Research Council, 1983). The risk assessment process has always been a part of the priority setting activity in environmental health, although it has become increasingly formalized and explicit in recent years, particularly in the case of cancer endpoints. To the extent that either the public or government agencies sense an environmentally associated chemical threat, some form of risk assessment is sure to follow. Elsewhere in this volume such an assessment is described in detail for the principal chemical constituents of drainwater that are of health concern (Klasing, 1991). In this chapter some of the generic aspects of that process will be reviewed with the object of gaining insight into its inherent limitations and speculation will be offered on new directions that the future may bring to the risk assessment process as applied to low level exposures to environmental contaminants.

## THE SINGLE AGENT PARADIGM

Epidemiology is the study of the distribution of disease in human populations. Its primary objective is to gain an understanding of those risk factors that determine this distribution. Clearly, epidemiology is a post hoc activity in that cases of disease must be present to study. Contemporary American society is wealthy enough, and increasingly committed enough, to support risk assessment and prevention programs as part of environmental planning and industrial and agricultural development. The goal is to be proactive and intervene early enough to avoid the need to confront failure by the subsequent enumeration of cases of disease. If, for example, a municipality is faced with the construction of a domestic waste incinerator it is now common practice to forecast the potential number of cancer cases its emissions may cause in the adjacent population over the lifetime of the facility. Sometimes there is human data to assist in this assessment and sometimes not. In either case, the generic process that must be undertaken is depicted in figure 1.

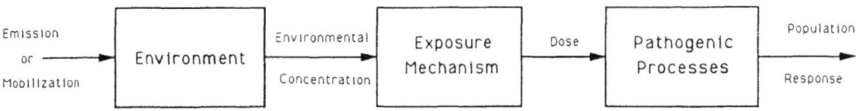

Figure 1. Causal linkages between emission and population response to environmental contaminants.

In the context of agricultural drainage, the process begins where irrigation water is introduced into the "environment," and some chemical, selenium for example, is mobilized and transported by one or more mechanisms to a point where it interfaces with the human population. At that point, an exposure mechanism operates which mediates between the concentration in the environment and the dose delivered to individuals in the population. Perhaps selenium is taken up into the food chain and bioaccumulates in animal or fish species to a point where health effects may ensue if these species are consumed by humans. A description of the exposure mechanism, in this case, requires quantitative estimates of the frequency of consumption of the contaminated foodstuff and the amount. The characterization of the environmental aspect of this process may involve a prediction of the selenium levels at various points in the environment and in the food chain. In cases where the system is already in place and operational, direct measurements of selenium levels are possible at various points in the system. A complete exposure assessment must take into consideration all potential exposure routes. In practice, there is usually considerable uncertainty in quantitative estimation of exposure, particularly when attempting to estimate historical exposures.

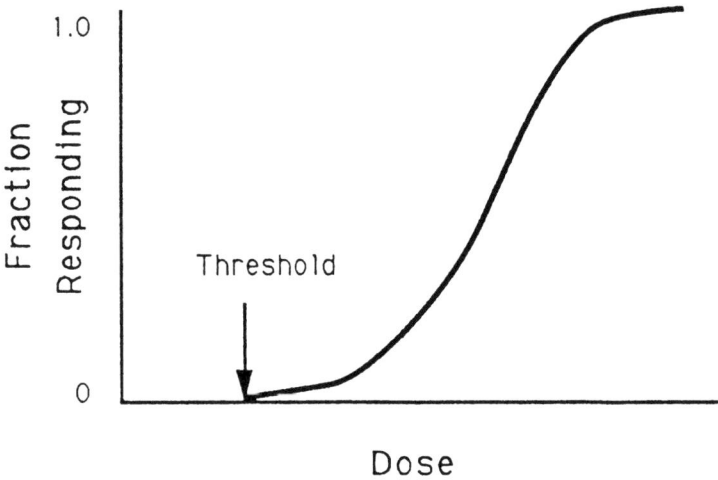

Figure 2. The dose-response function.

The final step in the assessment requires the estimation of the dose-response relationship for the particular chemical and the particular response in question. Because toxic response to chemicals is often sensitive to very small differences in molecular configuration, the vast majority of toxicological data pertain to single chemical species, usually in a highly purified form. The dose-response concept lies at the heart of toxicology and at the heart of quantitative risk assessment. In principal, the concept is straightforward as depicted in figure 2. As the dose increases, the fraction of an exposed population who manifest the response increases also, i.e., "the dose makes the poison." (Alternatively, the curve can be interpreted as giving the response probability at any dose for a randomly selected member of the population.) With the exception of carcinogens, it is generally assumed that there is a threshold dose below which no response occurs. At the other end of the scale there is assumed to be a dose above which all members of the population will respond.

The dose-response curve is most easily illustrated in the context of animal testing for acute toxins. The endpoint is death and the dose-response relation is established from a set of experiments in which different doses are administered and the fraction of animals dying at each dose is observed. These data are often used to estimate the LD50, the dose which kills one-half the animals. The LD50 is the most elementary index of acute toxicity in common use. Clearly, there are endless variations on the dose-response theme; animal cancer testing in which the dose is commonly the cumulative lifetime dose administered to the animal; subtle behavioral responses evoked by either acute or chronic exposures, etc. The difficulty in the risk assessment context is, of course, that there is very little dose-response data in human populations and extrapolation from animal results is fraught with difficulty (Ames, 1989).

For the moment, let us rise above the fact that we seldom have dose-response data useful for predicting human response and move ahead to see what could be done if such data were available. If the dose-response curve quantifies the biological variability in response to the agent in question, then the missing link is a description of the variability in dose across the exposed population. A moment's reflection should suffice to convince anyone that there is likely to be great variability in human exposure and delivered dose to virtually all environmental chemicals. This issue has been studied most extensively in the workplace where one might expect the variability in exposure to be modest since the location is fixed and workers can be identified who are engaged in more or less similar tasks. Figure 3 shows exposure data for a group of workers in an alkyl lead manufacturing facility (Cope et al., 1979). These data show the variability in average daily concentration in the breathing zone of the workers. This concentration is directly proportional to daily inhaled dose. As can be seen, among this group of workers the average daily concentration varies over a thirtyfold range despite the fact that they are engaged in the same tasks in the same place from day to day.

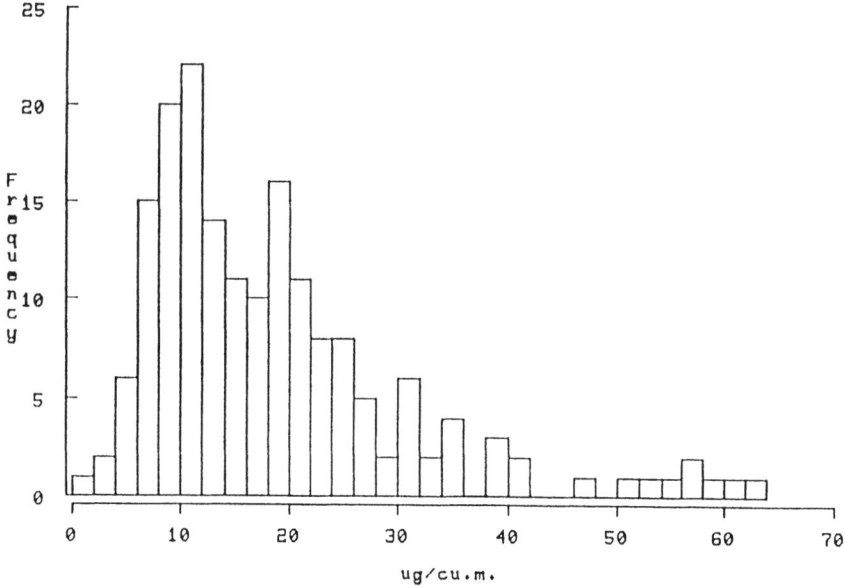

Figure 3. Time-weighted average exposures to airborne alkyl lead in a manufacturing facility as reported by Cope et al., 1979.

As in the foregoing example, exposure to environmental chemicals is often described by statistical distributions of daily exposure concentration or daily dose. Almost universally these distributions are assumed to be log-normal, i.e., the logarithms of the concentration values are normally distributed. A common parameter used to describe the variability in these data is the geometric standard deviation which tends to range from about 1.5 to 5 or more for day-to-day variations in workplace exposure concentrations. There is very little data on the variability in exposure to environmental chemicals outside the workplace. However, for any chemical exposure of public health concern, the exposure variability across the population is likely to be at least as great as encountered in occupational settings.

So, in concept, the dose-response curve reflects population variability in response at a given dose and the exposure or dose distribution reflects the variability in the dose received by individuals comprising this population. Putting these two pieces of information together, as is conceptually indicated in figure 4, would allow the prediction of the distribution of the probability of response, which will be called risk, across this population. That is, each individual in the population receives a dose from the dose distribution and, according to the dose-response curve, has then a certain probability of response. The collection of those probabilities is the population distribution of risk as indicated in the figure.

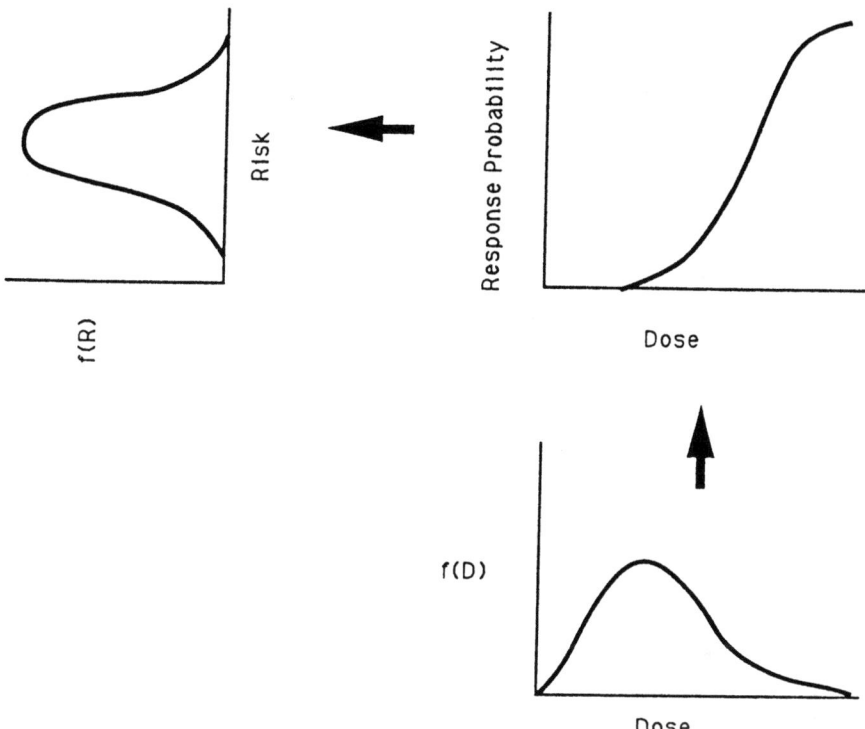

Figure 4. The distribution of disease risk as a function of the variability in dose and in the dose-response probability.

It is a particularly interesting fact that, if the largest risk in the population is less than about 0.01 (which is quite high in the general context of environmental exposures), then the distribution of the number of cases of disease is approximately Poisson with a parameter that depends only on the product of the average risk and the population size (Bogen and Spear, 1987). The fact that the distribution of cases is a function only of the average risk, or equivalently the average dose in the case of the linear dose-response function, underscores the fact that it is very difficult to detect the effects of environmental exposures in general population studies or from commonly available health statistics. This is because, with few exceptions, the exposure of most people to a particular environmental contaminant is very low by most conventional criteria and only subpopulations, occupational groups for example, may be exposed at levels high enough to lead to significant risk. This is also why risk assessments have as one of their primary objectives the search for the highly exposed subpopulations, for example, the subsistence gardener mentioned by Klasing (1991) in discussing selenium exposure.

To trace a chemical contaminant from its source in the environment through the complex web of exposure pathways with the objective of quantifying population dose distributions, and then moving on to risk estimation in any quantitative way, is clearly a very complex challenge and one fraught with inherent uncertainty. It is, in many ways, the ultimate reductionist challenge. In a particular case one might require specific knowledge from field and laboratory scientists as diverse as geochemists, hydrologist, wildlife experts, fisheries experts, nutritionists, toxicologists, and physicians. In such a case, each of these experts may be able to contribute only a small piece of the needed information, but it may be a crucial piece and it is often difficult to predict in advance just which piece this may be. Risk assessments are typically carried out by generalists relying on the literature for this specific input, but in important cases involving litigation, in particular, the array of expert witnesses can be extensive. As is easily seen, the inherent uncertainty in both the structure of the linkages between environment and toxic response, notwithstanding the uncertainty leading to quantitative predictions of risk, contain the seeds of transscientific arguments between competing experts and an erosion of public confidence in the process, as well as in its quantitative outcome. It is not the purpose here to discuss risk management, other than to point out that many of its challenges emanate from the changing focus of concern from acute health effects to chronic and delayed responses which cannot be usefully studied by conventional epidemiological methods, at least not in the context of environmental management and land use planning decisions.

## NEW DIRECTIONS IN RISK ASSESSMENT

The inherent difficulty with the set of procedures falling under the general rubric of quantitative risk assessment is that its predictions are, in general, unverifiable. There will be no one who actually does live for 70 uninterrupted years at the point of highest exposure downwind from a domestic waste incinerator that discharges trace amounts of dioxin. And if there were, there would be no way to isolate this individual from all other environmental exposures that might be responsible for his or her cancer or its absence, let alone could one expect to have a good estimate of the individual's dioxin exposure. There is much to be said for the epidemiological approach in that its results are based on observed facts rather than predictions. Yet, waiting for cases of chronic disease or cancer to occur is simply not an acceptable means of guiding preventive health policies or deciding on the need for environmental interventions. Fortunately, the revolution in biology holds promise of reinvigorating the epidemiological approach by providing indices of biological response to chronic and delayed toxins in human populations well in advance of the onset of clinical disease or disability. These methods do not offer a solution to the problem of forecasting the effects of new activities, but they hold promise for new and efficient ways to address the potential hazard of current exposures to chronic or delayed toxins which, if successful, would make forecasting a less crucial activity.

The idea is straightforward and has a long history in occupational health in which it has been common practice, for example, to monitor workers' blood lead levels or blood levels of the enzyme cholinesterase which is important to the mechanisms of toxicity of the organophosphate pesticides. Taking the lead example, epidemiological studies and clinical investigations have established blood lead levels below which no clinical disease is likely to develop in adults and this level can be used in risk management through the routine screening of high risk populations. Workers with levels above this cutoff are removed from exposure.

This same approach can be used in the investigation of hazardous community exposures where there are measurable biological markers of response. Indeed, the use of biological monitoring has not been possible for most diseases arising from chronic or delayed toxic mechanisms because of the absence of reliable preclinical endpoints. But now the possibilities are promising and, perhaps, best exemplified in the context of carcinogenesis. If a substance is to initiate processes in people which eventually are manifested as cancer, the substance presumably must interact with DNA. It is now possible to detect DNA modified by the adduction of various environmental chemicals or their metabolites. Take the case of styrene, a common chemical used extensively in reinforced plastics fabrication. DNA adducts arising from styrene exposure have been measured in workers [6]. Styrene has not been shown to be a human carcinogen and just what the implications of measurable styrene adducts are for the eventual cancer risk is quite unclear. However, the presence of styrene adducts is unequivocal evidence of exposure and reaction with biologically important macromolecules. It may be that these adducts will be repaired over time and that the cessation of exposure will lead to their eventual disappearance. If one were to know the level of cancer risk predicted by adduct levels, as we know for lead-related diseases and blood lead, then the entire issue of assessing exposure risks to the human population is restored to the realm of measurements and facts rather than of calculation and speculation in the presence of unresolvable uncertainty.

Note that DNA adducts remain part of the single agent paradigm of traditional toxicology in that they are compound specific. There are other biochemical and cytogenetic endpoints which respond to a broad range of chemical exposures, but which share the unfortunate attribute of adducts of being of unknown utility as predictors of disease risk. Sister chromatid exchange (SCE) is a cytogenetic endpoint which can be measured in peripheral blood and which shows a dose-response relationship for at least some chemicals. SCE's are induced by many animal carcinogens (Mason et al., 1990). Although the clinical significance of this test remains unknown, the observation of SCE's in worker populations has played a role in Federal regulatory decisions relating to safe standards of exposure (Yager, 1984). The micronucleus test is another such method which is receiving much current attention (Yager, 1990).

There is little doubt that the study of biomarkers of human exposure and response to environmental chemicals will be the focus of a great deal of research in the coming years. While their application is likely to also remain largely in the research arena for the foreseeable future, it is already clear that they are attractive to plaintiff attorneys who are litigating toxic tort actions. Quantitative risk assessment methods also have found their place in the courts. In both cases this has led to considerable uneasiness among scientists in the relevant disciplines who have become aware that different standards of evidence operate in legal and regulatory matters than in the scientific journals. On the other end of the spectrum, the politicians continue to respond to the environmental concerns of the electorate and the regulators, placed squarely in the middle, are faced with responding to this mandate. At least, quantitative risk assessment provides a relatively structured framework for the regulators to attempt to meet the demands of their various constituents. In this regard, it can be anticipated that the general factors motivating the development of quantitative risk assessment for carcinogens will be extended to other biological endpoints.

Even with the tools of modern biology the assessment of the health risk of exposure to drainage water, either directly or via the food chain, will remain an uncertain endeavor. To go back to the beginning of the discussion, the toxicologist, for example, would immediately ask "what is drainwater" by which he or she means "what exactly is the chemical composition of this material." It immediately becomes a difficult issue because drainwater is not well defined chemically and, as an entity, it certainly does not fit the single chemical paradigm of toxicology. That problem will not disappear even with the new tools of biology because, even if one were to find no adducts, SCE's or other indices of response in one locality, we cannot assume that the composition of drainwater in the next county will be toxicologically identical.

So, for the foreseeable future, science will not be able to give the public the unambiguous assurances of absolute safety that it demands and the management of environmental health risks will continue to be played out in the labyrinth of administrative procedures and spill over into the political arena and the courtroom. Lowrance (1976) has given a definition of safety which is instructive in this context. "A thing is safe if its attendant risks are judged to be acceptable." While there will remain uncertainty in the technical aspects of risk assessment, much of the argument will continue to be focused on the question of acceptability and, in particular, acceptability to whom.

## REFERENCES

Ames, B. N., 1989. Mutagenesis and Carcinogenesis: Endogenous and Exogenous Factors, *Environ. Mutagenesis,* 13.

Bogen, K. T. and Spear, R. C., 1987. Integrating Uncertainty and Interindividual Variability in Environmental Risk Assessment, *Risk Analysis,* 7.

Cope, R.; Pancamo, B.; Rinehart, W.; and Ten-Haar, G., 1979. Personal Monitoring for Tetra-Alkyl Lead Exposure in the Workplace, *American Industrial Hygienists Assn. Journal,* 40.

Klasing, S. A., 1991. Consideration of the Public Health Impacts of Agricultural Drainage Water Contamination. This Volume.

Liu, S. F.; Rappaport, S. M.; Pongracz, K.; and Bodell, W. J., 1988. *Detection of Styrene Oxide-DNA Adducts in Lymphocytes of a Worker Exposed to Styrene, Methods for Detecting DNA Damaging Agents in Humans.* International Agency for Research on Cancer (IARC) Scientific Publication 89.

Lowrance, W. F., 1976. *Of Acceptable Risk.* William Kaufman Inc., Los Altos, CA.

Mason, J. M.; Langenbach, R.; Shelby, M.; Zeiger, E.; and Tennant, R. W., 1990. Ability of Short-Term Tests to Predict Carcinogenesis in Rodents, *Ann. Rev. Pharmacol, Toxicol.,* 30, pp. 149-168.

National Research Council, 1983. *Risk Assessment in the Federal Government: Managing the Process.* National Academy Press, New York, NY.

Yager, J. W., 1990. Development of Cytogenetic Indices for Exposure Assessment: Micronuclei in Peripheral Lymphocytes. In: Thomas, V. and Ogata, M. (Eds.), *American Conference of Industrial Hygienists, Inc., Cincinnati, OH.*

Yager, J. W., 1984. Testimony before the Department of Labor, Occupational Safety and Health Administration Hearings on Occupational Exposure to Ethylene Oxide, *Federal Register,* 49(122), pp. 25746-25748.

# 17 CONSIDERATION OF THE PUBLIC HEALTH IMPACTS OF AGRICULTURAL DRAINAGE WATER CONTAMINATION

Susan A. Klasing, Health Officers Association of California

## ABSTRACT

While drainage water contamination can have severe impacts on the agricultural economy and wildlife in an affected region, the potential costs to humans from a public health perspective cannot be overlooked. Some trace elements found in drainage waters are necessary for adequate human nutrition; however, often a relatively narrow margin exists between safe and toxic levels. Natural evaporative processes as well as bioaccumulation and/or biomagnification through the food chain may occur during irrigation and subsequent disposal of drainage water. This can greatly increase the ultimate exposure of humans to these elements. Selenium is used as an illustration of this problem in the western San Joaquin Valley (Valley).

## INTRODUCTION

With the recognition of waterfowl deaths, deformities, and reproductive failures at Kesterson Reservoir in 1983, questions began to arise regarding potential human health impacts resulting from exposure to agricultural drainage water contamination. Although humans are not expected to have continuous, direct contact with drainage contaminants as can occur in the case of fish and many wildlife, significant human exposure pathways are possible.

Almost immediately following reports of toxicity to wildlife at Kesterson Reservoir, the Merced County Health Department initiated a survey of possible health problems in Kesterson area residents. The survey was limited in scope, but evidence was not found to indicate that adverse health effects were occurring as a result of proximity to Kesterson Reservoir (Merced County Health Department, 1985). Similarly, results from repeated medical examinations of Kesterson National Wildlife Refuge personnel showed traditional biochemical indicators of selenium exposure to be within normal limits (see Klasing and Pilch, 1988, for further discussion).

This chapter summarizes the findings of 3 years of research on this issue to evaluate the toxicity, metabolism, and environmental fate of the primary drainage water constituents, as well as potential routes of human exposures to these compounds. The need for additional information such as appropriate indicators of excessive exposure to these substances for human population groups was also identified.

## SUBSTANCES OF CONCERN IN AGRICULTURAL DRAINAGE WATERS

Although adverse effects in wildlife exposed to drainage waters in the western Valley were generally attributed to selenium, the potential toxicity of other contaminants also required investigation. A preliminary evaluation of the scientific literature was conducted to assess the potential human toxicity of approximately 20 substances considered to be elevated in drainage waters.

Selenium, boron, and molybdenum were deemed by most agencies to comprise the most serious threat to either plant, wildlife, or human health. Selenium was immediately chosen for further human health evaluation because of its known human toxicity and previously documented effects in wildlife of the region. While boron had been considered primarily a plant toxicant, concentrations of the element in drainage waters were sufficiently high to warrant evaluation from a public health perspective. Additionally, recent investigations by the U.S. Fish and Wildlife Service (FWS) had indicated that boron, in levels commonly found in drainage waters, was capable of causing reproductive toxicity in ducks (FWS, 1986). Although less data were available to document molybdenum levels in various exposure media in the Valley, published reports had suggested that naturally elevated concentrations of molybdenum had caused human toxicity in India (Deosthale and Gopalan, 1974) and Russia (Kovalsky et al., 1961). The National Research Council (NRC) had also noted that molybdenum toxicity from natural sources was more likely to occur in human populations than deficiency syndromes (NRC, 1980).

In general, much less data were available regarding concentrations of arsenic, mercury, nitrates/nitrites, uranium, and vanadium in potential exposure sources in the region. However, for arsenic, mercury, and nitrates, the scientific literature was replete with toxicity information. Arsenic is a known human carcinogen with an established Ambient Water Quality Criterion of 0.002 p/b set by the U.S. Environmental Protection Agency (EPA, 1989a). Nitrates are a well recognized contaminant of ground water in many agricultural regions. Nitrates can cause potentially fatal methemoglobinemia in infants (WHO, 1984) and, possibly, sensory impairments in older children (Petukov and Ivanov, 1970). Mercury contamination of water results in accumulation of an extremely toxic form of the element in fish (Lu, 1974; Jensen and Jernelov, 1969; and NRC, 1977); this process has caused severe human illness and death in at least one country (Study Group of Minamata Disease, 1968). While much less is published regarding the human toxicity of uranium and vanadium, discussions with many scientists conducting various types of drainage water research led to the conclusion that concentrations of these trace elements were potentially problematic in local drainage waters, warranting further study. Toxicological and preliminary exposure assessments for each of the above elements were published in Klasing and Pilch (1988) and Klasing et al. (1990).

## POTENTIAL HUMAN EXPOSURE ROUTES TO DRAINAGE CONTAMINANTS

In almost any geographic area, various forms and concentrations of the typical elemental constituents in agricultural drainage waters will be found. Many of these elements are required for adequate human or plant nutrition and occur routinely in air, ground water, surface water, soil, plants, and animals. Although these elements exist naturally, their elevated concentrations in some waters, sediments, and biota in certain areas of the western San Joaquin Valley can be directly related to agricultural practices. The irrigation and drainage of highly mineralized soil mobilizes soil constituents; subsequent transport of the drainage waters can lead to accumulation of high concentrations of these constituents in areas where they would otherwise not be expected. It is in these cases, where levels of a particular substance grossly exceed normal "background" concentrations, that excessive human exposures are most likely to occur.

The possible human exposure routes of drainage contaminants can be classified as direct or indirect. Direct routes would include consumption, dermal contact with, or inhalation of contaminated surface water, ground water, soil, or air. Constituents can occur in air in particulate, aerosol, or gaseous (volatilized) forms. Selenium, for example, is volatilized naturally from soil, sediments, plants, and (when the element is consumed in excess) from animals (Abu-Erreish et al., 1968; Chau et al., 1976; McConnell and Roth, 1966; and Zieve and Peterson, 1984). Indirect routes of exposure would include the consumption of fish, wildlife, cultivated or wild plants, and livestock (or livestock products) that have been exposed to drainage water either directly or indirectly. Some plants and animals selectively accumulate certain minerals to levels several orders of magnitude above that in the waters to which they are exposed (a process known as "bioconcentration"). In this way, indirect exposure routes can be much more significant from a human health perspective than direct routes. It is for this reason that ambient water quality criteria for the protection of human health can be more stringent than drinking water standards for the same element.

## POPULATIONS AT RISK

Determination of the populations at risk from exposure to agricultural drainage water contaminants relates directly to the toxicity of the particular substance of concern and the degree of human exposure. A thorough examination of the available toxicological data base for a substance usually permits the development of an acceptable daily intake (ADI) from all sources. From the ADI, acceptable concentrations of each substance can be determined for different exposure media (e.g., soil, air, water), given typical human exposure rates to each source (e.g., the amount of water consumed in one day, or the amount of air inhaled in one day). Various standards and criteria have been developed by State and Federal agencies, which may be enforceable or merely advisory in nature, to limit concentrations of certain substances in different exposure media. For other substances, particularly many of those that are naturally occurring, restrictive criteria or standards do not exist and predicted safe concentrations must be determined on an ad hoc basis.

To calculate the total human exposure to a particular drainage contaminant, the concentrations must be known for all potential exposure routes. From these results, the most likely routes of significant exposure can be determined and, thus, which population groups are likely to be at special risk from overexposure. In the western San Joaquin Valley, concentrations of many substances of concern in agricultural drainage waters have been surveyed in area soils, ground and surface waters, sediments, and biota. Additionally, selenium has been analyzed in approximately 125 crop samples from the region, in the air surrounding Kesterson Reservoir, and in several liver and milk samples from bovine raised in the area.

The management of drainage water from the irrigation of arid (saline) lands has led to relatively unique potential exposure scenarios for the agricultural industry. Transport of drainage waters through surface sloughs and rivers as well as disposal of such waters in evaporation ponds has led to the accumulation of drainage contaminants in aquatic biota (Moore et al., 1989). In particular, fish and aquatic birds residing in and around affected waters have, in some cases, concentrated sufficient quantities of some contaminants in bodily tissues to cause death or serious illness to the animal. Human consumers of these contaminated fish and game are also exposed to elevated concentrations. Thus, hunters and fishermen are at special risk from overexposure to drainage contaminants.

Cultivated crops can accumulate above-average concentrations of some elements when grown in soils or irrigated with water containing high levels of those constituents. However, because of the typical food distribution patterns in this country, it is unlikely that a person would receive repeated exposure to the same produce source (Cook, 1988). An exception to this would be people who meet the bulk of their dietary needs from a single geographic area. Subsistence gardeners, for example, are at special risk from both deficiency and toxicity of constituents found lacking or elevated in their soil or irrigation water.

In some regions, cultural or economic factors may result in foraging of nontraditional food items from the environment (Campbell and Christensen, 1989). Foragers may consume wild plants with a greater propensity for accumulation (such as selenium accumulator species) and/or greater contact with drainage contaminants (such as plants growing in and along drainage canals and sloughs). Foragers may also consume significant quantities of fish and game from contaminated regions, thus increasing their overall risk of overexposure.

Agricultural workers are also at special risk from exposure to drainage water contaminants. The most likely occupational exposures are dermal contact with water, soil, or sediment and inhalation of particulate, aerosol, and volatilized contaminants occurring in water, soil, or sediments.

The above discussion describes population subgroups that may be exposed to greater than average concentrations of drainage contaminants and are therefore at greater risk of suffering adverse health outcomes than the population as a whole. Other population subgroups may be at special risk from toxicity to these contaminants because they are more highly susceptible to their effects. For example, infants are more sensitive to nitrate exposure than are older children or adults (WHO, 1984). Similarly, mercury compounds are particularly toxic to the fetus (Berlin, 1986). In adults, poor nutrition or the genetic absence of an enzyme necessary for the detoxification of a chemical can lead to unexpected toxicity from exposure to a given dose of that chemical.

In summary, individuals may be at special risk from toxicity to drainage contaminants for two reasons: (1) elevated exposure and (2) increased susceptibility to their toxic effects.

## THE EXAMPLE OF SELENIUM AS A PUBLIC HEALTH RISK FROM DRAINAGE WATER CONTAMINATION

As mentioned previously, samples of soil, air, water, sediment, crops, fish, wildlife, and some bovine tissues from the western Valley have been collected and analyzed for selenium by various State and Federal entities. Klasing and Pilch (1988) collated and converted the available data to appropriate forms for public health analyses. For foodstuffs, typical portion sizes were obtained from U.S. Food and Drug Administration and U.S. Department of Agriculture data (Pennington and Church, 1985 and Watt and Merrill, 1975). For other media (such as drinking water), U.S. Environmental Protection Agency figures were used to estimate total daily consumption.

Table 1 presents selenium concentrations in cow's milk and liver, as well as in a variety of fish and game tissues obtained from different regions within the western Valley. The data, which were obtained from many different studies and sources, are presented as a range of mean concentrations and as the maximum concentration observed. Additionally, levels have been converted to the number of micrograms of selenium present in a single serving (based on typical portion sizes); these are presented as either a typical range or "worst" case for each foodstuff.

Table 2 provides similar data for representative vegetable crops grown in the western Valley. In the case of many food items, National averages for nutrient concentrations are available from various surveys (see, for example, Pennington and Church, 1985). Where available, these averages have been included in table 2 to provide comparisons to values of local products. It must be noted, however, that because of wide geographic variability, as well as inconsistencies in different analytical methodologies, National trace element averages may not be reliable (Pennington and Church, 1985).

Information from the scientific literature also has been used to evaluate the relative availability of selenium from different exposure routes. This is especially important in cases where a particular exposure route may strongly influence the rate of selenium absorption (e.g., dermal exposure). This information has been described more fully in Appendix A of Klasing and Pilch (1988).

Based on the information provided in tables 1 and 2, exposure routes have been ranked for their potential contribution to selenium exposure of humans in the local area and are discussed in the following pages and summarized in table 3. It is important to note, that unless specifically stated, rankings do not necessarily imply concomitant health risk (i.e., a ranking of 3 does not imply a threefold greater health risk of exposure from this route than from a route with a rank of 1). The criteria used to rank exposure routes were as follows: 1 was used when levels of the element from this route generally were within Nation-

# PUBLIC HEALTH IMPACTS

Table 1. Examples of selenium concentrations in and potential human exposures from consumption of livestock, fish, and wildlife in the western San Joaquin Valley[a].

| Source | Tissue | Geographic Area[b] | Se Concentration (p/m)[c] Range of Means | Maximum | Se Exposure (ug/serving)[d] Typical Range | "Worst" |
|---|---|---|---|---|---|---|
| Bovine | Milk | A | 0.013-0.023 | 0.041 | 3.05-5.59 | 10 |
|  | Liver | A | .2982-0.3522 | 0.510 | 29.8-35.2 | 51 |
| Calif. ground | Liver | K | 0.243-0.546 | 0.870 | 24-55.6 | 87 |
| squirrel | Liver | VWA | 0.174-0.330 | 0.33 | 17-33 | 33 |
| Cottontail | Thigh m. | K | 0.504-0.528 | 0.792 | 50-53 | 79 |
| rabbit | Thigh m. | VWA | 0.027-0.029 | 0.029 | 2.7-2.9 | 3 |
| Muskrat | Liver | K,SLD | 0.510-9.63 | 27.6 | 51-963 | 2760 |
|  | Liver | VWA | 0.105-0.570 | 0.570 | 11-57 | 57 |
| American | Liver | K | 21.3 | 41.8 | 2130 | 4180 |
| coot | Liver | GWD,VWA,LHR, TLD,STW | 0.92-6.70 | 11.0 | 92-670 | 1100 |
|  | Egg | K | 8.96 | 21.5 | 448 | 1075 |
|  | Egg | VWA | 0.22 | 0.32 | 11 | 16 |
|  | Breast m. | TLD, STW, LHR | 0.26-2.50 | 5.8 | 26-250 | 580 |
| Cinnamon | Egg | K | 2.12-4.19 | 11.47 | 106-210 | 574 |
| teal | Egg | GWD, VWA | 0.65-2.02 | 2.08 | 33-101 | 104 |
|  | Breast m. | TLD,STW,LHR | 0.58-1.50 | 2.8 | 58-150 | 280 |
|  | Liver | TLD,STW,LHR,WF | 2.5-5.7 | 11.0 | 250-570 | 1100 |
| Mallard | Liver | GWD | 0.99-1.60 | 3.40 | 99-160 | 340 |
|  | Egg | K | 3.22-4.71 | 9.61 | 161-236 | 481 |
|  | Egg | GWD, VWA | 0.37-1.13 | 1.86 | 19-57 | 93 |
| Black bullhead | Whole | MS | 1.7-2.5 | 2.5 | 170-250 | 250 |
| Bluegill | Whole | MS,VW,SJR, SS,HC | 0.22-4.60 | 4.60 | 22-460 | 460 |
| Channel | Whole | CD | 0.95-1.20 | 1.20 | 95-120 | 120 |
| catfish | Whole | SLD | 13.5 | 17.0 | 1350 | 1700 |
|  | Muscle | SJR,SS,MS,CD | 0.24-0.75 | 0.90 | 24-75 | 90 |
| Common | Whole | SLD | 34.3 | 60.0 | 3430 | 6000 |
| carp | Whole | AC,HC,SJR,SLC, CD,LB,MS,SS,VWA | 0.08-3.79 | 3.79 | 8-379 | 379 |
| Green sunfish | Whole | SS,HC,MS,LB, GWD,CD,MC | 1.2-5.5 | 5.5 | 120-550 | 550 |
| Sacramento blackfish | Whole | SLD | 16.2 | 33.0 | 1620 | 3330 |
|  | Whole | CD,HC,LC,SS,SJR | 0.18-3.9 | 3.9 | 18-390 | 390 |
| Striped Bass | Whole | LB,MS,SS,VW, MC,SJR,HC | 0.30-2.27 | 2.35 | 30-227 | 235 |
| White | Whole | MS,LB | 0.37-1.30 | 1.30 | 37-130 | 130 |
| catfish | Muscle | SS,SJR | 0.20-0.29 | 0.34 | 20-29 | 34 |
| Yellow bullhead | Whole | PC,GLC | 0.344-0.387 | 0.452 | 34.4-38.7 | 45 |
| Bent clam | Muscle | MS | 0.50 | 0.50 | 50 | 50 |
| Freshwater clam | Muscle | SFC,CD,MS, SS,VWA | <0.2-4.0 | 4.0 | <20-400 | 400 |

[a]Data were obtained from the SJVDP Biological Residue Data Base and the State Water Resources Control Board's Selenium Verification Study.
[b]A, bordered by Fresno Slough and San Joaquin River on the east, I-5 on the west, Fresno-Kings County Line on the south, and San Joaquin County on the north; AC, Agatha Canal; CD, Camp 13 Ditch; GLC, Goose Lake Canal; GWD, Grassland Water District; K, Kesterson Reservoir; LB, Los Banos Creek; LHR, Lost Hills Ranch; MC, Main Canal; MS, Mud Slough; PC, Poso Creek, Kern Co; SFC, Santa Fe Grade Canal; SJR, San Joaquin River; SLC, San Luis Canal; SLD, San Luis Drain; SS, Salt Slough; STW, Semitropic Water Storage District Ponds; TLD, Tulare Lake Drainage District Ponds; VWA, Volta Wildlife Area; WF, Westfarmers Evaporation Ponds.
[c]Concentrations are based on fresh weights; fresh weight conversion factors were obtained from either the respective studies or Pennington and Church (1985).
[d]Exposures are based on intake (ug) per serving; typical serving sizes were obtained from Pennington and Church (1985) and Watt and Merrill (1975).

Table 2. Examples of selenium concentration in and potential human exposure from consumption of agricultural crops in the western San Joaquin Valley[a].

| Source | Tissue | Size Sample | Se Concen. (p/m)[b] Geometric Mean | Maximum | Se Exposure (ug/serving)[c] Average | "Worst" | U.S. Average[d] |
|---|---|---|---|---|---|---|---|
| Bell pepper | Fruit | 9 | 0.0078 | 0.0132 | 0.78 | 1.32 | |
| Broccoli | Head | 14 | 0.0530 | 0.1106 | 5.30 | 11.06 | |
| Cabbage | Head | 30 | 0.0136 | 0.0281 | 1.36 | 2.81 | 2.2 |
| Cantaloupe | Fruit | 9 | 0.0128 | 0.0303 | 2.05 | 4.85 | 0.4 |
| Carrot | Root | 4 | 0.0407 | 0.0652 | 4.07 | 6.52 | 2.2 |
| Cauliflower | Head | 15 | 0.0385 | 0.0877 | 3.85 | 8.77 | 0.6 |
| Corn | Grain | 7 | 0.0036 | 0.0328 | 0.58 | 5.25 | 0.4 |
| Garlic | Root | 4 | 0.0381 | 0.0739 | 3.81 | 7.39 | 24.9 |
| Lettuce | Head | 6 | 0.0023 | 0.0081 | 0.23 | 0.81 | 0.8 |
| Lima bean | Seed | 5 | 0.0685 | 0.0796 | 6.85 | 7.96 | |
| Onion | Bulb | 5 | 0.0070 | 0.0082 | 0.70 | 0.82 | 1.5 |
| Sugar beet | Root | 5 | 0.0030 | 0.0068 | 0.30 | 0.68 | |
| Tomato | Fruit | 12 | 0.0034 | 0.0072 | 0.34 | 0.72 | |

[a]Data were obtained from Burau et al. (1988).
[b]Se concentrations are based on fresh (wet) weight; fresh weight conversation factors were obtained from Pennington and Church (1985).
[c]Exposures are based on intake (ug) per serving; typical serving sizes were obtained from Pennington and Church (1985) and Watt and Merrill (1977).
[d]National averages were obtained from Pennington and Church (1985).

Table 3. Potential relative contribution of different routes of increased selenium exposure in the western San Joaquin Valley.

| Route | Rank | Comments |
|---|---|---|
| WC | 5 | Certain forms of wildlife in the drainage area contain sufficient levels of Se that consumption of these animals, particularly organs which accumulate Se, should be limited. Allowable consumption rates will depend on predicted total dietary Se loads. |
| FC | 5 | As is the case with animals, certain fish in the drainage area contain excessive Se levels. See comments above for consumption restrictions. |
| PC | U | Field and laboratory studies of plant Se levels in the drainage area have been limited. Proposed drainage water recycling techniques may increase plant Se levels. Further monitoring of plant Se levels is warranted. |
| APC | U | Preliminary evidence from bovine milk and liver samples does not suggest excessive Se in these tissues. Sampling has been limited, however, and more data are necessary to assure acceptable Se levels in animal products. |
| GWC | 1 | Currently, most ground water in the area does not exceed federal drinking water guidelines for Se. With the closure of the San Luis drain, it is possible this situation could change. Ground water Se levels should be closely monitored in the drainage area. |
| SWC | 1 | Surface water Se levels outside Kesterson Reservoir and the San Luis Drain generally do not exceed 35 ppb. Direct consumption of this water would probably not adversely impact human health, but could contribute to higher Se levels in fish and wildlife. |
| BDV | U | Preliminary tests of airborne particulate Se have indicated that levels are in the acceptable range. Concentrations of volatilized Se compounds are largely unknown, however. |
| DC | UL | Dermal absorption of environmental Se appears to be poor. Hand-to-mouth contact could occur in small children playing in seleniferous soils. |

WC = wildlife consumption
FC = fish consumption
PC = plant consumption
APC = animal and animal product consumption
GWC = ground water consumption
SWC = surface water consumption
BDV = breathing dusts and volatilized compounds
DC = dermal contact
U = unknown
UL = unlikely
1 = least exposure potential
5 = greatest exposure potential

ally expected norms; 2 was used when levels of the element routinely were higher than National norms, but the route would not necessarily provide a significant portion of the daily intake; 3 was used if the route provided higher than normal levels and could provide a significant portion of the diet; 4 was not used, but would indicate a route currently being evaluated for some regulatory or advisory action; 5 was used when health advisories limiting exposure via this route were in effect.

Studies of fish and wildlife in the Grassland Water District have indicated that selenium has accumulated to such a degree that consumption of animals from some areas should be limited in the general population or avoided altogether by certain subgroups of the population (e.g., children and pregnant women). Health advisories issued by local authorities and the California Department of Health Services have notified persons entering affected areas of these restrictions. More recent data (Moore et al., 1989) have shown that aquatic birds found near evaporation ponds primarily in the Tulare Basin region of the western Valley also have selenium concentrations that may warrant health-based consumption restrictions. Given present information, consumption of fish and wildlife is considered the largest non-occupational source of potential human exposure to selenium.

Selenium analyses of agricultural crops in the area have been limited, but, so far, do not indicate a potential health threat for persons consuming commercial produce grown in the western Valley. It has been questioned whether persons practicing subsistence gardening or foraging free-growing vegetation in seleniferous regions may be at greater risk of excessive selenium intake. Investigations to date have not been sufficient to rule out this possibility; however, preliminary ethnographic surveys of Southeast Asians indicated that foraging meets only a small portion of the total family diet for this ethnic group (Campbell and Christensen, 1989). Analyses of bovine tissues obtained from cows raised in the drainage area have not shown elevated tissue selenium levels. The degree to which these animals may have been exposed to excessive dietary selenium, however, is unknown. Additional data are necessary to completely assess concentrations of selenium existing in local domestic animal products.

Selected ground waters along both sides of the California coastal range have been found to contain selenium in levels that exceed the EPA-proposed Maximum Contaminant Level (MCL) of 50 p/b (Federal Register 40 CFR, Parts 141, 142, and 143, May 22, 1989). In some cases, these levels are not known to be related to agricultural drainage or industrial practices and appear to exist naturally. In the western Valley, ground water used for drinking has not been found to exceed drinking water standards, and is therefore presumed not to be a health risk at this time (the highest selenium concentrations in domestic wells in the Kesterson vicinity was 5 p/b, U.S. Bureau of Reclamation, 1987). However, concern has been expressed that, because of the closure of the San

Luis Drain (which collected drainage waters along the west side of the Valley for deposition into Kesterson Reservoir), water containing high levels of selenium may begin to percolate down to ground-water aquifers and contaminate presently safe drinking water sources. Numerous wells now exist to monitor local drinking water aquifers for such an occurrence.

With the exception of evaporation ponds and the now closed Kesterson Reservoir and San Luis Drain, surface waters in the study area do not usually exceed 35 p/b selenium (U.S. Bureau of Reclamation, 1987). These and somewhat lower values have been found in Mud and Salt Sloughs, which drain seleniferous lands. Direct consumption of this water, although not expected, would increase an individual's daily selenium intake and impinge upon the safety margin for this element. A few canals in Merced County have been found to contain considerably higher selenium concentrations on an irregular basis (as high as 100-200 p/b selenium). These canals are used interchangeably to move both irrigation and drainage waters, and thus, selenium concentrations may vary fifty-fold depending on the date of sampling. Because canal water is not believed to be consumed directly by humans, these waters are not expected to serve as a regular source of selenium exposure for human populations. Surface waters in the study area that provide drinking water for municipal uses have not been found to exceed established EPA limits.

Although most surface waters in the drainage area are not used for drinking water, they may provide an important indirect source of selenium exposure in humans. Bioaccumulation of selenium may occur in plants or wildlife found in these waters when selenium concentrations exceed very low levels. Consumption of such wildlife must be considered a potentially significant source of selenium.

As mentioned previously, selenium is naturally volatilized from soil, sediments, plants, and, when consumed in excess, from animals. The predominant form of volatilized selenium is believed to be dimethylselenide. This form of selenium is of lower acute toxicity than many other forms of selenium (Raabe et al., 1988); however, the effects of long-term inhalation of this compound have not yet been investigated.

Other forms of selenium may exist in the air as particulate matter. These forms of selenium are considered to predominate in urban (as a result of industrial processes), rather than rural environments and have not been shown to contribute significantly to the average daily intake of selenium (Medinsky et al., 1985 and Wilbur, 1980). Airborne movement of particulate selenium from seleniferous soils could possibly increase exposure of nearby residents, but preliminary testing has shown concentrations of airborne particulate selenium to be within acceptable occupational limits within Kesterson Reservoir itself (air quality data are complied in the U.S. Bureau of Reclamation Kesterson Program EIS report, 1986). Decommissioning of evaporation ponds may present significant risks from inhalation of particulate or volatilized selenium compounds, although carcinogenic minerals (e.g., arsenic and hexavalent chromium) may provide greater hazards under those circumstances.

Humans may come in dermal contact with selenium from soil, water, and air. Very few data exist regarding dermal absorption of or adverse reactions to dermal contact with nonindustrial selenium compounds; this suggests that environmental forms of selenium are poorly absorbed through intact skin. Dermal exposure to seleniferous soils could lead to increased selenium ingestion via hand-to-mouth contact, particularly in certain population groups such as small children. Based on maximum soil selenium concentrations outside Kesterson Reservoir or evaporation pond sediment, even daily consumption of a few grams of soil would not increase selenium intake levels significantly in these children. Health risks from this exposure route may be higher in occupational settings.

Health correlates for different selenium exposures are presented in table 4. These examples are described in more detail in Klasing and Pilch (1988). Clearly, some levels of selenium consumption several-fold above those recommended for safety and adequacy have failed to cause discernible toxicity symptoms. Accordingly, there appears to be a safety margin between required levels of selenium and those causing overt toxicity that is difficult to exceed by normal dietary means. However, there is the possibility that previously unrecognized effects could result from exposure to selenium within this margin. Recent data have suggested that somatomedin C (a secondary growth-promoting factor) levels can be depressed in rheumatic disease patients receiving 256 $\mu$g/day supplemental organic selenium. Although these results warrant further study, the authors caution that final conclusions cannot yet be drawn on this subject (Thorlacius-Ussing et al., 1989).

## CONCLUSIONS AND RECOMMENDATIONS

Clearly, irrigation of arid, saline lands under certain conditions has led to the mobilization of soil constituents and their release in drainage waters. As a consequence of environmental transfer, the concentration of these substances in some areas has increased in air, surface waters, ground waters, plants, and animals. At least in the case of selenium, and possibly with other compounds, contaminant concentrations in some environmental media are sufficient to cause potential adverse impacts on human health when consumption is unrestricted. Although cases of human selenium (or other drainage component) intoxication have not been identified in local human populations, data are not sufficient at this time to rule out the occurrence of previously undefined chronic health effects.

Table 4. Human health effects associated with selenium exposures.

| Dose | Source | Comments |
|---|---|---|
| 45 ug/d | Infant Formula | Recommended upper limit for selenium in infant formula (Levander, 1989) |
| 55, 70 ug/d | Total Diet | Recommended dietary allowance for adult female and male, respectively. (NRC, 1989) |
| 100-250 ug/d* | Water | No increase in prevalence of 85 abnormal health status indicators (Tsongas and Ferguson, 1977) |
| 256 ug/d | Supplement | Lowered somatomedin C levels in rheuatic disease patients compared to paired controls provided organic selenium from yeast for 6 months (Thorlacius-Ussing et al., 1989) |
| 500 ug/d | Total Diet | Estimated consumption in extreme fish-eating Japanese populations; recommended as the tentative maximum permissible intake (Sakurai and Tsuchiya, 1975) |
| 590 ug/d | Total Diet | No evidence of toxicity in population from seleniferous region of U.S. (Longnecker et al., 1987) |
| 600 ug/d | Diet + Suppl. | No evidence of toxic effects in one person after 18 months (Schrauzer and White, 1978) |
| 2000 ug/d | Supplement | Consumption of NaSO3 for 2 years caused thickened, fragile nails and garlic odor of dermal excretions in one 62 year old male (Yang et al., 1983). |
| 3500 ug/d** | Supplement | Caused in apparent toxic effects when given to Finnish neuronal ceroid lipofuscinosis patients for 1 year (Westermarck, 1977) |
| 3200-6690ug/d | Total Diet | Estimated selenium intake in families in seleniferous region of China (Yang et al., 1983) |
| 700-14,000** | Total Diet | 76 percent of those interviewed in ug/d seleniferous region of U.S. had mild, nonspecific symptoms (Smith and Westfall, 1937) |
| 18,000 ug/d* | Water | Consumption for 3 mo. caused hair loss, weakened nails, listlessness in Ute Indian family in Colorado (Anon., 1962) |
| 350,000 ug | Supplement | Single dose induced vomiting, diarrhea, cramps, paresthesias (Hogberg and Alexander, 1986) |
| 27,000 ug/d | Supplement | From 1 tablet to 77 tablets over 2½ months caused nausea, nail and hair changes, peripheral neuropathy, garlic breath odor, fatigue, irritability (Helzlsouer et al., 1985) |

* Assumed consumption of 2 liters water/day.
** Converted to 70 kg man equivalent.

*Sources:* Burau et al. (1988), Pennington and Church (1985), and Watt and Merrill (1975).

Many factors impinge upon the ability to completely assess the risk to human health from exposure to agricultural drainage water contaminants. Most notably are the comparatively limited toxicity and exposure data related to drainage components.

The chronic human toxicity of many chemicals found in agricultural drainage water is not well understood. Bioaccumulation of chemicals through the food chain may produce as yet undefined chemical forms with similarly unidentified human health effects. Additionally, few data are available to assess interactions between the many chemicals that occur in drainage water. Environmental interactions between chemicals can greatly effect the subsequent human toxicity; interactions may render chemicals unavailable for absorption (and thus lessen the toxicity) or they can potentiate toxicity by a variety of synergistic mechanisms.

Even when the chronic human toxicity of a drainage contaminant is known or can be estimated, quantification of human exposure to contaminants is difficult and expensive. Determination must be made of the concentration of a substance in a specific exposure source (food, water, soil, air) and also the degree to which humans ingest or inhale that particular source. Additionally, biological indicators of excessive exposure (quantification of the trace element in blood, urine, hair, or other body tissues) can be used to estimate human body burdens of the compounds. Unfortunately, methodologies to obtain such data are usually expensive, technically difficult, and, in the case of biological measurements, frequently involve invasive techniques. Moreover, as is the case with selenium, biological indicators are often only useful for the determination of overtly toxic or deficient states; smaller deviations in assessment measures may have unknown physiological or toxicological significance (Levander, 1985).

In any attempt to acquire information regarding exposure of certain population groups to specific chemicals in the environment, great caution must be exercised to obtain valid information. Samples must be obtained in a statistically acceptable manner and analyses must include strict quality assurance/quality control procedures. A complete discussion of exposure assessment is outside the scope of this chapter; however, the reader is referred, for example, to the *Environmental Factors Handbook* (EPA, 1989b).

Currently, many alternatives are being studied to solve drainage and drainage-related problems in the Valley (SJVDP, 1989). Possible options have included source control, ground-water management, drainage water treatment, drainage water reuse, drainage water disposal, fish and wildlife measures, and institutional changes. It is unlikely that any single option will serve to solve a significant portion of drainage problems; undoubtedly, many options will be combined to form an integrated approach to drainage management.

Most alternatives for solving drainage-related problems are not without potential public health impacts. The use of evaporation ponds to store and dispose drainage water, for example, may present significant human hazards through bioaccumulation in the food chain. Drainage water treatment techniques may generate entirely new contaminants (such as bacterial sludge or volatilized forms of chemicals) and exposure scenarios.

Environmental and public health must be primary considerations during all phases of irrigated agriculture. History has shown that relatively moderate human intervention in arid climates can significantly decrease the quality of water and some foodstuffs. Careful planning and monitoring can help alleviate these problems in the future.

## REFERENCES

Abu-Erreish, G. M.; Whitehead, E. I.; and Olson, O. E., 1968. Evolution of Volatile Selenium from Soils, *Soil Science*, 106, pp. 415-420.

Berlin, M., 1986. Mercury. In: Friberg, L.; Nordberg, G. F.; Vouk, V. B. (Eds.), *Handbook on the Toxicology of Metals*, Vol. II. Elsevier, Amsterdam, pp. 387-445.

Burau, R.G.; McDonald, A.; Jacobson, A.; May ,D.; Grattan, S.; Shennan, C.; Swanton, B.; Sherer, D.; Abrams, M.; Epstein, E.; Rendig, V. 1988. Selenium in Tissues of Crops Sampled from the West Side of the San Joaquin Valley, California. In: Tanji, K.K. (Ed.), *Selenium Contents in Animals and Human Food Crops Grown in California*. University of California, Division of Agriculture and Natural Resources, Publication 330, pp. 61-66.

Campbell, M. and Christensen, L. C., 1989. Foraging in Central Valley Agricultural Drainage Areas, *California Agriculture*, 43, pp. 23-25.

Chau, Y. K.; Wong, P. T. S.; Silverberg, B. A.; Luxon, P. L.; and Bengert, G. A., 1976. Methylation of Selenium in the Aquatic Environment, *Science*, 192, pp. 1130-1131.

Cook, R., 1988. Organization of the Fruit and Vegetable Marketing System and Implications for the Distribution of Fresh Produce with High Selenium Levels. In: Tanji, K. K. (Ed.), *Selenium Contents in Animal and Human Food Crops Grown in California*. University of California, Division of Agriculture and Natural Resources, Publication 330, pp. 85-88.

Frankenberger, W. T., Jr. and Karlson, U., March, 1988. *Microbial Volatilization of Selenium at Kesterson Reservoir*. Interim Report. Prepared for the U.S. Department of Interior, Bureau of Reclamation, Contract No. 7-FC-20-05240.

Deosthale, Y. G. and Gopalan, C., 1974. The Effect of Molybdenum Levels in Sorghum (Sorhum vulgare Pers.) on Uric Acid and Copper Excretion in Man, *Br. J. Nutr.*, 31, pp. 351-355.

Jensen, S. and Jernelov, A., 1969. Biological Methylation of Mercury in Aquatic Organisms, *Nature*, 223, pp. 753-754.

Klasing, S. A. and Pilch, S. M., 1988. *Agricultural Drainage Water Contamination in the San Joaquin Valley: A Public Health Perspective for Selenium, Boron, and Molybdenum.* Prepared for the San Joaquin Valley Drainage Program under U.S. Bureau of Reclamation Cooperative Agreement No. 7-FC-20-04830, 41 p. (plus Appendices).

Klasing, S. A.; Wisniewski, J. A.; Pilch, S. M.; and Anderson, S. A., 1990. *Agricultural Drainage Water Contamination in the San Joaquin Valley: A Public Health Perspective for Arsenic, Mercury, Nitrates/Nitrites, Uranium, and Vanadium.* Prepared for the San Joaquin Valley Drainage Program under U.S. Bureau of Reclamation Cooperative Agreement No. 7-FC-20-04830, 64 p. (plus Appendices).

Kovalsky, V. V.; Yarovaya, G. A.; and Shmavonyan, D. M., 1961. The Change in Purine Metabolism of Humans and Animals Under the Conditions of Molybdenum Biogeochemical Provinces, *Zh. Obshch. Biol.*, 22, pp. 179-191.

Levander, O. A., 1985. Considerations on the Assessment of Selenium Status, *Fed. Proc. Fed. Am. Soc. Exp. Biol.*, 44, pp. 2579-2583.

Lu, F. C., 1974. Mercury as a Food Contaminant, *WHO Chronicle*, 28, pp. 8-11.

McConnell, K. P. and Roth, D. M., 1966. Respiratory Excretion of Selenium, *Proc. Soc. Exp. Biol. Med.*, 123, pp. 919-921.

Medinsky, M. A.; Cuddihy, R. G.; Griffith, W. C.; Weissman, S. H.; and McClellan, R. O., 1985. Projected Uptake and Toxicity of Selenium Compounds from the Environment, *Environ. Res.*, 36, pp. 181-192.

Merced County Health Department, 1985. *Proceedings from the Scientific/Medical Committee to Review the Impact of Kesterson Contamination on Human Residents.*

Moore, S. B.; Detwiler, S. J.; Winckel, J.; and Weegar, M. D., 1989. *Biological Residue Data for Evaporation Ponds in the San Joaquin Valley, California.* San Joaquin Valley Drainage Program.

National Research Council (NRC), Safe Drinking Water Committee, 1977. *Drinking Water and Health.* National Academy of Sciences, Washington, DC., pp. 270-279.

National Research Council (NRC), Food and Nutrition Board, 1980. *Recommended Dietary Allowances.* 9th Rev. National Academy Press, Washington, DC., 185 p.

Pennington, J. A. T. and Church, H. N., 1985. *Food Values of Portions Commonly Used*. 14th Ed. Harper & Row, New York, NY., 257 p.
Petukhov, N. I. and Ivanov, A. V., 1970. Investigation of Certain Psychophysiological Reactions in Children Suffering from Methemoglobinemia due to Nitrates in Water, *Hyg. Sanit.*, 35, pp. 29-32.
Rabbe, O. G. and Al-Bayati, M. A., 1988. *Toxicity of the Inhaled Dimethylselenide in Adult Rat*. Final Report. Prepared under U.S. Department of the Interior Cooperative Agreement No. 7-FC-20-05240.
Study Group of Minamata Disease, 1968. *Minamata Disease*. Kumamoto University, Japan, 330 p.
Thorlacius-Ussing, O.; Flyvbjerg, A.; Tarp, U.; Overvad, K.; and Orskov, H, 1989. Selenium Intake Induces Growth Retardation through Reversible Growth Hormone and Irreversible Somatomedin C Suppression. In: Wendel, A. (Ed.), *Selenium in Biology and Medicine*. Springer-Verlag, New York, NY., pp. 126-129.
U.S. Bureau of Reclamation, 1987. *Water Quality Analyses: West Side San Joaquin Valley*.
U.S. Environmental Protection Agency, 1989a. IRIS. *Inorganic Arsenic*, September 20, 1989.
U.S. Environmental Protection Agency, 1989b. *Exposure Factors Handbook*. EPA 600/8-89, 43 p.
U.S. Fish and Wildlife Service, Patuxent Wildlife Research Center. *Effects of Irrigation Drainwater Contaminants on Wildlife*. Annual Report, Fiscal Year 1986. Prepared for the San Joaquin Valley Drainage Program under Intra-Agency Agreement No. 6-AA-20-04170.
Watt, B. K. and Merrill, A. L., 1975. *Composition of Foods*. Agricultural Handbook No. 8. Consumer and Food Economic Institute, Agricultural Research Center, U.S. Department of Agriculture, 189 p.
Wilbur, C. G., 1980. Toxicology of Selenium: A Review, *Clin. Toxicol.*, 17, pp. 171-230.
World Health Organization, 1984. Recommendations. In: *Guidelines for Drinking-Water Quality*, Vol. 1. World Health Organization, Geneva.
Zieve, R. and Peterson, P. J., 1984. Dimethylselenide - An Important Component of the Biogeochemical Cycling of Selenium, *Trace Sub. Environmental Health*, 18, pp. 262-267.

# 18 CONTAMINANTS IN DRAINAGE WATER AND AVIAN RISK THRESHOLDS

Joseph P. Skorupa, U.S. Fish and Wildlife Service and Harry M. Ohlendorf, CH2M HILL, Sacramento, California

## ABSTRACT

The toxicity of selenium to avian embryos is one of the most restrictive constraints on options for managing agricultural drainage water. Although selenium in eggs strongly predicts embryotoxicity, waterborne selenium (on a total recoverable basis) often is an unreliable predictor of average realized selenium in eggs. For the San Joaquin Valley, however, the algebraically derived equation Log (Mean Egg Se) = 3.66 + 0.57 Log (Waterborne Se) is a good predictor of the maximum potential for selenium bioaccumulation in avian eggs. Using eared grebes (Podiceps nigricollis) as an indicator species for bioaccumulation potential, the average absolute difference between observed and predicted mean selenium in eggs was only 6 percent for test cases at waterborne concentrations of 2.8, 15, 126, 176 p/b (total recoverable) selenium. Various estimates of biologically important thresholds indicate that it would be prudent to consider drainage water with 3 to 20 p/b selenium as peripherally hazardous to aquatic birds (i.e., hazardous to some species under some environmental conditions) and drainage water with more than 20 p/b selenium as widely hazardous to aquatic birds (i.e., hazardous to most species under most environmental conditions). To prevent most avian toxicity, a reasonable goal for chemical or biological decontamination technologies would be concentrations of waterborne selenium < 10 p/b. Likewise, to minimize avian contamination, a reasonable goal of purity would be waterborne selenium < 2.3 p/b. When these water standards are technically or financially unattainable, actions to significantly reduce avian use of contaminated drainage water are necessary.

## INTRODUCTION

Over the past four decades, many water projects made it possible to irrigate large tracts of otherwise nonarable land in the arid western United States. For example, irrigated croplands increased in the Central Valley of California by 43 percent between 1959 and 1975 (Shelton, 1987). A substantial portion of this land, however, requires artificial drainage of shallow ground water to maintain crop productivity (Letey et al., 1986). In California more than 500 million cubic meters of this subsurface agricultural drainage water (drainage water) are already discharged annually (Ohlendorf and Skorupa, 1989) to surface aquatic ecosystems, primarily the San Joaquin River, its west-side tributaries, the Delta-Mendota Canal, evaporation ponds in the Tulare Basin, or the Salton Sea and its principal tributaries.

Concurrent with agricultural and other development (including pre-1959), more than 90 percent of the Central Valley's historic wetlands have been lost (Moore et al., 1990). Remnant wildlife populations have been concentrated onto the remaining wetlands, including those receiving drainage water. In at least one area, the Tulare Basin of California's southern San Joaquin Valley, ponds for evaporative reduction of drainage water (evaporation ponds) are typically the most common type of wetland available to wildlife during the spring (Moore et al., 1990). The shallow and nutrient-enriched waters of evaporation ponds lead to high primary and secondary productivity (Euliss, 1989) and provide the ready source of proteinaceous foods required by breeding birds. Accordingly, the ponds are particularly attractive to breeding waterbirds (Schroeder et al., 1988) and provide a pathway for wildlife exposure to contaminants in drainage water.

Although environmental exposure to drainage-water contaminants is documented for amphibians, reptiles, birds, and mammals, the impairment of avian reproduction is the most pronounced adverse biological effect documented for wildlife (Ohlendorf and Skorupa, 1989). It is this effect that will most likely impinge on the economics of drainage-water management because, under the Federal Migratory Bird Treaty Act (16 U.S.C. Sections 703-712), migratory birds are legally protected from human-caused poisoning (Olive and Johnson, 1986). The cost of drainage-water treatment, for example, depends on the standard of purity for treated water. The legal mandate that requires management of drainage water to be protective of migratory birds (including their embryos) is apparently the most restrictive constraint on acceptable standards of purity and acceptable methods for disposal of drainage water--treated or untreated. Therefore, this chapter attempts to clarify some of the biological constraints on drainage-water management by focusing principally on aquatic birds and on the toxicity of drainage-water contaminants to avian embryos (i.e., embryotoxicity as indicated by the overt deformity or death of an embryo).

Nearly a dozen inorganic constituents in drainage water are of toxicological concern (CSWRCB, 1987). Many of these constituents are found in tissues of wildlife sampled at evaporation ponds including arsenic, boron, cadmium, mercury, molybdenum, selenium, and strontium (Moore et al., 1989). Selenium, however, is the only constituent commonly found at embryotoxic concentrations in the eggs of aquatic birds (Ohlendorf and Skorupa, 1989). Experimental studies (Heinz et al., 1989 and Hoffman and Heinz, 1988) confirmed that the toxic effects of selenium alone are sufficient to explain most adverse effects on avian reproduction observed at evaporation ponds.

Boron, molybdenum, and strontium also have been detected at elevated levels in bird eggs from evaporation ponds. Elevated concentrations of boron and molybdenum (Ohlendorf and Skorupa, 1989 and Skorupa et al., unpubl. data) are usually well below known thresholds for avian embryotoxicity (Smith and Anders, 1989 and Eisler, 1989). The authors are unaware of critical threshold values for strontium-induced avian embryotoxicity, but eggs with elevated levels of strontium (i.e., > 75 p/m) are rare. (All tissue concentrations of contaminants cited in this chapter are on a dry-weight basis.)

In addition to the individual toxicity of drainage-water contaminants, chemical interactions can result in magnification or reduction of a contaminant's embryotoxicity. Also, noninteractive additive effects can cause cumulative toxicity, even though all the individual contaminants are below embryotoxic thresholds. The potential for interactive embryotoxic effects was evaluated in two experimental studies. Smith and Heinz (1990) found that the embryotoxic effects of boron and selenium seemed to be neither synergistic nor additive. Another study (USFWS, 1990) focused on the interaction between selenium and arsenic and found that 400 p/m dietary sodium arsenate reduced the embryotoxicity of 10 p/m dietary selenomethionine. In nature, however, the aquatic invertebrates that constitute the dietary staple of aquatic birds at evaporation ponds (Euliss, 1989) rarely exceed 25 p/m arsenic (Moore et al., 1989). Arsenic was below the limit of detection (ca. 0.4 p/m) in all bird eggs sampled from evaporation ponds (Moore et al., 1989 and Skorupa et al., unpubl. data). Although evaluation of the potential for interactive effects should be continued, current evidence is not compelling for important interactive or additive embryotoxic effects in the field. Therefore, as a matter of parsimony, the contaminant focus of this chapter will be on selenium toxicity.

The objective here, within the overall theme of biological constraints on drainage-water management, is to review and provide new syntheses of the results of field and laboratory studies of selenium embryotoxicity in birds. This chapter will emphasize what is known about significant thresholds and then discuss the general implications for the management of drainage water.

For this chapter, "avian contamination" is defined as mean selenium in eggs (mean egg selenium) above normal (background) concentrations, and "avian toxicity" is defined as mean egg selenium above embryotoxic thresholds. Avian contamination per se warrants the separate consideration given here because so little is known about subtle nonlethal adverse effects of selenium on avian embryos or about secondary hazards to predators of avian eggs.

To maintain a standard of best available information, unpublished data are cited occasionally in this review. When the unpublished data are the authors', the data are presented in appropriate detail. When they are not the authors', the details have been considered (usually from the raw data), but are not presented here. Results from both population-level analyses and individual-level analyses are discussed. It is stressed that these levels of analyses are not interchangeable.

## SELENIUM AND THE KESTERSON SYNDROME

Selenium is an essential trace element in animal diets, but the range between nutritional requirements and toxic levels is narrow (Ganther, 1974). In areas with seleniferous soils, selenium toxicosis was documented in poultry and livestock more than 50 years ago (e.g., Poley et al., 1937). Few studies, however, were conducted before the 1980's to examine selenium toxicity in wildlife (Ohlendorf, 1989).

Toxicity in wildlife was first observed at Kesterson Reservoir (Kesterson), a drainage-water evaporation pond system in the northern San Joaquin Valley. Field and controlled experimental studies identified selenium as the principal cause of embryotoxicity among birds at Kesterson (Ohlendorf, 1989). The drainage water discharged to Kesterson Reservoir during 1983-85 averaged about 300 p/b selenium (Presser and Barnes, 1984 and Saiki and Lowe, 1987). This extremely high concentration of selenium in the water (concentrations are normally < 1 p/b; e.g., Schroeder et al., 1988) was bioaccumulated to levels in avian foods, such as aquatic plants and insects, that were typically more than 30 times the normal concentrations for these taxa (Ohlendorf, 1989).

The extreme conditions at Kesterson provided little opportunity to assess thresholds for selenium toxicity to aquatic birds (but see Ohlendorf et al., 1986). However, two major research schemes, one directed by the U.S. Department of Interior National Irrigation Water Quality Program (Sylvester et al., 1989) and one directed by the U.S. Fish and Wildlife Service Patuxent Wildlife Research Center (USFWS, 1990), have recently expanded the basis for understanding avian exposure to selenium and the thresholds for toxicity.

## REFERENCE VALUES FOR SELENIUM IN EGGS OF WILD BIRDS

As of the early 1980's when Eisler (1985) reviewed selenium hazards to fish, wildlife, and invertebrates, little information was available to set quantitative guidelines for normal selenium concentrations in eggs of wild birds (i.e., in eggs of birds not exposed to selenium-enriched environments). By the mid-1980's slightly more information was available, and based on that information Ohlendorf (1989) suggested that normal concentrations averaged about 1 to 3 p/m selenium. Three dozen reference values for mean egg selenium in wild birds were available by the late 1980's, allowing Ohlendorf and Skorupa (1989) to estimate the reference interquartile boundaries as 1.4 and 2.7 p/m. This agreed with Ohlendorf's (1989) original estimate of normal concentrations. More recently, the reference data for wild birds inhabiting nonmarine wetlands have expanded to 74 sample means that allow a detailed percentile table to be constructed (table 1).

Table 1. Percentile values for mean selenium concentrations in samples of bird eggs from uncontaminated nonmarine wetlands (N = 74 sample means).

| Percentile | Mean Selenium Concentration[a] (p/m, dry weight) |
|---|---|
| 10th | 1.0 |
| 20th | 1.3 |
| 25th | 1.4 |
| 30th | 1.4 |
| 40th | 1.6 |
| 50th (Median) | 1.9 |
| 60th | 2.0 |
| 70th | 2.3 |
| 75th | 2.4 |
| 80th | 2.5 |
| 85th | 2.8 |
| 90th | 2.9 |

[a]The extreme sample means were 0.6 and 7.8 p/m. Sample means were typically based on samples of 2 to 9 individual eggs. Thus, the percentile values are approximate and apply only to means from small samples of eggs. As per the central limit theorem (e.g., DeGroat 1975:227), however, the median is valid for comparison to individual eggs or means from any size sample. At background concentrations, arithmetic and geometric means are practically equivalent, however, this table is best suited for comparison against geometric means from contaminated sites.

*Sources:* Haseltine et al. (1981,1983), Henny and Herron (1989), Hothem et al. (unpubl. data), Kepner et al. (unpubl. data), K. King (pers. comm.), Lambing et al. (1988), Ohlendorf et al. (unpubl. data), Ohlendorf and Marois (1990), Ohlendorf and Skorupa (1989), S. Schwarzbach (pers. comm.), Skorupa et al. (unpubl. data), USFWS (1989).

Significantly, the reference interquartile boundaries have changed very little (from 1.4-2.7 to 1.4-2.4 p/m selenium) with a doubling of the available data base and an increase in the taxonomic and geographic coverage. This suggests that the current reference interquartile boundaries are widely applicable taxonomically and geographically. More than 90 percent of all reference sample means are below 3 p/m selenium (table 1). Thus, > 3 p/m mean egg selenium seems to be a reasonable indicator threshold for avian contamination in nonmarine environments. In the Tulare Basin, avian contamination (i.e., mean egg selenium > 3 p/m) is associated with evaporation ponds containing as little as 1 to 3 p/b waterborne selenium (tables 2 and 3).

## TOXIC CONCENTRATIONS OF SELENIUM IN EGGS OF WILD BIRDS

Selenium toxicity, as indicated by abnormally high rates of teratogenesis (i.e., embryo deformity, particularly multiple overt deformities; Hoffman et al., 1988 and Hoffman and Heinz, 1988) or embryo death, was observed in several populations of waterbirds at Kesterson and at evaporation ponds in the Tulare Basin (Ohlendorf and Skorupa, 1989 and Skorupa et al., unpubl. data). Teratogenic populations averaged from about 15 to 80 p/m egg selenium. Assessments of average egg selenium and embryo status at Kesterson (northern San Joaquin Valley; Ohlendorf and Skorupa, 1989 and Ohlendorf et al., unpubl. data), in the Grassland Water District (northern San Joaquin Valley; R. L. Hothem et al., U.S. Fish and Wildlife Service, unpubl. data), in the Tulare Basin (southern San Joaquin Valley; Ohlendorf and Skorupa, 1989 and Skorupa et al., unpubl. data), and outside the San Joaquin Valley (Stephens et al. 1988 - Utah; Henny and Herron, 1989 - Nevada; S. G. Schwarzbach et al., U.S. Fish and Wildlife Service, unpubl. data - California/Oregon; D. U. Palawski et al., U.S. Fish and Wildlife Service, unpubl. data - Montana; P. Ramirez et al., U.S. Fish and Wildlife Service, unpubl. data - Wyoming) yield a clear dose-response relationship (figure 1).

A distinct dose-response relationship is evident in figure 1 (Spearman rank correlation = 0.943; N=6; p < 0.05; Siegel, 1956) despite a relatively coarse (but unambiguous) measure of contaminant response (presence or absence of overt deformities in a sample of embryos), uneven embryo sampling effort, multiple bird species, and the diversity of chemical environments represented, all of which are expected to weaken the dose-response graph. This dose-response relationship generated from field sampling (figure 1) suggests a teratogenesis threshold between 13 and 24 p/m mean egg selenium. One experimental study that exposed game-farm mallards (Anas platyrhynchos) to dietary selenomethionine, a form of selenium that seems to be an excellent model for environmental exposure (Hamilton et al., 1990), suggests that the teratogenesis threshold lies between 12 and 37 p/m mean egg selenium (Hoffman and Heinz, 1988 and Heinz et al., 1989). Mean egg selenium as high as about 25 p/m is associated with waterborne selenium as low as 10 to 20 p/b in the Tulare Basin (table 2).

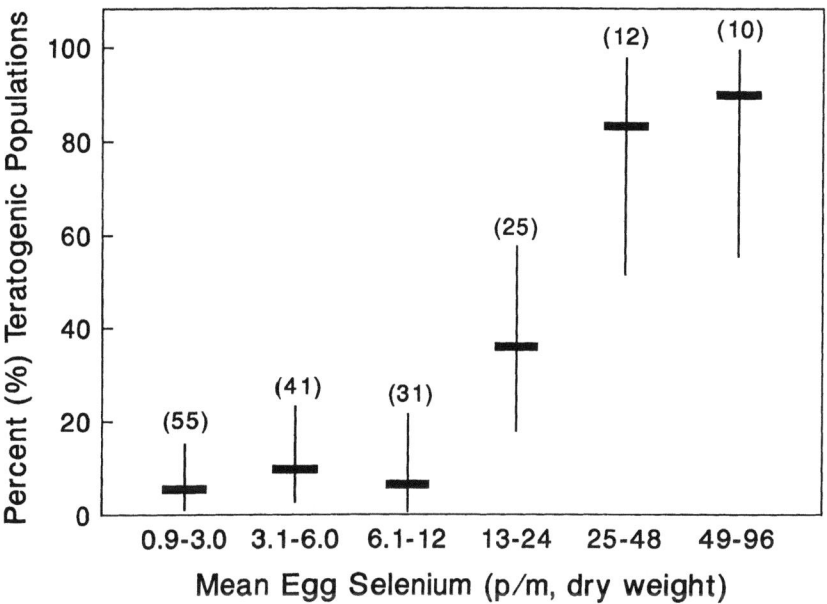

Figure 1. Dose-response relationship between mean egg selenium and teratogenic classification of aquatic bird populations.

*Dose intervals were delineated so that the first interval encompasses normal concentrations of mean egg selenium, and the succeeding intervals form a geometric progression. For each dose interval the observed percent of populations classified as teratogenic is plotted along with 95 percent binomial confidence intervals. Sample sizes (number of populations assessed) for each dose interval are listed above the response plots. Note, this plot is a population level analysis and cannot be used to infer the probability of teratogenesis in individual eggs of known selenium content.*

Another response variable for embryotoxicity is egg hatchability (e.g., see Ohlendorf et al., 1989). Hatchability is a more sensitive response variable than overt teratogenesis, but, in principle, it is also more ambiguous because of its equal sensitivity to noncontaminant-related perturbations (such as hen nutrition, unusual weather, observer disturbance, etc.). In practice, results of artificial incubation studies with eggs of black-necked stilts (Himantopus mexicanus) and American avocets (Recurvirostra americana) indicate that nearly all hatchability depression at evaporation ponds is contaminant-induced (Skorupa et al., unpubl. data).

Ohlendorf et al. (1986) related embryonic selenium exposure to embryo viability (= egg hatchability) for individual eggs of American coots (Fulica americana) and black-necked stilts at Kesterson and a reference site. The resulting regression for stilt eggs suggested that the minimum probability (i.e., lower 95 percent confidence band) of hatching failure started increasing sharply at about 10 p/m egg selenium. A similar evaluation of the regression for coot eggs is not possible because of the lack of low-selenium samples.

Preliminary population-level data from the Tulare Basin suggest that significantly reduced hatchability is associated with average selenium concentrations of about 8 p/m or greater (Skorupa et al., unpubl. data). This preliminary threshold value is based on monitoring the reproductive performance of 17 black-necked stilt and American avocet breeding aggregations during 1987 and 1988. Low hatchability was documented in eight of nine populations with mean egg selenium > 8 p/m, but in only two of eight populations with mean egg selenium < 8 p/m. Note that the Tulare preliminary analysis is a population-level analysis, and that the populations averaging 8 p/m or more egg selenium include individual eggs with > 10 p/m selenium (Skorupa et al., unpubl. data). Thus the individual-level analysis of Kesterson data and the population-level analysis of Tulare data seem compatible. The lowest concentration of waterborne selenium associated with populations of stilts or avocets over the 8 p/m mean egg selenium threshold is 10 p/b. Eggs of snowy plovers (Charadrius alexandrinus) and eared grebes (Podiceps nigricollis), however, have averaged 7 to 8 p/m selenium at ponds in the Tulare Basin with as little as 2 to 3 p/b waterborne selenium (table 3 and Skorupa et al., unpubl. data).

## SELENIUM BIOACCUMULATION: FROM WATER TO THE AVIAN FOOD CHAIN

Studies at evaporation ponds in the Tulare Basin and at lakes and ponds in Colorado and Wyoming demonstrated strong correlations between concentrations of selenium in the water and in aquatic plants and insects (Birkner, 1978 and Shelton et al., 1990). Data for waterborne selenium and food-chain selenium from the Tulare Basin yield statistically significant correlation coefficients of 0.91 to 0.98 for widgeon grass (<u>Ruppia maritima</u>), water boatmen (Corixidae), brine shrimp (<u>Artemia franciscana</u>), midge fly larvae (Chironomidae), and damselflies (Zygoptera) (J. Shelton et al., California Department of Water Resources, unpubl. data). At typical bioaccumulation factors of 1,000 to 5,000 (Birkner, 1978 and Schuler, 1987) for normal concentrations of waterborne selenium (i.e., < 1 p/b; Schroeder et al., 1988), samples of uncontaminated aquatic invertebrates should usually average < 4 p/m selenium (Ohlendorf, 1989). Results summarized in table 2 suggest that corixids, a common aquatic insect in evaporation ponds, begin to bioaccumulate selenium to concentrations averaging > 4 p/m at a waterborne selenium concentration between about 2 and 10 p/b.

## SELENIUM BIOACCUMULATION: FROM THE DIET (I.E., AVIAN FOOD CHAIN) TO THE EGG

Ohlendorf (1989) reported that bird eggs generally contain concentrations of selenium that are 1 to 3 times the dietary exposure of breeding females. Studies relating egg selenium to precisely verified levels of dietary exposure in the field have not been conducted. Heinz et al. (1989) experimentally exposed game-farm mallards to selenomethionine and demonstrated that egg selenium is closely related to a hen's dietary exposure. This has also been reported in the poultry literature (see citations in Heinz et al., 1989 and Ohlendorf, 1989). In the mallard experiment, average egg selenium varied from about 2.5 to 4.0 times the dietary exposure (dry weight basis). If biologically incorporated organoselenium consumed in the wild is assimilated with similar efficiency as dietary supplements of selenomethionine in the lab, a dietary intake averaging roughly 5 p/m organoselenium leads to an average egg selenium of about 15 p/m, the lowest mean concentration of egg selenium associated with embryo teratogenesis at Kesterson.

Much lower diet-to-egg bioaccumulation factors of 0.10 to 0.18 have been experimentally demonstrated for diets supplemented with inorganic forms of selenium (Heinz et al., 1987). However, evidence suggests that the selenium content of natural foods is predominantly in the form of organoselenium (Boyum and Brooks, 1988 and Hamilton et al., 1990). The diet-to-egg bioaccumulation factors of 1 to 3 implied by the field data presented in table 2 indicate substantial dietary exposure to organoselenium, although dietary exposure in the field likely includes a mixture of inorganic and organic forms of selenium.

A critical dietary threshold of about 5 p/m is consistent with the findings of Heinz et al. (1989) and Smith and Heinz (1990) for mallards. They found that the dietary threshold for elevated embryo teratogenesis (and reduced hatchability) was between 4 and 7 p/m of selenium as selenomethionine. If 80 percent of the selenium in natural foods is organoselenium (Boyum and Brooks, 1988), then toxic contamination of the food chain occurs between about 2 and 13 p/b waterborne selenium in Tulare evaporation ponds (estimated from unpublished regression equations for food-chain selenium available from John Shelton, California Department of Water Resources, Fresno, CA; for brine shrimp equation see figure 3).

## WATERBORNE SELENIUM AS A PREDICTOR OF EGG SELENIUM

Because measures of egg selenium are relatively precise indicators of the potential for adverse biological effects, identification of a quantitative relationship between waterborne selenium and egg selenium would be extremely desirable. However, waterborne selenium only determines the potential for selenium bioaccumulation in bird eggs (hereafter cited as "potential egg selenium"). Many variables are interposed between waterborne selenium and egg selenium (figure 2) that can alter the actual bioaccumulation of selenium (hereafter cited as "realized egg selenium"). Consequently, waterborne selenium is often an imprecise predictor of realized egg selenium (table 2).

The four sites listed in table 2 exhibit distinctly separated concentrations of waterborne selenium. Even though between-site separation in mean corixid (food-chain) contamination is distinct, only the lowest selenium site (TLDD-N) can be separated clearly from other sites on the basis of mean selenium concentrations in bird eggs (i.e., on the basis of "realized egg selenium"). Essentially, overlap in the spread of species' means for realized egg selenium is substantial when waterborne selenium (on a total recoverable basis) is anywhere between about 10 and 350 p/b (table 2).

Data for corixids (table 2) are consistent with the general finding (previously cited) that waterborne selenium strongly predicts food-chain selenium. Thus, in figure 2, the variables between step 1 (water selenium) and step 4 (food-chain selenium) must be fairly constant within the San Joaquin Valley and must not be responsible for the confounding results for realized egg selenium. Likewise, within species, variables between step 5 (avian exposure) and step 7 (egg selenium) should be constant. Hence, the variable between step 4 and step 5, avian behavioral ecology, may be the primary source of confounding variation.

Ecologically mediated behavioral characteristics such as degree of residency, home-range size, habitat preferences, and food preferences are very flexible between and within species. These variables may determine whether a site's potential for selenium bioaccumulation, based on waterborne selenium,

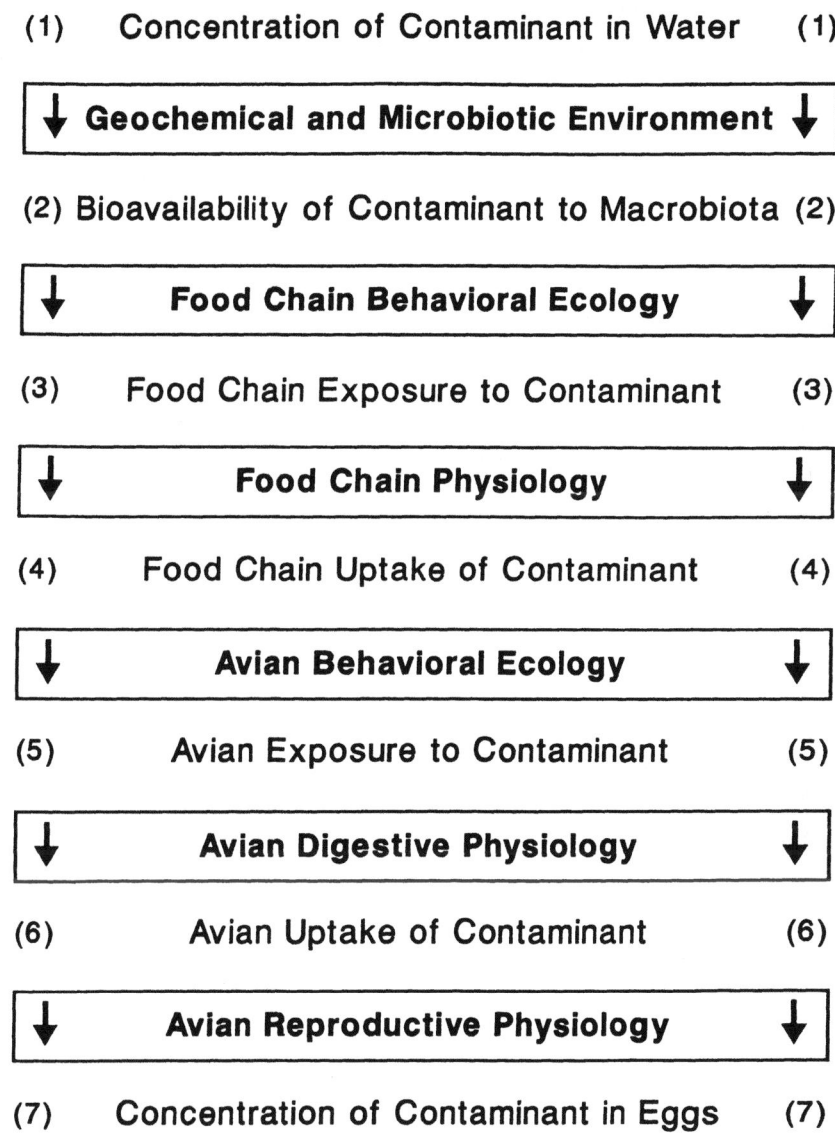

Figure 2. Major variables potentially confounding the relationship between waterborne selenium and egg selenium.

*In this simplistic representation of a water-to-egg contaminant pathway, movement between each step of the path is potentially influenced by an interposed variable (bold type enclosed by boxes).*

Table 2. Geometric mean selenium concentrations (and number of samples analyzed) of corixids (an aquatic insect) and bird eggs relative to waterborne selenium at four evaporation pond systems in the San Joaquin Valley, California.

|  | TLDD-N[a] | TLDD-S[b] | Kesterson[c] Reservoir | WFarm[d] |
|---|---|---|---|---|
| Water *(total recoverable, p/b)* | 1.1 - 2.5 | 9.8 - 23 | (65 - 225) | 140 - 345 |
| Corixids (p/m, dry wt.) | 3.4 ( 9) | 13 ( 6) | 22 (13) | 38 ( 6) |
| Eared Grebe Eggs *(p/m, dry wt.)* |  | 23 ( 9) | 70 (18) | 79 ( 5) |
| American Coot Eggs |  |  | 32 (17) |  |
| Waterfowl Eggs: |  |  |  |  |
| Gadwall | 2.9 (17) | 20 ( 9) | 20 (22) |  |
| Mallard | 1.8 (21) | 15 ( 3) | 12 (21) |  |
| Cinnamon Teal | 1.9 (31) | 20 ( 7) | 11 (12) |  |
| Northern Pintail | 2.6 ( 6) | 25 ( 3) | 13 ( 1) |  |
| Redhead | 3.4 ( 6) | 26 ( 4) |  |  |
| Ruddy Duck |  | 13 ( 1) |  |  |
| Canvasback |  | 10 ( 4) |  |  |
| Shorebird Eggs: |  |  |  |  |
| Black-necked Stilt | 2.6 (15) | 13 (20) | 32 (124) | 24 (39) |
| American Avocet | 3.7 (13) | 12 (10) | 19 (60) | 22 (40) |
| Snowy Plover |  | 23 (12) | 21 ( 1) | 25 ( 1) |
| Killdeer |  |  | 41 (32) |  |
| Range of Species Means for Egg Selenium | 1.8 - 3.7 | 10 - 26 | 11 - 70 | 22 - 79 |

*Note:* The National median for mean selenium concentration in samples of bird eggs from uncontaminated reference sites is 1.9 p/m (table 1). Medians for all taxonomic and geographic subgroups within the reference data are in the range 1.0 to 3.0 p/m (Skorupa et al., unpubl. data).

[a]<u>Tulare Lake Drainage District - North:</u> Waterborne selenium is for June, 1987 (Westcot et al., 1988a). Corixid selenium is for September, 1988 (Moore et al., 1989). Bird egg selenium is for April-July, 1987 and/or 1988 (Ohlendorf and Skorupa, 1989 and Skorupa et al., unpubl. data).

[b]<u>Tulare Lake Drainage District - South:</u> Waterborne selenium is for June, 1987 (Westcot et al., 1988a). Corixid selenium is for June 1987 (Moore et al., 1989). Bird egg selenium is for April-July, 1987 and/or 1988 (Ohlendorf and Skorupa, 1989 and Skorupa et al., unpubl. data).

[c]<u>Kesterson Reservoir:</u> Waterborne selenium is for May, 1983 (Saiki and Lowe, 1987) and May, 1984 (Schuler 1987) with an appropriate conversion from dissolved basis to approximate total recoverable basis (see footnote g in table 3). Corixid selenium is for May, 1983 (Saiki and Lowe, 1987), May 1984 (Schuler, 1987) and April-June, 1985 (Hothem and Ohlendorf, 1989). Bird egg selenium is for April-June, 1983 and/or 1984 and/or 1985 (Ohlendorf and Skorupa, 1989 and Ohlendorf et al., unpubl. data), except for snowy plover which is for April-June 1986 (F.L. Paveglio, unpubl. data).

[d]<u>Westfarmers:</u> Waterborne selenium is for June, 1987 (Westcot et al., 1988a).Corixid selenium is for June 1987, and June 1988 (Moore et al., 1989). Bird egg selenium is for April-July, 1987 and/or 1988 (Ohlendorf and Skorupa, 1989 and Skorupa et al., unpubl. data).

will be fully or only partially realized. For example, the counter-intuitive finding that waterfowl eggs from TLDD-S were equally or more contaminated than waterfowl eggs from Kesterson (table 2) is probably due to ecologically mediated behavioral variation. TLDD-S is isolated within an intensively developed agricultural landscape mostly devoid of nondrainwater wetlands during the spring. Kesterson was in a landscape with abundant neighboring wetlands that contained considerably lower concentrations of selenium (Ohlendorf et al., 1987). Thus, ducks at Kesterson had opportunities to use habitat that would reduce exposure to drainage-water contaminants whereas ducks at TLDD-S did not. This interpretation is supported by the results (table 2) for eared grebes (a very sedentary forager during the breeding season) that suggest duck eggs at TLDD-S were representative of local contaminant conditions, whereas duck eggs at Kesterson may have realized only 15 to 30 percent of the site potential for bioaccumulating selenium.

Because of eared grebes' long residency time (they are usually the latest breeders; C. J. Henny, U.S. Fish and Wildlife Service, pers. comm.; pers. obser.), localized foraging range (most foraging on evaporation pond systems is done in the same cell as the nest colony; pers. obser.), and stereotyped food preferences (for aquatic invertebrates; Johnsgard, 1987), grebes may consistently come the closest to realizing the full potential for selenium bioaccumulation in eggs at any site (i.e., realized egg selenium may often equal potential egg selenium). Eared grebes probably come close to meeting the special circumstances required for a one-to-one correspondence between steps 4 and 5 (in figure 2). This correspondence, in turn, best meets the special condition for predicting egg selenium from waterborne selenium:

[1]     If,   Log (FCS) $= a + b$ Log (WS)
[2]     and,  Log (MES) $= c + d$ Log (DS)
          and,  FCS = DS (the special condition)
          then, Log (MES) $= c + d[a + b$ Log (WS)$]$
                          $= (c + da) + db$ Log (WS)
[3]                                 $= e + f$ Log (WS)
    where,  DS   = p/b dry weight dietary selenium
                FCS = p/b dry weight food-chain selenium
                MES = p/b dry weight arithmetic mean egg selenium
                WS   = p/b total recoverable waterborne selenium
                a-d   = fitted regression parameters
                e      = (c + da)
    and,    f      = db.

Based on different taxa of aquatic invertebrates, Shelton et al. (unpubl. data) calculated four estimates of equation [1] for evaporation ponds in the Tulare

Basin. An estimate of equation [2] can be calculated from Heinz et al.'s (1989) data for game-farm mallards. This results in the following four solutions for equation [3]:

[4]  Log (MES) = 3.86 + 0.57 Log (WS)  (based on corixids)
[5]  Log (MES) = 3.66 + 0.57 Log (WS)  (based on brine shrimp)
[6]  Log (MES) = 4.07 + 0.72 Log (WS)  (based on midge larvae)
[7]  Log (MES) = 3.81 + 0.67 Log (WS)  (based on damselflies)

Predicted (from equations [4]-[7]) and observed mean egg selenium for eared grebes in the San Joaquin Valley can be compared (table 3). The performance of equation [5] is particularly encouraging, because the average absolute difference between predicted and observed mean egg selenium was only 6 percent. More importantly, the differences between predictions and observations were < 10 percent in the critical lower range of waterborne selenium (i.e., < 20 p/b) that is likely to embrace important biological thresholds.

Although brine shrimp are a highly preferred food of eared grebes in saline environments (Jehl, 1988), brine shrimp apparently do not occur at the nesting sites listed in table 3 (Hothem and Ohlendorf, 1989 and D. A. Barnum, U.S. Fish and Wildlife Service, pers. comm.); thus, there is no obvious reason for the brine-shrimp-based regression equation [5] to perform so well. Perhaps the bioavailable (for transfer to bird eggs) organoselenium concentrations biologically incorporated into macroinvertebrate tissues do not vary much between species (within a pond), and measures of total recoverable selenium from brine shrimp most closely estimate the bioavailable organoselenium fraction. Unlike corixids, midges, and damselflies, brine shrimp do not have a well-developed chitinous exoskeleton to which confounding fractions of inorganic selenium can become externally adsorbed (Krantzberg and Stokes, 1988 and Newman and McIntosh, 1989). The fact that all the other equations tend to overestimate mean egg selenium is consistent with this interpretation. Or perhaps brine shrimp are very representative of the modal type of aquatic invertebrate (i.e., nonchitinous, water column dwelling) preferred by eared grebes in saline environments even where brine shrimp are not available (Mahoney and Jehl, 1985). Future studies will have to further test the reliability of the brine-shrimp-based predictive model and, if it continues to prove reliable, focus on elucidating exactly why it performs so well.

One of the biological thresholds of inherent interest is the contamination threshold, that is, the concentration of waterborne selenium associated with a potential for mean egg selenium of about 3 p/m (the threshold between background and contaminated eggs). Ideally, the management goal for all wetlands is to keep waterborne selenium under the contamination threshold.

Table 3. Comparison of observed and predicted mean egg selenium for eared grebes nesting on evaporation ponds in the San Joaquin Valley, California.

| Site | Waterborne Se[a] (p/b) | Mean Egg Se (p/m) | | | | |
|---|---|---|---|---|---|---|
| | | Observed (N) | Predicted[b] | | | |
| | | | corix | brshp | mdlve | damfy |
| Lost Hills Ranch | 2.8[c] | 8.5[d] ( 7) | 13 | 8.2 | 25 | 13 |
| TLDD - South | 15[e] | 23[f] ( 9) | 34 | 21 | 83 | 40 |
| Kesterson Reservoir | (126)[g] | 75[h] (13) | 114 | 72 | 382 | 165 |
| Westfarmers | 176[i] | 81[j] ( 5) | 138 | 87 | 486 | 206 |
| Average Absolute Difference from Observed: | | | 56% | 6% | 341% | 100% |

[a]On a total recoverable selenium basis.
[b]Predicted values (arithmetic means) are from equations 4-7 of text which were based on food-chain data for corixids (corix), brine shrimp (brshp), midge fly larvae (mdlve), and damselflies (damfy). The observed values are also arithmetic means and therefore do not always match the geometric means reported from the same data in table 2.
[c]Measured in pond 1 during June, 1988 (Westcot et al., 1988b).
[d]Measured in eggs from pond 1 during June and July, 1988.
[e]Measured in pond 4 during June, 1988 (Westcot et al., 1988b).
[f]Measured in eggs from pond 4 during June and July, 1988.
[g]Saiki and Lowe (1987) measured 68 p/b dissolved selenium in pond 11 during May, 1983. That measurement has been multiplied by a factor of 1.85 to convert it to an approximate total recoverable selenium basis. Fujii (1988) reported an average ratio of 1.85 for total recoverable selenium to dissolved selenium in a Tulare Basin evaporation pond system. Moore et al. (1990) reported an aggregate ratio of 1.98 for Kesterson water analyses, but that is not based on a matched set of split samples as are Fujii's ratios.
[h]Measured in eggs from pond 11 during 1983.
[i]Measured in pond 1 during June, 1988 (Westcot et al., 1988b).
[j]Measured in eggs from pond 1 during June, 1988.

From equation [5] a concentration of about 0.5 p/b waterborne selenium has the potential to result in mean egg selenium of about 3,000 p/b ( = 3 p/m). This prediction can be compared to field data from Foxtail Lake and Carson Lake of the Stillwater Wildlife Management Area, Nevada. Eared grebe eggs collected from Foxtail Lake averaged 3.4 p/m selenium (N=10; C. J. Henny, unpubl. data) when waterborne selenium was < 1.0 p/b (R. J. Hoffman, U.S. Geological Survey, unpubl. data). Eared grebe eggs sampled from Carson Lake averaged 2.3 p/m selenium (N=11; C. J. Henny, unpubl. data) when waterborne selenium also was < 1.0 p/b (Hoffman et al., 1990). Thus, these field data suggest that eared grebe eggs cross over the 3.0 p/m mean selenium threshold between 0.0 and 1.0 p/b waterborne selenium which is consistent with the prediction generated from equation [5].

An estimate of uncertainty associated with the prediction of a contamination threshold at 0.5 p/b cannot be obtained through routine least squares estimates of variance because equation [5] was derived algebraically. A rough estimate of uncertainty, however, can be obtained by a graphical procedure (figure 3). Point B in figure 3 is derived from the lower 95 percent confidence band of the diet-to-egg regression equation, and it therefore is a rough estimate of the maximum mean dietary selenium consistent with a mean egg selenium of 3 p/m (point A). Similarly, point C in figure 3 is a rough estimate of the maximum waterborne selenium that can be linked with point A through point B. Thus point C is an estimate of the maximum waterborne selenium consistent with a mean egg selenium of 3 p/m, given the variation associated with the two empirical regression equations that equation [5] was algebraically derived from. Point C is estimated as 2.3 p/b waterborne selenium (figure 3). Consequently, the prediction of a contamination threshold at 0.5 p/b waterborne selenium is associated with a relatively narrow range of uncertainty ranging up to about 2.3 p/b.

A more direct approach to estimate the contamination threshold and its uncertainty is to derive an empirical least squares regression equation relating potential mean egg selenium to waterborne selenium directly from the four data points for eared grebes presented in table 3. This yields a regression equation of Log (MES) = 3.69 + 0.55 Log (WS) [R-squared = 0.997; p = 0.001] and a predicted contamination threshold of 0.4 p/b waterborne selenium with 95 percent confidence limits of 0.1 to 0.9 p/b (estimation of X from Y; Sokal and Rohlf 1981:496). The drawbacks of this approach are that the contamination threshold and its confidence limits are extrapolations outside the range of the four data points, and the regression from those four points is not as likely as the graphical approach of figure 3 to fully represent the variation embraced by San Joaquin Valley evaporation ponds. The graphical approach is based on larger sample sizes covering a wider range of environmental conditions (including the crucial threshold region). Both approaches, with low uncertainty, yield a maximum likelihood estimate of about 0.5 p/b waterborne selenium for the contamination threshold.

## IMPLICATIONS FOR DRAINAGE-WATER MANAGEMENT

Based on best available estimates of several critical thresholds (summarized in table 4), there is a fairly narrow range of about 0.5 to 20 p/b waterborne selenium between the minimum estimate for the contamination threshold (for eggs) and the maximum estimate for the embryotoxicity threshold. Many

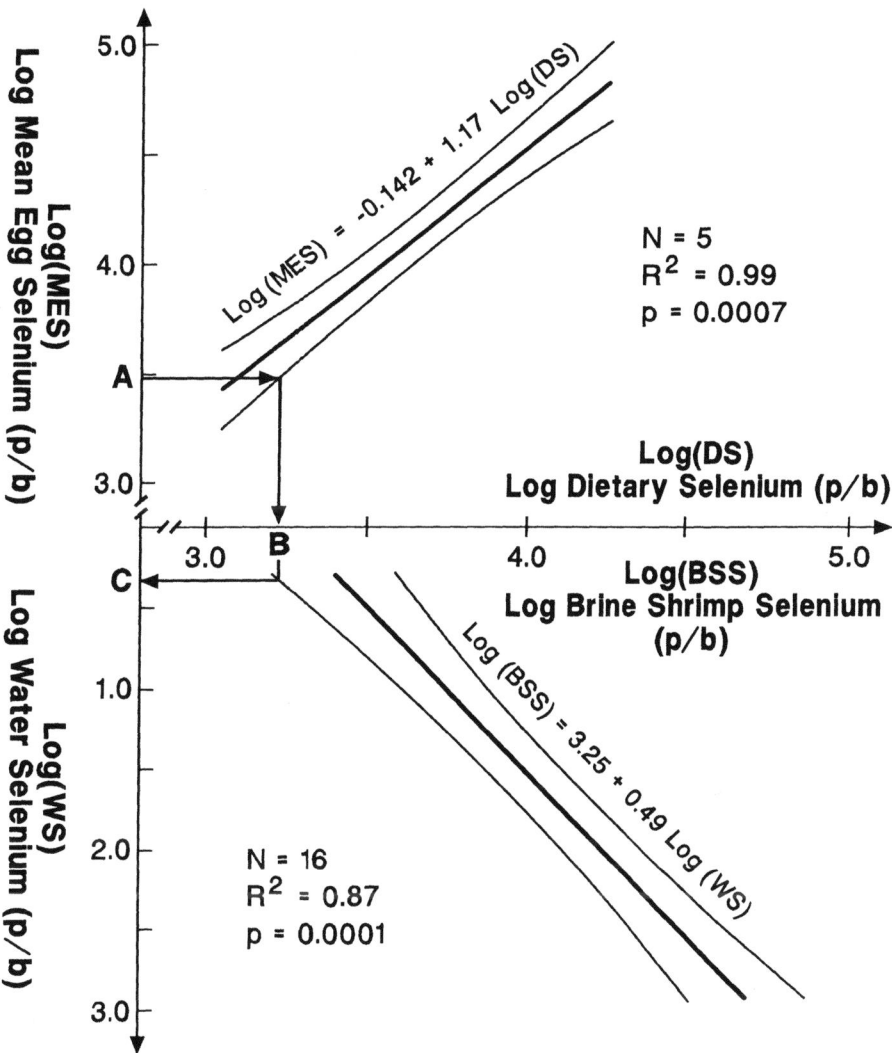

Figure 3. Graphical estimate of the uncertainty associated with predicting the avian contamination threshold for waterborne selenium through separate regressions for food chain uptake and avian uptake of selenium.

*In this figure, the lower 95 percent confidence bands of two regression equations are utilized to estimate the maximum concentration of waterborne selenium (point C) consistent with the bioaccumulation of 3 p/m mean egg selenium (point A) by waterbirds. The estimate of point C is 2.3 p/b waterborne selenium (total recoverable). See text for additional explanation.*

Table 4. Summary of estimated risk thresholds for selenium.

| Estimated Thresholds | Criterion |
|---|---|
| *Waterborne Selenium* (p/b total recoverable) | *Contamination Thresholds* |
| 0.5 | From equation [5] for eared grebe eggs. |
| < 1.0 | Observed for eared grebe eggs from Stillwater Wildlife Management Area, NV. |
| 1 - 3 | Observed for eggs of several species of aquatic birds from the Tulare Basin, CA. |
| | *Embryotoxicity Thresholds* |
| 2 - 13 | Based on critical dietary threshold of ca. 5 p/m organoselenium and empirically derived bioaccumulation curves for total selenium in food-chain items from Tulare Basin evaporation ponds. |
| 10 - 20 | Based on minimum waterborne selenium associated with mean egg selenium > 24 p/m in the Tulare Basin, CA. |
| *Egg Selenium* (p/m, dry weight) | *Contamination Threshold* |
| 3.0 | Upper boundary for normal <u>mean</u> egg selenium estimated from field sampling for various species of waterbirds at Nationwide reference sites. |
| | *Embryotoxicity Threshold* |
| 8.0 | Approximate lower boundary for <u>mean</u> egg selenium associated with populations of black-necked stilts and American avocets exhibiting impaired egg hatchability in the Tulare Basin, CA. |
| 10 | Approximate lower boundary for <u>individual</u> egg selenium associated with impaired embryo viability among black-necked stilts at Kesterson Reservoir, CA. |
| 13 - 24 | Threshold range for <u>mean</u> egg selenium associated with teratogenic populations of aquatic birds sampled in western and northern plains states. |
| 12 - 37 | Threshold range for <u>mean</u> egg selenium associated with impaired egg hatchability and elevated incidence of teratogenesis in mallard embryos when diets of mallard hens are supplemented with selenium in the form of selenomethionine. |

factors can influence whether the full potential for bioaccumulation of selenium in eggs (associated with any given concentration of waterborne selenium) will be realized. Although in many cases local site conditions and the idiosyncrasies of avian behavior may keep realized egg selenium below the site's full potential, it would be prudent to consider drainage water containing 3 to 20 p/b selenium as peripherally hazardous to aquatic birds (i.e., hazardous to some species under some environmental conditions) and drainage water containing more than 20 p/b selenium as widely hazardous to aquatic birds (i.e., hazardous to most species under most environmental conditions; table 4 and equation [5]).

Because impounded drainage water in the Tulare Basin averages roughly 50 p/b selenium (Moore et al., 1990), the protection of aquatic birds is dependent on management actions. Such actions should either reduce the concentrations of contaminants or reduce avian use of contaminated ponds. To prevent most avian toxicity, a reasonable provisional goal for chemical or biological decontamination technologies is purification of drainage water to < 10 p/b waterborne selenium. This goal will not, however, prevent avian contamination. To minimize contamination and the possibility of subtle nonlethal adverse effects and secondary hazards, a reasonable provisional goal is purification to < 2.3 p/b waterborne selenium. When these standards of purity cannot be met by decontamination technology, as is currently the case (Hanna et al., 1990), actions to significantly reduce avian use of contaminated drainage water are necessary.

## ACKNOWLEDGMENTS

The authors are particularly indebted to the many scientists providing access to unpublished data, especially John Shelton of the California Department of Water Resources (others are individually cited in text and tables). P. Albers, D. Barnum, C. Bunck, H. Coulombe, C. Hanson, R. Hothem, D. Lemly, S. Moore, E. Rockwell, and G. Rose offered review comments on earlier drafts of this chapter. W. Hohman, D. Roster, R. Stein, and D. Welsh assisted in collection and management of the Tulare Basin data. T. Charmley, D. Barnum, and their staffs provided logistic support at Kern National Wildlife Refuge. This work was funded by the U.S. Bureau of Reclamation through intra-agency Agreement No. 6-AA-20-04170 and by the California Department of Water Resources through Contract No. B-57626.

# REFERENCES

Birkner, J. H., 1978. *Selenium in Aquatic Organisms from Seleniferous Habitats*. Ph.D. Dissertation, Colorado State University, Fort Collins.

Boyum, K. W. and Brooks, A. S., 1988. The Effects of Selenium in Water and Food on Daphnia Populations, *Arch. Environ. Contam. Toxicol.*, 17, pp. 555-560.

California State Water Resources Control Board, 1987. *Regulation of Agricultural Drainage to the San Joaquin River*. Technical Committee Final Report on Order WQ 85-1.

DeGroat, M. H., 1975. *Probability and Statistics*. Addison-Wesley Publishing Co., Menlo Park, CA.

Eisler, R., 1985. Selenium Hazards to Fish, Wildlife, and Invertebrates: A Synoptic Review, *U.S. Fish and Wildlife Service Biological Report 85(1.5)*.

Eisler, R., 1989. Molybdenum Hazards to Fish, Wildlife, and Invertebrates: A Synoptic Review, *U.S. Fish and Wildlife Service Biological Report 85(1.19)*.

Euliss, N. H., Jr., 1989. *Assessment of Drainwater Evaporation Ponds as Waterfowl Habitat in the San Joaquin Valley, California*. Ph.D. Dissertation, Oregon State University, Corvallis.

Fujii, R., 1988. *Water-Quality and Sediment-Chemistry Data of Drainwater and Evaporation Ponds from Tulare Lake Drainage District, Kings County, California, March 1985 to March 1986*. U.S. Geological Survey Open-File Report 87-700.

Ganther, H. E., 1974. Biochemistry of Selenium. In: Zingaro, R. A. and Cooper, W. C. (Eds.), *Selenium*. Van Nostrand Reinhold Co., New York, NY., pp. 546-614.

Hamilton, S. J.; Buhl, K. J.; Faerber, N. L.; Wiedmeyer, R. H.; and Bullard, F. A., 1990. Toxicity of Organic Selenium in the Diet to Chinook Salmon, *Environ. Toxicol. Chem.*, 9, pp. 347-358.

Hanna, G.P., Jr.; Owens, L.P.; and Kipps, J. 1990. *Agricultural Drainage Treatment Technology Review: Executive Summary*. San Joaquin Valley Drainage Program.

Haseltine, S.D.; Heinz, G.H.; Reichel, W.L.; and Moore, J.F., 1981. Organochlorine and Metal Residues in Eggs of Waterfowl Nesting on Islands in Lake Michigan of Door County, Wisconsin, 1977-78, *Pestic. Monit. J.*, 15, pp. 90-97.

Haseltine, S.D.; Sutcliffe, S.A.; and Swineford, D.M., 1983. Trends in Organochlorine and Mercury Residues in Common Loon (Gavia immer) Eggs from New Hampshire. In: Yahner, R. H. (Ed.), *Trans. NE Sect., The Wildlife Society*, 40th NE. Fish and Wildlife Conf., West Dover, VT., pp. 131-141.

Heinz, G. H.; Hoffman, D. J.; Krynitsky, A. J.; and Weller, D. M. G., 1987. Reproduction in Mallards Fed Selenium, *Environ. Toxicol. Chem.*, 6, pp. 423-433.

Heinz, G. H.; Hoffman, D. J.; and Gold, L. G., 1989. Impaired Reproduction of Mallards Fed An Organic Form of Selenium, *Journal of Wildlife Management*, 53, pp. 418-428.

Henny, C. J. and Herron, G. B., 1989. DDE, Selenium, Mercury, and White-Faced Ibis Reproduction at Carson Lake, Nevada, *Journal of Wildlife Management.*, 53, pp. 1032-1045.

Hoffman, D. J. and Heinz, G. H., 1988. Embryotoxic and Teratogenic Effects of Selenium in the Diet of Mallards, *Journal Toxicol. Environ. Health*, 24, pp. 477-490.

Hoffman, D. J.; Ohlendorf, H. M.; and Aldrich, T. W., 1988. Selenium Teratogenesis in Natural Populations of Aquatic Birds in Central California, *Arch. Environ. Contam. Toxicol.*, 17, pp. 519-525.

Hoffman, R. J.; Hallock, R.J.; Rowe, T. G.; Lico, M. S.; Burge, H. L.; and Thompson, S. P., 1990. *Reconnaissance Investigation of Water Quality, Bottom Sediment, and Biota Associated with Irrigation Drainage in and near Stillwater Wildlife Management Area, Churchill County, Nevada, 1986-87*. U.S. Geological Survey Water-Resources Investigation Report 89-4105.

Hothem, R. L. and Ohlendorf, H. M., 1989. Contaminants in Foods of Aquatic Birds at Kesterson Reservoir, California, 1985, *Arch. Environ. Contam. Toxicol.*, 18, pp. 773-786.

Jehl, J. R., Jr., 1988. Biology of the Eared Grebe and Wilson's Phalarope in the Nonbreeding Season: A Study of Adaptations to Saline Lakes, *Studies in Avian Biology*, 12, pp. 1-74.

Johnsgard, P. A., 1987. *Diving Birds of North America*, University of Nebraska Press, Lincoln.

Krantzberg, G. and Stokes, P. M., 1988. The Importance of Surface Adsorption and pH in Metal Accumulation by Chironomids, *Environ. Toxicol. Chem.*, 7, pp. 653-670.

Lambing, J. H.; Jones, W. E.; and Sutphin, J. W., 1988. *Reconnaissance Investigation of Water Quality, Bottom Sediment and Biota Associated with Irrigation Drainage in Bowdoin National Wildlife Refuge and Adjacent Areas of the Milk River Basin, Northeastern Montana, 1986-87*. U.S. Geological Survey Water-Resources Investigation Report 87-4243.

Letey, J.; Roberts, C.; Penberth, M.; and Vasek, C., 1986. *An Agricultural Dilemma: Drainage Water and Toxics Disposal in the San Joaquin Valley*. University of California Kearny Foundation, Soil Sciences, Special Publication 3319.

Mahoney, S.A. and Jehl, J.R., Jr., 1985. Avoidance of Salt-Loading by a Diving Bird at a Hypersaline and Alkaline Lake: Eared Grebe, *Condor*, 87, pp. 389-397.

Moore, S.B.; Detwiler, S.J.; Winckel, J.; and Weegar, M.D., 1989. *Biological Residue Data for Evaporation Ponds in the San Joaquin Valley, California*. San Joaquin Valley Drainage Program.

Moore, S.B.; Winckel, J.; Detwiler, S.J.; Klasing, S.A.; Gaul, P.A.; Kanim, N.R.; Kesser, B.E.; DeBevec, A.B.; Beardsley, K.; and Puckett, L.K., 1990. *Fish and Wildlife Resources and Agricultural Drainage in the San Joaquin Valley, California*. San Joaquin Valley Drainage Program.

Newman, M.C. and McIntosh, A.W., 1989. Appropriateness of Aufwuchs as a Monitor of Bioaccumulation, *Environ. Pollut.*, 60, pp. 83-100.

Ohlendorf, H.M., 1989. Bioaccumulation and Effects of Selenium in Wildlife. In: Jacobs, L.W., (Ed.), *Selenium in Agriculture and the Environment*, Special Publication 23, American Society of Agronomy and Soil Science Society of American, Madison, WI., pp. 133-177.

Ohlendorf, H.M. and Marois, K.C., 1990. Organochlorines and Selenium in California Night-Heron and Egret Eggs, *Environ. Monit. Assess.*, 15, pp. 91-104.

Ohlendorf, H.M. and Skorupa, J.P., 1989. Selenium in Relation to Wildlife and Agricultural Drainage Water. In: Carapella, S.C., Jr., (Ed.), *Proceedings, Fourth International Symposium on Uses of Selenium and Tellurium*. Selenium-Tellurium Development Assn., Inc., Darien, CT., pp. 314-338.

Ohlendorf, H.M.; Hothem, R.L.; Bunck, C.M.; Aldrich, T.W.; and Moore, J. F., 1986. Relationships Between Selenium Concentrations and Avian Reproduction, *Transactions of the North American Wildlife and Natural Resources Conference*, 51, pp. 330-342.

Ohlendorf, H.M.; Hothem, R.L.; Aldrich, T.W.; and Krynitsky, A.J., 1987. Selenium Contamination of the Grasslands, a Major California Waterfowl Area, *Sci. Tot. Environ.*, 66, pp. 169-183.

Ohlendorf, H.M.; Hothem, R.L.; and Welsh, D., 1989. Nest Success, Cause-Specific Nest Failure, and Hatchability of Aquatic Birds at Selenium-Contaminated Kesterson Reservoir and a Reference Site, *Condor*, 91, pp. 787-796.

Olive, S. W. and Johnson, R. L., 1986. Environmental Contaminants: Selected Legal Topics, *U.S. Fish and Wildlife Service Biological Report 87(1)*.

Poley, W. E.; Moxon, A. L.; and Franke, K. W., 1937. Further Studies of the Effects of Selenium Poisoning on Hatchability, *Poultry Science*, 52, pp. 1841-1846.

Presser, T. S. and Barnes, I., 1984. *Selenium Concentrations in Waters Tributary to and in the Vicinity of the Kesterson National Wildlife Refuge,*

*Fresno and Merced Counties, California.* U.S. Geological Survey Water-Resources Investigation Report 84-4122.

Saiki, M. K. and Lowe, T. P., 1987. Selenium in Aquatic Organisms from Subsurface Agricultural Drainage Water, San Joaquin Valley, California, *Arch. Environ. Contam. Toxicol.*, 16, pp. 657-670.

Schroeder, R. A.; Palawski, D. U.; and Skorupa, J. P., 1988. *Reconnaissance Investigation of Water Quality, Bottom Sediment, and Biota Associated with Irrigation Drainage in the Tulare Lake Bed Area, Southern San Joaquin Valley, California, 1986-87*, U.S. Geological Survey Water-Resources Investigation Report 88-4001.

Schuler, C. A., 1987. *Impacts of Agricultural Drainwater and Contaminants on Wetlands at Kesterson Reservoir, California.* M.S. Thesis, Oregon State University, Corvallis.

Shelton, J.; Hoffman-Floerke, D.; and Jacobsen, D., 1990. Bioaccumulation of Trace Elements in Agricultural Evaporation Pond Organisms in the San Joaquin Valley, California. Abstract from Joint Annual Meeting of the Northwest and Western Sections of The Wildlife Society, Sparks, NV.

Shelton, M. L., 1987. Irrigation Induced Change in Vegetation and Evapotranspiration in the Central Valley of California, *Landscape Ecology*, 1, pp. 95-105.

Siegel, S., 1956. *Non-Parametric Statistics.* McGraw-Hill Book Co., New York, NY.

Smith, G. J. and Anders, V. P., 1989. Toxic Effects of Boron on Mallard Reproduction: Implications for Agricultural Drainwater Management, *Environ. Toxicol. Chem.*, 8, pp. 943-950.

Smith, G. J. and Heinz, G. H., 1990. The Interaction of Selenium and Boron: Effects on Mallard Reproduction. Abstract from Fifth Annual Symposium on Selenium and Its Implications for the Environment, Department of Conservation Resources Studies, University California, Berkeley, and The Bay Institute of San Francisco, Berkeley, CA.

Sokal, R.R., and F.J. Rohlf., 1981. *Biometry.* W.H. Freeman and Co., New York, NY.

Stephens, D. W.; Waddell, B.; and Miller, J. B., 1988. *Reconnaissance Investigation of Water Quality, Bottom Sediment, and Biota Associated with Irrigation Drainage in the Middle Green River Basin, Utah, 1986-87.* U.S. Geological Survey Water-Resources Investigation Report 88-4011.

Sylvester, M. A.; Deason, J. P.; Feltz, H. R.; and Engberg, R. A., 1989. Preliminary Results of the Department of Interior's Irrigation Drainage Studies, *Proc. on Planning Now for Irrigation and Drainage.* IR Div/ASCE, Lincoln, NE., pp. 665-677.

U.S. Fish and Wildlife Service, 1989. *Baseline Study of Trace Elements in the Aquatic Ecosystem of the James River Garrison Diversion Unit, 1986-88.* Final Report to the U.S. Bureau of Reclamation, USFWS, Bismarck, ND.

U.S. Fish and Wildlife Service, 1990. *Effects of Irrigation Drainwater Contaminants on Wildlife.* Final Report on Intra-agency Agreement 6-AA-20-04170 with the U.S. Bureau of Reclamation, USFWS, Patuxent Wildlife Research Center, Laurel, MD.

Westcot, D.; Rosenbaum, S.; Grewell, B.; and Belden, K., 1988a. *Water and Sediment Quality in Evaporation Basins Used for the Disposal of Agricultural Subsurface Drainage Water in the San Joaquin Valley, California.* Central Valley Regional Water Quality Control Board.

Westcot, D.; Toto, A; Grewell, B.; and Belden, K., 1988b. *Uranium Levels in Water in Evaporation Basins Used for the Disposal of Agricultural Subsurface Drainage Water in the San Joaquin Valley, California.* Central Valley Regional Water Quality Control Board.

# 19 PRELIMINARY ASSESSMENT OF THE EFFECTS OF SELENIUM IN AGRICULTURAL DRAINAGE ON FISH IN THE SAN JOAQUIN VALLEY

Michael K. Saiki, Mark R. Jennings, and Steven J. Hamilton,
U.S. Fish and Wildlife Service

## ABSTRACT

Concentrations of total selenium were measured in whole-body samples of seven fishes from the Sacramento and San Joaquin River systems and the San Francisco Bay complex. Concentrations of selenium (up to 11 $\mu$g/g dry weight in whole-body composite samples) were highest in fish from canals and sloughs in the Grassland Water District (Grasslands) that received large inflows of subsurface agricultural drainage water. Slightly lower selenium concentrations occurred in fish from the San Joaquin River immediately downstream from tributaries draining the Grasslands. Although circumstantial evidence suggests that selenium-sensitive species such as bluegills and largemouth bass are being excluded from the Grasslands, conclusive evidence of selenium toxicity is still lacking. In response to earlier reports of high concentrations of selenium in several species collected from the Grasslands, the California Department of Health Services has urged people to limit consumption of fish from this region.

## INTRODUCTION

Selenium occurs naturally in high concentrations in soils along the west side of the San Joaquin Valley floor (U.S. Bureau of Reclamation, 1984). It is mobilized from soils and transported to surface waters (e.g., ponds, streams) by subsurface agricultural (tile) drainage water. It accumulates in aquatic organisms through uptake directly from tile drainage water, or indirectly from the consumption of contaminated food-chain organisms, or both (Ohlendorf et al., 1986 and Saiki and Lowe, 1987). Excessive concentrations of selenium in the tissues of adult aquatic birds at Kesterson Reservoir (Kesterson) and evaporation ponds elsewhere in the San Joaquin Valley are probably responsible for the high mortalities and deformities observed in the young (Ohlendorf et al., 1986 and Ohlendorf and Skorupa, 1989). By comparison, the effects of seleniferous tile drainage water on fish are poorly understood; however, studies conducted in selenium-polluted environments elsewhere in the United States suggest that selenium concentrations $\geq 12\,\mu g/g$ (dry weight basis) in the whole bodies of freshwater fishes may cause reproductive failure and other problems (Lemly and Smith, 1987). The health and well-being of humans who consume waterfowl, fish, and other aquatic organisms that inhabit seleniferous waters could also be affected. Although public health surveys near Kesterson have identified no unusual problems attributable to excessive selenium exposure (Fan et al., 1988), at least one study has shown that the ingestion of high-selenium vegetables and maize in China was responsible for selenosis in humans (Yang et al., 1983).

The objectives of this chapter are (1) To provide a brief overview of selenium concentrations in fish from the Central Valley (with emphasis on the San Joaquin River system) and San Francisco Bay and (2) to determine if the selenium concentrations have approached or exceeded concentrations known or suspected to be toxic to fish and to humans who consume the fish. Detailed documentation and discussion of selenium and other elements (e.g., arsenic, boron, chromium, and mercury) determined in fishes as part of this field investigation are reported elsewhere (e.g., Saiki and Palawski, in press).

## STUDY AREA AND METHODS

Surface waters in the Sacramento and San Joaquin Valleys and in San Francisco Bay are affected to differing degrees by tile drainage water (figure 1). Flows in the Sacramento and San Joaquin Rivers and in other tributaries originating in the Sierra Nevada (American, Merced, Stanislaus, and Tuolumne Rivers) are initially derived from snowmelt and rainfall. During the irrigation season (usually April to September in the Sacramento River basin, and March to October in the San Joaquin River basin), surface return flows (tailwater) from irrigated fields often contribute substantially to the discharge in downstream reaches of all rivers. However, only the lower reaches of the San Joaquin River are known to receive subsurface drainage water.

Except for canals and sloughs in the Grassland Water District (Grasslands) in western Merced County, the southern and western tributaries of the San Joaquin River derive nearly all of their discharge from surplus irrigation water and tailwater. Although irrigation water and tailwater contribute to flows in canals and sloughs of the Grasslands, the canals also carry wastewater from tile-drained fields located upslope from the Grasslands to Salt and Mud Sloughs for

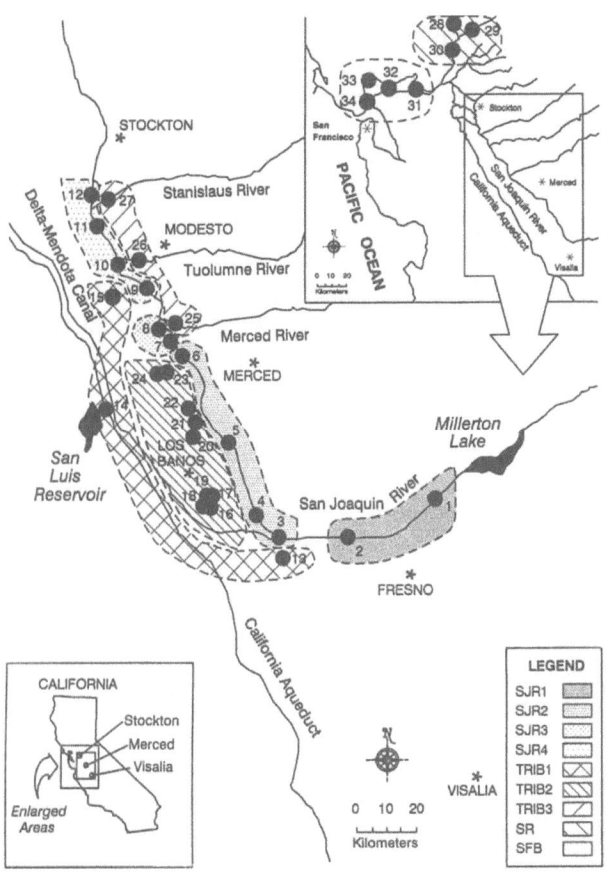

Figure 1. The study area, showing the general locations of broad geographic regions and the sampling sites within each region. Sampling sites in regions that subdivide the San Joaquin River are as follows: SJR1, (1) Fort Washington Beach Park, and (2) Highway 145; SJR2, (3) Mendota Pool, (4) Firebaugh, (5) Highway 152, and (6) Lander Avenue; SJR3, (7) Fremont Ford State Recreation Area and (8) above Hills Ferry Road; SJR4, (9) Crows Landing Road, (10) Laird County Park, (11) Maze Road, and (12) Durham Ferry State Recreation Area. Sampling sites in the remaining regions are as follows: TRIB1, (13) Fresno Slough at the Mendota Wildlife Area, (14) Delta-Mendota Canal at O'Neill Forebay, and (15) Orestimba Creek at Highway 33; TRIB2, (16) Helm Canal, (17) Agatha Canal, (18) Main Canal, (19) Camp 13 Ditch, (20) Mud Slough at Los Banos Wildlife Area, (21) Salt Slough at Hereford Road, (22) Salt Slough at the San Luis National Wildlife Refuge, (23) Mud Slough at Gun Club Road, and (24) Los Banos Creek at Gun Club Road; TRIB3, (25) Merced River at George J. Hatfield State Recreation Area, (26) Tuolumne River at Shiloh Road, and (27) Stanislaus River at Caswell State Park; SR, (28) Sacramento River at Knights Landing, (29) American River at Discovery Park, and (30) Sacramento River at Clarksburg; and SFB, (31) Honker Bay, (32) Suisun Bay, (33) San Pablo Bay, and (34) San Francisco Bay near Alcatraz Island.

disposal in the San Joaquin River. Slightly over 60 percent (115.9 million $m_3$) of the water delivered annually to the Grasslands consists of a variable mixture of tailwater and seleniferous tile drainage (collectively referred to as "agricultural drainage water"); the only requirement is that the concentrations of Total Dissolved Solids (TDS) and boron in water entering the Grasslands must not exceed 2,500 mg/L and 6.0 mg/L, respectively (Jones and Stokes Associates, 1985). The State Water Resources Control Board (1987) estimated that, in 1984-85, subsurface tile drainage composed about 39 percent (42.1 million $m_3$) of the agricultural drainage water entering the Grasslands.

Water quality surveys have revealed longitudinal (upstream to downstream) patterns for several physiochemical variables in the San Joaquin River that suggest progressively increasing environmental degradation caused by inflows of agricultural drainage water (Saiki, 1984 and Saiki and Palawski, in press). In particular, total alkalinity, total hardness, TDS, and conductivity increase at downstream sampling sites that receive the most concentrated flows of agricultural drainage water. In 1985, Salt and Mud Sloughs collectively supplied only 12 percent of the total flow in the San Joaquin River between Lander Avenue and its confluence with the Merced River; however, these sloughs contributed about 81 percent of the selenium, 69 percent of the boron, 44 percent of the molybdenum, and 46 percent of the dissolved salts occurring in this reach of the river (State Water Resources Control Board, 1987).

San Francisco Bay receives its inflow primarily from the Sacramento and San Joaquin Rivers. Minor contributions are made by rainfall in local watersheds and by municipal and industrial discharges. Although San Francisco Bay is enriched with selenium from the San Joaquin River and from oil refineries near Carquinez Strait, Cutter (1989) reported that the concentrations of dissolved selenium in water from the bay were within the ranges found in other estuaries, and far lower than concentrations typically measured in the San Joaquin River.

## Sample Collection and Handling

Samples from 27 sites on selected reaches of the San Joaquin River and its tributaries, 3 sites in the Sacramento River system, and 4 sites in the San Francisco Bay complex (SFB) (figure 1) yielded a total of 7 species of fish: bluegills (Lepomis macrochirus), chinook salmon (Oncorhynchus tshawytscha), common carp (Cyprinus carpio), largemouth bass (Micropterus salmoides), mosquitofish (Gambusia affinis), Sacramento blackfish (Orthodon microlepidotus), and striped bass (Morone saxatilis). All samples were collected between August 1986 and August 1987.

Fish were captured by electrofishing and gillnetting (Sacramento and San Joaquin River systems only); seining; trawling (SFB only); or a combination of these methods. Immediately after collection, the fish were rinsed with water at the respective sites, wrapped and bagged in polyethylene, then chilled on ice. Fish of similar total lengths were sorted into composite samples of about five fish or a minimum of 50 g, then rewrapped and bagged in polyethylene and frozen (-10 °C). Samples remained frozen until they were prepared for chemical analysis.

## Elemental Analyses and Quality Assurance

Samples of striped bass were analyzed at the U.S. Fish and Wildlife Service's Patuxent Analytical Control Facility, Laurel, Maryland; and samples of bluegills, chinook salmon, common carp, largemouth bass, mosquitofish, and Sacramento blackfish were analyzed at the National Fisheries Contaminant Research Center, Columbia, Missouri. The moisture content of samples was determined either by oven-drying overnight at 105 °C, or by lyophilization. For striped bass, aliquants of the samples were digested with a modified nitric acid procedure (hydrogen peroxide was added during the digestion process to enhance the solubilizing of tissue); total selenium was then quantified by graphite furnace atomic absorption spectrophotometry with Zeeman background correction (Krynitsky, 1987). Selenium concentrations were determined in samples of the six other species by digesting with a combined wet chemical (nitric and hydrochloric acids) and dry ash procedure, then quantifying the total selenium by using hydride-generation atomic absorption spectrophotometry (Brumbaugh and Walther, 1989).

Quality assurance measures included analyses of blind replicates, spiked samples, blanks, and reference materials from various sources: RM #50 albacore tuna from the U.S. National Institute of Standards and Technology (NIST, formerly referred to as the U.S. National Bureau of Standards); A-6 fish solubles from the International Atomic Energy Agency (IAEA); and an in-house reference material consisting of ground, whole striped bass (from the National Fisheries Contaminant Research Center). The relative standard deviation and relative percent difference, both of which estimate the precision of the "method" from blind replicates, ranged from 0.6 percent to 7.5 percent ($N = 20$). All determinations of selenium from digestion blank solutions were below the method detection limits (0.0023-0.0513 $\mu$g/g dry weight). Selenium concentrations measured in the reference fish samples were within their certified limits (IAEA RM-A-6 fish solubles, 3.07 ± 1.22 $\mu$g/g; NIST RM #50 albacore tuna, 3.60 ± 0.40 $\mu$g/g; in-house reference material, 2.26 ± 0.2 $\mu$g/g; values are dry weights). Mean recoveries, based on spiked samples, were 103 percent for 141 samples analyzed by the Patuxent Analytical Control Facility and 94.1 percent for 81 samples analyzed by the National Fisheries Contaminant Research Center; the concentrations of selenium in the samples were not adjusted for mean recovery efficiency.

## RESULTS AND DISCUSSION

A total of 337 fish samples were analyzed for moisture content and selenium concentration. Moisture content varied significantly ($P \leq 0.05$, one-way ANOVA) among species and sites. Overall, mean moisture content (percent) of the various species were as follows (minimum and maximum values in parentheses): bluegills, 73.0 (71-75); chinook salmon, 76.8 (70-80); common carp, 74.7 (64-82); largemouth bass, 73.6 (72-76); mosquitofish, 75.6 (72-78); Sacramento blackfish, 74.4 (73-76); and striped bass, 74.2 (69-84). To standardize the moisture content of the samples, all concentrations are reported on a dry-weight basis, unless clearly indicated otherwise. Wet weight concentrations of selenium can be estimated from dry weight concentrations by using the formula

[1] $\quad WW = DW \cdot (100 - MOIST) / 100$

where WW = wet weight, DW = dry weight, and MOIST = mean moisture content of the species.

Except for one sample of striped bass from Fresno Slough, detectable concentrations of selenium were measured in all samples of fish during this survey (table 1). The geometric mean concentrations of selenium seemingly varied in relation to inflows of tile drainage water; concentrations were relatively high ($>3.4 \mu g/g$) in samples from canals and sloughs in the Grasslands (TRIB2; see table 1, figure 2) and nearly as high in the reach of San Joaquin River adjacent to or immediately downstream from sloughs draining the Grasslands (SJR3). With few exceptions, mean selenium concentrations were uniformly low ($\leq 2 \mu g/g$) in fish collected from other areas of the Central Valley and San Francisco Bay that receive little or no direct flows of tile drainage water.

During the survey, selenium concentrations were $>5 \mu g/g$ in fish collected from the southern and north-central portions of the Grasslands (e.g., Agatha and Main Canals, Camp 13 Ditch, Mud Slough at the Los Banos Wildlife Area, and Mud Slough at Gun Club Road); concentrations were generally lower in fish from other localities. Similar geographic (spatial) patterns for selenium concentrations in fish from the Grasslands were noted in samples collected in 1984 and 1985 (Saiki, 1986 and 1989). The southern and north-central sites also contained some of the highest concentrations of dissolved selenium

Table 1. Total selenium concentrations (μg/g dry weight) in fishes from the Sacramento and San Joaquin River Systems, and the San Francisco Bay complex, 1986-87.

| Region and statistic | Bluegills | | | Chinook salmon | | | Common carp | | | Largemouth bass | | | Mosquitofish | | | Striped bass | | |
|---|---|---|---|---|---|---|---|---|---|---|---|---|---|---|---|---|---|---|
| | N | X̄ | MIN-MAX | N | X̄ | MIN-MAX | N | X̄ | MIN-MAX | N | X̄ | MIN-MAX | N | X̄ | MIN-MAX | N | X̄ | MIN-MAX |
| **San Joaquin River System** | | | | | | | | | | | | | | | | | | |
| *San Joaquin River* | | | | | | | | | | | | | | | | | | |
| SJR1 | 6 | 0.457 D | 0.38-0.53 | 0 | -- | -- | 4 | 0.567 D | 0.50-0.65 | 0 | -- | -- | 6 | 0.528 E | 0.45-0.63 | 0 | -- | -- |
| SJR2 | 14 | 1.53 B | 1.1-2.1 | 0 | -- | -- | 5 | 1.66 BC | 0.56-2.8 | 3 | 0.919 C | 0.88-0.94 | 12 | 1.54 BCD | 0.91-2.1 | 9 | 2.07 BC | 0.61-3.5 |
| SJR3 | 10 | 1.89 B | 1.2-2.4 | 2 | 2.95 A | 2.9-3.1 | 6 | 4.07 AB | 2.2-6.9 | 3 | 1.89 B | 1.7-2.0 | 6 | 2.82 AB | 2.0-4.0 | 5 | 4.29 AB | 2.7-5.8 |
| SJR4 | 14 | 1.62 B | 1.3-2.1 | 11 | 2.02 A | 1.2-3.2 | 10 | 1.75 C | 1.3-2.5 | 3 | 1.38 BC | 1.3-1.5 | 12 | 1.61 BC | 1.5-1.8 | 7 | 1.68 C | 0.66-2 |
| *Tributaries* | | | | | | | | | | | | | | | | | | |
| TRIB1 | 3 | 1.25 BC | 1.2-1.3 | 0 | -- | -- | 1 | 1.3 ACD | -- | 0 | -- | -- | 9 | 1.35 CD | 0.96-1.8 | 6 | 1.29 C | <0.41-2.8 |
| TRIB2 | 21 | 5.47 A | 1.9-9.4 | 0 | -- | -- | 15 | 4.28 A | 1.7-10. | 5 | 4.71 A | 2.8-9.7 | 27 | 3.44 A | 1.6-11.1 | 8 | 5.00 A | 2.7-7.9 |
| TRIB3 | 9 | 0.954 C | 0.76-1.1 | 9 | 1.07 B | 0.65-1.4 | 4 | 1.51 CD | 1.3-1.7 | 9 | 1.02 C | 0.90-1.2 | 9 | 0.924 DE | 0.66-1.4 | 1 | 3.2 AC | -- |
| **Sacramento River System** | | | | | | | | | | | | | | | | | | |
| SR | 3 | 1.83 B | 1.7-2.1 | 6 | 1.75 A | 1.5-1.9 | 3 | 1.35 CD | 1.2-1.5 | 3 | 1.22 BC | 1.1-1.3 | 3 | 1.46 BCD | 1.4-1.4 | 0 | -- | -- |
| **San Francisco Bay Complex** | | | | | | | | | | | | | | | | | | |
| SFB | 0 | -- | -- | 11 | 1.04 B | 0.66-2.1 | 0 | -- | -- | 0 | -- | -- | 0 | -- | -- | 11 | 1.73 C | 0.86-3.3 |
| F(df)[b] | 69.93** | | | 14.63** | | | 12.31** | | | 23.13** | | | 24.83** | | | 10.59** | | |

N, number of composite samples. X̄, mean (geometric mean if ≥2 samples analyzed). MIN-MAX (minimum-maximum).[c,d]
[a]See figure 1 for names and locations of regions.
[b]Values of df: bluegills, 7,72; chinook salmon, 4,34; common carp, 7,40; largemouth bass, 5,20; mosquitofish, 7,76; striped bass, 6,50; ** $P \leq 0.01$.
[c]For Sacramento blackfish, selenium concentrations were as follos (Region, N, X̄, MIN-MAX: Trib2, 3, 387, 3.7-4.1.
[d]Within a column, means with the same capital letter are not significantly different ($P > 0.05$, Tukey-Kramer hsd test).

(45-64 µg/L) reported by Presser and Barnes (1985), presumably due to major inflows of tile drainage waters into the Agatha, Helm, and Main Canals, and Camp 13 Ditch. By comparison, the existing National Ambient Water Quality criterion for dissolved selenium, established by the U.S. Environmental Protection Agency (EPA, 1987) to protect freshwater aquatic organisms, is only 5.0 µg/L (4-d average concentration that should not be exceeded more than once every 3 years).

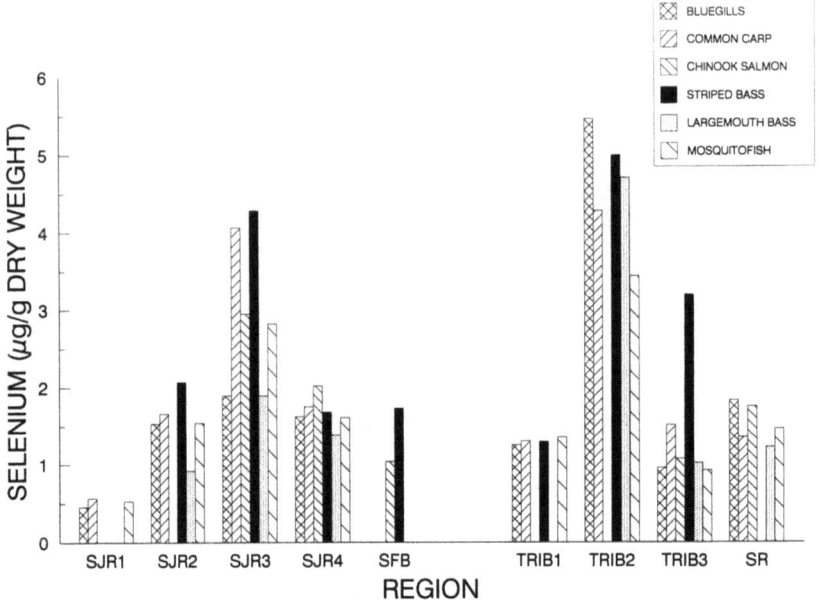

Figure 2. Mean concentrations of total selenium in whole-body samples of fishes from selected regions in the Sacramento and San Joaquin River basins, and San Francisco Bay. Regions subdividing the San Joaquin River (SJR1, SJR2, SJR3, and SJR4) and San Francisco Bay (SFB) are arranged in longitudinal (upstream-downstream) order; the remaining regions--tributaries of the San Joaquin River (TRIB1, TRIB2, and TRIB3) and the Sacramento River system (SR)- -are also arranged in longitudinal order relative to the San Joaquin River and San Francisco Bay.

Mean selenium concentrations were usually lowest (<2 lg/g) in fish from reaches that received little or no tile drainage water, such as the San Joaquin River upstream from the Grasslands (SJR1 and SJR2; see table 1, figure 2), the southern and western tributaries exclusive of the Grasslands (TRIB1), the eastern tributaries (TRIB3), and the Sacramento River system (SR). A single composite sample of juvenile striped bass from the Tuolumne River (one of three eastern tributaries included in TRIB3) was exceptional in containing 3.2 lg/g of selenium, which was about threefold higher than in samples of other fish species from the eastern tributaries. According to Saiki and Palawski (in press), this concentration was similar to values measured in striped bass from nearby sites in the San Joaquin River, and possibly represented fish that had recently entered the Tuolumne River. Dissolved concentrations of selenium in these waters are generally <1 lg/L (Gilliom, 1986 and Shelton and Miller, 1988).

Fish collected from the San Joaquin River adjacent to or immediately downstream from the Grasslands (SJR3) contained mean selenium concentrations ranging from 1.2-fold to 7.2-fold (mean, 3.3-fold) higher than in fish collected farther upstream at SJR1 and SJR2. Most selenium in fish collected from SJR3 probably originated either directly or indirectly from tile drainage water flowing through Salt and Mud Sloughs (State Water Resources Control Board, 1987).

Mean concentrations of selenium in fish were generally lower in the lowest reach of the San Joaquin River (SJR4) and SFB than in fish from SJR3; however, the differences were not always statistically significant (table 1). Exposure of fish at SJR4 and SFB to selenium from tile drainage water is probably reduced by (1) Dilution with low-selenium waters from eastside tributaries (for SFB, this includes SR), (2) uptake of selenium by aquatic biota, and (3) deposition of selenium in the sediments (Saiki 1989 and Saiki and Palawski, in press).

## Implications for Fish

The selenium content of whole freshwater fish from throughout the United States is routinely monitored by the National Contaminant Biomonitoring Program (NCBP). The 85th percentile concentrations from the NCBP represent arbitrary values that are useful for identifying locations where concentrations of elements in fish are relatively high (May and McKinney, 1981), but do not necessarily imply an associated toxic effect. For selenium, concentrations associated with toxic effects in whole fish (e.g., reduced growth rates and survival; impaired reproductive success) are much higher than the NCBP 85th percentile concentrations (Hamilton and Wiedmeyer, in press; Lemly, 1985; and Lemly and Smith, 1987).

The 1976-84 NCBP 85th percentile values for selenium have ranged from 0.70 µg/g to 0.82 µg/g wet weight (about 2.7-3.2 µg/g dry weight, assuming 74 percent moisture; Schmitt and Brumbaugh, in press). In the present survey, mean selenium concentrations exceeding the NCBP 85th percentile values were measured in common carp, mosquitofish, and striped bass from the San Joaquin River at sites adjacent to, or immediately downstream from, the Grasslands (SJR3), and in bluegills, common carp, largemouth bass, mosquitofish, Sacramento blackfish, and striped bass from the Grasslands (TRIB2, see table 1).

Judging by laboratory studies of selenium uptake from either food or water, adverse effects on growth and survival of young chinook salmon may occur when the fish accumulate whole-body burdens of selenium >5-8 µg/g (Hamilton and Wiedmeyer, in press). Juvenile chinook salmon from the present survey contained a maximum concentration of 3.2 µg/g (table 1), suggesting that this species had not accumulated toxic concentrations of the element.

According to Sato et al. (1980), survival was not affected in common carp that accumulated about 23 µg/g of selenium (whole-body concentration, assuming 74 percent moisture content) after exposure to 1,000 µg/L of selenite for up to 84 d; however, carp exposed to higher concentrations ( ≥ 10,000 µg/L) of selenite died after accumulating >38 µg/g of selenium. During this survey, the highest concentration of selenium measured in carp was 10 µg/g (table 1), indicating that this element was probably not sufficiently elevated to directly affect survival.

Lemly and Smith (1987) reported that freshwater fishes such as centrarchids (e.g., bluegills and largemouth bass) may experience reproductive failure at whole-body concentrations of selenium ≥12 µg/g. Bluegills exposed to high concentrations of selenium typically produce larvae that exhibit teratological effects--e.g., edema, lordosis, and lower jaw gape--and poor survival (Gillespie and Baumann, 1986 and Woock et al., 1987). Although the maximum concentrations of selenium measured during the present survey were only 9.4 µg/g in bluegills and 9.7 µg/g in largemouth bass collected from Mud Slough at Gun Club Road, centrarchids collected from the Grasslands in 1984 and 1985 contained as much as 23 µg/g (Saiki, 1986 and 1989). Bluegills, largemouth bass, and other centrarchids are common-to-abundant in the San Joaquin River; however, they are rarely collected in the canals and sloughs that traverse the Grasslands (Saiki, 1984 and M. R. Jennings and M. K. Saiki, unpublished data). For example, between August and December 1986, the average catch of all centrarchids in the Grasslands (based on a standard "effort" of 30-min of electrofishing) was 3 fish compared with 5-23 fish in other regions of the San Joaquin River system (M. R. Jennings and M. K. Saiki, unpublished data). As judged by available data (e.g., Saiki, 1984 and Saiki and Palawski, in press), water quality variables--temperature, turbidity, dissolved oxygen, pH, total dissolved solids, etc.--measured from sites in the Grasslands are well within limits tolerated by centrarchids (Bennett, 1965; Buckley, 1975; and Emig, 1966). Therefore, the paucity of centrarchids in the Grasslands suggests that they are being excluded from this region, perhaps by selenium toxicity that adversely affects their reproduction.

A study conducted in 1984-85 on the reproductive success of viviparous mosquitofish collected from the San Luis Drain at Kesterson, where whole-body concentrations of selenium averaged >100 $\mu$g/g, indicated that as much as 30 percent of the fry produced by adult females were stillborn; in contrast, stillbirths accounted for ≤3 percent of the fry born to females from the nearby Volta Wildlife Area, where whole-body concentrations of selenium averaged <1.6 $\mu$g/g (M. K. Saiki, unpublished data). Studies have not been attempted with mosquitofish from seleniferous environments elsewhere in the Grasslands and the San Joaquin River, where body burdens of selenium are generally less than 1/10 the concentrations measured in fish from the San Luis Drain (table 1).

## Implications for Humans

Selenium is an essential trace element in human nutrition (National Research Council, 1989). In the United States, people generally receive adequate amounts of selenium in the foods they consume, primarily meats, poultry, grains and grain products, and seafood (Pennington and Church, 1985). Although documented cases of poisoning among humans from eating naturally grown foods containing excessive amounts of selenium are rare, available data suggest that the margin of safety between the required and toxic concentrations of this element is relatively narrow (National Research Council, 1989).

Except for a drinking-water standard of 10 $\mu$g/L (EPA, 1986), there is no current established action level for selenium in food for human consumption in the United States. In California, health advisories are considered or issued for fish and waterfowl harvested from a given locality when selenium concentrations in their flesh approach or exceed 2 $\mu$g/g wet weight (about 7.7 $\mu$g/g dry weight, assuming 74 percent moisture; Fan et al., 1988).

In the survey described here, selenium concentrations were reported for whole fish, whereas humans typically consume only the flesh (fillets). A close association exists between the concentrations of selenium in whole fish and their fillets. For two popular game fishes from the San Joaquin River system, these relations are described as follows (species, N, coefficient of determination, regression equation):

[2] bluegills, 15, 0.982, SF = 0.045 + 1.227 WB

[3] largemouth bass, 11, 0.996, SF = -0.388 + 1.322 WB

where SF and WB are the skinless fillet and whole-body concentrations ($\mu g/g$), respectively. From these equations, it is estimated that bluegills and largemouth bass contained as much as 12 $\mu g/g$ of selenium in their fillets. Moreover, the estimated geometric mean concentrations of selenium in fillets of fishes from three sites in the Grasslands--the Agatha and Helm Canals (bluegills only) and Mud Slough at Gun Club Road (bluegills and largemouth bass)--exceeded 7.7 $\mu g/g$, the threshold concentration reported by Fan et al. (1988) that normally elicits health advisories in California.

According to the Food and Nutrition Board of the National Research Council (1989), the recommended dietary allowance of selenium in adults is 70 $\mu g/d$ for men and 55 $\mu g/d$ for women; the allowance for infants, children, and adolescents are extrapolated from adult values on the basis of bodyweight, and a factor is added for growth. However, EPA (1984) considers 210 $\mu g/d$ to be the acceptable daily intake (ADI) for selenium; this consumption rate is derived from a fifteen fold safety factor applied to the lowest intake-concentration (3,200 $\mu g/d$) associated with human toxicity in China (Yang et al., 1983). From the ADI, Fan et al. (1988) advised that a permissible amount of a dietary item containing selenium can be calculated as follows:

[4]  (ADI-DI)/C = PI

where DI is the daily intake of selenium from all other sources, C is the concentration of selenium in the food item, and PI is the permissible intake of the food item. From this formula, people who consume fish from the Grasslands should eat no more than about 67 g/d of fillets (wet weight basis, assuming that the fillets contain 74 percent moisture) to remain within the "acceptable daily intake" for selenium.

Since 1986, the California Department of Health Services has recommended that people eat no fish caught in Kesterson, and that the consumption of fish from other areas in the Grasslands be limited to 56.7 g/week (wet weight basis; Fan et al., 1988). Pregnant women or those who might soon become pregnant, and children of age 15 and under, were further advised not to consume any fish from the Grasslands (Fan et al., 1988).

## CONCLUSIONS AND RESEARCH NEEDS

High concentrations of environmental selenium can adversely affect the reproduction, growth, or survival of fish, and require public health advisories for humans who eat affected fish. This survey demonstrated that fish from certain parts of the Grasslands and the San Joaquin River adjacent to or immediately downstream from the Grasslands have accumulated excessive concentrations of selenium. Although circumstantial evidence suggests that selenium-sensitive fishes may be excluded from surface waters in the Grasslands that receive concentrated tile drainage water, conclusive evidence of a cause-effect relation is still lacking.

To better understand and predict the consequences of discharging seleniferous tile drainage water into the San Joaquin River or its tributaries, scientists need additional information about several aspects of selenium bioaccumulation and its effects on fish. Five examples follow:

1. The chemical forms of selenium--selenate, selenite, elemental selenium, and selenide (including organic selenium)--that occur in water and forage organisms should be determined so that appropriate form(s) can be used when fish are exposed to this element in laboratory toxicity experiments.
2. Better knowledge is needed of selenium concentrations that occur in different components of the aquatic food chain as a result of exposure to various concentrations of waterborne selenium. Also, the biochemical and physiological processes that underlie the uptake and excretion of selenium at each trophic level in the food chain must be understood to enable the development of predictive models of selenium bioaccumulation.
3. Threshold waterborne, dietary, and fish-tissue concentrations of selenium that affect survival, growth, reproduction, and other biological processes of fish must be determined. Laboratory studies are needed to estimate the toxic threshold concentrations in several ecologically and economically important fish species throughout their complete life cycle, and preferably through several generations, because not all adverse effects might be expressed over a shorter time period. In addition, field studies are needed to verify that the toxic thresholds estimated from laboratory studies are applicable to fish in their natural habitat.
4. Although field and laboratory studies have documented the occurrence of mortality and teratogenic effects in fish exposed to high-selenium environments, there remains a need for better diagnostic indicators of selenium toxicity. These should include sublethal characteristics (e.g., behavior) that can be measured in live animals and also those that are diagnostic of selenium-induced mortality.
5. Knowledge of how selenium interacts with other contaminants must be gained if the ecological significance of selenium is to be accurately assessed. It is known that the toxic effects of selenium can be greatly

altered by other chemicals (such as heavy metals and other trace elements), or factors such as water quality variables, species of organism and its life stage, nutritional status, disease state, and level of stress.

In addition to the research needs listed above, considerably more information may be needed by regulatory and management agencies to ensure that the disposal of seleniferous tile drainage into surface waters does not adversely affect fish populations or lead to further restrictions on the consumption of fish.

## ACKNOWLEDGEMENTS

The authors thank field personnel from the U.S. Fish and Wildlife Service and the California Department of Fish and Game for assistance in collecting fish, and the Portland (Oregon) Regional Office of the U.S. Fish and Wildlife Service for funding the analysis of selenium in striped bass. All other support was provided by the San Joaquin Valley Drainage Program, a cooperative effort between the U.S. Department of the Interior and the State of California.

## REFERENCES

Bennett, G. W., 1965. The Environmental Requirements of Centrarchids with Special Reference to Largemouth Bass, Smallmouth Bass, and Spotted Bass. In: Tarzwell, C. M. (Ed.), Biological Problems in Water Pollution, Third Seminar, 1962. Public Health Service Publication No. 999-WP-25, *Environmental Health Series, Water Supply and Pollution Control*, U.S. Department of Health, Education, and Welfare, Cincinnati, OH. pp. 156-160.

Brumbaugh, W. G. and Walter, M. J., 1989. Determination of Arsenic and Selenium in Whole Fish by Continuous-Flow Hydride Generation Atomic Absorption Spectrophotometry, *Journal Assn. Off. Anal. Chem.*, 72, pp. 484-486.

Buckley, R.V., 1975. Chemical and Physical Effects on the Centrarchid Basses. In: Clepper, H. (Ed.), *Black Bass Biology and Management*, Sport Fishing Institute, Washington, DC., pp. 286-294.

California State Water Resources Control Board, 1987. *Regulation of Agricultural Drainage to the San Joaquin River*. Technical Committee Report on Order WQ 85-1.

Cutter, G. A., 1989. The Estuarine Behaviour of Selenium in San Francisco Bay, *Estuarine Coastal Shelf Science*, 28, pp. 13-34.

Emig, J. W., 1966. Bluegill Sunfish. In: Calhoun, A. (Ed.), *Inland Fisheries Management*, California Department of Fish and Game, pp. 375-392.

Fan, A. M.; Book, S. A.; Neutra, R. R.; and Epstein, D. M., 1988. Selenium and Human Health Implications in California's San Joaquin Valley, *Journal Toxicol. Environmental Health*, 23, pp. 539-559.

Gillespie, R. B. and Baumann. P. C., 1986. Effects of High Tissue Concentrations of Selenium on Reproduction in Bluegills, *Trans. Am. Fish. Soc*, 115, pp. 208-213.

Gilliom, R. J., 1986. *Selected Water-Quality Data for the San Joaquin River and its Tributaries, California, June to September 1985*. U.S. Geological Survey Open-File Report 86-74, p. 12.

Hamilton, S. J. and Wiedmeyer, R. H. Concentration of a Mixture of Boron, Molybdenum, and Selenium in Chinook Salmon, *Trans. Am. Fish. Soc.*, 119. In Press.

Jones and Stokes Associates, Inc., 1985. *Ecological and Water Management Characterization of Grassland Water District*. Draft Report (dated December 5, 1985), Prepared for the Grassland Water District, Los Banos, CA., 49 p. (plus Appendices).

Krynitsky, A. J., 1987. Preparation of Biological Tissue for Determination of Arsenic and Selenium by Graphite Furnace Atomic Absorption Spectrometry, *Anal. Chem*, 59, pp. 1884-1886.

Lemly, A.D., 1985. Toxicology of Selenium in a Freshwater Reservoir: Implications for Environmental Hazard Evaluation and Safety, *Ecotoxicol. Environmental Safety*, 10, pp. 314-338.

Lemly, A. D. and Smith, G. J., 1987. *Aquatic Cycling of Selenium: Implications for Fish and Wildlife*. U.S. Fish Wildlife and Service, Fish Wildlife Leafl. 12, p. 10.

May, T. W. and McKinney, G. L., 1981. Cadmium, Lead, Mercury, Arsenic, and Selenium Concentrations in Freshwater Fish, 1976-77 -- National Pesticide Monitoring Program, *Pest. Monit. Journal*, 15, pp. 14-38.

National Research Council, 1989. *Recommended Dietary Allowances*, 10th Ed. National Academy Press, Washington, DC., p. 284.

Ohlendorf, H. M.; Hoffman, D. J.; Saiki, M. K.; and Aldrich, T. W., 1986. Embryonic Mortality and Abnormalities of Aquatic Birds: Apparent Impacts by Selenium from Irrigation Drainwater, *Sci. Total Environ.*, 52, pp. 49-63.

Ohlendorf, H. M. and Skorupa, J. P., 1989. Selenium in Relation to Wildlife and Agricultural Drainage Water. In: Carapella, S. C. Jr. (Ed.), *Proceedings of the Fourth International Symposium on Uses of Selenium and Tellurium*, Selenium-Tellurium Development Assn., Darien, CT., pp. 314-338.

Pennington, J.A.T. and Church, H. N., 1985. *Bowes and Church's Food Values of Portions Commonly Used*, 14th Ed., J. B. Lippincott Co., Philadelphia, PA., p. 257.

Presser, T. S. and Barnes, I., 1985. *Dissolved Constituents Including Selenium in Waters in the Vicinity of Kesterson National Wildlife Refuge and the West Grassland, Fresno and Merced Counties, California*. U.S. Geological Survey Water-Resources Investigation Report 85-4220, p. 73.

Saiki, M. K., 1984. Environmental Conditions and Fish Faunas in Low Elevation Rivers on the Irrigated San Joaquin Valley Floor, *California Fish Game*, 70, pp. 145-157.

Saiki, M. K., 1986. A Field Example of Selenium Contamination in an Aquatic Food Chain. *Symposium on Selenium in the Environment*. Publication CAT1/860201, California Agricultural Technology Institute, California State University, Fresno, pp. 67-76.

Saiki, M. K., 1989. Selenium and Other Trace Elements in Fish from the San Joaquin Valley and Suisun Bay, 1985. In: Howard, A. Q. (Ed.), *Selenium and Agricultural Drainage, Implications for San Francisco Bay and the California Environment (Selenium IV)*. The Bay Institute of San Francisco, Tiburon, CA, pp. 35-49.

Saiki, M. K. and Lowe, T. P., 1987. Selenium in Aquatic Organisms from Subsurface Agricultural Drainage Water, San Joaquin Valley, California, *Arch. Environ. Contam. Toxicol.*, 16, pp. 657-670.

Saiki, M. K. and Palawski, D. U. Selenium and other Elements in Juvenile Striped Bass from the San Joaquin Valley and San Francisco Estuary, California, *Arch. Environ. Contam. Toxicol.*, 19. In Press.

Sato, T.; Ose, Y.; and Sakai, T., 1980. Toxicological Effect of Selenium on Fish, *Environmental Pollution*, (Series A) 21, pp. 217-224.

Schmitt, C. J., and Brumbaugh, W. G. National Contaminant Biomonitoring Program: Concentrations of Arsenic, Cadmium, Copper, Lead, Mercury, Selenium, and Zinc in U.S. Freshwater Fish, 1976-1984, *Arch. Environ. Contam. Toxicol.*, 19. In Press.

Shelton, L. R., and Miller, L. K., 1988. *Water-Quality Data, San Joaquin Valley, California, March 1985 to March 1987*. U.S. Geological Survey Open-File Report 88-479, 210 p.

U.S. Bureau of Reclamation, 1984. *A Status Report on the San Luis Unit Special Study*. Information Bulletin 3.

U.S. Environmental Protection Agency, 1984. *Health Effects Assessment for Selenium (and Compounds)*. EPA/540/1-86/058, Environmental Criteria and Assessment Office, Office of Research and Development, Cincinnati, OH., 52 pp.

U.S. Environmental Protection Agency, 1986. *Quality Criteria for Water, 1986*, EPA 440/5-86-001, U.S. Government Printing Office, Washington, DC.

U.S. Environmental Protection Agency, 1987. *Ambient Water Quality Criteria for Selenium, 1987*. EPA-440/5-87-006, Environmental Research Laboratories, Duluth, MN., and Narragansett, RI., 121 pp.

Woock, S. E.; Garrett, W. R.; Partin, W. E.; and Bryson, W. T., 1987. Decreased Survival and Teratogenesis During Laboratory Selenium Exposures to Bluegill, Lepomis macrochirus, *Bulletin Environmental Contam. Toxicol.*, 39, pp. 998-1005.

Yang, G.; Wang, F.; Zhou, R.; and Sun, S., 1983. Endemic Selenium Intoxication of Humans in China, *American Journal Clinical Nutrition*, 37, pp. 872-881.

# Five: VALUING NON-AGRICULTURAL BENEFITS FROM WATER USE

# 20 MEASURING THE BENEFITS OF FRESHWATER QUALITY CHANGES: TECHNIQUES AND EMPIRICAL FINDINGS

Richard T. Carson, University of California, San Diego and
Kerry M. Martin, Harvard University

### ABSTRACT

This chapter gives an overview of the techniques used to value the nonmarket benefits of water-related public goods and of the major empirical studies in this area. Travel cost, hedonic pricing, and contingent valuation are described; special emphasis is placed on the problems and limitations of implementing these methods to value changes in the quality and quantity of water-related amenities. Major empirical efforts to value National and regional water quality improvements, water-based recreation, ecosystem preservation, instream flows, ground-water protection, and water supply reliability are discussed.

### INTRODUCTION

Water is the quintessential multiattribute good. It supports fish and birds. Humans drink it, cleanse with it, grow food with it, recreate on it, maintain landscaping with it, and generate electricity from its flow. Policy decisions have been made from the time of recorded history about how water should appear, when it should appear, and how it should be used. This chapter deals with valuing policy changes which would result in changing some attribute of water; in particular, this chapter focuses on placing a dollar value on attributes of water which are not directly priced by the marketplace.

Much of modern benefit-cost analysis has roots in the valuation of water projects (Krutilla and Eckstein, 1958). Some of the services provided by these water projects, such as electricity, had readily available market prices. Other benefits, for example, flood control, were also presumed to be measured by market prices although major problems with this presumption were immediately apparent. Water-based recreation, another class of benefits, had no readily available, even if imperfect, market price. Valuing water-based recreation was the initial focus of nonmarket valuation and has remained one of its mainstays.

Water-based recreation takes many forms. The building of a reservoir creates new boating and fishing opportunities and, in some instances, new swimming, picnicking, and camping opportunities. The building of dams alters riverflows, which in turn influences rafting opportunities. In some instances, a water project indirectly affects recreation. For example, draining a wetland may create valuable commercial real estate but may at the same time destroy nesting grounds for ducks prized by hunters or birds sought by birdwatchers.

Enhancing recreation was a key component of the Clean Water Act of 1972, which set its objectives in terms of achieving fishable, swimmable quality water. In order to determine the benefits of upgrading sewer treatment plants, installing industrial pollution control equipment, and reducing agricultural runoff, dollar values must be assigned to improved recreational opportunities at National, regional, and local levels.

The environmental legislation of the 1970's and 1980's introduced the issue of nonuse values, such as the stewardship of resources and the desire to bequest resources to future generations. Nonuse values have received increased attention as policymakers have confronted drinking water contamination, groundwater contamination, oil spills, and toxic chemicals in rivers and streams -- all problems which cannot be readily remedied. Finally, the increased demand for water from all sectors of the economy has created the potential for water shortages, particularly in the Western States, prompting researchers to address the question: What is the value of a reliable water supply?

Table 1 is a list of the aspects of water currently being valued and cites representative studies valuing each aspect. In later sections an illustrative study is discussed from each category.

## THEORETICAL BASIS OF BENEFITS

For goods which are readily bought and sold on the open market, price serves as a satisfactory indicator of value. For nonmarket goods, an alternative measure of value must be used. The preferred measure for benefits is usually the willingness to pay (WTP) Hicksian consumer surplus. This measure is the maximum amount of money that the consumer would be willing to pay for a good and have utility remain at some specified level. Less frequently the willingness to accept (WTA) Hicksian consumer surplus measure is used. Both WTP and WTA are derived from the Hicksian compensated demand curve. WTP is referred to as compensating variation and WTA as equivalent variation. These two Hicksian welfare measures differ in their assumptions about the consumers' property rights with respect to the good in question, the WTP measure assuming the agent does not have the right to the good and must therefore buy it, the WTA measure assuming the agent has the right to the good and can sell it. The ordinary (Marshallian) consumer surplus, which is the area below the demand curve and above the price paid by the consumer, falls between the two Hicksian measures and is often used as an approximation of those measures.[1]

Table 1. Representative water valuation studies[a].

| Aspect of Water Valued | Author(s) |
|---|---|
| (1) Water Reliability | Carson (1989a) |
| (2) Water Flows | Daubert and Young (1981) |
| | Bishop, Brown, Welsh, and Boyle (1990) |
| (3) Water Quality | Gramlich (1977) |
| | Greenley, Walsh, and Young (1982) |
| | Sutherland and Walsh (1985) |
| | Smith and Desvousges (1986) |
| | Carson and Mitchell (1988) |
| (4) Water-Based Recreation (e.g., boating, fishing, swimming, beach activity) | McConnell (1977) |
| | Hanemann (1978) |
| | Bowes and Loomis (1980) |
| | Russell and Vaughan (1982) |
| | Sellar, Stoll, and Chavas (1985) |
| | Cameron and James (1987) |
| | Carson, Hanemann, and Wegge (1989) |
| (5) Ground Water | Edwards (1988) |
| | Mitchell and Carson (1989a) |
| (6) Ecosystem Preservation | Hammack and Brown (1974) |
| | Bishop and Boyle (1985) |
| | Walsh, Sanders, and Loomis (1985) |
| | Loomis (1987) |
| (7) Water-Enhanced Amenities | Brown and Pollakowski (1977) |
| | Blomquist (1983) |
| (8) Toxic Contamination | Freeman (1987) |
| | Hanemann, Kanninen, and Loomis (1990) |

[a] A fairly comprehensive bibliography of travel cost and contingent valuation studies of outdoor recreation appears in Walsh, Johnson, and McKean (1988).

## TYPOLOGY OF POSSIBLE BENEFITS

A comprehensive assessment of the benefits accruing from a change in the level of a public good should include consideration of all types of benefits which would result from that change. The typology in figure 1 illustrates how the possible types of benefits from a change in freshwater quality might be categorized. This typology, from Mitchell and Carson (1989), while certainly not the only way to categorize the different types of benefits, is reasonably exhaustive of the possible benefits of freshwater quality improvements.[2] As figure 1 indicates, the major division in this classification of benefits is that between use and nonuse or existence values. The classification of benefits to be valued into these two categories often dictates the appropriate technique by which these benefits should be measured. While throughout this chapter water quality is used as an illustrative water-related public good, the broader classification in figure 1 also applies to other public goods. Before examining the techniques typically used to measure the value of water-related public goods, the different types of benefits are introduced in more detail.

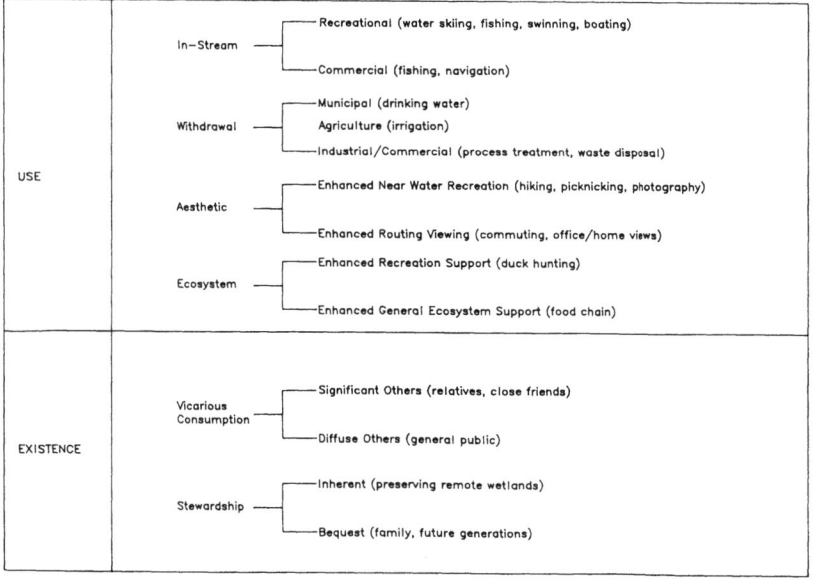

Figure 1. Typology of possible benefits.

## Use Values

The use class of benefits consists of all the current direct and indirect ways in which an agent makes physical use of water quality. Direct use benefits may arise as a result of recreational or commercial activities that occur on the water; or they may arise from withdrawal activities such as agricultural irrigation, cooling or washing operations in industrial processes, or drinking water. This class of use benefits also has an indirect dimension created when the water body's characteristics enhance nearby activities. Figure 1 lists two types of indirect use benefits: those occurring because water quality is a vital component of an ecosystem or habitat that supports certain types of recreation, such as, hunting or birdwatching, and those occurring because water quality provides an esthetically pleasing setting for activities like picnicking or gazing at the scenery.

## Nonuse (Existence) Values

In contrast to use values, which exist because people are physically affected by an amenity, existence values, or more generally nonuse values, embody the notion that a person need not visit a physical site or use services from that site to gain utility from its maintenance or improvement.[3] The motives for existence values usually stem from vicarious consumption or stewardship concerns.

In the case of existence values induced by vicarious consumption, an individual's utility arises from the consumption of the good by others. These "others" may be generalized, or they may be particular individuals. The motivation behind vicarious consumption values may stem either from a sense of obligation to provide the good for use by others or from a sense of interdependent utility.

Stewardship values involve a desire to see public resources used in a responsible manner and conserved for future generations. Two types of stewardship values are distinguished here: bequest and inherent. Bequest values exist if the motivation lies in knowing that the preservation of an amenity will make that amenity available for others to use in the near or distant future, whether those users are family or others. The other kind of stewardship value, termed inherent, stems from the satisfaction that an amenity is preserved regardless of its eventual use.

## BENEFIT MEASUREMENT TECHNIQUES

The challenge confronting economists for the past several decades has been that of measuring the public's WTP or WTA compensation for the changes in the level of provision of a public good. Economists have developed several techniques for measuring the potential benefits of water quality changes, including travel cost, hedonic pricing, and contingent valuation. These methods differ greatly in their data requirements and in their assumptions about economic agents and the physical environment. Travel cost and hedonic pricing rely on data from actual market choices by consumers, such as deciding on a trip or buying a house. These two benefit measurement techniques are known as observed indirect methods because they rely on observed behavior with respect to an activity which is related to the amenity change to be valued. In contingent valuation, a hypothetical direct method, agents are queried directly about their valuation of particular hypothetical changes in amenity quality or quantity. These three general techniques differ in the types of values they are able to measure. For example, the travel cost and hedonic pricing techniques cannot measure existence values.

### *Travel Cost*

The travel cost method proposed by Hotelling in 1949 (Clawson and Knetsch, 1966) has been used extensively to value site-specific recreation benefits.[4] In the simplest type of travel cost model, concentric circles of different radii around a particular site are used in calculating the average number of per capita visits to the site by the residents in each distance zone, the area between two concentric circles. The travel information is obtained by survey interviews or site visitation records. These data are used to trace a site-specific, trips-per-capita demand curve as a function of distance. The per-capita demand curve is used to estimate price until the number of trips is driven to zero. If a monetary value is assigned to each mile from the site, consumer surplus is calculated by measuring the area below the demand curve.

One data set often discussed in the travel cost literature is that of 1977 river recreation permits on the Colorado River through Westwater Canyon in Utah. Some 211 trips were recorded and assigned into 28 zones of varying distance from Westwater Canyon. The trips per capita were calculated from the population for each respective zone. Bowes and Loomis (1980) were the first to publish results for this model; and Vaughan, Russell, and Hazilla (1982) later commented on the estimates reported by Bowes and Loomis. These papers point to the necessity of using the correct functional form to model the relationship between trips and travel costs. The following table combines information from both papers and shows the trips-per-capita demand curve estimates and the value of benefits resulting from each. The GLS functional form is the OLS regression weighted by the square root of $N^i$ where $N^i$ is the

| Regression Type | Equation | Total Benefits |
|---|---|---|
| OLS | 5.7192-.0218·cost | $78,000 |
| GLS | 1.0260-.0031·cost | $24,000 |
| Box-Cox | 2729-.0255·cost | $14,000 |

population of zone i; this method corrects the problem of heteroscedasticity. Vaughan et al. concluded the Box-Cox travel cost model with a heteroscedasticity correction was best supported by the data. Their results indicate that benefit estimates from even simple travel cost models can be very sensitive to the functional form by which they are specified.

Several other considerations make the simple travel cost model problematic for measuring the benefits of water quality changes. First, the model is a highly site-specific technique in which the substitutes are fixed; the model does not allow for varying the substitute sites available or for varying the characteristics of any particular site. Second, with only one site, the explicit incorporation of environmental quality into the travel cost model is not usually possible. People travel to a lake for a variety of reasons which relate in various tenuous ways to the lake's water quality. Some progress has been made in overcoming these two problems in recent generalized travel cost models which incorporate a first-stage participation estimation with water quality as an argument (Hanemann, 1978; Vaughan and Russell, 1982; Bockstael, Hanemann, and Strand, 1985; Caulkins, Bishop and Bowes, 1986; and Smith and Desvousges, 1986). The third problem with the travel cost method is the difficulty of handling the role of time: identifying the temporal elements to be interpreted as costs of recreation activity and assigning monetary values to those elements (Freeman, 1979). Although recent progress on this topic by Willman (1980) and, particularly, Bockstael, Strand, and Hanemann (1985) represents a substantial advance, this difficult question is far from resolved. The fourth and fifth problems are even less likely to yield solutions. Bockstael and McConnell (1980) have shown that benefit measures using the travel cost method are very sensitive to the functional forms used in the estimation.[5] Even more fundamentally problematic, for which no solution seems likely, is that travel cost models can only measure a limited range of benefit categories: the direct recreation benefits and some benefits in the esthetic and ecosystem-use categories (Hay and McConnell, 1979). No travel cost model can measure existence values.

Some of the problems with travel cost models discussed above have led researchers to consider a number of extensions to the simple one site travel cost model. The most straightforward of these was to generalize the basic travel cost model to multiple sites along the lines of Burt and Brewer (1971), Cicchetti, Fisher, and Smith (1976), and Vaughan and Russell (1982); the current state of the art is represented by Smith and Desvousges (1986).[6] Another direction explored was the estimation of gravity models, a concept taken from geographers and regional scientists, in which the attributes of sites attract potential recreationalists. How such models should be estimated and the set of conditions under which the gravity model is consistent with utility theory was pursued by Sutherland (1982). The hedonic travel cost model (Brown and Mendelsohn, 1984) was another attempt to overcome the difficulties of the single site travel cost model. A number of prominent resource economists have heavily criticized the gravity models and the hedonic travel cost models (Smith and Kaoru, 1986 and Bockstael, McConnell, and Strand, 1989).

The most popular direction for travel cost models appears to be the discrete choice, random utility models (RUM) proposed by McFadden (1974). These models allow the straightforward incorporation of multiple sites and of quality indicators for those sites. To a large degree the RUM models represent a switch from the aggregate data sets typically used by other types of travel cost models to data sets with observations on individual recreational trips. The earliest applications of this framework for valuing recreation are found in Hanemann (1978), which looks at Boston beaches, and Vaughan and Russell (1982), which looks at National fishing behavior. Both studies focused on the role of water quality in recreational behavior. These models have steadily increased in size and complexity as better microdata sets have become available.

The largest and most comprehensive discrete choice travel cost model was constructed by Carson, Hanemann, and Wegge (1989) in a study of recreational fishing behavior in Alaska. That study used a random utility model statistically implemented as a nested multinomial logit model with four different levels. In the first level, a fisherman chooses a particular species of fish, for example, king salmon, red salmon, silver salmon, or pink salmon, and, in the second, chooses among broader groupings, such as, salmon, freshwater, saltwater, and no target. The third level captures the number of times a fisherman wishes to fish in a given week; and, in the fourth level, a household decides whether or not to fish that season.

The model has 29 fishing areas and 13 types of fish. The quality of fishing varies each week at each area for each type of fish. In some weeks, certain types of fish, for instance, king salmon, are not available at some sites. The characteristics of sites may be changed in several ways, for example by adding cabins, reducing congestion, changing the quality of fishing for one or more types of fish, or closing down one or more sites to one or more types of fishing. The model can produce dollar estimates of the welfare effects of these types of changes. Tracing the effects of a single change, for example, closing the Kenai

River to king salmon on a particular week in July, helps explain how the model works.[7]

The first result of this closure is the redistribution of king salmon fishing that week to sites near the Kenai River with similar king salmon runs. The second effect is the reduction of the total number of king salmon trips that week, some of which are reallocated to other types of salmon. Some of the original Kenai king salmon fisherman switch from salmon to other types of fish, for instance, halibut, while others decide not to fish at all that week. The welfare economic effects put into play by closing the Kenai River to king salmon for that week was estimated by the model to be $482,000. This model is currently used by Alaska's Department of Fish and Game (DFG) to allocate the fixed fish stock between recreational and commercial fisherman and to ascertain the cost of closing down a particular site for biological management purposes.

## HEDONIC PRICING

The other major, observed, indirect method is hedonic pricing (Adelman and Griliches, 1961; Ridker and Henning, 1967; and Rosen, 1974).[8] This method assumes that the price of a marketed good is a function of its different characteristics. Letting X represent a commodity class and $P_i$ the price of some good i in commodity class X which has characteristics $Q_j$, results in

[1] $P_i = P(Q_1...Q_j...Q_n)$.

The implicit price of a particular characteristic $Q_j$ is found by differentiating $P_i$ with respect to $Q_j$. In a second stage analysis, this relationship is combined with other restrictions on the demand and supply of the characteristic of interest, and a demand function for that characteristic is estimated. Consumer surplus is calculated using that demand function.

The most common use of the hedonic pricing technique is to access property values which include the value of some environmental good such as local shorelines or air quality (Brown and Pollakowski, 1977; Harrison and Rubinfeld, 1978; Freeman, 1979b; and Brookshire, Thayer, Schulze, and d'Arge, 1982). Most studies have focused on residential real estate; however, an increasing number of studies focus on the quality of land, particularly in agriculture (Miranowski and Hammes, 1984 and Palmquist and Daniels, 1989). The quality of the land varies with characteristics such as potential groundwater contamination and salinity content. The other major area of interest for researchers has been hedonic wage studies (Thaler and Rosen, 1976) in which different risk levels are thought to be incorporated into the wages for different jobs.

An illustrative study on water-enhanced amenities is the 1977 Brown and Pollakowski study. Brown and Pollakowski examined the value associated with shoreline development, specifically, the value associated with the demand for open space and the welfare gains and losses resulting from a change in the amount of water-related open space. They used market sales data obtained from 1969 to 1974 for houses in three relatively homogeneous neighborhoods in Seattle. The data set has a comprehensive set of structural attributes for each unit. A function was estimated by regressing the housing bundle attributes on the selling price to derive marginal implicit prices for each attribute, in particular for proximity to the water and water-related open space.[9] The results indicated both variables of interest were statistically significant and important in explaining variation in the dependent variable. As expected, the sample was willing to pay a premium for open space and proximity to the waterfront; the premium decreased as proximity to the waterfront decreased.

While the hedonic pricing technique can, in principle, value all of the use categories, in practice, it suffers from serious problems. First, the data requirements for a valid hedonic pricing study are unusually exacting. All relevant characteristics--structural, neighborhood, and environmental--must be subject to control; if many resources or unique resources are already in public hands, this control may be impossible. Second, sufficient market data for reliable estimations are often difficult to obtain. Housing turnover, for instance, is relatively slow; and locating genuinely comparable houses in relevant neighborhoods is not easy. Third, the functional forms of the true underlying hedonic pricing equations are unknown; and the researcher is confronted with a number of competing formulations with quite different implications (Freeman, 1979b and Halvorsen and Pollakowski, 1981). Fourth, people must be aware of the actual physical differences in the levels of characteristics being valued. For example, to assume such awareness may be unreasonable when dealing with risk levels posed by chemicals or with colorless or odorless air pollutants. Fifth, expectations about changes in the good being valued and its relevant characteristics are generally unobservable; but presumably such expectations affect the determination of prices. In particular, such expectations may affect the prices of property. For example, people may incorporate their assumptions concerning future water quality in a given location into their WTP or WTA a given purchase price. Sixth, to value simultaneous changes in substitute sites is difficult, if not impossible. Finally, perhaps the most serious problem, identified by Brown and Rosen (1982), is that the standard method (Rosen, 1974) of identifying systems of hedonic pricing, supply and demand equations, is tautological and incorrect except in special and unlikely cases. Separately identifying the equations has been the subject of much recent work (Epple, 1987; Bartik, 1987a and 1987b; and

Mendelsohn, 1985) as identification is crucial to the estimation of benefits (Palmquist, 1988).

## CONTINGENT VALUATION

The Contingent Valuation Method (CVM) uses survey questions to elicit preferences for public goods by discovering what respondents would be willing to pay for specified improvements in those goods.[10] The method elicits WTP in dollar amounts and circumvents the absence of markets for public goods by presenting consumers with hypothetical markets in which they may buy the good in question. The hypothetical market may be modeled after either a private goods market or a political goods market. Because the elicited WTP values are contingent upon the particular hypothetical market described to the respondent, this approach came to be called the CVM.

Respondents are presented with material which consists of three parts. First, a detailed description of the good being valued and the hypothetical circumstances under which it would be made available to the respondent is presented. The researcher constructs a model market, often in considerable detail, which is communicated to the respondent in the form of a scenario read by the interviewer during the course of the interview. The hypothetical market, designed as plausibly as possible, typically describes the good to be valued, the baseline level of provision, the structure under which the good is to be provided, the range of the available substitutes, and the method of payment. To trace the demand curve for the good, respondents are often asked to value several levels of the provision.

Second, questions which elicit the respondents' WTP for the good being valued are asked. These questions are designed to facilitate the valuation process without themselves biasing the respondent's WTP amounts.

Third, questions about respondents' demographics, their preferences relevant to the good being valued, and their use of the good are asked. This information, some of which is usually elicited preceding and some following the reading of the scenario, is used in regression equations to estimate a valuation function for the good. Successful estimation using variables which theory identifies as predictive of WTP provides evidence for the reliability and validity of the results.

If the survey is well designed and carefully pretested, the respondents' answers to the valuation questions should represent valid WTP responses.[11] These amounts are used to develop a benefit estimate. If the sample is meticulously selected through random sampling procedures, if the response rate is high enough, and if the appropriate adjustments are made to compensate for participants who fail to respond and for those who give "bad" quality data, such as "protest zeros," the results can be generalized with a known margin of error to the population from which the respondents were drawn. Generalization is a powerful feature of the sample survey method; a thousand people can be used to estimate the responses that would be given by everyone in a river basin, a state, or region, or even the entire country.

The following is presented as an illustrative contingent valuation study. The goals of the Clean Water Act support certain forms of water-based recreation. Mitchell and Carson (1988) attempted to measure the value of achieving boatable, fishable, and swimmable quality water on a National level. They asked a large National sample how much they would be willing to pay to move from a baseline level below boatable (which would occur without current levels of expenditure on sewage treatment and industrial/agricultural pollution control) to boatable, fishable, and swimmable water quality levels.

The key features of this study included: a water quality ladder which maps recreation levels onto physical quality levels, a payment vehicle which facilitates the elicitation of values, a test for whether air quality values were included in the WTP for water quality, a look at the reduction in benefits by obtaining a somewhat less than uniform National level of water quality, and a prediction of Smith and Desvousges' (1986) estimates for the Monongahela River Basin.

The suggested annual benefits of a uniform swimmable quality water level were in the $20 billion range, an amount sufficient to justify most of the water pollution control activities to date, but not enough to justify the cost of obtaining a uniform National level of swimmable water quality, and certainly not enough to justify the Clean Water Act's second goal of "zero discharge."

## EMPIRICAL RESULTS OF REPRESENTATIVE STUDIES

The remaining studies discussed in this section employ the CVM to value regional water quality, ecosystem preservation, water flows, ground-water contamination, and water reliability.[12]

### Regional Water Quality

Greenley, Walsh, and Young (1982) attempt to measure use values, option values, and other preservation values (bequest and existence) associated with preserving water quality in the South Platte River Basin in Colorado.[13] This particular river basin is subject to potential irreversible degradation in water quality due to mining activity. Greenley et al. used a bidding game with a random sample of 202 households in Denver and Fort Collins to measure WTP for incremental changes in water quality to enhance recreational enjoyment.

Two alternative payment vehicles were used, a general sales tax and a residential water sewer fee.[14]

Approximately 80 percent of the households interviewed in the sales tax version actually engaged in water-based recreational activities in the river basin; this portion of the sample was willing to pay an average of $57 per year for water quality to enhance enjoyment of recreational activities. In addition, these households were willing to pay an additional $23 per year for the option to engage in recreational activities in the future. Nonuser values totaled $42, $27 existence and $17 bequest, annually for the 20 percent of the sample who did not engage in recreational activities in the basin and $67, $34 existence and $33 bequest, for the remainder of the sample.[15] Total annual benefits aggregated over the approximately 576,000 households residing in the river basin were estimated to be $61 million.

## Ecosystem Preservation

Bishop and Boyle (1985) conducted a study in the early 1980's to measure the benefits, both use and nonuse, associated with preserving and maintaining the Illinois State Nature Preserve. Due to erosion of a ridge of sand dunes, there was concern that Lake Michigan would invade and flood the Nature Preserve unless a series of offshore breakwaters were built. Bishop and Boyle mailed their survey to a stratified random sample of 600 Illinois heads of households; the response rate was 63 percent. The elicitation technique chosen was dichotomous choice, that is, take-it-or-leave-it, and the payment vehicle was annual membership to a private foundation that would effect the necessary measures to maintain the Nature Preserve. The average Illinois head of household responding to the survey placed a value of approximately $28 per year on the Nature Preserve. Extrapolating this estimate to the State of Illinois yields an annual value of about $60 million.

## Instream Flows

The value of instream flows to recreationists is often ignored in water allocation decisions, a disregard sometimes leading to suboptimal policy changes. Daubert and Young (1981) used the contingent valuation approach to estimate the value of instream flows for a sample of recreationalists on the Cache la Poudre River in Northern Colorado.[16] A total of 134 personal interviews of recreationists using the river (49 fisherman, 45 shoreline recreationists, and 40 whitewater enthusiasts) were conducted in the summer of 1978. The interviewers used color photographs of eight different instream flow rates at four different sites and presented corresponding physical stream characteristics. Each respondent was asked in a bidding game framework about his WTP for instream flows. The payment vehicles were either an increase in sales tax or an increase in a hypothetical entrance fee. The values from the sales tax version always exceeded the values from the entrance fee version. Benefit functions were estimated for three different activities: Trout fishing, whitewater boating, and streamside recreation (picnicking, camping, and hiking).

Instream flow quantity was identified as the key variable in determining WTP for fishing and whitewater boating, explaining over 40 percent of the variation in the regression equations. Waterflow was also significant in the shoreline benefit function, but to a lesser degree. In the entrance fee version, the maximum value per day for fishing was $30 (1978 dollars) for a flow of 500 cubic feet per second ($ft^3/s$) and the shoreline activity, $10 for a flow of 700 $ft^3/s$. The WTP estimates for whitewater recreation increased throughout the range of observations which varied from 100 to 900 $ft^3/s$. Daubert and Young concluded that during periods of low flows, the marginal value of instream flows was greater than the marginal value of water used for irrigation, suggesting a need for reexamination of allocation choices.

## Ground-Water Protection

Edwards (1988) conducted a contingent valuation study of WTP to prevent possible future nitrate contamination of a potable supply of ground water in Cape Cod.[17] Option price was chosen as the appropriate measure of economic value because of demand and supply uncertainty and the uncertainty surrounding the probabilities of actual contamination. Edwards used 10 questionnaire versions varying year of expected future contamination, probability of nitrate contamination, and price of bottled water. After conducting a pilot study of 200 households, the discrete choice referendum format was adopted; bids ranged from $10 to $2,000. One thousand random households were sampled with a mail survey instrument. Three telephone followups led to a respectable response rate of 78 percent; however, only 58 percent were deemed by Edwards to be usable in the final analysis.

The study results indicated respondents with strong bequest motivations held dramatically higher option prices relative to those who did not. Edwards found that income displayed a very strong, positive effect on WTP for ground-water protection and that the respondents' perception of their own probability of future demand for the ground water was positively related to their WTP. The results also indicated the WTP bids varied with the level of uncertainty of future contamination; thus policymakers who only explore the worst-case scenario, that is, who assume contamination is certain when making aquifer management decisions, are potentially overestimating benefits.

## Water Supply Reliability

The issue of water shortages in the West looms large. Carson (1989a) informed California voters of the likely prospect of water shortages in the coming years. Four different water shortage scenarios were valued with each respondent valuing two scenarios: One 10-15 percent shortage in a 5-year period, two 10-15 percent shortages in a 5-year period, one 30-35 percent shortage in a 5-year period, and one 30-35 percent shortage and one 10-15 percent shortage in a 5-year period. Respondents were told what changes in water consumption behavior the shortages of different magnitudes would likely entail.

Two thousand respondents were interviewed by telephone; random digit dialing was used. Each respondent was asked two binary discrete questions with randomly assigned dollar amounts for each of the two scenarios valued. An interval-censored survival analysis technique was used to evaluate the resulting data. Median annual household amounts ranged from $83 annually for the mildest shortage scenario to $258 for the most severe. Northern and southern California household responses were, for the most part, quite similar.

## CONCLUDING REMARKS

Nonmarket valuation techniques have made large strides during the past decade and have now reached the point at which they can be used in many circumstances for serious policy analysis and decisions. The work of the past decade reflects three important lessons. First, the data requirements to successfully implement nonmarket valuation techniques are much more demanding and expensive than first believed. Travel cost analysis now typically requires large microdata sets with individual trips to an array of potential substitute sites and quality variables for each of these sites. Hedonic pricing requires detailed data on land prices and typically requires such data across time periods or markets. Contingent valuation requires extensive instrument development. Focus groups, pretests, and pilot studies are often needed to ensure that the desired good is being valued and that the scenario is well understood and accepted by respondents. Furthermore, contingent valuation in many cases may require expensive in-person surveys with well-trained interviewers. The econometric skills needed for correct implementation of any of the nonmarket techniques have increased substantially relative to the practices of even a few years ago.

The second lesson is that a significant learning curve frequently hampers nonmarket valuation when dealing with a new type of good. The implications of this lesson are many. Agency funding is needed to conduct exploratory research in such cases, which need not be used immediately for policy purposes. Multiple researchers need to attack the problem in order to determine the best solutions and the common features of different solutions. Interaction between economists, survey researchers, physical scientists, engineers, and policymakers is necessary to define the issues so that research results will be eventually useful to policymakers.

The third lesson is that economists tend to rely too heavily on existing data sources and tend to overlook the potential limitations those data sources may impose. Collecting new data is an expensive undertaking; but the cost of data collection often pales in comparison to the magnitude of costs involved in the pending policy decision. Trying to construct benefit estimates from bad or inappropriate data can only serve to discredit nonmarket benefit measurement techniques.

## NOTES

[1] See Just, Hueth, and Schmitz (1982) for a comprehensive look at welfare economics. See Hanemann (forthcoming) for a discussion of the relationship between these welfare measures in the case of changes in the quantity or quality of a public good.

[2] This typology is widely, although not universally, accepted. Disagreements are usually over the terms used to describe nonuse values and whether and how to include different types of uncertainty (i.e., option and quasi-option values).

[3] A large portion of the millions of dollars in fees and voluntary contributions paid by members of environmental groups and the willingness of environmental activists to volunteer their time to lobby for such legislation as the Alaska Wilderness Bill can be cited as evidence for the reality of existence values for wilderness amenities. Referenda on environmental programs often receive very strong voter support even among voters whose communities are unaffected by the improvements.

[4] See Bockstael, McConnell, and Strand (1989) and Smith (1989) for recent reviews of the current state of the art of travel cost modeling.

[5] The work of Vaughan and Russell (1982) and Kling (1988) confirms this finding.

[6] See Kling (1988) for a critique of welfare estimates from these and other modern variants of the travel cost model.

[7] The Kenai River is perhaps the world's premiere king salmon fishery.

[8] See Palmquist (1989) for a review of the current state of the art of hedonic pricing models.

[9]Both the dimensions of the setback area (i.e., size of open space) and the distance to the waterfront appeared in log form in the equation as the relationship was expected to be a nonlinear one.

[10]Mitchell and Carson (1989b) provide an extensive discussion of most contingent valuation issues.

[11]The survey designer must take care to ensure that respondents are valuing the good intended to be valued. Focus groups are often used for this purpose.

[12]Hanemann, Kanninen, and Loomis (1990) recently conducted a study of toxic contamination in the San Joaquin/Sacramento Bay Estuary. A subsequent chapter in this book is devoted entirely to that study.

[13]See Smith and Desvousges (1986) or Gramlich (1977) for additional examples of contingent valuation studies on water quality in river basins.

[14]The results indicate WTP was quite sensitive to the payment vehicle used; WTP was about 75 percent less in the residential sewer fee version. This difference is not surprising as increasing the sewer fee only affects the residents whereas increasing the sales tax also affects tourists so the sales tax vehicle may have been perceived as more equitable. One consistent finding of the contingent valuation literature is that the public often has strong preferences over how they pay for a particular public good. Often policy dictates the payment vehicle which could actually be used.

[15]At present, whether WTP may be meaningful and uniquely divided into subcomponents is subject to considerable debate.

[16]A more recent study examining waterflows is Bishop, Brown, Welsh, and Boyle (1990).

[17]See Mitchell and Carson (1989a) for another example of a study examining potential contamination of ground-water supplies.

## ACKNOWLEDGEMENTS

The authors wish to thank Michael Conaway and Steve Waters for their helpful comments. We also wish to thank the University of California Water Resources Center Grant W-722 for financial support.

## REFERENCES

Adelman, I. and Griliches, Z., 1961. On an Index of Quality Change, *Journal of American Statistical Assn.*, 56, pp. 531-548.

Bartik, T. J., 1987a. The Estimation of Demand Parameters in Hedonic Price Models, *Journal of Political Economy*, 95, pp. 81-88.

Bishop, R. C. and Boyle, K. J., 1985. *The Economic Value of Illinois Beach State Nature Preserve*. Report to the Illinois Department of Conservation, Heberlein and Baumgartner Research Services, Madison, WI.

Bishop, R. C.; Brown, C. A.; Welsh, M. P.; and Boyle, K. J., 1990. *Grand Canyon Recreation and Glen Canyon Dam Operations: Economic Evaluation*. Draft Report.

Blomquist, G. C., 1983. Measurement of the Benefits of Water Quality Improvement. In: Tolley, G. S.; Yaron, D.; and Blomquist, G. C. (Eds.), *Environmental Policy: Water Quality*, Ballinger, Cambridge.

Bockstael, N. E. and McConnell, K. E., 1980. Calculating Equivalent and Compensating Variation for Natural Resource Facilities, *Land Economics*, 56, pp. 56-62.

Bockstael, N. E.; Hanemann, W. M.; and Strand, I. E., 1985. *Measuring the Benefits of Water Quality Improvements Using Recreation Demand Models*. Report to the Economic Analysis Division, U.S. Environmental Protection Agency.

Bockstael, N. E.; McConnell, K. J.; and Strand, I. E., 1985. Recreation. In: Baden, J. and Kolstad, C. (Eds.), *Measuring the Demand for Environmental Commodities*, North Holland, Amsterdam.

Bowes, M. D. and Loomis, J. B., 1980. A Note on the Use of Travel Cost Models with Unequal Zonal Populations, *Land Economics*, 56(4), pp. 465-470.

Brookshire, D. S.; Thayer, M. A.; Schulze, W. D., and d'Arge, R. C., 1982. Valuing Public Goods: A Comparison of Survey and Hedonic Approaches, *American Economic Review*, 72, pp. 165-178.

Brown, G. M., Jr. and Pollakowski, H. O., 1977. Economic Valuation of Shoreline, *Review of Economics and Statistics*, 59, pp. 272-278.

Brown, G. M., Jr. and Mendelsohn, R., 1984. The Hedonic Travel Cost Method, *Review of Economics and Statistics*, 66, pp. 427-433.

Brown, J. N. and Rosen, H. S., 1982. On the Estimation of Structural Hedonic Price Models, *Econometrica*, 50, pp. 765-768.

Burt, O. R. and Brewer, D., 1971. Estimation of Net Social Benefits from Outdoor Recreation, *Econometrica*, 39, pp. 813-828.

Cameron, T. A. and James, M. D., 1987. Efficient Estimation Methods for Use with 'Close-Ended' Contingent Valuation Data, *Review of Economics and Statistics*, 69, pp. 269-276.

Carson, R. T., 1989a. *The Value of Diamonds and Water: Water Supply Reliability in Southern California*. Paper Presented at Association for Public Policy Analysis and Management, Washington, DC.

Carson, R. T. and Mitchell, R. C., 1988. *Value of Clean Water: The Public's Willingness to Pay for Boatable, Fishable, and Swimmable Quality Water*. Discussion Paper 88-13, Department of Economics, University of California, San Diego.

Carson, R. T.; Hanemann, W. M.; and Wegge, T. C., 1989. *A Nested Logit Model of Recreational Fishing Demand in Alaska*. Paper Presented at the Annual Western Economic Association Conference, Lake Tahoe.

Caulkins, P. P.; Bishop, R. C.; and Bowes, N. W., Sr., 1986. The Travel Cost Model for Lake Recreation: A Comparison of Two Methods of Incorporating Site Quality and Substitution Effects, *American Journal of Agricultural Economics*, 68(2), pp. 291-297.

Cicchetti, C. J.; Fisher, A. C.; and Smith, V. K., 1973. Economic Models and Planning for Outdoor Recreation, *Operations Research*, 21, pp. 1104-1113.

Clawson, M. and Knetsch, J., 1966. *Economics of Outdoor Recreation*. Resources for the Future, Johns Hopkins University Press, Baltimore, MD.

Daubert, J. T. and Young, R. A., 1981. Recreational Demands for Maintaining Instream Flows: A Contingent Valuation Approach, *American Journal of Agricultural Economics*, 63(4), pp. 666-676.

Edwards, S. F., 1988. Option Prices for Groundwater Protection, *Journal of Environmental Economics and Management*, 15, pp. 475-487.

Epple, D., 1987. Hedonic Prices and Implicit Markets: Estimating Demand and Supply Functions for Differentiated Products, *Journal of Political Economy*, 95, pp. 59-80.

Freeman, A. M., III, 1979. *The Benefits of Environmental Improvement: Theory and Practice*. John Hopkins University Press, Baltimore, MD.

Freeman, A. M., III, 1987. *Assessing Damage to Marine Resources: PCBs in New Bedford Harbor*. Prepared for Presentation at the Meetings of the Association of Environmental and Resource Economists and American Economic Association, Chicago, IL.

Gramlich, F. W., 1977. The Demand for Clean Water: The Case of the Charles River, *National Tax Journal*, 30(2), pp. 183-194.

Greenley, D. A.; Walsh, R. G.; and Young, R. A., 1982. *Economic Benefits of Improved Water Quality: Public Perceptions of Option and Preservation Values*. Westview Press, Boulder, CO.

Halvorsen, R. and Pollakowski, H. O., 1981. Choice of Functional Form for Hedonic Price Functions, *Journal of Urban Economics*, 10(1), pp. 37-49.

Hammack, J. and Brown, G. M., Jr., 1974. *Waterfowl and Wetlands: Toward Bioeconomic Analysis*. Johns Hopkins University Press, Baltimore, MD.

Hanemann, W. M., 1978. *A Methodological and Empirical Study of the Recreation Benefits from Water Quality Improvement*. Ph.D. Dissertation, Harvard University, Cambridge, MA.

Hanemann, W. M.; Kanninen, B.; and Loomis, J., 1990. Estimation Efficiency and Precision of Benefit Estimates From Use of Double Bounded Dichotomous Choice Contingent Valuation. Paper Presented at a Joint Meeting of the Western Regional Science Association and the W-133 Regional Research Project, Molokai, HI.

Hanemann, W. M. Willingness to Pay and Willingness to Accept: How Much Can They Differ? *American Economic Review*. Forthcoming.

Harrison, D. and Rubinfeld, D. L., 1978. The Distribution of Benefits from Improvements in Urban Air Quality, *Journal of Environmental Economics and Management*, 5, pp. 313-332.

Hay, M. J. and McConnell, K. E., 1979. An Analysis of Participation in Nonconsumptive Wildlife Recreation, *Land Economics*, 55, pp. 460-471.

Just, R. E.; Hueth, D. L.; and Schmitz, A., 1972. *Applied Welfare Economics and Public Policy*. Prentice Hall, Englewood Cliffs, NJ.

Kling, C. L., 1988. Comparing Welfare Estimates of Environmental Quality Changes from Recreation Demand Models, *Journal of Environmental Economics and Management*, 15, pp. 331-340.

Krutilla, J. V. and Eckstein, O., 1958. *Multiple Purpose River Development*. Resources for the Future, John Hopkins University Press, Baltimore, MD.

Loomis, J. B., 1987. *An Economic Evaluation of the Public Trust Resources of Mono Lake*. Institute of Ecology Report No. 30, College of Agriculture and Environmental Sciences, University of California, Davis.

McConnell, K. E., 1977. Congestion and Willingness to Pay: A Study of Beach Use, *Land Economics*, 53, pp. 185-195.

McConnell, K. E. and Strand, I. E., 1981. Measuring the Cost of Time in Recreation Demand Analysis: An Application to Sport Fishing, *American Journal of Agricultural Economics*, 63, pp. 153-156.

McFadden, D., 1974. Conditional Logit Analysis of Qualitative Choice Behavior. In: Zarembka, P. (Ed.), *Frontiers in Econometrics*, Academic Press, New York, NY.

Mendelsohn, R., 1985. Identifying Structural Equations with Single Market Data, *Review of Economics and Statistics*, 67, pp. 525-529.

Miranowski, J. A. and Hammes, B. D., 1984. Implicit Prices of Soil Characteristics for Farmland in Iowa, *American Journal of Agricultural Economics*, 66, pp. 745-749.

Mitchell, R. C. and Carson, R. T., 1989a. *Existence Values for Groundwater Protection*. Draft Final Report to the U.S. Environmental Protection Agency by Resources for the Future.

Mitchell, R. C. and Carson, R. T., 1989b. *Using Surveys to Value Public Goods: The Contingent Valuation Method*. Resources for the Future, Johns Hopkins University Press, Baltimore, MD.

Palmquist, R. B., 1988. Welfare Measurement for Environmental Improvements Using the Hedonic Model: The Case of Nonparametric Marginal Prices, *Journal of Environmental Economics and Management*, 15, pp. 297-312.

Palmquist, R. B., 1989. Hedonic Methods. In: Baden, J. and Kolstad, C. (Eds.), *Measuring the Demand for Environmental Commodities*, North Holland, Amsterdam.

Palmquist, R. B. and Danielson, L. E., 1989. A Hedonic Study of the Effects of Erosion Control and Drainage on Farmland Values, *American Journal of Agricultural Economics*, 71, pp. 55-62.

Ridker, R. G. and Henning, J. A., 1967. The Determinants of Residential Property Values with Special Reference to Air Pollution, *Review of Economics and Statistics*, 49, pp. 246-257.

Rosen, S., 1974. Hedonic Prices and Implicit Markets: Product Differentiation in Pure Competition, *Journal of Political Economy*, 82, pp. 34-55.

Sellar, C.; Stoll, J. R.; and Chavas, J. P., 1985. Validation of Empirical Measures of Welfare Changes: A Comparison of Nonmarket Techniques, *Land Economics*, 61, pp. 156-175.

Smith, V. K. and Desvousges, W. H., 1986. *Measuring Water Quality Benefits*. Kluwer-Nijhoff, Boston, MA.

Smith, V. K. and Kaoru, Y., 1986. Modeling Recreation Demand Within a Random Utility Framework, *Economic Letters*, 22, pp. 395-399.

Smith, V. K., 1989. Household Production Functions. In: Baden, J. and Kolstad, C. (Eds.), *Measuring the Demand for Environmental Commodities*, North Holland, Amsterdam. Forthcoming.

Sutherland, R. J., 1982. A Regional Approach to Estimating Recreation Benefits of Improved Water Quality, *Journal of Environmental Economics and Management*, 9, pp. 229-247.

Sutherland, R. J. and Walsh, R. G., 1985. Effect of Distance on the Preservation Value of Water Quality, *Land Economics*, 61, pp. 281-291.

Thaler, R. H. and Rosen, S., 1976. The Value of Saving a Life, in Nestor E. Terleckyj, (Ed.), *Household Production and Consumption*. National Bureau of Economic Research, New York, NY.

Vaughan, W. J. and Russell, C. S., 1982. *Freshwater Recreational Fishing: The National Benefits of Water Pollution Control*. Resources for the Future, Washington DC.

Vaughan, W. J.; Russell, C. S.; and Hazilla, M., 1982. A Note on the Use of the Travel Cost Model with Unequal Zonal Populations: Comment, *Land Economics*, 58, pp. 400-440.

Walsh, R. G.; Sanders, L. D.; and Loomis, J. B., 1985. *Wild and Scenic River Economics: Recreation Use and Preservation Values*. Report to the American Wilderness Alliance, Department of Agriculture and Natural Resource Economics, Colorado State University.

Walsh, R. G.; Johnson, D. M.; and McKean, J. R., 1988. *Review of Outdoor Research Economic Demand Studies with Non-Market Benefit Estimates: 1968-1988.* Technical Report No. 54, Colorado Water Resources Research Institute.

Wilman, E. A., 1980. The Value of Time in Recreation Benefit Studies, *Journal of Environmental Economics and Management*, 7, pp. 272-286.

# 21 WILLINGNESS TO PAY TO PROTECT WETLANDS AND REDUCE WILDLIFE CONTAMINATION FROM AGRICULTURAL DRAINAGE

John Loomis, University of California, Davis;
Michael Hanemann and Barbara Kanninen,
University of California, Berkeley; and
Thomas Wegge, Jones and Stokes Associates,

### ABSTRACT

This chapter presents the results of a survey of the general population in California regarding their willingness to pay for alternative programs to protect and expand wetlands as well as reduce wildlife contamination in the San Joaquin Valley (Valley). The results of 803 completed interviews from 1,573 successfully contacted households indicate that Californians would pay $154 each year in higher taxes to purchase water to prevent a decrease in wetland acreage from 85,000 acres to 27,000 acres. This value rose to $254 to provide foran increase in wetland acreage to 125,000 acres with an associated 40 percent increase of bird populations. California households would pay $313 each year in additional taxes to implement agricultural drainage programs that would reduce waterbirds exposure to contamination from 70 percent to 20 percent exposure. The water management implication of these results is that Californians value clean water supplies for refuges at over $3 billion a year.

## WILDLIFE RESOURCES AT RISK

This chapter provides information on the economic value to society from resolving fish and wildlife resource problems associated with agricultural drainage in the Valley. As noted in preceding chapters, the Valley provides important wildlife habitat which supported an estimated 2 million birds during the mid-1970's (Jones and Stokes, 1989). About one-third of the entire Pacific Flyway's migratory waterfowl population winters in the Valley. Although only a portion of this represents residential breeding waterfowl, the Valley provides critical habitat for the flyway's migrating population, especially for certain species (Jones and Stokes, 1989). The Valley supports about 90,000 acres of seasonal and permanent wetlands, with a majority on private lands. These wetlands represent about 10 percent of the original wetland area in the Valley (Frayer et al., 1989).

Much of the remaining wetlands have only about 25 percent of the water required for optimum management (U.S. Bureau of Reclamation, 1987, Table S-2). Some of the water provided to Valley refuges and wildlife areas has included agricultural drainage, a common practice for irrigated agricultural areas in the West. As earlier chapters have demonstrated, agricultural drainage water may contain high levels of selenium, boron, arsenic, and other trace elements which are concentrated to hazardous levels. Discharge of agricultural drainage water to wetlands at Kesterson Reservoir (Kesterson) and Kesterson National Wildlife Refuge was halted by the Federal Government in 1985. Since that time, farmers have been increasing their use of onfarm evaporation ponds. Such evaporation ponds, which attract waterfowl, biomagnify trace elements in waterbird populations. As an attractive nuisance to wildlife, these evaporation ponds may become "population sinks" which attract birds that subsequently become incapable of reproducing successfully and may experience high levels of mortality (Jones and Stokes Associates, 1989).

Another critical habitat in the Valley is the San Joaquin River, which prior to the mid-1940's supported naturally spawning stocks of chinook salmon. The construction of Friant Dam and reduced riverflows have resulted in near elimination of the chinook salmon fishery in the San Joaquin River. This chapter presents the economic value to society which could result from reversing many of the adverse conditions experienced by fish and wildlife in the Valley.

## RESEARCH METHODOLOGY

Since the objective of this chapter is measurement of the economic value to society of fish and wildlife resources in the Valley, it is important to define the types of economic benefits to be measured. Loomis et al. (1984) relate the concept of Public Trust values and environmental values to Randall and Stoll's (1983) notion of "total economic value." In particular, total economic value is made up of five components: (1) Onsite recreation use of the resource; (2) commercial use of the resource; (3) an option demand from maintaining the potential to visit the resource in the future; (4) an existence value derived from simply knowing the resource exists in a preserved state; and (5) a bequest value derived by individuals from knowing that future generations will be able to enjoy existence or use of a resource.

To quantify "total economic value" requires measurement of an individual's maximum willingness to pay (WTP) or minimum willingness to accept for alternative levels of fish and wildlife in the Valley. Increasing fish and wildlife populations beyond current levels can be viewed as an increment in a person's well being or utility. As such, willingness to pay could be argued to be the appropriate measure.

## Techniques for Measuring Willingness to Pay

The Contingent Valuation Method (CVM) is the technique used to measure California residents' willingness to pay for different levels of wildlife management in the Valley. CVM has been used for valuing both recreation and nonmarketed benefits of environmental resources (see Cummings, Brookshire, and Schulze, 1986 for a review of CVM). CVM has been recommended twice by the U.S. Water Resources Council (1979 and 1983) as one of two preferred methods for valuing outdoor recreation in Federal benefit-cost analyses. Recently, the U.S. Department of Interior (1986) endorsed CVM as one of the two preferred methods for valuing natural resource damages. CVM is capable of not only measuring the value of outdoor recreation but is the only method available to measure other resource values such as option, existence, and bequest.

The basic concept of CVM is that a realistic but hypothetical market for "buying" use and/or preservation of a nonmarketed natural resource is described to an individual. Then the individual is told to use this market to express his or her valuation of the resource. Recently, a dichotomous choice or "referendum" approach has been developed where the respondents answer "yes" or "no" to one randomly assigned dollar amount chosen by the interviewer. For more details on these approaches see Cummings et al. (1986); Hanemann (1984); and Kriesal and Randall (1986).

There are several advantages to using the dichotomous choice referendum approach in this analysis. First, a pretest indicated that people perceived a need for social rather than individual action to correct the many threats to fish, wildlife, and their habitats in the Valley. Therefore, respondents felt a voter referendum format was more credible than a market format. The referendum format is similar to how State residents make decisions on many environmental programs such as clean water, pesticide control, or recreation development. This format is also similar to the successful Proposition 70 (wildlife, open space, and parks bond) on the June 1987 ballot. This proposition asked voters to approve a bond issue for the purchase of habitat and open space in

California, with the bond funding to be repaid from State tax monies. Additionally, the voter referendum format avoids making CVM surveys sound like a solicitation for charitable contributions.

As Kriesel and Randall (1986) indicate, the dichotomous choice approach to CVM is structured such that the individual's best response strategy is to tell the truth. Since the referendum format is a dichotomous choice, individuals must only determine if the value to them is greater than or less than the dollar amount they are asked to pay. This is simpler than having to specify their willingness to pay as an exact amount. Because it was necessary to ask a total of 11 WTP questions in the survey, the particular advantage to the respondent of the dichotomous choice format is especially apparent. It is only feasible to ask respondents to answer this many questions if a close-ended, yes-no format is used. Asking this many open-ended or iterative questions would place too much of a burden on respondents.

The means of paying for the benefit described in the survey must be realistic and as neutral as possible for the respondent. To improve realism, the payment vehicle should be appropriate for the resource and market constructed. Given the political "market" (the voter referendum), the use of additional taxes was realistic and credible. While some people may react emotionally to tax issues, this problem can be ascertained using a protest check question. In addition, the focus groups used to develop and pretest the survey indicated that additional taxes would be a realistic and acceptable payment vehicle.

## THEORY

Initially, the basic structure of the dichotomous-choice CVM method will be presented to represent the traditional, single-bound approach. This approach will then be extended to the double-bound approach used in this chapter. Using this framework the differences in the variance-covariance matrix of standard single and double bound are derived. Differences in the variance-covariance matrix translate into different levels of precision in the benefit estimates.

The general structure of a discrete-choice CVM survey involves asking an individual if he or she would pay $B to secure a given improvement in environmental quality. The probability of obtaining a "No" or a "Yes" response can be represented, respectively, by the statistical models:

[1] $\pi_n(B) = G(B;\theta)$

[2] $\pi_y(B) = 1-G(B;\theta)$

where $G(\cdot;\theta)$ is some statistical distribution function with parameter vector $\theta$. As pointed out in Hanemann (1984), this statistical model can be interpreted as a utility maximization response within a random utility context where $G(\cdot;\theta)$ is the cumulative density function (cdf) of the individual's true maximum willingness to pay.

Since utility maximization implies:

Pr{ No to $B} <==> Pr{ $B > maximum WTP}

Pr{ Yes to $B} <==> Pr{ $B $\leq$ maximum WTP}

In Bishop and Heberlein's (1979) study, $G(\cdot;\theta)$ is the log-logistic cdf:

[3] $G(B) = 1/[1 + e^{a+b(\ln \$B)}]$

where $\theta \equiv (a,b)$. Another alternative is the logistic cdf:

[4] $G(B) = 1/[1 + e^{a+b(\$B)}]$.

In both cases, equations [1] and [2] correspond to a form of the logit model. Similarly, if one were to substitute the lognormal or normal cdf's for [3] and [4], [1] and [2] would correspond to a probit model. Other distribution functions could readily be employed, although logit and probit models are by far the most common to date.

While there are other estimation techniques with equivalent asymptotic properties, it is convenient to focus on the maximum likelihood approach. In general, the participants in a CVM survey will be offered different bids. Suppose there are N participants, and let $B_i$ be the bid offered to the ith individual. Then, the log-likelihood function for this set of responses is:

[5] $\ln L(\theta) = \sum_{i=1}^{N} \ln L_i(\theta)$

$= \sum_{i=1}^{N} I(B_i) \ln (B_i) + (1-I(B_i))\ln(Bi)$

$= \sum_{i=1}^{N} (B_i) \ln [1-G(B_i;\theta)] + (1-I(B_i))\ln G(B_i;\theta)$

where the indicator function $(I(B_i)) = 1$ if the ith response is 'yes,' and zero otherwise.

Here $L_i(\theta)$ is the individual's contribution to the likelihood function, which is given by $\pi_y$ for those who responded "Yes" and $\pi_n$ for those who responded "No". For logit and probit models, McFadden (1974) and Haberman (1974) established the global concavity of the log-likelihood function; thus in these cases the matrices

$$\frac{\partial^2 \ln L_i}{\partial \theta \, \partial \theta'}$$

are negative definite for all i. The maximum likelihood estimator, denoted $\hat{\theta}$, is the solution to the equation

[6] $\quad \dfrac{\partial \ln L(\hat{\theta})}{\partial \theta} = 0.$

As proven, for example, in Amemiya (1981), this estimator is consistent (though it may be biased in small samples) and asymptotically efficient. Thus the asymptotic variance-covariance matrix of $\theta$ is given by the Cramer-Rao lower bound

[7] $\quad V(\theta) = \left[ -E \, \dfrac{\partial^2 \ln L}{\partial \theta \, \partial \theta'} \right]^{-1}$

## Double Bounded Logit

So far we have been describing a conventional-dichotomous choice CVM survey in which the participants are each presented with a single bid. Now consider an alternative format in which each participant is presented with two bids. The level of the second bid is contingent upon the response to the first bid. If the individual responds "Yes" to the first bid, the second bid (to be denoted $B^u_i$) is some (random) monetary amount greater than the first bid ($B_i < B^u_i$); if the individual responds "No" to the first bid, the second bid ($B^d_i$) is some amount smaller than the first bid ($B^d_i < B_i$). Thus the overall survey has four possible outcomes: (1) Both answers being "Yes", (2) both answers being "No", (3) a "Yes" followed by a "No", and (4) a "No" followed by a "Yes". The likelihood of these outcomes will be denoted respectively by $\pi_{yy}$, $\pi_{nn}$, $\pi_{yn}$, and $\pi_{ny}$. Under the assumption of a utility maximizing respondent, the formulas for these likelihoods are as follows. In the first case, we have $B^u_i > B_i$ and

[8] $\quad \pi_{yy}(B^u_i, B_i) = \Pr\{B_i \leq \max \text{WTP and } B^u_i \leq \max \text{WTP}\}$

$$= \Pr\{B_i \leq \max \text{WTP} \mid B^u_i \leq \max \text{WTP}\} \Pr\{B^u_i \leq \max \text{WTP}\}$$

$$= \Pr\{B^u_i \leq \max \text{WTP}\}$$

$$= 1 - G(B^u_i; \theta)$$

since with $B^u_i > B_i$, $\Pr\{B_i \leq \max \text{WTP} \mid B^u_i < \max \text{WTP}\} \equiv 1$.

Similarly, with $B^d_i < B_i$, $\Pr\{B^d_i \leq \max \text{WTP} \mid B_i \leq \max \text{WTP}\} \equiv 1$, hence,

[9] $\pi_{nn}(B_i, B^d_i) = \Pr\{B_i > \max \text{WTP} \text{ and } B^d_i > \max \text{WTP}\} = G(B^d_i, \theta)$.

When a "Yes" is followed by a "No", we have $B^u_i > B_i$ and

[10] $\pi_{yn}(B_i, B^u_i) = \Pr\{B_i \leq \max \text{WTP} \leq B^u_i\}$

$$= G(B^u_i; \theta) - G(B_i; \theta)$$

and when a "No" is followed by a "Yes" we have $B^d_i < B_i$ and

[11] $\pi_{ny}(B_i, B^d_i) = \Pr\{B_i \geq \max \text{WTP} \geq B^d_i\} = G(B_i; \theta) - (B^d_i; \theta)$.

The log-likelihood function for the double bounded approach takes the form:

[12] $\ln L^D(\theta) = \sum_{i=1}^{N} I(B_i) I(B^u_i) \ln \pi_{yy}(B_i, B^u_i) +$

$(1-I(Bi))(1-I(B^d_i)) \ln \pi_{nn}(B_i, B^d_i) +$

$I(Bi)(1-I(B^u_i)) \ln \pi_{yn}(B_i, B^u_i) +$

$(1-I(B_i))(I(B^d_i)) \ln \pi_{ny}(B_i, B^d_i)$

Where $I(\cdot)$ is the indicator function defined as before.

The maximum likelihood estimator for the double bounded approach, $\hat{\theta}^D$, is obtained by solving an equation analogus to [6] but for equation [12]. The asymptotic variance covariance of $\theta^D$ is given by the analog of [7]:

[13] $V(\theta^D) = \left[ \dfrac{-E \, \partial^2 \ln L^D}{\partial \theta \, \partial \theta'} \right]^{-1}$

With regard to the comparison between the estimators $\hat{\theta}$ and $\hat{\theta}^D$, the following result is established using the theorem in Hanemann et al.: If the likelihood function [5] is globally concave in $\theta$, it follows that $V(\theta) \geq V(\theta^D)$. (See Appendix 1.)

The implication is that the estimator $\hat{\theta}^D$ is asymptotically more efficient than the estimator $\hat{\theta}$. The reduction in variance can be translated into tighter confidence intervals for the WTP estimates by adapting an approach first suggested by Krinsky and Robb (1986) for elasticities.

## METHODS

To implement the double bounded approach which involved four log-likelihood functions, Generalized Optimization Program (GQOPT) was used. The four log-likelihood functions implicit in equation [12] were programmed into a Fortran subroutine. The subroutine reads each individual's responses, determines which log-likelihood function to apply, then calculates the individual log-likelihood function. Finally, the sum of all of the individual log-likelihood functions is computed. It is this resulting log-likelihood function that is maximized by GQOPT. This log-likelihood function was maximized by using a simplex algorithm to find starting values and then applying a Davidson-Fletcher-Powell (DFP) method to find the maximum.

## STATISTICAL METHODS

To estimate WTP from yes/no responses to different dollar amounts requires two steps. The first step is to maximize the likelihood function given in equation [12] using the GQOPT program described above. This results in equation

[14]    $\text{LOGODDS} = A_o - A_1(\$B_i)$ where

LOGODDS is the log of the probability of a yes response to $\$B_i$ divided by the probability of a no response to $\$B_i$.

The next step is to compute WTP from the logit equation. This is basically the area under the logit curve or the expected value of WTP (i.e., the probability a person would pay each dollar amount times the respective dollar amount). Following Hanemann, (1989), WTP is given by:

[15]    $\text{WTP} = (1/A_1) \cdot \ln(1 + e^{A_o})$

The specific logit equations will be presented in the results section of the chapter.

## SURVEY DESIGN

The basic survey booklet and interview involve an introductory set of questions about wildlife, followed by WTP questions, and ending with demographic questions about the respondent. The survey booklet mailed out was the result of three focus groups and a pretest.

There were four major sections to the survey. The first was entitled "Wildlife and You." This section included questions about the importance of fish and wildlife, whether the respondents had visited the Valley, and if so, for what purpose. Additional questions were asked to determine the respondents' familiarity about fish and wildlife issues in the Valley. Next was a series of attitude questions about threats to fish and wildlife and importance of wildlife species. These questions also provided some nonmonetary indicators of the importance of wildlife and prompted the respondents to begin thinking about wildlife issues. In addition inquiries were made about their motivations for protecting wetlands, wildlife, and fisheries resources.

The next major section of the questionnaire, entitled "Alternative Futures," described alternative fish and wildlife programs that could be implemented in the Valley. The issues were set up to elicit votes on a referendum regarding each resource issue and level of management. The individual was asked to vote on three programs, with two of the programs having two alternative levels of management intensity. Thus a person voted a total of five times. Lastly, a check question was asked to determine whether any "no, would not pay" responses were made to protest some feature of the referendum.

The specific programs people voted for included a Wetlands Habitat and Wildlife Program which provided three alternative levels (a no action, a maintenance, and an improvement level) of three key characteristics: (1) Wetland acreage, (2) resident waterbirds and wintering waterfowl, and (3) public viewing of wildlife. The three levels for wetlands involved a loss down to 27,000 acres in the Valley; a maintenance of the current 85,000 acres; and an expansion to 125,000 acres. The improvement level involved purchase of additional wetlands and required water supply. With regard to bird populations, the relative percentage decreases in resident and wintering species under the no-action alternative (approximately a 70-percent loss) and the percentage increase of both groups of species with the improvement program (+40 percent) were illustrated using bar charts.

The Wildlife Contamination Control Program provided three alternative levels for two key indicators: (1) Percentage of resident waterbird exposure to contaminated waters and (2) cases of reproductive failure in the Valley's nesting waterbirds. The no-action level involved 95 percent of the Valley's resident waterbirds being exposed to contaminated water. The maintenance program involved 70 percent exposure while the improvement program resulted in only 20 percent of the Valley's resident waterbirds being exposed to contamination. This was represented by bar charts in the survey.

The San Joaquin River and Salmon Improvement Program involved comparison of the no-action level and an expansion level with two key indicators: (1) Chinook salmon populations and (2) sport and commercial catch of chinook salmon. The two alternative levels were illustrated graphically.

The particular contamination and wildlife levels for the program were developed jointly by biologists with Jones and Stokes Associates, U.S. Fish and Wildlife Service, Bureau of Reclamation, and California Department of Fish and Game. For details see Jones and Stokes Associates (1989).

The exact wording of the question sequence for Wetland Maintenance was: "If the Maintenance program were the only program you had an opportunity to vote on, and it cost every household in California $B dollars each year in additional taxes would you vote for it? YES or NO."

The followup telephone questions were set up as follows: If the respondents said NO to $B dollars, they were asked "What if the cost were 1/2 $B, how would you vote then?" If they had said YES to $B, they were asked how they would vote at two times $B. The range of the dollar bid amounts was $30 to $130 for the Maintenance questions and $45 to $225 for the Improvement questions. The followup vote at 1/2 $B or twice $B involved a double-bounded logit which increased the precision of WTP significantly (Hanemann et al., 1990).

This basic question format was also asked for the Wetlands Improvement Program, Wildlife Contamination Control Maintenance and Improvement Programs, and a San Joaquin River and Salmon Improvement Program. Five votes were asked with this format. The resulting benefit estimates reflect annual household total WTP (Randall and Stoll, 1983 and Loomis et al., 1984).

## DATA COLLECTION AND DATA SOURCES

The data collection procedure used in this chapter involved a combination of a mailed survey booklet followed by the telephone interview. Specifically, the actual interview and data collection from the respondent were conducted over the telephone, with the respondents having a survey booklet in front of them at the time of the interview.

Initial phone calls were made to random samples of households in the Valley and other areas throughout California to solicit their participation in the study. A total of 1573 households were contacted and 991 were scheduled for interviews yielding a participation rate of 63 percent. Of these 991 households, 803 (227 in the Valley and 576 in the rest of California) completed the interview when called back after receiving the survey booklet. This represents an overall completion rate of 51 percent for both steps.

## RESULTS

### Reduced Protest Responses with Voter Referendum Format

As is normal for all CVM studies, the completed questionnaires were screened for protest responses to the willingness to pay question. The voter referendum format had a very low percentage of respondents protest the WTP questions. Only 4.5 percent of the respondents voted against all programs because they either felt that the referendum was unrealistic; the Government wastes money; they already pay enough in taxes; or that others (e.g., farmers or visitors) should pay. This protest rate was substantially below the 10 to 23 percent protest rates found by Walsh et al., (1982 and 1984) who used a payment-into-a-trust-fund approach in Colorado. This protest rate was also much lower than found by Loomis (1987) using both a trust fund and water bill for preservation of Mono Lake. Thus the voter referendum format seems to have a greater credibility with the general public than other approaches. However, more comparisons are needed before any final conclusions can be drawn.

Table 1 presents the logit equations estimated using the double-bound approach for both residents of the Valley and other areas of California. As table 1 illustrates, all of the slope coefficients in the logit equations are statistically significant at the 1 percent level.

Table 1 also presents the benefit estimates (net WTP) for both residents of the Valley and rest of California. These benefit estimates conform to economic theory. The net WTP for the improvement level programs is higher than for the maintenance level. In addition, the gain in WTP between Maintenance and Improvement is smaller than between No-Action and Maintenance level. There is, as theory would predict, diminishing marginal value of additional wildlife habitat improvements. The benefit estimates are about the same magnitude as a dichotomous choice CVM survey of California households for protection of wildlife habitat at Mono Lake (Loomis, 1987).

Table 1. Double-bound logit equations and benefits per household.

| Program/Location | Logit Equation | | Benefit Estimates | |
|---|---|---|---|---|
| | Intercept | Slope | Mean | 90% Confidence Interval |
| *Wetland Maintenance* | | | | |
| California | 3.77 | -0.0249 | $152 | 123-188 |
| | (16.74) | (-13.94) | | |
| San Joaquin Valley | 3.80 | -0.022 | $174 | 157-196 |
| | (9.88) | (-7.52) | | |
| *Wetland Improvement* | | | | |
| California | 3.042 | -0.0123 | $251 | 235-268 |
| | (17.73) | (-14.75) | | |
| San Joaquin Valley | 2.80 | -0.010 | $286 | 255-325 |
| | (10.08) | (-8.27) | | |
| *Contamination Maintenance* | | | | |
| California | 3.61 | -0.0194 | $187 | 177-199 |
| | (17.49) | (-14.57) | | |
| San Joaquin Valley | 3.65 | -0.0187 | $197 | 179-216 |
| | (12.05) | (-9.63) | | |
| *Contamination Improvement* | | | | |
| California | 2.87 | -0.0095 | $308 | 289-331 |
| | (17.74) | (-14.86) | | |
| San Joaquin Valley | 2.434 | -0.0070 | $360 | 317-415 |
| | (9.77) | (-8.14) | | |
| *Salmon Improvement* | | | | |
| California | 3.450 | -0.0192 | $181 | 171-193 |
| | (16.85) | (14.04) | | |
| San Joaquin Valley | 3.10 | -0.0156 | $202 | 180-231 |
| | (10.16) | (-7.81) | | |

*Note:* t-statistics shown in parenthesis

The confidence intervals were calculated using Park et al.'s, adaption of Krinksy and Robb (1986) technique for calculating confidence intervals for elasticities. This approach involves following three steps: (1) A multivariate normal distribution for the estimated parameters is constructed having as its mean the parameter estimates, and having its variance developed from the parameter's variance-covariance matrix; (2) a large number of draws (here 4,000) are made from the resulting multivariate normal distribution. At each draw, the resulting parameters are used to calculate WTP; and (3) the vector of WTP are ranked and 5 percent of the WTP estimates in each extreme are dropped to form a 90-percent confidence interval on WTP.

These confidence intervals demonstrate that benefits rise in a statistically significant manner as wildlife management moves from the Maintenance to the Expansion level. The confidence intervals also show the relatively high degree of precision in these benefit estimates. Nearly all of the 90-percent confidence intervals are within plus or minus 10 percent of the mean.

## POPULATION ESTIMATES OF WTP FOR EXPANDING FISH AND WILDLIFE

These average values per household must be increased upward to the number of households Statewide in California. The accuracy of increasing a sample to the total population is dependent upon the degree to which the sample is truly representative. While the original sample was a representative sample of California residents, the 51-percent response rate is somewhat lower than desirable. However, the sample appears to represent many of the key socioeconomic characteristics of the State population fairly well.

Therefore our best estimate of Statewide benefits is obtained by multiplying our sample value per household by the total number of households in California. This involves weighting households in the Valley by .09 and rest of California by .91, their respective representations in the population. Aggregate benefits are given in table 2 for the State of California. One could compute a lower bound benefit estimate from the figures in table 2, by assuming (we believe somewhat incorrectly) that the nonrespondents to the survey had a zero WTP. In essence the conservative lower bound values then would be half the numbers reported in table 2.

Table 2. Benefits to California residents from wildlife management.

| Program | Mean Value Per Household | Total Benefits (Millions) |
|---|---|---|
| Wetland Maintenance | $154 | $1,515 |
| Wetland Improvement | 254 | 2,501 |
| Contamination Maintenance | 188 | 1,849 |
| Contamination Improvement | 313 | 3,077 |
| Salmon Improvement | 183 | 1,800 |

As the results in table 2 indicate, the benefits are $3 billion for reducing the percentage of waterbirds exposed to contamination from 95 percent to 20 percent. The benefits of expanding wetlands from 27,000 acres to 125,000 acres and increasing waterbird populations by 40 percent is $2.5 billion. It is important to note the diminishing incremental benefits would apply to additional wetlands in excess of 125,000 acres. That is, the total benefits of increasing wetlands to 225,000 acres would not be $5 billion ($2.5 billion times two), but perhaps $3.75 billion. The benefits of restoring chinook populations to the San Joaquin River is worth $1.8 billion. It is also important to note that the total benefits for performing all three improvement programs is not the simple sum of these three benefit estimates. Research by Loomis et al., indicates there are statistically significant interaction effects between these programs. The aggregate benefits for performing all three improvement programs appear to be about half as much as the simple sum of the individual program benefits. This result is consistent with economic theory of benefit measurement (Hoehn and Randall, 1989).

## SOME APPROXIMATE BENEFIT-TO-COST COMPARISONS

Of course, these benefits would need to be compared to the costs of water and wildlife management necessary to increase wetlands and fisheries, as well as to reduce contamination. For example, to increase wetland acreage from the current 85,000 acres to 125,000 would require about 410,000 acre-feet of water annually (Jones and Stokes Associates, 1989, p. 16). Gibbons' (1986, p. 38) survey of irrigation values of water in the Valley shows the $40 per acre-foot associated with cotton and melons is about the highest value in the region. Updating this value to 1988 costs, the annual cost of 410,000 acre-feet would be $23 million. The annual conveyance, operation, maintenance, power, and annualized construction costs (if any) associated with delivering about half this water volume to the Valley's eight refuges and wildlife management areas as well as the Grassland Resource Conservation District has been estimated by the Bureau of Reclamation to be $1.53 million (U.S. Bureau of Reclamation, 1987). Even if the costs for delivering the full 410,000 acre-feet of water were twice this $1.53 million figure, the $2.5 billion in benefits of expanding wetlands to 125,000 acres would substantially outweigh the water and conveyance costs of $26 million per year.

The amount of water required to provide the minimum 150 cubic feet per second ($ft^3/s$) releases needed from October through January for spawning and adult migration of chinook salmon would be 44,000 acre-feet annually (Jones and Stokes Associates, 1989). Adding to this the expected value of supplemental flows for outmigration during dry years brings the total to 61,500 acre-feet.

If these water releases could not be used downstream at that time of year and reduced the amount of water available to agriculture, the cost would be $3.5 million using the same agricultural water values from Gibbons' as before. In addition, there may be some small loss in hydropower values as well since two of the irrigation canals have "run of the river" hydropower that generates power from irrigation releases. If water releases for fish in the river reduce irrigation releases in these canals, then there would be forgone hydropower. This is likely to be quite small however. Given these relatively small costs, it appears that the benefits to society outweigh the opportunity costs of providing the flows for salmon in the San Joaquin River.

## CONCLUSIONS

This research has demonstrated the acceptability of the voter referendum format as a useful mechanism to record society's willingness to pay for improving wetlands and wildlife in the Valley. This format had the lowest protest rate of any contingent valuation method approach reviewed. Only about 5 percent of the respondents rejected the simulated voter referendum as not being a credible or fair approach to solving environmental problems in the Valley.

The estimate of benefits per household was quite precise with the 90 percent confidence interval being within 10 percent of the mean willingness to pay. The best estimate of California's total willingness to pay to largely eliminate waterbird exposure to contamination is $3 billion. Increasing the amount of wetlands is worth $2.5 billion.

These values strongly suggest that Californian's are concerned about the loss of wetlands and the exposure of wildbirds to contamination. The benefits of correcting these threats to wildlife appear to substantially outweigh the cost of the control measures. Implicit in these comparisons is that the value of the first half million acre-feet of water needed to produce wetlands and wildlife free of contamination is worth more than the value that half million acre-feet of water could produce in agricultural production. Given society's rising value for wildlife, far too little water has gone to wildlife and far too much to agriculture. Since Western water law is based on the concept of beneficial use, it should be flexible enough to recognize the increasing social benefits of water used in wildlife enhancements. Wholesale changes in water use are not needed. Rather, an incremental reallocation of a few percentage points of agriculture's use of nearly 90 percent of California's available water to wildlife would restore the balance in the Valley.

## APPENDIX 1

*Theorem:* $V(\theta^D) \leq V(\theta)$

*Proof:* Since $V(\theta) = \left[ -E \dfrac{d^2 \ln L}{d\theta d\theta'} \right]^{-1} = \left[ I(\theta) \right]^{-1}$

where $I(\theta)$ is the information matrix, $V(\theta^D) \leq V(\theta)$ is equivalent to $I(\theta^D) \geq I(\theta)$. For simplicity, assume $B_i = B$ and $B_i^u = B^u$, $B_i^d = B^d$ for all i's.

from [12] we can find:

$$I(\theta^D) = \frac{1}{1 - G(B^u)} \left[ G_\theta(B^u) \right] \left[ G_\theta(B^d) \right]'$$

$$+ \frac{1}{G(B^d)} \left[ G_\theta(B^d) \right] \left[ G_\theta(B^d) \right]'$$

$$+ \frac{1}{G(B^u) - G(B)} \left[ G_\theta(B^u) - G_\theta(B) \right] \left[ G_\theta(B^u) - G_\theta(B) \right]$$

$$+ \frac{1}{G(B) - G(B^d)} \left[ G_\theta(B) - G_\theta(B^d) \right] \left[ G_\theta(B) - G_\theta(B^d) \right]$$

and from [5]:

$$I(\theta) = \frac{1}{1 - G(B)} \left[ G_\theta(B) \right] \left[ G_\theta(B) \right]'$$

$$+ \frac{1}{G(B)} \left[ G_\theta(B) \right] \left[ G_\theta(B) \right]'$$

$$I(\theta^D) - I(\theta) = \frac{1}{G(B^u) - G(B)(1-G(B))(1-G(B^u))} \Big[ G_\theta(B) - (G_\theta(B^u) -$$

$$G_\theta(B)G(B^u) + G_\theta(B^u)G(B) \Big] \cdot \Big[ G_\theta(B) - (G_\theta(B^d) -$$

$$G_\theta(B)G(B^d) + G_\theta(B^d)G(B) \Big]'$$

$$+ \frac{1}{G(B^d)\Big[G(B) - G(B^d)\Big]G(B)} \Big[ G_\theta(B^d)G(B) - G_\theta(B)G(B^d) \Big] \cdot$$

$$\Big[ G_\theta(B^d)G(B) - G_\theta(B)G(B^u) \Big]'$$

Both of these terms are positive semidefinite.
Therefore $I(\theta^D) \geq I(\theta)$ and $V(\theta^D) \leq V(\theta)$.

## NOTES

The likelihood function, statistical results and benefit estimates calculated using the log of bid amount in the logit equations yielded a empirical results qualitatively identical to what is reported in this paper for the linear logit. Results of the log of bid results are available from the authors.

## ACKNOWLEDGEMENTS

The authors wish to thank Dr. Michael King, California State University, Chico, for coordination and implementation of the telephone survey. Steve Moore and Craig Stroh of the San Joaquin Valley Drainage Program provided valuable guidance on the survey design.

## REFERENCES

Amemiya, T., 1981. Qualitative Response Models, *Journal of Economic Literature*, 19, pp. 1483-1536.

Cummings, R.; Brookshire, D.; and Schulze, W., 1986. *Valuing Environmental Goods: An Assessment of the Contingent Valuation Method.* Rowman and Allanheld, Totowa, NJ.

Frayer, W. E.; Peters, Dennis; and Pywell, Ross, 1989. *Wetlands of the California Central Valley: Status and Trends.* U.S. Fish and Wildlife Service, Portland, OR.

Gibbons, D., 1986. *The Economic Value of Water.* Resources for the Future, Washington, DC.

Haberman, S. 1974. *The Analysis of Frequency Data.* University of Chicago Press, Chicago, IL.

Hanemann, M., 1984. Welfare Evaluations in Contingent Valuation Experiments with Discrete Responses, *American Journal of Agricultural Economics*, 66(3), pp. 332-341.

Hanemann, M., 1989. Welfare Evaluations in Contingent Valuation Experiments with Discrete Reponse Data: Reply, *American Journal of Agricultural Economics*, 71(4), pp. 1057-1061.

Hanemann, M.; Loomis, J.; and Kanninen, B., 1990. *Estimation Efficiency and Precision of Benefit Estimates from Use of Double Bounded Dichotomous Choice Contingent Valuation.* University of California Paper Presented at the 1990 Regional Research Project of the Agricultural Experiment Station, Molakai, HI.

Hoehn, J. and Randall, A., 1989. Too Many Proposals Pass the Benefit-Cost Test, *American Economic Review*, 79(3), pp. 544-551.

Jones and Stokes Associates, 1989. *Alternative Scenarios for the Study of Environmental Benefits of the San Joaquin Valley Drainage Program.* Jones and Stokes Associates, Sacramento, CA.

Kriesel, W. and Randall, A., 1986. *Evaluating National Policy by Contingent Valuation.* Paper Presented at the Annual Meetings of the American Agricultural Economics Association, Reno, NV.

Krinksy, I. and Robb, A. L, 1986. On Approximating the Statistical Properties of Elasticities, *Review of Economics and Statistics*, 68, pp. 715-719.

Loomis, J.; Peterson, G.; and Sorg, C., 1984. *A Field Guide to Wildlife Economic Analysis.* Transactions of 49th North American Wildlife and Natural Resources Conference. Wildlife Management Institute, Washington, DC.

Loomis, J., 1987. Balancing Public Trust Resources of Mono Lake and Los Angeles' Water Right: An Economic Approach, *Water Resources Research*, 23(8), pp. 1449-1456.

Loomis, J.; Hoehn, J.; and Hanmeann, M., *Testing the Fallacy of Independent Valuation and Summation in Multi-Part Policies: An Empirical Test of Whether 'Too Many Proposals Pass the Benefit-Cost Test.'* Unpublished Paper, University of California, Davis.

McFodder, D. , 1974. The Measurement of Urban Travel Demand, *Journal of Public Economics,* 3, pp. 303-328.

Park, T.; Loomis, J.; and Creel, M., 1989. Confidence Intervals for Evaluating Benefit Estimates from Dichotomous Choice Contingent Valuation Surveys, *Land Economics.* Forthcoming.

Randall, A. and Stoll, J., 1983. Existence Value in a Total Valuation Framework. In: Rowe, R. and Chestnut, L. (Eds.), *Managing Air Quality and Scenic Resources at National Parks and Wilderness Areas,* Westview Press, Boulder, CO.

U.S. Bureau of Reclamation, 1987. *Refuge Water Supply Investigation.* Draft Report.

U.S. Department of Interior, 1986. Natural Resource Damage Assessments: Final Rule, *Federal Register,* 43, CFR Part 11, 51(148), pp. 27674-27753.

U.S. Water Resources Council, 1979. Procedures for Evaluation of National Economic Development (NED) Benefits and Costs in Water Resources Planning (Level C), *Federal Register,* 44-243, pp. 72892-72976.

U.S. Water Resources Council, 1983. *Economic and Environmental Principles for Water and Related Land Resources Implementation Studies.* U.S. Government Printing Office.

Walsh, R.; Loomis, J.; and Gillman, R., 1984. Valuing Option, Existence and Bequest Demands for Wilderness, *Land Economics,* 60(1), pp. 14-29.

Walsh, R.; Sanders, L.; and Loomis, J., 1985. *Wild and Scenic River Economics: Recreation Use and Preservation Values.* Department of Agricultural and Natural Resource Economics. Colorado State University, Fort Collins, CO.

# 22 VALUING ENVIRONMENTAL GOODS: A CRITICAL APPRAISAL OF THE STATE OF THE ART

H. S. Burness, Ronald G. Cummings, and Philip T. Ganderton, University of New Mexico, Albuquerque, and Glenn W. Harrison, University of South Carolina, Columbia

### ABSTRACT

This chapter develops a contingent valuation method for estimating nonmarket, typically environmental goods. Economists usually ignore critically important issues in the application of the methods. These issues define a research agenda to be developed here for the application of the method for deriving meaningful value estimates for recreation, environmental, and other nonmarket goods.

### INTRODUCTION

There are a wide range of effects associated with the management of agricultural drainage water which, while they impact human values and welfare, do not command a market price. These include effects such as environmental degradation or improvement (for example, deterioration or improvement of water or air quality) and effects on public health and safety. Economists view such effects in the context of a market analogy and refer to them as "goods" or "commodities." However, these goods generally do not have a market price since they are not typically traded in well-established markets.

Given the importance society places upon environmental goods, a great deal of effort has been expended in the economics profession over the last decade on methodologies to assign values to such goods. Of these methodologies, the Contingent Valuation Method (CVM) has received a considerable amount of research attention, given the wide range of environmental goods for which it could potentially be used. The CVM is a survey method whereby participating subjects are asked their contingent maximum willingness to pay (WTP) (their valuation) for a specified commodity (such as a 10-percent improvement in water quality). The contingency involved is the actual provision of the commodity which is necessarily hypothetical at the time the survey is conducted ("what would you pay if [contingent upon] water quality [BOD levels] was improved?").

At a conceptual level, the CVM may be viewed as a methodological tool with considerable potential for estimating values relevant to agricultural drainage, water quality, and recreational programs. It must be recognized, however, that considerable debate exists as to the extent to which the present state of the art of the CVM is sufficiently advanced to warrant its application to broad classes of public goods. This debate is particularly intense for applications of the CVM which aspire to estimate reasonably "precise" values. Perhaps the most important debate concerns the use of the CVM for estimating damages pursuant to Section 301 of the Comprehensive Environmental Response, Compensation and Liability Act of 1980 (for this specific legal context, see Phillips and Zeckhauser, 1989; for a more general overview of biases in CVM values, see Cummings, Brookshire, and Schulze, 1986; Bentkover, Covello, and Mumpower, 1986; and Mitchell and Carson, 1986).

In their assessment of CVM state of the art, Cummings, Brookshire, and Schulze (1986) express cautious optimism for the method's promise. They note that for some applications of the method, CVM value estimates are remarkably similar to value measures derived for the same goods using various other methods. The realization of this promise is an important issue if economists are to develop defensible methods for valuing nonmarket goods. Thus, this chapter will focus on issues which pose the greatest challenge to the economists' goal of deriving credible values for nonmarket goods with the CVM. These issues are captured in the question: *Which Households Place What Value on Which Nonmarket Goods?* This question will initially be approached in the following way. The next two sections will address the "which households" and "what value for which goods" issues. Both sections adopt a level of exposition which is hoped will be comprehensible to the nontechnical reader. In the fourth section, the exposition necessarily becomes a bit more technical as relevant results from empirical studies are briefly reviewed. A few implications for future research are suggested in the final section.

## THE "WHICH HOUSEHOLDS" ISSUE

This section focuses on the problem of identifying households to which CVM values can be defensibly assigned for some particular nonmarket good. This problem is relevant for all nonmarket goods, but its importance becomes eminently clear when attributing a nonuse value to nonmarket goods. Thus, for the purpose of clarity, the focus is limited to nonuse values in the discussions that follow.

For readers who are not familiar with the relevant literature, a brief overview of the substance of and rationale for nonuse values may be warranted. The concept of nonuse values arose in response to inquiries on possible differences between "social" and "private" values of resources such as unique National parks. Thus, following Weisbrod (1964), if user fees paid to the private owner of a park were insufficient to cover costs, the park would be closed. From a societal standpoint, such closure was argued to be nonoptimal given that individuals would be willing to pay to retain their "options" to use the park at some future date. Weisbrod (1964) correctly observed that a full social accounting of the benefits of deciding to keep the park open must "... recognize the existence of people who anticipate purchasing the commodity (visiting the park) at some time in the future, but who, in fact, never will purchase (visit) it... if these consumers behave as 'economic men' they will be willing to pay something for the option to consume the commodity in the future."

The "something" that consumers would be willing to pay is referred to as an "option" value. "Option" value and "nonuse" value were essentially used interchangeably until later writers posited other motivations, or arguments in an individual's utility function, which might give rise to a nonuse value. For example, a "bequest" value is said to reflect the desire to leave unsullied environments to future generations, and an "existence" value is said to represent values which one might place on simply knowing that a unique environmental resource exists, whether or not he/she visits or uses the resource.

The notion of an option demand, then, is simply the idea that environmental goods might enter positively into the utility function of prospective users. It was considerably enhanced by the seminal work of Krutilla (1967) concerning conservation issues. It is important, however, to recognize the restrictions imposed by Krutilla on the classes of environmental resources for which the notion of option demand would be most justifiable. His persuasive arguments for a general application of the option demand notion focused upon: unique attributes of nature--a geomorphologic feature such as the Grand Canyon; grand scenic wonders; a threatened species, or an ecosystem or biotic community essential to the survival of the threatened species; and natural environments which have no close substitutes (Krutilla, 1967).

Two related questions implied by this notion of a nonuse value have direct relevance for the "which households" question. The first, obvious question is: who are these individuals (which households) who could be expected to hold positive nonuse values for particular environmental resources? Krutilla's response was that they are individuals for whom a significant part of their welfare is the preservation and continued availability of grand scenic wonders or unique, fragile ecosystems. Krutilla (1967) identifies them as the "... spiritual descendants of John Muir, the present members of the Sierra Club, the Wilderness Society, National Wildlife Federation, Audubon Society, and others to whom the loss of a species or the disfigurement of a scenic area causes acute distress and a sense of genuine relative impoverishment." Note that there is no intention here to argue that every household in the United States holds an option/nonuse value for every environmental resource. The argument is that some households may hold an option/nonuse value for some environmental resources. Thus, it may be palatable to argue that a family in New York might indeed be motivated to "save" the Grand Canyon even if they have not visited it. But, absent strong supportive evidence, the credulity of this argument is stretched when the family in New York is supposed to be motivated to preserve, for instance, the Puerco River in Central New Mexico ("the what river, where?").

The second issue concerns the CVM response to the question of which households can be expected to hold nonuse values for particular environmental resources. The CVM response assumes that all households in the population hold a nonuse value for any (indeed, as is shown below, every) environmental good at issue in an application of the CVM. In this regard, consider the procedures used in any typical CVM study designed to estimate values for an environmental commodity. Suppose that 1,000 questionnaires are mailed, and only 200 are returned. These 200 households are then taken to be "representative" of the population (the State of Colorado, or the Western United States, or, in some cases, the entire United States). Estimates of the marginal willingness to pay for increments in the environmental good are then estimated as simple averages of the sampled responses, perhaps classified by certain household characteristics. These values, differentiated by household characteristics, are then assumed to apply to all households with the same demographic/income characteristics throughout the entire population.

The question may well be asked on what basis can the values offered by a few households who completed and returned the questionnaire be reasonably applied to all households in the population (ignoring the 800 households who did not respond). Within the context of Krutilla's arguments, this is tantamount to assuming that all households in the United States are members of the Sierra Club.

The rationale for this conclusion is as follows. Responding to a CVM questionnaire may be reasonably assumed to be costly for a randomly selected household ("costly" in terms of forgone leisure time). Accordingly, one would expect the active respondents to be a biased subsample of the sample of households originally sent the questionnaire. Only those households whose nonuse value exceeds the cost to them of responding will in fact (rationally) complete the survey. Those households whose nonuse value is close to zero will not perceive any net benefit in responding, will therefore not do so, and will be underrepresented in the subsample on which the CVM estimate is finally computed.

Herein lies the first methodological issue requiring attention if the CVM is to provide credible value estimates. On a conceptual basis, the idea that some individuals have nonuse values for an environmental commodity reflecting option-related motivations has intuitive appeal and stands on equal footing in economic theory with any other "commodity" assumed to be an argument in an individual's utility function. The issue here is not whether individuals might, conceptually, hold such values. The issue is: can they be reliably measured? Has the state of the art of the CVM advanced to the extent of identifying in a credible way those households in the population to which such a value can be attributed? In the absence of such an identification it is impossible to assess the size of the biases inherent in the voluntary CVM survey, although one can easily assess the direction of the biases (viz., to overstate nonuse values). The current practice of assigning such values to all households diminishes the meaning of such values for any context requiring precision or reliability.

## WHAT VALUE FOR WHICH NONMARKET GOODS?: THE "GOOD CAUSE" ISSUE

After many applications of the CVM, particularly when nonuse values account for a significant portion of the total value, one result becomes obvious: it doesn't matter what public good is being valued, the CVM will yield a nonuse value for the good or commodity of an amount that represents a relatively small part of the household income. When respondents are asked for an annual payment, the amount is typically around $30-$60/household/year. When subjects are asked for a monthly payment, the amount is typically around $5-$10/household/month. For example, if improvements in air quality at the Grand Canyon National Park are valued, responses will be $5-$10/month/household; if the CVM commodity is improvements in air quality in the Grand Canyon National Park and in 5 or 6 other specific National parks, responses will be $5-$10/month/household; if the CVM commodity is improvements in air quality in all National parks, responses will be $5-$10/month/household; if the CVM commodity is air quality throughout the Nation (not just in National parks) responses will be $5-10/month/household; or if the CVM commodity is improvements in both air and water quality throughout the Nation, responses will be $5-$10/month/household (see Burness and Cummings, 1983; Rae, 1983; and Schulze, Cummings, and Brookshire, 1983). How can a valuation process be rationalized that yields essentially the same value for (1) a "package" containing one environmental good, and (2) a "package" containing any number of environmental goods, including the good in the single package?

Regrettably, with the present state of the art in the CVM, no better response is available to this question than that households express a similar willingness to pay for any "good cause" stated in a CVM survey questionnaire. This anomaly is widely recognized by researchers who have been at the forefront of experimental efforts to develop the CVM. For example:

"An important concern in valuation studies is that surveys that value an environmental good... might be capturing more generalized values... There are many competing claims on our resources... charities, research, and other good causes... if twenty-five or more medical research or other good causes were lined up to receive a donation, most households would face a difficult decision." (Rowe, Dutton, and Chestnut, 1985, pp. 3-26.)

"The question as to whether CVM bids for a specific environmental improvement are disaggregative values or, in fact, more likely values associated with some broader, environment- (or good cause) related aggregative 'account' raises an issue of particular concern... no researcher would be willing to defend the summation of CVM values that have been obtained in various studies for many types of environmental effects..." (Schulze, Cummings, and Brookshire, 1983, p. 6.)

Scores of CVM studies exist, each of which is focused on a different nonmarket commodity (improvement or damage). Take, for example, 20 studies each suggesting that all households would pay $30-$60 per year for the commodity at issue in the particular CVM study. To seriously argue that households would pay this amount for each commodity would then imply that these values can be added. Household WTP (for 20 studies, although there are many more) would be assumed to range from $600 ($30 for each of the 20 commodities) to $1,200 ($60 for each of the 20 commodities) per year for these 20 commodities, plus amounts associated with any other CVM commodity which came under study.

This is clearly specious and, as suggested above, no researcher would be willing to defend such a summation (see, also, Mitchell and Carson, 1989, pp. 40-47), and thus emphasizes the importance of the question (which households place) *what* value on *which* environmental goods? As noted above, Schulze, Cummings, and Brookshire (1983) and others suggest a possible response to this dilemma: Values derived with the CVM reflect only that amount that households will offer to any worthwhile cause. Moreover, these values are to a large extent independent of the good (cause) being considered and are generally not additive.

## SOME RELEVANT EMPIRICAL EVIDENCE

Various aspects of the "which households, what value, which goods" question have been considered in both of the state-of-the-art reviews of the CVM which appear in the literature. Cummings, Brookshire, and Schulze (1986) touch on the good cause issue (what value, which goods), but their focus tends to be blurred by their preoccupation with the "accuracy" of CVM values. Mitchell and Carson (1989) consider the "what value, which goods" issue within the context of whether or not CVM values for specific environmental goods can be aggregated (summed). In this regard they conclude that "The basic implication... is that aggregation of independent estimates of WTP for each (environmental good) will not in general equal the correct WTP for all changes simultaneously" (pp. 45-46). It should be noted in passing that Mitchell and Carson's observations in this regard are not limited to public goods--they apply equally well to goods with market prices.

The "which households" issue may be considered within two contexts. The first, more general context asks what are the characteristics of households who select and value a particular environmental improvement (or a particular set of environmental improvements), when faced with a comprehensive set of possible environmental improvements (good causes). A related question would involve the issue of nonrespondents in a CVM survey: are values derived with the CVM appropriately representative of households who are nonrespondents? Mitchell and Carson acknowledge and address this second version of the "which households" question (Mitchell and Carson, 1989), and suggest two approaches for responding to it.

Notwithstanding the profession's recognition of the "which households, what value, which goods" problem, there is little in the way of empirical efforts designed to resolve the problem. One study which focuses on the nonrespondent context of the "which household" question by Edwards and Anderson (1987) tested for selection biases and nonresponse biases in a coastal water quality study. They employed a model of self-selection that is drawn directly from Heckman's (1979) work and employed his two-stage "Heckman's lamda" correction method and test for selectivity bias. Unfortunately, while this study suggests an innovative use of the Heckman correction methodology for measuring nonresponse bias, the analyses are flawed by the authors' apparent misinterpretation of the implications of the technique. They incorrectly conclude that while the inclusion of the inverse Mills ratio (lamda) provides a test for biases induced by zero and missing bids, it does not allow for a test of nonresponse bias. Perhaps this reflects some confusion on their part between the definitions of self-selection and nonresponse bias (for example, see Mitchell and Carson, 1986). Edwards and Anderson do not find the correction for self-selection to be statistically different from zero, but do argue that there exists nonresponse biases. Until the difference between these two biases (if any exists) is resolved, this result provides mixed evidence on the ability to assign CVM values to the population of households that the response sample is assumed to represent.

In terms of the "what value, which goods" question, two studies address the question within a "good cause" context. The first of these is the study by Schulze, Cummings, and Brookshire (SCB), 1983. Unfortunately, while this study asserts a test of the good cause hypothesis, the test used is based upon a misinterpretation of data and is therefore flawed. Experiment 1 of several conducted for the study involved CVM values from two sets of subjects for two commodity "bundles." One set of subjects valued the commodity $Z_1$, improved visibility in the Grand Canyon National Park. This value is denoted here as $V^1(Z_1)$. SCB asserted that a second set of subjects valued a bundle consisting of <u>two</u> commodities: $Z_1$, improved visibility in the Grand Canyon National Park, and $Z_2$, improved visibility in five <u>other</u> National Parks, denoted here as $V^2(Z_1,Z_2)$. Close examination of their experiment and questionnaires reveals that subjects in experiment 2 actually reported two <u>separate</u> values, one for improved visibility in the Grand Canyon and another for improved visibility in five other parks. Subjects do <u>not</u> value the joint commodity $(Z_1,Z_2)$; thus, the value reported as $V^2(Z_1,Z_2)$ is the SCB study should in fact be $V^2(Z_1)$, with $V^2(Z_2)$ not reported. SCB then stated that a test of the good cause hypothesis would take the form:

$H^o$: $V^2(Z_1,Z_2) = V^1(Z_1)$

with the alternative hypothesis

$H_a$: $V^2(Z_1,Z_2) > V^1(Z_1)$.

SCB found no statistical difference between the values and accepted the hypothesis $H^o$.

With the misinterpretation of values described above, this test is of course meaningless. What they actually tested, and show, is that $V^1(Z_1) = V^2(Z_1)$. That is, the average valuation of improved visibility at the Grand Canyon is not altered when respondents are subsequently asked to separately value increased visibility at five other parks. This does not provide a test of the good cause hypothesis.

The second study of direct relevance to the "what value, which goods" question is one by Kahneman and Knetsch (KK), 1990. KK set out to investigate what they considered the most serious shortcoming of the CVM: "... that the assessed value of a public good is demonstrably arbitrary, because WTP for the same good can vary over a wide range depending on whether the good is assessed on its own or embedded as part of a more inclusive package." Indeed, KK argue that for a potentially large class of environmental goods the value derived with the CVM is best interpreted as the purchase of "moral satisfaction," rather than as a value associated with the specific good valued with the CVM. It can be argued that this is little more than a restatement of the good cause response to the "what value, which goods" question.

The concern here is with those parts of the KK study which test hypotheses related to the "what value, which goods" question. KK cast this issue in the context of the question "to what extent are values for one particular good embedded in subject valuations for more general environmental goods which include the one particular good?" To address this question, they structured the following three groups of goods for which CVM values are obtained from telephone interviews with three different groups of individuals. KK's telephone script, paraphrased as follows, makes use of the notation introduced above; descriptions in brackets [ ] represent a summary of what respondents were seemingly asked to remember and value. Individuals in each group were given different information sets. Individuals in group 1 were given the information described below as set 1; individuals in group 2 were given information set 2; and individuals in group 3 were given information set 3. Superscripts on V below denote the groups (1, 2, and 3) from which values are derived.

## Information Set 1

a. Respondents were told that Federal and provincial governments provide a wide range of services which include education (E), health (H), police protection (P), roads (R), and environmental services (Z): [Government provides services S: {E,H,P,R,Z,......}].

b. Environmental services, Z in S, include: preserving wilderness areas ($Z_7$), protecting wildlife ($Z_6$), providing parks ($Z_5$), controlling air pollution ($Z_4$), ensuring water quality ($Z_3$), routine treatment and disposal of industrial wastes ($Z_2$), and preparing for disasters ($Z_1$). How much would you pay in higher taxes to improve Z? [Z = {$Z_7,Z_6,Z_5,Z_4,Z_3,Z_2,Z_1$,......}; what is V(Z) = V($Z_7,Z_6,Z_5,Z_4,Z_3,Z_2,Z_1$,......)].

c. Keeping in mind services in Z, including $Z_7,Z_6,Z_5,Z_4$, and $Z_2$, what part of V(Z) should go to $Z_1$? [Either Z = {$Z_7,Z_6,Z_5,Z_4,Z_2$,....}, or Z = {$Z_7,Z_6,Z_5,Z_4,Z_3,Z_2,Z_1$,......}, what part of V(Z) is V($Z_1$)].

d. Preparedness for disasters includes: Emergency services in hospitals ($Z_{14}$), maintenance of large stocks of medical supplies, food, fuel, and communication equipment ($Z_{13}$), preparing for cleanup of oil, toxic chemicals, or radioactive materials ($Z_{12}$), and ensuring the availability of equipment and trained personnel for rescue operations ($Z_{11}$). [$Z_1 = \{Z_{14}, Z_{13}, Z_{12}, Z_{11}, ...\}$, what part of $V(Z_1)$ should be $V(Z_{11})$].

## Information Set 2

c'. Consider preparedness for disasters ($Z_1$), such as earthquakes, floods, hurricanes, tornados, etc. What is $V^2(Z_1)$?

d'. Same as d in set 1.

## Information Set 3

d''. Consider the availability of equipment and trained personnel for rescue operations in the event of a natural disaster. What is $V^3(Z_{11})$?

KK then tested the following hypothesis:

H°: $V^1(Z) = V^3(Z_{11})$

Using median values to avoid distortions from outlying values, KK found no significant difference between the median WTP value for $V^1(Z)$ and $V^3(Z_{11})$ using a nonparametric Mann-Whitney test (KK, p. 11), and conclude that this provides strong evidence of embeddedness.

There are at least two problems with KK's test. First, and perhaps most obviously, it is difficult to see how participants in the telephone survey could be expected to mentally keep track of all of the components and alternatives relevant for the choices evaluated; this is particularly the case for group 1 which received (a) through (d).

Secondly, KK's data yield the following relationships.

(1) $V^1(Z) = V^2(Z_1) = V^3(Z_{11})$

(2) $V^1(Z_1) < V^2(Z_1)$

(3) $V^1(Z_{11}) < V^2(Z_{11}) < V^3(Z_{11})$

(4) $V^1(Z_{11}) < V^1(Z_1) < V^1(Z)$

These relationships appear to be the basis for KK's conclusion that CVM values are arbitrary. The term "arbitrary" suggests capriciousness, which does not seem to be demonstrated by the data for the following reasons. First, relationship (4) is certainly consistent with received theory. Secondly, when individuals are provided information on value choices unfamiliar to them the information provided may be reviewed as limiting the individual's choice set; thus, the sets of information provided to KK's three groups can be considered as being different choice sets. Third, and finally, KK's relationship (1) is consistent with the good cause, or mental accounts, notion(s). It can be argued, however, that while this outcome would not be predicted a priori by standard economic theory, it is consistent with received theory in the sense that it does not violate standard assumptions underlying the theory of choice.

## WHICH HOUSEHOLDS, WHAT VALUE, WHICH GOODS?: SUGGESTIONS FOR FUTURE RESEARCH

This section briefly describes the implications of earlier discussions on the "which households, what value, which goods" question for future research. The objective here is simply to provide a sense of the direction that experimental CVM research must go if the challenges posed by this question are to be met.

Beginning with the nonrespondent problem as relevant to the "which households" issue, the approach suggested by Edwards and Anderson (1987) provides an interesting point of departure. Survey nonresponse can be caused by factors related to the issues in question, either directly or indirectly, or by completely unrelated factors. Problems of bias are introduced unless nonresponse is purely random, in which case estimates derived from survey responses may have greater variance, but no bias. The direction for future research to address these issues would be to model the decision to respond, or not respond, to a survey as having explicit structure. This may involve a maximizing or a strategic model of behavior. A procedure that is based on maximizing behavior, and which relies on the statistical constructs contained in the selectivity bias literature, exhibits potential beyond that realized in the Edwards and Anderson paper.

A brief discussion of the estimation procedures commonly used in applications of the CVM may help in understanding the nature and effect of this correction method. Individual characteristics are obtained at the same time as valuations are elicited. These valuations are regressed upon a set of explanatory variables resulting in an estimated relationship between observable characteristics and valuations. The regression model used assumes that the dependent variable is a function of both observed and unobserved variables.

Obviously, however, the sample of respondents may differ from the population as a whole with respect to observed and/or unobserved variables. So long as the distribution of the unobserved influences in the sample is the same as that of the population, the standard regression equation can be used to predict the expected valuation of the good for any set of values for the explanatory variables, be they sample mean values, population mean values, or specific individual values.

Correction is required if the error distribution of the estimated equation, determined by the distribution of the unobservables in the sample, is systematically related to some set of variables, which may include some, or all, of the observable variables influencing the valuation. The correction procedure attempts to model this explicitly, assuming that there is a process by which people choose to respond and that this process and the valuation of the good may not be independent. An equation explaining the decision to respond (a simple binary dependent variable) as a function of a set of observable characteristics and a random error term is estimated simultaneously with the valuation equation, thereby allowing for any correlation between the error terms in both equations.

The procedure requires information about nonrespondents which, by definition, is not available from the CVM. Several alternatives exist for obtaining these needed data. Edwards and Anderson obtained this information by following up nonrespondents with a telephone survey. An alternative source is a parallel survey, using a sample frame selected on the same basis as the CVM. The parallel survey collects only demographic and attitudinal data--the same demographic and attitudinal data included in the (typical) information section of the CVM survey. Even though this survey will suffer from nonresponse also, the nonresponse will not be related to the issues of concern in the CVM. Other possible sources for information on nonrespondents may be independent surveys such as the Census of the Current Population Survey (CPS).

If the propensity of people to respond to CVM-type surveys can be systematically predicted by socioeconomic characteristics of respondents and nonrespondents, one can then determine the rule used by individuals in deciding whether or not to participate in the CVM. The method provides a weak test of the hypothesis that participation and valuation are independent and can be used to adjust estimates derived from the CVM as if they were obtained from a complete sampling of the population, not just those who replied to the survey.

Nonresponse may, however, be just a part--and possibly a very small part-- of the "which households" question; indeed, the usefulness of separating it from the more general "which households, what value, which goods" issue is questionable. Research to date appears to make clear ("clear" in the sense that there is generally no compelling contradictory evidence) that CVM surveys for one specific commodity yield values which cannot be defensibly attributed to that specific commodity. This follows from, among other considerations, the argument that these single-good values cannot be added. KK demonstrate that the value of, say, Good A derived from disaggregating a value for three goods, A, B, and C, will differ dramatically from the value derived when only A is valued. Tolley and Randall (1983) have demonstrated that the value of A derived from aggregating values from Good A, then for Goods A and B, then for Goods A, B, and C will result in different values for A, depending in large part upon where, in the sequence, the Good A is introduced (the value of A when A is introduced first in the sequence differs from the value of A when it is introduced second, or third, in the sequence).

At this point in time, two lines of inquiry, can be identified which might be productive in addressing these issues. The first approach is to simply present households in the CV survey with a wide range of possible environmental improvements (or a wide range of public goods). Households are asked to state their WTP for any or all of the alternatives presented to them. The potential strength of this approach is that it can avoid the problem caused by what can be called the "targeting" of commodities. In virtually all applications of the CVM, respondents are essentially told what commodities are "important"--they are important inasmuch as they are the only commodities mentioned in the questionnaire. As one example, KK explicitly instruct participants to focus on a subset of environmental goods. To argue that the amount of residual income offered for good causes is fixed, is not to argue that the amount of that residual which the individual would attribute to a specific commodity cannot be determined. This point is made clear by a comparison of the following structures for the choice question: (i) Here is a set of commodities, what is the maximum amount that you would pay for commodity i; (ii) here is a set of commodities; on which of these commodities would you place a value, and what value would you assign to those valued by you? The commodity of concern may not be chosen, or may be chosen by few subjects. But ceteris paribus, it can be asserted with some confidence that values associated with a specific commodity derive from: First, the individual assigning priority to the commodity; and secondly, the individual's assignment of value to the commodity within a context consistent with standard utility maximization behavior.

The obvious weakness of this approach is the lack of an objective means for identifying the public goods to be included in--and, of course, excluded from--the set presented to CVM subjects. It may be sanguine to assume that values derived for particular components in a given set of public goods would remain unchanged with in introduction of "other" public goods, or the substitution of one or more of those included in the original set; this remains as an empirical question.

A second line of inquiry for the "which households, what value, which goods" question which should be mentioned is admittedly speculative. This approach involves moving away from methods based upon choices resulting from individual maximizing behavior to those which focus upon social choice mechanism. That is, eliciting from individuals a valuation (willingness to pay/ accept) that is representative of the relevant group's valuation in a context where allocation and compensation decisions are made by group, or social choice, rules. There are a number of mechanisms that do not necessarily collapse under Arrow's Possibilities Theorem. For example, the choice elicitation procedures introduced by Becker, DeGroot, and Marschak (BDM), 1964 can be adapted for the context of a group decision. An uncertain prospect--one in which subjects have property rights which they are asked to surrender--is identified, and subjects are asked to determine a group selling price (or, when more of the good is acquired, an "offer" price). In operational terms, subjects may be asked to think of a process wherein a group or government is negotiating with other agents to determine their buying/selling price. BDM's incentive property does not require regularity on the probabilities of buyouts (or sellouts), simply that every possible buyout price has some positive probability of occurring. Payoff dominance is avoided by keeping the probability of each possible buyout price reasonably large, thus ensuring that selling price alternatives appear credible to subjects. The potential for adapting the CVM to social choice contexts is problematical at this point. This possibility is offered here as an alternative for dealing with the "which households, what value, which goods" question which has yet to be explored.

## ACKNOWLEDGEMENTS

The authors are grateful to the Waste Education Research Consortium of the U.S. Department of Energy for funding. Parts of this paper draw upon work by Cummings which was financed by attorneys for the Idarado Mining Company.

## REFERENCES

Becker, G. M.; DeGroot, M. H.; and Marschak, J., 1964. Measuring Utility by a Single-Response Sequential Method, *Behavioral Science*, 9, pp. 226-232.

Bentkover, J. D.; Covello, V. T.; and Mumpower, J., 1986. *Benefit Assessment: The State of the Art*. D. Reidel Publishing Co., Boston, MA.

Burness, H. S. and Cummings, R. G., 1983. Valuing Alternatives for Managing Hazardous Wastes. In: Schulze, W. D.; Cummings, R. G.; and Brookshire, D. S., *Methods Development in Measuring Benefits of Environmental Improvements*, Vol. II, Report to the U.S. Environmental Protection Agency, No. CR808-893-01.

Cummings, R. G.; Brookshire, D. S.; and Schulze, W. D., 1986. *Valuing Environmental Goods: An Assessment of the Contingent Valuation Method*. Roman and Allanheld, Totowa, NJ.

Edwards, S. F. and Anderson, G. D., 1987. Overlooked Biases in Contingent Valuation Surveys: Some Considerations, *Land Economics*, 62(2), pp. 168-178.

Kahneman, D. and Knetsch, J. L., 1990. Valuing Public Goods: The Purchase of Moral Satisfaction, *Journal of Economics and Environmental Management*. Forthcoming.

Krutilla, J. V., 1967. Conservation Reconsidered, *American Economic Review*, 57, pp. 777-786.

Mitchell, R. C. and Carson, R. T. 1986. *Using Surveys to Value Public Goods--The Contingent Valuation Method.* Resources for the Future, Washington, DC.

Phillips, C. V. and Zeckhauser, R. J., 1989. Contingent Valuation of Damage to Natural Resources: How Accurate? How Appropriate? *Toxics Law Reporter*, pp. 520-529.

Rae, D., 1983. The Value to Visitors of Improving Visibility at Mesa Verde and Great Smoky National Parks. In: Rowe, R. D. and Chestnut, L. G. (Eds.), *Managing Air Quality and Scenic Resources at National Parks and Wilderness Areas,* Westview Press, Boulder, CO.

Rowe, R. D.; Dutton, R. A.; and Chestnut, L. G., 1985. *The Value of Ground Water Protection.* Unpublished Manuscript, Energy and Resource Consultants, Inc., Boulder, CO.

Schulze, W. D.; Cummings, R. G.; and Brookshire, D. S., 1983. *Methods Development in Measuring Benefits of Environmental Improvements*, Vol. II, Report to the U.S. Environmental Protection Agency, No. CR808-893-01.

Tolley, G. S. and Randall A. et al., 1983. *Establishing and Valuing the Effects of Improved Visibility in the Eastern United States.* Interim Report to the U.S. Environmental Protection Agency (Final Report, 1985).

Weisbrod, B. A., 1964. Collective Consumption Services of Individual Consumption Goods, *Quarterly Journal of Economics*, 77, pp. 471-477.

# 23 ECONOMIC VALUE OF WILDLIFE RESOURCES IN THE SAN JOAQUIN VALLEY: HUNTING AND VIEWING VALUES

Joseph Cooper and John Loomis,
University of California, Davis

### ABSTRACT

This chapter quantifies the effects of agricultural drainage on the recreational demand for wildlife resources in the San Joaquin Valley (Valley). The current value of waterfowl hunting is $3.2 million annually at public refuges and $16.5 million for the entire Valley. The value of viewing birds in the Valley is $64.7 million annually. An estimate of the change in waterfowl hunting benefits at Kesterson National Wildlife Refuge (NWR) resulting from control of agricultural drainage water is made by combining information on wildlife response to selenium with a quality differentiated demand equation for waterfowl hunting. This simulation illustrates how a bioeconomic analysis of waterfowl hunting benefits from reducing wildlife contamination can be performed.

### INTRODUCTION

Wildlife have a variety of values. They have ecological values for other species of animals and plants and for the communities in which they dwell. They also have ecological and economic values for society at large. Recreational demand for wildlife is often the largest portion of the total economic value of wildlife. This chapter examines the effects of agricultural drainage as a wetland water supply on the recreational demand for wildlife resources in the Valley. An estimate of the change in waterfowl hunting benefits resulting from a change in the level of drainage is made by combining the methodology and results of studies done on the recreational demand for Valley wildlife with what is known about the effects of drainage on the wildlife populations.

This study focuses on the onsite recreational demand for wildlife. Waterfowl hunting and bird viewing are the primary onsite recreational uses of Valley wildlife that are affected by agricultural drainage and which have been examined in an economic context. The economic value to society from preserving wildlife in the Valley has recently been performed by Loomis et al. in another chapter of this volume.

## CONCEPTS OF ECONOMIC VALUE OF WILDLIFE

Unlike most commercial goods, wildlife species in the Valley are largely nonmarket environmental resources. Hence, the economic value of the wildlife is not readily apparent. A nonmarket good, as opposed to a market good, is one which is not readily traded on the open market. The dollar value of a market good, a packaged frozen fryer for example, is readily determined: it is the dollar value - determined through the interaction of the forces of supply and demand - charged for the good in the local grocery store. On the other hand, waterfowl taken in a wildlife refuge are nonmarket goods. The payment the hunter must make to use a public area is the cost of the hunting application. This payment, or fee, is administratively set by the Government, frequently with little consideration for the interaction of supply and demand for the animal. Hence, the fee is not a market-clearing price (i.e., the price at which the supply equals the demand for the good). To make an estimate of what the market-clearing price would be if there was a market for the good, some sort of nonmarket resource valuation techniques are needed. Possible techniques to measure economic values of recreation include the Hedonic Price Approach, the Contingent Valuation Method (CVM), and the Travel Cost Method (TCM).

In addition to economic values of onsite recreation use and commercial uses of wildlife, there are many offsite user values. These include option, existence, and bequest values, all of which can be held by the general population as well as recreationists. Option value can be thought of as an insurance premium people would pay to insure availability of wildlife recreation opportunities in the future. Existence value is the economic benefit received from simply knowing wildlife exist. Bequest value is the willingness to pay (WTP) for providing wildlife resources to future generations. While option and existence values may be present for manufactured consumer goods, Randall and Stoll (1983) claim those values are likely to be empirically insignificant in size compared with the value of certain scarce wildlife species. Since the focus of this chapter is recreation, offsite values are not quantified here. However, offsite values can be estimated through survey techniques such as the CVM, as discussed for the Valley in the chapter by Loomis et al.

To clarify the discussion of the economic benefits of hunting and viewing wildlife, several terms will be defined in this section.

Economic value is measured in terms of the consumer's net willingness to pay. Consumer surplus is the economist's term for the consumer's net willingness to pay, which is the maximum increase in price above current costs a person would be willing to pay to purchase a good or service. Consumer surplus represents the consumer's additional (net) willingness to pay for the opportunity to, for example, hunt at some specific site. It is net or additional willingness to pay beyond current expenditures. Examples of a "good or service" as related to wildlife would be a waterfowl hunting trip or the experience of viewing wild birds. Total or gross willingness to pay is the sum of net willingness to pay and the amount actually spent on the good. Since the amount actually spent is part of the cost of participation, the benefits (i.e., the net willingness to pay) are just the amount in excess of what people spend.

It has been suggested (U.S. Water Resources Council, 1983 and U.S. Department of the Interior, 1986) that economic values lost to society be measured in terms of net willingness to pay in assessing natural resource damage and mitigation measures. The net willingness-to-pay criteria has also been broadly used in textbooks on Benefit Cost Analyses (Sassone and Schaffer, 1978; Just, Hueth, and Schmitz, 1982).

The CVM and the TCM are the dominant methodologies used in estimating the recreational value of wildlife resources. CVM is sometimes referred to as "the bidding method." In essence, a hypothetical but realistic market is established for some type of nonmarket good, say a recreational trip to a particular site. In open-ended questions, the respondent is asked to specify the maximum amount he or she would pay for that trip, including access and use fees. In close-ended (or dichotomous choice) questions, the respondent is asked whether he or she would pay some amount stated in the question. This dollar amount varies from individual to individual. By evaluating the probability of the respondent stating "Yes I would pay the [specific dollar amount]," an expected value of willingness to pay can be computed. For a thorough discussion of the strengths and weaknesses of the CVM see Schulze et al. (1981) or Cummings et al. (1986, and this volume).

Research on the accuracy of CVM has been performed by Welsh (1986), who bought and sold 1-day deer hunting tags for the Sandhill Demonstration area. Two parallel markets were established: (1) A real market where the hunters surveyed could actually buy the deer tags for real money and (2) a CVM survey of hunters, identical to (1) with the exception that no cash changed hands. Comparison of the results from the two markets showed that CVM yielded a value 25 percent higher than the actual cash value of the deer hunting tag.

Loomis (1989) tested the reliability of the CVM using the test-retest approach. In surveys of both visitors and the general public, willingness to pay responses were not statistically different between the first survey and the second survey 9 months later.

The TCM statistically traces out a demand equation, using observations of travel distance as a measure of price and number of trips taken as a measure of quantity. The resulting first stage, or per capita demand equation allows the calculation of the additional amount a recreationist would pay over travel costs (i.e., consumer surplus) to have access to a particular wildlife site for viewing, hunting, or fishing. This calculation is made using a "second stage," or site, demand curve that relates added distance or added travel cost of, for example, trips to a particular hunting area. See Clawson and Knetsch (1966); Dwyer, Kelly, and Bowes (1977); Sorg and Loomis (1985); or Ward and Loomis (1986) for a discussion of the basic TCM approach.

## IMPORTANCE OF WILDLIFE RESOURCES IN THE SAN JOAQUIN VALLEY

### Waterfowl Hunting Statistics for the Valley

Table 1 presents waterfowl hunting statistics for California counties in the Valley for the years 1983 through 1985. Waterfowl considered include ducks, geese, and coots, which together form the vast majority of the waterfowl species hunted in the Valley. The table aggregates California Department of Fish and Game (DFG) data (various years) for San Joaquin, Stanislaus, Merced, Madera, Fresno, Kings, Tulare, and Kern Counties, which comprise the geographical area of the Valley.

Table 1. Waterfowl hunting use for the San Joaquin Valley (includes hunting on both public and private lands).

| Year | Take | Hunters | Hunter-days |
|---|---|---|---|
| 1985 | 468,508 | 37,779 | 265,727 |
| 1984 | 445,184 | 40,212 | 255,816 |
| 1983 | 567,226 | 39,100 | 308,016 |
| 1982 | 501,688 | 36,603 | NA |

Reports for years prior to 1983 do not present the hunter-days by county, and only 1982 is included here. Although the table exhibits a great deal of variability, hunter take and hunter days have exhibited a downward trend over the last few years. By the October 1987 through January 1988 hunting season (not included in table 1), only 27,603 hunters were recorded at the seven public refuges in the Valley.

## Wildlife Viewing Statistics

A mail survey of 3,000 randomly selected California households conducted in 1987 provided the data for the analysis. The survey was conducted by the University of California, Davis, using a population-weighted sample drawn by a professional survey research firm (Survey Sampling, Inc.). The survey asked questions about viewing birds and deer in California. After deleting the undeliverable questionnaires, the overall response rate to this survey was 44 percent. While a higher response rate would have been desirable, this response rate is acceptable and believed to be representative of Californians with focused interest in wildlife. It is equal to or greater than the response rates for similar CVM surveys conducted in California over the last few years.

The respondents, each representing a California household, were requested to answer questions on whether they saw any wild birds on any outdoor recreation trips during the 12 months prior to the date of the survey. Table 2 presents the summary statistics on all outdoor recreation trips taken in the Valley during 1987 for the primary purpose of viewing birds. Table 2 also presents the summary statistics for general purpose outdoor recreation trips taken in the Valley during 1987 in which the respondents viewed birds. This category includes the data on trips both for the primary purpose of viewing birds plus trips for all other recreational pursuits.

Table 2 is organized by county. Sample size is the total number of surveys returned specifying Valley counties as the trip destination. The table presents the total number of trips to each county and the sum of all trips to the region. Note that for primary purpose trips, data were available for only the Valley counties of San Joaquin, Merced, and Fresno.

Because only a percentage of the total California households were sampled, the data on trips must be expanded to an equivalent Statewide use level. The estimated total trips in table 2 expands the trips per region to account for the difference between the actual population size of California and the size of the sample.

Table 2. Estimated total general purpose recreational trips during which wild birds were seen and estimated total trips for the primary purpose of viewing birds by Californians in 1987.

| | Primary purpose trips | Sample size | General purpose trips | Sample size |
|---|---|---|---|---|
| San Joaquin | 23,430 | 3 | 487,344 | 8 |
| Stanislaus | | | 159,324 | 8 |
| Merced | 166,353 | 4 | 185,097 | 5 |
| Madera | | | 107,778 | 7 |
| Fresno | 23,430 | 2 | 149,952 | 13 |
| Kings | | | 65,604 | 5 |
| Tulare | | | 318,648 | 34 |
| Kern | | | 260,073 | 15 |
| Total | 213,213 | 9 | 1,733,820 | 95 |

*Data source:* Nonconsumptive Wildlife Use Bird Survey.
*Note:* Primary purpose trips data are available for only three Valley counties.

## APPLICATION TO SAN JOAQUIN VALLEY

### Benefits of Bird Viewing

The same survey used to estimate the number of bird viewing trips to the Valley was also used to estimate the willingness to pay for the experience of viewing wild birds. Close-ended CVM questions were used to estimate trip values.

Specifically, the average cost and the maximum willingness to pay for the most recent trip were estimated for all Californians. The respondents were asked: (1) What their approximate costs were for transportation, food, and lodging on their most recent trip when they saw wild birds and (2) if their annual expenses where $X higher, would they still visit that site?

Unfortunately, the sample for the Central and San Joaquin Valley counties is so small that no reliable inferences can be made about the value of viewing birds in this specific California region. Hence, the overall results for California must serve as a proxy for the San Joaquin Valley values.

The specific question asked is, "If your annual cost of visiting just this area [the area of the most recent trip where wild birds were seen] increased by $X would you still visit the site?" The $X amount is the bid amount written into each survey.

With close-ended willingness-to-pay questions, the calculation of expected willingness-to-pay is a two-step process. In the first step, a logistic regression, which is equivalent to an inverse demand function, is estimated with probability of a "Yes would pay $X" response as the dependent variable and the amount ($X) as the independent variable. The logit model is an econometric model in which the statistical equation has a limited dependent variable, i.e., the dependent, or left-hand side variable, consists only of zeros and ones. If a "yes" is assigned a value of 1 and a "no" a value of 0, the logit model can be used to perform regressions on willingness to pay questions that require a dichotomous "yes" or "no" answer. Once this logit curve is estimated, the area under that curve, which is the expected willingness-to-pay, is calculated. The area under a logit regression function is estimated by integration of the function. The vertical axis of this two-dimensional area is the probability that a particular increase in trip cost would be paid by the respondent.

The constant and the slope, or log of the bid amount, are entered in an integration program, which then calculates the expected value, or average willingness-to-pay for a trip, under each of three conditions: (1) Current conditions; (2) 1.5 times more birds seen than under the current conditions; and (3) twice as many birds seen.

The estimated model is

[1]  BCRPAY = f (BID, BIFL, INC, BSEEN, TRIPS)

where: BCRPAY is the dichotomous answer.

BCRPAY =  O no, will not visit the site.
          1 yes, will visit the site.

BID is the dollar amount of increased annual trip cost the outdoor recreationist was asked to pay to visit the most recent site visited where wild birds were seen.

INC is the recreationist's annual household income ($).

TRIPS is the number of recreational trips to the most recent area visited where wild birds were seen.

BSEEN is the number of birds seen during the receationist's most recent trip.

BIFL is the influence that the potential of seeing wildbirds at a site has on the choice of what sites to visit.

Table 3. Logit equation for maintenance of current conditions.

| Variable[a] | Coefficient | t-Statistic | Mean All[b] |
|---|---|---|---|
| Constant | -1.4734 | -0.95023 | 1 |
| Bid | -8.8507 | -8.1408 | $30.22 |
| Bird influence | 0.7495 | 2.7146 | 1.41 |
| Income | 0.4039 | 2.8310 | $36,791.00 |
| Birds seen | 0.2926 | 3.2388 | 28.43 |
| Trips | -0.03616 | -0.3425 | 3.04 |

Note: 370 Cases where BCRPAY = 1; 163 Cases where BCRPAY = 0
[a] Note: the independent variables are in natural log form in the regression.
[b] Note: means are of the untransformed variables.

Table 4 presents the willingness-to-pay estimates for the three potential levels of bird viewing. Since the number of birds seen was found to be positively related to willingness-to-pay, it is possible to calculate how WTP changes if the number of birds to be seen was increased. As the results indicate, the respondents are willing to pay more to see more birds.

As shown in table 4, trip benefits (economic values) do increase with the number of birds to be seen. However, the principle of diminishing marginal returns is evident here: each additional bird seen adds less additional enjoyment than the previous bird seen. For example, trip enjoyment increases by approximately $0.50 per additional bird seen up to a 50-percent increase and then about $0.20 more per bird up to double the population (100 percent more birds). Since each bird seen is a public good available for all the visitors to view, if there are 1,000 visitors a day viewing birds over a 10- to 20-day period, the aggregate benefits of additional birds could be several thousand dollars.

Table 4. Willingness to pay estimates for viewing birds in California under three different scenarios.

| | Annual Total WTP | Avg. No. of Trips Per Year | Net WTP Per Trip | No. Birds Seen Per Trip |
|---|---|---|---|---|
| Current conditions | $112.00 | 3 | $37.33 | 28 |
| 50% more birds | 135.00 | 3 | 45.00 | 42 |
| 100% more birds | 140.00 | 3 | 46.67 | 56 |

The estimated value of viewing birds is based on the total number of recreational trips in the Valley in which birds were seen. This number is 1,733,820 trips (table 2); total value per trip is $37.33. The total annual value for bird viewing in the Valley is then $64,723,500. Since data on trips for the primary purpose of viewing birds exist for only three Valley counties, an estimate of the total annual value of Valley trips for the primary purpose of viewing birds cannot be made.

## Demand for and Benefits of Waterfowl Hunting

Using hunter application data, TCM demand curves were estimated and net willingness-to-pay calculated for waterfowl hunting in Valley refuges for the 1987-88 hunting season. During the October 1987 through January 1988 hunting season, 27,603 hunters visited these seven refuges.

To estimate the demand for waterfowl hunting, a variation of the usual TCM model was estimated. The traditional TCM demand equation uses trips per capita from a given zone (e.g., county) of origin to a particular site as its dependent variable. However, one of the assumptions of TCM is that all recreationists at any given distance are able to visit as frequently as they desire. That is, observed visitation rates are supposed to reflect the desired level of consumption given the travel cost facing the hunter (Dwyer, Kelly, and Bowes, 1977). However, in the case of waterfowl hunting in the Central Valley and San Joaquin Valley refuges, there is excess demand for permits. As a result not all hunters desiring to go waterfowl hunting in the refuges at the current permit and travel price are allowed to do so. The excess demand is rationed by the California DFG by means of a lottery. As an approach to account for the real, underlying demand (rather than just that portion of demand actually realized as an outcome of the lottery) applications per capita is used rather than trips per capita as the dependent variable. Applications reflect the participation level that waterfowl hunters desire at current permit and travel prices. Thus, use of applications meets the assumptions of the TCM whereas trips, in this case, would not. For more details see Loomis, 1982.

In addition to the seven Valley refuges (listed in table 5), the data set for the TCM regression included five Central Valley refuges (Colusa, Delevan, Gray Lodge, Sacramento, Sutter). The estimated model is :

[2] $\ln(APPLICATONS_{ij}/POP_i) = -24.277 - 1.406[\ln(TWOWYDIST_{ij})]$
     (-7.77)  (-15.25)

  $+ 0.235[\ln(HVST_j)] + 0.733[\ln(AVINCOME_i)] + 1.301[\ln(WATER_j)]$,
   (2.53)               (2.33)                    (9.96)

$R^2 = 0.607$, $F = 102.23$, observations = 270.

Numbers in parenthesis are t-values

where,

APPLICATIONS/POP is the per-capita number of applications

TWOWYDIST is the two-way trip distance from the hunter's resident county i to the refuge. This variable is the price (in terms of distance traveled) of visiting a refuge.

HVST is the average of the monthly total waterfowl harvest in the previous season, i.e., in the 1986 season.

AVINCOME is average hunter income.

WATER is total water supplied (acre-feet) to the refuge's wetlands during the hunting season. This variable is a proxy for the amount of waterfowl habitat at a refuge.

The $R^2$ is quite high for a cross-sectional TCM regression. In addition, all the coefficients are of the expected sign, and all are significant at the 5-percent level or higher.

The equation was estimated in the double-log form for a number of reasons. The most important reason for choosing a log model is that past research has shown that taking the natural log of the applications per capita minimizes two problems that arise with a linear model. First, with the log model the possibility of predicting negative applications per capita from distant counties is eliminated. Second, heteroskedasticity associated with zones of different population sizes is minimized using the log of the dependent variable (Strong, 1983 and Vaughan et al., 1982). The double-log model was selected over the semilog form as it provided a better statistical fit to the data. Because the model is estimated in double-log form, the coefficients are elasticities. Except for the coefficients on distance and total water use, all the elasticities are inelastic.

From the per-capita demand equation, each site's second stage demand curve was calculated. Because the price variable in the per-capita demand equations is scaled in terms of miles instead of dollars, the area under the second stage demand curve represents willingness to "pay" by traveling additional miles. In order to calculate net economic values in dollars, the hunter's additional willingness to "pay" by traveling additional miles must be converted to willingness-to-pay in dollars. This involves multiplying the added distance by a cost per mile. This travel cost per mile is the sum of two components: vehicle operating cost per mile and value of travel time.

Converting the added willingness-to-pay from miles into dollars follows the approach suggested in the U.S. Water Resource Council procedures (1979, 1983) of using (1) one-third the wage rate as the opportunity cost of travel time and (2) variable automobile costs. For a midsize vehicle, the variable transportation costs per mile is $.172 for fuel and repair costs (Hertz, 1986). To account for the likelihood that there is more than one hunter per vehicle and that each hunter in the vehicle will pay his or her share of the vehicle operating costs, the $.172 per vehicle-mile is divided by the average number of hunters (passengers) per vehicle, which is assumed to be 2.41 hunters (Sorg, 1987).

The opportunity cost of travel time reflects the deterrent effect that longer drives have on visiting more distant sites, independent of the vehicle operation costs. For example, many higher income people could afford the extra $8 of gasoline to drive an additional 2 hours, but could not "afford" the additional time cost in terms of other activities forgone. The hourly wage is used as a proxy for the opportunity cost of time. This is based in part on work by Cesario (1976), which demonstrated that the opportunity cost of time in commuting studies equaled between one-fourth and one-half the wage rate. In the current study, U.S. Water Resources Council Principles and Guidelines (1983) were followed, with the opportunity cost of time calculated as one-third of average wage rate. The calculated opportunity cost per mile is $0.1282 for this data set. Total variable cost per mile per hunter is then $0.1282 + $0.172/2.41 = $.20.

With the double-log model, trips can never fall to exactly zero. To be conservative, the top of the second stage demand curve was truncated at the maximum observed trip distance, which was 1,000 round-trip miles. The area under this curve starting at the base value and ending at a distance of 1,000 miles is the net willingness to pay, or the amount the sampled waterfowl hunters are willing to pay above the actual amount paid.

The total consumer surplus for each of the seven Valley sites is the product of that site's consumer surplus per hunter day and the total number of hunter days (U.S. Department of Interior, 1987) at that site. The sum of the total consumer surplus across all seven sites is $3.2 million. Table 5 presents the consumer surplus per hunter day and the total consumer surplus per site for the San Joaquin NWR's and wildlife areas examined in the survey. As the demand equation [2] tends to underestimate total trips, the benefit estimates err on the conservative side.

From table 1, the total number of waterfowl hunting days in all private and public areas in the Valley is 265,727 (source: *Report of the Game Take Hunter Survey*, published annually by the California DFG.) If hunters at all wildlife areas are assumed to have a similar hunting experience as hunters at the public areas listed above, this figure can be multiplied by the average consumer surplus value of $55.41, yielding a total annual value for waterfowl hunting of $16,475,074. In the next section of this chapter, the model will be used to estimate how waterfowl hunting benefits change with water levels and contamination control.

Table 5. Consumer surplus per hunter day and total consumers for the 1987-88 season for selected San Joaquin Valley wildlife areas.

| Refuge | Consumer Surplus per Hunter-Day | Hunter Days | Total Consumer Surplus |
|---|---|---|---|
| Kesterson NWR | $37.19 | 3,900 | $145,041 |
| San Luis NWR | 51.11 | 9,000 | 459,990 |
| Merced NWR | 43.46 | 1,700 | 73,882 |
| Volta WA | 60.01 | 3,500 | 210,035 |
| Los Banos WA | 62.98 | 3,500 | 220,430 |
| Mendota WA | 63.74 | 31,723 | 2,022,024 |
| Kern NWR | 69.36 | 1,300 | 90,168 |
| Average | $55.41 | Total 54,623 | $3,221,570 |

## EFFECTS OF AGRICULTURAL DRAINAGE ON THE ECONOMIC BENEFITS OF WATERFOWL HUNTING

The primary harm to wildlife in Valley refuges from agricultural drainage is associated with the high concentrations of selenium in much of the drainage water. Although selenium is a necessary nutrient for life, high concentrations have been implicated in waterfowl deformities and death. At Kesterson NWR, which used agricultural drainage as a major source of water supply, high selenium levels were lethal to a large percentage of the waterfowl population. In general, most of the refuges listed in table 5 now receive little agricultural drainage, and correspondingly, have nonlethal levels of selenium.

Ohlendorf (1989) estimated the frequency of embryotoxicity (dead or deformed embryos or chicks) attributable to selenium levels in nesting aquatic birds at the Kesterson refuge for the period 1983-85. In 1983 (the only year for which coot data is available), 64.4 percent of coot nests had one or more dead or deformed embryos or chicks. For the period 1983-85, an average of 34.9 percent of duck nests had one or more dead or deformed embryos or chicks. Estimated reductions in these death and deformity figures are used to determine the increase in waterfowl hunting benefits at Kesterson associated with reducing selenium concentrations to nonlethal levels. To do this, the 1986 waterfowl harvest data used to estimate equation [2] was separated into duck, geese, and coot components, which were 94.4 percent, 2.0 percent, and 3.6 percent of total 1986 harvest, respectively, for the Valley refuges (DFG, *1986 Waterfowl Hunting Season Report*). Using DFG-estimated 1989 breeding population data for the Valley refuges (1986 data were not used as data were not collected that year for coots), the percent of harvested ducks and coots bred in the Valley (geese do not breed there) is estimated. The above figures suggest that 11.5 percent of the total winter duck population and 3.9 percent of the total winter coot population are bred there. It is reasonable to assume, therefore, that of the ducks and coots harvested in Kesterson, 11.5 percent and 3.9 percent, respectively, were bred there.

Using the embryotoxicity figures listed above, the increase in the number of harvested ducks and coots bred at Kesterson attributable to decreasing selenium levels to nonlethal concentrations is calculated. Without factoring the possibility of compensatory mortality (due to a lack of information on its magnitude), it is assumed that the 64.4 percent of dead or deformed coot embryos or chicks and the 34.9 percent of dead or deformed duck embryos or chicks would have survived at nonlethal selenium concentrations. For want of more detailed embryo or chick mortality data, the dead or deformity percentages, which are the percentages of all nests with one or more dead or deformed embryos or chicks, are assumed to be the total death or deformity percentages for a clutch of eggs. This plus the preceding assumption may lead to a liberal estimate of the increase in native waterfowl population due to a decrease to nonlethal levels of the selenium concentration. On the other hand, no adjustment is made for the possible decrease in reproductive ability of waterfowl that inhabit the refuge in winter but breed somewhere else as no data exist on this topic.

Using the figures cited above, of the 509 waterfowl harvested in Kesterson in 1986, 51 ducks and 1 coot were estimated bred there. With a reduction in the selenium level to a nonlethal concentration, 538 waterfowl (a 5.7-percent increase) would have been harvested there. Substituting this harvest figure into equation [2], yields a 1.4-percent increase in Kesterson hunting applications. This percentage increase translates into an increase of 55 hunter days in the sample expansion of Kesterson hunter visitation figures from table 5. With this increase in hunter visitation, the total consumer surplus increases by $2,030. Assuming a 100-year horizon for this increased surplus and an 8-percent discount rate used by Federal water resources agencies, the present value of this increase in consumer surplus is $25,400. Note that this is the value only to waterfowl hunters visiting Kesterson. It is provided as an example of how the preceding valuation technique can be applied rather than as a definitive value.

An increase in the total economic benefits of bird viewing at Kesterson resulting from a decrease in selenium concentration to nonlethal levels should be added to this figure. A lack of Kesterson bird viewing data makes this addition difficult at this time. However, Loomis et al.'s chapter in this book quantifies the option and existence values to all members of society of Valley wetlands and of reducing Valley contamination.

## CONCLUSIONS

The Valley is heavily used for waterfowl hunting and wildlife viewing. While these recreational activities at National wildlife refuges and State wildlife management areas are nonmarket goods, the economic values have been quantified in this chapter using the travel cost and the contingent valuation methods. Waterfowl hunting at the seven public refuges and wildlife areas is worth $3.2 million annually. This value was found to be statistically related to waterfowl take, which in turn can be impacted by habitat contamination. By linking reductions in contamination to increases in waterfowl breeding populations at Kesterson NWR, an estimate of added benefits to waterfowl hunters can be computed for reductions in contamination. The same basic linkages apply to estimating the added wildlife benefits to viewers, but lack of viewing data for Kesterson prevented such a calculation. However, the benefits to other members of society of wildlife throughout the Valley is quantified by Loomis et al., in a subsequent chapter of this book. Even though some values for Kesterson wildlife could not be quantified, this chapter demonstrated how recreational use related to wildlife could be quantified and linked to agricultural contamination issues. More precise estimates of the economic effects await better biological data of onsite and offsite contamination effects on migratory birds.

## ACKNOWLEDGEMENTS

The authors thank the California Department of Fish and Game, especially Bruce Deuel and Sandra Wolfe, for providing access to data; Howard Hirahara and Gary Bedker at the Bureau of Reclamation; and Thomas Wegge of Jones and Stokes Associates for their assistance and suggestions in estimating the demand models.

## REFERENCES

California Department of Fish and Game. *Results of the Game Take Hunter Survey*. Administrative Report, Annual 1982-85 Editions.

California Department of Parks and Recreation, 1987. *Public Opinions and Attitudes on Outdoor Recreation in California*.

Cesario, Frank. Value of Time in Recreation Benefit Studies. *Land Economics*, 52, pp. 32-41.

Clawson, M. and Knetsch J., 1966. *Economics of Outdoor Recreation*. John Hopkins University Press, Baltimore, MD.

Cummings, R.; Brookshire, D.; and Schulze, W., 1986. *Valuing Environmental Goods: An Assessment of the Contingent Valuation Method.* Roman and Allanheld, NJ.

Dwyer, J.; Kelly, J.; and Bowes, E. M., 1977. *Improved Procedures for Valuation of the Contribution of Recreation to National Economic Development.* Research Report No. 128, Water Resources Center, University of Illinois, Urbana.

Just, R.; Hueth, D.; and Schmitz, A., 1982. *Applied Welfare Economics and Public Policy.* Prentice Hall, NJ.

Loomis, J. B., 1982. Use of Travel Cost Models for Evaluting Lottery Rationed Recreation: Application to Big Game Hunting, *Journal of Leisure Research*, 14, pp. 117-124.

Ohlendorf, H. M., 1989. *Bioaccumulation and Effects of Selenium in Wildlife. Selenium in Agriculture and the Environment.* Soil Science Society of America and American Society of Agronomy, Madison, WI.

Randall, A. and Stoll, J., 1983. Existence Value in a Total Valuation Framework. In: Rowe, R. and Chestnut, L., *Managing Air Quality and Scenic Resources at National Parks,* Westview Press, Boulder, CO.

Sassone, P. and Schaffer, W., 1978. *Cost Benefit Analysis: A Handbook.* Academic Press, New York, NY.

Schulze, W.; D'Arge, R.; and Brookshire, D., 1981. Valuing Environmental Commodities. Some Recent Experiments, *Land Economics*, 57(2), pp. 151-172.

Sorg, C. and Loomis J., 1985. An Introduction to Wildlife Valuation Techniques, *Wildlife Society Bulletin,* 13, pp. 38-46.

Strong, E., 1983. A Note on Functional Form of Travel Cost Models with Unequal Populations, *Land Economics,* 59(3), pp. 342-349.

U.S. Department of the Interior, 1986. Natural Resource Damage Assessments: Final Rule, *Federal Register,* 43 CFR Part 11, Vol. 51(148).

U.S. Department of the Interior, Bureau of Reclamation, 1987. *Report on Refuge Water Supply Investigations, Central Valley Hydrologic Basin, CA.* Draft Report. Vol. 1.

U.S. Water Resources Council, 1979. Procedures for Evaluation of National Economic Development (NED) Benefits and Costs in Water Resources Planning: Final Rule, *Federal Register,* Vol. 44(242).

U.S. Water Resources Council, 1983. *Economic and Environmental Principles and Guidelines for Water and Related Land Resources.*

Vaughan, W. and Russell, C., 1982. Valuing a Fishing Day: An Application of a Systematic Varying Parameters Model, *Land Economics,* 58, pp. 450-463.

Ward, F. and Loomis, J., 1986. The Travel Cost Demand Model as an Environmental Policy Assessment Tool: A Literature Review, *Western Journal of Agricultural Economics,* 11(2), pp. 164-178.

Welsh, M. P, 1986. *Exploring the Accuracy of the Contingent Valuation Method.* Unpublished Ph.D. Dissertation, Department of Agricultural Economics, University of Wisconsin, Madison.

If you have any concerns about our products,
you can contact us on
**ProductSafety@springernature.com**

In case Publisher is established outside the EU,
the EU authorized representative is:
**Springer Nature Customer Service Center GmbH
Europaplatz 3, 69115 Heidelberg, Germany**

Printed by Libri Plureos GmbH
in Hamburg, Germany